MODERN BIOELECTRICITY

edited by

Andrew A. Marino

Louisiana State University
School of Medicine
Shreveport, Louisiana

Marcel Dekker, Inc. • New York and Basel

Library of Congress Cataloging in Publication Data

Modern bioelectricity.

1. Electrophysiology. 2. Electromagnetism--Physiological effect. 3. Electromagnetism--Therapeutic use. 4. Electromagnetism--Toxicology. I. Marino, Andrew A.
QP341.M63 1988 574.19'127 87-27433
ISBN 0-8247-7788-3

MARCEL DEKKER, INC.
270 Madison Avenue, New York, New York 10016

Current printing (last digit):
10 9 8 7 6 5 4 3 2 1

PRINTED IN THE UNITED STATES OF AMERICA

Preface

It has been almost 25 years since I was introduced to bioelectricity by my teacher, Robert Becker. The subject was then in its infancy and had no natural constituency because it did not fit easily within any of the orthodox scientific pigeonholes. In physiology, electricity usually meant action potentials, within engineering it related to microwave heating, in physics and medicine it was associated with X-rays and radiotherapy treatment of cancer, and in chemistry it was linked to electrode reactions. Against this backdrop, two important themes emerged. What is the nature of the system or process that controls the living organism? Some might hold it to the finger of God, and declare its inherent mechanics to be unknowable. The other extreme involves focusing on molecular minutiae in the belief that life can be defined at that level. Modern bioelectricity is a middle-of-the-road approach which began with a crystallizing perception that electrical interactions are more fundamental than biochemical reactions, and hence that they perhaps have a greater probability of explaining the physical basis of life and the processes that control and express it. Bioelectricity's other major theme -- environmental electromagnetic pollution -- became important beginning in the early 1970's.

Much has happened during the past two decades, and this book is a monument to that work. As with all new initiatives, many questions have been raised, and previously unrecognized problems have become manifest. But it is the business of science to uncover and solve these problems, and it is precisely this effort, which is taking place on a broad scale across many traditional scientific disciplines, that constitutes the chief development in bioelectricity during the past 20 years. To build a new science there must be a group of scientists that are equal to the task, as well as a need for the approach that it embodies. The reader can judge whether the authors of this book meet that test.

As Machiavelli observed more than 400 years ago, any new idea is inherently born in struggle because it has preexisting opponents who are the adherents of the old idea, whereas the idea's potential proponents are still inchoate. If bioelectricity is to become a science and serve as a tool for solving mankind's problems, then the public-health issue of environmental electromagnetic fields must be faced and solved. This issue, which ironically was one of the seminal rationales for bioelectricity, has now come to serve as an incubus on progress because the lion's share of research funding of bioelectrical studies comes from organizations that create the electromagnetic fields whose risk-causing propensity is at issue. This is not a good or even acceptable pattern of funding if our true interest is the unbridled search for truth.

Perhaps the reader is interested in the men who, at least in my view, are responsible for pioneering bioelectrical research. Robert Becker's contributions to bioelectricity span its entire domain -- more so than any other investigator. He is a bold innovator and thinker of new thoughts, an indomitable spirit and a great influence on this field. C. Andrew Bassett introduced bioelectricity into the scientific mainstream. The magnetic-field device he developed for clinical use was made available to a wide range of investigators, leading to many reports in the scientific and medical literature. Such reports were rare when Bassett began his studies in the 1960's; now they are commonplace. Zachary Friedenberg and Carl Brighton have a long involvement with bioelectrical studies, and their career-long commitment has done much to regularize bioelectricity as a science and to underscore its usefulness. As successive chairmen of a clinical department of a prestigious medical school, they have been instrumental in establishing the credibility of the new science. Their attention to methodology, quantitation, and detail has earned them widespread respect. Allan Frey was for many years a solitary witness to the scientific fact that electromagnetic fields are physiologically significant. The effects do not exist, he was told, or if they do they are trivial, and if not, they are classical and hence are no threat to the status quo. Hindsight shows us that Frey was correct, and his tenaciousness has enriched this field. Perhaps least known of the pioneers is Milton Zaret, a soft-spoken man who saw what he saw through his bimicroscope, wrote about it, and continued to write even though some found his reports displeasurable. It is the duty of the scientist to report on nature, and the duty of the citizen to contribute to the collective judgment of how the scientific facts are to be incorporated into the fabric of society. Perhaps the well-being of our society demands that we accept a specific prevalence of microwave-induced cataracts as the price for living in an affluent and militarily strong society. That is, however, an entirely different issue from whether microwave radiation causes cataracts.

Bioelectrical research in the Soviet Union antedates that done in the West but language and other difficulties have hindered widespread appreciation of the Soviet studies. Nevertheless, the work of Yu. Kholodov and A.S. Presman must be mentioned. Presman's book was one of the earliest systematic treatments of bioelectricity, and it had a profound influence on me and many others. Although I have never met them, I feel that they must have the same indomitable spirit as their American counterparts.

I have organized the chapters according to the general framework of bioelectricity. Chapters dealing with the biological significance of natural electrical signals are grouped following the introductory chapter. Part III deals with measurements and computations of electrical properties and characteristics of tissue. Laboratory studies of biological changes induced in living organisms are described in Part IV. The rationale for bioelectricity is what it tells us about life, health, and disease, and these topics are covered in the last two parts.

The King of Hearts advised "begin at the beginning, and go on till you come to the end: then stop." Practical considerations, alas, have necessitated a relatively arbitrary end to this book, and I cannot offer the reader a guarantee of completeness. But I can guarantee that it is an authoritative exposition of the major threads of modern bioelectricity.

Andrew A. Marino

Contributors

E. AARHOLT, Department of Electronic and Electrical Engineering, University of Salford, Salford, M5 4WT, United Kingdom

ROBERT O. BECKER, Erie Canal Road, Star Route, Lowville, New York 13367

HERMANN BERG, Academy of Sciences of the GDR, Central Institute of Microbiology and Experimental Therapy, Department of Biophysical Chemistry, DDR-69 Jena, German Democratic Republic.

I.S. BEZDOL'NAYA, Marzeev Scientific Research Institute of General and Communal Hygiene, 50 Popudrenko Str., Kiev-94, 252660, USSR

MARIAN BIELEC, Department of Biological Effects of Nonionizing Radiation, Center for Radiobiology and Radiation Safety, Warsaw, Poland

NANCY A. BLAKEMORE, Department of Microbiology, University of New Hampshire, Durham, New Hampshire 03824

RICHARD P. BLAKEMORE, Department of Microbiology, University of New Hampshire, Durham, New Hampshire 03824

MARTIN BLANK, Department of Physiology and Cellular Biophysics, Columbia University, 630 West 168th Street, New York, New York 10032

RUGGERO CADOSSI, Center for Experimental Haematology, Department of Internal Medicine II, University of Modena, Policlinics Via del Pozzo 71, I41100 Modena, Italy

GIOVANNI CECCHERELLI, Center for Experimental Haematology, Department of Internal Medicine II, University of Modena, Policlinics Via del Pozzo 71, I41100 Modena, Italy

A. DAL MONTE, Department of Pediatric Orthopedics, Instituto Ortopedico Rizzoli, Bologna, Italy

Yu D. DUMANSKIY, Marzeev Scientific Research Institute of General and Communal Hygiene, 50 Popudrenko Str., Kiev-94, 252660, USSR

GIOVANNI EMILIA, Center for Experimental Haematology, Department of Internal Medicine II, University of Modena, Policlinics Via del Pozzo 71, I41100 Modena, Italy

G. FONTANESI, Department of Orthopaedics and Traumatology, Arcispedale S. Maria Nuova, Reggio Emilia, Italy

RICHARD B. FRANKEL, Francis Bitter National Magnet Laboratory, M.I.T., Cambridge, Massachusetts 02139

ALLAN H. FREY, Randomline, Inc., County Line and Mann Roads, Huntingdon Valley, Pennsylvania 19006

WILLIAM GENSLER, Department of Electrical Engineering, Building 20, University of Arizona, Tucson, Arizona 85721

F. GIANCECCHI, Department of Orthopaedics and Traumatology, Arcispedale S. Maria Nuova, Reggio Emilia, Italy

EUGENE M. GOODMAN, Biomedical Research Institute, University of Wisconsin - Parkside, Kenosha, Wisconsin 53141

BEN GREENEBAUM, Biomedical Research Institute, University of Wisconsin - Parkside, Kenosha, Wisconsin 53141

FRANCIS X. HART, Department of Physics, The University of the South, Sewanee, Tennessee 37375

M. JABERANSARI, Department of Electronic and Electrical Engineering, University of Salford, Salford, M5 4WT, United Kingdom

A.H. JAFARY-ASL, Department of Physiology, Medical School, Shiraz University, Shiraz, Iran

SIDNEY B. LANG, Department of Chemical Engineering, Ben-Gurion University of the Negev, P.O. Box 653, Beer Sheva 84105, Israel

SLAWOMIR LIPSKI, Department of Biological Effects of Nonionizing Radiation, Center for Radiobiology and Radiation Safety, Warsaw, Poland

ANDREW A. MARINO, Department of Orthopaedic Surgery, LSU School of Medicine in Shreveport, P.O. Box 33932, Shreveport, Louisiana 71130-3932

MICHAEL T. MARRON, Office of Naval Research, Arlington, Virginia

P.N. MARSH, Department of Electronic and Electrical Engineering, University of Salford, Salford, M5 4WT, United Kingdom

ROCHELLE MEDICI, Analytical Services, 2220 El Molino Place, San Marino, California 91108

MICHAEL A. PERSINGER, Behavioral Neuroscience Laboratory, Department of Psychology, Laurentian University, Sudbury, Ontario, P3E 2C6, Canada

RONALD PETHIG, Institute of Molecular and Biomolecular Electronics, University College of North Wales, Dean Street, Bangor, Gwynned LL57 1UT, United Kingdom

ARTHUR A PILLA, Bioelectrochemistry Laboratory, Department of Orthopedics, Mount Sinai School of Medicine, One Gustave L. Levy Place, New York, New York 10029

G. POLI, Department of Pediatric Medicine, Instituto Ortopedico Rizzoli, Bologna, Italy

MARIA REICHMANIS, 1951 Bolin Road, North Augusta, South Carolina 29841

M.G. SHANDALA, Marzeev Scientific Research Institute of General and Communal Hygiene, 50 Popudrenko Str., Kiev-94, 252660, USSR

BETTY F. SISKEN, Department of Anatomy, University of Kentucky, Lexington, Kentucky 40506

C.W. SMITH, Department of Electronic and Electrical Engineering, University of Salford, Salford, M5 4WT, United Kingdom

STEPHEN D. SMITH, Department of Anatomy, University of Kentucky, Albert B. Chandler Medical Center, Lexington, Kentucky 40506

GRAZYNA SOKOLSKA, Department of Biological Effects of Nonionizing Radiation, Center for Radiobiology and Radiation Safety, Warsaw, Poland

WILLIAM E. SOUTHERN, Department of Biological Sciences, Northern Illinois University, DeKalb, Illinois 60115

J.A. SPADARO, Department of Orthopedic Surgery, State University of New York, Health Sciences Center, 750 East Adams Street, Syracuse, New York 13210

STANISLAW SZMIGIELSKI, Department of Biological Effects of Nonionizing Radiation, Center for Radiobiology and Radiation Safety, Warsaw, Poland

H. Ti TIEN, Membrane Biophysics Laboratory, Department of Physiology, Michigan State University, East Lansing, Michigan 48824

GUISEPPE TORELLI, Center for Experimental Haematology, Department of Internal Medicine II, University of Modena, Policlinics Via del Pozzo 71, I41100 Modena, Italy

MILTON M. ZARET, 1230 Post Road, Scarsdale, New York 10583-2030

JOZEF R. ZON, * Department of Physiology, Michigan State University, East Lansing, Michigan 48824

*_Permanent address_: Department of Theoretical Biology, The Catholic University of Lublin, Lublin, Poland

Contents

I

OVERVIEW

1
Electromagnetism and Life

ROBERT O. BECKER
Erie Canal Road
Star Route
Lowville, New York

The word "bioelectricity" appears to be a contradiction in terms, implying a connection between biology and electricity. The established textbooks of biology seem to ignore electricity while the electrical texts seem equally unaware that biology exists. To compound the confusion, there are now a number of competing terms used to describe this field: biomagnetics, electrobiology, bioelectromagnetics, magnetobiology, bioelectrochemistry, and so on. Quite apparently something is happening at the interface between the life sciences and electromagnetism that is new and exciting, and that is ignored by the scientific establishment. The situation is typical of what Kuhn called a revolution in scientific concepts (1), and it is my belief that this is exactly what is occurring. It is the purpose of this chapter to introduce this new scientific paradigm by providing a brief historical perspective. As such, it can only touch on what I perceive to be the significant links in the chain of events that have led to the development of this new scientific discipline.

Bioelectricity, defined broadly, is the study of the electromagnetic forces generated by living organisms, and the effects of external electromagnetic forces and fields upon living organisms. This new term is compounded of two Greek roots, and it is fitting that bios comes before elektron, since in the historical perspective most of the major turning points can be attributed to biological scientists motivated by a recognition of the deep uncertainties in their understanding of life. The history of bioelectricity can be divided into three epochs, the first and second separated by the monumental contributions of Galvani, the second divided from the third by the scientific explosion that accompanied the global conflict of World War II, which has led to our modern technological world.

Any history of electricity and magnetism must begin with the publication in 1600 of DeMagnete by William Gilbert, physician to Queen Elizabeth. For the first time these two forces were removed from the realm of mystery, clearly separated from each other and subjected to logical experimentation. While static electricity was the only kind then known, the following century saw the development of methods for generating, measuring, and storing this energy, as well as the appearance of possibly the greatest collection of scientific minds ever existing in one century - Bacon, Harvey, Descartes and Newton. Nevertheless, biology and medicine were still enmeshed in the longest running scientific dispute on record: the question of vitalism versus mechanism. This controversy, dating back to the early Greek philosophers, pitted the vitalists who believed that living things possessed an "anima" or vital spirit unaccessible to physical analysis, against the mechanists who thought that this was simply nonsense and that living entities were merely more complex than non-living things. Electricity became the prime candidate for the "anima," and when Galvani published his findings in 1791 he thought he had identified the vital spirit as "animal electricity." What Galvani actually had discovered was another much more useful form of electricity, direct current, constantly generated by the apposition of two dissimilar metals in a conducting solution. Volta's attack upon Galvani, which occurred two years later, clearly demonstrated this fact and appeared to conclusively exclude electricity as being produced by living tissues. Galvani's answer (later published anonymously), demonstrated that under certain conditions living tissue could produce electricity without the intervention of metallic contacts. However, it was largely ignored in the excitement over Volta's electrical pile which produced a continuous supply of electrical current. Thirty years later, Matteucci demonstrated that what Galvani actually had shown in his later work was the production of electrical current by injured tissue, the "current of injury." In the meantime, Oersted had demonstrated that direct current flowing in a wire produced a magnetic field that extended out in space from the wire, and Davy had shown that the electricity produced by the Voltaic pile was the result of chemical events in the two metals and the conducting solution. Following Matteucci's observations, Du Bois-Raymond demonstrated that stimulation of a nerve produced an electrically measurable impulse that traveled along the nerve from the point of stimulation, and concluded that he had confirmed "the identity of the nervous principle with electricity." However, this was conclusively disproven by von Helmholtz in 1850 when he measured the speed of the nerve impulse at about 30 meters per second, a figure far below that previously obtained by Cavendish and others using a Leyden jar discharged into a wire more than 12,000 feet long. While the nerve impulse could be measured electrically, it could not be electrical in nature. What was it?

The answer was provided by Julius Bernstein, one of Du Bois-Raymond's students. Bernstein published his hypothesis in 1868, proposing that the nerve membrane is polarized, with the interior having a different electrical polarity than the exterior, based upon some mechanism that enabled the membrane to selectively admit ions of only one sign. The nerve impulse was then postulated to be a localized breakdown of this polarity which then propagated along the nerve fiber. Thus, while the nerve impulse could be measured with electrical instruments, it was definitely not the actual passage of electrical current longitudinally along the nerve fiber. This "Bernstein hypothesis" of cell membrane polarization has become the cornerstone of modern electrophysiology and is invoked to explain all electrically measurable phenomena in living organisms, including the current of injury.

In 1864, Maxwell postulated mathematically the existence of a continuous spectrum of electromagnetic fields arranged on a scale of increasing frequency and decreasing wave length. About 20 years later Hertz demonstrated the reality of such fields and their ability to transmit signals through space without wires. Less than 10 years later, Roentgen, experimenting with electrical discharges in partially evacuated tubes, discovered X-rays.

By the turn of the century, Hertz's discovery was being used to transmit messages across the Atlantic ocean, and

Roentgen's discovery was being used by the medical profession in the diagnosis and reduction of fractures in humans. During the same time the position of the vitalists had been slowly whittled away as electricity was excluded from more and more physiological functions until, by the beginning of the 20th century, only the transmission of the nerve impulse across the synaptic gap was left. In 1920, this last straw was removed by Otto Loewi's demonstration of the production of the chemical acetylcholine by the arrival of the impulse at the nerve termination. There was no "vital spirit" and no place for electricity as such in living organisms, they were simply complex chemical machines, nothing more.

The mechanistic-reductionist philosophy that the total organism was simply the sum of its parts, and that one could isolate and study any part with the resultant data applicable to the whole organism, took firm control of all biological thinking. This approach has led to such advances in all aspects of biology and medicine that it has gained the status of unimpeachable dogma.

One result of this triumph of mechanistic concepts was the total debunking of electromedicine. This method of therapy had arisen not long after Galvani's observations, developing in a completely empirical fashion until by the late 1800's it was in common use for the treatment of a wide variety of ailments. In view of the lack of basic knowledge in both biology and physics during this time, it is not surprising that much of this was sheer nonsense, and medicine was better off for having discarded it. We now know, however, that in at least one instance a valuable method of treatment was also discredited. The method in use at that time for treatment of non-unions of human bone fractures with electrical currents was remarkably similar to that employed and accepted today.

It appeared that the last word had been said with Loewi's demonstration in 1920, but this was actually not the case. In 1929 Berger reported the human electroencephalogram (EEG), with a characteristic frequency range and pattern. It was then, and it still is now, impossible to relate this electrical phenomenon entirely to the nerve impulse. Other neurophysiologists, chiefly, Gerard and Libet, were convinced that the simplistic concept of the nerve impulse being the sole neural mechanism was inadequate to explain the complex functions of the brain. During the 1930's and 40's they reported evidence for actual electrical currents flowing outside of the nerve cells proper in the brain. They demonstrated that such currents influenced the way in which the neurons operated, but their work evoked little interest, not only because it flew in the face of established dogma but also because the measurement of direct currents of such small magnitude was extremely difficult. Thus, few neurophysiologists were interested in exploring what appeared to be a nonproductive backwater. During the same period of time, several biologists, primarily Burr and Lund, similarly convinced that the whole was greater than the sum of the parts, continued to search for evidences of electrical forces playing a role in the actions and processes of the total living organism. Their observations consisted chiefly of measuring the DC electrical potentials from a wide variety of living organisms, and correlating them with functional changes. They interpreted their observations as indicating a total body "bioelectrical field," a simple dipole oriented along the head-tail axis of the body. While they both published their data, their reports elicited no interest from the great majority of scientists and their work was criticized on the basis of inadequate instrumentation, with considerable truth to the latter point. The measurement tools available to them certainly introduced major errors into whatever data was obtained primarily because of their low impedance.

While reductionist biology was consolidating and strengthening its gains, enthusiastic and expanding usage was being made of the discoveries in electricity and electromagnetism. Electrical power was being made available in increasing amounts and was accompanied by increasing development in the use of radio transmissions. Progress in biology and medicine reinforced this development, providing both reassurances that this energy had nothing whatsoever to do with living organisms, and, further evidences of the utility of these discoveries in the form of high-frequency heating for clinical use (diathermy), and the use of X-rays for treatment of a wide variety of conditions ranging from acne to cancer. It was not until well into the second quarter of the 20th century that this development first demonstrated some undesirable side effects in the form of the production of cancers, particularly in those physicians who had made great use of X-rays. This was rapidly attributed to the ability of radiation higher in frequency than light to produce ionization of the tissues and cell structures. Since radiation that was below light in frequency lacked this property, it was concluded that it had no biological effects.

All controversy had ceased by the beginning of World War II. There were no biological effects of electrical currents below the level at which shock was produced, and living organisms made no use of electrical currents in their physiological functions. All actions of living organisms were best explained on the basis of chemical and molecular actions and polarized cell membranes. Electromagnetic fields had no biological effects, and this concept appeared to be borne out by our operating experience. We had been employing these fields at frequencies from the 60-Hz power frequency to short-wave radio with no obvious (or even demonstrable) effects occurring either in those persons most exposed or in the general population. The possibility that magnetic fields had any biological effect was assigned to the realm of charlatenism or worse. Classical concepts of physics simply did not allow for any meaningful interaction between any form of nonionizing electromagnetic radiation and living organisms. Wave lengths of the frequencies then available were far too long to produce resonance with any biological structure, and the energy delivered to the living system was far below kT. Reductionist and physiochemical concepts of biology were poised to deliver the answers to all of our questions about life, and to provide effective controls for all of our ills.

Yet disquieting problems remained. In the rush to embrace these new concepts that had emerged victorious after their long struggle with vitalism, any biological property or function that did not fit the established paradigm was either glossed over or ignored. On close examination it appeared that we did know a lot about those functions that could be well understood by chemical and molecular concepts, but we simply had no ideas about such basic issues as the nature of life, or even lesser ones such as what controlled growth, structure and embryonic development, the nature of cerebral activity, or the mechanisms involved in biological cyclic patterns. In 1941 a spokesman for this view appeared; Albert Szent Gyorgyi who had won the Nobel prize in medicine in 1937 for his work on biological oxidation and Vitamin C, presented the now famous Koranyi Lecture in which he proposed the radical concept of the transfer of energy within living cells by excited electrons moving within semiconducting matrices (2). Commenting on the failure of reductionist philosophy to explain basic problems, he said, "It looks as if some basic fact about life were still missing, without which any real understanding is impossible." Szent Gyorgyi went on to publish numerous other papers and books on the same theme, after the war, among which "An Introduction to Submolecular Biology," published in 1960, has become a classic. Yet his greatest achievement may well turn out to be that single lecture delivered in 1941. From the point of view of bioelectricity it must be considered a classical turning point of as great an import as Galvani's discoveries 150 years earlier.

The roots of the modern concepts of bioelectricity can be traced to the period following World War II when a great expansion in science occurred based upon the technological advances stimulated by the conflict. This had been the first major war ever fought in which electrical and electromagnetic forces played a major, perhaps pivotal, role in the outcome. Great advances in communications technology were made and rapidly put into use. The ability to generate higher and higher frequencies appeared and the development of the cavity magnetron enabled the first effective use of radar. To the public, this represented a new and novel form of energy, and as technology progressed, the power produced in a radar beam became far greater than that from any other type of field-generating equipment. This fact, coupled with some reports of health effects among radar workers, stimulated a debate beginning in the late 1950's over the possible biological effects of this new electromagnetic modality. The debate was thought to have been concluded in the early 1960's with the confirmation of the classical concept that heating effects were the only ones possible, and the subsequent adoption of the thermal effects standard for exposure of humans to microwave radiation. That this is very much not the case is clearly presented in Steneck's latest book, <u>The Microwave Debate</u> (3).

Also in the post-War period, Szent Gyorgyi's prescient concepts of electron flow were confirmed in the physical sciences with the development of the transistor and other solid state technology. Increasingly sophisticated electronic instrumentation required more sensitive and stable methods for measuring and recording electrical currents and potentials. It was the availability of such devices, which were much more sensitive and much less subject to artifact than those used by Burr and Lund, that effectively re-opened the controversy over the role of electricity in living systems and led to the modern era in the study of bioelectricity.

The initial application of this type of instrumentation occurred in 1960 with the reevaluation of the bioelectric field concept of Burr and Lund. A moist-skinned vertebrate, the salamander, was used, and rather than the simple head to tail dipole field previously described, a complex potential field was found that correlated spatially with the anatomical complexity of the central nervous system (4). Further evidence suggesting the involvement of the nervous system in the generation of these DC potentials was the observation that the strength of individual DC vectors in the field varied directly with the level of consciousness of the experimental animal. The same techniques were later applied to the measurement of the current of injury in animals capable of limb regeneration compared with a closely related species that did not have such capability. The results were significant from two points of view (5). First, the polarity of the potentials at the site of a limb amputation in a regenerating animal while initially positive became strongly negative, this polarity coinciding in time with the regrowth of the limb. The polarity at a similar site in a non-regenerating animal was initially positive and remained so throughout the period of healing by scarification. The observation was unambiguous, indicating a definite relationship between the electrical phenomenon and the type of healing that occurred.

The second observation was of equal import. Clearly measurable potentials existed in both species throughout the entire healing period of 3-4 weeks. The explanation of the current of injury, based upon the Bernstein hypothesis, was that damaged cell membranes became leaky and, whatever the mechanism was that produced the polarization, it then produced the externally measurable current. It was obviously not logical to conclude that such a mechanism could be operational over such a lengthy period of time. The same measurement techniques were next applied to bone tissue which was then found to be capable of producing electrical potentials when mechanically stressed (6). This capability was theorized to be related to the growth of bone in response to such stress. The relationship of the nerve to the DC potentials was further strengthened by the observation that a Hall effect was measurable from peripheral nerves (7) with the application of high-strength magnetic fields, and later by the observation that similar fields could produce major alterations in the pattern of the electroencephalogram (8). In this latter experiment it was further noted that the shift in the pattern was similar to that observed under deep anesthesia, and that the animals exposed to such fields demonstrated a similar behavioral change. While all of these observations had been suggestive of the existence of functionally significant, organized DC electrical currents existing within living organisms, the first substantiation of this concept from an experimental point of view was Smith's observation that simulation of the negative polarity of the current of injury in animals not normally capable of limb regeneration restored a significant measure of this growth (9). This seminal report stimulated Friedenberg's work, reported in 1970, in which he demonstrated the healing of a non-union of a human fracture by the application of a negative polarity direct current to the site (10). Friedenberg's technique, remarkably similar to that used in the 1880's and subsequently discarded as quackery, has evolved with minor changes into the clinically approved method in wide use today. The importance of this development is frequently overlooked. This was the first time since the beginnings of medicine that a method for <u>stimulating</u> growth based upon verifiable scientific concepts was made available to physicians. The fact that the method was derived from basic experimentation in bioelectricity did much to enhance the credibility of the entire concept that electrical forces <u>did</u> play a role in life processes.

During the 1960's magnetism also crept back into biology with the first report of the production of an actual, detectable magnetic field produced by the electrical activity of the heart

(11). This was shortly followed by Cohen's report in 1968 of a magnetic field produced by the activity of the brain (12). These reports had employed classical Helmholtz coils as detectors for the magnetic field, and the data obtained was minimal until the development of the SQUID magnetometer in 1969. The availability of this device has led to an explosive growth in the detection, measurement and analysis of magnetic fields produced by brain activity (the magnetoencephalogram). Despite the fact that this technology requires the existence of actual electrical current flow in the brain, thus far credit has yet to be extended to Libet and Gerard for their pioneering observations, and no re-evaluation of classical cerebral neurophysiological concepts in this light has taken place.

In 1971 two important conferences were convened both of which materially assisted the development of this discipline. The first was a small gathering at the Lamont-Dougherty Geological Observatory in which the evidence for a surprising observation was presented. While it had been known for some time that the Earth's magnetic field was subjected to occasional perturbations of a significant nature during which reversals of the Earth's magnetic poles occurred, the new evidence consisted of correlations of these events with major alterations in the total biota. During the 5,000 years required to effect the reversal, major extinction of the most evolutionarily advanced species occurred. The data base in both the physical and biological areas was inadequate at that time to reach any firm conclusions, however, it appeared highly unlikely that a reversal in the direction of the DC field would have any major effect upon any life form except those that migrated possibly via magnetic cues. The suggestion was made that perhaps the micropulsation frequencies changed during reversal periods and that organisms which derived some important information from such frequencies would suffer major physiological and behavioral alterations. This concept had been in part explored for some years by Brown who presented considerable evidence that the tidal fluctuations in the Earth's field could be the timer for the simultaneous cyclic fluctuations in all living organisms known as circadian rhythms or biological cycles (13). His work illustrates well the classic conundrum in science: Brown could demonstrate with careful experimentation that very small magnetic fields (often below 1 gauss) could unquestionably alter the cyclic behavior of various animals, yet, because the hold of dogma was so strong and there was no known linkage mechanism between such small fields and living organisms, his observations were simply not believed. However, the Lamont Conference did much to stimulate interest in the possible role that normal electromagnetic fields of the Earth played in biological processes. The second conference was a larger affair held at Princeton University under the aegis of the Electrochemical Society. This organized group had worked primarily in the area of classical electrochemistry, with no interest in its possible application to biological affairs. The Princeton conference represented the formation of a sub-area of electrochemistry, bioelectrochemistry, and provided both the mechanism for the formation of this group and a chance for many of the people working in the area of bioelectricity to exchange ideas. The bioelectrochemical group is not only alive and well but expanding in a most productive fashion. The latest observations and concepts will be detailed in other chapters.

In 1973 an even more significant international conference took place which substantiated bioelectricity as a valid scientific discipline. "Electrically Mediated Growth Mechanisms in Living Systems" was hosted by the New York Academy of Sciences and consisted of 47 formal papers, several panel discussions and an air of excitement over the progress made in this field over the preceding decade. It provided another important link in the chain of events that has led to the present status of this field.

Progress in bioelectricity over the past decade has been even more rapid. It has been characterized by the establishment of several scientific societies and excellent journals, and the one index that a subject "has come of age," the number of papers published per year, clearly indicates a level of maturity that is almost overwhelming. Progress in clinical applications is apparent, and understanding of the extent of the environmental hazards posed by abnormal electromagnetic fields is advancing rapidly and seems destined to become the environmental issue of this decade. In the present context of an introductory chapter, it is impossible to adequately review the progress that has taken place in the past decade and that task will be taken over by the remaining chapters in this volume. However, during the same period of time, great advances have been made in basic biological knowledge that have important consequences for the science of bioelectricity as a whole. These seem destined to provide a means to relate all living organisms to the normal electromagnetic fields of the Earth in a particularly important and vital fashion. They also have provided us with an answer to the question of the linkage mechanism which has prevented even more rapid progress and acceptance of the concepts of bioelectricity by the total scientific establishment.

As noted early in this chapter, the concepts of classical physics provided no mechanism by which low-strength electric or magnetic fields could have any impact upon a living system. Despite the advances made in all aspects of this field and the obvious evidence that low-strength fields do have important biological effects, this situation still prevails to some extent. However, these new biological discoveries appear to be capable of resolving this dilemma. In 1975, Blakemore first described the presence of deposits of magnetite mineral in certain magnetotactic bacteria (14), and proposed that it served a useful function in detecting magnetic north. Assisted by SQUID technology, subsequent progress in studying these deposits has been rapid. It is now known that animals, representative of the majority of phyla including the human, have such deposits and that they are organized in such a fashion as to be able to "resolve magnetic field direction to within a few seconds of arc, or magnetic field intensity differences of 1-100 nT" (15). Further, in all higher animals the deposits are intimately related to the central nervous system. The widespread occurrence of these deposits throughout the animal kingdom indicates that they are evolutionarily conserved and that they must serve a useful function for their hosts. What is being described and characterized is a true magnetic organ whose purpose appears to be the detection of the strength and direction of the Earth's field. The pineal gland has, within the past few years, also become identified as another magnetic organ that is responsive to slight changes in the Earth's normal field (16). In addition, the pineal has come to be recognized as probably the true master gland of the body, with its secretions regulating the activity of the pituitary, the thyroid, the adrenal and the reproductive organs. Its major secretion, melatonin, is the regulator for the biological cycles and a potent neurochemical agent acting on the brain (17). Thus, final confirmation of Brown's original thesis is at hand; the cyclic fluctuations of the Earth's normal magnetic field are the timer for circadian rhythms. Whether the action of this weak field upon the cellular systems involved in these two magnetic organs is explicable on the basis of classical physics or involves other mechanisms remains to be elucidated. However, the existence of such specifically designed organs in living systems should not surprise us. Consideration of the normal electromagnetic environment of the Earth (that which pre-existed the present abnormal environment resulting from our usage of this modality) contained two quasi-static components; the DC magnetic field with its associated micropulsation spectrum (0-30 Hz), and visible light. This environment had existed, with the naturally occurring magnetic reversals providing the only perturbations, for the entire period of the genesis and evolution of life. During that time living things had developed a non-classical organ for the detection of light to provide imaging of the environment. It should not then be surprising to find that specific organs were also developed for the detection of the DC and ELF magnetic field fluctuations to provide timing signals for biological cycles (18).

Bioelectricity thus provides us with a totally new scientific paradigm by which to understand the basic physiological mechanisms occurring within living organisms, as well as the basic mechanisms relating all living organisms to the natural electromagnetic field of the Earth. Further exploration of its intricacies will undoubtedly lead to major advances in medical science and an understanding of the basic relationship between all living things and environmental electromagnetic forces.

REFERENCES

(1) Kuhn, T.S.: The Structure of Scientific Revolutions, Univ. of Chicago Press, Chicago, 1970.

(2) Szent-Gyorgyi, A.: The study of energy levels in biochemistry, Nature 148:159, 1941.

(3) Steneck, N.H.: The Microwave Debate, The MIT Press, Cambridge, 1984.

(4) Becker, R.O.: The bioelectric field pattern in the salamander: Simulation by an electronic analog, IRE Trans. Med. Elec. ME-7:202-208, 1960.

(5) Becker, R.O.: The bioelectric factors in amphibian limb regeneration, J. Bone Joint Surg. 43-A:643-656, 1961.

(6) Bassett, C.A.L. and Becker, R.O.: Generation of electric potentials by bone in response to mechanical stress, Science 137:1063-1064, 1962.

(7) Becker, R.O.: Search for evidence of axial current flow in peripheral nerves of salamander, Science :101-102, 1961.

(8) Becker, R.O.: The neural semiconduction control system and its interaction with applied electrical current and magnetic fields, in Proc. XI Int'l. Cong. Radiol. #105, 1753-1759, Exerpta Medica, 1966.

(9) Smith, S.D.: Induction of partial limb regeneration in Rana pipiens by galvanic stimulation, Anat. Rec. 158:89-98, 1967.

(10) Friedenberg, Z.B., Harlow, M.C. and Brighton, C.T.: Healing of non-union of the medial malleolus by means of direct current: A case report, J. Trauma 11:883-885, 1971.

(11) Baule, B.M., Gerhard, M.S. and McFee, R.: Detection of the magnetic field of the heart, Am. Heart J. 66:95-96, 1963.

(12) Cohen, D.: Magnetoencephalography: Evidence of magnetic fields produced by alpha-rhythm currents, Science 161:784-786, 1968.

(13) Brown, F.A.: The "clocks" timing biological rhythms, Am. Scientist 60:756-766, 1972.

(14) Blakemore, R.: Magnetotactic bacteria, Science 190:377-379, 1975.

(15) Walker, M.M., Kirschvink, J.L. and Dizon, A.E.: A candidate magnetic sense organ in the yellowfin tuna, Thunnus albacares, Science 224:751-753, 1984.

(16) Semm, P., Schnieder, T. and Vollrath, L.: Effects of an earth-strength magnetic field on electrical activity of pineal cells, Nature 288:607-608, 1980.

(17) Preslock, J.P.: The pineal gland: Basic implications and clinical correlations, Endocrine Rev. 5:282, 1984.

(18) Becker, R.O.: Theory of the interaction between DC and ELF electromagnetic fields and living organisms, J. Bioelectricity 4:133-140, 1985.

II

ELECTROMAGNETIC ENERGY AND REGULATION OF LIFE PROCESSES

2
Bacterial Biomagnetism and Geomagnetic Field Detection by Organisms

RICHARD P. BLAKEMORE, NANCY A. BLAKEMORE, RICHARD B. FRANKEL*

Department of Microbiology, University of New Hampshire
Durham, New Hampshire, and *Francis Bitter
National Magnet Laboratory, M.I.T.
Cambridge, Massachusetts

INTRODUCTION

The Earth is a magnet. Its dipole character results from massive currents within the molten portion of its core. These currents, driven presumably by gravitational energy, induce, in the manner of a self-sustaining dynamo, a global dipolar magnetic field with a magnitude of roughly 0.7 gauss at the poles. Although the ancient Chinese were familiar with the polar alignment of magnetized needles, geomagnetism became science with the publication in 1600 of William Gilbert's classic exposition *De Magnete, Magneticisque Corporibus Et De Magno Magnete Tellure: Physiologia Nova, Plurimis & Argumentis & Experimentis Demonstrata*. Gilbert's predecessor, Peter Peregrinus de Maricourt in his *Epistola de Magnete* of 1269, had noted that a magnetized needle (compass) left free to float on water, merely rotates, coming to rest with its axis lying in the north-south plane, and is not pulled in a northward direction. He did not perceive that the source of the magnetism causing the compass deflection was the Earth itself. Other predecessors of William Gilbert had believed such magnetism was extraterrestrial or was due to some remote "magnetic mountains." Gilbert fashioned lodestone spheres which he called terrellas or little Earths; a term indicating his suspicion that the Earth itself was a magnet. By studying the interactions between his terellas and small bits of iron wire, he arrived at a novel and experimentally based philosophy of the attractive behavior or "coition" of ferromagnets, and presented in his book the first inductive rationale for the concept of terrestrial magnetism.

The direction of a magnetic field is, by convention, the direction in which the north-seeking end of a compass needle points. Gilbert showed that this is inward (downward) at the Earth's north geographic pole and outward (upward) at the south pole. Thus, the geomagnetic field inclines upward in the southern hemisphere, is totally horizontal at the magnetic equator, and is inclined downward in the northern hemisphere. Earth's magnetic field intensity has been more-or-less constant during the 3.5 billion year history of life on Earth. Evidence preserved in the paleomagnetic record of sediments indicates that changes of the geomagnetic field direction associated with meanderings and reversals of the magnetic poles have been gradual. Major dipole reversals require thousands of years.

It is not surprising that organisms have adapted to exploit geomagnetism as a directional cue for guidance in migration and homing. However, it was not until the 1970's that good experimental evidence was obtained that animals sense the Earth's magnetic field.

ANIMAL ORIENTATION AND HOMING

Keeton (1,2) demonstrated that small bar magnets, but not brass control bars, fastened near the heads of homing pigeons disrupted their homing ability under overcast conditions. (On sunny days pigeons use the sun for navigation.) Walcott et al. (3) extended and confirmed these field studies using small energized coils to produce uniform or homogeneous magnetic fields in the pigeon's head region. These scientists learned that pigeons could be disoriented with uniform magnetic fields directed antagonistically to Earth's. On overcast days pigeons flew directly away from the home loft direction if the experimentally contrived magnetic field in their head region was directed upward. Upward, as mentioned previously, is anti-parallel to the direction of the normal geomagnetic field in the northern hemisphere in which these studies were conducted. Pigeons released at sites of magnetic anomaly did not home well, either with or without attached magnets or coils, suggesting that birds may also rely upon a magnetic map in addition to a magnetic compass (2,4). Studies by Lindauer and Martin in the same decade revealed that honeybees incorporate information about the Earth's magnetic field into their tail-wagging dances to communicate direction of nectar sources (5,6). Their comb-building activities (7) and circadian rhythms are also examples of behavior influenced by the geomagnetic field (8).

In more recent studies, other migratory birds including the European robin (9) and bobolink (10) have been shown by means of conditioned behavior responses to detect the geomagnetic field. The list of organisms either known or highly suspected on the basis of behavior experiments, to be able to sense the geomagnetic field also includes planaria (11), mud snails (12), salamanders (13), elasmobranch fishes (14,15), yellowfin tuna (16), woodmice (17), and possibly humans (18,19). In the case of humans however, the results appear equivocal and controversial at the present time.

How do creatures detect the geomagnetic field? One strategy which appears to have evolved among elasmobranch fishes makes use of the Faraday effect. Kalmijn (15) showed that elasmobranchs could be trained to respond to changes in the geomagnetic field. He postulated that when swimming or drifting at right angles to the Earth's magnetic field at 100 cm/sec, these creatures could induce field gradients in their head region of 0.4 μV/cm. Fields of only 0.01 μV/cm were sufficient to elicit electrophysiological responses in these animals. Thus, sharks, skates and rays appear to detect the geomagnetic field by transducing magnetic information to electrical information which they can sense using special electroreceptive organs called ampullae of Lorenzini in their snouts. This electromagnetic inductive mechanism requires that the animal be in a highly conducting medium such as seawater. For aerial animals, the Faraday effect would require a circular electrically-conducting loop of millimeter dimensions within the animal's tissues (20). Hence, it is not a likely candidate for geomagnetic field detection in birds, insects or terrestrial life forms because no structural evidence for such conducting loops exists.

A second way in which creatures might sense geomagnetism is through direct magnetic dipole interactions with the Earth's field. To do so, organisms would need a permanently magnetic substance within their tissues. However, until 1975, the only known instance of strongly magnetic material in a biological system was the mineral capping of chiton teeth. The abrasion-resistant layer on each of the teeth of the rasping organ of these primitive marine mollusks had been shown by Lowenstam to consist of the dense, hard, mineral magnetite (21).

Impetus for considering that cells might be permanently magnetized came by surprise. In 1975 Blakemore reported a new taxis or behavior type in bacteria (22). The term "magnetotaxis" was used to denote the directed swimming of bacterial cells along magnetic field lines. This behavior is dramatic and unequivocal. When examined with a dark-field microscope at 40-100 X (inexpensive hand-held microscopes may be used), one can see in muds from marshes and lakes, millions of active, highly refractile bacteria. Some move toward one side of their water-drop world and their wobbling motion straightens into nearly uni-directional swimming when a magnetized object is brought near them. If the observer is looking at such magnetotactic bacteria present in northern hemisphere sediments they are observed to swim toward the pole of a magnet which attracts the north-seeking end of a magnetic compass, and away from the pole which attracts the south-seeking end. Magnetotactic bacteria in the southern hemisphere do the opposite; they swim toward the end of a magnet which attracts the south-seeking end of a compass needle (23,24). Although they cannot swim, dead magnetotactic bacteria are also aligned in a uniform magnetic field. Just as Peregrinus' magnetized needles, living magnetotactic bacteria passively align in the geomagnetic field and consequently swim preferentially along magnetic field lines by ordinary means using their flagella; they act as swimming compass needles. It is important to note that the bacteria are not pulled northward because the Earth has a uniform magnetic field.

THE MAGNETOSOME

The possibility that magnetotactic bacteria were permanently magnetized seemed likely because each of the dozens of cell types examined by electron microscopy contained cytoplasmic crystals containing iron (22,23). These regularly shaped, enveloped structures were later shown to consist of crystalline magnetite or lodestone, an iron oxide mineral (25,26). They were subsequently named "magnetosomes" (27). In forms in which they have been studied, magnetosomes are enveloped single crystals of the iron oxide magnetite (25,27,29,30). Each is a single magnetic domain with a crystal size approximately 400–1000 Å, depending upon the species. Consequently, individual magnetosomes are too small to be seen within the cells observed with the light microscope. Their high iron content, however, renders them quite impenetrable by electrons and they are easily visualized even in unstained cells by means of electron microscopy. Recently, magnetotactic algae were discovered in Brazil (31,32). Each of these single-celled eukaryotic microorganisms possesses thousands of magnetosomes arranged in rows along the long cell axis. Magnetosomes within a given strain or cell type are homogeneous in grain size, and are uniform in shape and arrangement within the cell. This species specificity argues for genetic control of biogenic magnetite formation. The maximum size of the magnetosome within a given bacterial species is limited by an unknown mechanism. The number of magnetosomes per cell, however, can vary in response to culture conditions including iron supply and dissolved oxygen. For instance, the average number of magnetosomes within cells of a magnetic spirillum species varied from 0–17 in response to culture oxygen tension, and optimal numbers were produced under microaerobic conditions (33).

Several morphologically distinct types of magnetosomes have been observed within various types of magnetotactic microorganisms. Magnetosomes within Aquaspirillum magnetotacticum are truncated octahedral prisms (28). Magnetosomes within coccoid cells studied by Mann et al. (29,34) as well as those within an unidentified cell from a pond in Japan (30) were truncated hexagonal prisms. The prismatic crystals of either hexagonal or octahedral type were oriented with their easy axes of magnetization along the chain axis (e.g., [111] faces adjacent). The crystal morphology of tear-drop or bullet shaped magnetosomes found in some bacterial species and in a magnetotactic algal species (see below) is completely unknown.

In some cell types the magnetosomes occur in clusters predominantly at one side of the cell. In others the magnetosomes occur as a string or chain of particles arranged along the axis of cell motility. The magnetosomes situated at ends of such chains are often smaller. This suggests that magnetosome chains grow bidirectionally along their long axis as iron newly transported into the cell is transformed into magnetite. At cell division, whether they exist in chains or not, magnetosomes appear to be partitioned between each daughter cell. Thus, bacteria and algae control the iron biomineralization process thereby determining the magnetosome crystal size, morphology, structure, chemical composition, arrangement and crystallographic orientation within the cell (28,35). This is a splendid example of natural selection as there are no readily apparent physical or chemical reasons why constraints on these unique features of biogenic magnetite should exist.

MAGNETOTAXIS

Magnetosomes are unequivocally responsible for the magnetotactic response of microorganisms. Mutants of magnetotactic bacteria have been obtained which do not synthesize magnetosomes. These are fully motile but not magnetotactic. With both bacteria and algae, the arrangement and volume of magnetite present within each cell is more than enough to align it passively in the Earth's field of 0.5 gauss. The ratio of magnetic to thermal energy ($\mu B/kT$) is greater than 10 for the bacteria and greater than 100 for the algae. Thus, each cell's magnetic moment easily overrides the effect of Brownian motion caused by thermal agitation which tends to randomize cell orientation in water (36,37). Moreover, the ability to remagnetize the cells by means of a brief, monophasic magnetic pulse of several hundred gauss and thereby instantaneously reverse their swimming direction without cell turning provided unequivocal proof that the magnetotactic behavior of these organisms is due to ferromagnetism (22,24).

The geomagnetic field over most of the Earth is inclined from the horizontal (e.g. it has an angle of dip). The vertical component of the local geomagnetic field exerts strong selective pressure on natural populations for cells with a direction of magnetization tending to direct them downward along the inclined field lines (23,24,36,38,39). This was first evident with northern hemisphere monopolarly flagellated forms which persistently swam forward and in the magnetic field direction (e.g. the direction indicated by the north-seeking end of a compass needle), and was further substantiated by field observations which revealed that cells in southern hemisphere natural populations were of opposite magnetic polarity to those in the northern hemisphere. Consequently, magnetotaxis tends to direct unidirectionally swimming cells downward in each hemisphere. Some magnetotactic bacteria are bipolarly flagellated and swim principally along the inclined geomagnetic field lines but in either direction. The direction actually taken at any instant depends not only upon magnetism but also upon other taxes. Aerotaxis, for instance, has been shown to override magnetotaxis in bipolarly flagellated magnetotactic spirilla (40). The observed effect of Earth's magnetic field in orienting cells so that they may swim preferentially downward is consistent with their observed natural distribution. They are found in sediments and in the sediment-water interface, not in surface films or the surface micro-layer.

FORMS OF IRON IN MAGNETIC BACTERIA

The most intensively studied magnetotactic organism is the bacterium *A. magnetotacticum* (41,42). This chemoheterotroph is a microaerophilic denitrifying (43,44) nitrogen fixer (45). On the basis of extensive spectroscopic analysis, cells of *A. magnetotacticum* are known to contain ferrous ions, a low-density hydrous-ferric-oxide, a high-density hydrous-ferric-oxide (ferrihydrite) and Fe_3O_4. Additional experiments with cell fractions show that ferrihydrite in the magnetotactic cells is associated with the magnetosomes (46). It has been proposed that *A. magnetotacticum* precipitates Fe_3O_4 in the sequence: Fe^{3+} quinate → Fe^{2+} low-density hydrous-ferric-oxide → ferrihydrite → Fe_3O_4. In non-magnetic cells the process stops with ferrihydrite. In cells of the cloned, nonmagnetotactic strain the process stops with low-density hydrous-ferric-oxide.

In the proposed sequence, iron enters the cell as Fe^{3+} chelated by quinic acid. Reduction to Fe^{2+} releases iron from the chelator. Fe^{2+} is reoxidized and accumulated as the low-density hydrous-iron-oxide. By analogy with the deposition of iron in the micellar cores of the protein ferritin, this oxidation step might involve molecular oxygen, which as noted previously, is required for Fe_3O_4 precipitation in A. magnetotacticum (33). Dehydration of the low-density hydrous-ferric-oxide results in ferrihydrite. Finally, partial reduction of ferrihydrite and further dehydration yield Fe_3O_4.

In high resolution TEM lattice imaging studies (28), no other crystalline phases in addition to Fe_3O_4 were detected. However, in some magnetosomes, noncrystalline material was found contiguous with the Fe_3O_4. This suggests that the hydrous-ferric-oxide phase is amorphous ferrihydrite, and that final crystallization of Fe_3O_4 occurs as a solution-reprecipitation process, possibly triggered by Fe^{2+} ions.

Additional experiments demonstrate that while the hydrous-ferric-oxide is primarily associated with magnetosomes, Fe^{2+} in the cell is very probably associated with the peptidoglycan wall layer of the cell (47). This association could occur during the conversion from the iron quinate complex outside the cell to ferric iron and ultimately to Fe_3O_4 within the cell.

Fe_3O_4 is thermodynamically stable with respect to hematite and ferrihydrite at low E_H and high pH (48). However, rapid transformation of ferrihydrite to magnetite appears to involve more than simple reduction and dehydration. While the degree of crystallinity of ferrihydrite can vary, in crystalline samples it has a structure related to hematite, with hexagonal close-packed oxygen atoms and Fe^{3+} octahedrally coordinated sites. Fe_3O_4 has a cubic, inverse spinel structure with Fe^{3+} in octahedral and tetrahedral sites, and Fe^{2+} in octahedral sites. This, plus the fact that the precipitation process requires spatial segregation of regions of differing E_H and possibly pH, suggests that the process falls into the biomineralization category described by Lowenstam (35) as "organic-matrix mediated." Thus the magnetosome envelope is probably an integral element in the precipitation process, functioning as a locus for enzymatic activities, compartmentalizing constituents, providing control of E_H and pH, as well as comprising a structural element anchoring the Fe_3O_4 particles to the remainder of the cell.

MAGNETITE IN EUKARYOTES

The unexpected finding that certain bacterial cells were geomagnetically responsive, were permanently magnetized and contained iron-rich structured particles (22), precipitated a search for permanent magnetic material in other organisms; particularly those known from behavioral studies to be able to sense geomagnetism. The results proved extremely rewarding. Gould et al. (49), using sensitive rock magnetometers, discovered magnetite in honeybees as did Walcott et al. in pigeons (50). Other groups of workers have located magnetic material in migratory birds such as bobolinks (10), buntings and sparrows (51), in Monarch butterflies (52), green sea turtles (53), yellowfin tuna (54), woodmice (17), dolphins (55), cetaceans (56), and humans (57,58). In the case of honeybees, it may be iron deposits discovered within abdominal cells which play a role in magnetic field detection (59). In the yellowfin tuna, single magnetic domain-sized magnetite particles similar to those of magnetotactic bacteria were found in the skull bone (54). Several of these fish were recently trained to discriminate in their swimming response between the presence of one as compared to two Earth-strength magnetic fields in their tanks (16). Despite these encouraging results, a direct connection between the presence of magnetite in animal tissues and geomagnetic responsiveness of animals has yet to be demonstrated as it has for magnetotactic bacteria and algae.

MAGNETOTACTIC ALGAE

TEM of magnetotactic saprozooic (non-photosynthetic) euglenoid algal cells magnetically separated from brackish sediments in Brazil (32) shows that they contain numerous Fe_3O_4 particles arranged in chains oriented more or less parallel to the long axis of the cell. Individual particles are arrowhead or tooth-shaped and are within the single magnetic domain size range for Fe_3O_4. Hence, each chain is a permanent magnetic dipole. If the moments of all the chains are oriented parallel to each other, a cell would have a geomagnetic dipole moment equal to the sum of the moments of all its particles. An estimate of the total magnetic moment M of algal cells gives $M = 5 \times 10^{-10}$ emu.

This is about 1000 times the moment of a typical magnetic bacterium, and corresponds to a total of about 3 x 10^3 aligned particles of the observed dimensions.

The biological significance of magnetotaxis in these algae (32,60) is not yet understood. However, highly ordered arrangement of the chains of particles in the cells suggests that they are chains of magnetosomes very much like the chains of magnetosomes in bacteria. Evidence for the presence of membranes enveloping the particles must await TEM of thin sections.

Thus, eukaryotic cells as well as prokaryotic cells can produce biogenic Fe_3O_4 in the form of single magnetic domains as an intracellular biomineralization product. It will be interesting to compare the biomineralization process and the role(s) of membranes in these fundamentally different types of organisms.

Recent discoveries of biogenic magnetite in deep sea sediments (61,62) are exciting, suggesting that these particles are the major contributors to the paleomagnetic record of sediments. Because magnetosomes appear to be formed only with O_2 available (33), they may also provide unique fossilized information concerning sedimentation processes which have occurred since the transition on Earth from an anoxic to aerobic atmosphere.

Obviously, much remains to be discovered concerning the manner in which unicellular and multicellular organisms sense, respond to, and use magnetite and the geomagnetic information in which they are constantly bathed. It is ironical though, that lodestone, the very substance used by the twelfth century Chinese to make compasses and also used by the Late Renaissance scholars to understand the magnetic character of Earth, may now help us understand how some living organisms use geomagnetism in their life activities.

ACKNOWLEDGEMENTS

RPB and NAB were partially supported by NSF Grant DMB 85-15540 and ONR Contract N00015-85-K-0502. RBF was partially supported by ONR Grant N000-85-K-0505.

REFERENCES

(1) Keeton, W.T.: Magnets interfere with pigeon homing, Proc. Natl. Acad. Sci. USA, 68:102-106, 1971.

(2) Keeton, W.T., Larkin, T.S. and Windsor, D.M.: Normal fluctuations in the earth's magnetic field influence pigeon orientation, J. Comp. Physiol. 95:95-103, 1974.

(3) Walcott, C.: Anomalies in the Earth's magnetic field increase the scatter of pigeon's vanishing bearings, in Animal Migration, Navigation and Homing, K. Schmidt-Koening and W.T. Keeton, eds., Springer, Heidelberg, 1978.

(4) Gould, J.L.: The map sense of pigeons, Nature 296:205-211, 1982.

(5) Lindauer, M. and Martin, H.: Magnetic Effects on Dancing Bees, in Animal Orientation and Navigation, S.R. Galler et al., eds., U.S. Gov't. Printing Office, Washington, DC, 1972.

(6) Martin, H. and Lindauer, M.: Der Einfluss des Erdsmagnetfelds auf die Schwereorientierung der Honigbiene (*Apis mellifica*), J. Comp. Physiol. 122:145-187, 1977.

(7) De Jong, D.: The orientation of comb-building by honeybees, J. Comp. Physiol. 147:495-501, 1982.

(8) Dyer, F.C. and Gould, J.L.: Honey bee orientation: A backup system for cloudy days, Science 214:1040-1041, 1981.

(9) Wiltschko, W. and Wiltschko, R.: Magnetic compass of European robins, Science 176:62-64, 1972.

(10) Beason, R.C. and Nicholls, J.E.: Magnetic orientation and magnetically sensitive material in a transequatorial migratory bird, Nature 309:151-153, 1984.

(11) Brown, A., Jr.: Responses of the planarium, *Dugesia*, and

the protozoan, Paramecium, to very weak horizontal magnetic fields, Biol. Bull. 123:264-281, 1962.

(12) Brown, F.A. Jr., Webb, H.M. and Barnwell, F.H.: A compass directional phenomenon in mud-snails and its relation to magnetism, Biol. Bull. 127:206-220, 1964.

(13) Phillips, J.B.: Use of the Earth's magnetic field by orienting cave salamanders (Eurycea lucifuga), J. Comp. Physiol. 121:272-288, 1977.

(14) Kalmijn, Ad.J.: The electric and magnetic sense of sharks, skates, and rays, in Oceanus, 1977.

(15) Kalmijn, Ad.J.: Experimental evidence of geomagnetic orientation in elasmobranch fishes, in Animal Migration, Navigation, and Homing, K. Schmidt-Koening and W.T. Keeton, eds., Springer, Heidelberg, 1978.

(16) Walker, M.M.: Learned magnetic field discrimination in yellowfin tuna, Thunnus albacores, J. Comp. Physiol. 155:673-679, 1984.

(17) Mather, J.G. and Baker, R.R.: Magnetic sense of direction in woodmice for route-based navigation, Nature 291:151-155, 1981.

(18) Baker, R.R.: Goal orientation by blindfolding humans after long-distance displacement: Possible involvement of a magnetic sense, Science 210:555-557, 1980.

(19) Gould, J.L. and Able, K.P.: Human homing: An elusive phenomenon, Science 212:1061-1063, 1981.

(20) Jungerman, R.L. and Rosenblum, B.: Magnetic induction for the sensing of magnetic fields by animals — An analysis, J. Theor. Biol. 87:25-32, 1980.

(21) Lowenstam, H.A.: Magnetite in denticle capping in recent

chitons (Polyplacophora), Geol. Soc. Amer. Bull. 73:435-438, 1962.

(22) Blakemore, R.P.: Magnetotactic bacteria, Science 190:377-379, 1975.

(23) Blakemore, R.P.: Magnetotactic bacteria, Ann. Rev. Microbiol. 36:217-238, 1982.

(24) Blakemore, R.P., Frankel, R.B. and Kalmijn, Ad.J.: South-seeking magnetotactic bacteria in the southern hemisphere, Nature 286:384-385, 1980.

(25) Frankel, R.B., Blakemore, R.P. and Wolfe, R.S.: Magnetite in freshwater magnetotactic bacteria, Science 203:1355, 1979.

(26) Towe, K.M. and Moench, T.T.: Electron-optical characterization of bacterial magnetite, Earth Planet Sci. Lett. 52:213-220, 1981.

(27) Balkwill, D.L., Maratea, D. and Blakemore, R.P.: Ultrastructure of a magnetotactic spirillum, J. Bacteriol. 141:1399-1408, 1980.

(28) Mann, S., Frankel, R.B. and Blakemore, R.P.: Structure, morphology and crystal growth of bacterial magnetite, Nature 310:405-407, 1984.

(29) Mann, S., Moench, T.T. and Williams, R.J.P.: A high resolution electron microscopic investigation of bacterial magnetite: Implications for crystal growth, Proc. Royal Soc. (London) 221:385, 1984.

(30) Matsuda, T., Endo, J., Osakube, N., Tonomura, A. and Arii, T.: Morphology and structure of biogenic magnetite particles, Nature 302:411, 1983.

(31) Lins de Barros, H.G.P., Esquival, D.M.S., Danon, J. and de Oliveira, L.P.H.: Magnetotactic algae, in Acad. Brasileria Notas de Fisica DBPF-NF-48, 1981.

(32) Torres de Araujo, F.F., Pires, M.A., Frankel, R.B. and Bicudo, C.E.M.: Magnetite and magnetotaxis in algae, Biophys. J., in press.

(33) Blakemore, R.P., Short, K.A., Bazylinski, D.A., Rosenblatt, C. and Frankel, R.B.: Microaerobic conditions are required for magnetite formation with Aquaspirillum magnetotacticum, Geomicrobiol. J. 4:62, 1984.

(34) Moench, T.T. and Konetzka, W.A.: A novel method for the isolation and study of a magnetotactic bacterium, Arch. Microbiol. 119:203, 1978.

(35) Lowenstam, H.A.: Minerals formed by organisms, Science 211:1126, 1981.

(36) Frankel, R.B.: Magnetic guidance of organisms, Ann. Rev. Biophys. Bioeng. 13:85, 1984.

(37) Frankel, R.B. and Blakemore, R.P.: Navigational compass in magnetic bacteria, J. Mag. Mag. Mater. 15-18:1562-1564, 1980.

(38) Blakemore, R.P. and Frankel, R.B.: Magnetic navigation in bacteria, Sci. Am. 245:58-65, 1981.

(39) Frankel, R.B., Blakemore, R.P., Torres de Araujo, F.F., Esquivel, D.M.S. and Danon, J.: Magnetotactic bacteria at the geomagnetic equator, Science 212:1269, 1981.

(40) Spormann, A.M. and Wolfe, R.S.: Chemotactic, magnetotactic and tactile behavior in a magnetic spirillum, FEMS Lett. 22:171-177, 1984.

(41) Blakemore, R.P., Maratea, D. and Wolfe, R.S.: Isolation and pure culture of a freshwater magnetic spirillum in chemically defined medium, J. Bacteriol. 140:720, 1979.

(42) Maratea, D. and Blakemore, R.P.: Aquaspirillum magnetotacticum sp. nov., a magnetic spirillum, Int. J. Syst. Bacteriol. 31:452, 1981.

(43) Escalente-Semerena, J.C., Blakemore, R.P. and Wolfe, R.S.: Nitrate dissimilation under microaerophilic conditions by a magnetic spirillum, Appl. Environ. Microbiol. 40:429, 1980.

(44) Bazylinski, D.A. and Blakemore, R.P.: Denitrification and assimilatory nitrate reduction in Aquaspirillum magnetotacticum, Appl. Environ. Microbiol. 46:1118, 1983.

(45) Bazylinski, D.A. and Blakemore, R.P.: Nitrogen fixation (acetylene reduction) in Aquaspirillum magnetotacticum, Curr. Microbiol. 9:305, 1983.

(46) Frankel, R.B., Papaefthymiou, G.C., Blakemore, R.P. and O'Brien, W.: Fe_3O_4 precipitation in magnetotactic bacteria, Biochim. Biophys. Acta 763:147, 1983.

(47) Ofer, S., Nowick, I., Bauminger, E.R., Papaefthymiou, G.C., Frankel, R.B. and Blakemore, R.P.: Magnetosome dynamics in magnetotactic bacteria, Biophys. J. 46:57, 1984.

(48) Garrels, R.M. and Christ, C.L.: Solution, Minerals and Equilibrium, Harper and Row, New York, 1965.

(49) Gould, J.L., Kirschvink, J.L. and Deffeyes, K.S.: Bees have magnetic remanence, Science 201:1026-1028, 1978.

(50) Walcott, C., Gould, J.L. and Kirschvink, J.L.: Pigeons have magnets, Science 206:1027-1029, 1979.

(51) Ueda, K., Kusunoki, M., Kato, K., Kakizawa, R., Nakamura, T., Yaskawa, K., Koyama, M. and Maeda, Y.: Magnetic remanences in migratory birds, J. Yamashina Inst. Ornith. 14:166-172, 1982.

(52) Jones, D.S. and MacFadden, B.J.: Induced magnetization in the monarch butterfly, Danaus plexippus (Insecta, Lepidoptera), J. Exp. Biol. 96:1-9, 1982.

(53) Perry, A., Bauer, G.B. and Dizon, A.E.: Magnetite in the green turtle, EOS Trans. Amer. Geophys. Union 62:850, 1981.

(54) Walker, M.M., Kirschvink, J.L., Chang, S.B.R. and Dizon, A.E.: A candidate magnetic sense organ in the yellowfin tuna, Thunnus albacares, Science 224:751-753, 1984.

(55) Zoeger, J., Dunn, J.R. and Fuller, M.: Magnetic material in the head of the common Pacific dolphin, Science 213: 892-894, 1981.

(56) Bauer, G.B., Perry, A, Fuller, M., Dunn, J.R., Zoeger, J. and Herman, L.M.: Biomagnetic studies of cetaceans, in Abstracts 5th Biennial Conference on the Biology of Marine Animals, Soc. Marine Mammals, Boston, MA, 1983.

(57) Baker, R.R., Mather, J.G. and Kennaugh, J.H.: Magnetic bones in human sinuses, Nature 301:78-80, 1983.

(58) Kirschvink, J.L.: Ferromagnetic crystals (magnetite?) in human tissue, J. Exp. Biol. 92:333-335, 1981.

(59) Kuterbach, D.A., Walcott, B., Reeder, R.J. and Frankel, R.B.: Iron-containing cells in the honey bee (Apis mellifera), Science 218:694-697, 1982.

(60) Farina, M., Lins de Barros, H., Esquivel, D.M.S. and Danon, J.: Ultrastructure of a magnetotactic microorganism, Biol. Cell 48:85, 1983.

(61) Peterson, N., Dobeneck, T.V. and Vali, H.: Fossil bacterial magnetite in deep sea sediments from the South Atlantic Ocean, Nature 320:611-615, 1986.

(62) Stolz, J.F., Chang, S-B.R. and Kirschvink, J.L.: Magnetotactic bacteria and single-domain magnetite in hemipelagic sediments, Nature 321:849-851, 1986.

3
The Earth's Magnetic Field as a Navigational Cue

WILLIAM E. SOUTHERN
Department of Biological Sciences
Northern Illinois University
DeKalb, Illinois

INTRODUCTION

It is well known by biologists, paraprofessionals and lay persons that many birds and a variety of other vertebrates (e.g., fishes, marine mammals) migrate annually. This means that these species make an annual round trip between their breeding and nonbreeding ranges. In many instances the distances traversed can be measured in thousands of kilometers and the goals at each end of the migration route frequently are precise, with individual birds, for example, returning to the same hectare of habitat year after year. Numerous studies involving marked animals, either banded or otherwise individually recognizable, have documented the distances traveled, the seasonal ranges of particular species and individuals, and the seasonal periodicity of such movements (1,2). The migratory ability of birds and the associated direction-finding methodologies have been studied more thoroughly than similar behavior in other vertebrates. Most of this review, therefore, will be devoted to the state of our knowledge about avian orientation and navigation.

Perhaps two-thirds of temperate North America's bird species make annual round trips between their seasonal ranges which average 2000-6000 km (3). Individuals of some species travel much farther, and the transequatorial journeys of some shorebirds exceed 20,000 km annually. These seasonal migratory flights are weather dependent, and often the route is broken into a series of discrete segments. Usually the journey will take several weeks but considerable inter- and intraspecific variability exists.

Some avian species migrate nocturnally, others are diurnal migrants, and some are capable of migrating during day or night. The altitude at which migrating birds fly varies greatly and is related to species and weather conditions. Radar evidence indicates that most migrants fly at elevations below 3000 meters but some, such as shorebirds and geese, may migrate at altitudes up to about 10,500 meters (3).

Because of the nature of these feats, it follows that migratory birds must possess some mechanism for determining and maintaining a given direction of travel. They should possess, in other words, an internal compass which uses one or more of the available environmental cues (e.g., geomagnetism, stars, sun) to know and follow a particular heading. It appears equally necessary that migrants possess some means of recognizing where they are spatially, and the relationship between a given location and the preferred destination; that is, for a bird to accurately navigate between two distant points a map component seems essential. Investigators have used a variety of approaches over the last 35 years or so to determine which of the available sources of environmental information are used as navigational cues, how they are used, and when (seasonally or developmentally) particular cues are used.

It is worth emphasizing that all species do not have the same direction-finding needs and abilities. Even within a species, orientational ability may vary from season to season, day to day, and bird to bird. For example, the orientation mechanisms used during local foraging flights are probably different from those used during long-distance migratory travel.

Griffin (4) described three types of homing ability which require increasingly sophisticated mechanisms. The following homing types of Griffin (4) discern between piloting, compass orientation, and true navigation: Type 1, or piloting, is the finding of a goal by referring to familiar landmarks during random or systematic searches; Type 2, or compass orientation, is the ability to orient in a given compass direction without use of landmarks; and Type 3, or true navigation, is the ability to select a direction toward a goal or home, thus necessitating a map and a compass.

During the 1950's and 1960's there was a tendency among researchers to seek a single solution to the orientation puzzle;

that is, there was a tendency to believe that a common denominator existed and that all or most birds used the same mechanism for orienting or navigating. When I entered the field in the early 1960's, the findings of Matthews (5) and Sauer (6) were heralded as solutions to the avian navigation riddle. Matthews' sun arc hypothesis was envisioned as the answer to how diurnal migrants navigated, and Sauer's findings about the use of stellar cues were considered the answer to the question of how nocturnal migrants navigated. This is far from being the consensus of opinion today.

Later studies revealed that the problem was much more complex than originally believed. It became apparent that some or all avian migrants, as well as homing pigeons, possess the ability to use more than one kind of environmental information for orientational or navigational purposes. A system of multiple cues enables birds to navigate under a variety of environmental conditions and in different geographical areas. It is likely that a redundancy of cues exists which provides birds with improved accuracy as well as flexibility. It appears that the array of cues birds may use for navigational purposes (e.g., landmarks, wind direction, sun, stars, geomagnetism, polarized light) are hierarchically arranged in the bird's repertoire. Those used most regularly, possibly because they result in higher accuracy or are more dependably available, appear to be at the top of the hierarchy.

Before it was generally recognized that birds had backup systems for direction finding, investigators were besieged by contradictory evidence from similarly designed studies. For example, controversy resulted from the inability of any other investigator to replicate the findings of Yeagley (7), which showed that pigeons were using magnetic cues during homing flights. This brought most studies dealing with magnetism to an end, and it was not until about 24 years later that Keeton (8) repeated part of Yeagley's experiment and found that pigeons used backup cues (e.g., the sun) when the Earth's magnetic field was disrupted. Thus, latency in our ability to comprehend that birds might be able to use more than a single type of environmental information for direction-finding purposes resulted in most hypotheses dealing with geomagnetic cues being placed on the shelf by all but a few tenacious investigators. The revival of the hypothesis that birds possessed a magnetic compass occurred in the late 1960's, and it came into its own during the next decade.

In this chapter, I review the studies that have contributed to our understanding of the avian magnetic compass, and attempt to place this method of orientation within the hierarchy of cues we recognize as being available to birds. Although I will generalize frequently and refer to birds as a group, it must be recognized that we have experimental data for only a few migrant species and variations on this theme are to be expected.

This is not the first review of this subject as several authors (5,9-19) have included a discussion of magnetic cues as part of reviews of avian orientation or as papers. To my knowledge only one other review has been devoted solely to the subject of bird orientation and geomagnetic cues (20). This chapter provides an updated review, elaborates on some of the studies, and places in chronological order a majority of the contributions leading to the present, although cautious, acceptance of the hypothesis that birds possess a magnetic compass. Evidence for a magnetically-based map is still circumstantial, although experiments clearly show that birds are able to solve navigational-type problems.

HISTORICAL PERSPECTIVE

Although the homing and navigational abilities of birds were not tested experimentally until well into the twentieth century, the importance of a compass to migrants was apparent to several nineteenth-century investigators. The likelihood that geomagnetism served as the source of directional information was proposed by Middendorff in 1859 (21), Viguier in 1882 (22), and Thauzies in 1898 (23). Viguier suggested that birds could detect and measure magnetic intensity, inclination, and declination which would provide a complex grid of the isolines. This contention was pursued by other investigators during the next three or four decades (24-26).

Opposition to the notion that birds could sense and use the Earth's magnetic field for orientational purposes developed during the 1920's and 1930's, and the results of experiments supported such views. Casamajor (27) and Wodzicki et al. (28)

attached magnets to the heads of homing pigeons and storks and found that this had no effect on their homing ability. Rochon-Duvigneaud and Maurain (29) also argued against the possibility, but on theoretical rather than experimental grounds. The opposing viewpoints were forcefully presented and as a consequence the contention that birds possessed a magnetic compass was shelved until Yeagley (7) addressed the subject.

Yeagley's (7) first experiment was conducted in 1943. At that time he attached hyflux-chrome magnets (0.172 Oersted) to the underside of both wings of the experimental pigeons. The control birds carried similarly attached copper plates of comparable mass. Twenty pigeons (10 experimental and 10 controls) were released singly on the same date at a site about 104 km from their loft. Eight (80%) of the controls returned within two days following release whereas only two (20%) of the pigeons bearing magnets returned. Notable differences also were recorded at the time each bird was released in the homing trial. The controls showed better initial orientation with their departure bearings deviating only 10-50 degrees from a direct line between the release site and the loft. In contrast, the experimental birds showed deviations in headings ranging between 45 and 180 degrees from the homeward bearing. In 1944, Yeagley (7) released 122 pigeons in similar experiments but different methods were used for presenting results. Rather than indicating the number of birds returning and not returning, he summed the vectors for the pigeons used in each release. Six (75%) of the total flight vectors supported the hypothesis that pigeons used the Earth's magnetic field during homing.

Yeagley (30) postulated that three factors were essential to pigeon navigation: 1) sensitivity to the effect of flying through the vertical component of the Earth's magnetic field; 2) sensitivity to the magnitude of the Coriolis effect which results from a natural relationship between the Earth's rotational speed and the speed of an object moving over the Earth's surface; and 3) visual sensitivity to velocity over the Earth's surface (ground speed). Yeagley proposed that by correlating the results of the first and third of these sensitivities, a bird could detect its magnetic latitude. By a similar correlation of the second and third sensitivities, the bird would be able to detect the true latitude of its location.

In spite of his early success, Yeagley's contention that pigeons were capable of using the Earth's magnetic field for orientational purposes soon encountered serious problems. Other investigators (31,32,33) working with pigeons and even Yeagley himself (34) were unable to reproduce his earlier results. In addition, studies with other species also produced negative results (35,36) and attempts to detect magnetic sensitivity in birds were unsuccessful (37-47).

The preponderence of negative results during this period contributed to the subject being tabled by all but a few researchers. Persistent work by biologists on invertebrates repeatedly produced results indicating that various species, ranging from planarian worms to insects, responded to weak magnetic fields (48-52). During the 1950's and early 1960's, the only proponents of the magnetic compass in birds were Merkel and his associates in Germany (53-56). These authors consistently found that European robins (Erithacus rubecula), during periods of migratory unrest (Zugunruhe), showed preferences for their natural migratory direction (SW in fall, NE in spring) when tested in orientation cages in the absence of celestial cues. Birds placed in an all-steel chamber, which had a shielding effect, were unable to maintain these bearings. The directional choice of the birds was reinstated when the investigators generated a magnetic field of the Earth's intensity in the steel chamber with Helmholtz coils. The findings of these investigators were not readily accepted because (a) the negative results produced by attempts to reproduce Yeagley's (7) early results convinced most researchers that birds could not perceive magnetic stimuli, (b) a receptor for magnetic stimuli had not been identified in birds, and (c) other investigators (57,58) were unable to duplicate the results reported by Merkel, Wiltschko, and their colleagues. Later, however, it was shown that methodological problems were responsible for the inability of some researchers to duplicate the results of Merkel and Wiltschko (55,59,60). The persistence of Merkel, Roswitha, and Wolfgan Wiltschko and their co-workers forced other investigators to seriously reconsider the possibility that birds, other than the European robin, were using a magnetic compass during orientation.

During 1962-1966, I obtained the first evidence suggesting that natural disturbances in the Earth's magnetic field might disrupt the orientational ability of free-flying wild birds (61). A total of 429 adult herring gulls (Larus argentatus) and ring-billed gulls (L. delawarensis) were transported in darkened containers to locations at various directions and distances from their breeding colony. Releases were made under a variety of environmental conditions (e.g., clear and overcast skies) in an attempt to determine which factors influenced homing success and speed of return. Some individuals were radio-tracked following release. The results showed that these experienced migrators were capable of homing under environmental conditions that were believed to render particular, supposedly essential, cues unavailable to them (e.g., heavy overcast obscuring solar cues) (62-66). This prompted the hypothesis that adult gulls, because of experience gained from using a variety of cues during previous migratory trips, possessed the ability to use more than one type of environmental information for direction-finding purposes. If a particular type of directional information was unavailable (e.g., the sun), they used the next most accurate method available to them. To test this hypothesis, 56 juvenile ring-billed gulls were released at sites 11-29 km from their breeding colony in 1964 and 1965. This was the maiden flight for each of the young gulls as they had never flown outside the confines of the colony site. The departure bearings of these individuals was principally to the east and southeast, which was not the case for adults of the same species (61,67). Such headings corresponded with the direction ring-billed gulls would take to reach their major winter range (68,69,70). If fledgling gulls (35+ days old) expressed preference for their future migrational bearing, it seemed possible that even younger gulls might respond similarly. This was tested by placing ring-billed gull chicks (about 3-20 days old) in a circular cage and plotting the route they walked when exposed to selected environmental conditions. The results from an initial set of trials with 294 chicks showed that they possessed an innate ability for selecting a course suitable for reaching the primary winter range of the population (67,70). There was no evidence that solar cues were being used as directional information by the gulls and so the possible effect of naturally-occurring geomagnetic disturbances was examined.

The National Geomagnetic Observatory's K-indices were used as a measure of magnetic disturbance. These values, ranging from 0-9, reflect the amount of disturbance in the Earth's field that is caused by solar flares, solar storms, and related phenomena. K-values occurring at release times, or the mean value for the preceeding 12 hours, were compared with the responses of gull chicks to determine if the bearings selected by the birds varied in accordance with different levels of geomagnetic disturbance. All but 8 of the initial trials were conducted during low levels of disturbance, and it was not clearly evident that modifications in the geomagnetic field were influencing directional responses. Additional tests were conducted and the results from these 333 trials were combined with the earlier data for analysis. During the second series of tests, K-values ranged between 0 and 7, with 7 representing a moderately severe magnetic storm. At this time, it was clearly demonstrated that although ring-billed gull chicks were able to select a preferred bearing of SE during minor disturbances (0-3 K) in the Earth's field, they were unable to do so during higher intensity storms (4-7 K)(71). Similarly tested herring gull chicks (a more sedentary species than ring-billed gulls) from an adjacent colony did not show as much disorientation as ring-billed gulls (72). There was, however, a difference between the mean directional preferences of the experimental and control birds, and the experimental group also showed greater variance in the direction chosen. Moore (72) concluded that magnetic stimuli altered but did not completely disrupt herring gull chick orientation. Later ring-billed gull experiments involving magnets and a magnetically-shielded room (Earth's field reduced by a factor of 25) also produced disorientation. Results from trials using Ruben's coils to simulate the Earth's field within the shielded chamber were inconclusive because of lighting and ventilation problems within the apparatus (73).

Keeton et al. (74) examined the directional tendencies of pigeons during periods of naturally-occurring magnetic disturbances. The results showed a significant inverse correlation with K-values. Since most of the magnetic fluctuations during their releases were less than 70 gamma, Keeton et al. (74) concluded that the sensitivity of pigeons to magnetic cues probably approached that already demonstrated for honeybees

(0–300 gamma)(75). In contrast, ring-billed gulls did not exhibit disorientation prior to disturbances measuring about 500 gamma (K=4)(73).

As the body of evidence grew in support of birds being able to derive directional information from the Earth's magnetic field, other investigators became interested. Old approaches were retested, and new methodology was applied. Keeton (8,76) replicated Yeagley's (7,30) experiment by placing magnets on the backs (instead of the wings as done by Yeagley) of experimental pigeons and brass weights on the controls. In contrast to Yeagley, however, Keeton also considered sky condition at the time of release. He found that both the experimentals and controls oriented homeward when released under sunny conditions. But when the experimental and control birds were released under overcast conditions, the controls oriented homeward but the experimentals usually departed randomly. This was the first solid experimental demonstration of the possible redundancy of cues, i.e., that birds could use both solar and magnetic cues. Young untrained pigeons wearing magnets were unable to orient homeward under either clear or overcast conditions, suggesting that (a) young pigeons are dependent upon a magnetic compass, and (b) the sun compass develops through experience. Keeton and Gobert (77) concluded that inexperienced pigeons required both sun cues and magnetic cues to orient homeward. This also may be the case with young ring-billed gulls (78,79).

Other studies during the 1970's supported the contention that birds indeed must be capable of perceiving the Earth's magnetic field and using it as a compass. Walcott (80) equipped pigeons with two small Helmholtz coils, one glued to the top of the bird's head and the other around its neck. The power supply was attached to the bird's back. This arrangement produced a magnetic field of approximately 0.1 gauss between the coils in the vicinity of the bird's head. Experimental birds exhibited vanishing bearings under sun that were usually more scattered than bearings of control birds wearing an unenergized set of coils. Larkin and Keeton (81) compared the responses of pigeons with attached magnets and those with brass weights during periods of natural magnetic disturbance. Both magnets and natural disturbances caused pigeons to shift their bearings slightly to the left. Following these results, Larkin and Keeton concluded there was a direct cause-and-effect relationship between fluctuations in the Earth's magnetic field and the variations in initial bearings of pigeons.

Particularly important were the results from the continuing and increasingly refined studies of W. Wiltschko and his colleagues. These studies indicated that European robins and Old World warblers (Sylvia spp.) showed oriented nocturnal migratory activity (Zugunruhe) in a seasonally appropriate direction even when deprived of vision of the natural environment (e.g., stars or sun), provided they were exposed to a magnetic field comparable to that of the Earth (82-93). Working in cooperation with the Wiltschko's, Emlen et al. (60) found that indigo buntings (Passerina cyanea) were able to orient in the appropriate migratory direction when exposed to a minimum of visual cues and normal geomagnetic stimuli. When the horizontal component of the magnetic field was deflected clockwise 120° with Helmholtz coils, the buntings shifted their orientation clockwise. Emlen et al. (60) concluded from these results that indigo buntings are able to detect the Earth's magnetic field and to use the resultant information in determining their migratory direction.

It is particularly difficult to design studies that will test for the effects of naturally occurring geomagnetic disturbances on free-flying migrants. Such studies, however, are extremely important. Moore (94) provided the first direct visual evidence that the orientation of free-flying nocturnal migrants was affected by natural fluctuations in the geomagnetic field. He found that the variability in flight directions of nocturnal migrants was significantly correlated with increasing geomagnetic disturbance as measured by both the K-index and various components of the Earth's magnetic field. As with all studies involving modifications in the geomagnetic stimuli birds receive (superimposing other fields with magnets or coils), these results do not show conclusively that birds use geomagnetism as an orientational cue. They do show, however, that disturbances in the Earth's field are in some way detected by birds and that in response they alter their direction of travel. Intuitively, it would make sense for migrant birds, which are specialized for long-distance travel, to have evolved filters for weak background disturbances (such as magnetic fluctuations measuring up to perhaps 1000 gamma) that affected the efficiency of migration. The fact that they have not done so suggests that this type of environmental information is important to them rather than being solely disruptive. The evidence available makes attractive the assumption that the Earth's magnetic field provides information essential to the avian compass or to the apparently necessary navigational map.

Ontogenetic studies involving young inexperienced migrants or homing pigeons have the potential for showing the innate ability of birds to orient or navigate. Such studies are based on the premise that young birds have a programmed directional choice or they develop an early attachment for the home loft. Studies have shown that the young of a number of avian species show apparently innate preferences for seasonally appropriate migratory bearings, although the goals at the ends of such routes must be learned (9,61,70,95,96). I demonstrated that the directional preferences expressed by young ring-billed gulls (preflight chicks and first-flight juveniles) corresponded to the bearing appropriate for reaching the winter range of the population of gulls studied (70). Selection of these bearings was influenced by changes in the Earth's magnetic field (magnetic storms) and by induced fields (magnets), and I proposed that young ring-billed gulls first used a magnetic compass and then developed an ability to use other environmental information to establish redundancy in the system and to increase accuracy (70,71,78,79,97). The relationship between the magnetic compass and cues learned during development were presented as a model (97).

Wiltschko et al. (98) have shown that the magnetic compass is involved in the learning process associated with establishment of the sun compass, but as yet they have not been able to unravel completely the complex relationship that apparently exists. Existence of the sun compass is well documented (15); the mechanism, which is based on a relationship between sun azimuth, time, and geographic direction, is learned rather than being innate. The magnetic compass is used by young pigeons before the sun compass is established, and it apparently provides the basis for the sun compass' geographic component which develops through experience (99-103). A similar mechanism has been reported for night-migrating birds wherein the star compass is influenced by, and possibly calibrated by, information obtained from the Earth's magnetic field (104). The evidence produced by R. Wiltschko and W. Wiltschko (105) demonstrates that the most important orientation mechanism during the first flights by young pigeons is the magnetic compass. Their studies show that young inexperienced pigeons use route reversal to find their way between an unfamiliar release site and the home locality. This means the birds use the magnetic field to determine the direction of the outward journey, and for determining the home direction upon release. Pigeons transported in a distorted magnetic field apparently were unable to obtain the necessary information while en route. Magnets attached to the pigeons upon release also caused disorientation.

The magnetic compass enables birds to establish and maintain a course but it does not provide them with the "map" component of a navigational system. This apparently is based upon learned information (106). Studies by Beck and Wiltschko (107,108) with pied flycatchers (Ficedula hypoleuca) show that the magnetic compass develops independent of any exposure to the sky, and that it provides an adequate mechanism for selecting the migratory direction. Selection and maintenance of a migratory bearing are two separate and independent processes. Determination of a particular bearing is dependent upon a compass, in this case the magnetic compass, but maintenance of a direction strongly depends on the presence of other cues such as stars in the case of nocturnal migrants (93). The stars do not contain directional information in themselves, but they are secondary sources of orientation when information from the magnetic field has been transferred to them (88). In situations where information provided by stars and experimental magnetic fields was contradictory, the garden warbler (Sylvia borin) selected its bearing according to information provided by the altered magnetic field (96). This suggests that the magnetic compass is of primary importance to this species. Rabol (109), on the other hand, considered any tendency for sylvilid warblers to use the Earth's magnetic field for orientation in the absence of stellar cues to be weak at best. It seems that we are far from being able to weigh the various cue systems with respect to their relative importance within a species, let alone among species.

A considerable amount of evidence indicates that at least some birds apparently possess a magnetic compass. But if they do, work by W. and R. Wiltschko (86) suggests that the avian compass is functionally different from the mechanical device familiar to humans. The magnetic compass of the European robin does not use the polarity of the Earth's magnetic field for detecting north. Instead, the robins derive north direction from interpreting the inclination of the axial direction of the magnetic field lines in space, and then taking the direction on the magnetic north-south axis for north where the field lines and the gravity vector form the smaller angle. The involvement of gravity (or some other secondary reference point) in the system is essential if a bird is to determine where north is along the north-south axis of the magnetic field lines. Such a mechanism would provide birds with a highly flexible direction-finding system as it would be able to adjust to the varying intensity ranges encountered geographically in the Earth's field.

The magnetic compass described by Wiltschko (82) is not consistent with what would be expected on the basis of the findings of Southern (71,78,79), Keeton (10), Moore (72), and Wagner (110). Magnetic storms and magnetic anomalies involve only small distortions in the total magnetic field, yet both events have been shown to affect orientation by pigeons, gulls, and migrating passerines. Wiltschko (82) contends that changing the intensity of the magnetic field by as much as 10% had no effect on the orientation of captive European robins. Since the field changes reported during geomagnetic storms or at anomalies were less than 10%, the possibility exists that the effects reported by myself (61,67) and the others were not the result of the magnetic compass being disrupted (82). If such high sensitivity to geomagnetism is not an essential component of the avian magnetic compass, what role does magnetism play in the orientation or navigation process? This brings us to the possibility that geomagnetic information also may play a role in the experimentally elusive map that is essential to true navigators.

Walcott (111) suggested that the Earth's magnetic field could provide at least some information about position and hence contribute to the avian map component (i.e., a grid of coordinates). Objections to such a contention were raised in response to Yeagley's (30) conclusion that pigeons used a grid of latitudinal and longitudinal lines produced by the geomagnetic field and Coriolis effect. Similar objections also have been raised more recently by Kreithen and Keeton (112), Griffin (113), and Schmidt-Koenig (15). Several studies, however, have made the possibility of a map based on geomagnetic information worthy of further consideration (114). Much of the disagreement about this possibility has centered on the relative lack of physiological evidence indicating that pigeons or other birds actually are sensitive to magnetic stimuli. To date, a magnetic receptor has not been identified in birds and most of the data pointing to the existence of such a capability are from behavioral studies.

Because pigeons and some migratory birds apparently respond to small changes in the Earth's magnetic field (see studies dealing with K-values and anomalies), Walcott (111) suggested that these effects possibly were the result of the navigational map having a magnetic component. If this is correct, these effects provide a measure of avian magnetic sensitivity. The angular deviations reported by myself (67,73,79) and Keeton et al. (74), and the magnetic anomaly results suggest that a change in field strength of about 10 gamma is detectable by birds (111). This would provide the map resolution (1-mile accuracy) proposed by Schlichte and Schmidt-Koenig (115) and Schmidt-Koenig and Walcott (116). Walcott (111) also pointed out that there is roughly a 10 gamma/mile change in the Earth's magnetic field strength which could provide positional information. Perception of these subtle differences in field strength indeed could contribute to the map component but the results of studies designed to test for such ability have done little to resolve the problem.

Another, but even less convincing, possibility has been proposed by Wallraff and Foa (117). They concluded that olfaction was an integral part of the pigeon's navigation mechanism, but anosmic birds that showed rudimentary homeward orientation relied on magnetic cues. There is, however, only circumstantial evidence showing that birds, other than the domesticated homing pigeon, may be capable of navigating over long distances by olfactory cues. Furthermore, the evidence for pigeons navigating by this means is far from convincing (118). Some birds, however, use olfaction during foraging activities (e.g., vultures, shearwaters) and possibly for recognizing their home locality (e.g., petrels)(119,120).

GEOMAGNETIC SENSITIVITY AND THE SEARCH FOR A RECEPTOR

The results from four studies at the organ or cellular level suggest that pigeons may be able to perceive magnetic stimuli, but studies reporting negative results also exist (44,47,58,121,122). Reille (123) reported successful cardiac conditioning to a field intensity of 0.8 Oersted. Yakovleva and Medvedeva (124) also reported conditioning of the heart but at much higher intensity magnetic fields (520 Oe), which makes their results of questionable relevance to the orientation question. Bookman (125,126) used a different approach to test for sensitivity to Earth-intensity field strengths. A Y-maze was installed in a metal room which reduced the natural field intensity to about 0.02 Oe. Pigeons were trained to travel a flight tunnel and select the compartment containing food. The absence or presence of food was linked with a 0.5 Oe magnetic field produced by Helmholtz coils. Some birds discriminated between the presence and absence of the magnetic field. These individuals fluttered rather than walked down the tunnel, and correctly selected the part of the maze with an Earth-intensity magnetic field.

The results from these types of studies have not shown conclusively that birds possess a sensitivity to magnetic fields because: (a) only a small proportion of the sample tested responded positively (e.g., Bookman's study), and (b) other investigators (e.g., Kreithen and Keeton)(112) have failed in their attempts to replicate the results from some of the conditioning experiments (e.g., Reille's). One piece of evidence, however, suggests that with improved methodologies investigators may be able to determine conclusively whether or not magnetic sensitivity exists in potential receptor sites. Semm et al. (127,128) reported that the firing rates of some cells in the pigeon's pineal organ were altered when exposed to

Earth-strength magnetic fields. The remaining evidence pointing to a sensitivity to magnetic stimuli is indirect, being based on results from studies wherein birds were subjected to fields produced by magnets or coils (8,71,78,79,129,130), naturally-occurring magnetic anomalies (110,131-135), or man-made electromagnetic disturbances (136,137).

Many of the objections to the idea that birds can use geomagnetic cues for orienting or navigating stem from the fact that a receptor for magnetic stimuli has not been described. Studies of some non-avian species, however, have shown that the existence of such a receptor in birds is possible. Three methods of magnetic-field detection by living organisms have been described (138): (a) induction; (b) the presence of some type of paramagnetic material that will react to the geomagnetic field; (c) the existence of permanent magnets, such as magnetite particles, that will align themselves with the earth's magnetic field.

Induction occurs in some marine fishes. Kalmijn (139-142) described how sharks locate prey through electroreceptors (143) which register current flow generated by the prey (a conductor) passing through the Earth's magnetic field. The ability of skates to sense magnetic stimuli per se was suggested by Kalmijn's successfully training individuals to select a hiding place in a laboratory tank on the basis of stimuli from the Earth's magnetic field (141,142). These studies do not prove that elasmobranchs are using the induction process for orientation, however, as they do not rule out the possibility of some yet unknown magnetic detector system (111). The induction strategy represents a difficult approach for terrestrial vertebrates since they are not immersed in saltwater which serves as the return current path for sharks (134).

Search for an avian magnetic receptor was revitalized by the discovery that mud bacteria contain small magnetic inclusions of magnetite (Fe_3O_4) which act as single magnetic domains and result in the organisms being magnetotactic (144-149). These bacteria contain a chain-like alignment of magnetite particles (referred to as magnetosomes) which function as a compass. The cellular inclusions cause the organism to orient in the same direction as the lines of force of the geomagnetic field. Torque exerted on this biomagnetic compass by the magnetic field passively steers the swimming bacterium, and the field's inclination directs it down into the mud substrate. In the southern hemisphere, the bacteria's magnetic polarity is reversed, thereby producing the same relationship between the compass and the behaviorally relevant direction of orientation (149,150,151). The earlier report of magnetite in the radula of chitons (152), however, also should alert investigators to the possibility that this dense substance may be used within organisms for functions totally unrelated to orientation (153).

Positive results have been reported from some of the searches for iron-rich particles, magnetite or precursors thereof (e.g., hydrous iron oxides) in other animals. Gould et al. (138) and Kuterbach et al. (154) reported such substances in honeybees, and similar discoveries have been made in vertebrates, such as pigeons (155,156) and dolphins (157). The magnetite-containing tissue of pigeons was found, by using a magnetometer, on the inner surface of the dorsal cranium in an area just posterior to the orbits (155), but a later effort to expand upon this discovery produced only negative results (158). Changes in methodology again produced positive results. Stained serial sections of the pigeon's head were examined from the beak posteriorly. This approach thus far has revealed three major sites in the head with iron-containing tissue. The locations of the sites are: (a) the harderian gland which is positioned medial to each eye within the orbits; (b) the base of the beak; and (c) a sheet of cells more centrally positioned in the brain near the olfactory lobes (158).

The eye pecten, an intraocular pleated and highly vascular structure, has been suggested as a possible magnetic sensor (159), but this possibility has been ruled out in at least one species by detailed histological examination of the pecten's ultrastructure and the absence of magnetite-containing cells (160,161).

The search goes on to identify structures that may serve as the receptor of geomagnetic stimuli in birds. At this time, the search is concentrating on the head region and some positive results to date justify this approach. The finding of magnetite in pigeons has provided the first solid evidence that a potential component of a magnetic compass exists in vertebrates. Still lacking, however, are physiological data showing that these structures are indeed functioning as receptors. The results from mud bacteria suggest that magnetite-containing cells could function as a magnetic compass in birds, but the map component of the navigational system remains as elusive as ever.

MAGNETISM AND OTHER VERTEBRATES

A number of studies with fishes, in addition to those previously mentioned, have demonstrated responses to magnetic fields (162-167). Amphibians also have shown an ability to perceive magnetic stimuli. Phillips (168) trained salamanders to respond directionally according to geomagnetic stimuli and Phillips and Adler (169) documented magnetic sensitivity in two species of salamander. There is also evidence that reptiles are sensitive to magnetic fields. Rodda (170-172) concluded that alligators are not only true navigators, but that they use a geomagnetic map to select homeward directions when displaced. A few reports also indicate that mammals, ranging from rodents (173,174) to humans (175,176), are capable of perceiving and orienting by geomagnetic cues. The results reported for humans have been subjected to criticism as other investigators have been unable to duplicate the findings (e.g., Gould and Able) (177). Such controversy is not new, however, as it has existed in the bird literature for decades. As has been clearly shown for birds, it may be counterproductive to immediately dismiss evidence showing an effect in a carefully designed experiment simply because another investigator has been unable to reproduce the findings. Methodologies appear to be extremely important in this field, and subtle changes in apparatus or treatment of subjects have been causes for some of the disagreement.

The amount of evidence for these non-avian groups is small compared to what is available for birds. It is impossible, therefore, to pass judgment on the importance of magnetism in the orientation and navigation of representatives of these groups. It is interesting, however, that the apparent ability to perceive and use geomagnetic information appears to be so widespread. This suggests the ability may have been essentially universal among vertebrates at some point in their evolutionary history, but that the degree to which it is used today is contingent upon their orientational requirements. Long-distance migrants may have evolved increasing dependence upon supplemental cue types (e.g., sun, stars), particularly those which enhance efficiency, thereby obscuring the underlying sensitivity to geomagnetism and the magnetic compass.

CONCLUDING REMARKS AND SUMMARY

Almost 40 years have passed since Yeagley's (7) first publication on the subject of pigeons navigating by geomagnetic cues. We have gone full circle since then with respect to our willingness to accept the possibility that organisms can perceive and use geomagnetic information. Critical examination of the literature on the subject, with its contradictions and ambiguities, leads one to conclude that the road to scientific truth often is serpentine.

In the foregoing sections I have emphasized the positive results and provided the picture of a developing consensus favoring: (a) a great variety of organisms being able to perceive the Earth's magnetic field, (b) representatives from several major taxa having a magnetic compass, and (c) the magnetic field, in some way, possibly influencing the map component of the avian navigational system. Much of the available evidence strongly supports the first two possibilities, with the third being less well documented. The riddle of how birds navigate and the actual role of geomagnetism in the process is far from being answered (16). There are reasons for being cautious about accepting some of the more prevalent views until more evidence becomes available.

The consistent results which the Wiltschkos have produced over the years would appear, at first glance, as a breakthrough toward solving the avian orientation riddle. The findings, however, have not been readily accepted by ornithologists, either justly or unjustly. The cautionary approach of investigators is based on: (1) the difficulty others have had in replicating the results; (2) the statistical methods that are required to show positive effects; (3) the absence of conclusive physiological evidence showing that birds have a magnetic receptor; and

(4) the tendency of the Wiltschkos to place the magnetic compass in the dominant position in the hierarchical arrangement of orientational cues. As a result, the mood appears to be that of tolerance rather than acceptance of the contention that birds are navigating by geomagnetic cues. Ornithologists are closer to agreeing on the existence of a magnetic compass which would establish a bird's general direction of travel during migration (i.e., orientation) than they are to accepting geomagnetism as the basis for a true navigational ability in birds (i.e., the basis for both the compass and map components). But until more comparative studies are conducted, so as to provide a broader species base for the apparent ability to orient by geomagnetic cues, the uncertainties will persist.

A brief discussion of some of the concerns that have been raised appears in order. Firstly, when tests similar to the Wiltschko's are conducted in other than their exact cage design, or if the perch arrangement within the cage is different, negative results usually are obtained (57,178). Positive results appear possible only if the cage is eight-sided and has radially aligned perches. If perch arrangement is tangential, for example, other researchers (179) have been unable to replicate the Wiltschko's findings. Why should cage design be so important in such tests unless the cage itself is contributing to the responses obtained? This question has been raised repeatedly, and no acceptable answer is available although other investigators who have switched to Merkel and Wiltschko's cage design or collaborated with the Wiltschkos have obtained comparable results at other locations.

The second, and potentially more serious problem, pertains to the statistical procedures necessary to show positive effects in Wiltschko-type experiments. It is necessary to pool data and the practice used is to calculate nightly means for birds tested, and then to group these means and calculate a grand mean (3). This procedure of using second-order statistics obscures how individual birds respond, and high variance around such grand means is usually ignored. It is difficult to envision directional preferences (i.e., orientation) which are so weakly expressed that they can only be documented in this manner (i.e., second-order means) being useful to migratory birds. Evidence pertaining to the effects of naturally-occurring magnetic disturbances (e.g., storms and anomalies), however, has relied upon first-order means, suggesting that naturally occurring changes in the Earth's magnetic field have a more pronounced effect on orienting birds than the methods used by the Wiltschkos. Until the magnetic compass they propose is documented by other methods and in other species, it must be accepted with reservation.

Other types of problems are associated with some of the research approaches, and these have been reviewed by others (20). One group of problems relates to magnetic-field characteristics. Ossenkopp and Barbeito (20) aptly point out that over the years investigators have used two approaches; one assumes that the stronger the stimulus, the stronger the response, whereas the other attempts to simulate the Earth's field characteristics as closely as possible and looks for more subtle effects. It is possible that high-intensity fields may stress systems other than, or in addition to those associated with orientation thereby camouflaging any possible effect on direction-finding ability. The point is that the right stimulus must be used to test for a specific response or effect. Another set of problems is associated with the physiological or psychological state of the organism being tested. Emlen (180) and others have shown that orientation behavior changes according to the endogenous state of the organism. Failure to take this into account can lead to tests and results that do not accurately address the question being asked. Similarly, I found that the homing tendencies of individual gulls was influenced by their reaction to the test procedures themselves (61). Some individuals that were trapped and released, but not transported away from their nesting colony, showed tendencies similar to those usually used as measures of homing success. For example, some immediately abandoned their nest sites (this would equate with unsuccessful homers), others disappeared for several hours (slow homers), and some returned immediately to their nests as if nothing had happened (rapid homers). It can be misleading, therefore, for investigators to assume that the quantifiable responses of their test subjects are solely in reaction to the variable for which they want to test. In other words, control subjects are required. These types of design problems were more frequent in the past than they are now, but they must be placed in perspective when reviewing and evaluating the published accounts.

If we accept the precautionary stance which seems appropriate at this time, where does this leave us with respect to understanding the role of geomagnetic cues in vertebrate orientation? In the case of birds, the answer seems to be that it leaves us in a state of turmoil (16). It is now widely accepted that birds are capable of using a multiplicity of environmental information for orientational purposes. This realization was a significant advancement, but it also made it more difficult to design ways of testing for the importance of specific cues. The list of types of environmental information that could provide for redundancy in the system is lengthy, including the sun, stars, wind, odors, magnetic field, infrasound and landmarks (16). Not only is it possible that such cues may be used alternately for direction-finding purposes, but the use of two or more could be integrated in some way so as to improve accuracy. The difficulties that have been encountered have greatly slowed advances in this field during the 1980's. A possible contributor to the slowdown may have been our overreaction to the possibility that multiple cues are used by some birds. It may be that we have lost sight of the forest because of the trees.

The results from homing and migration studies indicate that an array of avian species is capable of solving navigational-type problems. At this point in time, however, it is impossible to describe the complete mechanism that enables any bird to accomplish such feats (for reviews of the role of various cues see 3,16). The likelihood that magnetism plays an integral role in the orientation of some avian species is supported by a number of studies discussed in this review. It remains to be discovered whether this apparent ability to sense and use geomagnetism is widespread among species and, if so, how they actually perceive and process such information.

REFERENCES

(1) Dorst, J.: The Migration of Birds, Houghton Mifflin Co., Boston, 1962.

(2) Gauthreaux, S.A. Jr.: Animal Migration, Orientation, and Navigation, Academic Press, New York, 1980.

(3) Emlen, S.T.: Migration: Orientation and Navigation, in Avian Biology, Vol. V, D.S. Farner, J.R. King and K.C. Parkes, eds., Academic Press, New York, 129-219, 1975.

(4) Griffin, D.R.: Bird navigation, in Recent Studies in Avian Biology, A. Wolfson, ed., University of Illinois Press, Champaign, 154-197, 1955.

(5) Matthews, G.V.T.: Bird Navigation, Cambridge University Press, Cambridge, 1955.

(6) Sauer, E.G.F.: Die sternenorientierung nachlich ziehender grasmucken (Sylvia atricapilla, borin und curruca), Z. Tierpsychol. 14:29-70, 1957.

(7) Yeagley, H.L.: A preliminary study of a physical basis of bird navigation, J. Appl. Phys. 18:1035-1063, 1947.

(8) Keeton, W.T.: Magnets interfere with pigeon homing, Proc. Nat. Acad. Sci. U.S.A. 68:102-106, 1971.

(9) Matthews, G.V.T.: Bird Navigation, Second Edition, Cambridge University Press, Cambridge, 1968.

(10) Keeton, W.T.: The orientational and navigational basis of homing in birds, in Advances in the Study of Behavior, Vol. 5, D.S. Lehrman, R. Hinde and E. Shaw, eds., Academic Press, New York, 47-132, 1974.

(11) Keeton, W.T.: Avian orientation and navigation, Ann. Rev. Physiol. 41, 353-366, 1979.

(12) Keeton, W.T.: Avian orientation and navigation: A brief overview, Brit. Birds 72:451-470, 1979.

(13) Keeton, W.T.: Pigeon navigation, in Neural Mechanisms of Behavior in the Pigeon, A.M. Granda and J.H. Maxwell, eds.,

Plenum Publishing Corp., New York, 5-20, 1979.

(14) Schmidt-Koenig, K. and Keeton, W.T., eds.: Animal Migration, Navigation, and Homing, Springer-Verlag, New York, 1978.

(15) Schmidt-Koenig, K.: Avian Orientation and Navigation, Academic Press, New York, 1979.

(16) Able, K.P.: Mechanisms of orientation, navigation, and homing, in Animal Migration, Orientation, and Navigation, S.A. Gauthreaux, Jr., ed., Academic Press, New York, 283-373, 1980.

(17) Papi, F. and Wallraff, H.G.: Avian Navigation, Springer-Verlag, New York, 1982.

(18) Baker, R.R.: Bird Navigation: The Solution to a Mystery?, Holmes & Meier Publ. Incl, New York, 1984.

(19) Wallraff, H.G.: Migration and navigation in birds: A present-state survey with some digressions to related fish behaviour, in Mechanisms of Migration in Fishes, J.D. McCleave, G.P. Arnold, J.J. Dodson and W.H. Neill, eds., Plenum Publishing Corp., New York, 509-544, 1984.

(20) Ossenkopp, K. and Barbeito, R.: Bird orientation and the geomagnetic field: A review, Neurosci. Behav. Rev. 2:255-270, 1978.

(21) Middendorff, A.: Die isepipetesen russlands. Grundlagen zur erforschung der zugzeiten und zugrichtungen der vogel russlands, Mem. L'Acad. Imp. Sci., St. Petersburg, Sixth Series 10, 8:1-144, 1859.

(22) Viguier, C.: Le sens d'orientation et ses organes chez les animaux et chez l'homme, Rev. Phil. 14:1-36, 1882.

(23) Thauzies, A.: L'orientation, Rev. Sci. (4th Series) 9:392-397, 1898.

(24) Stresemann, E.: Haben die Vogel einen Ortssinn?, Ardea 24:213-226, 1935.

(25) Daanje, A.: Haben die vögel einen sinn für den erdmagnetismus, wie deklination, inklination, und intersitat?, Ardea 25:107-111, 1936.

(26) Daanje, A.: Heimfindeversuche und erdmagnetismus, Vogelzug 12:15-17, 1941.

(27) Casamajor, J.: Le mysterieux "sens de l'espace", Rev. Sci. 65:554-556, 1927.

(28) Wodzicki, K., Puchalski, W. and Liche, H.: Untersuchungen über die orientation und geschwindigkeit des fluges bei vögeln. V. Weitere versuche an storchen, J. Ornithol. 87:99-114, 1939.

(29) Rochon-Duvigneaud, A. and Maurain, C.: Enquete sur l'orientation du pigeon voyageur et son mechanisme, Nature (London) 51:232-238, 1923.

(30) Yeagley, H.L.: Final Scientific Report on Bird Navigation, Unpublished report, Department of the Army, Signal Corps Engineering Lab, File No. 3664-PH-47-91 (756), 1949.

(31) Gordon, D.A.: Sensitivity of the homing pigeon to the magnetic field of the Earth, Science 108:710-711, 1948.

(32) Matthews, G.V.T.: The experimental investigation of navigation in homing pigeons, J. Exp. Biol, 28:508-536, 1951.

(33) van Riper, W. and Kalmbach, E.R.: Homing not hindered by wing magnets, Science 115:577-578, 1952.

(34) Yeagley, H.L.: A preliminary study of a physical basis of bird navigation. II. J. Appl. Phys. 22:746-760, 1951.

(35) Matthews, G.V.T.: An investigation of homing ability in two species of gulls, Ibis 94:243-264, 1952.

(36) Bochenski, Z., Dylewska, M., Gieszczykiewicz, J. and Sych, L: Homing experiments on birds. XI. Experiments with swallows Hirundo rustica L. concerning the influence of earth magnetism and partial eclipse of the sun on their orientation, Zesz. Nauk, W.J. Zool. 5:125-130 (In Polish, English summary), 1960.

(37) Griffin, D.R.: Homing experiments with leach's petrels, Auk 57:61-74, 1940.

(38) Griffin, D.R.: Bird navigation, Biol. Rev. Cambridge Phil. Soc. Phil. Soc. 27:359-400, 1952.

(39) Henderson, G.H.: Physical basis of bird navigation, Science 107:597-598, 1948.

(40) Clark, C.L., Peck, R.A. and Hollander, W.F.: Homing pigeon in electromagnetic fields, J. Appl. Phys. 19:1183, 1948.

(41) Kramer, G.: Über richtungstendenzen bei der nächtlichen zugunruhe gekäfigten vögel, in Ornithologie als biologische Wissenschaft, E. Mayr and E. Schuz, eds., Winter, Heidelberg, 269-283, 1949.

(42) Kramer, G.: Orientierte zugaktivität gekäfigter singvögel, Naturwissenschaften 37:188, 1950.

(43) Wilkinson, D.H.: Some physical principles of bird orientation, Proc. Linn. Soc. London 160:94-99, 1949.

(44) Orgel, A.R. and Smith, J.C.: Test of the magnetic theory of homing, Science 120:891-892, 1954.

(45) Neville, J.R.: An Experimental Study of Magnetic Factors Possibly Concerned with Bird Navigation, Ph.D. Thesis, Stanford University, Palo Alto, CA, 1955.

(46) Fromme, H.G.: Untersuchungen über das orientierungsvermögen nächtliche ziehender kleinvögel (Erithacus rubecula, Sylvia communis), Z. Tierpsychol. 18:205-220, 1961.

(47) Meyer, M.E. and Lambe, D.R.: Sensitivity of the pigeon to changes in the magnetic field, Psychon. Sci. 5:349-350, 1966.

(48) Becker, G.F: Ruheeinstellung nach der himmelsrichtung eine magnetfeld-orientierung bei termiten, Naturwissenschaften 50:455, 1963.

(49) Brown, F.A. and Park, Y.H.: Phase-shifting a lunar rhythm in planarians by altering the horizontal magnetic vector, Biol. Bull. 129:79-86, 1965.

(50) Lindauer, M. and Martin, H.: Die schwereorientierung der bienen unter dem einfluss des erdmagnetfieldes, Z. Vergl. Physiol. 60:219-243, 1968.

(51) Bennett, M.F. and Huguenin, J.: Geomagnetic effects on a circadian difference in reaction times in earthworms, Z. Vergl. Physiol. 63:440-445, 1969.

(52) Brown, F.A.: Some orientational influences of nonvisual terrestrial electromagnetic fields, Ann. N.Y. Acad. Sci. 188:224-241, 1971.

(53) Merkel, F.W. and Fromme, H.J.: Untersuchungen über das orientierungsvermögen nächtlich ziehender rotkehlchen (Erithacus rubecula), Naturwissenschaften 45:499-500, 1958.

(54) Merkel, F.W., Fromme, H.J. and Wiltschko, W.: Nichtvisueles orientierungsvermögen bie nächtlich zugunruhigen rotkehlchen, Vogelwarte 22:168-173, 1964.

(55) Merkel, F.W. and Wiltschko, W.: Magnetismus und richtungsfinden zugunruhiger rotkehlchen (Erithacus rubecula), Vogelwarte 23:71-77, 1965.

(56) Wiltschko, W. and Merkel, F.W.: Orientierung zugunruhiger rotkehlchen im statischen magnetfeld, Verh. Deut. Zool, Ges. Jena, 1965, 362-367, 1966.

(57) Perdeck, A.C.: Does navigation without visual clues exist in robins?, Ardea 52:91-104, 1963.

(58) Emlen, S.T.: The influence of magnetic information on the orientation of the indigo bunting, Passerina cyanea, in Animal Orientation and Navigation, S.R. Galler, K. Schmidt-Koenig, G.J. Jacobs and R.E. Belleville, eds., NASA SP-262, 191-210, 1970.

(59) Wallraff, H.G.: Nichtvisuelle orientierung zugunruhiger rotkehlchen (Erithacus rubecula), Z. Tierpsychol. 30:374-383, 1972.

(60) Emlen, S.T., Wiltschko, W., Demong, N.J., Wiltschko, R. and Bergman, S.: Magnetic direction finding: Evidence for its use in migratory indigo buntings, Science 193:505-508, 1976.

(61) Southern, W.E.: The role of environmental factors in ring-billed and herring gull orientation, Ph.D. Dissertation, Cornell University, Ithaca, 1967.

(62) Southern, W.E.: Gull orientation behavior: Influence of experience, sex, age, and group releases, Jack-Pine Warbler 47:34-43, 1969.

(63) Southern, W.E.: Sky conditions in relation to ring-billed and herring gull orientation, Trans. Ill. State Acad. Sci. 62:342-349, 1969.

(64) Southern, W.E.: Influence of disturbances in the Earth's magnetic field on ring-billed gull orientation, Am. Zool. (Abstract) 10:23, 1970.

(65) Southern, W.E.: En route behavior of homing herring gulls as determined by radio-tracking, Wilson Bull. 82:189-200, 1970.

(66) Southern, W.E.: Orientation in gulls: Effect of distance, direction of release, and wind, Living Bird 9:75-95, 1970.

(67) Southern, W.E.: Orientation behavior of ring-billed gull chicks and fledglings, Condor 71:418-425, 1969.

(68) Southern, W.E.: Colony selection, longevity, and ring-billed gull populations: Preliminary discussion, Bird-Banding 8:52-60, 1967.

(69) Southern, W.E.: Seasonal distribution of great lakes region ring-billed gulls, Jack-Pine Warbler 52:154-179, 1974.

(70) Southern, W.E.: Migrational orientation in ring-billed gull chicks, Auk 93:78-85, 1976.

(71) Southern, W.E.: Influence of disturbances in the Earth's magnetic field on ring-billed gull orientation, Condor 74:102-105, 1972.

(72) Moore, F.R.: Influence of solar and geomagnetic stimuli on the migratory orientation of herring gull chicks, Auk 92:655-664, 1975.

(73) Southern, W.E.: The effects of superimposed magnetic fields on gull orientation, Wilson Bull. 86:256-271, 1974.

(74) Keeton, W.T., Larkin, T.S. and Windsor, D.M.: Normal fluctuations in the Earth's magnetic field influence pigeon orientation, J. Comp. Physiol. 95:95-103, 1974.

(75) Lindauer, M. and Martin, H.: Magnetic effect on dancing bees, in Animal Orientation and Navigation, NASA SP-262, S.R. Galler, K. Schmidt-Koenig, G.J. Jacobs and R.E. Belleville, eds., 559-567, 1972.

(76) Keeton, W.T.: Effects of magnets on pigeon homing, in Animal Orientation and Navigation, NASA SP-262, S.R. Galler, K. Schmidt-Koenig, G.J. Jacobs and R.E. Belleville, eds., 579-594, 1972.

(77) Keeton, W.T. and Gobert, A.: Orientation by untrained pigeons requires the sun, Proc. Nat. Acad. Sci. U.S.A. 65:853-856, 1970.

(78) Southern, W.E.: Magnets disrupt the orientation of juvenile ring-billed gulls, BioSci. 22:476-479, 1972.

(79) Southern, W.E.: Orientation responses of ring-billed gull chicks: A re-evaluation, in Animal Migration, Navigation, and Homing, K. Schmidt-Koenig and W.T. Keeton, eds., Springer-Verlag, New York, 311-317, 1978.

(80) Walcott, C.: The navigation of homing pigeons: Do they use sun navigation?, in Animal Orientation and Navigation, NASA SP-262, S.R. Galler, K. Schmidt-Koenig, G.J. Jacobs and R.E. Belleville, eds., U.S. Govt. Printing Office, Washington, DC, 283-292, 1972.

(81) Larkin, T.S. and Keeton, W.T.: Bar magnets mask the effect of normal magnetic disturbances on pigeon orientation, J. Comp. Physiol. 110:227-231, 1976.

(82) Wiltschko, W.: The influence of magnetic total intensity and inclination on directions preferred by migrating European robins (Erithacus rubecula), in Animal Orientation and Navigation, NASA SP-262, S.R. Galler, K. Schmidt-Koenig, G.J. Jacobs and R.E. Belleville, eds., 569-578, 1972.

(83) Wiltschko, W.: Der magnetkompass der gartengrasmücke (Sylvia borin), J. Ornithol. 115:1-7, 1974.

(84) Wiltschko, W.: Der magnetkompass der zugvögel und seine biologische bedeutung, Vogelwarte 29:76-82, 1977.

(85) Wiltschko, W. and Merkel, F.W.: Zugorientierung von dorngrasmucken (Sylvia communis) im erdmagnetfeld, Vogelwarte 26:245-249, 1971.

(86) Wiltschko, W. and Wiltschko, R.: Magnetic compass of European robins, Science 176:62-64, 1972.

(87) Wiltschko, W. and Wiltschko, R.: Grasmucken benutzen den magnetkompass auch bei sternsicht, Naturwissenschaften 12:1-3, 1973.

(88) Wiltschko, W. and Wiltschko, R.: The interaction of stars and magnetic field in the orientation system of night migrating birds, Z. Tierpsychol. 37:337-355, 1975.

(89) Wiltschko, W. and Wiltschko, R.: Die bedeutung des magnetkompasses für die orientierung der vögel, J. Ornithol. 117:362-387, 1976.

(90) Wiltschko, W. and Wiltschko, R.: Interaction of magnetic compass and star orientation in night-migrating birds, J. Comp. Physiol. 109:91-99, 1976.

(91) Wiltschko, W. and Wiltschko, R.: A theoretical model for migratory orientation and homing in birds, Oikos 30:177-187, 1978.

(92) Wiltschko, W. and Hock, H.: Orientation behavior of night-migrating birds (European robins) during late afternoon and early morning hours, Wilson Bull. 84:149-163, 1972.

(93) Wiltschko, R. and Wiltschko, W.: Relative importance of stars and the magnetic field for the accuracy of orientation in night-migrating birds, Oikos 30:195-206, 1978.

(94) Moore, F.R.: Geomagnetic disturbance and orientation of nocturnally migrating birds, Science 196:682-684, 1977.

(95) Bingman, V.P.: Savannah sparrows have a magnetic compass, Anim. Behav. 29:962-963, 1981.

(96) Wiltschko, W.: The migratory orientation of garden warblers, in Avian Navigation, F. Papi and H.G. Wallraff, eds., Springer-Verlag, Berlin, 50-58, 1982.

(97) Southern, W.E.: Comparative distribution and orientation in North American gulls, in Behavior of Marine Animals, Vol. 4, J. Burger, B.L. Olla and H.E. Winn, eds., Plenum Publishing Co., New York, 449-498, 1980.

(98) Wiltschko, W., Wiltschko, R., Keeton, W.T. and Madden, R.: Growing up in an altered magnetic field affects the initial orientation of young homing pigeons, Behav. Ecol. Sociobiol. 12:135-142, 1983.

(99) Wiltschko, W., Wiltschko, R. and Keeton, W.T.: Effects of a "permanent clock-shift" on the orientation of young homing pigeons, Behav. Ecol. Sociobiol. 1:229-243, 1976.

(100) Wiltschko, R. and Wiltschko, W.: The development of sun compass orientation in young homing pigeons, Behav. Ecol. Sociobiol. 9:135-141, 1981.

(101) Wiltschko, R., Nohr, D. and Wiltschko, W.: Pigeons with a deficient sun compass use the magnetic compass, Science 214:343-345, 1980.

(102) Walcott, C.: Magnetic orientation in homing pigeons, IEEE Trans. Magnetics MAG 16:1008-1013, 1980.

(103) Viehmann, W.: Interrelation of magnetic compass, star orientation, and the sun in the orientation of blackcaps and robins, in Avian Navigation, F. Papi and H.G. Wallraff, eds., Springer-Verlag, Berlin, 59-67, 1982.

(104) Wiltschko, W.: The Earth's magnetic field and bird orientation, Trends Neurosci. 1980:140-144, 1980.

(105) Wiltschko, R. and Wiltschko, W.: The development of the navigational system in young homing pigeons, in Localization and Orientation in Biology and Engineering,

Varju and Schnitzler, eds., Springer-Verlag, Berlin, 337-343, 1984.

(106) Wiltschko, R.: The ontogeny of orientation in young pigeons, Comp. Biochem. Physiol. 76A:701-708, 1983.

(107) Beck, W. and Wiltschko, W.: Trauerschnapper (Ficedula hypoleuca Pallas) orientieren sich nickvisuell mit hilfe des magnetfeldes, Vogelwarte 31:168-174, 1981.

(108) Beck, W. and Wiltschko, W.: The magnetic field as a reference system for genetically encoded migratory direction in pied flycatchers (Ficedula hypoleuca Pallas), Z. Tierpsychol. 60:41-46, 1982.

(109) Rabol, J.: Magnetic orientation in night-migrating passerines, Ornis Scandinavia 10:69-75, 1979.

(110) Wagner, G.: Das orientierungsverhalten von brieftauben im erdmagnetisch gestorten geviete des chasseral, Rev. Suisse Zool. 83:883-890, 1976.

(111) Walcott, C.: Is there evidence for a magnetic map in homing pigeons, in Avian Navigation, F. Papi and W.G. Wallraff, eds., Springer-Verlag, Berlin, 99-108, 1982.

(112) Kreithen, M.L. and Keeton, W.T.: Attempts to condition homing pigeons to magnetic stimuli, J. Comp. Physiol. 129:355-362, 1974.

(113) Griffin, D.R.: The sensory physiology of animal orientation, Harvey Lect. Ser. 71:133-172, 1978.

(114) Moore, F.R.: Is the homing pigeon's map geomagnetic?, Nature (London) 285:69-70, 1980.

(115) Schlichte, H.J. and Schmidt-Koenig, K.: Zum heimfindevermogen der brieftaube bei erschwerter optischer Wahrnehmung, Naturwissenschaften 58:329-330, 1971.

(116) Schmidt-Koenig, K. and Walcott, C.: Tracks of pigeons homing with frosted lenses, Anim. Behav. 26:480-486, 1978.

(117) Wallraff, H.G. and Foa, A.: The roles of olfaction and magnetism in pigeon homing, Naturwissenschaften 69:504-505, 1982.

(118) Phillips, J.B. and Waldvogel, J.A.: Reflected light cues generate the short-term deflector-loft effect, in Avian Navigation, F. Papi and H.G. Wallraff, eds., Springer-Verlag, Berlin, 190-202, 1982.

(119) Grubb, T.C. Jr.: Olfactory navigation to the nesting burrow in leache's petrel (Oceanodroma leucorrhoa), Anim. Behav. 22:192-202, 1974.

(120) Wenzel, B.M.: Chemoreception in seabirds, in Behavior of Marine Animals, Vol. IV, J. Burger, B.L. Olla and H.E. Winn, eds., Academic Press, New York, 41, 1980.

(121) Orgel, A.R. and Smith, J.C.: A test of the magnetic theory of homing in pigeons, J. Genet. Psychol. 88:203-210, 1956.

(122) Beaugrand, J.P.: An attempt to confirm magnetic sensitivity in the pigeon, Colulmba livia, J. Comp. Physiol. 110:343-355, 1976.

(123) Reille, A.: Essai de mise en evidence d'une sensibilite du pigeon au champ magnetique a l'aide d'une conditionnement nociceptif, J. Physiol. Paris 60:85-92, 1968.

(124) Yakovleva, M.I. and Medvedeva, M.V.: Conditioned control of cardiac activity and respiration and morphological changes in the brain of pigeons under the effect of a constant magnetic field, Zh. vyssh. nerv. Deyat. I.P. Pavlova 22, 288-293 (NTIS No. AD-784-798), 1972.

(125) Bookman, M.: Sensitivity of the homing pigeon to an Earth-strength magnetic field, Nature (London) 267:340-342, 1977.

(126) Bookman, M.: Sensitivity of the homing pigeon to an Earth-strength field, in Animal Migration, Navigation and Homing, K. Schmidt-Koenig and W.T. Keeton, eds., Springer-Verlag, Berlin, 127-134, 1978.

(127) Semm, P., Schneider, T. and Vollrath, L.: Effects of an Earth-strength field on electrical activity of pineal cells, Nature (London) 288:607-608, 1980.

(128) Semm, P., Schneider, T., Vollrath, L. and Wiltschko, W.: Magnetic sensitive pineal cells in pigeons, in Avian Navigation, F. Pappi and H.G. Wallraff, eds., Springer-Verlag, Berlin, 329-337, 1982.

(129) Walcott, C. and Green, R.: Orientation of homing pigeons altered by a change in the direction of an applied magnetic field, Science 184:180-182, 1974.

(130) Visalberghi, E. and Alleva, E.: Magnetic influences on pigeon homing, Biol. Bull. 156:246-256, 1979.

(131) Graue, L.C.: Initial orientation in pigeon homing related to magnetic contours, Am. Zool. (Abstract) 5:704, 1965.

(132) Talkington, L.: Bird navigation and geomagnetism, Amer. Zool. (Abstract) 7:199, 1967.

(133) Frie, U. and Wagner, G.: Die anfangsorientierung von brieftauben im erdmagnetisch gestören gebiet des mont jorat, Rev. Suisse Zool. 83:891-897, 1976.

(134) Walcott, C.: Anomalies in the Earth's magnetic field increase the scatter of pigeon's vanishing bearings, in Animal Migration, Navigation and Homing, K. Schmidt-Koenig and W.T. Keeton, eds., Springer-Verlag, Berlin, 143-151, 1978.

(135) Walcott, C.: Homing-pigeon vanishing bearings at magnetic anomalies are not altered by bar magnets, J. Exp. Biol. 86:349-352, 1980.

(136) Southern, W.E.: Orientation of gull chicks exposed to project Sanguine's electromagnetic field, Science 189:143-145, 1975.

(137) Larkin, R.P. and Sutherland, D.J.: Migrating birds respond to project Seafarer's electromagnetic field, Science 195:777-778, 1977.

(138) Gould, J.L., Kirschvink, J.L. and Deffeyes, K.S.: Bees have magnetic remanence, Science 201:1026-1028, 1978.

(139) Kalmijn, A.J.: The electric sense of sharks and rays, J. Exp. Biol. 55:371-383, 1971.

(140) Kalmijn, A.J.: The electric and magnetic sense of sharks, skates and rays, Oceanus 20:45-52, 1977.

(141) Kalmijn, A.J.: Electric and magnetic sensory world of sharks, skates and rays, in Sensory Biology of Sharks, Skates and Rays, E.S. Hodgson and R.F. Mathewson, eds., U.S. Govt. Printing Office, Washington, DC, 507-528, 1978.

(142) Kalmijn, A.J.: Experimental evidence of geomagnetic orientation in elasmobranch fishes, in Animal Migration, Navigation and Homing, Springer-Verlag, Berlin, 347-355, 1978.

(143) Lissmann, H.W.: On the function and evolution of electric organs in fish, J. Exp. Biol. 35:156-191, 1958.

(144) Blakemore, R.P.: Magnetotactic bacteria, Science 190:377-378, 1975.

(145) Kalmijn, A.J., and Blakemore, R.P.: Geomagnetic orientation in marine mud bacteria, Proc. Int. Union Physiol. Sci. 13:364, 1977.

(146) Kalmijn, A.J. and Blakemore, R.P.: The magnetic behavior of mud bacteria, in Animal Migration, Navigation, and Homing, K. Schmidt-Koenig and W.T. Keeton, eds., Springer-Verlag, Berlin, 354-355, 1978.

(147) Frankel, R.B., Blakemore, R.P., and Wolfe, R.S.: Magnetite in freshwater magnetotactic bacteria, Science 203:1355-1356, 1979.

(148) Frankel, R.B., Blakemore, R.P., Torres de Araujo, F.F., Esquivel, D.M.S. and Dannon, J.: Magnetotactic bacteria at the geomagnetic equator, Science 212:1269-1270, 1981.

(149) Blakemore, R.P. and Frankel, R.B.: Magnetic navigation in bacteria, Sci. Am. 245:58-65, 1981.

(150) Blakemore, R.P., Frankel, R.B. and Kalmijn, A.J.: South-seeking magnetotactic bacteria in the southern hemisphere, Nature (London) 256:384, 1980.

(151) Kirschvink, J.L.: South-seeking magnetic bacteria, J. Exp. Biol. 86:345-348, 1980.

(152) Lowenstam, H.A.: Magnetite in denticle capping in recent chitons (Polyplacophora), Geol. Soc. Am. Bull. 73:435-438, 1962.

(153) Maugh, T.H. II.: Magnetic navigation an attractive possibility, Science 215:1492-1493, 1982.

(154) Kuterbach, D.A., Walcott, B., Reeder, R.J. and Frankel, R.B.: Iron-containing cells in the honey bee (_Apis mellifera_), Science 218:695-697, 1982.

(155) Walcott, C., Gould, J. and Kirschvink, J.: Pigeons have magnets, Science 215:1027-1029, 1979.

(156) Presti, D. and Pettigrew, J.D.: Ferromagnetic coupling to muscle receptors as a basis for geomagnetic field sensitivity in animals, Nature (London) 285:99-101, 1980.

(157) Zoeger, J. and Fuller, M.: Magnetic material in the head of a dolphin, Trans. Am. Geophys. Union 61:225, 1980.

(158) Walcott, B. and Walcott, C.: A search for magnetic field receptors in animals, in Avian Navigation, F. Papi and H.G. Wallraff, eds., Springer-Verlag, Berlin, 338-343, 1982.

(159) Danilov, V., Demirochaglyan, G., Avetysan, Z., Aelakhnyerdyanm, M., Grigoryan, S. and Saribekhyan, G.: Possible mechanisms of magnetic sensitivity in birds (in Russian), Biol. Zh. Arm. 23:26-34, 1970.

(160) Hanzely, L., Southern, W.E. and Molsen, D.: Ultrastructure of the ring-billed gull eye pecten, Cytobios 12:191-201, 1975.

(161) Southern, W.E., Hanzely, L., Bailey, R.L. and Molsen, D.V.: Is the avian eye pecten a magnetic sensor?, in Avian Navigation, F. Papi and H.G. Wallraff, eds., Springer-Verlag, Berlin, 344-351, 1982.

(162) Branover, G.G., Vasil'yev, A.S., Gleyzer, S.I. and Tsinober, A.B.: A study of the behavior of cells in natural and artificial magnetic fields and an analysis of its reception mechanism, J. Ichthyol. 11:608-614, 1971.

(163) Ovchinnikov, V.V., Gleyzer, S.I. and Alaktionov, G.Z.: Features of orientation of the European eel (Anguilla anguilla) (L.) at some stages of migration, J. Ichthyol. 13:455-463, 1973.

(164) Vasil'yev, A.S. and Gleyzer, S.I.: Changes in the activity of the freshwater eel (Anguilla anguilla)(L.) in magnetic fields, J. Ichthyol. 13:322-324, 1973.

(165) Andrinanov, G.N., Brown, G.R. and Ilyinsky, O.B.: Responses of central neurons to electrical and magnetic stimuli of the ampullae of Lorenzini in the Black Sea skates, J. Comp. Physiol. 93:287-299, 1974.

(166) Tesch, F.-W.: Influence of geomagnetism and salinity on the directional choice of eels, Helgolander Wiss. Meeresunters 26:382-395, 1974.

(167) Kalmijn, A.J.: Electric and magnetic field detection in elasmobranch fishes, Science 218:916-918, 1982.

(168) Phillips, J.B.: Use of the Earth's magnetic field by orienting cave salamanders (Eurycea lucifuga), J. Comp. Physiol. A121:273-288, 1977.

(169) Phillips, J.B. and Adler, K.: Directional and discriminatory responses of salamanders to weak magnetic fields, in Animal Migration, Navigation, and Homing, K. Schmidt-Koenig and W.T. Keeton, eds., Springer-Verlag, Berlin, 325-334, 1978.

(170) Rodda, G.H.: The orientation and navigation of juvenile alligators: Evidence of magnetic sensitivity, J. Comp. Physiol. A154:649-658, 1984.

(171) Rodda, G.H.: Homeward paths of displaced juvenile alligators as determined by radiotelemetry, Behav. Ecol. Sociobiol. 14:241-246, 1984.

(172) Rodda, G.H.: Navigation in juvenile alligators, Z. Tierpsychol. 68:65-77, 1985.

(173) Mather, J.G. and Baker, R.R.: Magnetic sense of direction in woodmice for route-based navigation, Nature 291:152-155, 1981.

(174) Baker, R.R. and Mather, J.G.: A comparative approach to bird navigation: Implications of parallel studies on mammals, in Avian Navigation, F. Papi and H.G. Wallraff, eds., Springer-Verlag, Berlin, 308-312, 1982.

(175) Baker, R.R.: Goal orientation by blindfolded humans after long-distance displacement: Possible involvement of a magnetic sense, Science 210:555-557, 1980.

(176) Baker, R.R.: Human Navigation and the Sixth Sense, Hodder and Stoughton, London, 1981.

(177) Gould, J.L. and Able, K.P.: Human homing: An elusive phenomenon, Science 212:1061-1063, 1981.

(178) Wallraff, H.G.: Versuche zur frage der gerichteten nachtzug-aktivaitat von gekafigten singvögeln, Verh. dt. zool. Ges. Jena 339-356, 1965.

(179) Howland, H.C.: Orientation of European robins to kramer cages, Z. Tierpsychol. 33:295-312, 1973.

(180) Emlen, S.T.: Bird migration: Influence of physiological state upon celestial orientation, Science 165:716-718, 1969.

4
NMR Conditions and Biological Systems

E. AARHOLT, M. JABERANSARI, A.H. JAFARY-ASL,* P.N. MARSH, C.W. SMITH

Department of Electronic and Electrical Engineering
University of Salford, Salford, United Kingdom
and *Department of Physiology, Medical
School, Shiraz University, Shiraz, Iran

INTRODUCTION

Nuclear magnetic resonance (NMR) is a phenomenon associated with atomic nuclei having an odd number of either protons or neutrons. The first demonstration of NMR was made by Rabi in 1938 using molecular beams. The first observations in condensed matter were made independently and almost simultaneously by Purcell, Torry and Pound in Cambridge, and by Block and Hansen at Stanford. The Nobel Prize was awarded jointly to Purcell and Bloch in 1952.

The physical fundamentals and applications of NMR have been the subject of many texts (1-8). The two major NMR applications involve chemistry and medical imaging. The chemical applications arise because the magnetic field at a nucleus is never exactly equal to the externally applied magnetic field, but depends in many ways on the magnetic properties of the molecular structures surrounding an atom. These give rise to characteristic shifts in the nuclear magnetic resonance conditions of the order of parts per million.

The possibility of using NMR for in vivo medical imaging of body structures arises because the relaxation times for the return to an equilibrium condition following the application of a nuclear magnetic resonance perturbation to a biological system vary not only among different kinds of normal tissues, but more importantly between healthy and diseased tissues. Application of electronics and computing techniques have enabled living systems to be scanned while the relaxation times are being measured. This results in a map or image of their spatial distribution in such detail that it correlates with the anatomical landmarks, and shows the distribution of physiological and biochemical parameters.

BASIC THEORY OF NUCLEAR MAGNETIC RESONANCE

NUCLEAR MAGNETIC MOMENT

The magnetic moment μ of a coil in a magnetic field B is the torque Γ per unit magnetic field and is equal to the product of the current i and the area of the coil S,

$$\Gamma/B = \mu = i \cdot S \tag{1}$$

Scaling things down to nuclear dimensions, a current i can be considered equivalent to a charge e rotating at $\omega/2\pi$ revolutions per second where ω is the angular velocity in radians per second.

$$i = e\,\omega/2\pi \tag{2}$$

Taking an idealized nucleus as an annulus of mass M, radius r and charge e rotating about the principal axis with angular velocity ω, the magnetic moment μ will be from Equations (1) and (2):

$$\mu = e\omega r^2/2 \tag{3}$$

The ratio of the magnetic moment to the angular momentum P, is the gyromagnetic ratio;

$$\gamma = \mu/P = e\omega r^2/2M\omega r^2 = e/2M \tag{4}$$

A fundamental postulate of theoretical physics is that the total angular momentum of an isolated particle can only take on certain discrete values. It is said to be quantized, and can only have values which are integral multiples of the quantity $h/2\pi$, often written $\hbar$, where h is Planck's constant (6.626176×10^{-34} J.sec). The nuclear energy levels are quantized as a direct result of the quantized nature of the nuclear angular momentum. It takes on a series of values corresponding to unity changes in a quantity known as the spin quantum number I, which can range from +I to -I. Since the proton has I = 1/2, only the two values corresponding to I = ±1/2 are allowed. The value of I depends on the particular nucleus. Isotopes with equal numbers of protons and neutrons have I = 0 and isotopes having odd mass numbers tend to have 1/2 integral spin values. The highest value is for the isotope $^{176}{}_{71}Lu$, for which I = 6.

NUCLEI IN A MAGNETIC FIELD

Remembering that magnetic moments are generated by the equivalent of rotating charges, a non-zero angle between an applied magnetic field and the magnetic moment of a nucleus is the equivalent of the classical Larmor precessional motion of a gyroscope with its axis at an angle to the gravitational field. Force is given by the rate of change of momentum, torque is given by the rate of change of angular momentum. If a system with angular momentum P is precessing at an angular velocity Ω, then the change of angular momentum in time dt which can give rise to the torque is

$$dP = P\,\Omega \sin\theta\, dt \tag{5}$$

The torque is given by

$$\Gamma = \mu B \sin\theta = \gamma P B \sin\theta \tag{6}$$

Comparison of Equations (1), (4), and (6) gives the angular velocity of the precession

$$\Omega = \gamma B \tag{7a}$$

or in terms of the Larmor precessional frequency, ν_o

$$\nu_o = \gamma B/2\pi \tag{7b}$$

POPULATION OF SPIN STATES

In most cases, NMR transitions between the energy levels will be the result of stimulated emission or absorption of radiation. Irradiation of a specimen with an intense radiofrequency field would rapidly equalize the populations of the energy levels setting up a dynamic equilibrium in which no changes in emission or absorption would be detected. In matter in general, the population distribution between two energy states can be described by the Boltzmann factor,

$$N_2/N_1 = \exp(-\Delta E/kT) \tag{8}$$

where N_1 and N_2 are the number of nuclei in energy states 1 and 2 respectively, the energy state 1 is less than that of 2, k is the Boltzmann constant and T is the absolute temperature.

In a magnetic field of strength B, the separation of the energy levels ΔE is

$$\Delta E = 2\mu B \tag{9}$$

$$N_2/N_1 = \exp(-2\mu B/kT) \approx 1 - (2\mu B/kT) \tag{10}$$

if $2\mu B << kT$.

The observation of NMR through some physical process depends on being able to detect a net absorption of energy by the small excess population defined by Equation (10). Living systems are, by definition, systems far from thermal equilibrium. Their population distributions are determined by stimulated emission and absorption, and will not be Boltzmann if the system is alive and active (29).

SPIN-LATTICE AND SPIN-SPIN RELAXATIONS

1. NMR Relaxation Times

If a system is perturbed from a condition of stable equilibrium and can eventually return to the initial stable state, the return may take the form of an exponentially damped oscillatory response, such as the motion of a pendulum swinging in air, or the exponential return of a pendulum in a viscous oil. The relaxation rate is expressed by the constant of the exponent.

The concept of relaxation times for assemblies of magnetic dipoles is of great importance in the applications of NMR techniques. By arranging detector coils in various directions relative to the applied field, the magnetizations and relaxations in these directions can be determined. When a radiofrequency pulse is applied, the magnetic dipoles begin to precess at the Larmor frequency about a direction defined by the resultant of the steady and oscillatory magnetic fields. The magnetization nutates, or nods, away from the direction defined by the applied steady field. The nutation stops when the radiofrequency magnetic field is switched off. If this occurs when the angle of nutation reaches 90°, the magnetization would be left to relax in a transverse plane. A 180° pulse is one that reverses the direction of the magnetization. Experimentally, the NMR signal is usually detected by a coil placed perpendicular to the steady magnetic field and resonant at the radiofrequency. The decrease in the induced voltage measured after the end of the radiofrequency pulse is a measure of the relaxation of the magnetization as the nuclei return to their equilibrium.

2. Spin-Lattice Relaxations

The interaction between the nuclear spin and the nuclear environment is small but finite, and given enough time and an absence of stimulation, the nuclei will eventually come into thermal equilibrium with the lattice. Absorption of radiofrequency energy reduces the population of the lower energy state, and the flow of thermal energy opposes this process. Since spontaneous emission is insignificant, the NMR relaxation is only by stimulated downward transitions from those lattice magnetic fields that happen to be at the Larmor frequency. For protons in biological tissues, a typical time constant for this process is 0.05-3 sec (9).

3. Spin-Spin Relaxations

Each nuclear magnetic moment experiences not only the applied magnetic field, but the resultant local interaction field arising from the neighboring nuclear magnetic fields. The local field may have both static and oscillatory components, and it can result in a broadening of the energy levels. The lifetime of a nuclear spin state in the absence of an applied radiofrequency field, is an approximate definition of the spin-spin relaxation time. In biological systems, the spin-spin relaxation times are typically 0.04-2 sec (9).

4. Saturation

In the absence of radiation to stimulate transitions between different states, the relaxation processes outlined will occur. When a strong radiofrequency field is applied, the spin system becomes saturated. In the case of a nucleus for which I = 1/2, the populations of the upper and lower states equalize, leaving no magnetization in any direction.

OBSERVATION OF NMR EFFECTS

To observe a nuclear magnetic resonance it is necessary to perturb the system from the condition of steady-state precession. Magnetic resonance in bulk matter can be excited in several ways (10). The present concern is less with the observation of magnetic resonances by physical measurements as such, because the application of magnetic-resonance techniques to biological measurements is well documented (11,12). Rather, it is to seek ways in which the excitation of NMR can result in the perturbation of a living biological system which can in turn be investigated by physical measurements.

The three methods commonly used for the observation of NMR are described elsewhere (2).

MEASUREMENTS OF BIOLOGICAL SYSTEMS UNDER NMR CONDITIONS

DIELECTROPHORESIS

Dielectrophoresis is the movement of electrically neutral particles in a nonuniform electric field due to their polarization (13-15). The dielectrophoretic yield is the number of particles collected at the electrodes per unit time. The small dielectrophoresis cell used in our earlier studies was constructed on a standard microscope slide and consisted of a pair of spherical platinum electrodes (diameter, 0.8 mm) sealed into a shallow well that contained a suspension of yeast cells. We now use evaporated metal film electrodes to achieve short optical working distance at high magnification (45). The electrodes were connected to an audio-frequency oscillator.

The yeasts used in these experiments (16,40) were _Saccharomyces cerevisiae_ (normal diploid strain). They were grown as pure colonies in a suitable medium, harvested, and suspended in

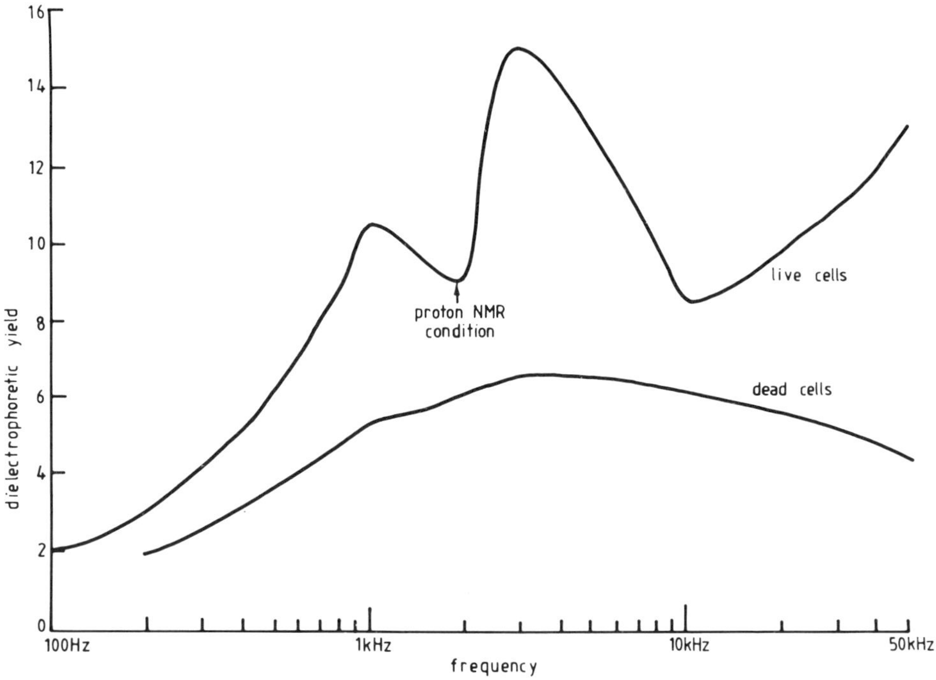

FIGURE 1. Dielectrophoretic yield spectra of live and dead yeast cells as a function of the frequency of the electric field.

an ion-free isotonic solution. Cell concentration was measured by standard dilution and plating techniques.

The clean, dry electrode chamber was filled with 0.5 ml of the cell suspension, mounted on the microscope stage, and the electrodes were connected to the oscillator. When the selected voltage and frequency were applied, the cells migrated toward the electrodes and attached themselves in pearl-chain formations. After 3 minutes the oscillator was turned off and the average chain length was measured with the microscope eyepiece graticule.

Figure 1 shows the dielectrophoretic yield as a function of frequency for both live yeast cells and cells that had been killed by exposing them to ultraviolet light (254 nm). The anomaly in the region of 2 KHz is clearly seen in the curve for the live cells. Figures 2a and 2b show that the dielectro-

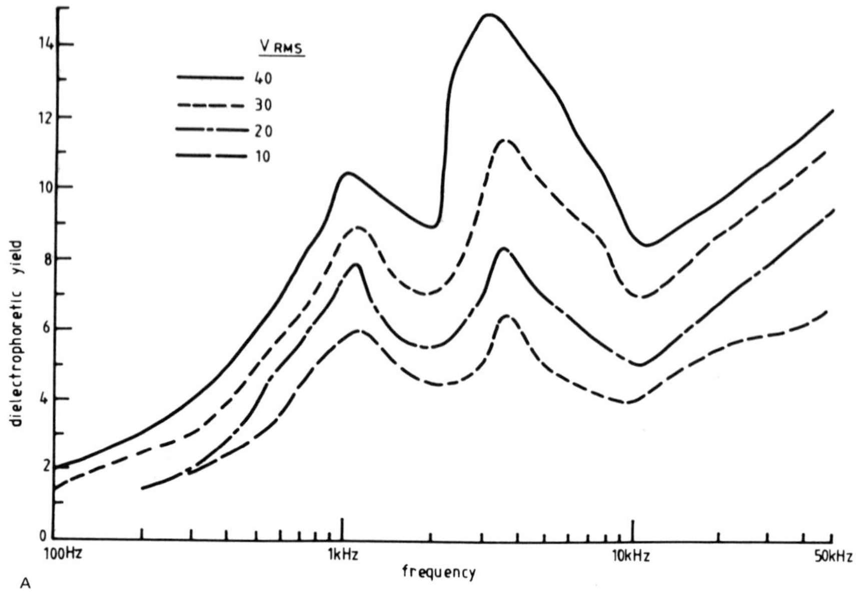

FIGURE 2a. Dielectrophoretic yield spectra of live yeast cells as a function of the frequency of the electric field.

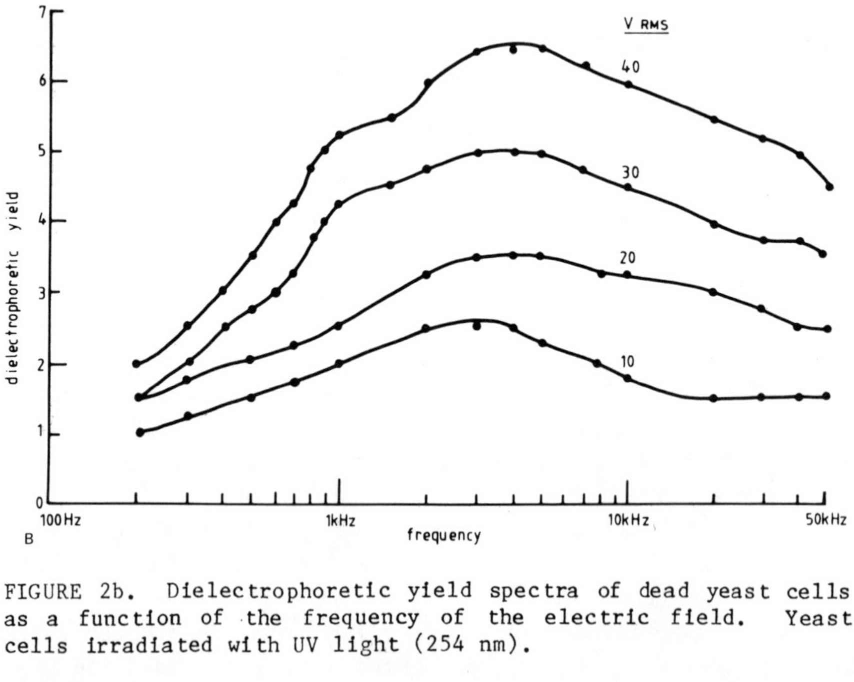

FIGURE 2b. Dielectrophoretic yield spectra of dead yeast cells as a function of the frequency of the electric field. Yeast cells irradiated with UV light (254 nm).

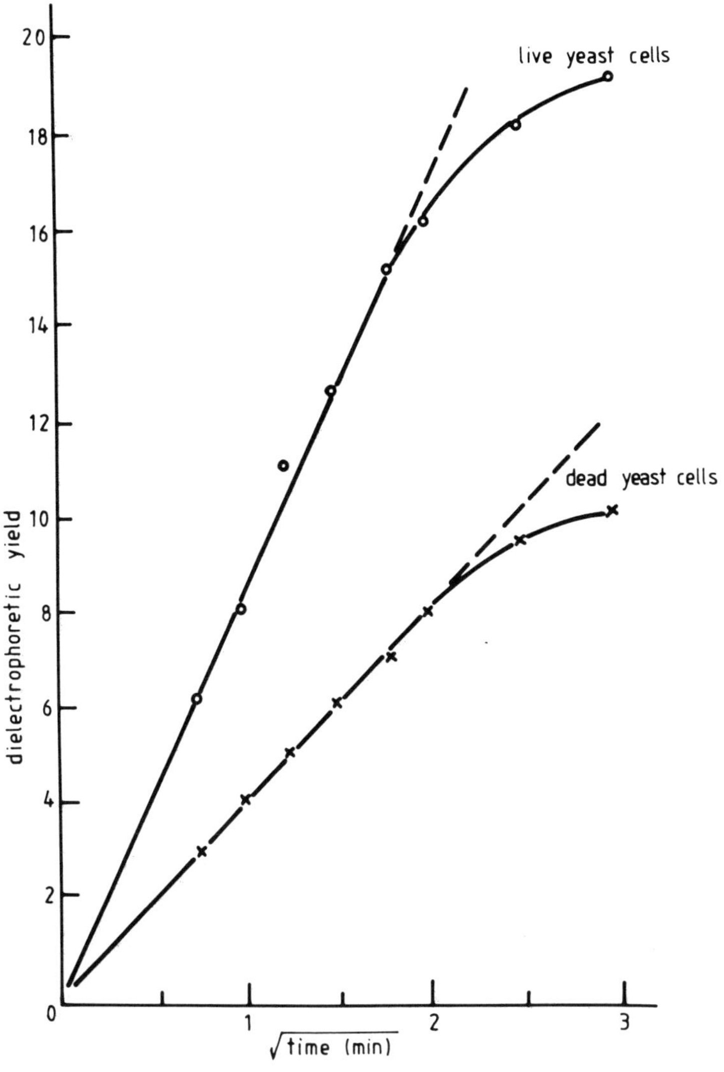

FIGURE 3. Variation of dielectrophoretic yield with the square root of the length of time the field is applied. V = 40 V r.m.s., frequency = 3 KHz.

phoretic yields were proportional to the applied voltage, and that they remained at the same frequency. Figure 3 shows that the yield of live and dead cells differed.

Figure 4 shows the results of measurements on live cells from 0.5 to 5 G (50 μT to 500 μT) normalized by plotting KHz/G as the abscissa. The anomaly then shows up clearly as a sharp resonance at exactly the proton magnetic resonance condition of 4.26 KHz/G from results taken over a 10:1 range of magnetic field strengths. Figure 5 shows the difficulty in repeating these measurements under nominally the same conditions. The variation in the positions of the curves is due to the variation of the laboratory magnetic field at the microscope specimen stage. These experiments took place on three successive days using yeast cells from the same culture.

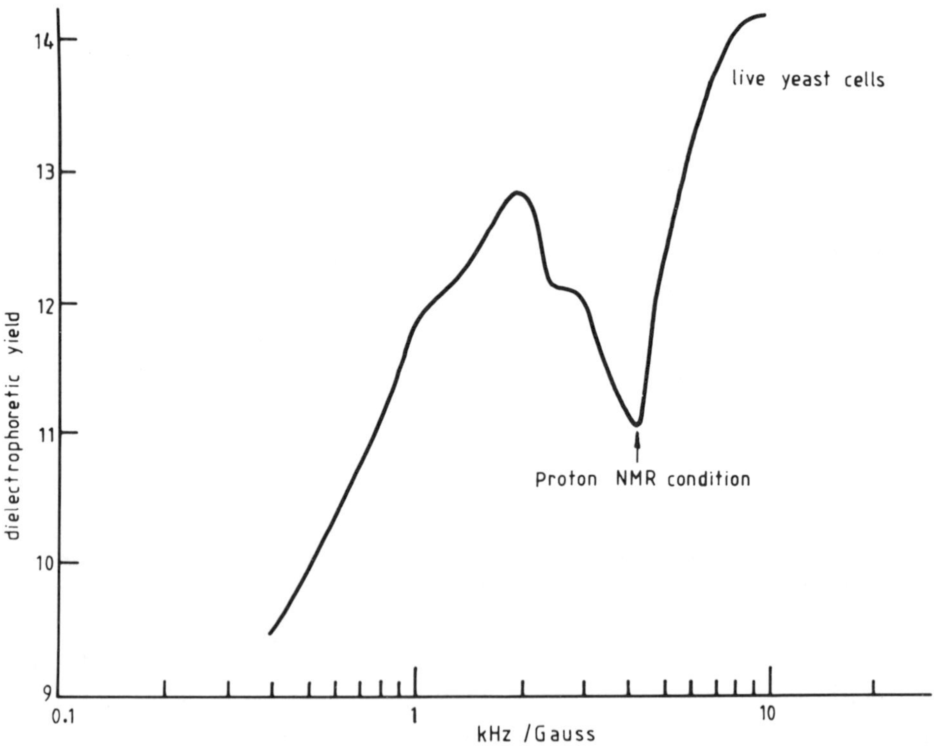

FIGURE 4. Measurements on live cells from 0.5 to 5 G normalized by plotting the abscissa as KHz per gauss.

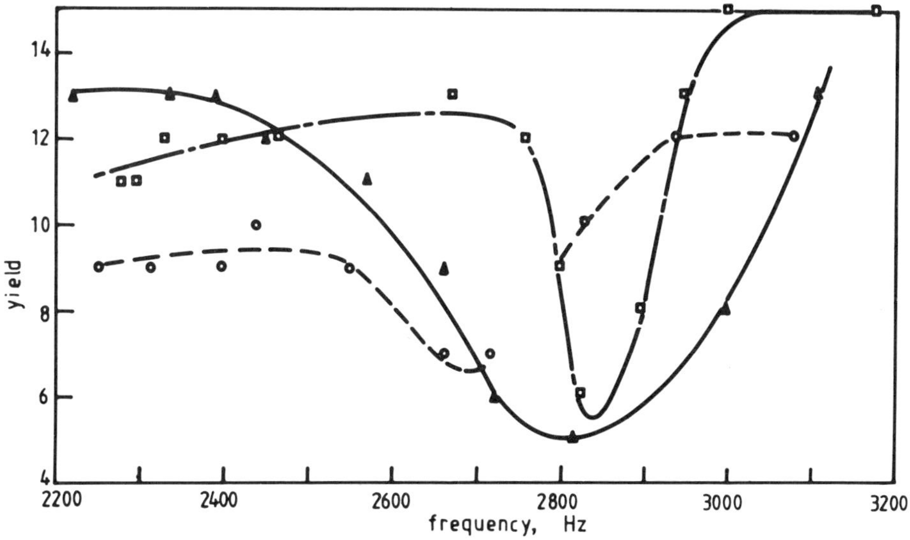

FIGURE 5. The difficulty in repeating measurements under nominally identical conditions is shown. The variation in the positions of the curves is due to the variation of the laboratory magnetic field at the microscope specimen stage. These experiments took place on 3 successive days using yeast cells from the same culture.

DIELECTRIC CONSTANT AND LOSS

The permittivity and dielectric loss of live yeast cells were measured for a suspension containing 1.9 x 10^6 cells/ml at room temperature and in magnetic fields of 0.5-5 Gauss (16,40). Figure 6 shows the very sharp resonances that were observed in the dielectric loss. Figure 7 shows values of the frequency and magnetic field for the dielectric loss peaks; the line represents the proton magnetic resonance condition.

More detailed measurements showed peaks in the dielectric loss corresponding to the NMR conditions of ^{1}H, ^{31}P, ^{23}Na, ^{35}Cl, and ^{39}K (Figure 8a). Figure 8b shows these peaks to be absent in the case of dead yeast cells. The magnitudes of these resonances were temperature dependent, as shown in Figures 9-11, where the points were taken at 0.5-Hz intervals and represent the limit of accuracy of the apparatus rather than the shapes of the resonances.

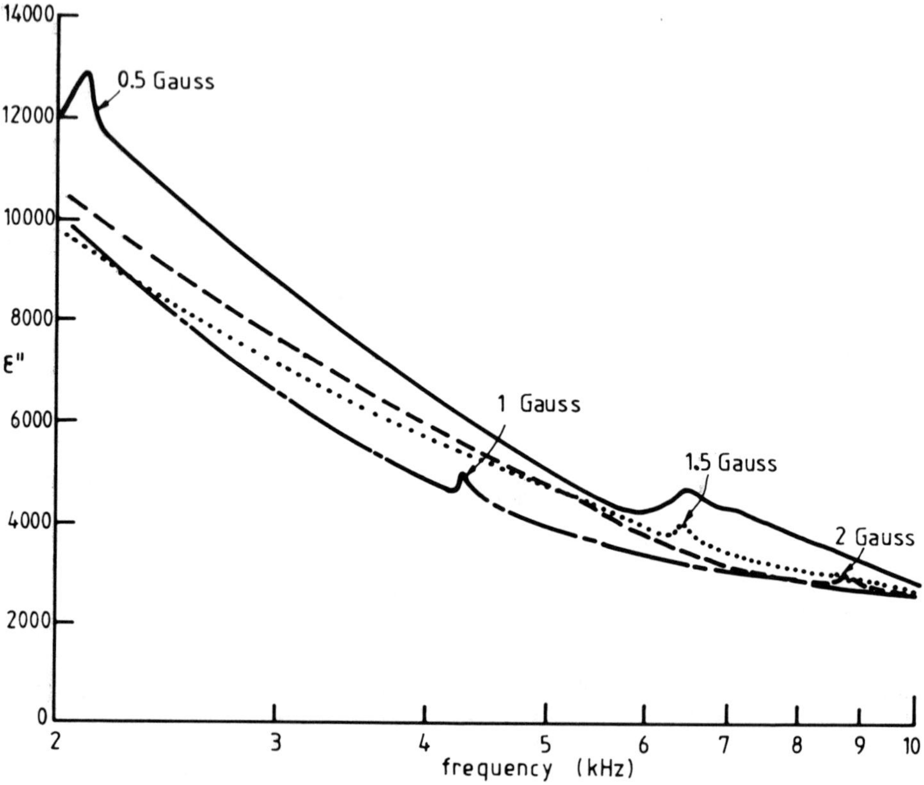

FIGURE 6. Dielectric loss peaks at NMR field conditions of a 1% by weight bakers' yeast in deionized water at 24°C.

There is not enough evidence to determine the specific process responsible for the observed effects on the dielectric constant and loss at NMR conditions when the applied Larmor frequency arises from an oscillating electric field. There is an interaction at proton NMR conditions with the dielectric properties of proton conducting pH meter glass (45). The basis of interactions between electric and magnetic susceptibilities has been given by Van Vleck (17).

GROWTH OF *E. COLI* IN NMR CONDITIONS

E. coli cultures were grown at 39.0°C (18), which is 1.5°C above the value at which the multiplication rate is a maximum. Any additional heating of the test cultures due to the current

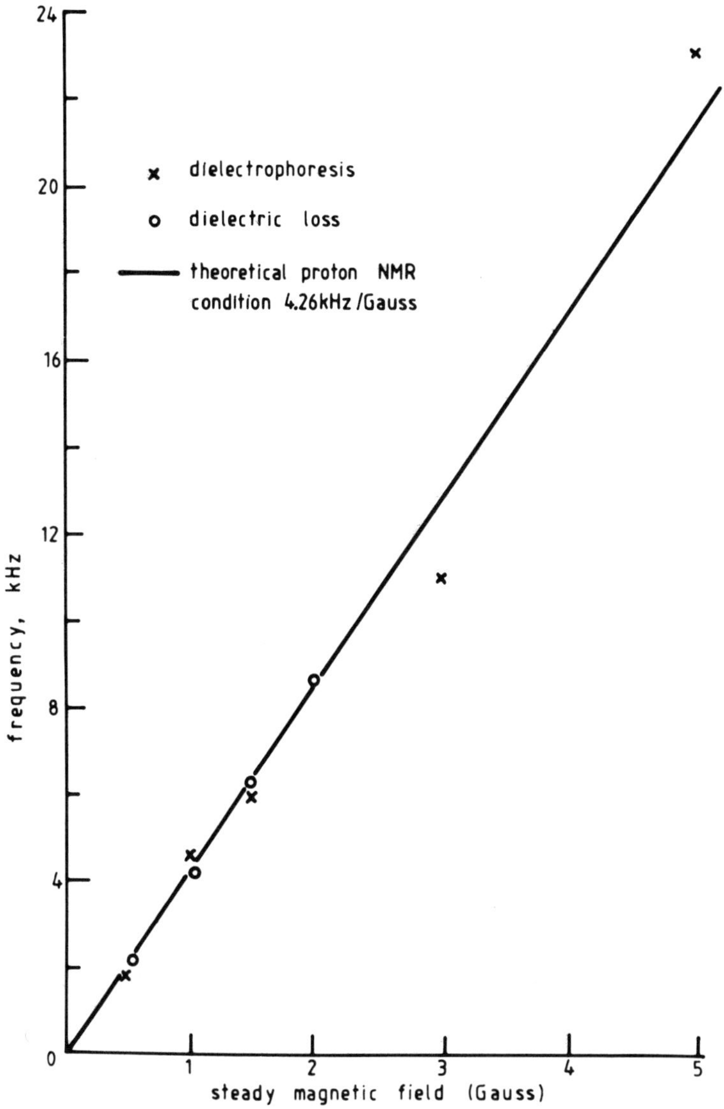

FIGURE 7. Variation of the proton NMR and dielectric loss peaks with frequency with applied magnetic field.

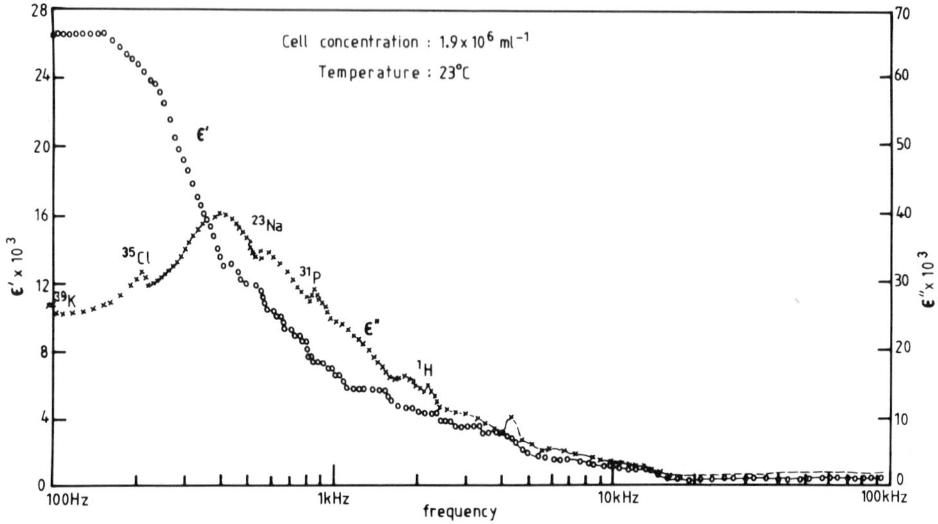

FIGURE 8a. The real and imaginary parts of the complex permittivity of the live yeast (S. cerevisiae) as a function of frequencies. The laboratory ambient-magnetic field was 0.5 G.

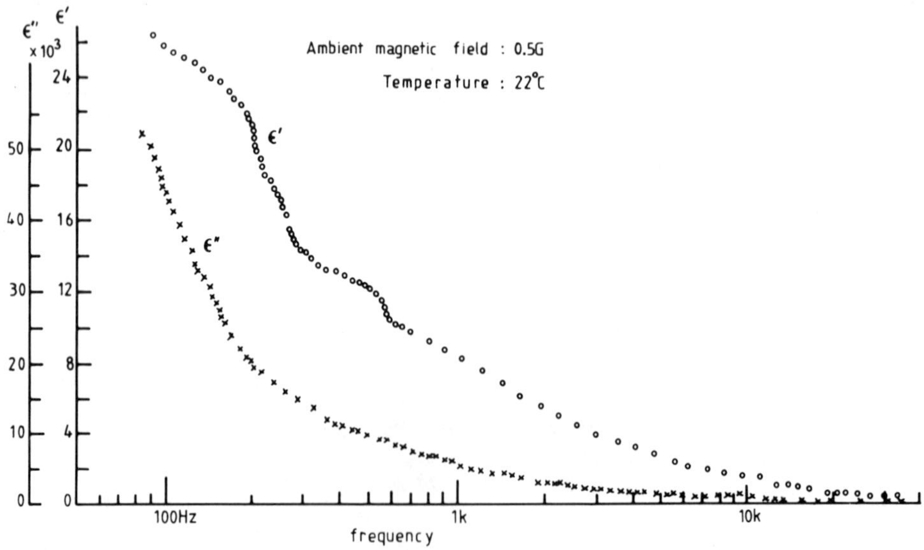

FIGURE 8b. The real and imaginary parts of the complex permittivity of the dead yeast (S. cerevisiae) as a function of frequencies. The laboratory ambient-magnetic field was 0.5 G.

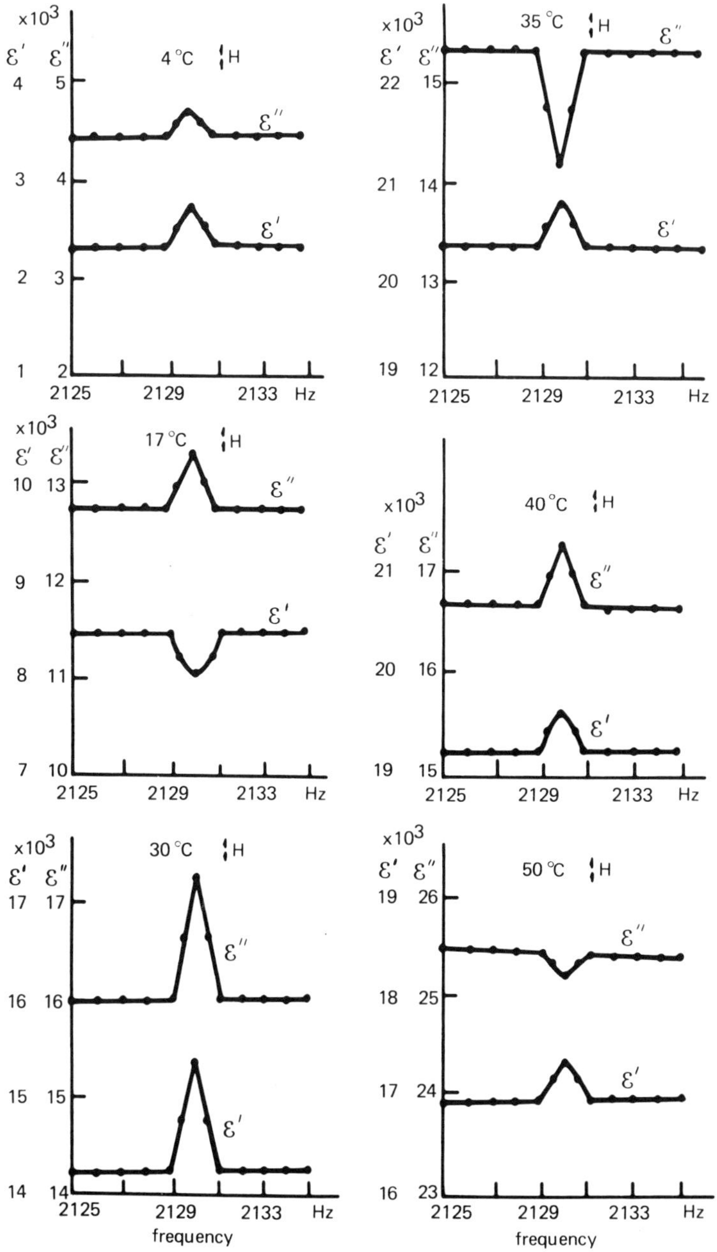

FIGURE 9. Dielectric constant and dielectric loss of live yeast cell suspensions (1.9 x 10^6 cells/ml) in a laboratory ambient field of 0.5 G (electric field strength of the order of 20 kV/m). Proton NMR conditions.

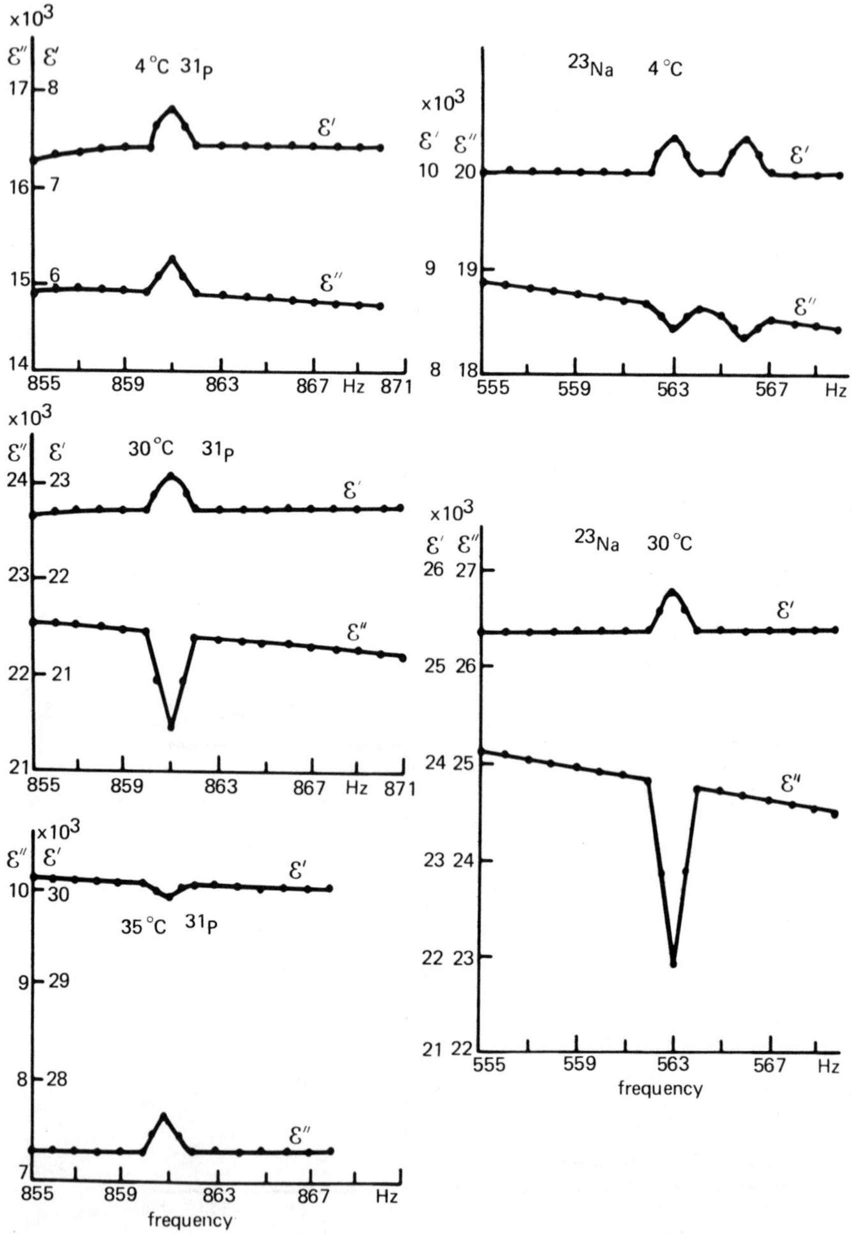

FIGURE 10. Dielectric constant and dielectric loss of live yeast cell suspensions (1.9 x 10^6 cells/ml) in a laboratory ambient field of 0.5 G (electric field strength of the order of 20 kV/m). Phosphorus and sodium NMR conditions.

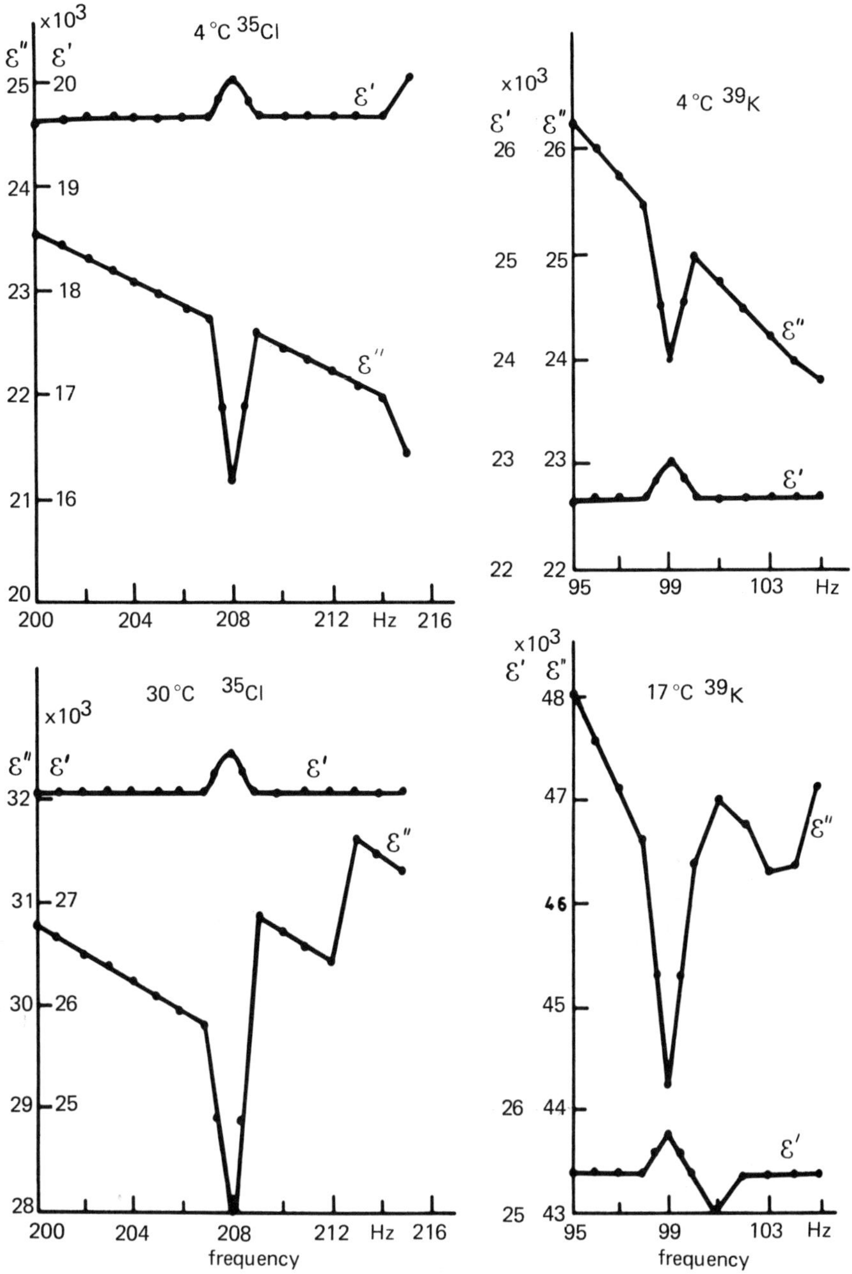

FIGURE 11. Dielectric constant and dielectric loss of live yeast cell suspensions (1.9×10^6 cells/ml) in a laboratory ambient field of 0.5 G (electric field strength of the order of 20 kV/m). Chlorine and potassium NMR conditions.

flowing in the energizing coil of the electromagnet would therefore be expected to decrease the growth rate.

Each culture was grown in a plastic spectrophotometry cuvette. Each experiment involved thirty cuvettes: eighteen cultures were grown in the field, nine were grown inside a closed mu-metal box (giving effectively zero-field conditions), and three were left uninoculated as a check on the sterility of the culture medium.

Approximately 10^4 bacterial cells were used for each inoculation. Each culture was rapidly transferred to the incubator and grown in darkness without aeration for 10-12 hours. This period represents approximately 14 cell divisions at a mean generation time (MGT) of 0.85 hours per division. An electromagnet, the test cultures, the control cultures inside their mu-metal enclosure, and the uninoculated sterility control cuvettes, were all situated inside the incubator. The standard deviation of the MGT between cultures of the same batch was approximately 0.5%.

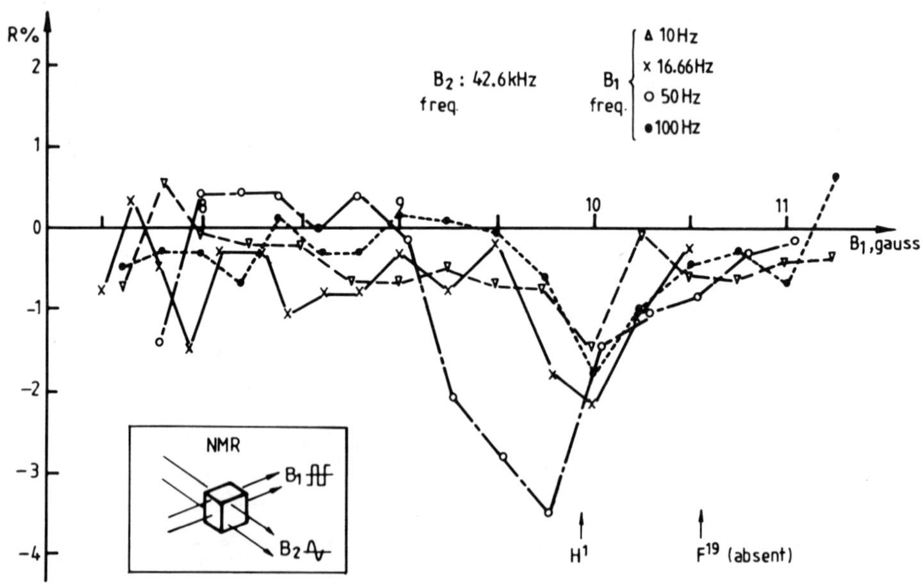

FIGURE 12. Percentage difference in the MGT of the test and control cultures as a function of the magnetic field strength.

After incubation, cell density was determined turbidimetrically at 650 nm using a spectrophotometer (19). A square-wave magnetic field was used at 10, 16.66, 50 and 100 Hz, with a perpendicular sinusoidal magnetic field at 42.6 KHz. The results are presented in Figure 12, where the relative difference in MGT between test and control cultures is plotted as a function of magnetic-field strength (41).

There was a marked drop in MGT when the magnetic field strength satisfied the proton resonance condition. The effect was most marked for 50-Hz magnetic fields, giving a decrease in the MGT of about 3.5% compared with the zero-field controls. The effect also occurred at 16.66 Hz (about 2.25%), and 10 Hz and 100 Hz (1.5%); the standard deviation of the control cultures was 0.7%. No resonance for F^{19} was observed.

In these experiments it was demonstrated that when satisfying the NMR conditions, which involves the injection of small amounts of energy into protons within the _E. coli_, the bacteria responded by increasing the rate of division by only a few percent. Effects on enzyme activity were also investigated (20,40).

BOVINE EYE LENSES

Considering all the known biological effects of microwave and radiofrequency radiation, cataract induction appeared until recently to be the only irreversible effect of such an exposure with the notable exception of thermally lethal doses (21-35). We investigated the possible effects of low power density (lower than 10 mW/cm^2) microwaves on the bovine eye lens _in vitro_. The overall aim was to determine the effects of electromagnetic irradiation and NMR conditions on possible cataractogenesis.

A total of 720 bovine eye lenses were studied. An incision was made approximately 1 cm below the cornea, and the eye lens was removed via the vitreous humor and placed in modified Krebs Phosphate Ringer media. The lens could be maintained for up to 240 hours at 35°C using this medium without deterioration.

The microwave radiation studies were performed with the lens on a glass ring in the incubation vessel which was placed within a wire coupling loop. In most cases, the dimensions of the loop were small compared with the wavelengths used.

No significant changes in the sodium and potassium concentrations of the eye lenses occurred following microwave radiation by the time a cataract was visible. It was considered that the microwave radiation had produced an effect on the bovine eye lens if the formation of a cataract in the posterior cortex of the lens could be observed within 24 hours of the commencement of irradiation.

Experiments using the NMR conditions for sodium, chlorine and phosphorus did not result in the production of microwave cataracts in the bovine eye lens. However, the proton NMR conditions at 2000 MHz and modulation frequency of 2.13 KHz in a magnetic field of 0.5 Gauss gave highly developed microwave cataracts in the posterior cortex of the bovine eye lens. The maximum power density of the radiation was 1-2 $\mu W/cm^2$, although there were uncertainties due to absorptions in the nutrient medium. The length of irradiation was 18-20 hr, and all cataracts obtained were located in the posterior cortex of the eye lens, and all were subcapsular.

Experiments were performed in which the various parameters such as power, density, time of irradiation and physiological conditions of incubation remained constant, while the frequency of microwave radiation was the only variable. Microwave cataracts in the bovine eye lenses were then produced at 55-64 MHz; above and below this immediate range no cataracts were obtained. The spectral line-width was about 50 Hz. When the bandwidth was increased to 150-200 Hz under otherwise identical conditions, the cataractogenic effect disappeared. All these microwave cataracts were subcapsular and located in the posterior cortex of the lens. By maintaining the frequency at 55 MHz and starting at a power level that was known from the literature to cause cataracts, the power relationship to microwave cataractogenesis was studied using fixed calibrated attenuators to reduce the microwave power output by known ratios. These experiments determined that microwave cataracts at 55 MHz could be produced over the range of 0.54 mW down to 0.0075 μW (a ratio of 70,000:1 in power).

Microwave cataracts were also produced over the range 900-2000 MHz. The time for development of these microwave cataracts was 1-20 hours depending upon the power density. At low power densities, the length of time of irradiation required to produce a cataract increased to 20 hours. The spectral line width of the 900 MHz radiation was 3 KHz.

The frequency range of 70 MHz-800 MHz was investigated, but microwave cataracts were not obtained in any of the exposed bovine eye lenses. The spectral bandwidths of the oscillators used in this range were up to 1 MHz, and this could account for the observed lack of cataract formation. The above results are summarized in Table 1.

That highly coherent radiation can produce biological effects at very low incident power densities is in agreement with the theoretical predictions of Fröhlich concerning the effects of coherent excitations in biological systems (36). Weak microwave radiation seems able to exert a cataractogenic effect on in vitro bovine eye lenses only under two conditions. Proton NMR conditions must be satisfied by a combination of modulated microwaves and ambient magnetic fields such that the microwaves act as a carrier for the modulation frequency which is induced throughout the eye lens on absorption of the microwaves. Also, specific conditions must be satisfied involving frequency, power density, duration of microwave exposure, and coherence of microwave radiation. All these conditions can be met in the context of ambient geomagnetic and other environmental magnetic fields, taking due account of magnetic flux quantization conditions (37,38,40). The properties of the water involved must also be taken into account (48).

DISCUSSION

Although it happened that the first indications that a biological system might be sensitive to NMR conditions were seen in the results of dielectrophoresis measurements, this technique is not sufficiently precise to stand on its own. It is difficult to work with a chain longer than 10 cells, which immediately limits the accuracy to 10%. In addition, the reproducibility of the magnetic field strength at the microscope stage was only 5-10%. Observation of the distortion of pearl chains subjected to NMR conditions however, does give a qualitative demonstration

TABLE 1. Bovine Eye Lenses Exposed to Microwave Radiation

FREQUENCY OF IRRADIATION (MHz)	NO. OF CONTROLS	NUMBER OF EXPOSED EYES	
		Cataracts	No. Cataracts
50-110	35	7	28
110-800	21	0	21
810-890	10	0	21
900	10	3	7
910-950	4	2	2
960	59	9	50
1000-1900	15	2	13
2000	72	8	64
2000	7	7	0
2100-4000	10	0	4
3145	4	0	4
3580	27	8	19
9373	12	0	12
10116	8	1	7

of the existence of an NMR effect (Figure 13).

With the best dielectric measuring apparatus available to us, it was only just possible to detect the small peaks obtained when NMR conditions were satisfied by the bridge frequency and the ambient magnetic field. Experimental points were taken at 0.5 Hz intervals. If these peaks have widths corresponding to NMR relaxation times in biological materials, then readings should be taken at 0.1 Hz to 0.01 Hz intervals. The majority, and largest, effects were found at the proton NMR condition.

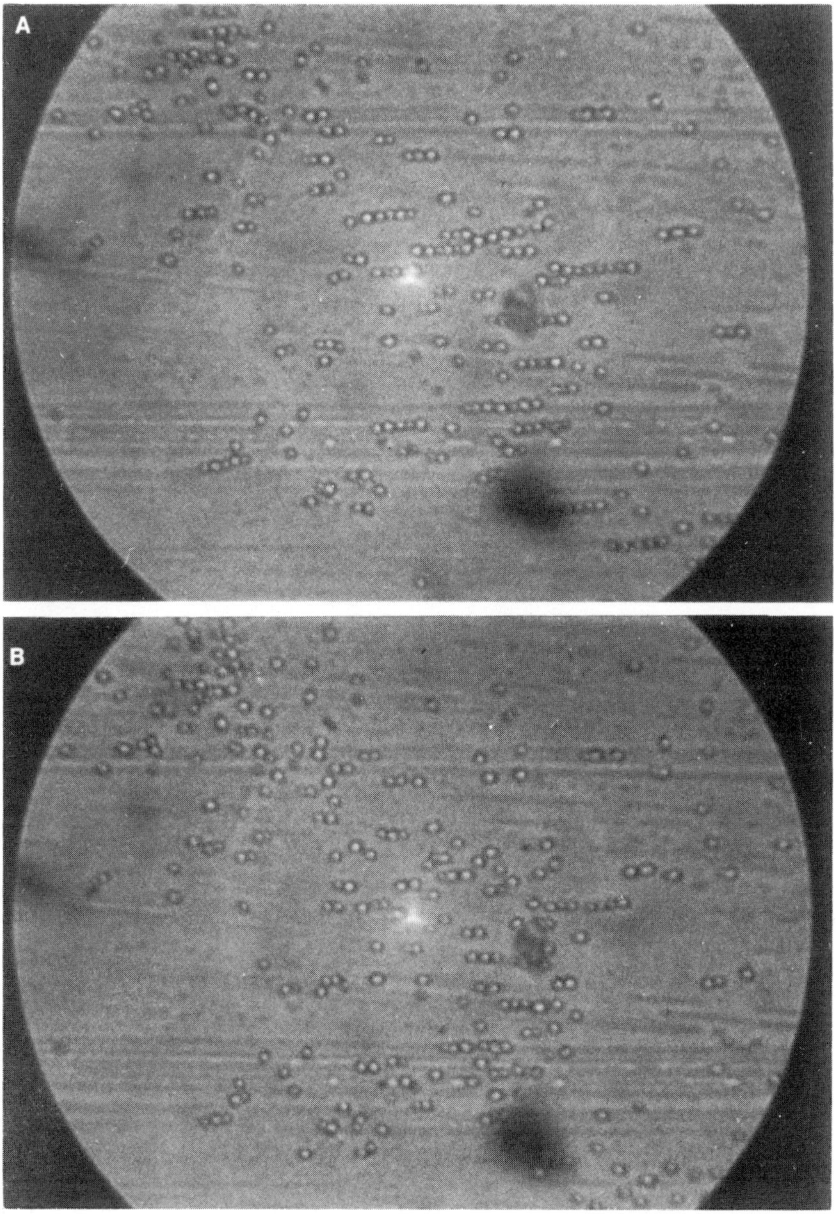

FIGURE 13. A, dielectrophoretic pearl chain of yeast cells. B, conditions identical, except for satisfying proton NMR conditions.

Water forms 70% of body weight and unlike most physical systems, the electrical conduction processes are largely protonic rather than electronic (48).

MGT of E. coli cultures grown in low-frequency magnetic fields showed a periodicity that had a probability of less than one in two million of having arisen by chance (18,38). However, the basic process whereby an alternating magnetic field between 10-100 Hz can affect the MGT by a few percent remained unexplained.

The formation of cataracts was observed after irradiating bovine eye lenses with microwaves, modulated so that the proton NMR conditions were satisfied in the ambient magnetic field. The cataracts were posterior, subcapsular, and developed within 24 hours.

The posterior cataract is the most important glare-causing cataract because it scatters light rays already nearly focussed by the rest of the lens. The sequence of events that characterize this type of cataract include the disorganization of cells at the equator, some of which subsequently migrate to the posterior capsule where they increase considerably in size to give enlarged nucleated cells called bladder cells or Wedl cells. It is possible that dielectric changes due to the NMR conditions in these cells speed their migration. Alternatively, osmotic imbalance rather than the posterior migration of equatorial cells may account for some kinds of posterior subcapsular cataract (42).

The major metabolic pathways for the lens are controlled by the enzymes, hexokinase, phosphofructokinase, and pyruvate kinase. Future experiments should test the sensitivity of these enzymes to NMR conditions.

Although the presence of high molecular weight (HMW) proteins in a cataract may not be a sufficient condition for opacity (42), water insoluble aggregates of protein can form and reach 0.5 μm in diameter as determined using polyacrylamide-gel electrophoresis. The bands holding the HMW proteins in the normal aging eye show weak non-covalent linkages, whereas the cataractous protein bands are not dissociated by detergents, but only by reduction. Electrophoresis without and with NMR conditions might be worth trying to see whether bond stabilization occurs. There is an indication that cataracts may also have an abnormally high phosphorous content (42). There are several reductase and hydrogenase enzymes in the major metabolic pathways which could give rise to this if they were inhibited, and which might be tested for sensitivity to proton NMR conditions.

In general, there are many biomedical systems that have exhibited sensitivities to low-level electromagnetic fields (36), and which should be tested for sensitivity to NMR conditions. Many of these are given by Becker and Marino (43) in a well referenced source book. It seems that the larger effects are obtained with biological systems which are under stress (39); this suggests that the various electromagnetic field effects involve but one of many redundant control pathways in a system which is normally under good homeostatic control (48). The various pulsed magnetic field therapies may be using NMR effects to generate highly coherent oscillations through spin echoes (49) to mediate in the healing process.

It is possible that the classical waveforms and experiments of electrophysiology may be to biology the equivalent of telegraph telecommunications in the days of Samuel Morse, and that nature is really in the pulse-modulated microwave communications business. Most experiments may only have looked at the modulation envelope of what happened to get detected and demodulated. The most likely spectral region for the carrier frequencies is 100 GHz to 1 THz (29), although optical frequencies should not be excluded from consideration (44). For the ELF region, recent work indicates that frequencies may need to be specified with a precision up to 1-10 mHz if results are to be meaningful; this implies possible latency periods up to hundreds of seconds.

ACKNOWLEDGEMENTS

The writers gratefully acknowledge many years of advice and encouragement from Professor H. Fröhlich, F.R.S. Much of the work was carried out with facilities and resources freely available within the University of Salford until 1981. The work on bovine eye lenses was supported during 1983-84 by the CAMPUS Venture & Enterprise Fund of the University of Salford which is gratefully acknowledged.

The assistance and cooperation of colleagues in the University of Salford is also acknowledged with thanks and especially, Dr. E.A. Flinn in respect of the work on mean generation time and cell growth, Dr. L.R. Croft in respect of the work on bovine eye lenses, and Dr. H.A. Foster in respect of much advice on the microbiological aspects of this work over many years.

REFERENCES

(1) Abragam, A.: The Principles of Nuclear Magnetism, Clarendon Press, Oxford, 1969.

(2) Emsley, J.W., Feeney, J. and Sutcliffe, L.H.: High Resolution Magnetic Resonance Spectroscopy, Vol. 1, Pergamon Press, New York, 1965.

(3) Andrew, E.R.: Nuclear Magnetic Resonance, University Press, Cambridge, 1969.

(4) Lynden-Bell, R.M. and Harris, R.K.: Nuclear Magnetic Resonance Spectroscopy, Nelson, London (UK), 1969.

(5) Farrar, T.C. and Becker, E.D.: Pulse and Fourier Transform NMR, Introduction to Theory and Methods, Academic Press, New York, 1971.

(6) Akitt, J.W.: NMR and Chemistry, Chapman & Hall, New York, 1983.

(7) Pople, J.A., Schneider, W.G. and Bernstein, H.J.: High Resolution Nuclear Magnetic Resonance, McGraw Hill, New York, 1959.

(8) Gunther, H.: NMR Spectroscopy - An Introduction, Wiley, New York, 1980.

(9) Bottomley, P.A.: Nuclear magnetic resonance: beyond physical imaging, IEEE Spectrum 20:32-38, 1983.

(10) Bloch, F., Hansen, W.W. and Packard, M.: The nuclear induction experiment, Phys. Rev. 70:474-485, 1946.

(11) Berliner, L.J. and Reuben, J., eds.: Biological Magnetic Resonance, 3 Vols., Plenum Press, New York, 1978, 1980, 1981.

(12) Lasker, S.E. and Milvy, P., eds.: Electron spin resonance and nuclear magnetic resonance in biology and medicine and magnetic resonance in biological systems, Ann. N.Y. Acad. Sci. 222, 1973.

(13) Pohl, H.: Dielectrophoresis: The Behaviour of Neutral Matter in Nonuniform Electric Fields, University Press, Cambridge, 1978.

(14) Pethig, R.: Dielectric and Electronic Properties of Biological Materials, Wiley, New York, 1979.

(15) Schwan, H.P.: Electrical properties of tissue and cell suspensions, in Advances in Biological and Medical Physics, J.H. Lawrence and C.A. Tobias, eds., Vol. 5, pp. 148-209, Academic, New York, 1957.

(16) Jafary-Asl, A.H.: The Interaction of Electric and Magnetic Fields with Biological Materials, Ph.D. Thesis, Salford University, Salford, U.K., 1983.

(17) Van Vleck, J.H.: The Theory of Electric and Magnetic Susceptibilities, University Press, Oxford, 1932.

(18) Aarholt, E., Flinn, E.A. and Smith, C.W.: Effects of low-frequency magnetic fields on bacterial growth rate, Phys. Med. Biol. 26:613-621, 1981.

(19) Gerhardt, P.: Manual of Methods for General Bacteriology, Am. Soc. Microbiol. ISBN 0-914826-29-8, 1981.

(20) Shaya, S.Y. and Smith, C.W.: The effects of magnetic and radiofrequency fields on the activity of lysozyme, Collective Phenomena 2:215-218, 1977.

(21) Zaret, M.M. and Snyder, W.Z.: Cataracts in aviation environments, Lancet, 484-485, 1977.

(22) Tyler, P.: Overview of the biological effects of electromagnetic radiation, IEEE Trans. Aerosp. & Electron. Syst. AES-9:225-228, 1973.

(23) Milroy, W.C. and Michaelson, S.M.: Biological effects of microwave radiation, Health Phys. 20:567-575, 1971.

(24) Kalant, H.: Physiological hazards of microwave radiation: a survey of published literature, Can. M.A.J. 81:575-582, 1959.

(25) Schwan, H.P. and Foster, K.R.: RF field interactions with biological systems: electrical properties and biophysical mechanisms, Proc. IEEE 68:104-113, 1980.

(26) Lerner, E.J.: The drive to regulate electromagnetic fields, IEEE Spectrum 21:63-70, 1984.

(27) Lerner, E.J.: Biological effects of electromagnetic fields, IEEE Spectrum 21:57-69, 1984.

(28) McRee, D.I.: Review of Soviet/Eastern European research on health aspects of microwave radiation, Bull. N.Y. Acad. Med. 55:1133-1151, 1979.

(29) Fröhlich, H.: Biological effects of microwaves and related questions, Adv. Electron. & Electron. Phys. 53:85-152, 1980.

(30) Milroy, W.C. and Michaelson, S.M.: Microwave cataractogenesis - critical review of literature, Aerospace Med. 43:67-76, 1972.

(31) Cleary, S.F.: Microwave cataractogenesis, Proc. IEEE 68:49-55, 1980.

(32) Duke-Elder, W.S.: Pathological action of light upon eye; Action on outer eye; Photophthalmia, Lancet 1:1137-1140, 1188-1191, 1250-1254, 1926; Lancet 2:16-19, 1926.

(33) Swerdlow, A.J.: Epidemiology of eye cancer in adults in England and Wales, 1962-1977, Am. J. Epidemiol. 118:294-300, 1983.

(34) World Health Organization, Environmental Health Criteria 35, Extremely Low Frequency (ELF) Waves, WHO, Geneva, 1984.

(35) World Health Organization, Environmental Health Criteria 16, Radiofrequency and Microwaves, WHO, Geneva, 1981.

(36) Fröhlich, H. and Kremer, F., eds.: Coherent Excitations in Biological Systems, Springer, New York, 1983.

(37) Bloch, F.: Simple interpretation of the Josephson effect, Phys. Rev. Lett. 21:1241-1243, 1968.

(38) Aarholt, E.: Biological Effects of Electromagnetic Fields, Ph.D. Thesis, Salford University, Salford, U.K., 1982.

(39) Aarholt, E., Flinn, E.A. and Smith, C.W.: Magnetic fields affect the lac operon system, Phys. Med. Biol. 27:603-610, 1982.

(40) Jafary-Asl, A.H., Solanki, S.N., Aarholt, E. and Smith, C.W.: Dielectric measurements on live biological materials under magnetic resonance conditions, J. Biol. Phys. 11:15-22, 1983.

(41) Jafary-Asl, A.H., Solanki, S.N., Aarholt, E. and Smith, C.W.: The interaction of fields satisfying NMR conditions with biological materials, Dielectrics Society Conf. on Dielectric Properties of Molecular Liquids and Biological Systems, Cambridge, March 30-April 1, 1982.

(42) Maisel, H., ed.: The Ocular Lens, Structure, Function and Pathology, Marcel Dekker, New York, 1985.

(43) Becker, R.O. and Marino, A.A.: Electromagnetism and Life, SUNY Press, Albany, New York, 1982.

(44) Popp, F-A. and Becker, G., eds.: Electromagnetic Bioinformation, Urban and Schwarzenberg, Baltimore, 1979.

(45) Jaberansari, M.: Dielectrophoretic, Electrodynamic and Magnetic Resonance Phenomena in Yeast Cells, MSc Thesis, Salford University, Salford, U.K., 1985.

(46) Smith, C.W., Marsh, P.N. and Croft, L.R.: Microwaves - new evidence, Salford Univ. Cave Suppl., 8-9, Sept. 1985.

(47) Marsh, P.N.: The Biological and Biochemical Effects of Microwave and Radiofrequency Radiation on the Bovine Eye Lens In Vitro, MSc Thesis, Salford University, Salford, U.K., 1986.

(48) Smith, C.W., Choy, R. and Monro, J.A.: Water - friend or foe, Lab. Pract. 34:29-34, 1985.

(49) Smith, C.W.: A proposed physical mechanism for pulsed electromagnetic field effects in biological systems, B.R.A.G.S., Utrecht, Netherlands, 1986.

5
Apoplastic Electropotentials in Plants: Measurement and Use

WILLIAM GENSLER

Department of Electrical and Computer Engineering
University of Arizona, Tucson, Arizona

INTRODUCTION

The object of this Chapter is to describe the characteristics and use of the electropotentials obtained by in vivo invasive sensors placed directly in plant tissue under laboratory or field conditions. First, the existence and possible origin of different types of electropotential variations will be described. Then, a specific use of the potential variations in agriculture will be presented.

BASIC TECHNIQUE AND TYPE OF POTENTIAL VARIATIONS

The basic technique is to place a palladium electrode invasively in the stem, petiole, or peduncle of the plant, and a reference electrode in the root environment (Figure 1). The measuring electrode consists of a palladium rod 150 μm in diameter and 10 mm long. The reference electrode is a conventional silver-chloride electrode, modified for long-term burial in the field. The two electrodes and the plant form a galvanic cell which yields a DC electrical potential which is coherent, reproducible, and which changes with environmental conditions. The fundamental problem is to explain the origin of the potential and the physiological reasons for the changes.

Several definitive changes in potential are observed. Immediately after the insertion of the probe, the potential rises towards more positive values. The change is termed a healing potential and is similar to the response in human tissue that occurs under the same circumstances (1,2). In virtually

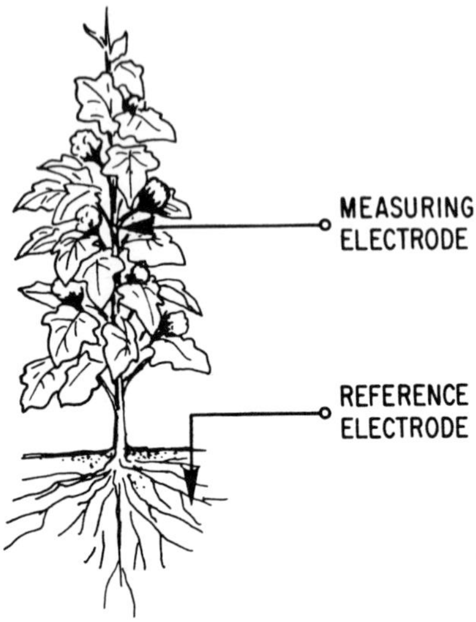

FIGURE 1. The basic measurement technique.

hundreds of electrode insertions, the author has never seen this direction of potential change violated. This suggests that the mechanism that causes the potential reaction is exceedingly basic. The timing of the rise varies widely. Tomato plants in environmental growth chambers take three days to reach a steady-state potential value (3), but cotton plants under field conditions take only a few hours.

After healing, disturbance of the probe causes a rewounding and a repeat of the insertion response. The potential under such circumstances drops precipitously and then rises in a roughly exponential manner to the original level (2). The rewounding potential is similar in form but smaller in magnitude than the original healing potential.

The steady-state or homeostatic potential level is 100-400 mV relative to the saturated silver-chloride reference electrode in the root zone. There is considerable variation from electrode to electrode, but the average value of the potential is similar in different plant types and in different soil conditions. The values obtained with a hydroponic root environment are the same as with soil under normal field

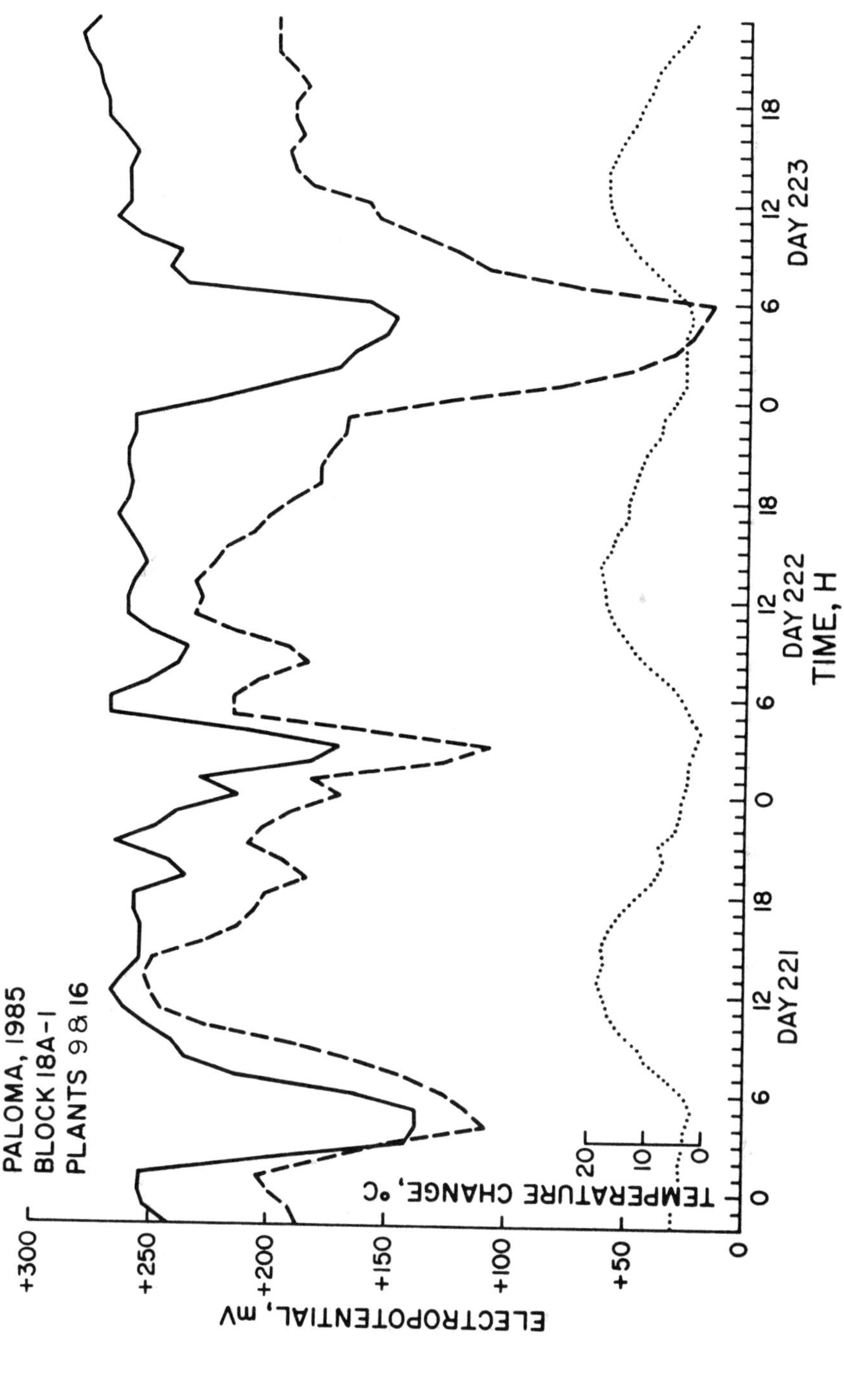

FIGURE 2. Cotton nocturnal potential. Electrode 9 and 16, Paloma Ranch, Gila Bend, Arizona, PDL90, 1985.

FIGURE 3a. Multiple electrode placement in one plant.

conditions. The origin of the potential variation between individual electrodes is an unsolved problem.

The homeostatic electropotential value in cotton changes during the 24-hour cycle, but under field conditions it settles in the mid-afternoon at approximately +230 mV. The subsequent variations are basically movements around this level.

A recent study by Ladezma-Rascon indicated that the plant attempted to maintain a homeostatic potential even in the presence of external attempts to change the level (4). He applied external loads to the measuring electrode to perturb the potential from equilibrium. The plant deviated from the homeostatic level while the disturbance was applied, but returned

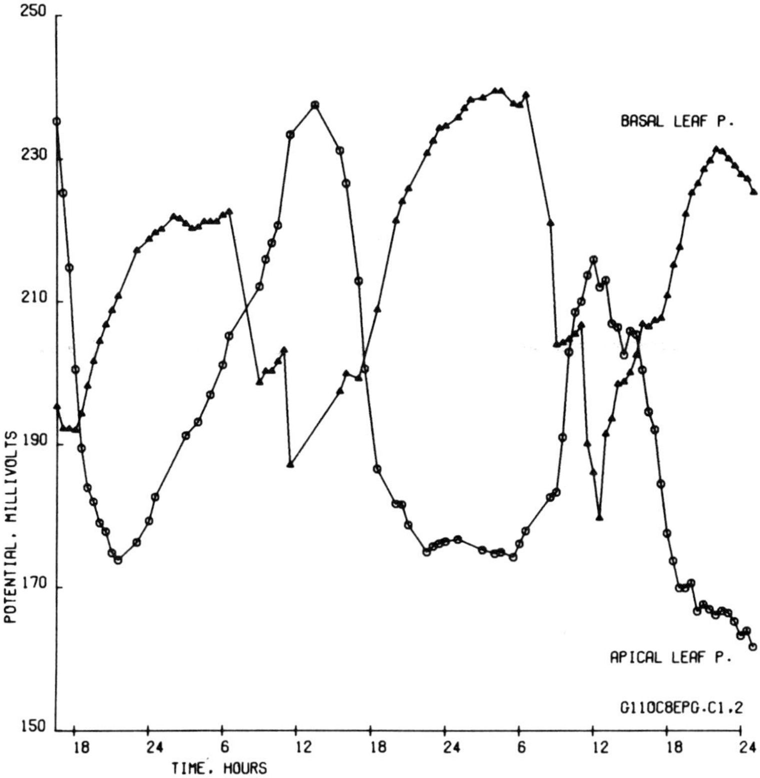

FIGURE 3b. Simultaneous electropotential variations in apical and basal petioles of the same side branch.

when the disturbance was removed. This same situation ensued if the plant potential was perturbed by the addition of external charge. The plant would accept the charge and then return slowly to the homeostatic level.

Internal plant processes, however, can readily change the homeostatic level. In water-stressed cotton under field conditions, there is a potential drop of as much as 300-400 mV upon irrigation or rainfall (5). If the drop occurs in the late afternoon, the potential remains low until about midnight when it rises for about an hour and then drops again until dawn. The potential remains high until the next late afternoon, at which time the entire sequence is repeated. This phenomenon repeats for three or four days with decreasing amplitude. Simultaneous with the drop in potential, there is an expansion of the main stem of the plant.

When cotton is not under water stress, the potential rises rapidly in the morning, and then exhibits a small drop around 0930 hours. This midmorning adjustment is not present if the rise is more moderate. Examples of this phenomenon are shown in Figure 6. The plant potential also changes at night (Figure 2).

By placing several probes in various points along a single branch, it is possible to observe systemic relations between the various regions of the plants. Figure 3a shows an example of the probe placement. Figure 3b shows the observed simultaneous potential changes. Additional experiments indicate that there is a simultaneity of potential change between the bottom and the top of the plant main stem. Whether this simultaneity implies some form of communication has not been determined. The timing of the potential change and the distance involved requires that the mode of transmission be rapid. Such rapidity is possible by hydraulic or electrochemical information transfer.

LOCATION OF THE POTENTIAL

The question that immediately arises when a galvanic circuit of this complexity is formed is the exact location of the potential changes. The DC level is an algebraic sum of the potentials of the string of interfaces that occur between the two pieces of wire that terminate the galvanic cell. It is not possible to determine the value of any individual contributor to this sum. The changes in potential can be localized, however. If one places two electrodes in the stem of the plant in a vertical direction and if the electrodes are relatively close, 1 cm, for example, then one can assume that there exists a common path from both measuring electrodes back to the reference electrode. If this is the case, a change in this common path will show up as a simultaneous change in the potential of both measuring electrodes. A change in the reference electrode itself would be such a disturbance. Non-simultaneous changes in potential can then be attributed to the interface of the measuring electrode. Additional evidence to support this contention occurs when one purposely causes a mechanical disturbance at one of the measuring electrodes. A rewounding potential occurs as described above. The conclusion is that the changes of potential that occur arise from the interface between the measuring electrode and the fluid in which it bathes. The fluid is extracellular or apoplastic which means that the measurement itself is extracellular.

The location of the interface of the measuring electrode and the plant tissue is an important question. The electrode penetrates into the plant tissue 2-3 mm. In most plant tissue, this means it passes through the dermal and vascular layers, and into the pith. Since the surface of the electrode is equipotential, electron transfer can occur anywhere along the length of penetration. It is therefore not possible to isolate the location of the potential-determining region. Electron and light microscopy studies have indicated that there is normal vascular tissue within 20 μm of the electrode surface (6). Studies of the anatomical reaction of plant tissue to probe penetration indicate that if the probe is placed at the upper shoulder of the S-shaped growth curve, almost no scar tissue forms (7). If the probe is placed in juvenile tissue, the reaction is such that scar tissue will form as an invagination of the outer dermal layers (8).

PLANT ACCEPTANCE

The question of plant acceptance of the electrode must also be considered from a materials viewpoint. Glass, plastic, silicon, silicon dioxide and tungsten carbide are not biocompatible, and result in necrosis of the tissue contiguous to the probe. Titanium, palladium, aluminum, carbon and stainless steel are readily accepted. The acceptance takes the form of healing and subsequent sealing off of the electrode entry point. It is essential that sealing occurs. This insures the presence of a normal fluid at the surface of the measuring electrode.

The tissue reaction has been extensively studied in tomato, cotton, and pecans (9-12); it appears to be similar in tomato and cotton. In pecans, the hardness of the tissue permits placement of the electrode only during the spring, in the region of new buds. If placement occurs at that time, the bud expands and substantial hardening occurs within a five-week interval. At that same time, scar tissue, referred to as a bolsa or pocket, forms around the electrode. When the electrode is removed for inspection, the pocket comes with it, indicating a substantial adhesion between the tissue and the electrode. This indicates that the scar-tissue layer renders the potential changes suspect, and further investigation is required before definitive conclusions can be reached. Nocturnal potentials in pecan have been seen, but the compartmentalization of the electrode obscures their meaning.

ORIGIN OF THE POTENTIAL

The electrochemical origin of the potential has been examined both theoretically and experimentally. Goldstein listed three possibilities for the origin of the potential: a redox couple or couples, image charge on the electrode due to charge in the contiguous tissue, and unbalanced charge in the region of the electrode (13). The first possibility is electrochemical while the two other possibilities are physical. Silva-Diaz used cyclic voltammetry to determine the presence of redox couples in the apoplast fluid, and found that no redox couples were discernible (13). The study did indicate that oxygen may be a factor in setting the potential. It also pointed out the main experimental limitation of in vivo cyclic voltammetry, namely, solution resistance. It was not possible to place the reference electrode physically close enough to the measuring electrode to achieve a low resistance in the path between the two electrodes. A low resistance is necessary to discern the presence of weak couples, and to facilitate interpretation of the voltamograms. Ledezma-Rascon compared carbon and palladium electrodes in vivo in cotton subjected to active and passive loads (4), and found that carbon electrodes had a consistently lower homeostatic potential. He also observed differences in the transient and steady-state response of the two electrode types. The responses were analyzed in terms of the circuit model in which the interface is characterized by a parallel resistance and capacitance, R_I and C_I

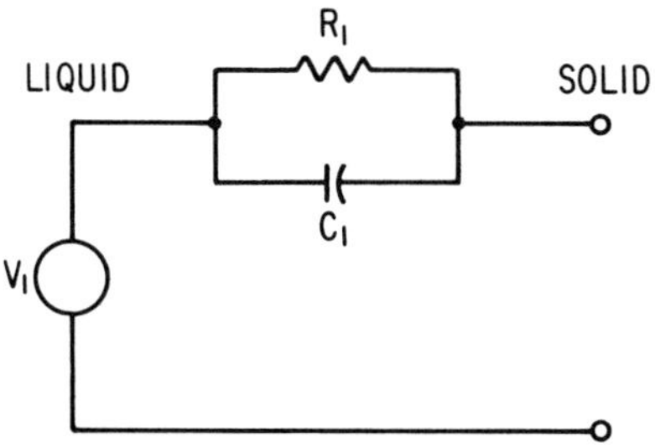

FIGURE 4. Single energy storage model of the electrode-apoplast fluid interface.

respectively (Figure 4). The resistance is the conventional charge-transfer resistance derived from the Butler-Volmer equation for small perturbations about the Nernst potential. The capacitance is a result of the double-layer of charge at the interface. The carbon electrode likely has a large active electrochemical surface area per unit of geometrical surface area, leading to a large surface capacitance and a small surface resistance. The opposite situation prevails for the palladium electrode in which there is a relatively small electrochemically active surface area per unit of geometrical surface area. The results were in agreement with the experimentally observed response, suggesting that the origin of the potential is electrochemical, and that surface characteristics of the electrode figure prominently in the change in the potential of plants subject to external disturbances.

The modified coulostatic response of carbon versus palladium also lends insight into the origin of the potential. This response is obtained by discharging a capacitor into two palladium or carbon electrodes placed within 1 cm of each other in the stem of the cotton plant. The carbon pair yields a much lower magnitude of response, and it recovers more quickly. This is in agreement with the theoretical model of carbon involving a relatively high interfacial capacitance and a relatively low charge-transfer resistance.

The difference in the steady-state potential between carbon and palladium electrodes also supports the thesis that the potential is electrochemical and not physical in nature. In the latter case, the potential would arise as a response to external charge.

If a palladium electrode and a reference electrode are placed in a beaker of tap water and the solution sparged with nitrogen, the oxygen is driven off and replaced by nitrogen, and the potential drops precipitously. This implicates oxygen in the origin of the potential, and thus also supports the electrochemical theory of origin.

These results have led to the hypothesis that the potential changes are due to varying concentrations of oxygen in the tissue (15). A low potential is associated with a low oxygen concentration, and a high potential with a high oxygen concentration. The healing potential can be explained in terms of this hypothesis by suggesting that the wounding and healing result in an initial precipitous consumption of oxygen and then a gradual increase in the oxygen concentration in the healing site. Under this hypothesis, the potential drop associated with the change in stem diameter would be a result of an expenditure of energy and concomitant consumption of oxygen. This energy is required to implement the transfer of water associated with the expansion of the stem. The potential rise in the morning is a result of an increase in the oxygen level in the tissue over the level caused by diffusion. The drop of potential under water-stress conditions is a result of a net decrease in the oxygen level in the tissue during the morning hours. The nocturnal potentials are a result of a sudden increase in metabolic activity on the part of the plant.

The fact that the carbon electrode produces a homeostatic potential and acts basically similar to the palladium electrode in its response would require oxygen to be involved in a complex form. A possibility is that adsorbed oxygen is presented at the surface of the electrode in a non-charge neutral complex.

ELECTRODE LOCATION AND PLANT ARCHITECTURE

The location of the electrode in the plant has received considerable attention. There are four possibilities: the main stem, side branches, petioles, and peduncles. Experimental results indicate the electropotentials from the petioles pass through a specific sequence as the leaf itself moves from juvenility to senescence. For example, a senescent leaf maintains an almost constant petiole potential until a change occurs in the water status of the root zone. At that time, the potential changes vigorously for a few days before returning to the predisturbance level. The peduncle potential is relatively stable compared to the potentials in the other parts of the plant. At the termination of the fruit development process in cotton, the potential of the peduncle drops precipitously. For long-term sensing of plant electrical activity, the main stem is the most desirable location. There is evidence that the age of the tissue has an influence on the level of potential variations, but in cotton over an entire growing season, the decrease in activity has not been significant.

The electrode placement described above is amenable to plants in which an anatomical region of sufficient size to insure long-term undisturbed placement is present. Cotton, tomatoes, soybeans, safflower, alfalfa, grapes, and potatoes are examples of plants with a stem architecture that permits placement of the probe. Sugar beets, lettuce, carrots, and cabbage are examples of plants not easily amenable to electrode placement. Several commercially important monocotyledons such as corn, wheat and rice are also not easily amenable to electrode placement. A distinct petiole is not present and the main stem in wheat consists of a series of concentric tubes which move relative to each other in the course of growth. An electrode placed in the stem passes through several of these tubes simultaneously. The relative movement then causes a tearing of the tissue around the probe.

REFERENCE ELECTRODE

The reference electrode is shown in Figure 5. It is basically a silver-chloride electrode with a connection to the root zone consisting of a polyacrylamide layer and a porous ceramic plug. The purpose of the latter two items is to prevent loss of the filling solution and maintenance of a constant junction potential. One of the major electrode design problems is to seal the wire connection from water penetration. The

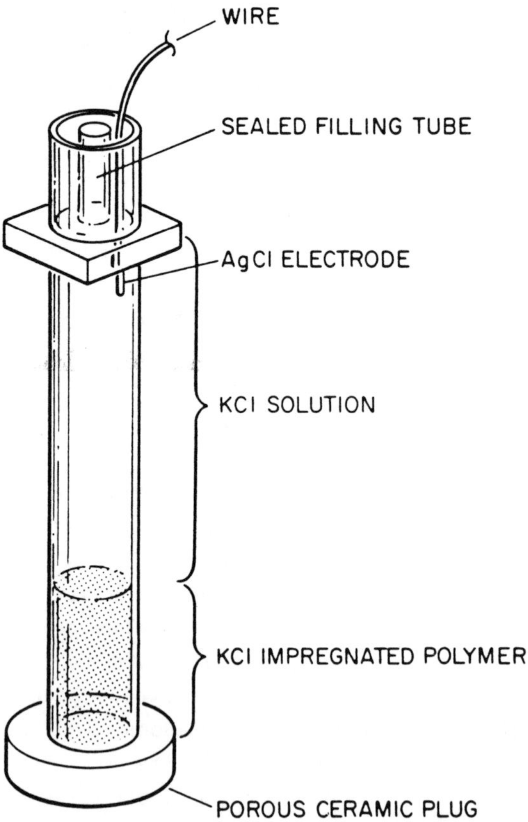

FIGURE 5. Electrical model of the reference electrode.

present method is to use a barrier of fiber-impregnated tar. This material will wet the polyvinylchloride or polyethylene insulation covering the copper wire, and thus prevent water encroachment into the soldered junction.

An auxiliary reference electrode is always employed to test the junction of the main reference electrode. The two electrodes are placed side by side in the root zone and form a concentration cell. A change in the potential of this concentration cell indicates a change in the reference electrode potential. A constant potential indicates that the potential between the bulk root zone and the wire leading from the main reference electrode is constant. Electrodes of the design shown in Figure 5 are stable to within a few millivolts for months at a time.

The distance between the reference electrode and the plant under measurement has not been a problem. The furthest separation employed has been about thirty meters. In addition, a common set of reference electrodes can be transported from site to site and used to determine absolute potential levels. For example, cotton electropotentials in fields 125 km apart have been compared by this technique.

APPLIED ASPECTS OF THE PHYTOGRAM TECHNIQUE

A major emphasis has been directed towards relating the variations in potential to the water status of the cotton plant under field conditions (16). The application begins with the analog plots of potential versus time (phytograms). An analysis of the phytogram patterns combined with an empirical knowledge of the water status of the field has led to general conclusions concerning the potentials that one can expect when a crop is in a water-stressed or non-stressed condition. A field not under stress will yield an average potential that will rise in the morning. A field under mild stress will yield a potential that rises only moderately. A field under serious water stress will yield a potential that falls in the morning. The method of using the potentials to determine water status is to quantify the average potential variation from 0500 hours to 1300 hours. The present method of operation is to place the electrodes in the main stems of the cotton plant at approximately node nine. Forty-four plants are monitored over an area approximately 31 meters by 20 meters. Data is transferred by wire from the electrode to a central acquisition canister (pod), buried in the field. The pod contains the electronics required to time the acquisition, multiplex the electrodes, process and store the data, and transfer the data to a central site by wireless telemetry.

The voluminous electropotential data is processed through the use of a cardinal point. The cardinal point is defined as the potential of each probe at 0500 hours each day. The potential is measured at later times and an increment formed by

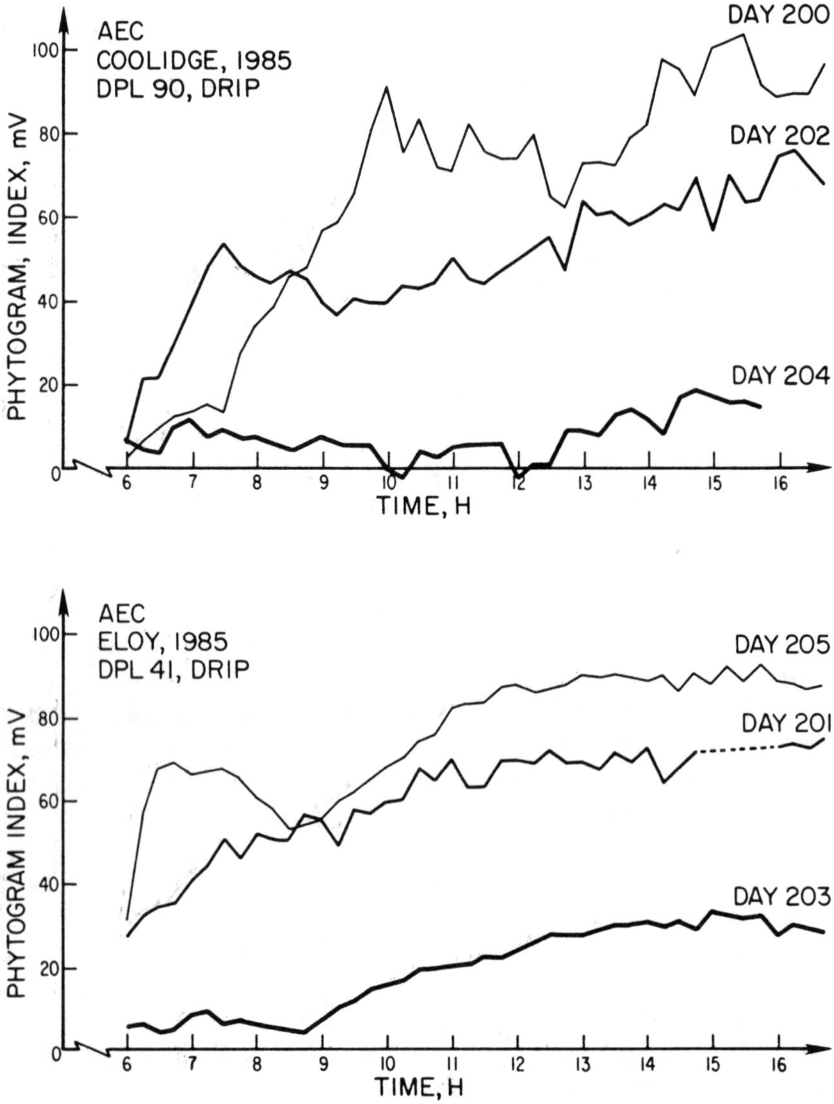

FIGURE 6. Daily trend in the phytogram index in cotton, Regal Farms, Eloy, AZ, and Sundance Farms, Collidge, AZ, 1985. Numbers are in Julian days.

algebraically subtracting the value of the cardinal point from the potential at these particular times. These incremental values are then summed and averaged to yield a potential increment. This average potential increment is termed a phytogram index. The index can be plotted at various times after 0500 hours to yield a trend in the potential between sunrise and solar noon. An example of this daily trend is plotted in Figure 6.

The phytogram index can be used as the essential measurement in a complete network wherein the bioelectrochemical status of the crop is monitored every 15 minutes throughout the 24-hour period. This status is transferred by radio telemetry to the farm headquarters where it is processed into a daily numerical index. The index is used to set a water application rate for the field.

ACTIVE BIOELECTROCHEMICAL MEASUREMENTS

The previous discussion has centered on the measurement and use of an electropotential obtained by purely passive means. No external energy was employed to make the measurement. Active bioelectrochemical measurements are also possible wherein energy is supplied externally to achieve a particular assay of some desired bioelectrochemical condition. For example, in vivo cyclic voltammetry can be performed by placing electrodes in the plant and sweeping the potential of the electrode above and below the homeostatic potential level in the same manner as in vitro voltammetry (14). Various other microelectroanalytical methods can be similarly applied. One method that circumvents the problem of solution resistance encountered in cyclic voltammetry is to use coulostatic impulses. In this method a pair of electrodes are placed very near each other in the main stem of the plant and electrical charge is precipitously transferred into and out of the tissue adjacent to the electrode. The potential changes that occur relative to the third reference electrode can be used to assay the condition of the plant electrolyte.

The active methods described above can use conventional macroelectrodes such as round rods placed invasively in the

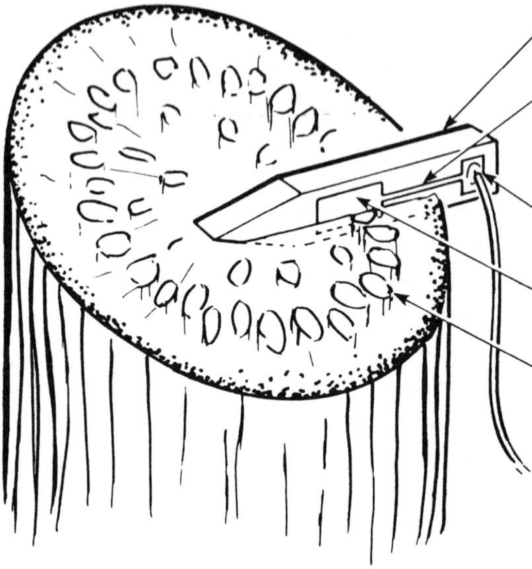

FIGURE 7. Invasive planar microelectronic electrode (not to scale).

tissue. Alternately, planar microelectronic electrodes can be employed in the manner shown in Figure 7. The electrode consists of a rigid substrate, a surface or window to achieve the assay, a channel to transfer the signal out of the plant and a bonding pad to connect the electrode to a lead wire. These electrodes are fabricated in the same manner as conventional chip production. A stainless steel rather than a silicon substrate is used to achieve biocompatibility, and polymide layers are used for insulation. Multiple-window electrodes with specific ion-sensitive layers can also be produced. There are myriad possibilities for in vivo electrochemical analysis of the plant electrolyte.

SUMMARY

This discussion has considered the basic method of determining the in vivo electropotential of the plant apoplast. Various characteristics of this potential have been presented.

The physiological interpretation and use of the patterns of potential variation have just begun. Determination of crop water status is an initial application of these studies. Active electrochemical techniques can also be employed using both conventional macroelectrodes and planar microelectrodes.

REFERENCES

(1) Gensler, W.: Tissue electropotentials in Kalanchoe blossfeldiana during wound healing, Am. J. Bot. 65:152-167, 1978.

(2) Gensler, W.: Electrochemical healing similarities between animals and plants, Biophys. J. 27:461-466, 1979.

(3) Gensler, W.: Bioelectric potentials and their relation to growth in higher plants, Ann. N.Y. Acad. Sci. 238:280-299, 1974.

(4) Ledezma-Rascon, E.A.: Modeling of the bioelectric system formed by palladium and carbon electrodes inserted in cotton (Gossypium hirsutum) plants, Master's thesis, University of Arizona, 1985.

(5) Gensler, W. and Diaz-Munoz, F.: Simultaneous stem diameter expansions and apoplastic electropotential variations following irrigation or rainfall in cotton, Crop Sci. 23:920-923, 1983.

(6) Goldstein, A.: Interface between cotton tissue and a penetrating noble metal probe, Am. J. Bot. 49:513-518, 1982.

(7) Ruggenstein, S.: Tissue response to palladium microprobe, as observed in Gossypium hirsutum, L. (Malvaeceae), Am. J. Bot. 64:519-528, 1982.

(8) Gensler, W.: Transition electropotentials and growth in Lycopersicon esculentum, 4th International Symposium on Bioelectrochemistry, Woods Hole, MA, October, 1977.

(9) Gensler, W.: An electrochemical instrumentation system for agriculture and the plant sciences, Electrochemistry Society Meeting, Seattle, WA, May, 1978.

(10) Gensler, W.: Transition electropotentials and growth in Lycopersicon esculentum, Biochem. Bioenergetics 5:152-167, 1978.

(11) Gensler, W.: An electrochemical instrumentation system for agriculture and the plant sciences, J. Electrochem. Soc. 127:2365-2370, 1980.

(12) Gensler, W.: Apoplastic electropotentials in cotton under variable water stress, Ann. Meeting Am. Soc. Pl. Physio., Urbana, IL, June, 1982.

(13) Goldstein, A. and Gensler, W.: Physiological basis for electrophytograms, Part I: Theoretical considerations, Bioelectrochem. Bioenergetics 8:645-659, 1981.

(14) Silva-Diaz, F., Gensler, W. and Sechaud, P.: In vivo cyclic voltammetry in cotton under field conditions, J. Electrochem. Soc. 30:1464-1468, 1983.

(15) Gensler, W.: Advanced agricultural instrumentation, design and use, Proceedings of NATO Advanced Studies Institute, Martinus Nijhoff, Dordrecht, 1986.

(16) Gensler, W.: Method and apparatus for electrically determining plant water status, U.S. Patent No. 3,967,198, 1976.

III

ELECTRICAL PROPERTIES OF TISSUE

6
Electrical Properties of Biological Tissue

RONALD PETHIG
Institute of Molecular and Biomolecular Electronics
University College of North Wales
Bangor, United Kingdom

INTRODUCTION

Increasing attention is being given to investigations of the electrical properties of biological materials. There are several reasons for this, which include an increasing awareness of the possible physiological effects associated with the absorption by tissues of electromagnetic fields (1-5). Studies of the ways in which tissues interact with electromagnetic energy are also important in the continuing developments in radiofrequency and microwave hyperthermia (6,7), impedance pneumography (8), impedance plethysmography (9), electrical impedance tomography (10,11), the thawing of cryogenically preserved tissue (12), and in the use of pulsed electric and magnetic fields to aid tissue and bone healing (13).

Measurements of the electrical properties of biological materials have provided important contributions to the biophysical and physiological sciences. For example, Höber (14) measured the electrical impedance of red blood cell suspensions up to 10 MHz and, finding that their impedance decreased with increasing frequency, concluded that the cells were surrounded by a poorly conducting membrane and that they contained a cytoplasm of relatively low resistivity. Among the first indications of the ultra-thin nature of this membrane were those provided by Fricke (15) who obtained a value of 0.81 $\mu F/cm^2$ for the red blood cell membrane. On assuming a value of 3 for the dielectric constant of the membrane material, Fricke derived a membrane thickness of 3.3 nm. Quantitative details of the molecular size, shape and hydration content of protein molecules were provided by the dielectric measurements of Oncley (16), and these were extended in the laboratories of Hasted (17), Grant (18), and Schwan (19) to provide more details of the physical nature of protein hydration. As examples of how electrical and dielectric studies continue to provide new knowledge to the biophysical and physiological sciences, we can quote the studies of the diffusional motions of lipids and proteins in cell membranes by Kell and Harris (20), the field-induced cell rotation studies of Zimmermann's laboratory (21), studies of the influence of hydration on enzyme activation and molecular mobility (22), and of protonic and ionic charge transport processes in protein structures (23,24).

In this Chapter we shall be concerned with the parameters that influence the way in which biological materials interact with nonionizing electromagnetic (EM) radiation. Since EM radiation is composed of electric and magnetic fields, we need in principle to consider not only the electrical parameters such as conductivity and permittivity, but also the magnetic permeability. However, apart from materials of the form of magnetite, which is known to be associated with orientational and navigational sensing abilities of some bacteria, insects and mammals (25), most biological materials have a permeability close to that of free space. Magnetic fields are also able to interact with the nuclear magnetic moments or with unpaired electrons associated with paramagnetic ions or free radical species. Such interactions are highly specific and in general have very little influence on bulk electrical properties. In this way, we shall be primarily concerned with the dielectric and electrical conduction properties of biological materials, and in particular with variations of the relative permittivity and conductivity as a function of frequency. A full historical account of this field of study cannot be attempted here, but may be derived from several texts (26-28). A full description of the electrical properties of cells and tissues should ideally include details associated with active ion transport and membrane potentials, but we shall mainly be concerned here with the so-called passive electrical parameters. Relevant aspects of other tissue electrical properties can be found elsewhere (29-31). Finally, valuable tabulations of the dielectric properties of various tissues and biomaterials have already been formulated by Geddes and Baker (32) and by Stuchly and Stuchly

(33). An update of these last two reports is attempted here, together with an overview of the present theories which have been formulated in an effort to understand the electrical properties in terms of the underlying physical and physiological processes involved.

DIELECTRIC THEORY: A SUMMARY

The electrical properties of a material held between two parallel electrodes of area A and separation d can be completely characterized by its electrical conductance G and capacitance C, as defined in the following two equations:

$$G = \sigma A/d \qquad \text{and} \qquad C = A\,\varepsilon_0\,\varepsilon/d \qquad (1)$$

The conductivity σ is the proportionality factor between electric current and electric field, and is a measure of the ease with which delocalized charge carriers can move through the material under the field's influence. For biological materials, the conductivity arises mainly from the mobility of hydrated ions. The factor ε_0 is the dielectric permittivity of free space (8.854×10^{-12} F/m) and ε is the material's permittivity relative to free space, sometimes referred to as the dielectric constant. The permittivity is the proportionality factor between electric charge and the electric field, and it reflects the extent to which localized charge distributions can be distorted or polarized under the influence of the field. For biological materials such charges are mainly associated with electrical double layers occurring at membrane surfaces or around solvated macromolecules, or with polar molecules which by definition possess a permanent electric dipole moment. Examples of electrical double layers at a cell membrane surface and around a globular protein, as well as of a molecular dipole, are shown in Figure 1. The simplest molecular dipole consists of a pair of electrical charges +q and −q separated by a vector distance s, and in this case the dipole moment m is given as m = qs. For a protein molecule such as that shown in Figure 1, positive and negative charges arise from the presence of ionizable acidic and basic amino-acid side chains in the protein structure, and these will give rise to a comparatively large

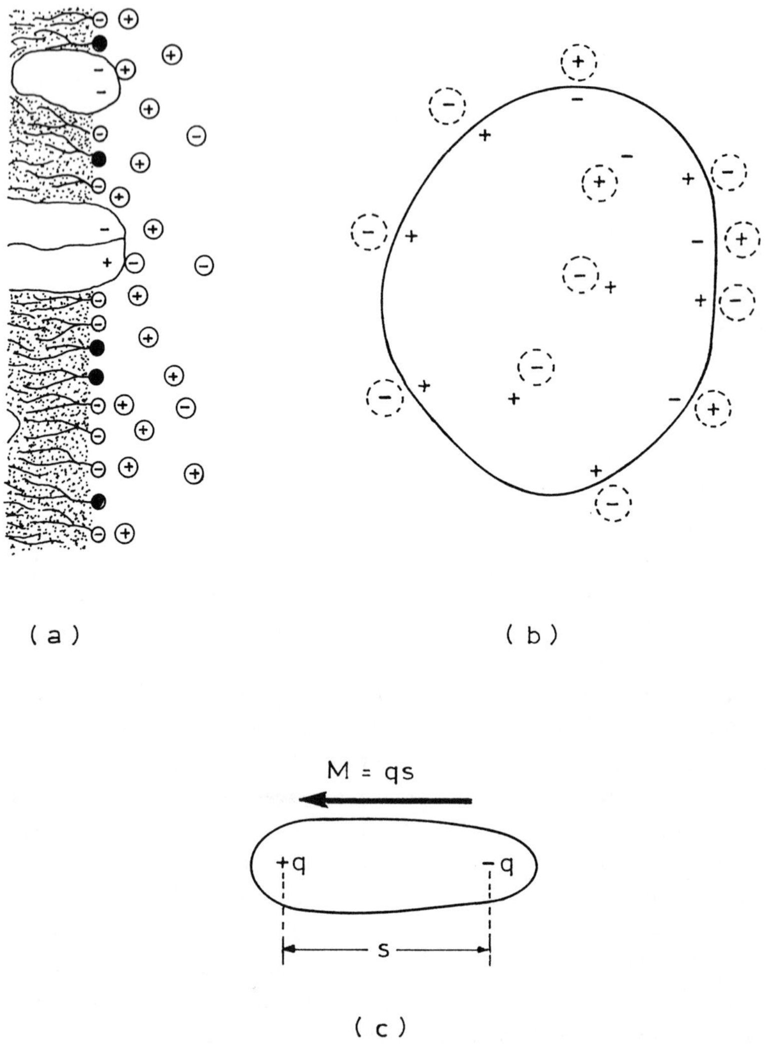

FIGURE 1. (a) Electrical double layers formed at the surface of a membrane and (b) around a globular protein. (c) A simple polar molecule of dipole moment M.

dipole moment whose value will vary with pH and molecular conformation.

Each type of polarizable entity will exhibit its own characteristic temporal response to an imposed electric field and this is mathematically handled by describing the relative permittivity as a complex function of the form:

$$\varepsilon^*(\omega) = \varepsilon_\infty + (\varepsilon_s - \varepsilon_\infty)/(1 + i\omega\tau) \qquad (2)$$

in which ε_∞ is the permittivity measured at a sufficiently high frequency for the polarizable entity to be unable to respond to the electric field, ε_s is the limiting low frequency permittivity where the polarization effect is fully realized, ω is the angular frequency, i is $\sqrt{-1}$, and τ is the characteristic response or relaxation time. The real and imaginary components of the relative permittivity can be expressed in the form:

$$\varepsilon^* = \varepsilon' - i\,\varepsilon''$$

in which the real part ε', corresponding to the permittivity parameter in Equation (1), is given by:

$$\varepsilon'(\omega) = \varepsilon_\infty + (\varepsilon_s - \varepsilon_\infty)/(1 + \omega^2\tau^2) \qquad (3)$$

The imaginary component ε'', corresponding to the dissipative loss associated with the polarizable charges moving in phase with the electric field, is given by:

$$\varepsilon''(\omega) = (\varepsilon_s - \varepsilon_\infty)\omega\tau/(1 + \omega^2\tau^2) \qquad (4)$$

and takes the form of a loss peak as shown in Figure 2 for water. The loss factor ε'' can also be defined in terms of a frequency-dependent conductivity as:

$$\varepsilon'' = \sigma(\omega)/\omega\varepsilon_0 = (\sigma_0 + \sigma_d(\omega))/\omega\varepsilon_0$$

where σ_0 is the steady-state conductivity arising mainly from mobile ions, and $\sigma_d(\omega)$ is the frequency-dependent conductivity arising from dielectric polarization losses.

A more useful form of the above equations can be derived from defining the magnitude of the dielectric dispersion as:

$$\Delta\varepsilon = \varepsilon_s - \varepsilon_\infty$$

From Equations (3) and (4) we then obtain the relationships:

$$\varepsilon'(\omega) = \varepsilon_\infty + \Delta\varepsilon/(1 + (f/f_r)^2) \tag{5}$$

and

$$\sigma(\omega) = \sigma_s + 2\pi\varepsilon_o f^2 \Delta\varepsilon/f_r(1 + (f/f_r)^2) \tag{6}$$

in which f_r is the relaxation frequency ($f_r = 1/2\pi\tau$). The factor σ_s is the low-frequency limit of the conductivity that includes the steady-state conductivity and dielectric losses associated with any polarization processes having relaxation frequencies well below that defined by f_r above. For frequencies very much greater than f_r, from Equation (6) we have:

$$\Delta\sigma = \sigma_\infty - \sigma_s = 2\pi\varepsilon_o f_r \Delta\varepsilon \tag{7}$$

This shows that the increment in conductivity is directly proportional to the permittivity change and can be used as a check on the validity of experimental data. We can also write Equation (6) in the form:

$$\tau = \varepsilon_o \Delta\varepsilon/\Delta\sigma$$

and this is a relationship that holds reasonably well for the case where the polarizable entity exhibits a spread of relaxation times.

A simple model, which can be used to describe dielectric relaxation processes that involve dipolar molecules, is one that considers the dipoles to be spheres whose rotation is opposed by the viscosity of the surrounding medium. The relevant relaxation time for such a sphere is given by:

$$\tau = \xi/2kT$$

where ξ is a molecular friction constant that relates the torque exerted on the dipole molecule by the external electric field to the molecule's angular velocity, k is Boltzmann's constant and T is the absolute temperature. Assuming the dipole molecule to be equivalent to a rigid sphere of radius a turning in a hydrodynamic fluid of macroscopic viscosity η, then using Stokes' law we have:

$$\xi = 8\pi\eta a^3$$

This gives the relaxation time as:

$$\tau = 4\pi\eta a^3/kT \qquad (8)$$

Considering the simple nature of the model used, Equation (8) gives surprisingly good results, even for the case of water molecules rotating in bulk water. The distance between adjacent oxygen molecules in bulk water is 0.28 nm, so we can take the molecular radius of water to be one-half of this, namely 0.14 nm. At 293°K the viscosity of water is 10^{-3} kg/msec and the value for the relaxation time derived from Equation (8) is 8.5×10^{-12} sec, in close agreement with the experimental value of 9.3×10^{-12} sec. This relaxation time is equivalent to a relaxation frequency of 17 GHz, and the frequency-dependencies of ε' and ε'' for bulk water at 293°K are shown in Figure 2(a). The corresponding increase in the conductivity of pure water, from its low-frequency value of 5×10^{-6} mho/m, is shown in Figure 2(b). The low-frequency conductivity of biological electrolytes (equivalent to 150 mM saline solution) is around 2 mho/m, and above 2 GHz their conductivities exhibit approximately the same frequency dependencies as the curve shown in Figure 2(b).

Following the work of Kirkwood (34,35) we can relate the dielectric dispersion strength to the dipole moment m and molecular weight M of the polar molecule according to the relationship:

$$\Delta\varepsilon = NCgm^2/2\,\varepsilon_0 MkT \qquad (9)$$

where N is Avogadro's number and C is the concentration (kg/m^3) of the polar molecule in the solvent. The parameter g was

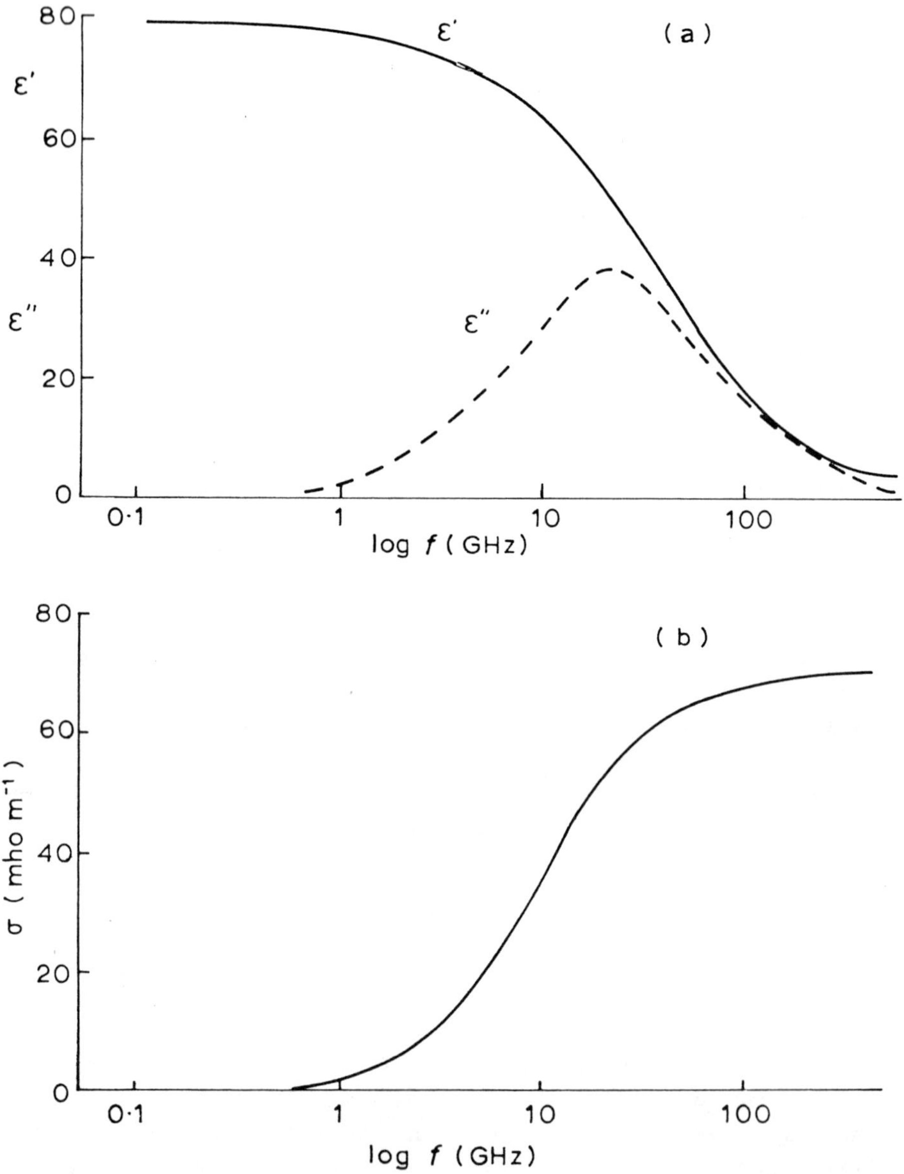

FIGURE 2. (a) The dielectric dispersion and loss peak for pure water at room temperature. (b) The associated increase in the conductivity of pure water from its low frequency value of 5×10^{-6} mho/m.

introduced by Kirkwood to take account of molecular associations and correlation effects between the solute and solvent molecules. Such effects are particularly large for hydrogen-bonded liquids such as water, where the rotation of one water molecule disrupts the local hydrogen bonding and requires, as compensation, the correlated reorientation of four or more neighboring water molecules. For water at room temperature, g is 2.8, and it drops to around 2.5 at 100°C. A value for g of unity implies no molecular associations, and this situation is usually approached with increasing molecular weight of the polar solvent molecule. In the next section we shall see that for α-amino acids g has an estimated value of around 1.2. For globular proteins in aqueous solution it is usually assumed that g has a value of unity. Sometimes the polar molecules are rotationally hindered, and this can give rise to a value for g of less than unity. This effect can occur in different cases including polar side-chains in protein structures, proteins partially immobilized in membranes, or water strongly associated (bound) with protein or membrane surfaces.

Hopefully, this brief outline of dielectric polarization processes and theory will aid those whose specializations lie outside the subject to more fully appreciate the dielectric topics described in this Chapter. More comprehensive treatments of dielectric theory of relevance to biological materials can be found in Grant et al. (27) and Pethig (28).

AMINO ACIDS, PROTEINS AND DNA

AMINO ACIDS

The term amino acid could be used to refer to any compound that contains amino and acidic groups, but it is usually considered to mean the α-amino carboxylic acids isolated from natural sources. Well over 100 of these have been isolated, but only 20 are commonly obtained upon hydrolysis of proteins. These acids contain an amino group in a position alpha to a carboxyl group and can be represented by the following chemical structure:

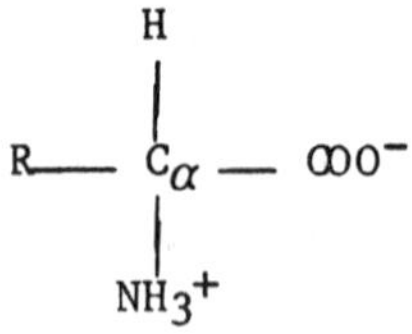

where R is the variable side chain characterizing the particular amino acid. The side chains largely determine the dielectric properties of proteins. The structure above is shown in the form that exists in aqueous solutions at neutral pH, in which the carboxylic acid and amino groups are ionized. One consequence of this is that the process of solvation of amino acids in water is accompanied by a strong negative volume change, resulting from the electrostatic forces of attraction between the polar water molecules and the two charged groups. Also, because the ionized form (zwitterion) represents a large dipole, neutral solutions of amino acids have high permittivities. Amino acids in neutral solution absorb infra-red radiation at 1580 cm^{-1}, which is characteristic of the COO^- carboxylate ion, and not at 1710 cm^{-1} as would be the case for a nonionized carboxyl group. We shall refer to this again when considering the properties of protein-associated water.

The simplest amino acid is glycine, for which the side chain R is one hydrogen atom. From a straightforward structural model of this molecule, the distance d between the positive charge on the nitrogen atom of the amino group and that of the negative charge shared between the oxygen atoms of the carboxyl group should be about 0.32 nm. The effective dipole moment should then have a value given by:

$$m = qs = (1.6 \times 10^{-19}) \times (3.2 \times 10^{-10}) = 5.1 \times 10^{-29}\ C \cdot m = 15.3\ \text{Debye units.}$$

This value compares reasonably well with the early estimate by Wyman (36) of 20 Debye units obtained from dielectric measurements. Dunning and Shutt (37) found that the permittivity of glycine solutions is constant from pH 4.5 to pH 7.5, but falls sharply on both sides of this pH range. This is another consequence of the dipolar ionic form of α-amino acids. At the lower pH values the carboxyl group will tend not to be ionized, and likewise for the amino group at the high pH range, so that the polar form will disappear in acid or alkali solutions. In its polar, zwitterion, form the dipole moment per unit volume of an α-amino acid is larger than that of a water molecule, and amino acid solutions therefore exhibit a greater permittivity than that of pure water, as can be seen for glycine in Figure 3. At room temperature the dielectric dispersion for glycine solutions is centered around 3.3 GHz (27), and it overlaps with the dispersion for water (Figure 3). Because of the large electrostatic interactions that occur between the charged carboxyl and amino groups and the surrounding water dipoles, the model used to derive Equation (8) is not valid (a relaxation frequency of 12.5 GHz is predicted from Equation (8)). In fact, no straightforward theoretical interpretation of the dielectric dispersion of amino acid solutions has been forthcoming. The molecular polarizabilities of amino acid solutions have often been expressed in the form:

$$\varepsilon = \varepsilon_1 + \delta c \quad (10)$$

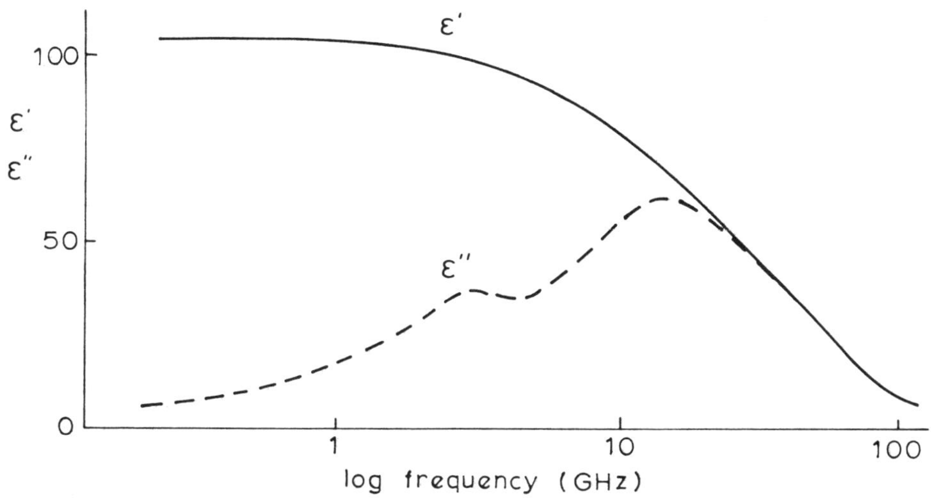

FIGURE 3. The dielectric dispersion of a 1 M aqueous glycine solution. The low-frequency permittivity is greater than that of pure water and the dispersions for glycine and water overlap.

where ε and ε_1 are the permittivities of the solution and pure solvent respectively, and c is the molar concentration of the solute. The specific dielectric increment δ is given by:

$$\delta = \Delta\varepsilon/c$$

and quantifies the increase in polarizability. For low concentrations δ is usually found to be constant, so that the measured permittivity varies linearly with concentration. For aqueous solutions of α-amino acids at 25°C, δ is 26-28 for frequencies up to 1 GHz and concentrations up to 2.5 M. Taking the dipole moment of 15.3 D obtained for glycine as being typical, a value for g of around 1.2 can be estimated (from Equation (9)) for aqueous α-amino acid solutions.

POLYPEPTIDES AND PROTEINS

The way in which linear chains of amino acids, called polypeptide chains, are built up by the formation of peptide bonds is shown in Figure 4(a). These peptide bonds take the form of an amide linkage between α-amino and α-carboxyl groups of adjacent amino acids. For each linkage formed there is the loss of one water molecule, so polypeptide chains are in fact formed of amino acid residues. As the polypeptide increases in length, the distance between the terminal carboxyl and amino charges also tends to increase and along with it the effective dipole moment and observed dielectric increment. For polypeptide chains up to glycine heptapeptide, the specific dielectric increment varies linearly according to the relationship:

$$\delta = 14.51(n) - 5.87$$

where n is the number of chemical bonds between the terminal carboxyl and amino group (28).

Protein molecules are composed of one or more polypeptide chains, and the three-dimensional structure of proteins is influenced by the way some of the peptide units link with one another to form regions of α-helical or pleated β-sheet configurations. The N-C bond in the peptide units has partial double-bond character and as a result the six atoms $C_\alpha NHCOC_\alpha$ are

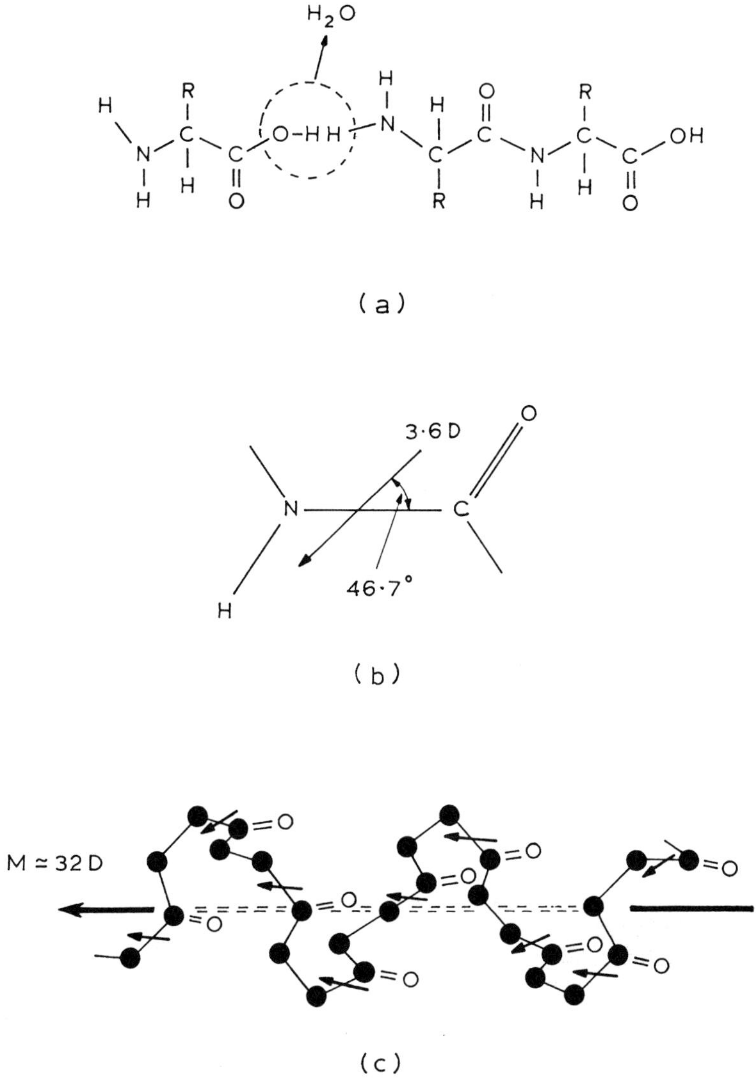

FIGURE 4. (a) The formation of peptide bonds to form a polypeptide chain. (b) The dipole moment of a peptide unit. (c) The individual peptide moments are additive in an alpha-helix.

coplanar. Also, the C-O bond is polar and this is largely responsible for the peptide unit possessing a permanent dipole moment. A relatively straightforward quantum mechanical calculation (28, pp. 44-49) gives the magnitude of this dipole moment as being around 3.6 D and directed at an angle of 46.7° to the C-N bond axis (Figure 4(b)). Since each peptide unit possesses a permanent dipole moment, then polypeptide chains effectively take the form of strings of connected dipoles. The contribution that a dipole of moment m makes to the polarizability of a medium is proportional to m^2. For a completely rigid linear polymer chain of n regularly spaced dipolar units of moment m, fixed and directed normal to the chain, the total contribution to the polarizability will either be zero or of the order of (nm^2), depending upon whether the dipoles vectorially cancel or are additive. If these dipolar units are perfectly free to rotate, the contribution will be nm^2. For most polypeptide systems, n will typically be 10^2-10^3, so that the peptide unit dipole contribution to the overall dipole moment value of the protein molecule, and hence to the permittivity of an aqueous protein solution, could be extremely sensitive to the polypeptide configuration.

A good example of where the peptide group moments are totally additive is the extended α-helix configuration depicted in Figure 4(c). The axis of a rigid polypeptide α-helix is directed at about 56° to the C-N bond of the constituent peptide units. Since the peptide dipole moment is directed at 46.7° to the C-N bond, the peptide contribution to the total moment parallel to the helix will be:

$$m_{11} = 3.6n \, \mathrm{Cos}(9.3°) = 3.6n \text{ Debye units}$$

The α-helix configuration contains 3.6 peptide units per turn of the helix, so even the relatively small helix of not quite three turns shown in Figure 4(c) will have a dipole moment of about 34 D. Such a helix dipole moment is equivalent to there being half a positive electronic charge at the N-terminus and half a negative charge at the C-terminus. The electrostatic field associated with helix dipole moments can be expected to aid in the binding of ions to proteins. Also, dipole-dipole interactions between neighboring α-helical segments of the main polypeptide backbone should influence the overall protein conformation (38).

The effective dipole moment value for small globular proteins, obtained from measurements of the dielectric increment and the use of Equation (9), is typically of the order of several hundred Debye units. Examples of the dielectric dispersions exhibited by protein solutions are shown in Figure 5 for bovine serum albumin (BSA) and myoglobin. The fall in the relative permittivities from a value of around 90 to 70 arises from the rotational relaxation of the protein molecules and is referred to as the β-dispersion. BSA is a larger molecule than myoglobin and hence its relaxation time (c.f. Equation 8) is greater than that for myoglobin. For frequencies below 10 KHz for BSA and below 1 MHz for myoglobin the rotational motion of the protein molecules can contribute fully to the polarizability of the solutions, and as the polarizability per unit volume for the proteins is greater than that of a water molecule, the resultant permittivity is greater than that of pure water. As the frequency is raised above the rate at which the protein molecules can rotationally reorientate, the protein molecules effectively act as polarizable "dead space", and the permittivity drops below the value of around 80 expected for bulk water. In estimating the dielectric dispersion strength $\Delta\varepsilon$ to be used in Equation (9), this dielectric decrement should be added to the dielectric increment defined in Equation (10). Because the dielectric dispersion for glycine overlaps with that of water, the magnitude of this dielectric decrement is difficult to estimate for glycine solutions. Consequently, it was not included in the derivation of the correlation parameter g in the last section. However, the dielectric decrement for amino acids can be expected to be relatively small and so the underestimation in g will also be relatively small.

Careful examination by Essex et al. (39) of their results obtained for BSA revealed two components of the relaxation process, arising from the fact that the BSA molecule takes the form of a spheroid of axial ratio 1:3 and is capable of rotation about both its major axes. Essex et al. (39) were able to determine that the small fall in permittivity (the so-called δ-dispersion) in the frequency range from around 10 MHz to 200 MHz could also be resolved into two separate processes. The

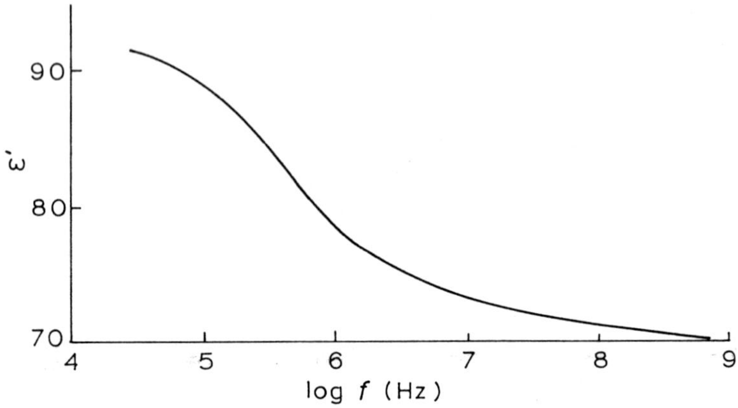

(a)

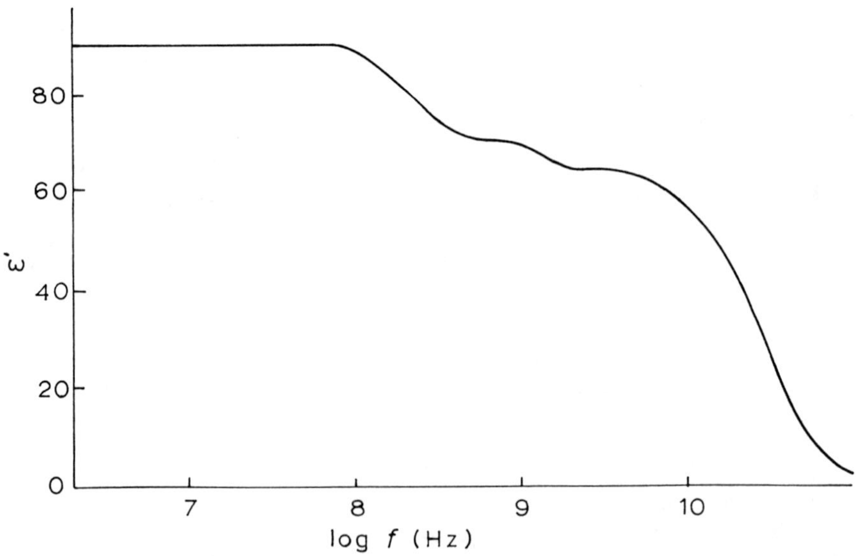

(b)

FIGURE 5. (a) The dielectric dispersion for a 100.8 mg/ml solution of bovine serum albumin (derived from reference 39). (b) The dielectric dispersion for a 10% solution of myoglobin (from reference 27).

process occurring at the higher frequency range was concluded to arise from relaxations of water molecules bound to the protein molecules. In Figure 5 the δ-dispersion can also be clearly seen at a frequency of around 100 MHz for myoglobin. Although no firm conclusions could be made, the effect of fluctuating proton locations on the protein surface, and the relaxation of amino-acid side chains in the protein structure were considered to be the two most probable contributions to the other component of the δ-dispersion for BSA. Other processes have also been considered to influence the dielectric properties of protein solutions, including the effect of the relaxation of the diffuse double layer formed around the protein molecule, and conduction processes associated with the movement of ions bound to the protein. Such effects have been described in detail elsewhere (27,28) but in general they are considered to contribute little to the dielectric properties of protein solutions. Finally, the dispersion centered around 20 GHz for both the BSA and myoglobin solutions is due to the relaxation of water molecules which are not strongly associated with the protein molecules.

It is generally found to be the case that the relaxation times for proteins, as derived from Equation (8), lead to molecular radii that are larger than the values which can be obtained directly from X-ray diffraction studies. The accepted way of interpreting this is to assume that there is a layer or two of water molecules (the so-called hydration sheath) so strongly associated or bound to the protein molecules, that they remain attached to the protein as it rotates. Water molecules that experience strong interactions with the protein structure can be expected to exhibit different relaxation processes compared to normal bulk water. There is supporting evidence to indicate that the δ-dispersions described above are associated with relaxations of such protein-associated water molecules. We shall return to this when considering the dielectric properties of protein-associated water.

DNA

DNA, the molecule that contains the genetic library of living systems, is composed of two counter-opposed alpha-helical chains of polynucleic acids and can take the form of linear rods or supercoiled rings. These macromolecules can have molecular weights of 10^6-10^9. Since the two helical chains point in opposite directions, the dipole moments of each chain cancel one another and this dipolar antiparallelism contributes towards the energy stabilization of the double helix structure. Although the DNA macromolecule has no net helix-contributed dipole moment, early dielectric measurements (40-42) provided conclusive evidence for DNA molecules possessing large dipole moments directed along the axis of the double helix. This dipole moment arises from the fact that the DNA molecule is predominantly negatively charged when in aqueous solution, so that it will be surrounded by an atmosphere of counter-cations. Under the influence of an applied electric field these counter-ions will tend to be displaced along the surface of the DNA macromolecule and so give rise to a large induced dipole moment. The resulting dielectric dispersion will depend on the effective mobility of the ions along the macromolecule surface, and for rod-shaped macromolecules the relaxation time will be given by:

$$\tau = \pi\varepsilon L^2/2uzq^2$$

where ε is the effective permittivity of the surrounding ionic atmosphere of z ions per unit length, u is the counter-ion surface mobility and L is the length of the DNA macromolecule. Such an L^2 dependence for the dielectric relaxation time has been observed (43). The dielectric relaxation times are commonly found to be of the order 1 msec and dielectric dispersion strengths $\Delta\varepsilon$ are large and usually greater than 1000.

Apart from dielectric dispersions arising from the induced dipole moment associated with the electrical double layer around the macromolecule, a smaller dispersion occurs between 1-50 MHz (44,45). The variation of the relative permittivity of an aqueous 1% DNA solution is shown in Figure 6 for the range from 200 KHz to 10 GHz. The high-frequency tail of the induced dipolar dispersion is clearly evident, as is the dispersion around 20 MHz which is considered to arise from polar groups in the DNA molecule (45). The high-frequency dispersion arises from the relaxation of the unbound solvent water molecules and does not differ markedly from normal bulk water.

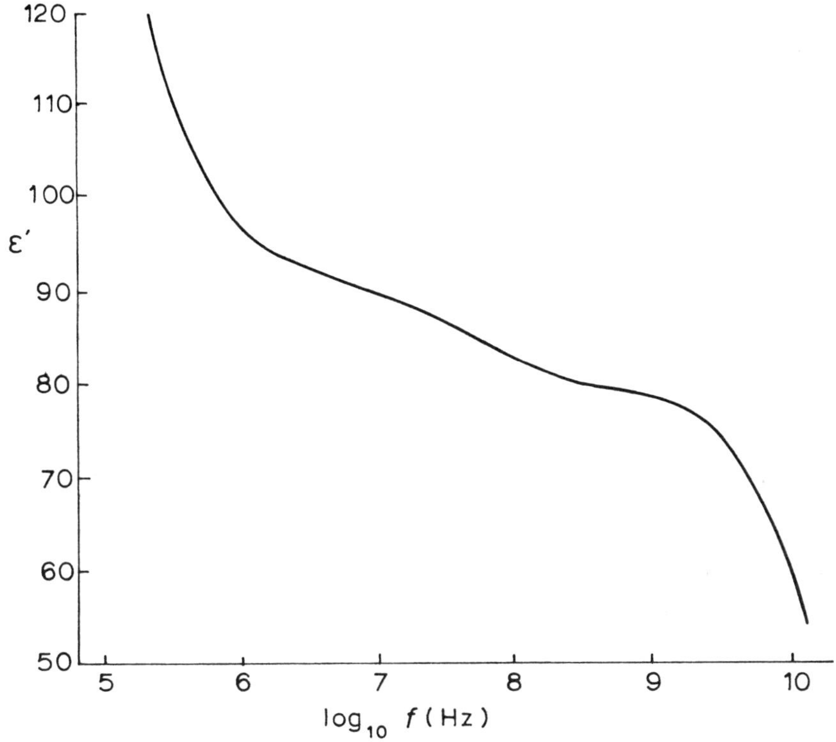

FIGURE 6. The dielectric dispersion for a 1% aqueous DNA solution (from reference 45).

These observations by Takashima et al. (45) that the dielectric absorption of an aqueous solution of DNA does not differ appreciably from normal bulk water at 1-10 GHz is of great interest in view of reports of large resonance absorptions in DNA solutions at these frequencies. Swicord and Davis (46,47) and Edwards et al. (48,49) have provided detailed evidence for the possibility that aqueous solutions of DNA molecules at room temperature can resonantly absorb microwave energy, and that the mechanism for this is associated with the coupling of the electric field to acoustic modes in the macromolecular structure. If such resonance effects can be confirmed they will have significant implications, such as the possibility that biochemical processes could be influenced by highly selective microwave frequencies and that acoustic modes could possibly provide a mechanism for transporting coherent energy over large molecular distances.

BIOLOGICALLY BOUND WATER

Although most of the hydrophilic amino-acid residues of a protein molecule will be situated on the outside surface of the protein in contact with the aqueous environment, significant areas of the protein surface will be made up of hydrophobic regions. In lysozyme, for example, about one half of the exposed protein structure has been found to consist of non-polar groups (50). The water molecules surrounding such hydrophobic groups will be forced to form hydrogen-bond networks with each other in a way that differs from normal bulk water structure. Computer simulations (51) indicate that such water-structure modifications can extend at least 10 Å into the bulk liquid away from a hydrophobic surface. Dielectric studies by Hallenga et al. (52) of the effect of hydrophobic solutes in water have shown that the dielectric relaxation times are increased, suggesting that there is an increased degree of hydrogen-bonding in the immediate neighborhood of the hydrophobic solutes. Some water molecules will form hydrogen bonds with the protein structure, and others will experience strong electrostatic interactions with charged groups such as glutamic acid and lysine. Water molecules near charged membrane surfaces will be similarly affected. All of these factors lead to the concept of there being a layer or two of water molecules near biomacromolecular surfaces having differing physical and dielectric properties from normal bulk water.

The existence of protein-associated or bound water in biological systems which as a result of its having hindered rotational mobility exhibits dielectric properties that differ from normal bulk water was first indicated by studies of the dielectric properties of protein solutions at microwave frequencies (53). On extrapolating the results to frequencies below where relaxation for normal water occurs, it was found that the relative permittivity value was lower than that for normal bulk water. This difference was assumed to be mainly due to the existence of water "irrotationally" bound to the protein molecules. Depending on the molecular shape assumed for the protein, the amount of such bound water was calculated to be 0.2-0.4 wt% of the protein. The identification of the δ-dispersion in terms of such bound water was first described by Schwan (19) and Grant (18) in studies of hemoglobin and albumin solutions.

By making measurements of the dielectric properties of hydrated protein powders, effects associated with rotations of the protein molecules and relaxations of electrical double layers are avoided. This enables the dielectric properties of the protein-bound water to be investigated directly. Measurements by Bone et al. (54,55) of the microwave dielectric properties of protein powders as a function of hydration indicate that 7-8% of the water molecules associated with a fully hydrated protein are bound, and are unable to contribute to the dielectric properties at 10 GHz. This is within the frequency range at which normal bulk water exhibits a strong dielectric absorption. Later investigations (56) of the δ-dispersion for hydrated lysozyme powders indicate that about 36 water molecules are tightly incorporated into each lysozyme molecule and form an integral part of the overall protein structure. It was also concluded that vibrational motions of the protein structure contribute to the overall δ-dispersion, especially at the higher hydration levels where the water appears to act as a plasticizing agent for protein structures (57).

The physical state of water associated with DNA molecules has also been investigated dielectrically. Microwave studies (58) on herring sperm DNA at 90-300°K have shown that a considerable proportion of the bound water, corresponding to around 280 water molecules per DNA helix turn, exhibit significantly different properties from those of normal bulk water in terms of dielectric relaxations below 273°K. An effective freezing point depression of 138°K was observed, indicating that a loss of rotational mobility of the DNA-associated water molecules with decreasing temperature is a much more gradual process than occurs in the normal water-ice transition. The collective nature of the interactions that can exist between water molecules and DNA has been clearly demonstrated by the computer simulations of Clementi (59) who showed that for hydrations up to around 270 water molecules per DNA helix turn the ensemble of water molecules is structured so as to reproduce a global image of the DNA electric field.

The presence of fluctuating proton populations will be of relevance to the dielectric properties of structural and membrane-bound proteins, and also of possible significance concerning proton conduction pathways. In an important series of measurements (60), significant details were obtained of the protein-water interactions for lysozyme. Measurements of the optical absorbance at 1580 cm^{-1} (which was earlier mentioned as the characteristic absorbance of the COO^- carboxylate ion) and the specific heat as a function of hydration, indicated that the first water molecules to bind to lysozyme interacted with the ionizable carboxylic and basic groups. At around 5 wt % redistribution of the proton population appeared to occur, together with transitions in the protein-water and water-water hydrogen-bonding networks. Proton transport effects as a function of hydration for protein powders have been described by Gascoyne et al. (61) and Behi et al. (62) and these measurements indicate that protons are able to migrate relatively freely in protein structures. Cyclodextrins have found use as artificial enzymes, and studies by Bone and Pethig (22) have shown that the flip-flop hydrogen bond networks that occur in the cyclodextrin hydrates provide a useful model system for studying proton conductivity.

BIOLOGICAL ELECTROLYTES

Mammals have a total body water content of around 65 to 70%, and apart from the effects of dissolved biomacromolecules and membrane surfaces, the dielectric properties of this water are influenced by dissolved ionic salts. The physical effects of such dissolved ions on the dielectric permittivity of biological fluids arise from more than simply the volume effect of replacing polar water molecules by non-polar ionic particles. The strong local electric field around each dissolved ion has the effect of orientating the water molecules, thereby reducing the way they can rotate in response to an applied electric field. In the case of amino acids in solution, an equation of the form of Equation (10) can be used to describe the permittivity (ε) of dilute electrolytes and that (ε_1) of the pure aqueous solvent in terms of a dielectric decrement as follows:

$$\varepsilon = \varepsilon_1 - \delta c$$

where δ is the sum of the decrements arising from the cation and anion and is given by:

$$\delta = (\delta^+ + \delta^-)$$

Values of δ^+ and δ^- for various ions in water are given in Table 1 for concentrations, c, of less than 1 molar. To estimate the extent to which, for example, KCl will reduce the permittivity of water, then from Table 1 the total decrement value for K^+ plus Cl^- is $\delta = 11$, giving the room temperature relative permittivity for aqueous molar KCl solution as ε=68 (for pure water, $\varepsilon_1 = 79$).

Besides lowering the relative permittivity of the aqueous solvent, dissolved ions generally increase the relaxation frequency. One exception to this is the effect of protons in decreasing the relaxation frequency. To a first approximation this effect can be considered to result from the solvated ions disrupting the hydrogen bond structure of normal water. For

TABLE 1. Dielectric Increment Values for Some Ions in Aqueous Solution

CATION	$\delta^+(\pm1)$	ANION	$\delta^-(\pm1)$
Na^+	8	Cl^-	3
K^+	8	F^-	5
Li^+	11	I^-	7
H^+	17	SO^{2-}	7
Mg^{2+}	24	OH^-	13

concentrations of less than 1 molar this effect can be expressed in terms of a relaxation frequency increment by the equation

$$f = f_1 + c\,\delta f$$

where

$$\delta f = (\delta f^+ + \delta f^-)$$

represents the effect of the cation and anion. Values for f^+ and f^- are given in Table 2 and can be used to estimate how far the relaxation frequencies for various electrolytes differ from the value of 17.1 GHz for pure water at 20°C.

The overall AC conductivity of a biological material is given by

$$\sigma(f) = \sigma_o + 2f\,\varepsilon_o\,\varepsilon''$$

where σ_o is the conductivity arising from electric-field induced motions of the various ions in the electrolyte, and has the value:

$$\sigma_o = q\sum_i z_i n_i u_i \quad (11)$$

In this equation, i denotes an ion species with valency z, concentration n and electrical mobility u_i. The mobility values

TABLE 2. Relaxation Frequency Increment Values for some Ions in Aqueous Solution

CATION	ΔF^+(±0.2)GHz	ANION	Δf^-(±0.2)GHz
H^+	-0.34	OH^-	0.24
Li^+	0.34	Cl^-	0.44
Na^+	0.44	F^-	0.44
K^+	0.44	SO^{2-}	1.20
Mg^{2+}	0.44	I^-	1.65

TABLE 3. The Electrical Mobility of Some Ions in Dilute Aqueous Solution at 25°C

CATION	MOBILITY ($10^{-4}cm^2/V.sec$)	ANION	MOBILITY ($10^{-4}cm^2/V.sec$)
H^+, H_3O^+	36.3	OH^-	20
K^+	7.6	CH^-, I^-	7.7
Na^+	5.0	F^-	5.4
NH_4^+	7.6	Br^-	7.8
Ca^{2+}	6.2	$SO^2{}_4^-$	8.3

for some biologically relevant cations and anions are listed in Table 3. A pH of 7.0 at 24°C is equivalent to there being 10^{-7} moles of H^+ and OH^- ions in a liter of water. This corresponds to a concentration of $6.03 \times 10^{19}/m^3$ of H^+ and OH^- ions. The conductivity of pure water at 24°C can then be calculated from Equation (11), using mobility values for H^+ and OH^- given in Table 3, as 5.4×10^{-6} mho/m. In fact, H^+ ions as such do not exist to a significant extent in aqueous solution since they rapidly become hydrated to the H_3O^+ ion. The mobility of the H_3O^+ ion (Table 3) is the same as that of the proton. This is because the apparent mobility of the hydronium ion is equal to the rate at which protons transfer between neighboring hydrogen-bonded water molecules. The positive charge can move a considerable distance (creating a hydronium ion whenever it resides on a water molecule) with little movement of the water molecules themselves.

It is often more convenient to deal with the effective molar conductivities of ions, and these values are given in Table 4 for dilute solutions. The conductivity of a dilute electrolyte can then be found from the equation:

$$\sigma = \sum_i m_i \sigma_{mi} \qquad (12)$$

where m_i is the molar concentration of ion species i having a molar conductivity σ_{mi}. The most dominant ions in human tissue

TABLE 4. Molar Conductivity of Some Ions in Dilute Aqueous Solution at 25°C

CATION	CONDUCTIVITY ($mho.cm^2/mol$)	ANION	CONDUCTIVITY ($mho.cm^2/mol$)
H^+, H_3O^+	350	OH^-	193
K^+	73	Cl^-, I^-	74
Na^+	48	F^-	52
NH_4^+	73	Br^-	75
Ca^{2+}	119	$SO^2{}_4^-$	160

fluids are those of sodium and chlorine, each of concentration around 150 mM per liter, which is equivalent to a 0.9 wt % saline solution. From Table 4 and Equation (12) the conductivity of such a solution at 25°C can be calculated to be 1.83 mho/m. The average water content of various tissues is listed in Table 5. For liver and muscle the water content is

TABLE 5. Water Content Values for Various Tissues and Organs

TISSUE	WEIGHT % WATER CONTENT	TISSUE	WEIGHT % WATER CONTENT
Bone	44-55	Lung	80-83
Bone Marrow	8-16	Muscle	73-78
Bowel	60-82	Ocular Tissues	
		Choroid	78
Brain		Cornea	75
White matter	68-73	Iris	77
Grey matter	82-85	Lens	65
		Retina	89
Fat	5-20	Skin	60-76
Kidney	78-79	Spleen	76-81
Liver	73-77		

about 75%. To a first approximation then, the conductivity should be around 1.4 mho/m. As described later in this Chapter (see Table 7) this accurately predicts the conductivity at frequencies around 1 GHz, where the electrical properties of tissues are determined by the electrolyte content and not by the tissue structures themselves, and before frequencies are attained where the dielectric relaxation of water molecules becomes the dominant effect.

MEMBRANES AND CELLS

As we have already discussed, cell membranes are ultrathin structures of thickness around 6 nm. They are composed mainly of long-chain hydrocarbon (lipid) molecules and proteins. The relative permittivity of a close packed assembly of lipids and proteins, such as exists in membrane structures, can be expected to be of the order 3. An approximate value for the effective capacitance per unit area of such a membrane can be found using the basic equation:

$$C = \varepsilon_o \varepsilon / d$$

Assuming a thickness d of around 6 nm and a relative permittivity ε of 3.0, then the membrane capacitance can be estimated to be 0.44 $\mu F/cm^2$, which compares favorably with the range 0.5-1.3 $\mu F/cm^2$ commonly observed for a wide range of biological membranes (63). Although cell membranes are largely hydrophobic, it is known that some water molecules are incorporated into membrane structures. Their presence, together with the possibility that the membrane-associated proteins can exhibit some degree of lateral diffusion and rotational freedom, will increase the effective overall polarizability of the membrane and increase the capacitance value estimated above to be within the observed range of values. Being largely composed of hydrocarbon-based lipid and protein molecules, the electrical resistivity of cell membranes can be expected to be large.

The capacitive and resistive components of a cell membrane can be represented as an equivalent electrical circuit of the form of Figure 7(a), where the membrane capacitance C_m is shown

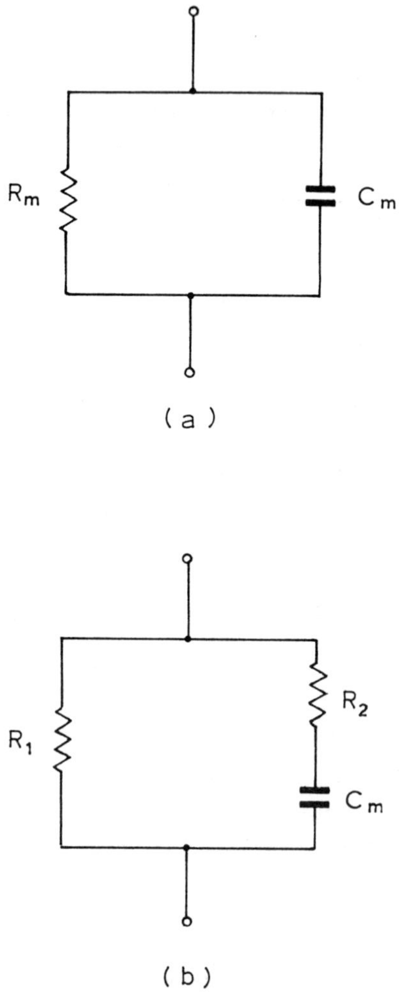

FIGURE 7. (a) Equivalent electrical circuit for a cell membrane, where C_m and R_m are the membrane capacitance and resistance, respectively. (b) An equivalent circuit for a cell, where R_1 is the resistance of the extracellular electrolyte, C_m is the membrane capacitance, and R_2 is a function of the membrane and cytoplasm resistance.

in parallel with the membrane resistance R_m. A feature of this circuit is that with increasing frequency the resistance R_m is increasingly short-circuited by the reactance $(1/\omega C_m)$ of C_m. The consequence of this as far as the electrical properties of a cell in aqueous suspension is concerned is shown schematically in Figure 8. At low frequencies (Figure 8a) the resistance of the cell membrane insulates the cell interior (cytoplasm) from an external electric field, and no electrical current is induced within the cell interior. In dielectric terms, the cell appears to take the form of an insulating spheroid, and it increases the effective resistivity of the aqueous suspension. At higher frequencies (Figure 8b) the short-circuiting effect of the membrane capacitance allows the electric field to penetrate into the cell until, at a sufficiently high frequency, the effective membrane resistance becomes vanishingly small and the cell appears dielectrically as a spheroid composed of the cytoplasmic electrolyte. The effective permittivity and resistivity of a cell suspension will therefore fall with increasing frequency and this gives rise to a dielectric dispersion (the β-dispersion) of roughly the same form as the β-dispersion exhibited by tissues shown in Figure 9. An analysis of this dispersion for cell suspensions can lead to values for the membrane capacitance and resistance, as well as for the cytoplasm resistivity. The representation of a completely insulating membrane, as depicted in Figures 8(a) and 8(b), is only an approximation. Ion conducting channels are known to exist in cell membranes and these could act as localized areas of relatively low resistance in the membrane. Klee and Plonsey (64) have shown how the existence of such low-resistance patches can alter the field pattern across the cell membrane, and an example of this is shown in Figure 8(c). The possible consequences of this on the dielectric properties of cell suspensions have not been addressed in the literature, and there could also be implications for studies of electric field induced cell-to-cell fusions (65).

An equivalent electrical circuit that is often used to represent a cell is shown in Figure 7(b), where R_1 represents the resistance of the extracellular medium, C_m is the membrane capacitance and R_2 is a function of both the membrane resistance and cytoplasm resistance. A full discussion of the analysis of

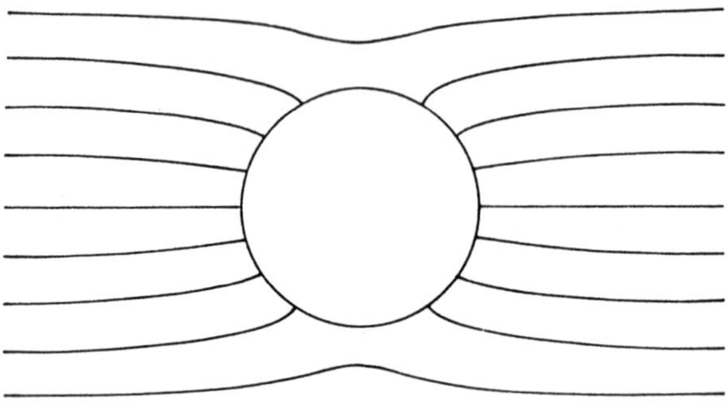
(a)

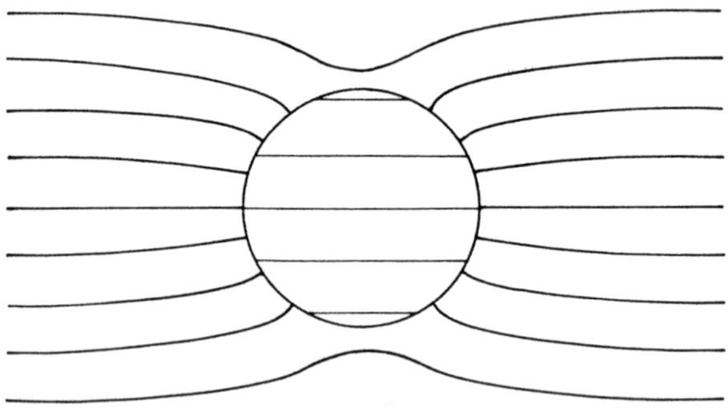
(b)

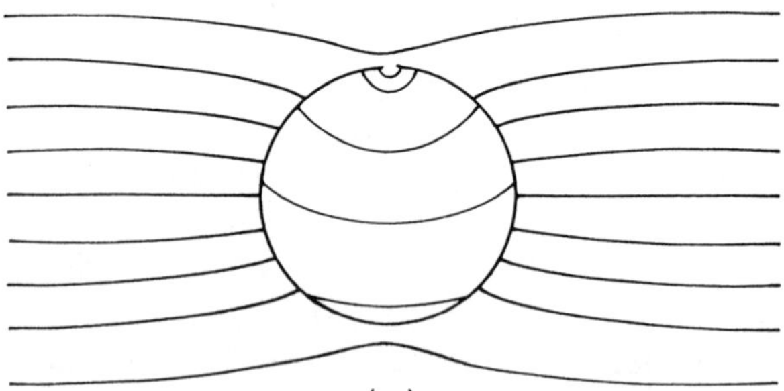
(c)

FIGURE 8. (a) At low frequencies the membrane resistance shields the cell interior from applied electric fields. (b) At higher frequencies the membrane resistance is progressively shorted-out by the membrane capacitance. (c) The effect of a conducting pore in the membrane is to change the voltage stress across the membrane compared to the commonly accepted picture of (a) above. (From reference 64.)

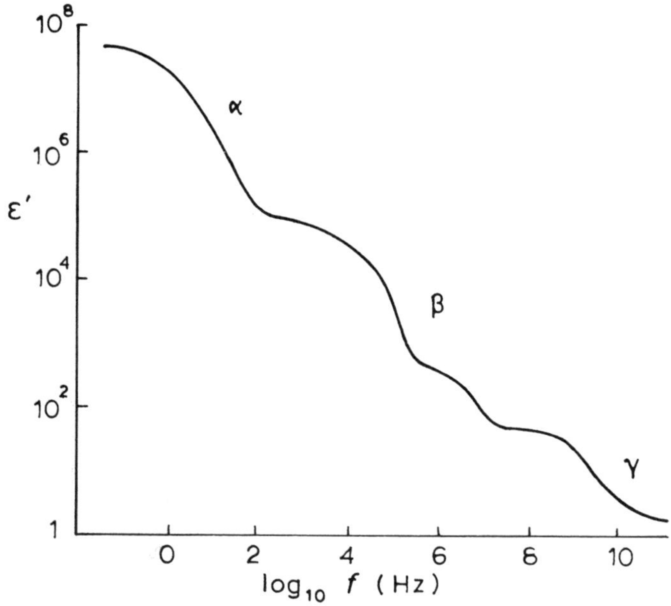

FIGURE 9. An idealized representation of the way in which the relative permittivity of a typical biological tissue varies with frequency.

the equivalent electrical circuits for cells and biological tissues is given by Schanne and Ceretti (31). These authors, together with Cole (30), also provide a comprehensive account of the values that have been obtained for the electrical properties of a variety of cell and bacterial membranes and cytoplasms. Values for cell membrane resistance of 10^3-10^5 ohm cm^2 and cytoplasm resistivity of 0.3-7 ohm/m have been obtained.

Cell membranes contain ionizable acidic and basic groups, although for most cells so far studied the acidic groups are dominant and the membranes carry a net negative charge. At the immediate interface between a cell membrane and its aqueous environment there will therefore be an electrical double layer formed by the charged groups on the membrane and the counter-ions in the electrolyte, as depicted in Figure 1(a). As for the case of DNA, we shall expect to find for cell suspensions a dielectric dispersion associated with ionic surface conduction effects and relaxations of the double layer, and in fact such a dispersion is found (the α-dispersion) of the same form as the α-dispersion shown in Figure 9. Similar dispersions can be observed for colloidal suspensions of glass or polystyrene spheres in electrolytes and also as a result of electrical polarizations at electrode-electrolyte interfaces. Recent studies (20,66) of the dielectric properties of bacterial suspensions have provided convincing evidence that effects associated with the diffusional motions of lipids and proteins in the membranes also contribute to the α-dispersion. Field induced diffusions of membrane proteins have been observed microscopically (67), and such effects are relevant to possible physiological effects of electrical fields.

TISSUES

FREQUENCY DEPENDENCE OF DIELECTRIC PROPERTIES

The relative permittivity of biological tissue typically decreases with increasing frequency in three major steps which are designated as the α-, β-, and γ-dispersion, and an idealized representation of this is shown in Figure 9. The α-dispersion is generally considered to be associated with interfacial polarizations associated with electrical double layers and surface ionic conduction effects at membrane boundaries. The β-dispersion in Figure 9 is shown to have two components and, although such fine detail is not commonly found in reality, the purpose of representing it in this way is to indicate that both capacitive shorting-out of membrane resistances and rotational relaxations of biomacromolecules can contribute to this dispersion. The γ-dispersion arises from the relaxation of free water in the tissue.

Comparatively little work has been reported for the dielectric properties of tissue in the frequency range up to a few KHz, but a good example of the α-dispersion has been obtained by Singh et al. (68) for _in vitro_ measurements on freshly excised kidney and _in vivo_ measurements using external electrodes for normal breasts and breasts with malignant tumors. Some of these results are shown in Figure 10 and it can be seen that malignancy appears to influence the observed dielectric properties. The possible cause for the difference in

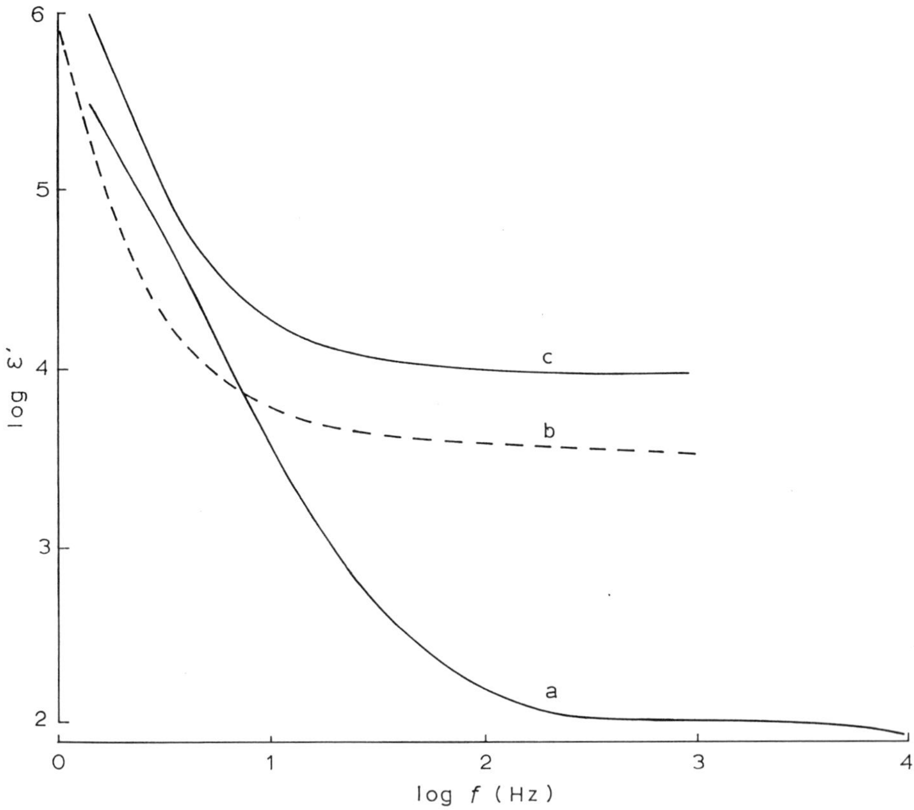

FIGURE 10. Frequency variation of the relative permittivity of (a) human kidney, (b) normal breast and (c) breast with malignant tumor (from reference 68).

the dielectric properties of normal and cancerous tissue will be discussed later in this Chapter. Another good example of the α-dispersion can be found in the results of Kosterich et al. (69) for freshly excised rat femurs, where the permittivity was observed to fall from 10^4 at 10 Hz to around 100 at 1 MHz. In the early work by Schwan (70), the α-dispersion was clearly demonstrated for muscle tissue and an important aspect of this work was the observation that the permittivity and resistivity measured at 1 KHz decreased steadily with time after excision of the tissue. This effect is consistent with the steady loss of integrity and physiological viability of the cell membranes.

The dependence of the β-dispersion upon the integrity of cell membranes was clearly shown (Figure 11) by Pauly and Schwan (71), who measured the effect of digitonin in lysing the fiber membranes of bovine eye lens. Until now we have considered the β-dispersion to be associated with the short-circuiting of the membrane resistance by the membrane capacitance. Whereas this is now the accepted interpretation, it does not completely make clear that the essential effect is one of interfacial polarizations associated with the heterogeneous structure inherent in membrane-electrolyte structures. At the interface between two dissimilar dielectrics there is a build-up of charge and this gives rise to interfacial, or Maxwell-Wagner, polarizations. The magnitudes of these polarizations are

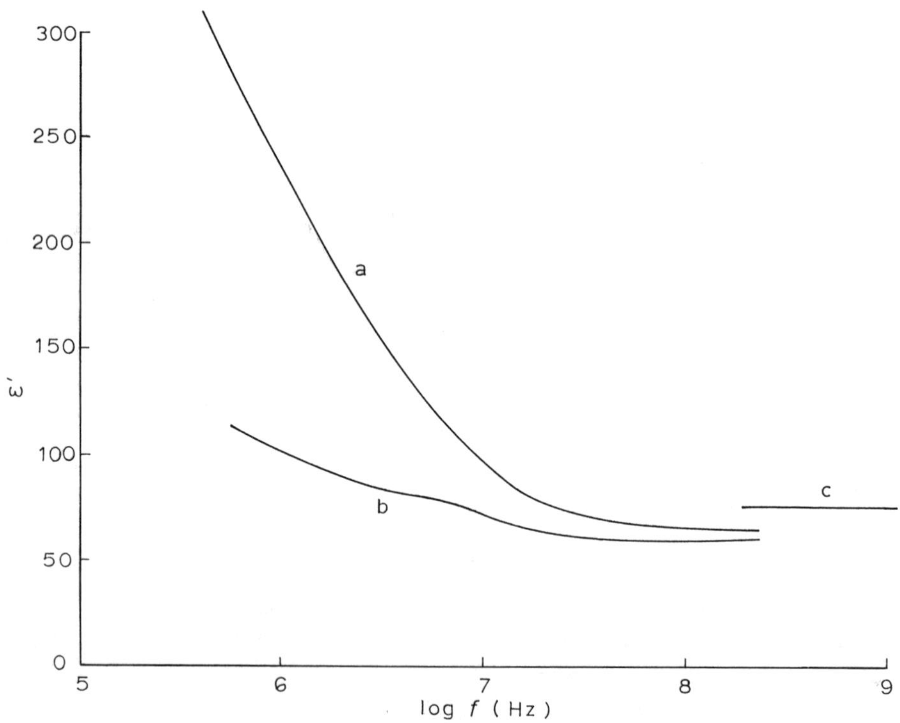

FIGURE 11. (a) Frequency variation of the relative permittivity of bovine eye lens fiber suspension and (b) the effect of digitonin lysis of the fiber membrane (71). (c) Permittivity of 0.9% NaCl solution at 27°C.

dependent on the conductivity, permittivity and geometry of the separate components of the heterogeneous structure. With increasing frequency, the more resistive components are neutralized by their associated parallel capacitances. The structure therefore becomes progressively more (electrically) homogeneous. A full account of the theories of interfacial polarizations of relevance to biological materials has been given elsewhere (28).

Also shown in Figure 11 is the permittivity of 0.9% sodium chloride solution, which corresponds to the ionic strength of cellular fluid. It can be seen that above 100 MHz the dielectric properties of tissue become largely independent of the membrane structures. The permittivity of tissue in the region 100 MHz to 1 GHz is a little less than that of a 100% concentration of the aqueous cellular fluids due to the presence of the non-polar membrane materials and other chemicals of low polarizability.

For frequencies higher than about 100 MHz, where the capacitive shorting-out effects of the cell membrane resistances begins to take effect, the dielectric characteristics of tissues can be expected to reflect the properties of the inter- and intra-cellular electrolytes and, in particular, to exhibit a dielectric dispersion associated with the relaxation of water dipoles. This can be seen in Figure 12, where the relative permittivity and resistivity values of brain, fat and muscle above 100 MHz are shown alongside those exhibited by 0.9% saline solution. Resistivity values, rather than conductivity, are used in Figure 12 as this aids in showing the differences in the electrical properties at the higher frequencies. A feature that can also be deduced from this figure is that the dielectric properties of tissues are influenced by their tissue water content. Muscle, for example, which in man can have water contents of 73-77.6 wt%, exhibits a much larger permittivity and conductivity compared to those of fat whose water content is 5-20 wt%. The water contents of various tissues are given in Table 5. In their studies of liver, muscle and skin, Schwan and Foster (72) concluded that the microwave dielectric properties of these tissues could be attributed directly to their free water contents, and also to the dispersion expected for normal bulk water. A similar conclusion was obtained by Foster et al.

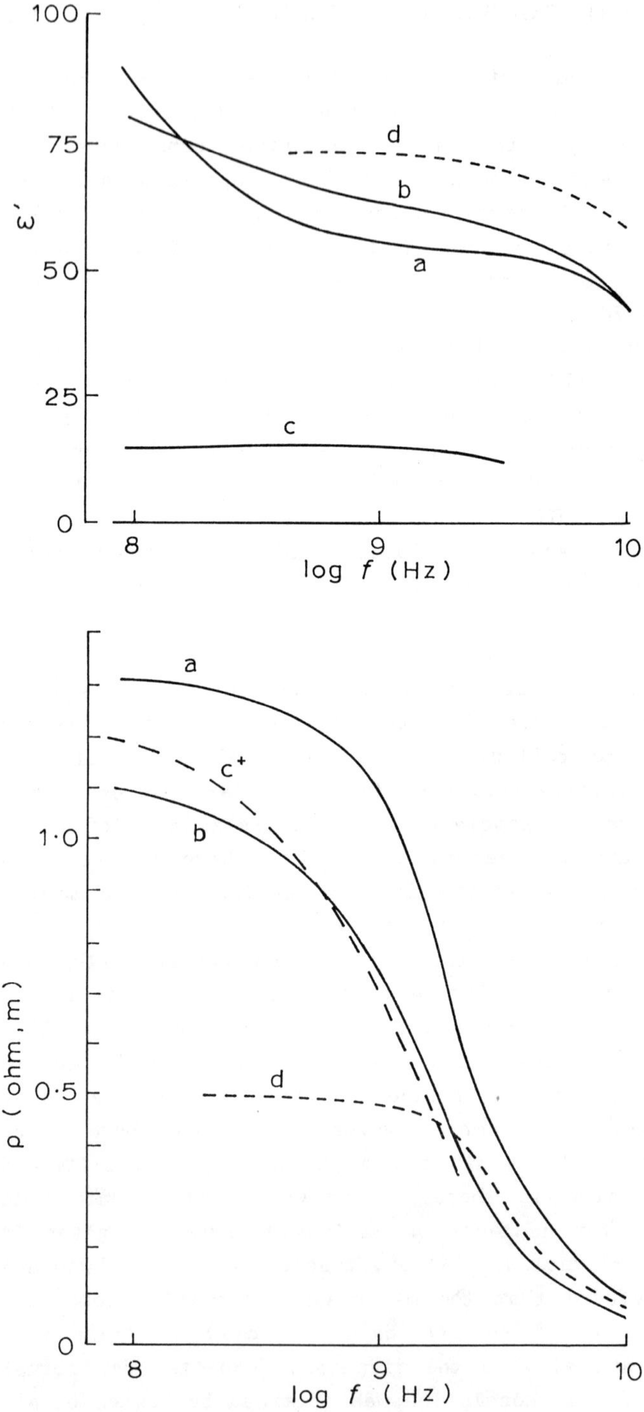

FIGURE 12.

(73) for the microwave dielectric properties of canine brain tissue. These studies show that for frequencies above 100 MHz the dielectric properties of tissue are consistent with those expected for a suspension of low conductivity, low permittivity, particles (cells) in an aqueous electrolyte.

SKIN

The dielectric properties of skin are of relevance to a large number of therapeutic and diagnostic techniques which involve the application or monitoring of electrical signals. There is also interest within the cosmetics industry in using the dielectric properties of skin as an indicator of the efficacy of dermatological and cosmetic treatments. The electrical properties of skin are largely determined by the corneum which has a thickness of the order 15 μm and consists of layers of dead cells. As these dead cells, formed mainly of keratin and membrane material, are worn off from the skin surface they are replaced by underlying epidermal cells. The dielectric properties of skin show considerable regional variability over the body, with the impedance being lowest in areas such as the palms where sweat ducts are in abundance. Rosendal (74) measured the properties of wet, freshly excised, skin approximately 1 mm thick at 1 KHz and 10 KHz and obtained values for the effective capacitance and resistance of the corneum of 4.6 nF/cm^2 and 34.9 kΩ.cm^2, respectively. A value of 6.2 kΩ.cm^2 was obtained for the effective series resistance of the skin and underlying deep tissue. If the relative permittivity of dry keratinized membrane material is assumed to be around 10, then these results suggest that the capacitive-resistive element in skin has a thickness of the order 2 μm. This layer provides the protective barrier between the body tissues and the environment.

The average *in vivo* electrical properties of skin over the range 1 Hz to 1 MHz, determined by Yamamoto and Yamamoto (75), are shown in Figure 13(a). An understanding of these properties

---

FIGURE 12. *In vivo* permittivity and resistivity measurements obtained (101) for: (a) rat brain, (b) rat muscle and (c) rat fat (scaled down by one-fifth). (d) 0.9% saline solution.

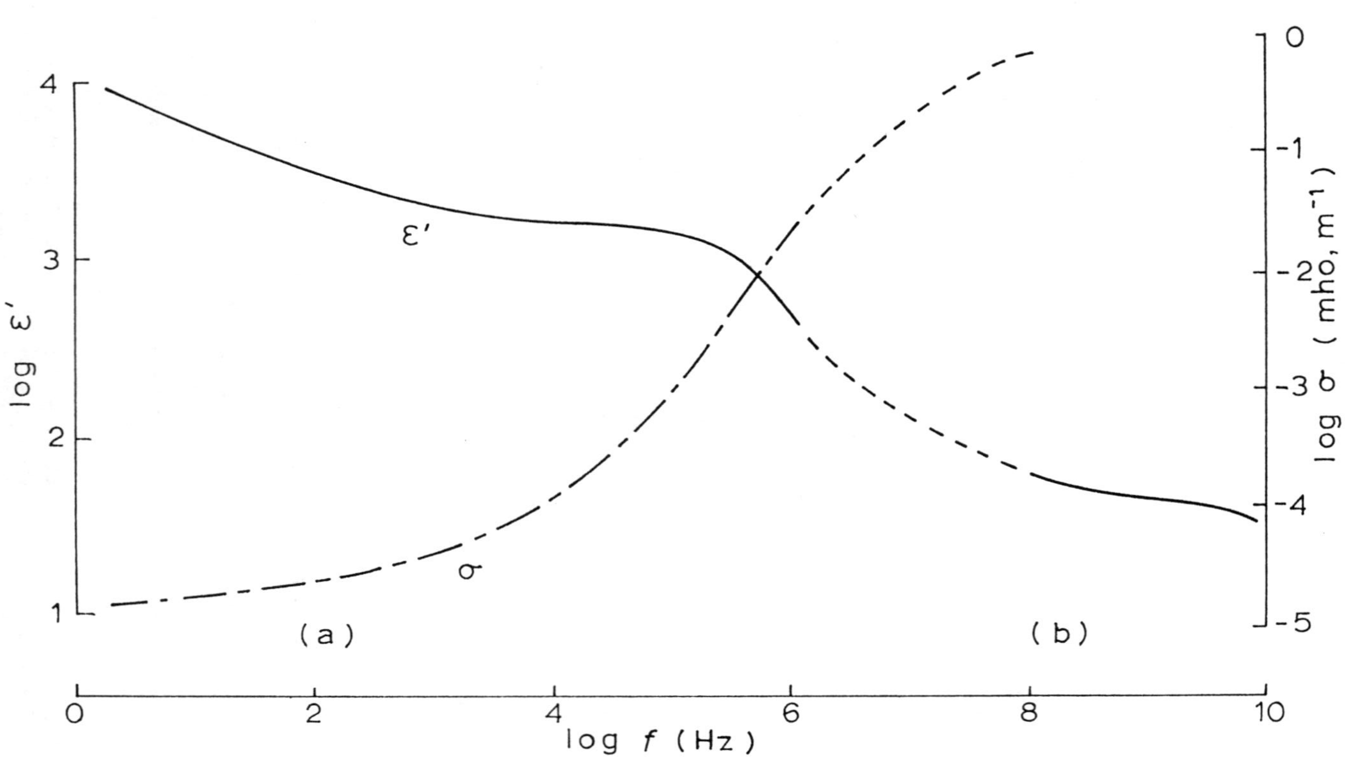

FIGURE 13. Frequency variation of the relative permittivity and conductivity for skin at 37°C (based on references 75 and 19).

was approached in terms of the inhomogeneous structure and composition of skin, and the way in which this varies from the skin surface into the underlying dermis and subcutaneous tissue. The appearance of a relatively weak α-dispersion for skin should be noted, and this lack of a significant dispersion in the frequency range from around 10 Hz to 10 KHz could be associated with the dead nature of the corneum. Clar et al. (76) found that the dispersion exhibited by normal skin in the range of 0.5 Hz to 10 KHz was characterized by two separate relaxation processes, centered around 80 Hz and 2 KHz, respectively. As a tentative proposal, the origin of these dispersions was considered to be located within the corneum and to be associated with relaxations of counterions surrounding the corneal cells. The dielectric response of psoriatic skin was found to differ significantly from that of normal skin. The dielectric properties determined by Schwan (19) for the microwave range of frequencies are shown in Figure 13(b), and from an interpolation of the two sets of results of Figure 9 the form of the β-dispersion for skin can be estimated.

OTHER TISSUES

The relative permittivity and conductivity of various tissues at 37°C are given in Tables 6 and 7, respectively, for several frequencies commonly used for therapeutic and diagnostic purposes. This data has been derived from the references cited, and although they represent the most reliable data presently available, they should be considered as representing average values. The tissue water contents given in Table 5 and the ranges for these can be used to estimate the likely spread of electrical properties expected for the various tissues. The data, mostly <u>in vivo</u>, was obtained for tissues of various animals (including man), but since the dielectric properties of the same tissue types from different animals are quite similar (77-79) this should not reduce the value of Tables 6 and 7.

TISSUE-BOUND WATER

The results of Figure 12 indicate that a large proportion of the water in tissues has rotational properties similar to those of normal bulk water. Schwan and Foster (72) obtained evidence to suggest that about 10% of the water in tissue is

TABLE 6. The Relative Permittivity of Biological Tissues at 37°C for Various Frequencies Commonly Used for Therapeutic Purposes

MATERIAL	13.56 MHz	27.12 MHz	433 MHz	915 MHz	2.45 GHz	REFERENCE
Artery	-	-	-	-	43	98
Blood	155	110	66	62	60	99
Bone (with marrow)	11	9	5.2	4.9	4.8	28
(in Hank's solution)	28	24	-	-	-	70
Bowel (plus contents)	73	49	-	-	-	100
Brain (white matter)	182	123	48	41	35.5	73
(grey matter)	310	186	57	50	43	73
Fat	38	22	15	15	12	104
Kidney	402	229	60	55	50	104,78
Liver	288	182	47	46	44	99,78
Lung (inflated)	42	29	-	-	-	99,100
(deflated)	94	57	35	33	-	99,100
Muscle	152	112	57	55.4	49.6	102
Ocular tissues (choroid)	240	144	60	55	52	80
(cornea)	132	100	55	51.5	49	80
(iris)	240	150	59	55	52	80
(lens cortex)	175	107	55	52	48	80
(lens nucleus)	50.5	48.5	31.5	30.8	26	80
(retina)	464	250	61	57	56	80
Skin	120	98	47	45	44	99, Fig. 13
Spleen	269	170	-	-	-	78

TABLE 7. The Conductivity (mho/m) of Biological Tissues at 37°C

MATERIAL	13.56 MHz	27.12 MHz	433 MHz	915 MHz	2.45 GHz	REFERENCE
Artery	-	-	-	-	1.85	98
Blood	1.16	1.19	1.27	1.41	2.04	99
Bone (with marrow)	0.03	0.04	0.11	0.15	0.21	78
(in Hank's solution)	0.021	0.024	-	-	-	78
Brain (white matter)	0.27	0.33	0.63	0.77	1.04	73
(grey matter)	0.40	0.45	0.83	1.0	1.43	73
Fat	0.21	0.21	0.26	0.35	0.82	104
Kidney	0.72	0.83	1.22	1.41	2.63	104,78
Liver	0.49	0.58	0.89	1.06	1.79	99,78
Lung (inflated)	0.11	0.13	-	-	-	99,100
(deflated)	0.29	0.32	0.71	0.78	-	99,100
Muscle	0.74	0.76	1.12	1.45	2.56	102
Ocular tissues (choroid)	0.97	1.0	1.32	1.40	2.30	80
(cornea)	1.55	1.57	1.73	1.90	2.50	80
(iris)	0.90	0.95	1.18	1.18	2.10	80
(lens cortex)	0.53	0.58	0.80	0.97	1.75	80
(lens nucleus)	0.13	0.15	0.29	0.50	1.40	80
(retina)	0.90	1.0	1.50	1.55	2.50	80
Skin	0.25	0.40	0.84	0.97	-	99, Fig. 13
Spleen	0.86	0.93	-	-	-	78

strongly bound and rotationally hindered, having a relaxation frequency 50-200 times lower than that for normal bulk water. A less pronounced lowering of the water relaxation frequency was observed by Gabriel et al. (80) for ocular tissues. The relaxation frequency of normal bulk water at 37°C is 25 GHz, whereas retina (89% water content) exhibited a relaxation frequency of around 21 GHz, and for the lens nucleus (65% water) this was reduced to around 9 GHz. The lowering of the relaxation frequency with falling water content was an interesting feature of these studies. Stuchly et al. (79) obtained in vivo measurements of the dielectric properties of skeletal muscle, brain cortex, spleen and liver of live cats and rats for 0.1-10 GHz. Above 1 GHz the dielectric properties correlated well with the various tissue water contents, and it was deduced that practically all the water in skeletal muscle was in the form of bulk water. For the other tissues, both free and bound water were deduced to be present, with spleen having a total tissue water content (volume basis) composed of 69% free water and 9% bound water. For liver, the total water content (80%) was found to consist of 62% free water and 18% bound water. Relaxations of such bound water produce a relatively weak dielectric dispersion, called the δ-dispersion, centered around 100 MHz and hence is located roughly midway between the β- and γ-dispersions. Most of the strongly bound water in tissues will be incorporated directly into the overall structure of the globular and membrane-associated proteins. It could be that perturbing the strongly bound water molecules by subjecting them to 100 MHz electromagnetic radiation induces changes in those physiological processes that depend on the functional behavior of the substrate proteins. Such a possibility should be included by those who wish to consider the harmful effects that might arise from deliberate or accidental exposure to electromagnetic energy of 50-500 MHz.

TUMOR TISSUE

The results shown in Figure 10 for breast cancer indicate that the permittivity of cancerous tissue is significantly greater than that of normal tissue. Such a trend was first noted by Fricke and Morse (81). This feature is also indicated by the data presented in Table 8, and is especially evident from

TABLE 8. Relative Permittivity (ε) and Resistivity (ρ ,ohm.m) Values for Some Tumor Tissues at 37°C

MATERIAL	13.56 MHz		27.12 MHz		433 MHz		915 MHz		2450 MHz		REFERENCE
	ε	ρ	ε	ρ	ε	ρ	ε	ρ	ε	ρ	
Hemangiopericytoma	136	0.91	106	0.88	57	0.73	55.4	0.62	50	0.35	102
Intestinal Leiomysarcoma	309	1.2	183	1.1	62	0.81	60	0.67	54	0.38	102
Splenic Hematoma	297	1.56	243	1.35	54	1.08	52	0.94	49	0.53	102
Rat Hepatoma D23	305	1.35	178	1.15	-	-	-	-	-	-	82
Normal Rat Liver	167	2.05	110	1.90	-	-	-	-	-	-	82
Canine Fibrosarcoma	48	5.6	29	5.3			(Before Hyperthermia)				100
	30	1.55	19	1.51			(After 43°C Hyperthermia)				100
Mouse KHT Tumor	-	-	135*	-	61	0.89	60	0.62	54	0.39	83
normal muscle	-	-	90*	-	56	1.08	56	0.66	48	0.38	83

* Extrapolated data

the work of Bottomley and Andrew (82) for normal and cancerous rat liver, and from the measurements of Rogers et al. (83) on normal and tumor mouse muscle. The conductivity of cancerous tissue also appears to be greater than that of normal tissue, and this property, noted also by Foster and Schepps (84), could be of use in further developments of the radiofrequency and microwave hyperthermic treatment of cancer.

As summarized by Hazlewood et al. (85) the relaxation times in malignant tissue are larger than those in normal tissue, indicating that a significant increase in the motional freedom of water has occurred. The dielectric results may also relate to the observation that the water content and sodium concentration in tumor cells is higher than in normal cells (86,87), and also to the facts that cancer cells have significantly reduced membrane potentials (88-90) and an altered ability to absorb positive ions (91,92). By incorporating these known differences in the water and ionic compositions of normal and cancerous tissues, Grant and Spyrou (93) have successfully been able to model the interfacial polarizations to predict the dielectric differences that were observed by Bottomley and Andrew (82) for normal and cancerous rat liver.

It would also be of interest to investigate the possibility that the dielectric properties relate to the cellular electrochemical potentials. Tumor cells are more electronegative than normal cells (91), and cancerous tissues are more electronegative than normal tissues (94,95). Such results suggest that structurally less differentiated tissues are more electronegative than normal tissues and it is of interest to note that regenerating tissue, which is another type of less differentiated tissue, is also relatively electronegative (96). Dielectric measurements may usefully complement such studies as those summarized here and provide new insights at the molecular level in the study of the cancer problem.

REFERENCES

(1) Sheppard, A.R. and Eisenbud, M., eds.: Biological Effects of Electric and Magnetic Fields of Extremely Low Frequency, New York University Press, New York, 1977.

(2) Konig, H.L., Drueger, A.P., Lane, S. and Sonnig, W.: Biologic Effects of Environmental Electromagnetism, Springer-Verlag, Heidelberg, 1981.

(3) Becker, R.O. and Marino, A.A.: Electromagnetism and Life, State University of New York Press, Albany, 1982.

(4) Fröhlich, H. and Kremer, F., eds.: Coherent Excitations in Biological Systems, Springer-Verlag, Heidelberg, 1983.

(5) Adey, W.R. and Lawrence, A.F., eds.: Non-linear Electrodynamics of Biological Systems, Plenum Press, New York, 1984.

(6) Hahn, G.M.: Hyperthermia and Cancer, Plenum, New York, 1982.

(7) Storm, F.K., ed.: Hyperthermia in Cancer Therapy, G.K. Hall, Boston, 1983.

(8) Baker, L.E.: Electrical impedance pneumography, in Non-Invasive Physiological Measurements, P. Rolfe, ed., Academic, London, 1979.

(9) Hill, D.W.: The role of electrical impedance methods for the monitoring of central and peripheral blood flow changes, in Non-Invasive Physiological Measurements, P. Rolfe, ed., Academic, London, 1979.

(10) Pryce, L.R.: Electrical impedance tomography (ICT): A new CT imaging technique, IEEE Trans. Nucl. Sci. NS-26:2736-2739, 1979.

(11) Barber, D.C. and Brown, B.H.: Applied potential tomography, J. Phys. [E] 17:723-733, 1984.

(12) Burdette, E.C., Wiggins, S., Brown, R. and Karow, A.M.: Microwave thawing of frozen kidneys: A theoretically based experimentally-effective design, Cryobiology 17:393-402, 1980.

(13) Barker, A.T. and Lunt, M.J.: The effects of pulsed magnetic fields of the type used in the stimulation of bone fracture healing, Clin. Phys. Physiol. Meas. 4:1-27, 1983.

(14) Höber, R.: Eine methode die elektrische leitfahigkeit im innern von zellen zu messen, Arch. Gesamt. Physiol. Mensch. 133:237-259, 1910.

(15) Fricke, H.: The electric capacity of suspensions with special reference to blood, J. Gen. Physiol. 9:137-152, 1925.

(16) Oncley, J.L.: The electrical moments and relaxation times of proteins as measured from their influence on the dielectric constant of solutions, in Proteins, Amino Acids and Peptides, E.J. Cohen and J.T. Edsall, eds., Reinhold, New York, 543-568, 1943.

(17) Haggis, G.H., Buchanan, T.J. and Hasted, J.B.: Estimation of protein hydration by dielectric measurements at micro-wave frequencies, Nature 167:607-608, 1951.

(18) Grant, E.H.: The structure of water neighboring proteins, peptides, and amino acids as deduced from dielectric measurements, Ann. N.Y. Acad. Sci. 125:418-427, 1965.

(19) Schwan, H.P.: Electrical properties of bound water, Ann. N.Y. Acad. Sci. 125:344-354, 1965.

(20) Kell, D.B. and Harris, C.M.: On the dielectrically observable consequences of the diffusional motions of lipids and proteins in membranes, Eur. Biophys. J. 12:181-197, 1985.

(21) Arnold, W.M., Wendt, B., Zimmermann, U. and Korenstein, R.: Rotation of a single swollen thylakoid vesicle in a rotating field. Electrical properties of the photo-synthetic membrane and their modification by ionophores, lipophilic ions and pH, Biochim. Biophys. Acta 813:117-131, 1985.

(22) Bone, S. and Pethig, R.: Cyclodextrins as model systems for the study of proton transport, Int. J. Quantum Chem; Quantum Biol. Symp. 10:133-141, 1983.

(23) Careri, G., Garaci, M., Giansanti, A. and Rupley, J.A.: Protonic conductivity of hydrated lysozyme powders at megahertz frequencies, Proc. Natl. Acad. Sci. USA 82:5342-5346, 1985.

(24) Morgan, H. and Pethig, R.: Protonic and ionic conduction in lysozyme, J. Chem. Soc. Faraday I 82:143-156, 1986.

(25) Gould, J.L.: Magnetic field sensitivity in animals, Annu. Rev. Physiol. 46:585-598, 1984.

(26) Schwan, H.P.: Electrical properties of tissue and cell suspensions, Adv. Biol. Med. Phys. 5:147-209, 1957.

(27) Grant, E.H., Sheppard, R.J. and South, G.P.: Dielectric Behaviour of Biological Molecules in Solution, Clarendon, Oxford, 1978.

(28) Pethig, R.: Dielectric and Electronic Properties of Biological Materials, Wiley, Chichester, 1979.

(29) Almers, W.: Gating currents and charge movements in excitable membranes, Rev. Physiol. Biochem. Pharmacol. 82:96-190, 1978.

(30) Cole, K.S.: Membranes, Ions and Impulses, University of California Press, Berkeley, 1972.

(31) Schanne, O.F. and Ceretti, E.R.P.: Impedance Measurements in Biological Cells, Wiley, New York, 1978.

(32) Geddes, L.A. and Baker, L.E.: The specific resistance of biological material: A compendium of data for the biomedical engineer and physiologist, Med. Biol. Eng. 5:271-293, 1967.

(33) Stuchly, M.A. and Stuchly, S.S.: Dielectric properties of biological substances - Tabulated, J. Microwave Power 15:19-26, 1980.

(34) Kirkwood, J.G.: Theory of solutions of molecules containing widely separated charges with special application to zwitterions, J. Chem. Phys. 2:351-361, 1932.

(35) Kirkwood, J.G.: The dielectric polarization of polar liquids, J. Chem. Phys. 7:911-919, 1939.

(36) Wyman, J.: Dielectric constants of polar solutions, J. Am. Chem. Soc. 56:536-544, 1934.

(37) Dunning, W.J. and Shutt, W.J.: The dielectric constants of zwitterions and polar molecules as related to pH, Trans. Faraday Soc. 34:479-485, 1938.

(38) Sheridan, R.P., Levy, R.M. and Salemme, F.R.: Alpha-helix dipole model and electrostatic stabilization of 4-alpha-helical proteins, Proc. Natl. Acad. Sci. USA 79:4545-4549, 1982.

(39) Essex, C.G., Symonds, M.S., Sheppard, R.J., Grant, E.H., Lamote, R., Soetweey, F., Rosseneu, M.Y. and Peeters, H.: Five component dielectric dispersion in bovine serum albumin solution, Phys. Med. Biol. 22:1160-1167, 1977.

(40) Takashima, S.: Dielectric dispersion of deoxyribonucleic acid, J. Mol. Biol. 7:455-458, 1963.

(41) Hanss, M.: Mesure de la dispersion dielectrique en tres basse frequence de solution de DNA, Biopolymers 4:1035-1042, 1966.

(42) Takashima, S.: Dielectric dispersion of deoxyribonucleic acid II, J. Phys. Chem. 70:1372-1377, 1966.

(43) Takashima, S.: Mechanism of dielectric relaxation of deoxyribonucleic acid, Adv. Chem. Ser. 63:232-244, 1967.

(44) Mandel, M.: Dielectric properties of charged linear macromolecules with particular reference to DNA, Ann. NY Acad. Sci. 74:303-311, 1977.

(45) Takashima, S., Gabriel, C., Sheppard, R.J. and Grant, E.H.: Dielectric behavior of DNA solution at radio and microwave frequencies (at 20°C), Biophys. J. 46:29-34, 1984.

(46) Swicord, M.L. and Davis, C.C.: Microwave absorption of DNA between 8 and 12 GHz, Biopolymers 21:2453-2460, 1982.

(47) Swicord, M.L. and Davis, C.C.: An optical method for investigating the microwave absorption characteristics of DNA and other biomolecules, Bioelectromagnetics 4:21-42, 1983.

(48) Edwards, G.S., Davis, C.C., Saffer, J.D. and Swicord, M.L.: Resonant microwave absorption of selected DNA molecules, Phys. Rev. Lett. 53:1284-1287, 1984.

(49) Edwards, G.S., Davis, C.C., Saffer, J.D. and Swicord, M.L.: Microwave-field-driven acoustic modes in DNA, Biophys. J. 47:799-807, 1985.

(50) Lee, B. and Richards, F.M.: The interpretation of protein structures: Estimation of static accessibility, J. Mol. Biol. 55:379-400, 1971.

(51) Lee, C.Y., McCammon, J.A. and Rossky, P.J.: The structure of liquid water at an extended hydrophobic surface, J. Chem. Phys. 80:4448-4455, 1984.

(52) Hallenga, K., Grigera, J.R. and Berendsen, H.J.C.: Influence of hydrophobic solutes on the dynamic behaviour of water, J. Phys. Chem. 84:2381-2390, 1980.

(53) Buchanan, T.J., Haggis, G.H., Hasted, J.B. and Robinson, B.G.: The dielectric estimation of protein hydration, Proc. R. Soc. Lond. A213:379-391, 1952.

(54) Bone, S., Gascoyne, P.R.C. and Pethig, R.: Dielectric properties of hydrated proteins at 9.9 GH, J. Chem. Soc.: Faraday Trans. 1 73:1605-1611, 1977.

(55) Bone, S., Eden, J. and Pethig, R.: Electrical properties of proteins as a function of hydration and NaCl content, Int. J. Quantum Chem.; Quantum Biol. Symp. 8:307-316, 1981.

(56) Bone, S. and Pethig, R.: Dielectric studies of the binding of water to lysozyme, J. Mol. Biol. 157:571-575, 1982.

(57) Bone, S. and Pethig, R.: Dielectric studies of protein hydration and hydration-induced flexibility, J. Mol. Biol. 181:323-326, 1985.

(58) Cross, T.E. and Pethig, R.: Microwave studies of the interaction of DNA and water in the temperature range 90-300K, Int. J. Quantum Chem; Quantum Biol. Symp. 10:143-152, 1983.

(59) Clementi, E.: Structure of water and counterions for nucleic acids in solution, in Structure and Dynamics: Nucleic Acids and Proteins, E. Clementi and R.H. Sarma, eds., Adenine Press, New York, 321-364, 1983.

(60) Careri, G., Gratton, E., Yang, P.H. and Rupley, J.A.: Correlation of IR spectroscopic, heat capacity, diamagnetic susceptibility and enzymatic measurements on lysozyme powder, Nature 284:572-573, 1980.

(61) Gascoyne, P.R.C., Pethig, R. and Szent-Györgyi, A.: Water structure dependent charge transport in proteins, Proc. Natl. Acad. Sci. USA 78:261-265, 1981.

(62) Behi, J., Bone, S., Morgan, H. and Pethig, R.: Effect of deuterium-hydrogen exchange on the electrical conduction in lysozyme, Int. J. Quantum Chem; Quantum Biol. Symp. 9:367-374, 1982.

(63) Tien, H.T. and Diana, A.L.: Biomolecular lipid membranes:

A review and a summary of some recent studies, Chem. Phys. Lipids 2:55-101, 1968.

(64) Klee, M. and Plonsey, R.: Extracellular stimulation of a cell having a non-uniform membrane, IEEE Trans. Biomed. Eng. BME-21:452-460, 1974.

(65) Zimmermann, U.: Electric field-mediated fusion and related electrical phenomena, Biochim. Biophys. Acta 694:227-277, 1982.

(66) Harris, C.M. and Kell, D.B.: On the dielectrically observable consequences of the diffusional motions of lipids and proteins in membranes, Eur. Biophys. J. 13:11-24, 1985.

(67) Poo, M-m.: In situ electrophoresis of membrane components, Ann. Rev. Biophys. Bioeng. 10:245-276, 1981.

(68) Singh, B., Smith, C.W. and Hughes, R.: In vivo dielectric spectrometer, Med. Biol. Eng. Comput. 17:45-60, 1979.

(69) Kosterich, J.D., Foster, K.R. and Pollack, S.R.: Dielectric permittivity and electrical conductivity of fluid saturated bone, IEEE Trans. Biomed. Eng. BME-30:81-86, 1983.

(70) Schwan, H.P.: Die elektrischen eigenschaften von muskelgewebe bei niederfrequenz, Z. Naturforsch. 9b:245-251, 1954.

(71) Pauly, H. and Schwan, H.P.: The dielectric properties of bovine eye lens, IEEE Trans. Biomed. Eng. BME-11:103-109, 1964.

(72) Schwan, H.P. and Foster, K.R.: Microwave dielectric properties of tissues: Some comments on the rotational mobility of tissue water, Biophys. J. 17:193-197, 1977.

(73) Foster, K.R., Schepps, J.L., Stoy, R.D. and Schwan, H.P.:

Dielectric properties of brain tissue between 0.01 and 10 GHz, Phys. Med. Biol. 24:1177-1187, 1979.

(74) Rosendal, T.: Concluding studies on the conducting properties of human skin to alternating current, Acta Physiol. Scand. 9:39-46, 1945.

(75) Yamamoto, T. and Yamamoto, Y.: Dielectric constant and resistivity of epidermal stratum corneum, Med. Biol. Eng. 14:494-499, 1976.

(76) Clar, E.J., Cambrai, M. and Sturelle, C.: Study of skin horny layer hydration and restoration by impedance measurement, Cosmetics & Toiletries 97:33-40, 1982.

(77) Lin, J.C.: Microwave properties of fresh mammalian brain tissues at body temperature, IEEE Trans. Biomed. Eng. BME-22:74-76, 1975.

(78) Stoy, R.D., Foster, K.R. and Schwan, H.P.: Dielectric properties of mammalian tissues from 0.1 to 100 MHz: A summary of recent data, Phys. Med. Biol. 27:501-513, 1982.

(79) Stuchly, M.A., Kraszewski, A., Stuchly, S.S. and Smith, A.M.: Dielectric properties of animal tissues _in vivo_ at radio microwave frequencies: Comparison between species, Phys. Med. Biol. 7:927-936, 1982.

(80) Gabriel, C., Sheppard, R.J. and Grant, E.H.: Dielectric properties of ocular tisses at 37C, Phys. Med. Biol. 28:43-49, 1983.

(81) Fricke, H. and Morse, S.: The electric capacity of tumors of the breast, J. Cancer Res. 10:340-346, 1926.

(82) Bottomly, P.A. and Andrew, E.R.: RF magnetic field penetration, phase shift and power dissipation in biological tissue: Implications for NMR imaging, Phys. Med. Biol. 23:630-643, 1978.

(83) Rogers, J.A., Sheppard, R.J., Grant, E.H., Bleehen, N.M. and Honess, D.J.: The dielectric properties of normal and tumour mouse tissue between 50 MHz and 10 GHz, Br. J. Radio 56:335-338, 1983.

(84) Foster, K.R. and Schepps, J.L.: Dielectric properties of tumor and normal tissues at radio through microwave frequencies, J. Microwave Power 16:107-119, 1981.

(85) Hazlewood, C.K., Chang, D.C., Nichold, B.L. and Woesner, D.E.: Nuclear magnetic resonance transverse relaxation times of water protons in skeletal muscle, Biophys. J. 14:583-606, 1974.

(86) Damadian, R. and Cope, F.W.: NMR in cancer. V. Electronic diagnosis of cancer potassium nuclear magnetic resonance: Spin signatures and T1 beat patterns, Physiol. Chem. Phys. 6:309-322, 1974.

(87) Pool, T.B., Cameron, I.L., Smith, N.K.R. and Sparks, R.L.: Intracellular sodium and growth control: A comparison of normal and transformed cells, in The Transformed Cell, I.L. Cameron and T.B. Pool, eds., Academic, New York, 1981.

(88) Cone, C.D.: Electro-osmotic interactions accompanying mitosis initiation in sarcoma cells _in vitro_, Trans. N.Y. Acad. Sci. 31:404-427, 1969.

(89) Cone, C.D.: Variation of the transmembrane potential level as a basic mechanism of mitosis control, Oncology 24:438-470, 1970.

(90) Cone, C.D. and Tongier, M.: Control of somatic cell mitosis by simulated changes in the transmembrane potential level, Oncology 25:168-182, 1971.

(91) Ambrose, E.J., James, A.M. and Lowick, J.H.B.: Differences between the electrical charge carried by normal and homologous tumor cells, Nature 177:576-577, 1969.

(92) Purdom, L., Ambrose, E.J. and Klein, G.: A correlation between electrical surface charge and some biological characteristics during the stepwise progression of a mouse sarcoma, Nature 181:1586-1587, 1958.

(93) Grant, J.P. and Spyrou, N.M.: Complex permittivity differences between normal and pathological tissues: Mechanisms and medical significance, J. Bioelectricity 4:419-458, 1985.

(94) Schaubel, M.K. and Habal, M.B.: Electropotentials of normal tissue, J. Surg. Res. 9:513-515, 1969; Electropotentials of tumor tissue, J. Surg. Res. 9:517-520, 1969.

(95) Schaubel, M.K. and Habal, M.B.: Electropotentials of surgical specimens, Arch. Pathol. 90:411-415, 1970.

(96) Becker, R.O.: The bioelectric factors in amphibian-limb regeneration, J. Bone Joint Surg. 43A:643-656, 1961.

(97) Altman, P.L. and Dittmer, D.S.: Biology Data Handbook (2nd Ed.), Vol. 1, Federation of American Societies for Experimental Biology, Bethesda, MD, 392-298, 1972.

(98) Brady, M.M., Symons, S.A. and Stuchly, S.S.: Dielectric behavior of selected animal tissues in vitro at frequencies from 2 to 4 GHz, Trans. Biomed. Eng. BME-28:307, 1981.

(99) Schwan, H.P.: Biophysics of diathermy, in Therapeutic Heat and Cold, S. Licht, ed., Waverly Press, Baltimore, 63-125, 1965.

(100) Hahn, G.M., Kernahan, A., Martinez, A., Pounds, D. and Prionas, S.: Some heat transfer problems associated with heating by ultrasound, microwaves or radio frequency, Ann. N.Y. Acad. Sci. 335:327-346, 1980.

(101) Burdette, E.C., Cain, F.L. and Seals, J.: In vivo probe

measurement technique for determining dielectric properties of VHF through microwave frequencies, IEEE Trans. Microwave Theory & Techn. MIT-28:414-427, 1980.

(102) Schepps, J.L. and Foster, K.R.: The UHF and microwave dielectric properties of normal and tumour tissues. Variation in dielectric properties with tissue water content, Phys. Med. Biol. 25:1149-1159, 1980.

7
Electronic Properties of Natural and Modeled Bilayer Membranes

JOZEF R. ZON* and H. TI TIEN
Department of Physiology, Michigan State University
East Lansing, Michigan

INTRODUCTION

Natural membranes are characterized by high selectivity, specificity, and control processes whose mechanisms of operation can be reduced to molecular levels. Many of these processes involve self-assembly of macromolecules such as lipids, proteins, and DNA. In 1941 Szent-Gyorgyi suggested that proteins and other materials may possess some of the solid-state electronic properties known to occur in amorphous materials and organic polymers (1). Since then, electronic processes in living systems have been extensively studied, especially in connection with coupled redox reactions in mitochondria, energy transfer and conversion in chloroplasts, and sensory transduction in visual systems. Bilayer lipid membrane-based structures responsible for carrying out the aforementioned processes may be the transducers. Whether we can construct similar membrane devices by emulating natural systems remains to be seen, but the approach is a viable one. The purpose of this chapter is to review data and concepts involving the study of electronic properties of both natural membranes and experimental lipid bilayer systems such as planar bilayer lipid membranes (BLMs) and spherical liposomes. Although the emphasis will be on the BLM systems, it is appropriate to deal first with natural membranes and then proceed to their models because the study of BLMs originated from attempts to better understand processes in

---

*Permanent address: Department of Theoretical Biology, The Catholic University of Lublin, Lublin, Poland

natural membranes. Jointly studying natural membranes and artificial lipid bilayers, leads to both a better understanding of natural membranes, and the possibility of constructing bilayer systems capable of performing basic biological functions such as photosynthesis (2-5).

Planar and spherical bilayer lipid membranes can be made from constituents of natural membranes, combinations of natural membranes and artificially synthesized components, or completely synthetic components (6).

The problem of electronic conductivity in both biological membranes and experimental lipid bilayers has a diverse background. One related area involves the study of the semiconductive properties of basic classes of chemical compounds of living organisms, paralleled by the theoretical study of the quantum chemistry of biomolecules. Another area involves attempts to link knowledge of physiological functions with electronic models of living processes (7,8). Neither of these related fields will be dealt with here, but they are treated elsewhere (1,9-11). Results of investigations of electronic properties of other models of biological membranes such as thick lipid films and multi-lamellar layers will also not be dealt with here. BLMs and liposomes have proven to be, in many respects, the most adequate models of biological membranes.

ELECTRONIC PHENOMENA IN MEMBRANES

Electrical current may arise as a result of movement of any kind of charged species, or from a change of polarization. In biological materials, currents of ions, electrons, and polarization phenomena are so highly interrelated that it is difficult to separate the individual components. But special techniques to do so have recently become available. Moreover, there is a group of specific phenomena that may be adequately explained only in terms of electron transfer and redox reactions.

MOLECULAR AND SUPRAMOLECULAR ASPECTS OF ELECTRON TRANSLOCATION

Every reaction in which transfer of an electron occurs from a reductant species, R, to an oxidant one, O, is called a redox reaction. However, the physical conditions under which the transfer occurs and the mechanisms of the electron transfer may vary. The factors that determine the mechanism of electron translocation are the ability of one molecule to donate or to accept an electron (redox potential), the geometry of the system in which the molecules and/or atoms are located (liquid or liquid crystal phase, solid crystal-like or noncrystalline phase), and various externally changeable factors such as temperature, pressure, and illumination. Redox reactions may be divided into five subgroups: (1) redox reactions occurring through collisions in solution or at a solution/solid interface; (2) charge-transfer reactions in which the electrons are shared between the components of the charge-transfer complex; (3) translocation of electrons in a solid phase where electrons overcome the relatively low energy barriers between molecules or atoms by activated hopping over a barrier; (4) quantum-mechanical tunneling of electrons between sites separated by a relatively high-energy barrier; (5) conduction of electrons along common energy bands. Taking into account the physical situation of biological membranes, one may confine the first two mechanisms to the solution/membrane interface, and the last three mechanisms to the membrane itself.

1. Collision Redox Reactions

The transfer of whole numbers of electrons takes place according to the scheme:

$$R + O \rightarrow R^{+} + O^{-} \tag{1}$$

When R and O are reductant and oxidant species, respectively, R^{+} is the oxidized reductant and O^{-} is the reduced oxidant. The result of these redox reactions depends on the redox potential difference between species, and the frequency of collisions determined by the temperature, and the concentration of both chemicals. The redox potential (or oxidative ability) of molecules of biological significance has been given (12). Equation (1) may be expressed in terms of electron acceptor (A) and donor (D). In this case we have reduced acceptor (A^{-}) and oxidized donor (D^{+}).

2. Charge-Transfer Reactions

If two chemical species form a charge complex, they share an electron in a common orbital. The probability of finding the electron either on the donor side or the acceptor side may differ, depending mainly on the difference between the ionization potential of the donor and the electron affinity of the acceptor. In the excited state of the complex, the shifting of the electron is much stronger. When an electron is strongly shifted towards the acceptor side, dissociation of the charge-transfer complex occurs, and the result may be regarded as a redox reaction.

3. Hopping

This occurs when two electron energy states exist that are isolated from each other by an energy barrier. Commonly, the mobility of the electrons and the conductivity of the material (a solid) are both low. However, if the height of the barrier is lower than the energy gap between the valence and conduction bands, activation of electrons by either an electric field or temperature will increase charge-carrier mobility, but not density.

4. Tunneling

The quantum-mechanical probability of tunneling across an energy barrier depends on its thickness and shape, and on the energy and mass of the particle. In biological systems, tunneling may be a mechanism of electron translocation over distances of tens of Angstroms (13). Tunneling of electrons in biocompounds was discovered by Chance, DeVault, and their co-workers (14). The probability of tunneling may be altered by variations in temperature or by applying an electric field, because these factors influence the shape and width of the energy barrier as well as the density of electrons and their effective masses. The macroscopic consequence of these changes are reflected in the conductivity of the material.

5. Conduction Along Energy Bands

In inorganic crystalline materials, common energy bands arise as a result of periodicity and overlapping of wave functions of electrons of a great number of atoms forming the solid. However, biological material is not periodic and homogeneous, therefore the formation of extensive common-energy bands is doubtful. Yet, one may expect the formation of localized common-energy bands which may be generated as a result of aggregation of a sufficient number of molecules having π-type bonds. The width of the conduction band (and thereby the effective mass of the charge carriers moving along them) is determined by the degree of overlapping of the π-electron systems of the molecules. The value of the forbidden energy gap, which is the crucial factor that determines the number of the charge carriers in the conduction band, depends primarily on the electronic structure of the atoms forming the given material, and on their proximity and periodicity. The common energy bands may extend through the whole solid or occur only locally, being isolated from other similar regions by energy barriers.

Temperature is the main factor governing the population of electrons that are free to move in the conduction band. The conductivity span of materials classified as semiconductors is 10^{-10}–$10^{1}(\Omega\,m)^{-1}$. However, the magnitude of the conductivity does not uniquely determine either the nature of the charge carriers or the mechanisms underlying the conductivity. Many ionic conductors and even insulators also have conductivities falling into this range. Therefore, a specific set of criteria is usually applied to identify the electronic nature of the semiconductivity of a material.

The conductivity of a semiconductive material should vary with temperature according to the formula:

$$\sigma(T) = \sigma_0 \exp(-E_g/2kT) \qquad (2)$$

where (T) is the value of conductivity at a given temperature, σ_0 is a constant (dependent primarily on the mobility and effective masses of the carriers, E_g is the energy gap separating the valence and conduction bands, k is the Boltzmann constant, and T is the absolute temperature. The value of E_g is usually given in eV; in semiconductors, it is 1-3 eV.

Intrinsic semiconductors are particularly sensitive to the presence of doping substances, which can change the value and character of the conductivity (holes or electrons). An especially important feature of the investigations on extracted biological materials is that they contain impurities which substantially influence the conductivity. The presence of ions and water (which decrease E_g by increasing the dielectric constant of the material, and act as electron donors) adds to the complications in characterizing the nature of the semiconductivity of biological materials (15). Another characteristic of some semiconductive materials (those in which E_g is lower than the energy of impinging light quanta) is the rise in conductivity that they exhibit upon illumination. This property is manifested in the photovoltaic effect. When one type of current carriers predominates, the change of conductivity upon illumination of the material is given by:

$$\Delta\sigma = e(u_+ \Delta C_+ + u_- \Delta C_-) \tag{3}$$

where $\Delta\sigma$ is the change of conductivity, e is the elementary charge, u_+, u_- are the hole and electron drift mobilities respectively, and ΔC_+, ΔC_- are the increases in concentration of the holes and electrons, respectively. The photovoltage generated in open circuit conditions, E_{op}, is given by:

$$E_{op} = (kT/e)\ \ln(1 + \sigma_1/\sigma_d) \tag{4}$$

where σ_1 and σ_d are the conductivities in light and in dark, respectively.

The thermoelectric effect is a phenomenon that occurs at a junction between two materials having different electrical and thermal conductivities, and different temperatures, T_1 and T_2. In this system a voltage, E_{11}, is generated that depends on the temperature difference and the relative thermoelectric power, E_{12} (defined as a voltage generated across the junction when the temperature changes by one degree). E_{12} has a positive value when the current generated flows from the conductor 2 to the conductor 1.

$$E_{11} = \int_{T_1}^{T_2} E_{12}\ dT \tag{5}$$

The thermopower generated in systems containing semiconductors is a very sensitive function of the level of impurities present. The magnitude of E_{12} in systems based on contact of a semiconductor and a metal usually exceeds by more than an order of magnitude that found in systems based on metal/metal contacts.

The presence of a junction between p-type and n-type conducting materials also gives rise to some characteristic conductivity phenomena. One of them is rectifying action and its derivative -- Zener-diode behavior. In a rectifier, the current has a much smaller resistivity in one direction than in the other, and in the Zener-diode, after the voltage across the junction reaches a certain value, it maintains a steady level.

Thus, all the above possible mechanisms of the translocation of the electrons create a spectrum extending from the most simple mechanism (collisions), through relay mechanisms to those involving large aggregates of atoms or molecules (Figure 1). In a broad meaning of the term "redox reactions," all of these mechanisms may give rise to interfacial, intramembrane, and transmembrane processes of reduction and oxidation. In both

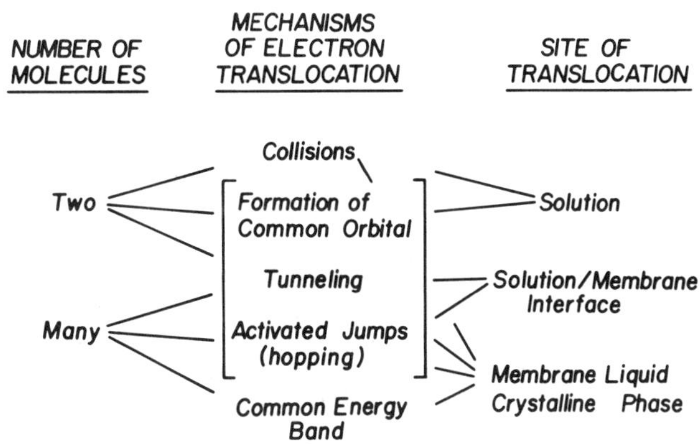

FIGURE 1. Possible mechanisms of electron translocations in natural and artificial bilayer lipid membranes. The bracketed mechanisms, when operating on molecules or atoms, may give rise to long-distance electron translocations. The conduction brought about by the existence of donor or acceptor levels located near the conduction or the valence band respectively, may be treated as intermediate between the hopping and band mechanisms. Thermally or electrically activated injection of electrons from the donor levels into the conduction band, or creation of holes in the valence band by the presence of acceptor levels, constitute the current flowing in these bands.

natural and artificial lipid bilayers immersed in an aqueous solution, it is possible that many of these mechanisms are operating at the same instant, with one of them being the prevailing or limiting factor.

ORIGIN OF CONDUCTIVITY

Ionic currents undoubtedly play an important role in both natural and artificial bilayer lipid membranes. However, experimental evidence of electronic conductivity in materials extracted from biological structures, as well as information gathered on the biochemical and biophysical processes taking place in the course of physiological activity of living membranes, lend support to the idea that electrons and/or holes must also play an important role in life processes. In this connection, a question arises of how to distinguish between electronic and ionic conductivity either in natural or artificial membranes. In principle, it is possible to tell the difference using methods that allow detection based on any of the following: (1) differences in mass; (2) structure of the material determining the behavior of the carrier; (3) energy requirements for initiating the conduction process; (4) specific reactions at the interface (electrostenolysis)(15).

1. Mass of Charge Carriers

An electron is nearly 2000 times lighter than a proton and much lighter than simple ions or molecular ions, and this difference manifests itself in many ways. The mobilities of electrons in solids are higher than those of ions, but the existence of energy barriers and the narrowness of the conduction bands (high effective masses) decrease the mobilities of electrons and holes. High-frequency techniques made it possible to overcome difficulties arising from the existence of energy barriers at the boundaries of microstructures in biological materials.

Another difference arising from that inequality of masses of ions and electrons is the time required for the development of a voltage or current response to an external factor such as a light pulse.

If a magnetic field is used to induce circular motion of electrons around the magnetic field lines, two kinds of reactions to the field may be observed: (1) voltage development is slower in the case of ions than electrons; (2) the cyclotron frequency of electrons is higher than that of ions.

When charge translocation occurs, the increase in weight of the part of the system where the ionic carriers are deposited may be observed. With the transfer of ions will be a concommitant polarization arising from the ions deposited in that part of the system.

2. Structure of the Material

A generally accepted structure of a biomembrane is depicted in Figure 2. One unique feature to be noted, as a result of the ultrathinness of the structure, is the field effect. If a potential as low as 10 mV exists across the membrane, electric field gradients on the order of 10^6 V/cm will be generated. If the membrane is asymmetrical, the current-voltage curve may be non-linear because of at least three mechanisms, each of which is analogous to well-known mechanisms in solid-state physics:

(1) Charge injection. When an insulating layer is interposed between two conducting layers, an applied voltage will effectively be dropped across the insulating layer. Because of the interplay of thermal diffusion and the electric fields (electrical double-layer theory), the space-charge distribution is a function of the applied voltage. The result is that the greater the applied voltage, the higher the majority-carrier concentration and, therefore, the higher the charge injection.

(2) The image-force effect. Image forces are an expression of the polarization of the dielectric environments of charged species. The forces exist near the membrane/electrolyte interface, and lead to a strong attraction of any charged species dissolved in the lipid bilayer. The effect is greater at higher voltages.

(3) The Wien effect (16). The applied electric field affects the equilibrium constant of dissolved charged species, leading to a net increase in dissociation, and therefore to more charge carriers.

In well-organized materials such as membranes, liquid crystals, and thin films, it may be expected that the conductivity will be brought about by electrons or holes in the conduction or valence bands (5,17,18). Therefore, crystallographic studies of biological structures may give some hints regarding the expected nature of charge carriers. The thermal behavior of the conductivity in such structures is described by Equation (2). To explore the electronic nature of the conductivity, additional tests may be carried out such as studying the effect on conductivity of infrared irradiation, and measuring the energy localization of traps by studying the time and temperature dependence of the delayed luminescence.

3. Energy Requirements

When transfer of charge occurs by ionic motion across a membrane where at least one very high energy barrier must be overcome, considerable activation energies of conductivity should be involved. According to the Born expression (if dielectric constant of the bathing solution is much higher than that of the membrane, ε_m):

$$E = q^2/2r\varepsilon_m \tag{6}$$

where E is the energy spent overcoming the barrier at the interface, and q is the electronic charge on the ion. According to Equation (6), the energy necessary to transfer an ion of radius of 10^{-10} m from water to a lipid medium is about 22.6 eV. Much less energy is needed to form a charge-transfer complex (0.02-0.54 eV per complex).

In tunneling processes it is not necessary that the energy be supplied externally, but external energy may change the height, shape or width of the barrier thereby altering the efficiency of the process.

For both intrinsic and extrinsic semiconductors, the conductivity will increase with increasing number of charge carriers in the conduction band (or holes in the valence band), and thereby increase their mobility. The factor determining these parameters is the temperature, and the energy expenditure necessary to bring about a certain value of conductivity in a given material may be estimated from Equation (2). However, in the case of biomembranes and BLMs, the conductivity may also vary with composition because of water dissociation (15).

4. Electrostenolytic Effect

This process involves reactions of oxidation and reduction at the opposite ends of an electronically conducting but high

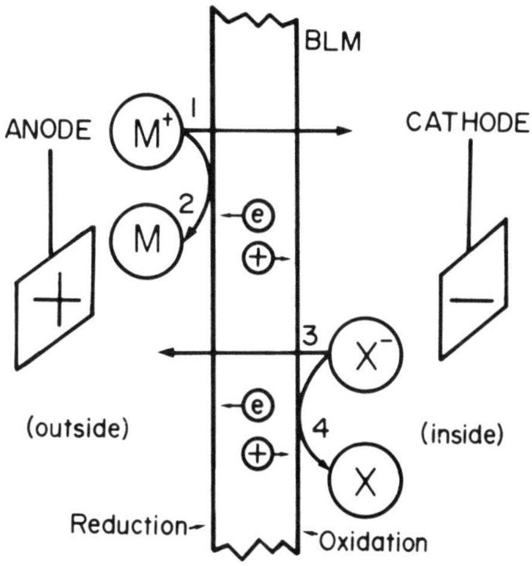

FIGURE 2. Mechanism of redox reactions across an ultrathin high-resistance membrane. BLM, bilayer lipid membrane; M^+, metal ion or electron acceptor; M, reduced atom, X^-, anion or electron donor; X, oxidized atom, e, electron, +, hole. 1 and 3 denote ion conduction; 2 and 4 denote redox reactions (15).

ion resistivity path (Figure 2). The process of the electrostenolysis across BLMs has been described previously (19). One of the most spectacular methods of demonstrating that electrons are transferred across the membrane is the formation of the highly reflective mirror, composed of reduced copper ions, at one of the BLM/bathing solution interface. Any reduction or oxidation taking part at the BLM/solution interface may be interpreted as electron conduction involvement in the process.

ELECTRONIC PROPERTIES OF BIOLOGICAL MEMBRANES

INTRODUCTION

In reviewing experimental data, mainly investigations on whole cells will be considered. But since the process described will be primarily or entirely dependent on the cell membranes, they will be reviewed here.

Among other suggestions made by Szent-Gyorgyi in 1941, ushering in the solid-state physics approach to living processes, was that the cell membrane creates an outer boundary of the common energy levels in the cell (7). This hypothesis has not been proved in its general formulation, however increasingly more data have been accumulated to support the notion that solid-state electronic processes can be ascribed to biomembranes. Strong support for the concept is provided by the results obtained in BLM research, which will be dealt with at length later.

Tunneling of electrons between membrane-bound, charge-transfer species and between molecules inside the membrane is one mechanism of electron transfer in biological membranes. It is considered an alternative mechanism to charge and energy transfer along the common energy bands (13,14).

The earliest suggestion that redox reactions may be the cause of transmembrane potential differences was made by Lund in 1928 (15,20). However, the idea was discarded on the basis of the argument that there are no electronically conductive paths in biological membranes. A new momentum was given to the idea when it was discovered that certain kinds of biologically significant compounds can be semiconductive. In 1962, Jahn renewed the hypothesis, suggesting that the electron-conductive paths across the membrane may consist of molecules that possess conjugate bond systems such as astacene, retinene, carotene, and vitamin A. The electromotive force moving electrons across the membrane was suggested to be due to differences in electron pressures in two electron systems placed across the membrane. The electron source was hypothesized to be a redox system in which the electron was released by action of ATP (20). Simultaneously with electron transport along the conjugate-bonded molecule, proton transport was assumed to take place along small diameter water pores. In Jahn's theory, both electronic and protonic flow across the membrane were considered to be coupled to the secretion of Cl^-.

MEMBRANES ACTIVE IN THE DARK

1. Secretory Cells and Erythrocytes

On the basis of the relationship between current flowing across the membranes of various secretory cells of many types of animals and the applied voltage, Mandel (21) concluded that the best approximation of the nature of processes taking place in the membrane is electron flow between a solution and semiconductive membrane. The hypothesis that electrons may be charge carriers passing through secretory cell membranes of the salivary gland of the Lone Star tick was tested by Pohl and Sauer (22). Using a suitable redox reaction indicator (Nyle Blue A), they were able to show that if a potential difference was created across the membranes (inside of the gland being negative) a color change from blue to red of the outside solution should take place. This result was interpreted as an indication of transfer of electrons across the membranes.

The flow of electrons across the erythrocyte membrane was shown in the following experiment by Marinov (23). Erythrocytes were isolated from the blood of rats, and after the hemoglobin inside them was converted into oxidized form (by incubation of the erythrocyte in the solution containing hydroxylamine), they were put into a solution containing eosin and NADH. After illumination of the suspension with visible light, the shift in the Soret band and the appearance of an absorption band with λ_{max} at 570-580 nm, indicative of reduction of hemoglobin, was observed. Simultaneously, the shift of the Soret band of extracellular NADH-eosin complex, and the appearance of an absorption band in the region of 580 nm was seen. These redox reactions must have been coupled by the membrane of the erythrocyte. On the basis of additional measurements, the mediation of membrane mobile electron carriers was excluded. The only possibility which remained was that protein (spectrin) spanning across the membrane was able to transfer electrons from the donor to the acceptor located on opposite sides of the membrane.

2. Muscle

Semiconductive electronic processes may be manifested as increases in the conductivity, thermoelectric force or photocurrent generation. All of these effects were observed in muscle membrane. A temperature-related increase in conductivity was demonstrated in the frog sartorius muscle in the region 6-24°C (24). Above this temperature, up to 42°C, there were no changes in conductivity, and above 42°C the changes were exponentially related with the temperature as predicted by Equation (2). The calculated value of E_g was about 2 eV. No measurements aimed at a more specific determination of the nature of charge carriers (electrons/holes) were carried out.

The differential thermoelectric power induced by the change in temperature was 230 + 40 μV/deg (25a). Below 19°C the dominating charge carriers were positive, and above 23°C they were negative. The ability of the muscle to respond thermoelectrically was abolished by treating it with chloroform. In contrast to the experiments described earlier, the measurements were done along the muscle cell membrane.

In another series of experiments, the heart of the frog was used (25b). Its beating was stopped by deleting K^+ from the bathing solution. If eosin (a light sensitizer) was added and the muscle was illuminated, the beating action reoccurred. Although a small (about 1°C) rise in temperature during illumination was observed, it apparently was not the reason for the effect observed. It was hypothesized that the light acted on the membranes of the muscle cells, exciting electrons in the eosin, which then migrated along the network of π bonds in the membrane. The same mechanism of changes of the level of excitation seen in the sartorius muscle of the frog was suggested. If the muscle membranes were stained with eosin and subsequently irradiated either with ultraviolet or visible light, the threshold of stimulation decreased (25c).

3. Nerve

In 1956 Ernst (26a) tried to establish links between the responses of nerve and semiconductors to various physical factors such as temperature and presence of impurities. He hypothesized that the frequency modulation of nerve impulses may be due to some specific current effect in the membrane. The fact that procaine and thioguanine can arrest the action potentials initiated by Pacinian corpuscles was explained as a result of the interaction between π electron systems of these compounds and the nerve membrane (26b). In these terms, and including also a charge-transfer interaction, he explained the blocking of the action potentials by veratrine and novacaine. Moreover, on the basis of formal similarities between generation of high-frequency oscillation by the Gunn diode and frequency modulated nerve impulses, Ernst pointed out that impulse modulation in the nerve might be due to an electronic mechanism similar to that involved in the Gunn diode.

Photovoltage and photocurrent generation was found in stained giant axons of _Sepia_ (27). During his experiments,

Chalazonitis took precautions to eliminate the infrared component of the incident light. The sign of the generated photovoltage (V) was dependent on the nature and concentration of the used pigment. The rate of photodepolarization one second after illumination was taken as a measure of the interaction between the neuron and the incident light. The rate was dependent on the intensity of the incident light of specific wavelength (I_λ), temperature (T), the membrane potential (E_r), the partial pressure of oxygen (pO_2), and the time lapse (t),

$$dV/dt = A \exp(kt) \tag{7}$$

where A and k are constants dependent on I_λ, T, E_r, and pO_2. The rise of I_λ above a certain level led to the generation of an oscillatory firing pattern of the nerve. The surface density of the photocurrent flowing across the membrane was estimated as being of the order 60 mA/m^2.

Electron excitation and electron transfer phenomena were believed to be the crucial mechanisms involved. It was pointed out that after exciting the electrons in the dye-molecule aggregates, trapped or freely moving electrons or holes may appear in the membrane. These free electrons (and also protons), before they reached their final acceptors, contributed to current flow across the membrane and played the main role in depolarizing it. The photovoltage was thought to be generated at the illuminated donor-acceptor junction between photoactivated hemoprotein and carotene-protein molecules. Cope showed that the conductance changes found by Chalazonitis obeyed the Elovich kinetic equation, and this was additional supportive evidence for the idea that semiconductivity was involved in the photoresponses (10).

An influence of light on the spiking activity of stained (eosin, neutral red, and Bengal rose) nerve membranes from several species was found by Lakatos (25a). Although the spiking activity was evoked by illumination in less than 50% of the stained nerves, no spiking was observed either in illuminated unstained nor in non-illuminated but stained neurons. Similarly, as in the sequence of events suggested by Chalazonitis, it was assumed that after excitation of the electrons in the dyes, migration of charge and energy along electron conductivity paths in the excited membrane occurred. The final event was the depolarization of the membrane.

Ultraviolet radiation (260, 280, 313 nm) and visible light (405 nm) were shown to be effective in eliciting the firing activity of unstained neurons (28). However, if the neurons were maintained at 20-22°C, no spiking was brought about by light; decreasing the temperature to 6-10°C again made the axons susceptible to the light.

Generation of a thermoelectric voltage in the sciatic nerve of the frog has been reported (25). The thermoelectric coefficient was found to be 45 + 15 μV/deg, and it decreased with time of preservation of the nerve and vanished completely after 9 days.

4. Mitochondria

Over 90% of all energy coupling in aerobic organisms involves redox reactions intimately associated with lipid membranes of mitochondria. At present, there is not one system for which the detailed mechanism of the electron-transfer and coupling process is known at a molecular level (29). Some transmembrane structure is required for a system that channels electrons in a directional manner whereby charge separation and proton movement are linked to the formation of high energy molecules such as ATP. Notwithstanding the uniqueness of the photosynthetic thylakoid membrane (discussed below), what is particularly striking is the similarity to other electron-transfer chains, such as that in mitochondria. The cristae membrane respiratory electron transport chain may have evolved from the photosynthetic system (Figure 3).

In normal conditions, the flow of electrons is coupled to ATP formation which is known to be a universal biological energy carrier and storage unit. But before mitochondria were discovered and identified as the high-energy bond synthesizing apparatus in cells of aerobic organisms, Szent-Gyorgyi (7), referring to a previous suggestion made by Jordan in 1938, pointed out that energy transfer in cells may involve a solid-state electronic mechanism. Specifically, Jordan hypothesized that in protein complexes insoluble in water, common energy levels of the outermost electrons may develop, and that an electron in such a band could move around in the band and drop to the ground level, releasing its energy in a place where it is in demand. This

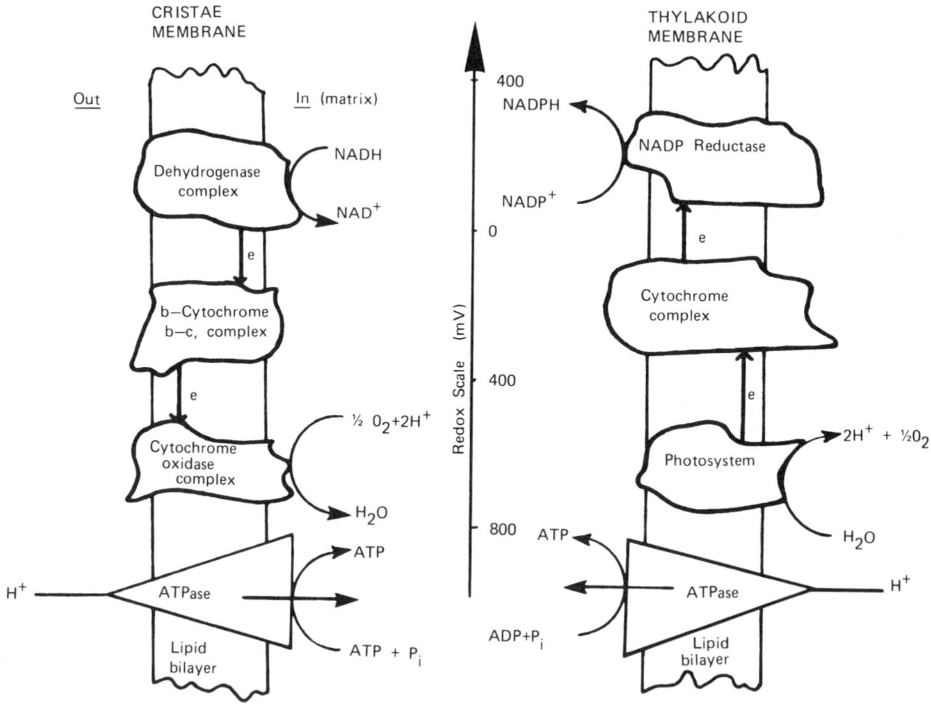

FIGURE 3. The bioenergetics of life processes in terms of the thylakoid membrane of the chloroplast and the cristae membrane of the mitochondrion. The energy-transducing membranes consist of two coupled redox reactions of photosynthesis and respiration, in which the products of one process are consumed by the other and vice versa (15).

suggestion gave rise to an extensive study of electronic properties of biological molecules, both their semiconductive properties and their donor or acceptor abilities. Here, attention will be paid only to whole mitochondrial structures and their electronic properties.

Since it is difficult to investigate electronic processes in mitochondria performing their functions in an intact state, many simplified approaches have been developed. One of them is the theoretical investigation of the feasibility of tunneling between components of the respiratory chain, based on measurements of electron transfer chain components of simpler biological systems (13,14).

Free-charge mobilities in dried samples of mitochondria have been estimated (9). The free charges were generated by a pulsed electron beam incident on a mitochondrial layer. Measurement of the speed of movement of the charges in an electric field interposed between two points on the layer allowed estimation of the electron mobility ($\mu \simeq 5 \times 10^{-6}$ m^2/V S). If small quantities of water (less than 1%) were present in the sample, the value increased by one order of magnitude.

Using the Hall microwave mobility measurement technique, Eley et al. observed the movement of electronic charges in lipid extracts that contained components of the respiratory chain of rat liver mitochondria (1). The Hall mobility of electrons was about 5×10^{-4} $m^2/V \cdot S$. If correction were made for the volumes involved, the mobility was calculated to be an order of magnitude higher. On the basis of thermoelectric force generation in pressed discs of a mixture of intact mitochondria and submitochondrial particles, it was shown that the majority charge carriers may be electrons (below the transition temperature, 342°K) or holes (above this temperature).

5. Plant Cells

Green plant photosynthesis is the process by which solar electromagnetic radiation is converted into chemical energy. This phenomenon is unique in that the light energy absorbed by pigment molecules embedded in a bilayer lipid membrane is converted eventually to stable energy-rich compounds such as ATP. The mechanism by which nature accomplishes this feat is not understood, and is being actively investigated by a number of approaches (31). Energy transduction by plant-cell membranes has broader significance beyond the process of photosynthesis itself. The mechanism of charge generation, separation and transport, and the coupling of separated charges to the production of ATP, is the central process of energy metabolism in all living systems. Thus, studies on algal cells and on pigmented lipid membranes, to be discussed in a later section, are of importance not only in their own right, but as a model for many similar processes, such as those taking place in the cristae membrane of the mitochondrion (Figure 3).

As an accidental observation made during measurements of the impedance of simple algal cells, a decrease in the longitudinal resistance of about 50% was observed when the part of the cell on which the measurements were carried out was immersed in water (32). The possibility that these changes of conductivity might be due to an outward current flow from the cell was excluded. A hypothesis that the applied current was electronic was put forward, and experiments involving the measurement of Hall voltage were suggested. The Hall phenomena may be observed both in liquid and solid conductors and semiconductors. However, due to the low mobility of charge carriers in electrolytes and most organic semiconductors, the Hall voltage is very low. In experiments carried out on the internodal cell walls and membrane of Nitellopsis obtusa using magnetic fields of 0.8-2 T and applied transmembrane currents of 1.3-2.2 μA, the Hall voltage ranged from 100-650 μV. In dead cells, no Hall voltage was observed.

Changing the resting potential in plant cells under the influence of light is a well-known phenomenon (33,34). However, these responses are not necessarily connected with the direct illumination of chloroplasts. The possibility that light may directly influence the resting potential of a plant cell membrane must be considered because this kind of interaction has been observed in membranes of rhizoids of some algae where only extremely small quantities of chlorophyll may be found (35).

To test the hypothesis that the light may also act directly upon algae membranes, ultraviolet radiation was used (32). After switching on the light, fast changes (much less than 1 sec.) of the membrane resting potential were recorded. The possibility that the observed changes may be due to UV absorption by lipids and proteins present in the membrane, and subsequent changes of its conductivity, was discussed and considered to have only minor importance.

MEMBRANES ACTIVE IN THE LIGHT

In spite of the minuteness of biologic photoelectric converters, rod sacs and thylakoids, some data has been obtained using microelectrodes. In studying the electronic properties of both types of organelles, the following methods have proved useful: (1) study of the dependence of the conductivity on temperature; (2) study of the temperature dependence of the delayed luminescence; (3) high-frequency conductivity measurements; (4) measurements of photopotentials evoked in visual receptors and chloroplasts; (5) electrochromic shift measurements.

1. Visual Receptors

As in the case of the thylakoid membrane, to be considered in the next section, the molecular mechanisms of light transduction by visual receptor membranes is obscure, but they appear to function as photon detectors, converting light quanta into electrical signals to trigger an action potential. How the absorption of light in the pigmented sac membrane is coupled to the electrical event in the plasma membrane is not known with certainty, although a hypothesis has been suggested (36). If photoreceptors are excited by a short flash, very fast photopotentials known as early receptor potentials (ERP) are generated. The ERP is usually biphasic, consisting of a corneal positive R_1 phase followed by a corneal negative R_2 phase. The R_2 of ERP's has been shown to be temperature dependent. Thus far, the ERP has not been detected in aqueous suspension of rhodopsin, or with aqueous suspension of rods and cones. This suggests that the generation of ERP requires the presence of pigments in an intact sac membrane, and has prompted a number of investigators to study dried rods and tissues. Some of the results obtained have been compared with the ERP of visual receptors and are reviewed in the following paragraph.

Trukhan and others (37) have shown that dried rods of the visual receptor of the sheep obey Equation (2), which gives a calculated value 2.3 eV. The rods exhibited a photocurrent 2-3 times greater than the dark current. Action spectra of the rods were similar to those of rhodopsin. More sophisticated experiments, avoiding some difficulties arising from use of the steady current, were also carried on the dried pigment epithelium of the frog's eye. By using a high-frequency non-contact method it was possible to show that the photoconductivity rose with increasing light intensity, the photoconductivity and dark conductivity increased with water content of the sample (3-40%), and that the charge carriers were holes with a mobility on the order of 1.5×10^{-3} $m^2/V \cdot S$. Protons were an alternative to the charge carriers responsible for the observed behavior of the sample (37b).

Electron conductivity was taken into account as an alternative for apparent proton conductivity along the membranes of the outer segments of the frog's retina (38). In this experiment, the rods' admittance changes (15 Hz-60 KHz) were tested in various solutions, including some that had a water content close to normal.

The ERP was discovered in 1964 in monkey retina (39), and has been found in the pigments in the visual receptor in invertebrates and in vertebrates (39-41). It is believed that this kind of electrical response is a common feature of structures containing pigments.

The ERP consists of three phases, the last and slowest being identified with the α-wave of the electroretinogram. Both the fast-positive and the fast-negative phases have proved to be independent of ionic movements across the pigmented membrane on the basis of the following experimental findings: (1) they are not affected by anoxia; (2) changes in composition of the ionic environment do not influence the occurrence of either phase; (3) lowering the temperature does not abolish the positive phase but does reversibly abolish the negative phase (increase of the temperature of the retina to -85°C and subsequent thawing changes the organization of the pigment molecules in the membrane and the fast signals do not appear); (4) fixation with formaldehyde or glutaraldehyde modifies only the shape of the fast phases of the ERP, but it abolishes the ion-dependent third phase; (5) after dehydration of the pigmented cells, only the fast-positive phase remains.

When the fast reaction of the retina to the pulse of light was compared with the response of a silicon cell connected with a circuit composed of resistance and capacitance representing the passive electrical properties of the biological material, the responses were indistinguishable (15,42).

This apparently non-ionic nature of the first and second phase of the ERP led to consideration of the mechanisms of its generation in terms of solid-state electronics. In this perspective, the first act of generation of the visual sensation is the shift of electronic charges on the rhodopsin and possibly also neighboring molecules, followed by conformational changes that trigger the subsequent event of the visual excitation.

2. Chloroplasts

Photosynthesis can be viewed as coupled redox reactions; water is oxidized and carbon dioxide is reduced. The driving force for the reaction is, of course, solar radiation mediated by chlorophylls embedded in the thylakoid membrane of the chloroplast. Van Niel (43) first proposed this idea in terms of the oxidant, OH and reductant, H. In terms of solid-state physics, these entities are positive holes and negative electrons (44). If one assumes that chlorophylls are in an ultrathin liquid-crystalline-like lipid bilayer, absorption of light excites an electron to the conduction band and leaves a hole in the valence band. According to Katz (44), electrons and holes are free to move to effect reduction and oxidation, respectively. In other words, chlorophyll aggregate or chlorophyll dispersed in a lipid bilayer acts as a semiconductor (15). The photogenerated electron can be transferred to an electron acceptor. Similarly, the photogenerated hole can be combined with an electron donor. That chlorophylls and their host, the chloroplast, behave as a semiconductor has long been suggested by many authors, including Arnold and Sherwood (45). Recently, owing to the interest in semiconductor electrochemical photocells for solar energy conversion (46), parallels between natural photosynthesis and semiconductor photoelectrolysis have been noted (1-5,31,46).

Considerations that energy conversion and transfer in chloroplast may involve conduction of electrons along common energy bands were first presented in 1938 (47). Consistent with the idea were the results on the temperature dependence of the conductivity of dried films of chloroplast (48), where conductivity changes were shown to obey Equation (2). In similar experiments, the photoconductivity of chloroplasts was found by Ichimura (49) to be also detectable in moist chloroplast films; however, the current induced by illumination of these samples was smaller than in the dry ones. In contrast to inorganic semiconductors and pure chlorophyll layers, the rising phase of the current was relatively slow (minutes), and the conductivity response was biphasic.

Using a condenser method for photoconductivity detection, McCree (50) was not able to detect any signal from dried chloroplasts or from chlorophyll monolayers, dried proteins, green algae, and leaves. Because the method used made it possible to detect photoconductivity signals 10^3 times weaker than those generated by inorganic photoconductors, the author concluded that, if photoconductivity actually occurs in plant material, it could not be an efficient mechanism of photosynthesis.

Measurements of photocurrents generated in dried layers of chloroplasts of various plants (using a sensitive electrometer) have shown that the action spectrum of photoconductivity corresponds to an absorption spectrum with two exceptions (48): (1) in the region of the maximum absorption of chloroplasts, the action spectrum revealed a peak that was explained as due to increased surface recombination of carriers; (2) in some cases, the photocurrent was also generated when chloroplasts were illuminated with infrared light. The maxima of the infrared action spectrum were at 950, 1040, 1260 and 1550 nm. The presence of maxima in the long-wave region was interpreted as evidence of the existence of aggregation of chlorophyll molecules. It was also found that photoconductivity spectrum coincided with the absorption spectrum of the main pigments present in chloroplast membranes. Dark conductivity and conductivity in the illumination phase, in the temperature range of -20 to +20°C, obeyed Equation (2) with conductivities and activation energy 10^{-13}–10^{-14} $(\Omega\ m)^{-1}$ and 1.72 eV, and 10^{-12}– 10^{-13} $(\Omega\ m)^{-1}$ and 0.05–0.2 eV for dark and illumination conditions, respectively. The mobility of current carriers was estimated to be 10^{-5}–10^{-3} $m^2/V \cdot sec$.

The process of generation of free charges in chloroplasts and in the membranes of photosynthetic bacteria has also been investigated by means of measurement at microwave frequencies of the dielectric loss factor, which is equivalent to conductivity losses in the sample. In the study by Blumenfeld et al. (51), the dependence of the loss factor on wavelength of incident radiation of leaves and chloroplasts extracted from _Vicia faba_ and _Sorghum sudanese_ was investigated. The action spectrum of photoconductivity was similar to the absorption spectrum of chlorophylls a and b and, probably of β-carotene. Although the action spectra of both leaves and chloroplasts were qualitatively similar, the action spectra of the leaves were higher by a factor of 5-7. When the chloroplasts were fresh and intact, a bipolar response was observed; when they were old or damaged (by heating, for example), the response was monophasic (51). A signal similar to that of chloroplasts was obtained in a model chloroplast system (52).

Bogomolni and Klein carried out combined measurements of photoconductivity at microwave frequencies, Faraday rotation, and electron-spin resonance in dried films composed of chromophores of photosynthetic bacteria (53). The photoconductivity signal was generated by both negative and positive carriers; the former seemed to originate in thermal releasing of electrons from the primary acceptors of electrons, and the latter in hole movement in dimer chlorophyll cation radicals. The Hall mobilities of the charge carriers were of the order of 10^{-4} $m^2/V{\cdot}sec$. As in the previously described experiments, the action spectrum of photoconductivity was in agreement with the absorption spectrum. The data did not support a band-conduction model of conductivity in the photosynthetic structure. Instead, they seemed compatible with a model where photoliberated charges migrated by a hopping or tunneling mechanism (see Figure 4).

Microwave conductivity in reaction centers of photosynthetic bacteria has been shown to consist of two components having different rise-times (54). One, which may originate in the migration of electrons between quinones, rose rapidly (much less than 1 sec), and a second component which rose more slowly (about 20 sec). The amplitude of the fast component depended strongly on the hydration level, and it rose when the hydration increased. Other explanations for the origin of the fast microwave photoconductivity signal, considering its hydration dependence, is that water may influence the mobility of electrons in reaction centers in proteins, or that water may make the liberation of electrons to the conduction band easier. On the other hand, the slow component of the response may be interpreted as reflecting an accumulation of mobile electrons in the electron transport chains, or changes of the charge distribution of the photosynthetic apparatus (54).

The emission of light, different from fluorescence and ordinary phosphorescence, from chlorella suspensions was discovered in 1951 (45). One-tenth of a second after illumination, the intensity of the so-called delayed light was about 10^{-3} of the intensity of the fluorescent light from the plant, and about 10^{-6} of the intensity of the absorbed light. Even after several hours, it was possible to detect this extremely weak light by means of light-sensitive photographic material. That the shortest living components of the delayed light were of purely physical nature, and did not involve chemical reactions,

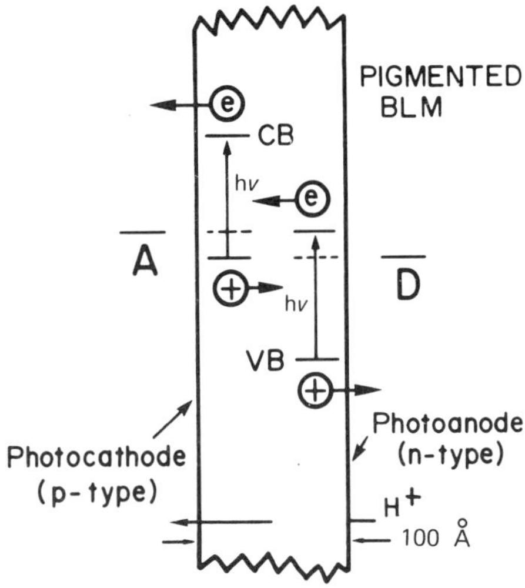

FIGURE 4. Electron tunneling in photoactive membranes. The operation of the double Schottky (or p-n junction) BLM cell is as follows: at each side of the BLM/solution interface, a space charge layer exists, which serves to separate photogenerated electron-hole (exciton) pairs. At the left-hand side, electrons move to the acceptor (A) at the interface where they effect a reduction, and the holes move to the interior of the lipid bilayer. At the right-hand side, separated holes move to the interface to cause an oxidation of the donor (D), and the electrons move toward the interior. Transmembrane electron movement is assumed to occur by tunneling (74). CB, conduction band; VB, valence band.

was shown in a series of experiments. In experiments with chloroplasts where the intensity of the delayed light as well as its decay times were investigated down to -140°C, free radical reactions, and excitation and decay of triplet states were ruled out. This left, as the only plausible explanation, a mechanism involving formation of trapped electrons in a quasi-crystalline lattice of chloroplasts, followed by emission of light after detrapping and recombination with holes (55). The relatively fast rate of the decay of delayed light at -120°C was also interpreted as ruling out the involvement of enzymatic processes in generation of that light.

The depth of particular charge traps in the forbidden energy gap of a photosynthetic preparation may be revealed by investigating the dependence of the delayed light intensity on the temperature, or (at stable temperature) the time distribution of the emitted light (45,48). In a general semiconductor approach to the mechanism of the generation of the delayed light, the following sequence of events is suggested by the evidence. After absorption of light quanta of sufficient energy, the electrons are transferred from the valence band to the conduction band. As a result of imperfections in the chloroplast-chlorophyll crystalline lattice, empty electron traps exist in the forbidden energy band. A quantum of delayed light is emitted if, due to thermal vibration, a trapped electron is released to the conduction band and eventually to the valence band where it recombines with a hole.

This simple picture of the mechanism of delayed light generation has been modified by taking into account hole traps, discharging of the energy accumulated in traps by formation of radicals, and stable chemical species, and by placing the scheme in the electrochemical context of photosynthesis (45). Although not discarded, this solid-state approach to the mechanism of delayed light generation is regarded by some as an oversimplification, and competitive approaches are reviewed by Malkin (56). Another approach to electronic conduction occurrence in chloroplast membranes is provided by experiments in which transient electric field gradients are brought about by fast movements of electrons across the membranes. These processes last less than 20 nsec and are accompanied by shifts in the absorption spectra of all photosynthetically active pigments (electrochromism) (57,58). If this mechanism operates in normal physiological conditions, one must take into account instantaneous flow of electrons from the inside of the thylakoid, where donors of electrons are located in its outer space. In addition, if Mitchell's mechanism of generation of ATP is correct, one would have to consider the flow of electrons along the thylakoid membrane transverse to the ion flow (59).

It has been shown that an outward current of electrons from chloroplasts was responsible for the transient induced electric field. In these experiments, chloroplast suspensions were illuminated with light flashes and small voltage changes were measured between two electrodes placed on the illuminated and dark sides of the chloroplasts. The risetime of the transient current induced in chloroplasts by the flash was less than

1 μsec, indicative of its direct relation to primary events of photosynthesis. The electrode on the side of the light source was polarized negatively (60).

The results obtained in the above experiment are in qualitative agreement with observations of chloroplast electrophoretic mobility changes after illumination (61). When illuminated with continuous light, the chloroplasts became negatively charged and their mobility increased by an average of 15%. Chemicals that inhibited oxygen evolution abolished the light-induced electrical mobility rise, but uncouplers of photophosphorylation and cyclic electron carriers did not affect the increase of mobility during illumination. The data obtained may be interpreted as evidence for continuous electron flow across the membrane of chloroplasts during their photosynthetic activity. The importance of membrane integrity for the occurrence of the photo-induced increase of electrical mobility of the chloroplasts was also evidence for the possible involvement of solid-state charge transport across the membrane.

The very fast electric field induction after a strong light flash was observed in the leaves of the gout weed (*Aegopodium podagrayia*). The response was identical with the ERP found in visual pigments, as noted earlier (42).

The direct measurement of transmembrane potential difference generation by light in chloroplasts of *Pepperonia metallica* reported by Bulychev et al. (62) is consistent with the experimental findings described above. The rising phase of the field was smaller than 0.01 sec, and the potential generated was negative on the outside of the chloroplast. Although it is questionable whether a transthylakoid membrane potential was actually measured, the experiments demonstrate at least provisionally that a photovoltaic effect exists in the chloroplast, as has been observed in a different system by Luttage and Pallagy (33).

ELECTRONIC PHENOMENA IN BLM SYSTEMS

INTRODUCTION

The electronic properties of planar bilayer lipid membranes (BLM) may be studied either in the dark or the light. In the first case, the changes of conductivity of the membranes and the nature of the current carriers are of importance. These problems will be dealt with in the first part of this section. The phenomena evoked in BLMs by light include the generation of photopotentials, photocurrents, and changes of conductivity.

Owing to the original idea underlying the study of the properties of BLMs, the photoelectric phenomena in them are closely related to vision and photosynthesis. But a number of experiments have shown photoelectric responses in BLMs that were not dependent on specific pigments, but on the ionic composition of the bathing solution. There is also a group of experiments in which ultraviolet radiation was the factor eliciting the photoresponse.

Attempts were made to study the electrical responses resulting from illumination of model membranes consisting of lipids and pigments known to occur in biological membranes. Other studies involved membranes containing molecules mimicking biological molecules. This area of study has developed to the extent that spherical bilayers (liposomes) made of completely synthetic components now constitute an independent field. By achieving the photosynthesis-like energy conversion, it could be seen as a remote consequence of the studies that were begun on the reconstitution of chloroplast membranes in 1968 (63).

ELECTRONIC PROPERTIES OF BLMs IN THE DARK

1. The BLM System

The usual picture of a BLM interposed between two aqueous solutions consists of a liquid hydrocarbon phase sandwiched between two hydrophilic regions. The electrical properties of BLMs have been extensively investigated, which usually entails the measurements of membrane resistance (R_m, or conductance, $G_m = 1/R_m$), capacitance (C_m), potential (E_m), dielectric breakdown voltage (V_b), and current/voltage (I/V) characteristics. Unmodified BLMs (i.e., BLMs formed from phospholipids or oxidized cholesterol dissolved in an n-alkane solvent in 0.1 M KCl solution) have typical intrinsic values of R_m greater than 10^8 ohms, C_m = 5000 pF, E_m = 0, V_b < 200 mV, and I/V curves obeying Ohm's Law. With a few exceptions, the interpretation of the results of these measurements has treated the BLM as electrically equivalent to an ionic resistor connected in parallel with a capacitor. The structure of the BLM is considered to be a thin slab of liquid crystals in two dimensions, having a fluid hydrocarbon core about 50 Å thick. The liquid-crystal portion of the BLM is an excellent insulator, but its electrical conductance can be drastically altered by incorporating a variety of compounds such as iodine, valinomycin, 2,4-dinitrophenol, chlorophyll and its related compounds, and dyes (15,64,65). Of interest in this connection is a theoretical paper on solitary waves and solitons in BLMs (69).

2. Electronic Conduction Across BLMs

One of the most striking changes of BLM electrical properties was observed when I_2 and I^- were added to the bathing solution. The usually very high electrical resistivity of the BLM, which was in the range 10^{16}–10^{17} Ω/m^2, dropped by several orders of magnitude (15,66,67). One possible explanation was that the membrane became more permeable to I^- or polyions of iodine (68). Another possibility was that the conductivity changes were due to electron conduction across the membrane (15). More specifically, conductivity changes due to charge-complex formation at the interface between the membrane and the bathing solution, and to the transfer of electrons across the membrane. After having measured capacitance and conductivity changes of BLM as a function of the frequency of the applied field and the concentrations of KI and I_2 in the bathing solution, Vodyanoy et al. (67b) concluded that $I^-{}_3$ was able to enter the BLM and act as an electron donor. It was also concluded that electrons in the membrane were transferred between donor and acceptor centers by hopping mechanisms. In similar experiments Boguslavskii et al. found that at the membrane-bathing solution interface, an exchange of electrons took place and that holes were the charge carriers inside the membrane (66).

To ascertain the role of ionic conduction in the observed drastic changes of conductivity of the BLMs under the influence I_2 and I^-, tests using ^{131}I were carried out. Jain et al. (67a) showed that conductivity changes of oxidized cholesterol-BLM were not accompanied by transfer of ^{131}I across the membrane. Moreover, the direction of the current of electrons was from the electron-donor side (I^- or thiosulphate) to the electron acceptor side (I_2). In a membrane of the same type of lipid, Karvaly et al. (70) also found that current transferred across the membrane was independent of ^{131}I (up to a specific flux level). From voltages generated by varying the iodine concentration in one of the compartments when $[I^-]$ was kept constant, and from I/V characteristics taken under the same conditions as with tracer measurements, Karvaly et al. concluded that, although some contribution of ionic conductivity could not be excluded, the current carriers across the membrane were electrons and that the membrane-bathing solution interface behaved as a semiconductive electrochemical electrode. The same conclusion was also drawn on the basis of voltages observed in systems where the concentration of the ferric ion in one of the compartments was also varied.

Feldberg et al. observed electronic conductivity across BLMs formed from glycerol mononucleate in n-hexadecane, and containing magnesium etiochlorin (71). The bathing solution contained a buffer and ferro/ferricyanide redox couples on both sides of the membrane. The value of the redox potential of the couples could be changed by varying the ratio between ferrocyanide and ferricyanide present in the bathing solution. One of the methods of detecting the electron current flowing across the membrane was by measuring the open-circuit voltage when the redox potentials of the couples on both sides of the membrane were changed. Another method consisted of measuring the open-circuit voltage decay after a very short current pulse was injected from one of the electrodes. The third method was by measuring the changes in a steady-state voltage across the membrane when different steady currents were applied to the membrane. All three methods showed that the predominant charge carriers of the current across the membrane were electrons. The density of the current flowing across the membrane was shown to be proportional to the magnesium etiochlorin in the membrane. When the magnesium etiochlorin concentration was kept constant and the redox potential of the couples was changed, the resulting current across the membranes followed the magnitude and sign of these changes.

Another phenomenon involving the movement of electrons is known as electrostenolysis (15). The phenomenon can be described as follows. When a direct current is passed through a membrane (or barrier) of high electrical resistance separating two aqueous solutions, coupled electrochemical reactions occur on opposite sides of the membrane. Oxidation takes place on the side facing the negative electrode. Implicit in these reactions is a transverse movement of electrons across the membrane.

A dramatic demonstration of electrostenolysis in BLMs was seen in the following experiment. If an oxidized cholesterol BLM is interposed between a solution of cupric nitrate and sodium sulphide, a shining mirror is observed to cover the entire BLM area in 3-10 minutes. The brightness of the mirror has been found to depend on several variables, such as the concentration of $Cu(NO_3)_2$, the BLM resistance, the pH of the bathing solution, and the duration and magnitude of the applied voltage. For example, using the cell arrangement consisting of aqueous solutions of Na_2S and $Cu(NO_3)_2$ separated by an oxidized cholesterol BLM, mirror formation was observed within 100 seconds. If a pair of calomel electrodes were used to monitor the potential difference across the BLM, a voltage of 0-50 mV was detected when the mirror became visible. This voltage rose to about 200 mV when the entire BLM surface was covered by the mirror. The brightest mirror observed was at about 350 mV. When the BLM resistance was lowered by the addition of tetraphenylborate to the bathing solution, the mirror began to form within 30 seconds after adding $Cu(NO_3)_2$ to one side of the BLM. Concurrently, the membrane voltage rose quickly to about 150 mV and eventually leveled off at about 300-350 mV (19). Similar reactions were demonstrated in the BLM when the time dependence of BLM conductivity (current) in the presence of 0.1 M KCl and 0.1 M KI at 60 mV was measured. In the case of the KCl solution, the BLM current displayed practically no time dependency. Thus, the results obtained can be explained in terms of electrostenolysis in which the BLM served as a bipolar electrode, just as in the case of the water/saturated KI solution interface. Upon passing a direct current, I^- was oxidized to iodine. The resulting product preferred the low dielectric BLM to the aqueous solution by a factor of 50. The enhanced conductivity of BLM under these circumstances has been attributed to the tendency of iodide to form polyiodides (68) which have a high solubility in the lipid phase, and which facilitate ion transport across the BLM.

3. Semiconductive Behavior of BLMs

As mentioned previously, an increase in conductivity with temperature may indicate semiconductivity as described by Equation (2). Membranes formed from oxidized cholesterol, both in the presence of substances forming charge-transfer complexes with membranes in bathing solution and without these substances, have been shown to obey Equation (2). Satisfactory agreement has been found in the temperature dependence of the conductivity between oxidized cholesterol BLMs and oxidized cholesterol in the solid state. A compilation of data for a variety of BLM systems using Equation (2) have been published (15).

PHOTOELECTRIC EFFECTS

Several comprehensive reviews on photoelectric effects in BLMs have been published (5,15,31,72,73). In this part of the review, some recently obtained data will be presented and emphasis will be put on the effects where electronic mechanisms seemed to be operating (6,65).

1. BLMs Containing Visual Pigments

The first observation of the photoelectric effect in lecithin/cholesterol BLMs containing various carotenoids (all-trans-retinol, 9-cis-retinal, all-trans-retinol, β-carotene) was reported in 1969 (74). The photovoltages evoked in these membranes ranged from tenths of a mV to several mV, with rise times of about 1 second. Depending on the external conditions of the membrane (applied voltage, pH), the response was biphasic or monophasic. In later experiments, it was found that when $FeCl_3$ was used as an acceptor of electrons, and a membrane containing all-trans-retinal was illuminated with a short-duration light flash (0.8-3 μs), then a very fast electrical response of the membrane, similar to the R_1 phase of the ERP was observed. As in experiments on the dependence of the R_1 phase of the ERP of natural systems on pH, the R_1 observed in the BLM could be masked by increasing the pH. However, the enhancement of R_1 in natural systems with lower temperature could not be tested at subzero temperatures. Decreasing the temperature of the BLM system to 9°C seemed to increase the R_1 phase (74). The possibility that this response was due to proton transport processes was rejected for the following reasons: (1) the sign of R_1 was totally dependent on the location of the electron acceptors (the acceptor side became negative upon illumination); (2) the magnitude of the voltage could be enhanced (up to 100 times) by addition of suitable electron acceptors; (3) in the presence of the high concentration of electron acceptors, the magnitude of R_1 was independent of the proton concentration gradient across the membrane; (4) at high buffer capacity, where the R_2 phase was completely abolished, R_1 could still be observed. Having excluded ionic mechanisms for the generation of the R_1, Kobomoto and Tien put forth an overall picture of the sequence of events in the development of the ERP in both model and natural membranes (15,73).

In an experiment where BLMs containing retinol were exposed on one side to a solution of $K_3Fe(CN)_6$, no biphasic response was found (75). Instead, a monophasic response was observed that depended on the relation of incident light, wavelength and absorption spectrum of retinol. A biphasic response was registered from BLMs containing vitamin A when the membrane was illuminated with a wavelength close to the absorption maximum of vitamin A. The photovoltage and photocurrent changes registered were on the order of a few seconds. Photo-oxidation of retinol and reduction of the Fe^{3+} present in solution was proposed as the explanation of the mechanism leading to photopotential development (75). This redox reaction has been shown capable of changing pH and of giving rise to the observed voltages. As the result of retinol oxidation, vitamin A acid is produced in the membrane and the response becomes biphasic.

The influence of light on the conductivity of BLMs formed of egg lecithin and cholesterol in heptane that contained rod outer segments, was studied by Fesenko and Lyubarskii (76). The time constant of the conductivity increase was about 30 msec after the light flash, and the rise was followed by a phase of increasing conductivity that was 16 times longer. Sometimes a photopotential of the order of 20 mV with a risetime of several msec was observed, but it was viewed as an artifact.

2. BLMs Containing Photosynthetic Pigments

This field of study is especially well developed. Motivation for this work comes both from attempts at understanding photosynthesis, and at constructing working models of efficient converters of light into electricity (2,3,72). In both cases, the efficiency of light transduction processes is the focal point of investigation.

The first observation of the photoelectric activity of BLMs containing chlorophyll and xantophyll pigments (both extracted from spinach leaves or purchased from commercial sources) was reported in 1968 (63). The observed value of photovoltage was of the order of few mV, and its rising phase developed in less than 0.1 sec. The value of the first phase of the photocurrent density was of the order 10^{-15} A/m^2. After the first phase reached its maximal value, it began to drop to a certain value and then kept increasing until the membrane broke.

The next step in approaching more closely the biological conditions was applying asymmetric conditions across the membrane by the presence of oxidizing or reducing compounds, either on one or both sides of the membrane. To avoid the artifacts arising from a potential induced by pH gradients, buffers were used in the bathing solutions. The presence of redox gradients across the BLM dramatically changed their photoresponses. Especially high open-circuit photovoltages (of the order of 100 mV or more) were observed when Fe^{3+} ions were present on one side of the membrane, and 1,4-dihydroquinone or ascorbic acid was on the other side. The sign of the voltage generated at the Fe^{3+} (acceptor) side was always negative.

Quantum efficiency of chlorophyll-containing BLMs has been calculated to be very low (less than 0.005%). The action spectrum of BLMs was identical with the absorption spectrum of chlorophyll in the bulk solution, which indicated that the chlorophyll in the investigated BLMs did not form crystal complexes, which may be found when there is sufficient concentration of chlorophyll in a membrane-forming solution (15,77).

In pursuing the direction of modeling the natural condition, the dependence of the photovoltage on the value of the redox gradient created across BLMs containing chloroplast extracts was studied, employing various concentrations of ceric/cerous ions, ferric/ferrous ions, and ascorbic/dehydroascorbic acid (78). It was found that the photovoltage generated across the BLM was proportional to the light intensity, and that at a given value of light intensity there was a limiting value of the photocurrent caused by increasing the redox potential gradient. The rise in temperature from 16°C to 25°C was followed by a drop in photovoltage, as shown earlier (15). However, the value of photoconductivity did not change with temperature, even when there were sufficient changes of the dark conductivity of the membrane. This experimental finding was interpreted as implying that the activation energy was zero or very small, and as a consequence, the photoconductive current could not be regarded as ionic in nature.

These observations, as well as more recent studies with cytochrome c-551 and flavine mononucleotide, were shown to be incompatible with an ionic mechanism of photoresponse generation. Instead, they appeared compatible with a model of electronic conductivity by the mechanism of quantum mechanical tunneling across the membrane (79).

In connection with electron tunneling, the following remarks are appropriate. Electron transfer and redox reactions in suitable membrane systems in which the energy storage is substantial, and for which the back reaction can be controlled, is a very important research area. One approach that has been pursued in our laboratory since 1968 is the pigmented bilayer lipid membrane system, in which the excited state transfers an electron to the acceptor through a tunneling mechanism. Figure 4 shows a simple diagram of a pigment on an ultrathin bilayer lipid membrane. The back reaction is prevented by the insulating lipid bilayer core. When the pigment is excited, the electron tunnels from the acceptor side to the donor side, overcoming the energy barrier.

It has been shown that at the interface between the BLM and the bathing solution, photoredox reactions take place in which electrons may be transferred from excited chlorophyll to oxidants present in the solution ($SmCl_3$ or $K_3Fe(CN)_6$ for example) or from reductants (ferrocytochrome c) to excited chlorophyll (80).

The ability of the BLM to respond photoelectrically has been shown to be dependent on a sufficient concentration of chlorophyll and carotenoids, and both pigments are regarded as essential in photocurrent generation. Chlorophyll is believed to act as a sensitizer, absorbing the light quanta and directing the electrons to carotenoids that span the membrane. Carotenoids provide a low electrical resistivity path for electrons moving across the membrane (81). However, this role of carotene pigment in BLM was questioned in experiments with liposomes, where electrons were shown to be transferred from chlorophyll across the lipid phase in the absence of carotenoids (82).

Experiments employing short impulses of light are especially useful in investigating the fast changes of the electrical event. Using different redox systems, pH gradients and applied external potentials, it has been found that the electrical response of the membrane after the light flash can be resolved into three components: the fastest one (risetime of 1-3 μsec), a slower component (about 20 msec), and the slowest part of the response (about 1 sec). Both the fastest and intermediary response were depicted as electronic in nature. The fastest response was interpreted as evidence of rapid charge separations; the intermediary response was explained as reflecting the process of exciton migration and transport of protons across the membrane. Electron mobility in the membrane was estimated on the basis of these experiments to be 10^{-5} $m^2/V \cdot sec$ (6,15,72).

Remarkably high photovoltages (155 mV) of the fast response (less than 10 μsec) were observed in a system consisting of a chlorophyll-BLM, and solutions of acetate chlorophyllin, and $(NH_4)Ce(NO_3)_6$. When $FeCl_2$ was added to the chlorophyllin-containing compartment, the photovoltage exceeded 200 mV (83).

Membrane-bound pigments of the photosynthetic bacterium, Halobacterium halobium, when incorporated in BLMs, make them photoelectrically responsive (84-87). Thus far, purple membranes of H. halobium has been investigated with the aim of testing the chemiosmotic theory of ATP generation, stressing the proton current across the membrane. However, electronic phenomena involving electron current transients across the membrane as a result of light absorption must be also operative in this system, as they are in the photosynthetic case (88).

3. Non-Pigmented BLMs

Photoelectric response of the lipid bilayer may be elicited in systems in which the bathing solution contains iodine and its derivatives. UV radiation from a nitrogen laser (337 nm) was shown to cause a fast (much less than 40 sec) drop of the conductivity across the membrane (89). The photovoltage induced across the membrane was small, about 0.2 mV, and the current drop reached about 7×10^{-10} A. Generation of about 60 mV was observed from BLMs formed of oxidized cholesterol between solutions of ferric chloride and pure water, and illuminated with 365-nm light. When 100 mV was applied across the membrane, a rise in conductivity of 150-200 times was observed (90).

The current across the membrane diminished when the concentration of $FeCl_3$ was increased beyond 8×10^{-3} M. This was interpreted as evidence against the possibility that the observed changes of conductivity were due to a thermal effect. If this was responsible for the observed changes of conductivity, the conductivity should have increased with concentration because of enhanced absorption by the ions (90). Biphasic photovoltage changes were obtained across oxidized cholesterol BLMs dividing $FeCl_3$ and I_2 solutions when the system was illuminated with violet (365 nm) and ultraviolet (254 nm) continuous light. Immediately after the illumination began, a very fast negative potential was developed which was followed by a slow positive phase reaching a saturation level. When $FeCl_3$ was present on one side of the BLM and iodide on the other side, a remarkably high (260 mV) photovoltage was generated. A significant difference was also found in the value of the negative current flowing across the membrane, depending on the polarity of the imposed potential difference. If the field was so directed that the negative charge carriers (probably electrons or hydrated ions) were expelled from the iodide-containing compartment to the one containing ferric chloride, the value of the current was significantly higher than if the field were oppositely directed (90).

4. BLMs Containing Various Molecules

Dyes embedded in BLMs or present in the bathing solution have been shown to determine the occurrence and the value of the photoelectric response. A biphasic photovoltage was seen in a system composed of egg lecithin, and bathed with cyanine dye from one side (91). The action spectrum was similar to the absorption spectrum of the dye, and the photovoltage was proportional to the light intensity until a saturation value was reached. A similar dependence of the photovoltage upon the dye concentration was found, however after exceeding a certain concentration the photovoltage decreased. An extensive study of the photoelectric responses of similar systems containing various cyanine dyes was carried out (92). Mechanisms involving intermolecular electron transfer in the photoresponse generation were assessed as unlikely for most of the cases. Instead, intermolecular charge shifts and translocations of dyes in membranes were suggested.

Using oxidized cholesterol-BLMs, and modifying the solution in one of the compartments either with methylene blue or rhodamine B, early investigators observed photopotentials of up to 6 mV (6,84,93). The potentials were dependent on the spectral composition of the incident light and on the addition of $KMnO_4$ to the compartment containing the dye. It was suggested that a redox reaction took place at the interface between the membrane and the dye/$KMnO_4$ solution. Because the solution with modifiers was charged negatively, it was concluded that potassium permanganate acted as an electron acceptor, and the membrane itself as a hole conductor.

Hong and Mauzerell investigated a system composed of lecithin and cholesterol BLMs modified with magnesium octaethylporphyrin, using buffered solutions of potassium ferro- and ferricyanide (73,89). They used a null current method which enabled them to differentiate between photovoltage dependent on redox gradients and photovoltage change arising from the movement of the porphyrin cation in the membrane. It was shown that continuous illumination gave rise to a voltage that depended on light intensity, applied voltage, and on the conductivity changes brought about by movement of the porphyrin cation. Using pulses of laser light and the voltage-clamp method in the same system, it was shown that the current risetime was shorter than 1 sec, and that photovoltage evolved exceeded 500 mV. A biphasic photocurrent was generated, and its first (extremely fast) phase depended on the capacitance established at the interface where a redox reaction between excited pigment and oxidant took place (72). At both interfaces, redox reactions occurred which were due to the charge carriers in the membrane and uncharged pigment molecules (73,89).

Recently, Heubner (94) by means of an apparatus able to register potential changes in the nanosecond range, was able to show that a BLM separating a mixture of chlorophyllin with Na_2HPO_4, and solution of $(Na_4)_2Ce(NO_3)_6$, illuminated with a flash of light, generates a potential of the order of 30 mV in much less than 20 nsec. Estimated current densities in both the BLM and chloroplast membranes were about 10^4 A/m^2 (58).

For efficient electron transfer and charge separation in green plant photosynthesis, a close proximity between a donor, such as chlorophyll, and an electron acceptor, such as quinone, is probably a prerequisite. To test this hypothesis, several investigators have synthesized covalently-linked porphyrin-quinone and porphyrin-carotene complexes as models for the initial photophysicochemical event in reaction centers of photosynthesis (95). Joshi et al. incorporated these newly available complexes into BLMs and found greatly enhanced photovoltage (302 mV) and photocurrent (22 nA), the highest reported values. Of all compounds studied, porphyrin-quinone (PQ) gave the most dramatic light response. The photoelectric action spectrum of the pigmented membrane closely followed the absorption spectrum of PQ, thereby providing strong evidence for the separation of electrons and holes in the lipid bilayer, and for the view that only the photons absorbed by the pigment were responsible for the observed light-induced redox reactions. To account for this finding, the pigmented BLM was considered to be an organic semiconductor separating two aqueous solutions. The membrane/solution contact was analogized to that of a Schottky barrier except that the BLM system had two interfaces. One side of the membrane acts as a photocathode (p-type) and the other side as a photoanode (n-type). At the membrane-electrolyte interface, the aqueous solution played the role of the metal. When the quinone structure, an electron withdrawing group, was covalently attached to porphyrin (an electron donating group) as in PQ, the energy gap between the ground and excited states of PQ was narrowed, facilitating a push-pull effect of electron transfer. Absorption of light in the presence of appropriate redox agents led to a reduction on one side and oxidation on the other side of the membrane. When porphyrin was simply mixed with quinone of β-carotene in the membrane-forming solution, photoelectric effects were also observed, but the magnitude was lower by a factor of four as compared with the membrane containing covalently-linked porphyrin-quinone complex. Thus, in the PQ compounds, a more favorable orientation and closer proximity were attained between the donor-acceptor pair for light-induced charge separation, rather than being dissipated by other pathways such as fluorescence (6,95).

In recent years, increasingly more attention has been paid to vesicular systems in the form of either liposomes or vesicles formed of surfactants (6). In one such system containing EDTA and methylviologen, a light-induced transmembrane reduction took place; EDTA in the inside of vesicles was oxidized and methylviologen-reduced (6). The reaction went against the redox gradient. In explaining this and other results, it was suggested that tunneling of electrons took place across the hydrocarbon-like core of the membrane (6,96,97).

In experiments involving egg lecithin membranes modified with various dyes and bathing solutions of reductants and oxidants, transmembrane photoreduction was observed (6,65). However, the mechanism of the charge movement across the membrane was depicted as based on the process of diffusion of the dye anions across the membrane (98).

Redox reactions across the interface between the membrane phase and solution were also investigated in surfactant vesicles. Laser radiation was used as a light source. In an experimental system where the dye pyrene was incorporated in the hydrophobic phase of negatively charged dihexadecylphosphate vesicles, it was found that a hydrophobic environment lowered the ionization potential of the dye, and that electrons were expelled by the electric gradient arising from polar groups on the surface of the membrane (99). An electron transfer reaction from a donor (N-Methylphenotiazine or N-dodecylphenothiazine), present in the hydrophobic phase at the positively charged dioctadecyldimethylammonium chloride vesicle, to an electron acceptor, a surfactant derivative of tris (2,2'-bipyridine) ruthenium perchlorate anchored on the surface of that vesicle, was observed by Infelta et al. (100). A more extensive study on the vesicles formed of the same surfactant, and tris (2,2'-bipyridine) ruthenium cation and methylviologen acting as donors and acceptors of electrons respectively, were carried out by Tunuli and Fendler (101). They investigated the efficiency of photoactivated electron transfer as a function of the location of the donor and acceptor on the membrane and found that efficient electron transfer occurred (with quantum efficiency 2.40×10^{-2})

if both donor and acceptor were attached to either the inner or the outer side of the vesicular membrane.

CONCLUSIONS AND SUGGESTIONS FOR FURTHER STUDY

The existence of free electrons in both natural membranes and experimental bilayer lipid membranes (planar BLMs and liposomes) is a question still under discussion, in spite of considerable variety of data accumulated in the last 20 years. There are several reasons for these difficulties. Firstly, BLMs are extremely thin structures, and the physical theory of ultrathin layers is still in its beginning phase. Secondly, BLMs formed even in simple electrolyte solutions are relatively complex physicochemical systems, and an adequate description of all the significant interrelations presents many difficulties. Thirdly, because of the thinness of both natural membranes and BLMs, it is difficult to apply many of the physical techniques that have been successfully applied to thicker samples (see below). In spite of these difficulties, data have been accumulated to justify treatment of natural and bilayer lipid membranes as systems in which electronic phenomena play an important role.

ARE ELECTRONIC PROCESSES IN NATURAL MEMBRANES ADEQUATELY MODELLED IN BLMs?

1. Electronic Phenomena Without Light

The adequacy between some features of the behavior of natural membranes in dark, which may be interpreted as due to electronic processes, and supposedly electronic behavior of BLMs in the same conditions, is more questionable than that in the case of light-evoked electronic phenomena (102). Some of the phenomena that are the basis of their phenomenological electronic-like behavior, such as rectification and negative resistance, may be explained in terms of active transport of ions across natural membranes. On the other hand, some seemingly electronic phenomena may be alternatively interpreted in terms of changes in the conductivity of the membrane for charged aggregates, or conduction due to the presence of the products of hydrolysis of water in the BLM (15). However, there is little doubt that electron translocation processes occur during normal activity of biological membranes. Therefore, the more accurate question is whether these electron translocations play any significant role in the physiology of membranes. In this connection, it must be said that electron translocations across the membranes of mitochondria play a crucial role in establishing transmembrane pH gradients leading to ATP synthesis. In this way, electronic conduction in mitochondrial and other proton-gradient-creating membranes appears to be connected with the basic process of bioenergetics.

With regard to information-transfer processes in biological membranes, it is possible that gating currents, discovered in nerve and muscle cell membranes, are electronic (103). Table 1 lists some experimental results on electronic behavior of natural and artificial BLMs obtained in the dark. To complete the overall picture of electronic effects, both in natural and artificial bilayer lipid membranes, estimations of the basic electronic parameters of mitochondrial, photosynthetic and artificial bilayer lipid membranes are compiled in Table 2.

2. Electronic Phenomena With Light

Table 3 presents a comparison of some of the features of the photovoltage generated in natural and artificial BLMs. On the basis of this comparison, it may be said that BLMs provide a realistic model of photoelectric phenomena in natural membranes. There are significant differences in the time course of the voltages. In particular, those measured in thylakoid membranes give much faster responses than those from BLMs. However, the main reason for this discrepancy lies in the limitation of the apparatus used in the experiments with the BLMs.

This apparent similarity of the photovoltage from natural membranes and BLMs may be regarded as an evidence that the response is generated by the same type of electronic mechanism in both systems. Besides the features of the photovoltage response enumerated in Tables 1A and B, other observations made on natural membranes (or intact organelles) are worth recalling here. These include the occurrence of ERP signals in chloroplasts at very low temperatures, light evoked electron transmembrane transfer in erythrocytes, and photoinduced microwave conductivity increases in chloroplasts and in photosynthetic bacteria (37).

TABLE 1A. Some Features of the Photovoltaic Response of Natural Membranes

PHOTOVOLTAGE FEATURES	NATURAL MEMBRANES			
	Thylakoid	References	Visual receptor	References
Amplitude (mV)	100	(105)	3	(36,39-41)
Latency (μs)	< 40	(106)	not detected (<10)	(39)
Rise-time (ms)	$< 2.10^{-2}$	(105)	0.2	(39-41)
Temperature dependence	Decrease of photovoltage with increasing temperature	(106)	Decrease of the photovoltage with increasing temperature	(39)
Adequacy between action and absorption spectre	Observed	(105,106)	Observed	
Time-course of the sign of the response	In the fast light absorption induced phase the outside of the membrane charges negatively	(105)	The retinal side of the R_1 of ERP changes after the light flash	(36,39-41)
Redox gradient dependence	Observed	(33,62)		

TABLE 1B. Some Features of the Photovoltaic Response of Artificial BLMs

PHOTOVOLTAGE FEATURES	Containing Photosynthetic Pigments	References	Containing Visual Pigments	References
Amplitude (mV)	< 100	(6,15,31,65)		(6,65,73,74)
Latency (μs)	Not observed (?)	(65,72)		(65,73,74)
Rise-time (ms)	10^{-2}	(31,65)		(72,89)
Temperature dependence	Photovoltage decreases with increasing temperature (b)	(6,65)	Photovoltage decreases with increasing temperature	(65) (79)
Adequacy between action and absorption spectre	Observed	(15,65)	Observed	(65,74)
Time-course of the sign of the response	Fast component negative on the oxidant side	(31,65)		(65,73,74)
Redox gradient dependence	Observed	(15,65)	Observed	(74)

TABLE 2. Some Representative Data on the Similarities of the Electronic Behavior of Natural Membranes and Artificial BLMs

EFFECT	NATURAL MEMBRANES	REFERENCES	ARTIFICIAL BLMs	REFERENCES
Increase of conductivity with rising temperature	Increase of 40–140% per deg 230 ± 40 μV deg (a)	(24)	Increase of 40–118% per deg	(6,15,65)
Sensitivity to Impurities	Electron donors and acceptors affect membrane conductivity	(1)	Observed increase of conductivity by a factor of about 10^3 when chloride ions in bathing solution are replaced by iodide ions (b)	(15)
Rectification	Observed rectification ratio 1:100 (c)	(1,55)	Observed rectification ratio 1:22 (d)	(15,31,65)
Negative resistance	Observed	(9)		
Thermoluminescence	Observed (e)	(55)		
Thermoelectric Power	Observed 230 ± 40 V/°C (a)	(18,25)	Observed 80 V (f)	(15,31)
Hall effect	Observed (g)	(1,32,53)		

(a) Cell membrane of the muscle of the frog, temp. range 0–35°C; (b) Phosphatidylcholine-BLM; (c) Squid axon membrane, ionic process; (d) Egg lecithin-BLM dividing solutions of ceric ammonium sulphate and ferrous chloride. Rectification ratio given for V_p=400 mV; (e) Spinach chloroplast and extracts from algae; (f) Chlorophyll-BLM, temp. range 15–40°C; (g) Net mitochondrial and in <u>Nitellopsis obtusa</u> cell membrane.

TABLE 3. Some Semiconductor Characteristics of Natural and Artificial BLMs

PARAMETER	NATURAL MEMBRANES				ARTIFICIAL BLMs	
	Chloroplast	References	Mitochondrial	References		References
Type of dominant charge carriers	Holes	(53)	Electrons	(9)	Electrons or holes	(5)
Density of dominant charge carriers (m^{-3})	10^{19} 8×10^{19}	(32,53)	10^{23}	(105)	6×10^{23}	(6,15,65)
Conductivity mobility	10^{-5}–10^{-3}	(9)	5×10^{-6}	(66)	10^{-6}	
Hall mobility of ($m^2V^{-1}\ s^{-1}$) charge carriers	10^{-4}	(9)	4.8×10^{-4}	(9)		
Life-times of the charge carriers (5)	$\leq 5.10^{-5}$	(9)	$< 10^{-6}$	(66)		
Specific Conductivity (Ωm^{-1})	10^{-15}–10^{-14}		10^{-12}	(9)	2.6×10^{-14}	(15)
Dominant mechanism of charge translocation	Tunneling and/or Hopping	(53)	Band conduction in macromolecules and tunneling between macromolecules	(1)	Charge transfer at the interface and implied band conduction in the membrane	(15)
Energy gap (ev)	1.7–1.72	(9)	2.6	(1)	1.1	(15)

One can distinguish two trends in the present stage of research on BLMs. The first is oriented at gaining more precise knowledge about natural membranes and mechanisms of their functioning, and the other is a technologically oriented approach. In this connection, it may be said that the photoresponses obtained in artificial BLMs surpass the biological membranes in many respects. Their quantum efficiencies (10^{-2}) are much closer to those of biological membranes (~ 0.5) than are the quantum efficiencies of BLMs containing chlorophyll ($< 5 \times 10^{-3}$). It seems quite probable that biomimetic BLMs will evolve much faster in the direction of higher efficiencies than BLMs containing biological materials. Since a significant role is attributed to proteins in natural membranes, BLMs modified with a pigment and a protein may be regarded as a more advanced model of natural membranes. Consequently, it may be expected that further development of BLM models containing different kinds of photoactive compounds will provide more support for the existence of electron currents in BLMs and, indirectly, in natural membranes. However, a word of caution is in order. The BLM experiments have shown the necessary and sufficient conditions for an electrical response from BLMs that is similar to that observed in natural systems. It does not necessarily mean that the same process must exist in biological membranes. It is highly probable that similar relationships exist in BLMs and natural systems, but this can be proved only by building photoactive systems with a complexity more like that of natural membranes.

Both BLMs and natural membranes have been shown to respond like organic semiconductors to illumination, temperature changes, and doping with electron donors or acceptors. On the basis of these properties, models of organization and electronic function of artificial lipid bilayers and of the thylakoid membrane have been put forward. Despite these investigations and models, there is a need for more quantitative data concerning electronic effects in both types of membranes.

SOME NEW POSSIBLE EXPERIMENTS

1. A Suggested Hall Experiment with BLMs

The electrical and electronic properties of BLMs have been investigated thus far mainly in the transmembrane direction.

The properties of the membranes in the plane of the membrane have been the subject of only a few studies. However, the properties in the plane of the membrane in natural systems are also very important. Firstly, in electron transfer chains of photosynthesis and oxidative phosphorylation, electrons are believed to be arranged across, as well as in, the plane of the membrane. Secondly, in the direction across the membrane, the path of an electron is limited to the membrane thickness, whereas in the plane of the membrane, electrons may move much greater distances, provided that there are conduction paths present (intrinsic proteins), and driving potential differences available.

If a method of measuring and applying the potential difference in the plane of the membrane is found, a new class of electronic phenomena in the BLMs may be studied. The Hall effect and its variations such as photomagnetoelectric voltages might be investigated. The Hall voltage, V_H (perpendicular to the direction of the magnetic induction B), induced in a sample of the thickness of t_m, (measured along the direction of B), is given by $V_H = R_H IB/t_m$, where R_H is the Hall coefficient of the material, and I is the current flowing through the sample perpendicular to B. If R_H of the BLM is taken to be 0.1 m^3/C, B = 1 T, I = 1 nA, the Hall voltage (measured in air) should be about 20 mV. V_H would be measured for polar groups of the BLM which, in air, should form the interior of the membrane.

Pursuing this line of consideration, it may be possible to investigate the magneto-electric effects and the cyclotron resonance of charge carriers in membranes. The data obtained in these kinds of experiments would allow differentiation between electrons and ions as charge carriers (the difference in the frequency of the resonance), and the evaluation of effective mass of electrons and the geometry of conduction bands in the BLMs.

Another possibility for discriminating between electronic and ionic conductivity of BLMs, is the measurement of the transport of mass that accompanies the transport of charge. The planar geometry of BLMs would be an advantage in such studies.

2. Biomolecular Devices

The following experiments are formulated specifically to obtain an understanding of the principles, at the molecular level, governing molecular junction effects in electron-conducting ultrathin bilayer lipid membranes, and to apply these principles to the design and construction of devices having a lipid bilayer as their central components (104). The focus on the lipid bilayer arises from the fact that all biomembranes possess such a structure as their key element. As has been demonstrated to some extent, ultrathin lipid membranes less than 100 Å thick, can function as gated channels, molecular diodes, photodetectors, and energy transducers. The choice of the compounds for incorporation into BLMs has been strongly motivated by the aim of modelling biomembranes. In the last few years, however, the interest generated by biomolecular electronic devices (BED) has provided new impetus. In particular, we believe that the bilayer lipid membrane system can be used as a tool in BED research.

To construct a molecular diode, for example, using a TCNQ-TTF (tetracyano-p-quinodimethane-tetrathiavulvalene) complex, the classic monolayer technique of Langmuir and Blodgett could be attempted (15), with a monolayer of TCNQ as the acceptor on one side and a monolayer of TTF on the other side.

The idea of biofuel cells based on glucose oxidation has been known for some time but, to our knowledge, no one has developed a biofuel cell based on the scheme shown in Figure 5. The key element is, again, an ultrathin electron-conducting BLM, which serves as a bipolar electrode. For the arrangement proposed, an open circuit voltage in excess of 600 mV at a current density of 100 mA/cm^2 is predicted. Both voltage and current density will depend on the concentrations of species involved.

3. Other Experiments

While the electrochemistry and photobiophysics of pigmented BLMs have been topics of study for many years, the interaction with microwaves has only recently been studied (65). Microwave energy has different effects than other forms of radiation, and it therefore can be expected to reveal different phenomena. One advantage is that the radiation damage to the system is extremely low. Thus, the dynamic properties of microwave interactions provide the possibility of probing states of motion in the membrane and its adjacent interfaces. Finally, voltammetric techniques should be applied to both BLMs and natural membrane

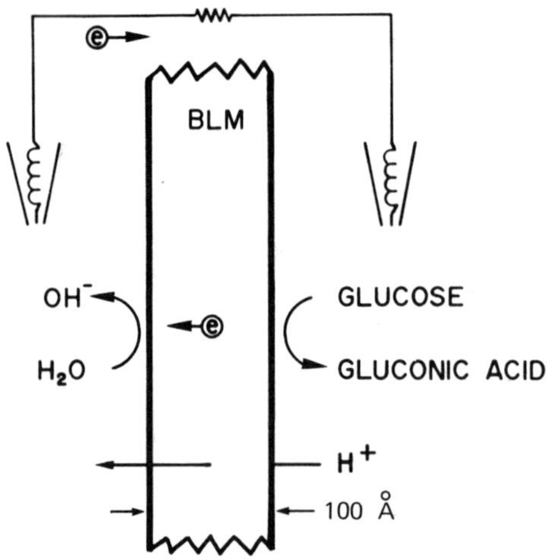

FIGURE 5. Biofuel cell using an electron-conducting BLM as bipolar electrodes (104).

systems to probe the mechanisms of electron transfer and bioredox reactions (102).

ACKNOWLEDGEMENTS

The preparation of this review was supported in part by an NIH grant (GM-14971) and a scholarship (to JZ) from the Kosciuszko Foundation, New York, NY. We thank Ms. Sharon Shaft for her excellent secretarial assistance and Ms. Carol Farmer for correcting the manuscript.

REFERENCES

(1) Eley, D.D., Lockhart, N.C. and Richardson, N.: Electrical properties and structural transitions in mitochondrion, J. Bioenerg. 3:271, 1977.

(2) Porter, G. and Archer, M.D.: In vitro photosynthesis, Interdiscipl. Sci. Rev. 1:119, 1976.

(3) Bolton, J.R., ed.: Solar Power and Fuels, Academic Press, New York, 167-225, 1977.

(4) Bolton, J.R. and Hall, D.O.: Photochemical conversion and storage of solar energy, Ann. Rev. Energy 4:353, 1979.

(5) Tien, H.T.: Semiconducting photoactive bilayer lipid membranes, in Solution Behavior of Surfactants: Theoretical and Applied Aspects, Vol. 1, K.L. Mittal and E.J. Fendler, eds., Plenum, New York, 229-240, 1982.

(6) Antonili, R., Gliozzi, A. and Gorio, A., eds.: Transport in Membranes: Model Systems and Reconstitution, Raven Press, New York, 57-75, 1982.

(7) (a) Szent-Gyorgyi, A.: Towards a new biochemistry: The study of energy-levels in biochemistry, Science 93:1941; Nature 148:157, 1941; (b) Weller, P.F., ed., Biology and semiconduction, in Solid State Chemistry and Physics, Vol. 2, Chapter 14, Dekker, Inc., New York, 1973.

(8) Sedlak, W.: Bioelectronics (In Polish), PAX, Warszawa, Poland, 1979.

(9) (a) Boguslavskii, L.I. and Vannikov, A.V.: Direct determination of the mobility of electrons and holes in mitochondria, Biophysics 14:444, 1969; (b) Cope, F.W., Evidence for semiconduction in Aplysia nerve membrane, Proc. Natl. Acad. Sci. (USA) 61:905, 1968.

(10) Cope, F.W.: A review of the applications of the solid state physics concepts to biological systems, J. Biol. Phys. 3:1, 1975.

(11) Kryszewski, M.: Semiconducting Polymers, Polish Scientific Publishers, Warsaw, 1980.

(12) Pullman, B. and Pullman, A.: Quantum Biochemistry, Interscience, New York, London, 1963.

(13) Hopfield, J.J.: Fundamental aspects of electron transfer in biological membranes, in Electrical Phenomena at the Biological Membrane Level, Vol. 1, E. Roux, ed., Elsevier, Amsterdam, 471-490, 1977.

(14) (a) Chance, B.: Electron Transfer Chains and Oxidative Phosphorylation, E. Quagliariello, et al., eds., North Holland Publ. Co., Amsterdam, 225-226, 1975; (b) DeVault, D., Parkes, J.H. and Chance, B.: Electron tunneling in cytochromes, Nature 215:642, 1967.

(15) Tien, H.T.: Bilayer Lipid Membranes (BLM), Theory and Practice, M. Dekker, New York, 1974.

(16) Blank, M., ed.: Surface Chemistry of Biological Systems, Plenum Press, New York, 135-154, 1970; Electrical Double Layers in Biology, Plenum, New York, 1986.

(17) Miyasaka, T. and Honda, K.: Photoelectrochemical observations on chlorophyll-carotene interactions in a lipid bilayer film, Thin Solid Films 102:173, 1983.

(18) Shimony, T.B. and Ishay, J.S.: Thermoelectric (Seebeck) effect of the cuticle of social wasps, J. Theor. Biol. 92:497, 1981.

(19) Tien, H.T., Karvaly, B. and Shieh, P.K.: Electrostenolysis in bilayer lipid membranes, J. Coll. Interf. Sci. 62:185, 1977.

(20) Jahn, T.L.: A theory of electronic conduction through membranes, and of active transport of ions, based on redox transmembrane potentials, J. Theor. Biol. 2:129, 1962.

(21) Mandel, L.J.: Active ionic transport across biological membranes: Possible role of electrons and protons, Nature 225:450, 1970.

(22) Pohl, H.A. and Sauer, J.R.: Redox reactions by indicator dyes, J. Biol. Phys. 6:118, 1978.

(23) Marinov, B.S.: Photoreduction of hemoglobin in erythrocytes, Mol. Biol. 13:81, 1979.

(24) Nagy, L.: Semiconductor property of frog sartorius muscle, Acta Biochim. Biophys. Acad. Sci. Hung. 5:341, 1970.

(25) (a) Lakatos, T. and Kollar-Morocz, A.: Acta Biochem. Biophys. Acad. Sci. Hung. 4:99, 1969; (b) Effect of visible light on stopped frog's heart, Acta Biochem. Biophys. Acad. Sci. Hung. 1:413, 1966; (c) Effect of visible light and ultraviolet rays on the stimulus threshold of frog muscle sensitized by eosin, Acta Biochem. Biophys. Acad. Sci. Hung. 2:191, 1967.

(26) (a) Ernst, E.: Semiconductors and the problem of nerve excitation: Halbleiter-Gesischtspunkte im Erregungspr problem, Biophysics 1, 296, 1956; Wiss. Z. Humboldt Univ. Berlin, Math.-Naturwiss. Rheie 5:167, 1955/56; (b) Ernst, E., Excitation as an electron process, Biochim. Biophys. Acta Acad. Sci. Hung. 1:321, 1966.

(27) Chalazoniis, N.: Light energy conversion in neuronal membranes, Photochem. Photobiol. 3:539, 1964.

(28) Ludvikovskaya, R.G. and Panglelova, T.K.: Light excitation of the giant axon of the earthworm, Biophysics 10:319, 1965.

(29) Skulachev, V.P. and Hinkle, P.C., eds.: Chemiosmotic Proton Circuits in Biological Membranes, Addison-Wesley, Reading, MA, 1982.

(30) Jordan, P.: Uber die physikalische structur organischen riesenmoleküle, Naturwissenschaften 26:693, 1938.

(31) Barber, J., ed.: Photoeffects in pigmented bilayer lipid membranes, in Photosynthesis in Relation to Model Systems, Elsevier/North Holland, New York, 115-173, 1979.

(32) Skierczynska, J., Bulanda, W., Gladyszewski, L., Kubacka,

Z., Kutnik, J., Ostrokolski, J. and Wardak, A.: Hall's phenomenon in cell membranes of Nitellopsis obtusa, Stud. Biophys. 63:225, 1977.

(33) Luttage, V. and Pallaghy, C.K.: Light triggered transient changes of membrane potentials in green cell in relation to photosynthetic electron transport, Z. Pfanzenphysiol. 61:58, 1969.

(34) Hansen, U.P., Warncke, J. and Keunecke, P.: Photoelectric effects in Characean cells. I. The influence of light intensity, Biophysik. 9:197, 1973.

(35) Adrianov, V.K., Kurella, G.A. and Litvin, F.F.: Influence of light on the electrical activity of Nitella cel' , Abhandl. Deutsch. Akad. Wiss. Berlin, Kl. Medizin 4a: , 1968.

(36) Knowles, A. and Dartnall, H.J.A.: Carotenoid pigments, in The Eye, H. Davson, ed., Vol. 2B, 425, 1977.

(37) (a) Trukhan, E.M., Perevozchikov, N.F. and Ostrovskii, M.A.: Photoconductivity of the pigment epithelium of the eye, Biophysics 15, 1090, 1970; (b) Rosenberg, B.: Photoconduction and semiconduction in dried receptors of sheep eyes, Arch. Biochem. Biophys. 93:395, 1961.

(38) Falk, G. and Fatt, P.: Conductance changes produced by light in rod outer segments, J. Physiol. (London) 198:647, 1968.

(39) Brown, K.T. and Murakami, M.: A new receptor potential of the monkey retina with no detectable latency, Nature 201:626; An early potential evoked by light from the pigment epithelium-choroid complex of the eye of the toad, Nature 204:739, 1964.

(40) Pak, W.L. and Ebrey, T.G.: Visual receptor potential observed at sub-zero temperatures, Nature 205:484, 1965.

(41) Cone, R.A. and Cobbs, W.H.: Rhodopsin cycle in the living eye of the rat, Nature 221:820, 1969.

(42) Adrian, R.H.: Charge movement in the membrane of striated muscles, Ann. Rev. Biophys. Bioeng. 7:85, 1978.

(43) Van Niel, C.B.: Comparative photosynthesis, Adv. Enzymol. 1:263, 1941.

(44) Katz, E.: Chlorophyll fluorescence as an energy flowmeter for photosynthesis, in Photosynthesis in Plants, Iowa State College Press, Ames, Iowa, 287, 1949.

(45) (a) Strehler, B. and Arnold, W.: Light production by green plants, J. Gen Physiol. 34, 809, 1951; (b) Arnold, W. and Sherwood, H.K.: Are chloroplasts semiconductors, Proc. Natl. Acad. Sci. (USA) 43:105, 1957.

(46) Gerischer, H.: Solar photoelectrolysis with semiconductor electrodes, Top. Appl. Phys. 31:115, 1979.

(47) Moglich, F. and Schon, M.: Zur Frage der Energiewanderung in Kristallen und Molekulkomplexen, Naturwissenschaften 26:199, 1938.

(48) Litvin, F.F. and Shuvalov, V.A.: Long afterglow of chlorophyll in plants and energy traps in photosynthesis, Dokl. Biophys. 181:86, 1968; Arnold, W.: Delayed light in photosynthesis, Ann. Rev. Biophys. Bioeng. 6:1, 1977.

(49) Ichimura, S.: The photoconductivity of chloroplast and the far red light effects, Biophys. J. 1:99, 1960.

(50) McCree, K.J.: Photoconduction and photosynthesis. II. Photoconduction in wet materials by the condenser method, Biochem. Biophys. Acta 102:96, 1965.

(51) Blumenfeld, L.A., Kafaleva, D.N., Livshits, A.W., Solovev, I.S. and Chetverikov, A.G.: Membrane phosphorylation, Dokl. Biophys. 193:700, 1979.

(52) Deryabkin, V.N., Kireev, V.B., Skachkov, M.P. and Trukhan, E.M.: Model membranes, Dokl. Biophys. 240:978, 1978.

(53) Bogolmoni, R.A. and Klein, M.P.: Mobile charge carriers in photosynthesis, Nature 258:88, 1975.

(54) Skachkov, M.P., Trukhan, E.M., Nokos, P.P., Kononenko, A.A. and Rubin, A.B.: Microwave effects on photosynthetic bacteria, Dokl. Biophys. 252, 1980.

(55) Tollin, G., Fujimori, E. and Calvin, M.: Delayed light emission in green plant material: Temperature dependence and quantum yield, Proc. Natl. Acad. Sci. (U.S.A.) 44:1035, 1958.

(56) Malkin, S.: Delayed luminescence, in Primary Processes in Photosynthesis, J. Barber, ed., Elsevier, Amsterdam, 1977.

(57) Wolff, Ch., Buchwald, H.-E., Ruppell, H., Witt, K. and Witt, H.T.: Rise time of the light induced electrical field across the function membrane of photosynthesis, Z. Naturforsch. 24b:1038, 1969.

(58) Witt, H.T.: The primary events in photosynthesis, Quart. Rev. Biophys. 4:365, 1971.

(59) Junge, W. and Witt, H.T.: On the ion transport system of photosynthesis, Z. Naturforsch 23b:244, 1968.

(60) Fowler, C.F. and Kok, B.: Direct observation of a light-induced electric field in chloroplasts, Biochim. Biophys. Acta 357:308, 1974.

(61) Noble, P. and Mel, M.: Electrophoretic studies of light-induced charge in spinach, Arch. Biochem. Biophys. 113:695, 1966.

(62) Bulychev, A.A., Adrianov, V.K., Kurella, G.A. and Litvin, F.F.: Photoinduction kinetics of electrical potential in

a single chloroplast as studied with micro-electrode technique, Biochim. Biophys. Acta 236:175, 1972.

(63) Tien, H.T.: Light-induced phenomena in black lipid membranes constituted from photosynthetic pigments, Nature 219:272, 1968.

(64) Antonov, V.F., Rovin, Y.G. and Trifmov, L.T.: A Bibliography of BLM: 1962-1975, All Union Institute for Scientific Information, Moscow, 1979.

(65) Tien, H.T.: Planar BLM, Prog. Surface Sci. 19, No. 3, 1985.

(66) Boguslavski, L.J., Bogolepova, F.J. and Lebiedev, A.V.: Conductivity mechanism of bimolecular phospholipid membranes in the iodine iodide system, Chem. Phys. Lipids 6:296, 1971.

(67) (a) Jain, M.K., White, F.P. and Cordes, E.H.: Electronic conduction across a black lipid membrane, Nature 227:705, 1970; (b) Vodyanoy, V.Y., Vodyanoy, I.Y. and Fedorovich, N.A.: On the mechanism of conduction of bimolecular phospholipid membranes, Physics. Solid State 12:2690, 1971.

(68) Finkelstein, A. and Cass, A.: Permeability and electrical properties of thin lipid membranes, J. Gen. Physiol. 52:145s, 1968.

(69) Joosten, J.G.H.: Soliatry waves and solitons in liquid films, J. Chem. Phys. 82:2427, 1985.

(70) Karvaly, B., Rosenberg, B., Pant, H.C. and Kemeny, G.: Contribution to the electrochemistry of bimolecular lipid membranes, Biophysik 10:199, 1973.

(71) Feldberg, S.W., Armen, G.H., Bell, J.A., Chang, C.K. and Wang, C.-B.: Electron transport across glycerol monoole-

ate bilayer lipid membranes facilitated by magnesium etiochlorin, Biophys. J. 34:149, 1981.

(72) Tien, H.T., ed.: Photoelectric BLM, Proceedings of the Symposium on Photoelectric Bilayer Lipid Membrane, The 3rd Annual Meeting of the American Society for Photobiology, Photochem. Photobiol. 24:95, 1976.

(73) Hong, F.T.: Displacement photocurrents in pigment-containing biomembranes: Artificial and natural systems, in Bioelectrochemistry, M. Blank, ed., ASC, Washington, D.C., 211-237, 1980.

(74) (a) Kobamoto and Tien, H.T.: Light-induced electrical effects in a retinal bilayer lipid membrane, Biochem. Biophys. Acta 241:129, 1971; Nature 224:1107, 1969; (b) Liu, Q.-Y. and Tien, H.T.: The photopotential of rhodopsin in association with a BLM, Photobiochem. Photobiophys. 10:1, 1985.

(75) Schadt, M.: Photoresponse of bimolecular lipid membranes pigmented with retinal and vitamin A acid, Biochim. Biophys. Acta 323:357, 1973.

(76) Fesenko, E.E. and Lyubarskii, A.L.: Effect of light on artificial lipid membranes modified by photoreceptor fragments, Nature 268:562, 1977.

(77) Csorba, I., Szabad, J., Erdei, L. and Fajszi, Cs.: Black lipid membranes BLM containing crystalline chlorophyll, Photochem. Photobiol. 21:377, 1975.

(78) Ilani, A. and Berns, D.S.: Photoresponse of chlorophyll-containing bileaflet membranes and the effect of phycocyanin as extrinsic membrane protein, J. Membrane Biol. 8:333, 1972.

(79) Alexandrowicz, G. and Berns, D.S.: Photovoltages in chloroplast extract bilayer membranes stimulated by micro-

molar amounts of oxidants and reductants, Photobioch. Photobiophys. 1:356, 1980.

(80) Tomkievich, M. and Corker, G.A.: Chlorophyll sensitized charge separation in phospholipid vesicles, Photochem. Photobiol. 22:249, 1975.

(81) Mangel, M., Berns, D. and Ilani, A.: Dependence of photosensitivity of bileaflet lipid membranes upon the chlorophyll and carotenoid content, J. Membr. Biol. 20:171, 1975.

(82) Kurihara, K., Toyoshima, Y. and Shukigara, M.: Photochemical reactions in bilayer lipid membranes, Biochim. Biophys. Acta 547:117, 1979.

(83) Huebner, J.S. and Tien, H.T.: Large amplitude photo-voltage transients of bilayer lipid membranes in the presence of chlorophyllin, Bioenerget. Biomembr. 4:469, 1973.

(84) Tien, H.T. and Karvaly, B.: Ultrathin barriers and solar energy conversion, in Solar Power and Fuels, J.R. Bolton, ed., Academic Press, New York, 167-225, 1977.

(85) Danzshazy, Z., Ormos, P., Drachev, L.A. and Skulachev, V.P.: Investigation by focused laser beam scanning of the photoelectric activity of bacteriorhodopsin-containing lipid bilayers, Biophys. J. 24:423, 1978.

(86) Lopez, J.R. and Tien, H.T.: Photoelectrospectrometry of bilayer lipid membranes, Biochim. Biophys. Acta 579:433, 1980; Reconstitution of photosystem I reaction center into bilayer lipid membranes, Photobiochem. Photobiophys. 7:25, 1984.

(87) Rayfield, G.W.: Kinetics of the light-driven proton movement in model membranes containing bacterio-rhodopsin, Biophys. J. 38:79, 1982.

(88) Oosawa, F. and Iami, H., eds.: Dynamic Aspects of

Biopolyelectrolytes and Biomembranes, Kodansha-Elsevier, Tokyo and Amsterdam, 445-456, 1982.

(89) D. Mauzerall: Photoinduced electron transfer at the water-lipid bilayer interface, in Light-Induced Charge Separation in Biology and Chemistry, H. Gerischer and J.J. Katz, eds., Dahlem-Konferenzen, Verlag, Chemie, Weinheim, 241-254, 1979.

(90) Karvaly, B. and Pant, H.C.: A new biphasic photoelectric response on bimolecular lipid membranes, Studia Biophys. 33:51, 1972.

(91) Ulrich, H.M. and Kuhn, H.: Photoelectric effects in bimolecular lipid-dye membranes, Biochim. Biophys. Acta 266:584, 1972.

(92) (a) Huebner, J.S., Arrieta, R.T. and Millar, D.B.: Aromatic amino acids and ultraviolet induced photo-electric effects in bilayer membranes, Photochem. Photobiol. 35:467, 1982; (b) Tien, H.T. and Kutnik, J.: Voltammetry of dye-sensitized BLM, Photobiochem. Photobiophys. 7:319, 1984.

(93) Olson, J.M. and Hind, G., eds.: Chlorophyll-Proteins, Reaction Centers and Photosynthetic Membranes, Brookhaven Sym. Biol. 28:105, 1977.

(94) Huebner, J.S.: Apparatus for recording light flash-induced membrane voltage transients with 10 ns resolution, Photochem. Photobiol. 30:233, 1979.

(95) Joshi, N.B., Lopez, J.R., Tien, H.T., Wang, C.-B. and Liu, Q.-Y.: Photoelectric effects in bilayer lipid membranes containing covalently linked porphyrin complexes, J. Photochem. 20:139, 1982.

(96) Ford, W.E., Otvos, J.W. and Calvin, M.: Photosensitized electron transport across phospholipid vesicle walls, Nature 274:507, 1978.

(97) Ford, W.F. and Tollin, G.: Direct observation of electron transfer across a lipid bilayer: Pulsed laser photolysis of an asymmetric vesicle system containing chlorophyll, methyl viologen and EDTA, Photochem. Photobiol. 35:809, 1982.

(98) Sudo, Y., Kawashima, T. and Toda, F.: The trans-membrane electron transport coupled with dye redox cycle in the liposome system, Chem. Lett. Jap., 355, 1980.

(99) Escabi-Perez, J.R., Romero, A., Lukac, S. and Fendler, J.H.: Aspects of artificial photosynthesis. Photoionization and electron transfer in dihexadecylphosphate vesicles, J. Am. Chem. Soc. 101:2231, 1979.

(100) Infelta, P.P., Gratzel, M. and Fendler, J.M.: Aspects of artificial photosynthesis. Photosynthesized electron transfer and charge separation in cationic surfactant vesicles, J. Am. Chem. Soc. 102:1479, 1980.

(101) Tunuli, M.S. and Fendler, J.H.: Aspects of artificial photosynthesis. Photosynthetized electron transfer across bilayer; charge separation and hydrogen production in anionic surfactant vesicles, J. Am. Chem. Soc. 103:2507, 1981.

(102) Tien, H.T.: Cyclic voltammetry of bilayer lipid membranes, J. Phys. Chem. 88:3172, 1984; J. Bioelectrochem. Bioenerg. 13:299, 1984.

(103) Petrov, E.G., Teslenko, V.I. and Kharkyanen, V.N.: Gating mechanisms in nerve and muscle cell membranes, Dokl. Biophys. 238:967, 1978.

(104) Mountz, J.M. and Tien, H.T.: Photoeffects of pigmented lipid membranes in a micropore filter, Photochem. Photobiol. 28:395, 1978.

(105) Zon, J.: The living cell as a plasma physical system, Physiol. Chem. Phys. 12:357, 1980.

8
Bioelectric Pyroelectricity

SIDNEY B. LANG
Department of Chemical Engineering
Ben-Gurion University of the Negev
Beer Sheva, Israel

INTRODUCTION

The pyroelectric effect has been observed in mineral crystals such as tourmaline since antiquity, and in hundreds of natural and artificial crystals, ceramics, and polymers during the past century (1). However, its existence in biological materials was unknown until 1966 when Lang (2) first reported measurements of it in animal bone and tendon. Since then, its presence has been demonstrated in many animal and plant tissues. Most of the data are qualitative, but they are sufficient to establish the generality of the phenomenon in living tissues. The unique association of pyroelectricity with a vectorial electric polarization, and the high probability of the existence of the long-range ordering characteristics of ferroelectricity suggest that the presence of pyroelectricity may have a profound influence on the processes of life.

This chapter is written to introduce the reader to the fundamentals of the pyroelectric phenomenon, to describe the research that has been carried out on biomaterials, and to present an overview and some suggestions for future research.

FUNDAMENTALS OF PYROELECTRICITY

The characteristics of pyroelectricity and related effects presented here are relevant to both nonbiological and biological materials. A more detailed treatment can be found elsewhere (1,3,4).

PYROELECTRICITY AND CONDITIONS FOR ITS EXISTENCE

Pyroelectricity is rigorously defined as the manifestation of the temperature dependence of the spontaneous polarization of certain anisotropic solids. The precise meaning of this definition will be clarified by means of a simple physical example which illustrates the character of the effect. Let us consider a thin parallel-sided sample of a pyroelectric material, such as a crystal of the mineral tourmaline or a disk of poled barium titanate ceramic, cut so that its crystallographic symmetry axis is perpendicular to the flat surfaces. The molecular subunits of the material have a dipole moment. The crystallographic symmetry of a pyroelectric material dictates that the unit dipoles pack in such a way that the components of these dipole moments in the direction normal to the flat surfaces are additive rather than self-cancelling. The dipole moment per unit volume of the material is called the spontaneous polarization P_s. This quantity is always nonzero in a pyroelectric material, and it exists in the absence of an applied electric field. The spontaneous polarization is equivalent to a layer of bound charge on each flat surface of the sample. The amount of charge per unit area is given by

$$Q/A = |P_s| \tag{1}$$

If the sample is suspended in the atmosphere, free charges (electrons or ions) will be attracted and adhere to the material, as illustrated in Figure 1(a).

Conductive electrodes may be applied to the flat surfaces and interconnected by an ammeter having zero internal resistance (Figure 1(b)). If the temperature of the sample is constant, P_s does not change and no current flows in the external circuit. A change in temperature will cause the spontaneous polarization to change. In most single crystals and ceramics, an increase in temperature causes the dipole moments and, consequently, the spontaneous polarization to decrease. (In pyroelectric polymeric materials such as polyvinylidene fluoride, an increase in temperature causes an increase in separation of centers of positive and negative charge, leading to an increase in the spontaneous polarization.) The quantity of bound charge then decreases, and the free charges are redistributed to compensate

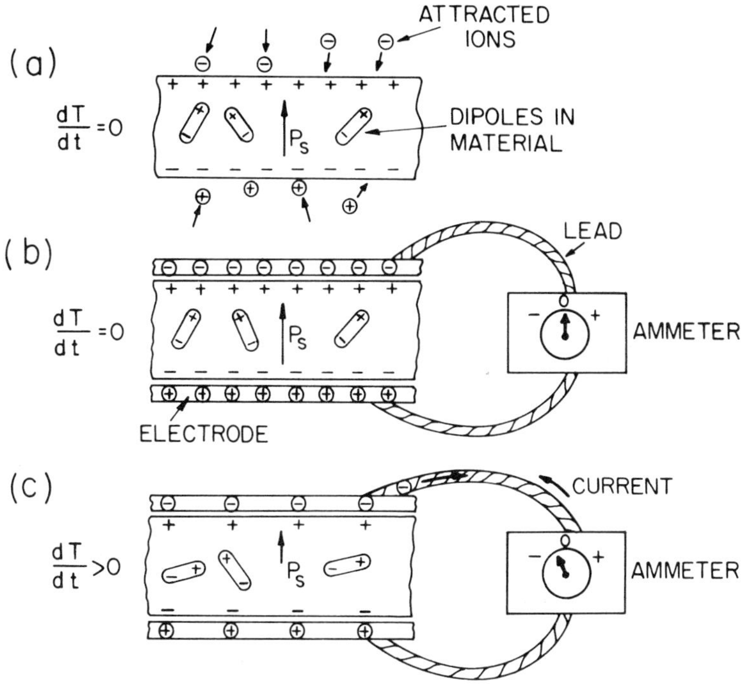

FIGURE 1. (a) Pyroelectric sample with spontaneous polarization and equivalent bound charges, and free charges attracted from its surroundings. (b) Sample with electrodes showing compensating free charges. (c) Charge distribution and current during increase in temperature.

for the change in bound charge resulting in a current flow in the external circuit (Figure 1(c)). The pyroelectric current I is given by

$$I = pA d\theta/dt \tag{2}$$

where $d\theta/dt$ is the rate of change of temperature, and p is the pyroelectric coefficient. Rigorously, the pyroelectric coefficient is defined by

$$\bar{p} = (\partial \bar{P}_s / \partial \theta) \tag{3}$$

The partial derivative in Equation (3) is taken under the constraints of constant electric field and constant elastic stress. Both $\bar{p}$ and $\bar{P}_s$ have the mathematical character of vectors or first-rank tensors. However, only the components of the vectors that are normal to the electrode surfaces will be considered in this chapter, and the quantities will therefore be written as scalars. If the sample had been cooled instead of heated, the signs of both the temperature derivative and the current in Equation (2) would have been reversed. It is important to note that the pyroelectric effect is only observable during the period in which the temperature is changing, either increasing or decreasing.

The existence of the pyroelectric effect in any solid material requires that three conditions must be satisfied: (a) the molecular structure must have a nonzero dipole moment; (b) pyroelectric substances must have no center of symmetry; and (c) pyroelectric substances must have either no axis of rotational symmetry, or have a single axis of rotational symmetry that is not included in an inversion axis.

Of the 32 crystal point-group symmetries (also called crystal classes), only ten permit the existence of pyroelectricity. These groups are: the triclinic group, 1; the monoclinic groups, 2 and m; the orthorhombic group, mm2; the tetragonal groups, 4 and 4mm; the trigonal groups, 3 and 3m; and the hexagonal groups, 6 and 6mm.

The symmetry characteristics of biological pyroelectric materials can often be described in terms of texture point groups, as originally developed by Shubnikov et al. (5). The textures considered here consist of aggregates of crystallites randomly arrayed in a plane, but possessing elements of order in a direction normal to that plane. Of the seven texture groups described by Shubnikov, two groups $(\infty)T$ and $(\infty \cdot m)T$, satisfy the symmetry requirements above, and thus may exhibit pyroelectricity. These groups possess, as elements of symmetry, an infinite-fold rotational axis and an infinite-fold rotational axis at the intersection of an infinity of mirror planes, respectively.

Bone is an example of a biological material describable by means of a texture group. The analysis is based on a consideration of both the molecular structures of the principal components and the histological character of osteons and interstitial lamellae. The major solid components of bone are the

FIGURE 2. Texture group (∞)T consisting of subunits having point group 3 symmetry (8). The figures are assumed to be black on their visible surfaces and white on their hidden surfaces.

crystals of hydroxyapatite and the reinforcing collagen protein fibers. Hydroxyapatite has the centrosymmetric point group 6/m (6) and is not pyroelectric. Collagen has the point group 3 which has a threefold rotational axis as its single element of symmetry (7), and thus it is pyroelectric. The long collagen fibers are oriented with their rotational axes parallel to the bone axis but rotated in a random fashion in the bone transverse plane. This pattern is depicted schematically in Figure 2, in which it is assumed that the symbols are colored black on their upper visible faces and white on their lower hidden faces. This structure corresponds to the texture group (∞)T.

PIEZOELECTRICITY

If an elastic stress is imposed upon certain crystals, they become electrically polarized or, equivalently, a distribution of bound charges appears on their surfaces. If electrodes are appropriately applied to the crystals, a voltage can be measured between the electrodes. This is the description of the direct piezoelectric effect. A converse effect also exists, whereby an imposed electric field gives rise to an elastic strain. Piezoelectric materials must have structures which satisfy the first two of the conditions required for pyroelectric materials. There are 21 point groups that have no center of symmetry and all are piezoelectric with the exception of point group 432. Three of the texture groups are piezoelectric, the two pyroelectric ones and (∞:2)T. It is readily apparent that all pyroelectric materials are piezoelectric, but the converse statement is not true.

Piezoelectricity is characterized by a quantity called the piezoelectric constant which has the mathematical properties of a third-rank tensor. This constant represents the proportionality constant between electric displacement $\bar{D}$ which is a vector (first-rank tensor) and stress $\overleftrightarrow{T}$ which is a second-rank tensor. For convenience, a simplified matrix notation has been developed in which the piezoelectric coefficient d is given by a 3 x 6 matrix and the stress by a 1 x 6 vector, as follows:

$$dD_i = d_{ij}\ dT_j \quad (i = 1,2,3;\ j = 1,2,\ldots,6) \tag{4}$$

The 1, 2, and 3 components of the stress are normal stresses parallel to the x_1, x_2, and x_3 crystallographic axes, and the 4, 5, and 6 components are shear stresses.

Piezoelectric effects in biological materials are beyond the scope of this chapter, with the exception of one special case. A hydrostatic piezoelectric response can be produced by placing a small electroded sample in an oil-filled cell which is closed by piston (Figure 3). The piston can be driven at a low

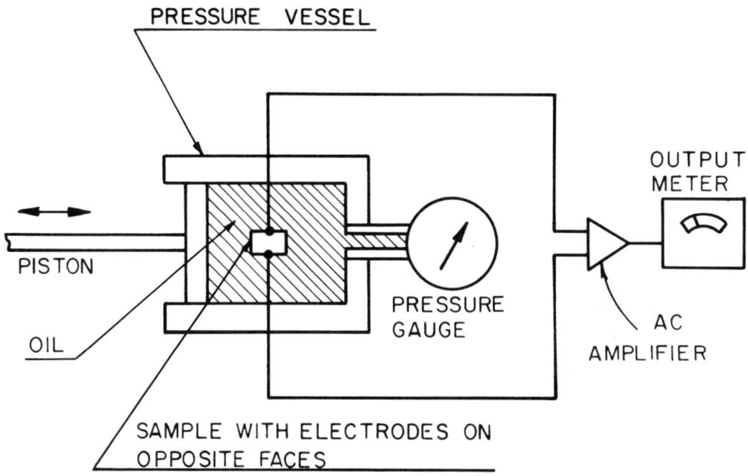

FIGURE 3. Apparatus for measuring hydrostatic piezoelectric coefficients.

frequency, say 20 Hz, to give pressure fluctuations. A hydrostatic pressure produces only normal stresses. Since stress, by convention, has a positive sign for tension and a negative one for compression, an incremental hydrostatic pressure dP^h is related to the normal stresses as

$$dP^h = -dT_1 = -dT_2 = -dT_3 \tag{5}$$

Equation (4) can then be written in the form

$$\begin{aligned} dD_1 &= -(d_{11} + d_{12} + d_{13})dP^h = -d_1{}^h dP^h \\ dD_2 &= -(d_{21} + d_{22} + d_{23})dP^h = -d_2{}^h dP^h \\ dD_3 &= -(d_{31} + d_{32} + d_{33})dP^h = -d_3{}^h dP^h \end{aligned} \tag{6}$$

The quantities $d_i{}^h$ are called the hydrostatic piezoelectric coefficients. It can be shown that they have the symmetry characteristics of a vector, just as do the pyroelectric coefficients. Thus a hydrostatic piezoelectric effect can only exist in a material having one of the ten polar point groups or two polar texture groups. A hydrostatic piezoelectric effect can be observed in tourmaline which is also pyroelectric, but not in quartz which is piezoelectric but not pyroelectric. The existence of a hydrostatic piezoelectric effect is proof that the material is pyroelectric.

ELECTRETS

An electret is a dielectric material that produces an external electric field which results from an ordering of molecular dipoles, or uncompensated surface or space charges. The dipole ordering or the introduction of charges may produce either permanent or transient structural changes. However, if the changes are transient, they must decay at a rate which is slow relative to the duration of the measurements made on the dielectric. The presence of the dipoles or charges will often permit the existence of pyroelectricity and piezoelectricity, so that pyroelectrics and piezoelectrics might be considered as subgroups of electrets. Electrets have been produced by a number of different techniques: dipole ordering by means of imposed electric fields at elevated temperatures, excitation of space charge by photoeffects and rearrangement of the charge distribution with an applied electric field, and implantation of electric charge by electron irradiation or corona charging processes. Several authors have explained biological pyroelectric effects by means of an electret model, as discussed later in this chapter.

FERROELECTRICITY

A polar material whose electric dipoles can be reversed in direction by means of an electric field is defined as ferroelectric. Ferroelectric materials form a subgroup of pyroelectric materials. Some evidence has been found for ferroelectricity in various biomaterials, as described later in this chapter.

Ferroelectric materials usually, but not always, exist in a nonpolar (paraelectric) state above a temperature known as the Curie point. Their dielectric constants are usually anomalously large and highly temperature dependent, especially in the region of the Curie point. At one time, ferroelectricity was thought to be quite rare, but it is now considered to be one of the commonest cooperative phenomena in nature. More than 1000 ferroelectric materials have been studied up to the present time. Several comprehensive books on nonbiological ferroelectrics have been published recently (9-11).

The most frequently used technique for demonstrating ferroelectricity in a material is to produce a ferroelectric hysteresis loop using a Sawyer-Tower circuit (11) similar to that shown in Figure 4. The test material is exposed to a high alternating voltage, typically at line frequency. The voltage drop across the test material is displayed on the x axis of an oscilloscope, and the voltage drop across a linear dielectric capacitor is displayed on the y axis. Results for a typical ferroelectric are illustrated in Figure 5. The spontaneous polarization is calculated by means of the construction shown. It is difficult, but possible, to determine p from the temperature dependence of P_s using Equation (3).

Ferroelectric hysteresis loops have been determined on several very interesting biomaterials, as discussed later. However, unless precautions are taken, it is possible to get results similar to those of Figure 5 using nonlinear, nonferro-

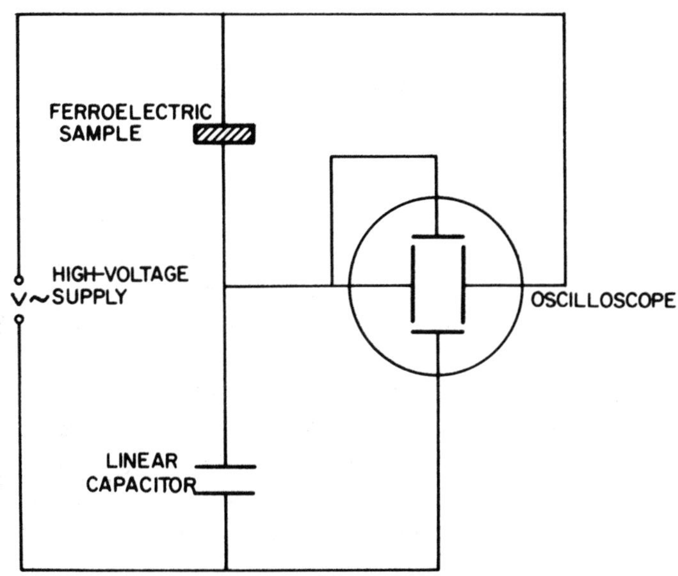

FIGURE 4. Sawyer-Tower hysteresis loop circuit.

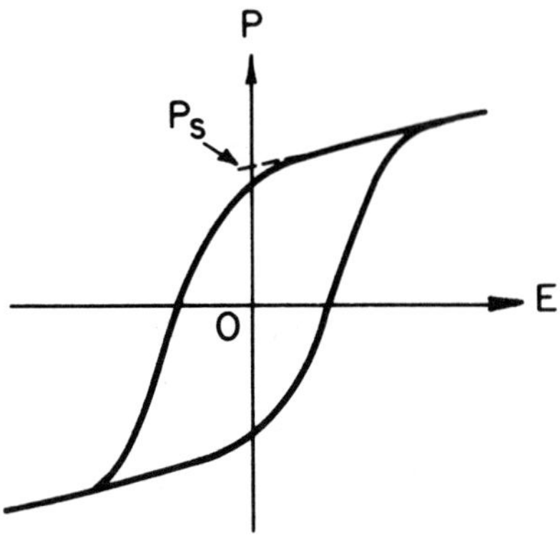

FIGURE 5. Polarization versus applied electric field for a ferroelectric, showing the calculation of the spontaneous polarization.

electric, conductive dielectrics. As a result, all claims of observed ferroelectricity in biomaterials are still quite controversial.

TECHNIQUES FOR MEASURING THE PYROELECTRIC EFFECT

A number of techniques have been used for quantitatively measuring the pyroelectric effect. The three described below have been used for testing biomaterials.

1. Static Method

The static method was used in studies of biomaterials by Lang (2) and Liboff and Furst (12). It is based on a technique developed by Lang and Steckel (13). It is the only one of the methods capable of giving absolute values of the pyroelectric coefficient without a calibration or additional measurement.

Samples are cut in the form of thin plates, and evaporated metal, silver paint, or colloidal graphite electrodes are placed upon the appropriate flat surfaces. The sample is mounted in an oven or a cryostat whose temperature can be varied either up or down, preferably linearly in time. As the temperature of the sample is varied, the voltage V that appears on the electrodes is measured with a high-impedance voltmeter or electrometer. The instrument is shunted with a resistor R whose resistance is less than that of the sample. The pyroelectric coefficient is found from

$$p = V/AR(d\theta/dt) \qquad (7)$$

where A is the area of an electrode and $d\theta/dt$ is the time derivative of the temperature change. Alternately, if the current I is measured, the following equation is used:

$$p = I/A(d\theta/dt) \qquad (8)$$

A third modification uses a charge amplifier as a measuring instrument. Then the charge Q produced during a change in temperature $\Delta\theta$ is related to p by

$$p = Q/A\Delta\theta \qquad (9)$$

The method is subject to two major sources of error: (1) steady or slowly varying voltages or currents are produced by many dielectrics, even when the temperature is held constant, and (2) voltages or currents can result from a varying nonuniform temperature distribution. The former error can be corrected by subtracting the voltage or current produced at constant temperature as measured either before or after a run. The latter error is known as the tertiary pyroelectric effect (1). It can be observed in any piezoelectric material, not only the pyroelectric ones. It can best be avoided by using thin samples, taking care in mounting them with respect to heating elements, and not using excessively large heating or cooling rates. The precautions of reversing the leads from the sample to the measuring instrument, making both heating and cooling runs, and trying several types of electrode materials are also advisable.

2. Rectangular Pulse Heating Method

The rectangular-pulse heating technique has been used extensively in recent years. It was first quantitatively analyzed by Shaulov and Simhony (14). In this method, samples cut in the form of thin plates are electroded and then connected to an electrometer or charge amplifier. A light beam from an xenon or argon lamp is focused on the sample, and the beam is broken into rectangular pulses by means of a photographic shutter or a light chopper. The output voltage of the amplifier may be displayed on an oscilloscope. Shutter speeds of 0.01-1 sec or chopping rates of 1-100 Hz are generally used. Rectangular pulse-heating of a pyroelectric material produces a characteristic pyroelectric signature, some examples of which are shown in Figure 6. The quasi-symmetric character of the heating and cooling curves should be noted.

The analysis (14) shows that the voltages rise at an exponential rate determined by the electrical time constant of the system, and fall at a rate set by the thermal time constant. Mathematically, the voltage is given by

$$V = V_o[\exp(-t/\tau_T) - \exp(-t/\tau_E)] \tag{10}$$

where

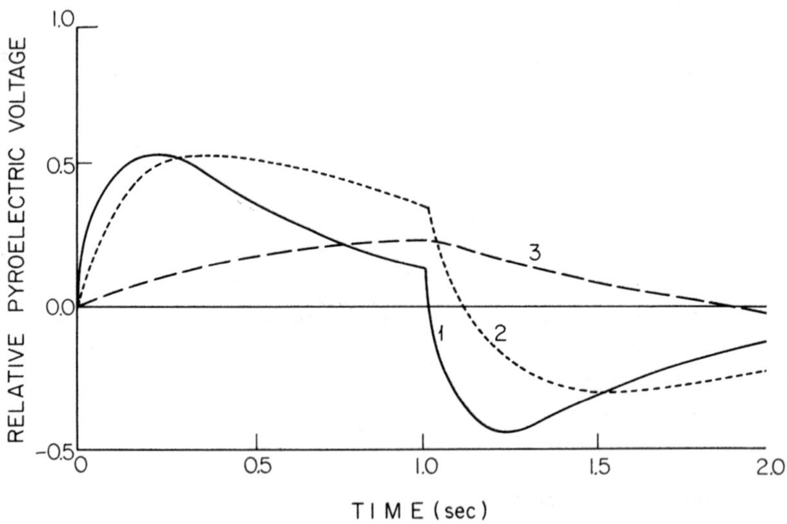

FIGURE 6. Calculated responses of a typical pyroelectric material to rectangular heat pulses (according to Equation (10)). Heat pulse was initiated at 0 seconds for a duration of 1 second. The thermal and electrical time constants (in seconds), respectively, were (0.5, 0.1), (1.0, 0.2), (2.0, 1.0) for curves 1-3 respectively.

$$V_o = pA^2F_o/CC_T[(1/\tau_E) - (1/\tau_T)]. \tag{11}$$

The parameters in Equation 10 are: τ_T, thermal time constant (= C_T/G_T); τ_E, electrical time constant (= RC); F_o, thermal heat flux absorbed; C, capacitance of sample and measuring circuit; C_T, thermal mass of sample; G_T, thermal conductance for heat loss of sample; R, electrical resistance of sample. The other parameters have been defined previously. The parameter of primary interest is the pyroelectric coefficient. The two time constants, although often measured, are not of fundamental importance. The electrical time constant is usually determined by the value of the amplifier shunt resistance, and the thermal time constant depends upon the mounting of the sample. It is often helpful to vary the electrical time constant with various shunt resistors, but only in order to verify that it is actually the pyroelectric effect that is being measured. The initial slope, maximum voltage, and the time to reach maximum voltage can be used to determine V_o, τ_E, and τ_T. These three values can be determined even more accurately by using a nonlinear least-squares curve-fitting technique to match Equation (10) to the experimental data. This is illustrated in Figure 7 from Lang and Athenstaedt (15).

This method yields only relative values of p unless the absorbed thermal flux is known. It can be calibrated by use of a test material whose properties have been previously measured such as tourmaline. However, corrections must be introduced for the differences in the thermal and electrical properties of the two materials.

3. Dielectric Heating Method

One of the difficulties in the two previous methods is that the temperature distribution in the sample may not be uniform. This can be largely overcome by a technique utilizing dielectric heating recently reported by Sussner et al. (16). He used a 500-kHz 20-msec pulse whose energy was uniformly absorbed in the sample, rapidly raising its temperature. The process is especially efficient in biological specimens because of their high electrical loss tangents.

If the loss tangent of the material is known independently as well as its density, heat capacity, and dielectric constant,

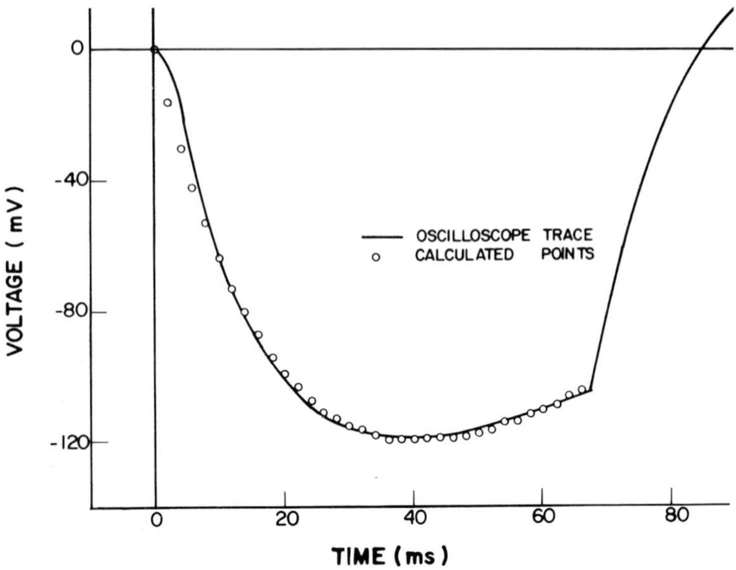

FIGURE 7. Oscilloscope trace of the response of a pyroelectric sample to a 1/15 second duration rectangular light pulse (15). The function which fits the data best in the sense of least squares was $V = -0.777 [\exp(-t/0.0328) - \exp(-t/0.05)]$.

it is possible to calculate the temperature increase. Thus, a quantitative value of the pyroelectric coefficient can be obtained without the need for temperature measurement. Furthermore, the rate of change of temperature is very great, yielding readily measureable voltages with small pyroelectric coefficients.

STUDIES OF BIOLOGICAL MATERIALS

PYROELECTRIC NATURE OF ELECTRICAL RESPONSES

A number of different artifacts may be confused with a true pyroelectric response, including dielectric absorption currents, thermally stimulated discharge currents, various electrochemical effects, and photocurrents. To prove that bovine tendon and the leaf of the Encephalartos villosus gave pyroelectric responses upon heating, Lang and Athenstaedt (15,17) performed the following experiment. Samples of tendon and Encephalartos, a tourmaline crystal, a copper-constantan thermocouple, and a silicon photodiode, were in turn exposed to rectangular pulses of filtered light from a xenon lamp. The filters were a series with 50% transmission points ranging from 305 to 1000 nm. The light was absorbed on each sample by a layer of silver paint. The electric charges produced by the tendon, Encephalartos, and tourmaline, and the voltages from the thermocouple and photodiode were measured. The relative responses are graphed versus the 50% transmission points in Figure 8. According to Equation (9), the responses of the pyroelectric materials should be proportional to the change in temperature, which is also the case with regard to the response of the thermocouple. The relative responses of these four items clearly diminished with increased filter opacity, whereas the photodiode, which responds to red and near infrared radiation, had a nearly constant output. The correspondence of the behavior of pyroelectric materials to the phenomenological theory was thus demonstrated.

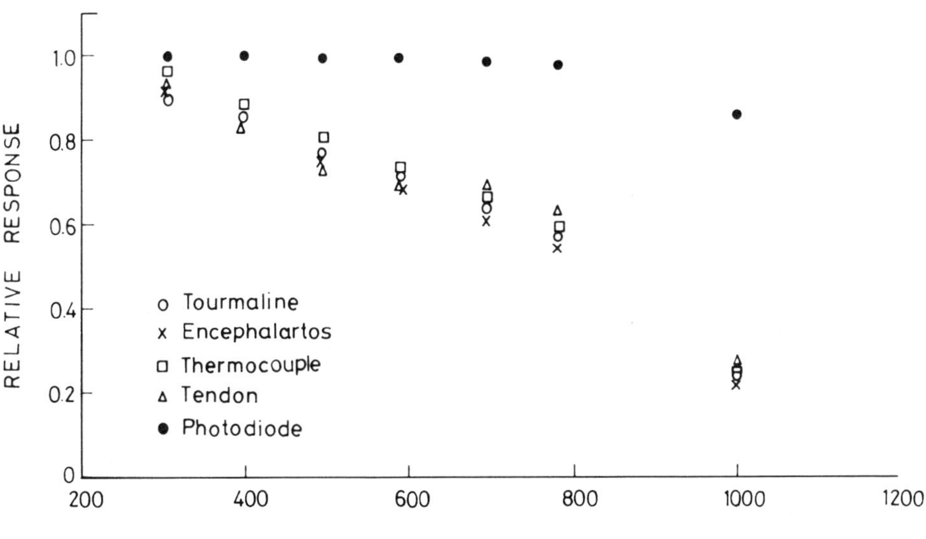

FIGURE 8. Effects on pyroelectric and photoelectric materials of optically absorbing the shorter wavelengths of light (15,17).

ANIMAL TISSUES

The first report of an observation of the pyroelectric effect in a biomaterial was made by Lang (2) in 1966. He examined bovine femur and phalanx bones and hoof tendon. The bone samples were degreased, and disk-shaped specimens 6 mm in diameter and 3 mm in thickness were prepared and oven-dried at 50-150°C. The tendon specimens were dehydrated in absolute alcohol and then dried. After drying, all samples had resistances greater than 10^{14} Ω. Several phalanx disks were demineralized in 2% HNO_3, which dissolved about 75 wt-% of the samples. Either silver paint or colloidal graphite electrodes were used. No effects due to differences in the electrodes were observed. The static method of measurement was used, utilizing a platinum resistance thermometer for temperature sensing and a vibrating-reed electrometer shunted with a 10^{10}-Ω resistor. The sample holder was heated or cooled at temperature rates as great as 14°K/min. The pyroelectric coefficients in samples cut with electrodes normal to the bone or tendon longitudinal axis are given in Table 1. No pyroelectric effect was observed in any sample cut with electrodes parallel to the axis. The pyroelectric coefficients were only slightly greater than the noise level of the system, accounting for the large standard deviations. The pyroelectric coefficient of the demineralized bone (based on the area of the sample before demineralization) had

TABLE 1. Pyroelectric Coefficients (at constant stress) of Bovine Bone and Tendon (2)

MATERIALS	TEMPERATURE RANGE (°C)	PYROELECTRIC COEFFICIENT AND STANDARD DEVIATION ($\mu Cm^{-2}K^{-1}$)
Phalanx	-35 to 55	0.0025 ± 0.0018
Demineralized phalanx	-35 to 60	0.0038 ± 0.0018
Femur	-25 to 60	0.0036 ± 0.0021
Hoof tendon	-35 to 85	0.0041 ± 0.0024

the same magnitude as those of the untreated specimens. The tendon, which is almost pure collagen, had a slightly larger pyroelectric effect. The results definitely indicate that pyroelectricity, like piezoelectricity, is due to the protein rather than the mineral component of bone.

Liboff and Furst (12) conducted similar experiments with bovine femur. In some tests, they found that the logarithm of the pyroelectric current varied inversely with the reciprocal of the absolute temperature, with a change in slope at the 200°C denaturation point. If the current obeys an Arrhenius relation of the form

$$I = I_0 \exp(-\Delta\varepsilon / k\theta) \tag{11}$$

where k is the Boltzmann constant, then the activation energy $\Delta\varepsilon$ was 0.6 eV below the denaturation point and 1 eV above it. Similar anomalous behavior was noted in the vicinity of the 160°C denaturation point of bovine Achilles tendon. In other experiments on bovine femur, pyroelectric currents that exhibited extremes and even reversals in sign were observed (Figure 9). No detailed explanations were given, although the possibility of electret behavior was suggested. A very brief

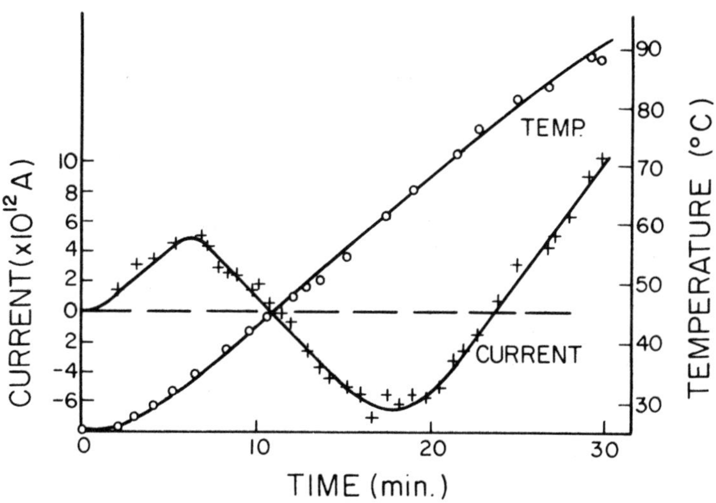

FIGURE 9. Reversal of sign of pyroelectric current (12).

description of a pulsed infrared CO_2 laser measurement technique was given. In a later abstract, Furst and Liboff (18) described successful application of the technique. Pulses of 250 nsec at a level of 0.5 J impinged upon samples of various collagenous materials. A rapid initial rise in electric signal due to the pyroelectric effect followed by an oscillatory ringing due to a piezoelectric effect was observed.

The origin of the pyroelectricity in a material can often be better understood by decomposing the pyroelectric coefficient into two parts. As defined in Equation (3), the change in spontaneous polarization is measured under the constraint of constant stress. Instead of maintaining constant stress while the temperature is changed, the material, in principle, could be rigidly clamped so that expansion or contraction is not possible (constant strain). The change in electric displacement per unit temperature change at constant strain is called the primary pyroelectric coefficient. Then the material is released and allowed to deform. The new change in displacement per unit temperature change is the secondary pyroelectric coefficient. The conventionally measured total pyroelectric coefficient is the sum of the two. Mathematically, the pyroelectric coefficient is decomposed as follows:

$$\underset{\text{Total}}{p_i^T} = \underset{\text{Primary}}{p_i^S} + \underset{\text{Secondary}}{d_{ijk}{}^{\theta} c_{jklm}{}^{E,\theta} \alpha_{lm}{}^{E}} \tag{12}$$

Here the subscripts are the tensor indices and the superscripts indicate constraints of constant stress (T), strain (S), temperature (θ), and field (E). Parameters not previously defined are the elastic stiffness coefficients (c) and the thermal expansion coefficients (α). Readers unfamiliar with the notation should consult Nye (3) or Mason (4).

The only separation of the primary and secondary components for a biological material was carried out by Lang (19). Equation (12) was expanded using the symmetry characteristics of texture group (∞)T to give

$$p_3^T = p_3^S + 2d_{31}\,(c_{11} + c_{12})\,\alpha_1 + 2d_{31}c_{13}\,\alpha_3 + 2d_{33}c_{13}\,\alpha_1 + d_{33}c_{33}\,\alpha_3. \tag{13}$$

The primary pyroelectric coefficient cannot be measured directly because rigid clamping is experimentally impossible. Instead it must be calculated from Equation (13). The experimental parameters used and the calculated results are listed in Table 2. The sign convention used was that the positive direction of the x_3 axis was toward the proximal end of the bone. It is interesting to note that the primary and secondary components have opposite signs but that the contribution of the secondary effect is the dominant one.

A comprehensive program of polarization studies on animal tissues was carried out by Athenstaedt, resulting in a series of more than 10 papers published in 1967-1974 (22-33). Tissues from a large number of animal species were examined. The major objective was to study the direction of spontaneous polarization

TABLE 2. Calculation of Primary and Secondary Pyroelectric Coefficients of Bovine Phalanx Bone

PARAMETER	VALUE	REFERENCE
p_3^T	0.0025 $\mu Cm^{-2}K^{-1}$	2
d_{31}	0.0033 pCN^{-1}	21
d_{33}	0.0033 pCN^{-1}	21
1	$30 \times 10^{-6}K^{-1}$	19
3	$18 \times 10^{-6}K^{-1}$	19
c_{11}	$2.12 \times 10^{10}Nm^{-2}$	20
c_{12}	$0.95 \times 10^{10}Nm^{-2}$	20
c_{13}	$1.02 \times 10^{10}Nm^{-2}$	20
c_{33}	$3.74 \times 10^{10}Nm^{-2}$	20
p_3^S (calculated primary)	-0.0092 $\mu Cm^{-2}K^{-1}$	19
$p_3^T - p_3^S$ (calculated secondary)	0.0117 $\mu Cm^{-2}K^{-1}$	19

in the tissues as a function of species and age of the animal. The orientation of the spontaneous polarization is uniquely related to the orientation of the pyroelectric vector as shown by Equation (3). Thus the pyroelectric vector clearly relates the orientation of the spontaneous polarization to the histology. However, Athenstaedt attempted to characterize the orientation of the spontaneous polarization by means of the hydrostatic piezoelectric coefficient. These two do not necessarily have the same sign, and, therefore, the vectors may be either parallel or antiparallel. As a result, the existence of a hydrostatic piezoelectric effect confirms that the spontaneous polarization is nonzero, but is ambiguous concerning its orientation. Because the term "electric polarization" in these papers (22-33) is not the same as "spontaneous polarization", we will refer to it in this chapter as "hydrostatic piezoelectric polarization" (d^h polarization).

Athenstaedt's results have some very interesting physiological implications. However, it is highly recommended that these data be verified by a more conventional measuring technique before further interpretation is made. The major results can be summarized as follows:

1. All the collagenous tissues tested exhibited a hydrostatic piezoelectric polarization in the axial direction and occasionally in a radial direction. Some chitin- and keratin-containing tissues also had a d^h polarization.
2. The directions of d^h polarization were different in the epiphyses (end sections) and the diaphyses (midshaft sections) of either the long bones of the appendicular skeleton (e.g., femur, tibia) or the vertebrae of the axial skeleton. The d^h polarization directions were reversed during the processes of growth so that the directions differed between the growth stage and the adult stage. The polarization in the diaphyses in old age had an antiparallel character so that large specimens had a small or zero net d^h polarization. These points are summarized in Figure 10, which shows typical results for higher mammals including humans.
3. The reversal of polarization with biological maturity occurred in all mammals as well as in birds. The bones of

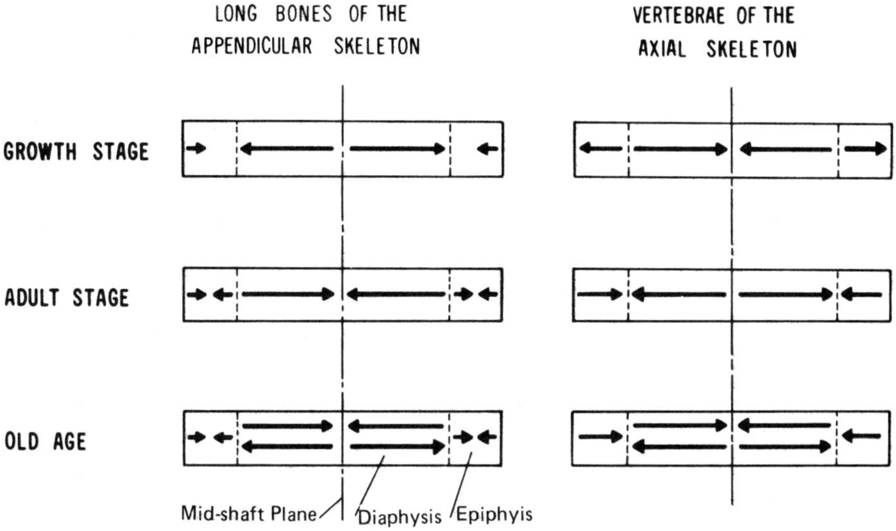

FIGURE 10. Directions of d^h polarization in different types of bones at various stages in the human life span (8). The arrows point to the ends which become positive under hydrostatic compression.

crocodiles exhibited a similar behavior, but those of simpler reptiles and all amphibians showed no reversal. The relative directions of polarization in the axial and appendicular skeletons of amphibians and simpler reptiles were alike, but they were opposite in higher animals.

4. Cartilage in higher animals exhibited a d^h polarization in the thickness direction. The polarization did not reverse after the cessation of growth, and it did not disappear in old age.

PLANT TISSUES

A large number of qualitative and a lesser number of quantitative studies on cells, tissues, and organs of plants were made by Athenstaedt (34). Many of the materials exhibited two directions of polarization, one parallel to a longitudinal axis and the second normal to the axis, usually in an inside-outside direction. This implies that the spontaneous polarization vector made an acute angle with the longitudinal axis.

Athenstaedt also reported some interesting results on grains of wheat and rye (35). Some measurements were made on the shell of the grain and others on intact single grains. The rectangular-pulse heating method was used, and it was made quantitative by using a tourmaline crystal for calibration. Pyroelectric coefficients were measured as a function of temperature over the range -20°C to 40°C. The surprising feature of the work was the striking difference between the temperature dependencies of the pyroelectric coefficient of the winter and spring species of the grain. The winter varieties showed a peak in p, at or below 0°C, whereas the pyroelectric coefficients of the spring varieties increased monotonically with temperature. Winter wheat forms ears only when exposed for a lengthy period of time after sowing to a temperature of about 0°C (up to a maximum of +3°C). For this reason, it is sown in autumn, before the beginning of winter with its low temperatures. If winter wheat is sown in spring, or if from the time of sowing in autumn the temperature remains constantly above 5°C, strong vegetation growth appears, but the reproductive phase does not follow. Spring wheat, however, shows a normal formation of ears when sown in spring. Athenstaedt emphasized the relationship between this physiological behavior and the differences in pyroelectric effect. He suggested that a pyroelectric test might serve as a simple means of distinguishing among the species.

Lang and Athenstaedt (15,17) examined the pyroelectric behavior of the leaves of the palmlike plant Encephalartos villosus. Samples of leaves were scraped with a scalpel until their thickness was 0.06-0.12 mm. They were air-dried briefly and then tested using the rectangular pulse method with a tourmaline calibration. Differences were observed in both the pyroelectric activity and the thermal time constant dependent on whether the scraped or the original external surface was heated. They suggested that only a thin layer in the epidermis close to the surface was pyroelectrically active. A room-temperature pyroelectric coefficient of 0.0129 $\mu Cm^{-2}K^{-1}$ was determined. The polarity was such that the external surface acquired a positive charge upon heating. Direct-current (DC) bias voltages used to determine if the epidermis was ferroelectric produced unexpected results. An example is shown in Figure 11. Without bias, a pyroelectric signal of 85 mV was

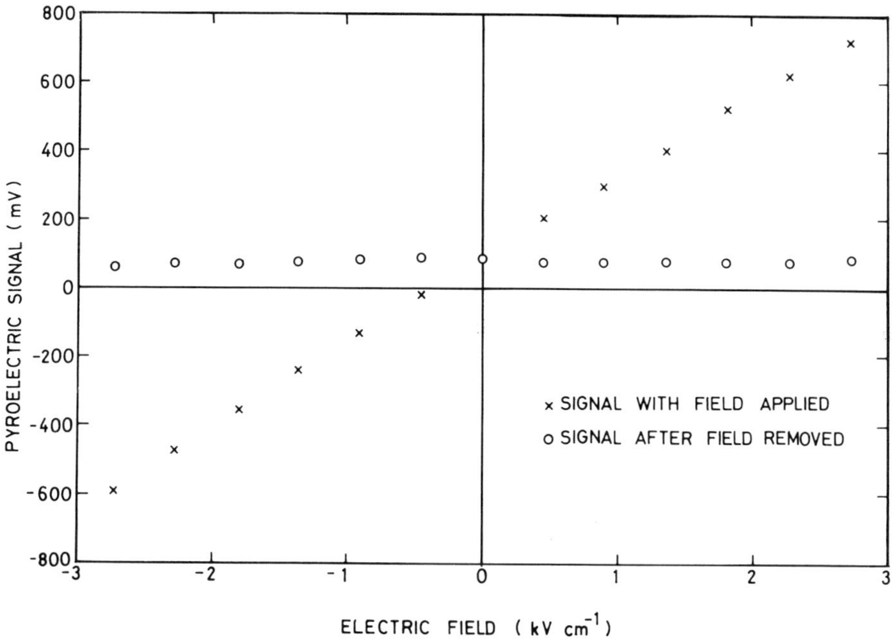

FIGURE 11. Dependence of pyroelectric response of Encephalartos villosus epidermis on bias electric field (15,17).

observed. Stepwise increase of the bias field to 2.7 kV/cm caused the pyroelectric response to increase linearly to 710 mV; a negative DC bias of 2.7 kV/cm caused the pyroelectric signal to decrease to -595 mV. In all cases, removal of the bias caused the pyroelectric signal to immediately return to its original value. On other samples, the same magnitude bias field caused the pyroelectric response to increase from 175 mV to 8.8 V, but the response was no longer linear with the bias field. Bias fields as large as 18.2 kV/cm were used in an attempt to reverse the polarization as an indication of ferroelectricity. No reversal could be achieved, suggesting that the material was not ferroelectric. These results were interpreted phenomenologically by considering that the pyroelectric coefficient could be expressed as a power series in electric field. A large term proportional to the electric field and a smaller one proportional to the cube of the field were present, but no effect proportional to the square of the field was observed. A

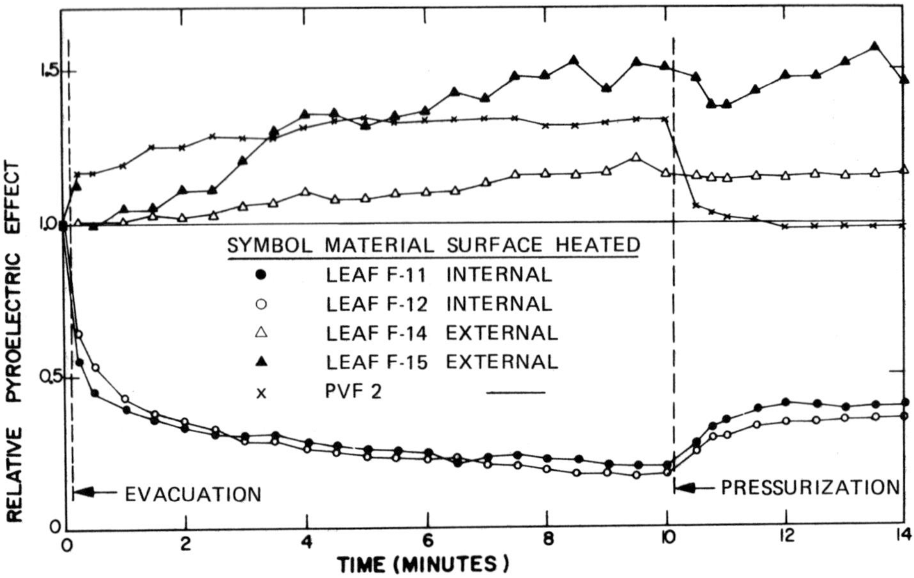

FIGURE 12. Relative pyroelectric coefficients of several *Encephalartos* and a polyvinylidene fluoride sample as functions of time after sample chamber evacuation and repressurization (12).

nonlinear stress-induced effect due to changes in hydrostatic pressure was also observed. Figure 12 shows the relative pyroelectric effects of several samples as functions of time after evacuation of the sample chamber to 0.05 torr followed by repressurization to atmospheric pressure. The pyroeffects increased for samples whose external surfaces were heated but decreased when the internal surfaces were heated. A similar effect was also observed in a sample of polyvinylidene fluoride. An explanation of these results in terms of a stress-dependent pyroelectric coefficient is given elsewhere (15).

The origin of the pyroelectric effect in plant materials is uncertain. Wax and lipid deposits have been suggested (15,17). Bazhenov (36), in his classic treatise on the piezoelectric properties of wood, found that the only nonzero piezoelectric coefficients of wood were d_{14}, d_{25}, and d_{36} (with $|d_{14}| > |d_{25}|$ and d_{36} extremely small. This suggests a texture group $(\infty:2)T$ (which, however, requires that $d_{14} = -d_{25}$ and $d_{36} = 0$) or a point group 222 (in which d_{14}, d_{25}, and d_{36} are all independent). In any event, neither group is pyroelectric. However, Preston (37) does not seem to rule out the possibility that cellulose may have a polar (pyroelectric) point group.

LIVING TISSUES

Recently, a series of papers by Athenstaedt and coworkers (38-41) appeared describing pyroelectric studies of living tissues. In order to make such measurements, a new technique was developed in which both electrodes were on the same surface of the sample. As described by Simhony and Athenstaedt (38), an adhesive tape strip containing two small holes was applied to the living specimen. Electrodes made from an aqueous suspension of graphite were painted through the holes onto the specimen. Contact to the electrodes was affected by means of aluminum foil leads attached to silver paste rings applied around the perimeters of the holes. In use, one of the holes was exposed to rectangular pulses of light from a xenon lamp and the other hole was shielded from the light source. The electrical responses showed the classical pyroelectric signature. A detailed theoretical analysis of a pyroelectric structure with such electrodes has not been published. The specimens studied were assumed to consist of a nonconductive thin pyroelectric layer overlying a thick nonpyroelectric layer in which ion conduction occurred. Presumably the lines of electric flux passed from one electrode through the pyroelectric layer, then the conductive layer, and through the pyroelectric layer again to the second electrode. This type of structure, with curved flux lines and layers having vastly different conductive and dielectric properties is not likely to behave in the same way as a homogeneous pyroelectric crystal. Although a true pyroelectric effect was apparently measured in these experiments, the results cannot be quantitatively interpreted by means of any present theory.

In these studies, three types of specimens were used: living materials, nonliving materials which had been excised within the previous several hours, and aged nonliving materials. Two electrodes on a single surface as described above were applied to the living specimens and the measurements were generally conducted in a Faraday cage. The nonliving ones had conventional electrodes and both rectangular-pulse and dielectric heating methods were used. The materials studied were leaves of the Rhododendron and Encephalartos, the insect *Periplaneta americana* (38), thorax or abdomen segments of the insect *Blaberus giganteus* (39), human skin (40), and skin specimens from a series of birds, mammals, reptiles, amphibians, insects, and others (41). Table 3 contains some of the quantitative results (values cited for living specimens are subject to uncertainty because of the lack of a theoretical model for the geometry).

The major conclusions which were drawn from these studies were:

1. The pyroelectric properties are due to the polar structures of the materials and not to the presence of a living state.
2. The living samples exhibited larger effects than fresh nonliving samples and far larger ones than dried nonliving specimens.
3. Pyroelectric activity in skin samples was due to the epidermis layer and not to the underlying corium. The outside surface invariably acquired a negative charge when the sample was heated and the inner surface, a positive sign. Insect integuments exhibited similar polarity. The pyroelectric axis was normal to the outside surface.

SENSORY ORGANS

Several authors have suggested that the pyroelectric effect may be of fundamental physiological importance in various sensory organs. All the evidence appears to be indirect. It is generally based on the similarity between the pyroelectric response to a rectangular pulse, as shown in Figure 6, and the physiological reaction of a stimulated sensory organ. Because the pyroelectric current is proportional to the rate of change of temperature (Equation (2)), a physiological response that is also related to $d\theta/dt$ would be expected to have a time dependence similar to that shown in Figure 6.

Cope (42,43) examined data on the frequency of impulses in single nerve fibers from the thermoreceptors in the ampulla of Lorenzini of fishes when they were subjected to temperature transients. A graph of rate of change of impulse frequency with time versus $d\theta/dt$ was linear, constituting preliminary quantitative evidence for pyroelectricity in this system.

TABLE 3. Pyroelectric Coefficients of Living Tissues

MATERIAL	PYROELECTRIC COEFFICIENT ($\mu Cm^{-2}K^{-1}$)	REFERENCE
Periplaneta americana		
Abdomen rings (scraped)	0.2	38
Thorax (*in vivo*)	3.5	38
Encephalartos leaves		
Upper epidermis (whole)	0.02	38
Upper epidermis (scraped layer)	0.005	38
Lower epidermis (whole)	0.06	38
Lower epidermis (scraped layer)	0.03	38
Rhododendron leaves		
Upper epidermis (whole)	0.03	38
Upper epidermis (scraped layer)	0.02	38
Lower epidermis (whole)	0.15	38
Blaberus giganteus		
In vivo	2.6 - 7.5	39
Dried	0.3 - 1.1	39
Human Skin		
In vivo	0.018 - 0.265	40
In vitro (fresh)	0.021 - 0.27	40

Other examples have been given by Athenstaedt (44). He pointed out that the electrical responses of many thermoreceptors, photoreceptors, electroreceptors, chemoreceptors, and mechanoreceptors can be considered as pyroelectric or piezoelectric signals. The evidence is based on the known pyroelectric properties of many integument tissues and the observation that the physiological responses are proportional to $d\theta/dt$, as shown by the similarity of their behavior to that of known pyroelectrics.

A good example of the similarity between some receptor responses and those of pyroelectric materials is shown by the results of Grundfest (45) in Figure 13. The recordings represent the responses of a single photoreceptor cell of the retina of a cat to light flashes of constant intensity but having durations which varied from 1.0 msec to 1.14 sec. It is possible to generate almost identical electrical responses by varying the electrical and thermal time constants of an inorganic pyroelectric material exposed to similar light flashes.

BIOLOGICAL FERROELECTRICITY

A very few experiments designed to determine if any biomaterials are ferroelectric have been carried out. Although

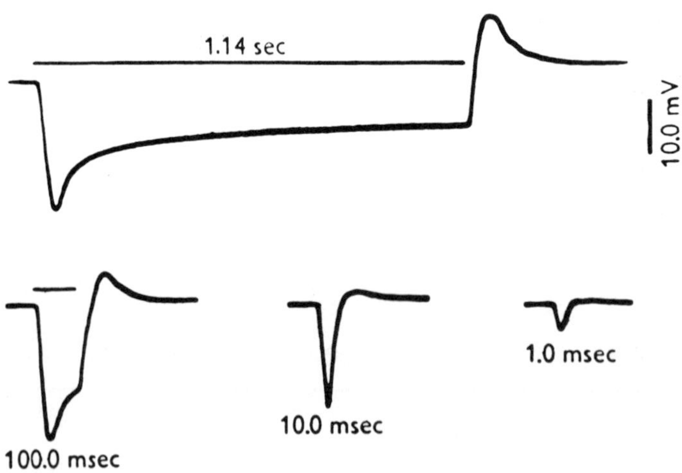

FIGURE 13. Responses of a single photoreceptor cell of the retina of a cat (45).

somewhat outside of the scope of this chapter, the studies are nevertheless interesting from the point of view of pyroelectricity.

In 1960, Duchesne et al. (46) proposed that the sodium salt of deoxyribonucleic acid (NaDNA) was piezoelectric. In the same year, Polonsky et al. (47) and Douzou et al. (48) measured ferroelectric hysteresis loops on NaDNA using a Sawyer-Tower circuit similar to that described in an earlier section. A frequency of 50 Hz and a maximum electric field of 1000 V/cm were used. Based on the hysteresis data, the Curie temperature appeared to be in the range of 50-60°C. The dielectric constant increased rapidly with temperature to a maximum near 70°C, although the Curie-Weiss law was not obeyed above the Curie point [$\varepsilon \approx 1/(\theta - \theta_c)$, where θ_c is the Curie temperature (9-11)]. Stanford and Lorey (49) observed hysteresis loops and anomalous dielectric constant behavior in the sodium salt of ribonucleic acid (NaRNA). They suggested, "RNA is seen to have the attributes of a ferroelectric substance at life temperatures and possesses the capability of storing information by a physical process, hysteresis."

However, O'Konski and Shirai (50) and Brot et al. (51) in experiments with NaDNA and Mascarenhas et al. (52) in studies of RNA disagreed with the ferroelectricity hypothesis. The latter groups all found evidence that the DNA and RNA behaved as nonlinear conductors, and the hysteresis loops were merely artifacts. O'Konski and Shirai explained the observed effects by means of electrolytic processes, and Mascarenhas et al. proposed an electret theory.

CONCLUSIONS AND SOME SPECULATIONS

We have two major objectives in this chapter: (1) to concisely summarize the physics of pyroelectricity and closely related effects and describe the techniques for measuring them; and (2) to present a brief discussion and some analysis of all the studies of biological pyroelectricity to date. Relatively little has been said of the physiological implications of the existence of pyroelectricity in biological tissues and biological systems. This is true largely because most of the research to date has been very exploratory in nature. It was first necessary to demonstrate that pyroelectricity and concomitant spontaneous polarization were fundamental and inherent properties of biomaterials. Although much of the data in the literature are qualitative and some require additional verification, it now seems certain that pyroelectricity is a widespread phenomenon in biomaterials.

The generality of materials in which pyroelectricity may exist is even further extended by studies such as the optical second harmonic generation measurements of Delfino (53) on 71 α-amino acids, 22 dipeptides, 6 tripeptides, 16 proteins, and 5 viruses. He found that all α-amino acids with the exception of α-glycine, all peptides containing at least one enantiomeric α-amino acid residue, and all proteins and viruses were non-centrosymmetric, permitting the existence of piezoelectricity and possibly pyroelectricity as well.

Since possible physiological significance is the raison d'etre for the study of pyroelectricity in biomaterials, it is appropriate to speculate on some areas wherein further research may be warranted. These speculations may not necessarily be valid and they certainly should not limit the scope of research; hopefully, they will be provocative (54).

1. What mechanism could produce a change in direction of polarization or a diminishing of polarization with biological growth and maturation (22-33)? Lang (55) has suggested that if this mechanism functions improperly, calcification might not occur where it should or might occur where it should not. For example, excessive calcium losses in bone and calcium deposits in cartilage are both pathological conditions. Is there a correlation between these conditions and behavior of polarization?
2. Why is there no polarization reversal in the simpler animals (28)? Is there some relationship to the ability of the simpler animals to spontaneously regenerate large organs?
3. Could some biomaterials be ferroelectric? The positive results on DNA and RNA have been challenged, but almost no other biomaterials have been tested for ferroelectricity. Matthias (56) has commented on the generality of the effect. Could ferroelectricity be the mechanism for the

apparently easy reversal in the polarization of bone during the growth stages? The hysteresis phenomenon in ferroelectrics has been used as a reversible information storage device in electronic systems. Fong (57) has suggested a brain memory mechanism based on biological electricity. Could ferroelectricity be related to this and other biological memory processes? Ferroelectrics are characterized by long-range ordering effects, and biological systems depend upon long-range order for their existence. Is there a correlation?

4. Mascarenhas (58) has discussed electret characteristics such as thermally stimulated currents in biomaterials. Are there possible relationships between pyroelectricity and the proposed electret properties?
5. Many biological membranes consist of highly oriented proteins such as collagen. If the protein is pyroelectric, the membrane itself would be expected to be pyroelectric and to exhibit spontaneous polarization. Could the large internal electric fields due to the polarization influence the mass transport of ions through the membrane?
6. According to the wheat studies of Athenstaedt (35), a subtle physiological difference manifested itself in significant differences in the temperature dependence of the pyroelectric effect. Are there other physiological factors that could be conveniently studied by means of their pyroelectric behavior?
7. Small bias DC fields were shown by Lang and Athenstaedt (15,17) to have a very large effect on the biological pyroelectricity of *Encephalartos* leaves. Pyroelectric effects in biomaterials are usually very small, perhaps too small to have biological significance. Very large DC fields exist naturally in biological systems, often due to the very small dimensions of relevant structures such as cells (59). Could these fields have a strong amplifying effect on the pyroelectricity?
8. Pyroelectricity may be responsible for various biological sensors and receptors. Lipinski (60) has suggested a relationship between an ancient Chinese therapeutic heating technique called moxibustion, which stimulates acupuncture loci, and the pyroelectric properties of tissues. Are

there mechanisms whereby pyroelectric responses can affect or control the nervous system?

In conclusion, it is suggested that pyroelectricity is an essential and universal property of all biological systems and that polarization homeostasis may be an important new biomechanism of living matter.

REFERENCES

(1) Lang, S.B.: Sourcebook of Pyroelectricity, Gordon & Breach Science Publishers, New York, 1974.

(2) Lang, S.B.: Pyroelectric effect in bone and tendon, Nature 212:704-705, 1966.

(3) Nye, J.F.: Physical Properties of Crystals, Clarendon Press, Oxford, 1957.

(4) Mason, W.P.: Crystal Physics of Interaction Processes, Academic Press, New York, 1966.

(5) Shubnikov, A.V., Zheludev, I.S., Konstantinova, V.P. and Silvestrova, I.M.: Etudes des Textures Piezoelectriques, Dunod, Paris, 1958.

(6) Müller, M.: The precipitation and the X-ray diffraction investigation of the mixed crystal system $Ca_{10}(Po_4)_6(OH)_2$ - $Pb_{10}(PO)_4)_6(OH)_2$, Helv. Chim. Acta 30:2069-2080, 1947.

(7) Ramachandran, G.N. and Kartha, G.: Structure of collagen, Nature 174:269-270, 1954.

(8) Lang, S.B.: Pyroelectric effect in biological materials, in Electronic Conduction and Mechanoelectrical Transduction in Biological Materials, B. Lipinski, ed., Marcel Dekker, New York, p. 157-200, 1982.

(9) Mitsui, T., Tatsugaki, I. and Nakamura, E.: An Introduc-

tion to the Physics of Ferroelectrics, Gordon & Breach Science Publishers, New York, 1976.

(10) Lines, M.E. and Glass, A.M.: Principles and Applications of Ferroelectrics and Related Materials, Clarendon Press, Oxford, 1977.

(11) Burfoot, J.C. and Taylor, G.W.: Polar Dielectrics and their Applications, Macmillan, New York, 1979.

(12) Liboff, A.R. and Furst, M.H.: Pyroelectric effect in collagenous structures, Ann. N.Y. Acad. Sci. 238:26-35, 1974.

(13) Lang, S.B. and Steckel, F.: Method for the measurement of the pyroelectric coefficient, DC dielectric constant, and volume resistivity of a polar material, Rev. Sci. Instrum. 36:929-932, 1965.

(14) Shaulov, A. and Simhony, M.: Pyroelectric voltage response to rectangular infrared signals in triglycine sulphate and strontium-barium niobate, J. Appl. Phys. 43:1440-1444, 1972.

(15) Lang, S.B. and Athenstaedt, H.: Anomalous pyroelectric behavior in the leaves of the palm-like plant Encephalartos villosus, Ferroelectrics 17:511, 1978.

(16) Sussner, H., Horne, D.E. and Yoon, D.Y.: A new method for determining the pyroelectric coefficient of thin polymer films using dielectric heating, Appl. Phys. Lett. 32:137-139, 1978.

(17) Lang, S.B. and Athenstaedt, H.: Pyroelectricity and induced pyroelectric polarization in leaves of the palmlike plant Encephalartos villosus, Science 196:985-986, 1977.

(18) Furst, M.H. and Liboff, A.R.: Pyroelectric effect in collagenous structures. Observed with pulsed 10.6 μm radiation, Biophys. J. 17:300a, 1977.

(19) Lang, S.B.: Thermal expansion coefficients and the primary and secondary pyroelectric coefficients of animal bone, Nature 224:798-800, 1969.

(20) Lang, S.B.: Ultrasonic method for measuring elastic coefficients of bone and results on fresh and dried bovine bones, IEEE Trans. Biomed. Eng. BME-17:101-105, 1970.

(21) Fukada, E.: Private communication.

(22) Athenstaedt, H.: Das pyroelektrische Verhalten und das permanente elektrische Moment von menschlichem und tierischem Sehnengewebe, Z. Zellforsch. Mikrosk. Anat. 81:62-73, 1967.

(23) Athenstaedt, H.: Permanent electric polarization and pyroelectric behaviour of the vertebrate skeleton. I. The axial skeleton of the vertebrates (excluding Mammalia), Z. Zellforsch. Mikrosk. Anat. 91:135-152, 1968.

(24) Athenstaedt, H.: Permanent electric polarization and pyroelectric behaviour of the vertebrate skeleton. II. The axial skeleton of the Mammalia, Z. Zellforsch. Mikrosk. Anat. 92:428-446, 1968.

(25) Athenstaedt, H.: Permanent electric polarization and pyroelectric behaviour of the vertebrate skeleton. III. The axial skeleton of man, Z. Zellforsch. Mikrosk. Anat. 93:484-504, 1969.

(26) Athenstaedt, H.: Permanent electric polarization and pyroelectric behaviour of the vertebrate skeleton. IV. The cranial bones of man, Z. Zellforsch. Mikrosk. Anat. 97:537-548, 1969.

(27) Athenstaedt, H.: Permanent electric polarization of the meninges of man, Z. Zellforsch. Mikrosk. Anat. 98:300-322, 1969.

(28) Athenstaedt, H.: Permanent longitudinal electric

polarization and pyroelectric behaviour of collagenous structures and nervous tissue in man and other vertebrates, Nature 228:830-834, 1970.

(29) Athenstaedt, H.: Permanent electric polarization and pyroelectric behaviour of the vertebrate skeleton. V. The appendicular skeleton of the vertebrates (excluding man), Z. Anat. Entwicklungsgesch. 131:1-20, 1970.

(30) Athenstaedt, H.: Permanent electric polarization and pyroelectric behaviour of the vertebrate skeleton. VI. The appendicular skeleton of man, Z. Anat. Entwicklungsgesch. 131:21-30, 1970.

(31) Athenstaedt, H.: Pyroelectric and piezoelectric behaviour of human dental hard tissues, Arch. Oral Biol. 16:495-501, 1971.

(32) Athenstaedt, H.: Pyroelectric behaviour of integument structures and of thermo-, photo- and mechanoreceptors, Z. Anat. Entwicklungsgesch. 136:249-271, 1972.

(33) Athenstaedt, H.: Pyroelectric and piezoelectric properties of vertebrates, Ann. N.Y. Acad. Sci. 238:68-94, 1974.

(34) Athenstaedt, H.: Pyroelectric polarization in cells, tissues and organs of plants, Z. Pflanzenphysiol. 68:82-91, 1972.

(35) Athenstaedt, H.: Pyroelectric properties of wheat, Ferroelectrics 14:753-759, 1976.

(36) Bazhenov, V.A.: Piezoelectric Properties of Wood, Consultants Bureau, New York, 1961.

(37) Preston, R.D.: X-ray analysis and the structure of the components of plant cell walls, Phys. Rep. 21C:183-226, 1975.

(38) Simhony, M. and Athenstaedt, H.: Measurement of the pyro-

electric coefficient and permittivity on rhodendron and encephalartos leaves and on the insect _Periplanta americana_, Biophys. J. 29:331-337, 1980.

(39) Athenstaedt, H. and Claussen, H.: Evidence for pyroelectric and piezoelectric sensory mechanisms in the insect integument, Biophys. J. 35:365-374, 1981.

(40) Athenstaedt, H., Claussen, H. and Schaper, D.: Epidermis of human skin: Pyroelectric and piezoelectric sensor layer, Science 216:1018-1020, 1982.

(41) Athenstaedt, H. and Claussen, H.: Evidence of inherent spontaneous polarization in the metazoan integument epithelia, Biophys. J. 41:359-366, 1983.

(42) Cope, F.W.: Piezoelectricity and pyroelectricity as a basis for force and temperature detection by nerve receptors, Bull. Math. Biol. 35:31-41, 1973.

(43) Cope, F.W.: A review of the applications of solid state physics concepts to biological systems, J. Biol. Phys. 3:1-41, 1975.

(44) Athenstaedt, H.: Pyroelectric sensors of organisms, Ferroelectrics 11:365-369, 1976.

(45) Grundfest, H.: Electrophysiology and pharmacology of different components of bioelectric transducers, Cold Spring Harbor Symp. Quant. Biol. 30:1-13, 1965.

(46) Duchesne, J., Depireaux, J., Bertinchamps, A., Cornet, N. and van der Kaa, J.M.: Thermal and electrical properties of nucleic acids and proteins, Nature 188:405-406, 1960.

(47) Polonsky, J., Douzou, P. and Sadron, C.: Mise en evidence de proprietes ferroelectriques dans l'acide desoxyribonucleique (DNA), C.R. Acad. Sci. 250:3414-3416, 1960.

(48) Douzou, P., Franco, J.-C., Polonsky, J. and Sadron, C.:

The electrical properties of sodium deoxyribonucleate, C.R. Acad. Sci. 251:976-978, 1960.

(49) Stanford, A.L. and Lorey, R.A.: Evidence of ferroelectricity in ribonucleic acid, Nature 219:1250-1251, 1968.

(50) O'Konski, C.T. and Shirai, M.: Electric properties of macromolecules. X. Conduction and the question of ferroelectricity in solid sodium deoxyribonucleate, Biopolymers 1:557-562, 1963.

(51) Brot, C., Lassier, B., Sharbaugh, A.H., Reynolds, S.I. and White, D.M.: On the nonlinear dielectric behavior of desoxyribonucleic acid fibers at low frequencies, J. Chem. Phys. 43:3603-3608, 1965.

(52) Mascarenhas, S., Onuchic, J.N. and Quezado, S.: On the non-linear dielectric properties and the alleged ferroelectricity of RNA, Ann. Acad. Bras. Cienc. 51:605-607, 1979.

(53) Delfino, M.: A comprehensive optical second harmonic generation study of the non-centrosymmetric character of biological structures, Mol. Crys. Liq. Cryst. 52:271-284, 1979.

(54) Lang, S.B.: Pyroelectricity: Occurrence in biological materials and possible physiological implications, Ferroelectrics 34:3-9, 1981.

(55) Lang, S.B.: Pyroelectricity research on biomaterials: Past, present, and some speculation, Presented at Joint XII International Conference on Medical and Biological Engineering and V International Conference on Medical Physics, Jerusalem, Israel, Aug. 19-24, Paper 80.4, 1979.

(56) Matthias, B.T.: Ferroelectricity, J. Appl. Phys. 38:928-930, 1967.

(57) Fong, P.: Ferroelectricity of RNA and brain memory, Bull. Ga. Acad. Sci. 30:13-23, 1972.

(58) Mascarenhas, S.: Bioelectrets: Electrets in biomaterials and biopolymers, Top. Appl. Phys. 33:321-346, 1980.

(59) Frohlich, H.: Ferroelectric excitations in biological systems, Presented at 6th Ann. International Meeting on Ferroelectricity, Kobe, Japan, Aug. 12-16, 1985.

(60) Lipinski, B.: Biological significance of piezoelectricity in relation to acupuncture, hatha yoga, osteopathic medicine and action of air ions, Med. Hypotheses 3:9-12, 1977.

9
Mathematical Modeling of Electromagnetic Interactions with Biological Systems

FRANCIS X. HART
Department of Physics
The University of the South
Sewanee, Tennessee

INTRODUCTION

FUNDAMENTAL PRINCIPLES

The interaction of electromagnetic fields with biological systems has attracted considerable interest in recent years. These fields are reported to have positive effects such as the enhancement of bone fracture healing and the use of hyperthermia as a cancer treatment. On the other hand there is concern regarding some potentially harmful effects of microwave and transmission line fields. Other chapters will discuss the details of many of these interactions. But in any such analysis it is important to relate the fields, currents and dissipated energy in the biological system to the externally applied fields. This chapter will survey the various types of mathematical models developed to determine that relationship.

These models can be technically quite complicated. This chapter will emphasize the basic physical principles underlying them. Its purpose is to provide a broad introduction to the area so that researchers entering it can determine what methods might be useful and where more detailed information regarding them can be found.

Space limitations prevent comparisons of the models with experimental systems, details of the mathematical solutions, and an exhaustive coverage of the many variations upon the basic models. Furthermore, there are no diagrams. Each model really merits its own particular illustration, but space is limited. Readers should consult the references for more information in these areas.

The basic principles of electromagnetic theory are well covered in undergraduate texts such as Reitz, Milford and Christy (1), or Lorrain and Corson (2) and in graduate texts such as Jackson (3). Of particular importance for the detailed understanding of the models used for biological systems is the classic work of Stratton (4). Rationalized MKS units will be used. Readers consulting other sources should be aware that different systems are sometimes used.

All of classical electromagnetic theory is contained in Maxwell's four equations. In differential form they are:

$$\nabla \cdot \bar{E} = \rho / K\varepsilon_0 \quad (1a)$$

$$\nabla \cdot \bar{B} = 0 \quad (2a)$$

$$\nabla X \bar{E} = -\partial \bar{B} / \partial t \quad (3a)$$

$$\nabla X \bar{B} = \mu_0 \bar{J}_c + \mu_0 K \varepsilon_0 \partial \bar{E} / \partial t. \quad (4a)$$

The electric field $\bar{E}$ and magnetic field $\bar{B}$ are functions of both position $\bar{r}$ and time t. It will be assumed in this article that the time dependence is sinusoidal, $e^{i\omega t}$, where $i = (-1)^{1/2}$. $\omega = 2\pi f$ where f is the frequency of the field. There is no loss of generality with this assumption. Any more complicated waveform such as a pulse can always be expressed using Fourier analysis as a linear combination of such sinusoidal functions. Thus $\partial \bar{E}(\bar{r},t)/\partial t = i\omega \bar{E}(\bar{r},t)$. The symbol ∇(del) represents a vector differential operator. In rectangular coordinates $\nabla = \bar{1}_x \partial/\partial x + \bar{1}_y \partial/\partial y + \bar{1}_z \partial/\partial z$ where $\bar{1}_x$, $\bar{1}_y$ and $\bar{1}_z$ are unit vectors in the x, y and z directions. Del has more complicated forms in other coordinate systems. $\nabla \cdot \bar{E}$ is called the divergence of $\bar{E}$; $\nabla x \bar{E}$, the curl of $\bar{E}$. Many authors express some of the electromagnetic equations in terms of $\bar{D} = K \varepsilon_0 \bar{E}$ and $\bar{H} = \bar{B}/\mu_0$. In this chapter only E and B will be used for consistency.

Equation 1a indicates that an electric field can be produced by a net charge density $\rho(\bar{r},t)$. Such a field diverges radially away from positive charges and converges toward negative charges. $\varepsilon_0 = 8.85 \times 10^{-12}\ C^2/Nm^2$ is the permittivity of free space. The dielectric constant K is a physical property of the medium in which the field exists. K serves as a relative permittivity in that one can associate with a medium a permittivity $\varepsilon = K\varepsilon_0$. According to Equation 2a there are no free magnetic charges in nature. One should never encounter a region of space from which there is a net divergence of magnetic field lines.

Equation 3a indicates that an electric field can be produced by a changing magnetic field. The electric field lines in this case form closed loops about the magnetic field lines. According to Equation 4a a magnetic field can be produced by an electric current or a changing electric field. $\bar{J}_c$, the conduction current density, represents the flow of free charge. $\mu_o = 1.25 \times 10^{-6}$ N/A^2 is the permeability of free space. For biological materials the relative permeability is essentially equal to 1, and is omitted from discussion in this article.

For computational purposes Maxwell's equations are often written in integral form as:

$$\oint_S \bar{E} \cdot \bar{n} \; d^2r = \int_V \rho d^3r / K \varepsilon_o \tag{1b}$$

$$\oint_S \bar{B} \cdot \bar{n} \; d^2r = 0 \tag{2b}$$

$$\oint_C \bar{E} \cdot \bar{dl} = - \partial / \partial t \int_s \bar{B} \cdot \bar{n} \; d^2r \tag{3b}$$

$$\oint_c \bar{B} \cdot \bar{dl} = \mu_o \int_s \bar{J} \cdot \bar{n} \; d^2r + \mu_o K \varepsilon_o (\partial / \partial t) \int_s \bar{E} \cdot \bar{n} \; d^2r. \tag{4b}$$

d^2r and d^3r are area and volume elements. In Equations 1b and 2b, S is the closed surface enclosing a volume v, and $\bar{n}$ is the outward directed unit normal to S. In Equations 3b and 4b, C is the closed curve enclosing the area S, and $\bar{dl}$ is a directed line element along C.

In principle, any problem in electromagnetics can be reduced to an application of Equations 1-4. Usually, however, it is more convenient to apply directly other equations that can be derived from them. Of particular importance is the low-frequency case in which the magnetic field is static or changes so slowly that the right side of Equation 3 is zero (quasi-static approximation). If $\nabla \times \bar{E} = 0$, then there exists a scalar function $V(\bar{r})$, the electric potential, given by:

$$\bar{E} = -\nabla V(\bar{r}). \tag{5}$$

∇V is called the gradient of V. Use of Equation 1a then yields Poisson's equation:

$$\nabla^2 V = -\rho/K\varepsilon_0. \qquad (6)$$

∇^2 is called the Laplacian of V. If there is no net charge density, then Laplace's equation holds:

$$\nabla^2 V = 0. \qquad (7)$$

Because $\nabla \cdot \bar{B} = 0$ for all situations, the magnetic field $\bar{B}$ can always be expressed in terms of a magnetic vector potential $\bar{A}(\bar{r})$:

$$\bar{B} = \nabla \times \bar{A}. \qquad (8)$$

Equations 2a and 3a can be combined to obtain a useful general expression for the electric field produced by both a charge distribution and a changing magnetic field:

$$\bar{E} = -\nabla V - \partial\bar{A}/\partial t = -\nabla V - i\omega\bar{A}. \qquad (9)$$

The conduction current term $\bar{J}_c$ on the right side of Equation 4 represents the motion of free charge. For linear (Ohmic) materials this motion is directly proportional to the local electric field.

$$\bar{J}_c = g\bar{E} \qquad (10)$$

where g is the conductivity of the medium. Conservation of charge relates the change in the free charge density inside a small volume to the conduction current moving into or out of that volume by:

$$\nabla \cdot \bar{J}_c = -\partial\rho/\partial t = -i\omega\rho. \qquad (11)$$

The $K\varepsilon_0\partial\bar{E}/\partial t$ term in Equation 4 acts analogously to $\bar{J}_c$ and is called the displacement current density $\bar{J}_d$. With the assumed $i\omega t$ time dependence the right side of Equation 4 can be expressed as a total current density $\bar{J}_t$ which is the sum of the conduction and displacement current densities:

$$\bar{J}_t = (g + i\omega K \varepsilon_o)\bar{E}. \tag{12}$$

When an electric field is applied to a material, in addition to the migration of free charge there can also be a net shift in the relative positions of the positive and negative bound charges. The resulting net polarization is given by:

$$\bar{P} = (K - 1)\varepsilon_o\bar{E}. \tag{13}$$

If the applied field varies with time, so will the polarization. Associated with the oscillation of charge is a polarization current density:

$$\bar{J}_p = \partial\bar{P}/\partial t = i\omega(K - 1)\varepsilon_o\bar{E} = i\omega\bar{P}. \tag{14}$$

Note that the displacement current depends on K whereas the polarization current depends on K - 1. For biological materials up to the radio frequency range $K \gg 1$ so that this difference is often unimportant.

Equations 1-4 can be used to obtain the boundary conditions relating the electric and magnetic fields at an interface between two media, 1 and 2. A subscript n refers to the field component normal or perpendicular to the surface; t, refers to the component that is tangential or parallel to the interface. One finds that:

$$K_1E_{1n} - K_2E_{2n} = \sigma/\varepsilon_o \tag{15}$$

$$B_{1n} = B_{2n} \tag{16}$$

$$E_{1t} = E_{2t} \tag{17}$$

$$B_{1t} - B_{2t} = \mu_o J_s \tag{18}$$

where σ is the induced surface charge density. J_s, the induced surface current density, is usually zero. For situations of biological interest the magnetic field will be continuous across an interface as will also the tangential component of $\bar{E}$. The difference in the normal component of $\bar{E}$ is related to the surface charge density induced at the interface and the dielectric constants of the two media. Equations 11 and 15 can be combined to yield a very useful additional boundary condition -- the continuity of the normal component of the total current density:

$$(g_1 + i\omega K_1 \varepsilon_o)E_{1n} = (g_2 + i\omega K_2 \varepsilon_o)E_{2n}. \tag{19}$$

It must be emphasized that a complete solution of a problem in electromagnetics requires that not only Maxwell's Equations 1-4, or their equivalents, but also the boundary conditions 15-19 be satisfied.

According to Equation 19 the term $\omega K \varepsilon_o$ acts as a conductivity. The electrical properties of a material, g and K, can be incorporated into a single complex-valued conductivity:

$$g^* = g + i\omega K \varepsilon_o. \tag{20}$$

In a similar fashion the permittivity can be generalized to a complex-value form:

$$\varepsilon^* = K \varepsilon_o - ig/\omega \tag{21}$$

In terms of the dielectric constant:

$$K^* = K - ig/\omega\varepsilon_o \tag{22}$$

K and g are functions of frequency for biological materials. The physical principles which determine these dielectric properties of biomaterials are discussed at length by Pethig (5).

During the passage of free charge associated with J_c, energy is dissipated in the form of heat. The time-averaged rate at which energy is dissipated per unit volume is given by the power density:

$$p = (J_c + J_p)E/2 = (g + i\omega(K - 1)\varepsilon_o)E^2/2 \tag{23}$$

For biological objects the conduction term generally dominates the polarization term and:

$$p \cong gE^2/2 \tag{24}$$

The Specific Absorption Rate (SAR) is the total power absorbed per unit mass (M) of the object. If d' is the density of the object:

$$\text{SAR} = p/d' = p(\text{volume}/M) \tag{25}$$

Whole-body SAR is the total energy absorbed by the entire object per unit mass per unit time. The local variation of the electric field within the object gives rise to a local SAR according to Equations 23 and 25.

PLANE WAVES

Far away from a time-varying charge or current source the electric and magnetic fields engage in a form of mutual self-propagation in which each field produces the other. In a medium with no net charges in the bulk, Maxwell's equations can be combined to yield an expression to be solved for $\bar{E}$:

$$\nabla^2\bar{E} + (\omega^2 K \varepsilon_o \mu_o - i\omega g \mu_o)\bar{E} = 0. \tag{26}$$

The propagation vector $\bar{k}^*$ has a magnitude defined by:

$$k^{*2} = \omega^2 K \varepsilon_o \mu_o - i\omega g \mu_o. \tag{27}$$

The complex-valued k* can be expressed as $k^* = k - ik_i$. The real part $k = 2\pi/\lambda$, where λ is the wavelength. The imaginary part k_i is the inverse of the penetration depth δ : $k_i = 1/\delta$. Far from the source the solution of equation 26 has the form of a plane wave with an exponentially decreasing amplitude. The direction of wave propagation is that of $\bar{k}$, and $\bar{E}$ is perpendicular to $\bar{k}$. An equation similar to 26 can be derived for $\bar{B}$. However, it is simpler to use Equation 3a to find $\bar{B}$ in terms of the solution to Equation 26:

$$\bar{B} = \bar{k} \times \bar{E}/\omega. \tag{28}$$

For this plane wave $\bar{B}$ is perpendicular to both $\bar{k}$ and $\bar{E}$. The electric and magnetic fields are completely coupled. Given $\bar{E}$, $\bar{B}$ is determined. For a plane wave propagating in the z direction:

$$\bar{E} = E_o\, e^{i(\omega t - kz)}\, e^{-z/\delta}\, \bar{1}_x \tag{29}$$

and

$$\bar{B} = B_o\, e^{i(\omega t - kz)}\, e^{-z/\delta}\, \bar{1}_y. \tag{30}$$

E_o and B_o are the amplitudes of the coupled waves. They are related by Equation 28:

$$E_o/B_o = \omega/k = v_p = (k\,\varepsilon_o\,\mu_o)^{-1/2} = \eta/\mu_o. \tag{31}$$

v_p is the phase velocity of the wave. $\eta = (\mu_o/K\,\varepsilon_o)^{1/2}$ has units of electrical resistance and is referred to as the impedance of the medium. The impedance of free space (K = 1) is 377 ohms.

The wave amplitude decreases to about 40% of initial value after traversing the penetration depth (skin depth) which is:

$$\delta = (2/\omega g\,\mu_o)^{1/2}. \tag{32}$$

The penetration depth decreases with increasing frequency and conductivity. Values of δ at various frequencies are presented by Johnson and Guy (6) for tissues with high and low water content. Beyond about 10 GHz (1 GHz = 10^9 Hz) plane waves do not penetrate very deeply into tissue.

The time-averaged energy flux (energy transmitted per unit area per unit time or Poynting vector) is given for a plane wave by:

$$\bar{S} = (K\,\varepsilon_o/\mu_o)^{1/2} E_o^2\bar{1}_k/2 = E_o^2\bar{1}_k/2\eta\,. \tag{33}$$

$\bar{1}_k$ is a unit vector in the direction of $\bar{k}$. Energy is transmitted in the direction of propagation, as expected.

CALCULATION OF THE INTERNAL FIELD

Polarization fields and currents are produced within an object exposed to an external electromagnetic field. These fields and currents then serve as sources of secondary electromagnetic fields. If $\bar{E}_o$ is the unperturbed field (the field which would exist at the site of the object if it were not present) and $\bar{E}^s$ is the secondary (at high frequencies, scattered) field produced by this polarization, then the internal field within the object is:

$$\bar{E}_i(\bar{r},t) = \bar{E}_o(\bar{r},t) + \bar{E}^s(\bar{r},t). \tag{34}$$

$\bar{E}^s$ may be expressed in terms of the induced potentials V^s and $\bar{A}^s$ using Equation 9. These potentials can, in turn, be related to the induced charge and current densities $\rho_i(\bar{r}')$ and $\bar{J}_i(\bar{r}')$ by:

$$V^s(\bar{r}) = 1/4\,\pi\,\varepsilon_o \int_v (\rho_i(\bar{r}')/\,|\bar{r}-\bar{r}'|\,)e^{ik\,|\bar{r}-\bar{r}'|}\,d^3r' \tag{35}$$

and $$\bar{A}^s(\bar{r}) = \mu_o/4\pi \int_v (\bar{J}_i(\bar{r}')/\,|\bar{r}-\bar{r}'|)\,e^{-ik\,|\bar{r}-\bar{r}'|}\,d^3r'. \tag{36}$$

$|\bar{r}-\bar{r}'|$ is the distance from a volume element of the source located at $\bar{r}'$ to the place where the potential is to be found, $\bar{r}$. The integration is carried out over v, the volume of the source. Some references use a $+ik\,|\bar{r}-\bar{r}'|$ factor, but then a $-i\omega t$ time dependence. Substitution of these expressions into Equation 34 yields:

$$\bar{E}_i(\bar{r}) + \nabla V^s(\bar{r}) + i\omega\bar{A}^s(\bar{r}) = \bar{E}_o(\bar{r}). \tag{37}$$

As will be noted V^s and $\bar{A}^s$ can be expressed in terms of $\bar{E}_i$. Hence Equation 37 is really an integral equation to be solved for $\bar{E}_i(\bar{r})$ with $\bar{E}_o(\bar{r})$ given. Many alternate forms are found in the literature.

For example, in some cases the total (conduction plus polarization) current density induced within the object is of primary concern:

$$\begin{aligned} \bar{J}_i = \bar{J}_c + \bar{J}_p &= (g + i\omega(K-1)\,\varepsilon_o)\bar{E}_i \\ &= i\omega(K^* - 1)\,\varepsilon_o\bar{E}_i \\ &= t^*\bar{E}_i. \end{aligned} \tag{38}$$

Substitution of Equation 11 into 35 yields:

$$V^s(\bar{r}) = i/4\pi\varepsilon_0\omega \int_V (\nabla'\cdot\bar{J}_i(\bar{r}')/(|\bar{r}-\bar{r}'|))e^{-ik|\bar{r}-\bar{r}'|}\, d^3r' \qquad (39)$$

Equations 39, 14, and 37 can be combined to give:

$$L^{op}(\bar{J}_i) + \bar{J}_i/i\omega(K^* - 1)\,\varepsilon_o = \bar{E}_o. \qquad (40)$$

where L^{op} is an integral operator obtained from Equations 36 and 39.

Usually, however, the integral equation is expressed in terms of the internal electric field. Equations 39 and 36 become:

$$V^s(\bar{r}) = (i/4\pi\varepsilon_0\omega)\int_V [\nabla'\cdot(t^*(\bar{r}')\bar{E}_i(\bar{r}'))/(|\bar{r}-\bar{r}'|)]e^{-ik|\bar{r}-\bar{r}'|}\, d^3r' \qquad (41)$$

and $$\bar{A}^s(\bar{r}) = \mu_o/4\pi\int_V (t^*(\bar{r}')\bar{E}_i(\bar{r}')/(|\bar{r}-\bar{r}'|))e^{-ik|\bar{r}-\bar{r}'|}\, d^3r'. \qquad (42)$$

The expression $ie^{-ik|\bar{r}-\bar{r}'|}/k|\bar{r}-\bar{r}'|$ happens to be [1] $H_o^{(2)}(k|\bar{r}-\bar{r}'|)$, the zeroth order spherical Hankel function of the second kind. It represents an outgoing spherical wave. Some authors write these equations in terms of Hankel functions. It should be noted that cylindrical Hankel functions are used for problems with cylindrical symmetry. For this reason, and also because some authors use different electromagnetic units, these equations assume a variety of forms in the literature.

Replacement of t^* by $i\omega(K^*-1)\,\varepsilon_o$ in Equations 41 and 42 and then substitution into 37 yields:

$$\bar{E}_o(\bar{r}) = \bar{E}_i(\bar{r}) - 1/4\pi \int_V (\nabla'\cdot(K^*-1)\bar{E}_i(\bar{r}'))\,\nabla((e^{-ik|\bar{r}-\bar{r}'|})/$$

$$(|\bar{r}-\bar{r}'|))d^3r' - k^2/4\pi \int_V (K^*-1)\bar{E}_i(\bar{r}')(e^{-ik|r-r'|}/|r-r'|)\, d^3r' \qquad (43)$$

K^* may be a function of $\bar{r}'$. The first integral is the ∇V^s term; the second, $i\omega\bar{A}^s$. k^2 replaces $w^2\mu_o\varepsilon_o$. This equation or a related version has been referred to (7) as the Free-Space Green's Function Integral Equation (FGIE).

A Green's function in electromagnetic theory can be regarded as giving the response of a system to a unit pulse source. For example, in electrostatics $G(\bar{r},\bar{r}')$ is a scalar function which represents the solution of Poisson's equation for a delta function charge source. Equation 6 becomes $\nabla'^2 G(r,r') = -\delta(\bar{r}-\bar{r}')/\varepsilon_o$ in free space. The potential $V(\bar{r})$ for a distributed source $\rho(\bar{r}')$ is the superposition of the solutions for an infinite number of pulse sources:

$$V(\bar{r}) = \int_V \rho(\bar{r}')G(\bar{r},\bar{r}')d^3r'/\varepsilon_o. \tag{44}$$

Comparison with Equation 35 indicates that for this situation $G(\bar{r},\bar{r}') = e^{-ik|\bar{r}-\bar{r}'|}/4\pi|\bar{r}-\bar{r}'|$ is a free space Green's function. Note that $G(\bar{r},\bar{r}')$ is symmetric, diverges as $\bar{r} \to \bar{r}'$, and can be expressed in terms of Hankel functions. Special problems then arise when $\bar{r}$ is within the object volume v.

More generally, the scattered electric field produced within an object by the induced currents can be considered to be the superposition of the fields generated by current pulses. One finds (8,9) that:

$$\bar{E}^s(\bar{r}) = PV \int_V \bar{J}_i(\bar{r}') \cdot \overleftrightarrow{G}(\bar{r},\bar{r}')d^3r' - \bar{J}_i(\bar{r}')/3i\omega\varepsilon_o \tag{45}$$

where $\overleftrightarrow{G}(\bar{r},\bar{r}') = -i\omega\varepsilon_o(\overleftrightarrow{I} + \nabla\nabla/k_o^2)\,\psi(\bar{r},\bar{r}')$

and $\psi(\bar{r},\bar{r}') = e^{-ik|\bar{r}-\bar{r}'|}/4\pi|\bar{r}-\bar{r}'| = G(\bar{r},\bar{r}')$ free space. $\overleftrightarrow{G}(\bar{r},\bar{r}')$ must be a tensor (dyadic) operator because it transforms one vector field $\bar{J}_i(\bar{r}')$ into a nonparallel field $\bar{E}^s(\bar{r})$. The PV (principal value) symbol indicates that special care is required in evaluating this integral because of the divergence. The unit tensor $\overleftrightarrow{I}$ factor originates with the $-i\omega\bar{A}$ term in Equation 9 and the $-\nabla\nabla$ factor with the $-\nabla V$ term. Use of Equations 34 and 38 yields the following integral equation for $\bar{E}_i(\bar{r})$:

$$\bar{E}_o(\bar{r}) = (1 + t^*(\bar{r})/3i\omega\varepsilon_o)\bar{E}_i(\bar{r}) - PV\int_V t^*(\bar{r}')\bar{E}_i(\bar{r}')\cdot\overleftrightarrow{G}(\bar{r},\bar{r}')d^3r'. \tag{46}$$

This important equation was introduced to the bioelectromagnetics literature by Livesay and Chen (9). Durney (7) refers to it as the Dyadic Green's Function Integral Equation (DGIE).

Equations 43 and 45 are volume integral equations (VIE) for the field. Alternatively, the field within a volume v can be expressed in terms of the field or currents induced on the surface S surrounding v. If the electric field has only one component (z), then a scalar solution for it is given by the Kirchhoff-Huygens principle (3,4) as:

$$E_z(\bar{r}) = \int_S (G(\bar{r},\bar{r}')\, \partial E_z(\bar{r}')/\partial n - E_z(\bar{r}')\, \partial G(\bar{r},\bar{r}')/\partial n)d^2r' \quad (47)$$

where $n(\bar{r}')$ is the outward directed normal to S at $\bar{r}'$ and G is the Green's function, not for free space, but for the homogeneous interior v. If the field has more than one component, a vector generalization must be used:

$$\bar{E}(\bar{r}) = -\int_S (i\omega(\bar{n} \times \bar{B}(\bar{r}'))G(\bar{r},\bar{r}') + (\bar{n} \times \bar{E}(\bar{r}'))\times \nabla G(\bar{r},\bar{r}')$$

$$+ (\bar{n}\cdot\bar{E}(\bar{r}')\, \nabla G(\bar{r},\bar{r}'))d^2r'. \quad (48)$$

The $\bar{n} \times \bar{B}$ and $\bar{n} \times \bar{E}$ factors act as effective surface electric and magnetic current density terms; $\bar{n} \cdot \bar{E}$ as a surface charge density term. Equations 47 and 48 yield surface integral equations (SIE) for the electric field. Similar expressions can be obtained for magnetic fields.

BASIC APPROACH OF THE CHAPTER

The mathematical models used to describe the interaction of electromagnetic fields with biological objects can be categorized according to the distance (d) of the object from the source of the fields, and according to the ratio of the wavelength of the field to the dimensions of the object (a). For $d \ll \lambda$, the object is in the "near zone" of the source. The electric and magnetic fields are decoupled. The interaction of the object with them must be treated separately according to the quasistatic approximation. For $d \gg \lambda$, the object is in the "far zone" or "radiation zone." The electric and magnetic fields are coupled. Plane-wave wave analysis may be used. The situation can be quite complicated in the "induction zone" ($d \sim \lambda$) where many of the simple relationships previously described no longer hold. Some authors use the term "near-field" to describe situations which are really induction zone. In many situations the analysis is simplified if $\lambda \gg a$. This "long wavelength" approximation can reduce a plane-wave or induction-zone problem to an application of the quasistatic approximation.

This chapter will discuss mathematical models used from static fields up to frequencies on the order of 50 GHz. At that point the applied fields no longer penetrate the object appreciably. In each section analytical models such as spheres will be discussed in the detail needed to convey the basic principles involved. References will be given for more complicated geometries. The details of the numerical methods will also be found in the references.

STATIC AND LOW FREQUENCY FIELDS

APPLIED ELECTRIC FIELDS

1. Analytical Models - Biological Object in Free Space

Suppose that a biological object is placed in a spatially uniform electric field $\bar{E}_o$. Suppose also that the surrounding medium is free space for which $g_e = 0$ and $K_e = 1$. The subscript e refers to external medium. A thorough discussion of this situation is presented by Kaune and Gillis (10).

For a static field ($\omega = 0$) and $g_e = 0$, no current can flow from external sources to the object. The internal charges will continue to move to the surface until $\bar{E}^s(\bar{r}) = -\bar{E}_o(\bar{r})$; that is, in Equation 34, $\bar{E}_i(\bar{r}) = 0$. Once this equilibrium is established there are no field, current or power dissipated within the object. It should be noted here that the external field at the surface of the object, $\bar{E}_s(\bar{r})$, differs from the unperturbed value $\bar{E}_o(\bar{r})$. Equation 17 implies that $\bar{E}_s$ is normal to the surface. From Equation 15 its magnitude is:

$$\bar{E}_s(\bar{r}) = \sigma(\bar{r})/\varepsilon_o. \tag{49}$$

For a time-varying field $\partial\bar{E}_s/\partial t \neq 0$. A free-space displacement current connects the object to the external source. According to Equation 19 the normal component of the total current density must be continuous across the interface. A current density and thus an electric field must then exist inside the object. Let g_i and K_i be the conductivity and dielectric constant of the object, which is assumed to be homogeneous. The subscript i refers again to internal medium. According to Equation 19:

$$i\omega\varepsilon_0\bar{E}_s(\bar{r})\cdot\bar{n} = (g_i\bar{E}_i + i\omega K_i\,\varepsilon_0\bar{E}_i)\cdot\bar{n}. \tag{50}$$

It will be noted later that $\bar{E}_i(\bar{r})$ is uniform. Hence the spatial variation of $\bar{E}_i$ will be omitted in the notation as in Equation 50. For biological materials at low frequencies $g_i \gg \omega K_i\,\varepsilon_0$ and:

$$\bar{E}_i\cdot\bar{n} = i\omega\varepsilon_0\bar{E}_s(\bar{r})\cdot\bar{n}/g_i. \tag{51}$$

For biological materials at low frequencies it must then also be the case that $g_i \gg \omega\varepsilon_0$. $\bar{E}_i$ is very small compared to $\bar{E}_s$. In fact, to a good first approximation $\bar{E}_i \sim 0$. Using the same arguments as for static fields, one may conclude that $\bar{E}_s$ is essentially normal to the surface and given by Equation 49. The current density induced inside a biological object placed in a low-frequency electric field is thus given to a good approximation by:

$$\bar{J}_c\cdot\bar{n} = i\omega\varepsilon_0 E_s(\bar{r}) = i\omega\sigma(\bar{r}). \tag{52}$$

Note that the current density is independent of the object's conductivity. The i factor indicates that the current is 90° out of phase with the external field and the induced surface charge.

In order to determine $\bar{E}_i$ and $\bar{J}_i$ it is first necessary to find $\bar{E}_s(\bar{r})$. At low frequencies the quasistatic approximation of Equations 5-7 may be used. The general procedure is to write down the solution to Laplace's equation for the external medium in a coordinate system corresponding to the shape of the biological object within. The solution will be an infinite series, the coefficients of which are determined by the boundary conditions.

This procedure will be illustrated for the simplest case -- spherical coordinates. The biological object is modelled as a sphere of radius a and conductivity g_i. The solution of Equation 7 outside the sphere is (1,2):

$$V_e(r, \theta) = \sum_m (A_m r^m + B_m r^{-(m+1)}) P_m(\cos \theta) \quad (53)$$

where θ is a colatitude, measured from the pole (z axis) to the equator, and the P_m are the Legendre polynomials of order m. If the unperturbed field is applied in the z direction, then far from the object the external field:

$$\bar{E}_e(\bar{r}) \rightarrow E_o \bar{1}_z \quad (54)$$

and

$$V_e(\bar{r}) \rightarrow -E_o z. \quad (55)$$

Because the object is a good conductor, its surface is an equipotential; that is, $V(a, \theta) = V_a$ for all θ. With these boundary conditions all $A_m = 0$ except $A_o = V_a$ and $A_1 = -E_o$, and all $B_m = 0$ except $B_1 = E_o a^3$. Thus with $P_1(\cos \theta) = \cos \theta$:

$$V_e(r, \theta) = V_a - E_o(1 - a^3/r^3) r \cos \theta \quad (56)$$

The external electric field is now found by applying Equation 5. The surface field $E_s = -\partial V/\partial r$ evaluated at $r = a$.

$$\bar{E}_s = 3E_o \cos \theta \; \bar{1}_r. \quad (57)$$

$\bar{1}_r$ is a unit vector in the radial direction. Hence $\bar{E}_s$ is normal to the surface as expected and decreases from a value of $3E_o$ at the poles to zero at the equator. The electric field at the surface of the biological object can be significantly greater than the unperturbed value. It will be noted later that inside the sphere $\bar{J}$ is uniform and parallel to the unperturbed field. One finds that (11):

$$\bar{J} = 3i\omega\varepsilon_o \bar{E}_o. \quad (58)$$

The internal field is then:

$$\bar{E}_i = 3i\omega\varepsilon_o E_o/g_i \bar{1}_z \tag{59}$$

The power dissipated is given by Equation 24. Both the internal field and current increase linearly with frequency. The field, but not the current, depends on the object's conductivity.

In some cases the object is not isolated in space, but is in direct contact with a grounded surface. Such situations can be treated by the method of images (1,2). The object is modelled as a hemisphere of radius a in contact with a flat plane for which V = 0. The image of the object beneath the plane forms the other half of the sphere. The problem is now reduced to that of an isolated sphere with surface potential $V_a = 0$. In more complicated situations the object may be insulated from ground or connected to ground through an impedance. The same approach is used, but V_a must be related to the external potential V_o and the system capacitances and impedances (10).

Many biological objects are modelled more realistically as isolated spheroids or as hemispheroids on a grounded plane. The earliest such model was developed in the classic work of Barnes et al. (12), who treated mice and men as conducting prolate spheroids. A prolate spheroid is formed by the rotation of an ellipse about its major axis. Laplace's equation is solved in spheroidal coordinates in the external region. Equation 55 again serves as a boundary condition along with $V = V_a$ on the surface of the spheroidal object. $\bar{E}_s$ is again normal to the surface and $E_s(\bar{r}) = \sigma(\bar{r})/\varepsilon_o$. The closed form expression for V_e is not as simple as Equation 53 because the geometry is more complicated. Laplace's equation can also be solved for the current density inside the spheroid. Both $\bar{J}$ and $\bar{E}_i = \bar{J}/g_i$ are uniform and parallel to the unperturbed field.

The enhancement of the surface field relative to the unperturbed field depends on a/b, the ratio of the spheroid's semimajor to semiminor axes. For $\bar{E}_o$ parallel to the major axis, $\bar{E}_s$ is greatest at the poles and decreases to zero at the equator. For $\bar{E}_o$ parallel to the minor axis, $\bar{E}_s$ is greatest at the equator and decreases to zero at the poles. Enhancement is greater in the former case because a > b. A more complete summary of the dependence of E_s/E_o on a/b and position on the spheroid is given by Sheppard and Eisenbud (13), who note that for humans E_s/E_o can be as high as 15.

A more general analytical shape for a biological object is an ellipsoid or a hemiellipsoid on a grounded plane. An ellipsoid has three perpendicular axes: $a > b > c$. Its cross-section normal to any one of these is an ellipse. Laplace's equation can be solved in ellipsoidal coordinates and the usual boundary conditions applied (14). The enhancement of the surface field relative to the unperturbed field is determined by a shape factor, which depends on a, b, and c and on an integral which must be evaluated numerically. These results are applied to a grounded hemiellipsoid by Kaune (15) and by Lattarulo and Mastronardi (16).

2. Analytical Models - Biological Object in a Lossy Medium

In some situations it is important to calculate the electric field produced inside a biological object which is embedded in a medium with a significant conductivity (lossy), such as a nutrient solution or soil. In that case it can no longer be assumed that $g_e = 0$, $K_e = 1$ or $E_i/E_o \sim 0$. $\bar{E}_s$ will no longer be perpendicular to the surface.

The boundary-value approach described in the previous section can, however, be extended to this case. Laplace's equation is solved separately for the potentials inside and outside the object: $V_i(\bar{r})$ and $V_e(\bar{r})$. The boundary conditions are Equations 19 and 55 and the continuity of the potential at the interface. For a sphere that would be:

$$V_e(a) = V_i(a). \tag{60}$$

In spherical coordinates the general solutions for $V_e(\bar{r})$ and $V_i(\bar{r})$ will both have the same form as Equation 53. Inside the sphere all B_m terms must vanish to prevent a divergence at the center. The internal electric field is found to be (11):

$$\bar{E}_i = [3(g_e + i\omega K_e \varepsilon_o)/((g_i + 2g_e) + i\omega\varepsilon_o(K_i + 2K_e))]\bar{E}_o. \tag{61}$$

As noted previously the internal electric field and thus the current density are uniform and parallel to the unperturbed field. Whether or not $E_i > E_o$ depends in general on the particular conductivities, dielectric constants and frequency. With free space as the external medium and $g_i \gg \omega K_i \varepsilon_o$ this expression reduces to Equation 59.

The more realistic prolate and oblate spheroid models can be treated in the same manner. Shiau and Valentino (17) present results for both shapes with the applied field either parallel to or perpendicular to the major axis. Fields at an arbitrary angle can be decomposed into perpendicular and parallel components to be analyzed separately. As for a sphere, both the internal field and current density are uniform and parallel to the unperturbed field. The actual expressions are not as simple as Equation 61 because of the more complicated geometry. The ratio E_i/E_o depends on g_i, g_e, K_i, K_e and ω in addition to the ratio of the axes a/b. As the ratio a/b increases, E_i/E_o increases for a field applied parallel to the major axis of a prolate spheroid, but decreases if the field is perpendicular to that axis. This difference in coupling can be understood by recalling from Equation 17 that the tangential component of $\bar{E}$ is continuous across an interface. With $\bar{E}_o$ parallel to the major axis, its tangential component increases with the ratio a/b whereas with $\bar{E}_o$ parallel to the minor axis its tangential component decreases with the ratio a/b.

Similar results are found for oblate spheroids. An oblate spheroid is formed by the rotation of an ellipse about its minor axis. The special case of an isolated, conducting oblate spheroid in air can be treated by using the general results (17) with $g_e = 0$ and $K_e = 1$.

An even more realistic model is provided by an ellipsoid surrounded by a lossy medium as discussed by Hart and Marino (18). The basic pattern of results is similar to that found for spheroids. The actual expressions for $\bar{E}_i$ and $\bar{J}$ are complicated by the presence of the shape factor, which must be evaluated numerically.

3. Analytical Models - Layered Objects

The electric-field interaction models considered so far have all assumed that the object is homogeneous. In some cases, such as a membrane surrounding a cell interior or bone surrounding brain material in the head, a more complicated, layered model is appropriate. Hart and Marino (19) have considered the object to be a sphere of radius a, conductivity g_3 and dielectric constant K_3 surrounded by a shell of thickness (b - a), conductivity g_2 and dielectric constant K_2. The object is embedded in a medium with conductivity g_1 and dielectric constant K_1. Laplace's equation is solved for each of the three regions. The boundary conditions are Equations 19 and 55 and the continuity of the potential at both interfaces. The interior electric field is found to be uniform and parallel to the unperturbed field. The shell field is more complicated. Penetration of the external field into the interior depends on the thickness and electrical properties of the shell.

Bernhardt and Pauly (20) developed a model to calculate the potential difference produced by an applied field across the membrane of an ellipsoidal cell. The external and internal media were assumed to be purely conductive whereas the membrane itself was purely capacitive and of negligible thickness. The potential difference depended strongly on ω with a relaxation frequency determined by the shapes of the ellipsoidal surfaces and the conductivities. No attempt was made to find the internal field or current density.

Marchesi and Parodi (21) devised a spherical shell model for a static applied field with the interesting variation that the membrane, though negligibly thick, behaves in a nonlinear fashion as a diode. Laplace's equation is solved for both the internal and external media. The normal component of the current density (conduction only, in this case) must be continuous across the membrane. In addition, this component is related to the potential difference across the membrane ΔV by $J = J_o(e^{\Delta V/C} - 1)$ where C and J_o are constants. The A_m and B_m coefficients in Equation 53 no longer vanish in general as before. Instead the series must be terminated after N terms. The boundary conditions are explicitly satisfied at N different positions on the surface of the sphere to yield N equations for the coefficients. A similar solution is obtained for a cylindrical system. For applied field strengths of practical interest this nonlinear model gives results similar to those of the usual linear approximation for spherical, but not for cylindrical systems. The numerical approach required here because of the nonlinear boundary condition serves as a transition to the next section.

NUMERICAL MODELS

Although the analytical solutions for the electric field provide valuable insights into how the internal field depends on such parameters as conductivity, dielectric constant, frequency and general shape, they cannot provide a detailed knowledge of the currents and fields inside a biological object because real objects possess more complicated, irregular shapes and are internally inhomogeneous. Various numerical models have been developed to make more realistic determinations of the internal fields and currents.

One general approach is a variation of the analytical method described in a previous section. The object is modelled as a good conductor. $\bar{E}_i \sim 0$, the object's surface is an equipotential, and $\bar{E}_s(\bar{r})$ is normal to the surface. The normal component of the current density can be found from Equation 52 and integrated over the surface of the object to give the total current. The difference is that $\bar{E}_s(\bar{r})$ is found numerically. The method can be extended by numerically solving Laplace's equation inside the object and using the appropriate boundary conditions.

Kobayashi et al. (22) photographed a mouse and digitized the resulting image to provide an accurate representation of its boundary. Because of the irregular geometry Laplace's equation cannot be solved in closed form. Instead a finite-difference method is used. The space surrounding the mouse is divided into a two-dimensional grid. The potential at each nodal point on the grid is expressed in terms of the potentials at the four neighboring points. The potential at the top of the grid is the applied potential. The digitized image of the grounded mouse occupies an irregular part of the bottom region of the grid. The potential of that part must be zero. At the left and right boundaries the horizontal component of the field is set equal to zero.

The successive over relaxation method is now applied. Potentials are assumed for those nodal points not specified by the boundary conditions. The potential at each node is calculated from the assumed values of the neighboring nodes. A new set of potentials for the interior nodes is thus generated. The process is now repeated to calculate each nodal potential from the new set of neighboring potentials and the previous value at the node. After a certain number of such iterations the difference in the successively calculated nodal potentials is less than a pre-assigned value and the process is terminated. The nodal potentials adjacent to the grounded object can be divided by the grid size to give the horizontal and vertical components of the field at the object's surface $\bar{E}_s(\bar{r})$ as in Equation 5. Although the authors do not extend this technique to obtain the internal current density and field, it can be done using the methods described earlier.

Such a two-stage approach is used by Kaune and McCreary (23) to calculate the electric field and current density inside a conducting cylinder resting on a grounded plane. The unperturbed field was applied parallel to the cylinder's axis. In the first stage the internal field was assumed to be negligible in comparison with the applied field. The cylinder's surface is thus an equipotential $V_s = 0$. The potential at any point in space is the unperturbed potential there ($-E_oz$) plus the contribution from the charges induced on all the surfaces. In particular, on the surface of the object itself:

$$V_s = -E_oz + (1/4\pi\varepsilon_o)\int_{S+S_1} (\sigma(\bar{r}')/|\bar{r}-\bar{r}'|)d^2r'. \qquad (62)$$

S is the surface of the object. S_i is the surface of the image cylinder used to account for the grounded plane.

The top and side surfaces of the cylinder are divided into small patches. In cylindrical coordinates:

$$V(r,\theta,z) = -E_oz + \sum_{i=1}^{N} \sigma_t(r_i)F_t(r_i) + \sum_{j=1}^{M} \sigma_s(z_j)F_s(z_j).$$

$\sigma_t(r_i)$ is the surface charge density on a top patch at a distance r_i from the axis; $\sigma_s(z_j)$ is the surface charge density on a side patch at height z_j above the plane. $F_t(r_i)$ and $F_s(z_j)$ are complicated functions of the patch geometry. This equation must hold in particular at each of the surface patches, where $V = 0$. Because of the symmetry of the problem many patches are equivalent. The resulting set of N+M equations is solved for the surface charges on the patches. The field at the surface of each conducting patch is found using $E_s = \sigma/\varepsilon_o$.

In the second stage the cylinder is no longer assumed to be a perfect conductor. Laplace's equation holds in the interior. The boundary conditions are V = 0 at the bottom ground plane, and current-density continuity at each of the top and side patches. The complicated boundary conditions require a numerical solution. A finite difference approach is used. The interior is divided into a cylindrical grid and the successive over relaxation method is applied. The internal electric field and from it the current density can be found from the difference in potential between the nodes of the grid according to Equation 5.

The similar method of spheres has been used by Di Placidio et al. (24) to obtain the surface field and induced current. The conducting object is replaced by a set of spheres such that the total surface area of the sphere equals the area of the object. In an external field surface charges are induced on each sphere. The total potential at any point in space is again the unperturbed potential plus contributions from the surface charge induced on each sphere. If the object is grounded, that potential is zero on each sphere. The resulting set of simultaneous equations is solved for the surface charges. Again $E_s = \sigma/\varepsilon_o$. If, instead, the object is isolated from ground, the potential has the same constant value for each sphere, and the total surface charge is zero.

Chiba et al. (25) use a very different variational approach. A human is modelled as a set of coaxial cylinders of different radii placed top to bottom on a grounded plane. The rate of change of electrical energy in the volume v can be expressed as:

$$p = \int_v g^* E^2 \, d^3r/2 = \int_v g^* (\nabla V)^2 \, d^3r/2.$$

One can imagine that the potential function within the object is allowed to redistiribute itself in many different ways. p will be a minimum for that potential configuration V'(r) for which charge conservation, Equation 11, is satisfied.

$$\partial/\partial V \int_v (g + i\omega K \varepsilon_o)(\nabla V)^2 d^3r = 0 \text{ for } V = V'.$$

This integral equation for V can be solved numerically by dividing the interior of the object into a grid. A matrix equation can be developed for the potentials V_i' at the various nodes i = 1....m. [C][V'] = 0. The matrix [C] depends on g, K and ω in addition to the details of the grid, which is curvilinear. After the nodal potentials are found, the internal field and current density are obtained as in the finite difference methods. The internal current densities are found to be insensitive to the choice of conductivity as expected from the analytical models.

The various integral equation formulations discussed previously can also be applied to low frequency fields. The $i\omega\bar{A}^s$ term is then small in comparison to the ∇V^s term and $e^{-ik|\bar{r}-\bar{r}'|} \sim 1$. Moment methods, which are described later in the section on plane wave numerical analysis, are used to obtain solutions to the resulting integral equations. These formulations are particularly useful for irregularly shaped, inhomogeneous objects.

In one approach Spiegel (26) applied Equation 40 to an object modelled as a three-dimensional array of cylindrical segments. He calculated the currents at various positions within a human standing on a grounded plane near a transmission line. Different choices for the array could be made depending on the person's posture. In a later paper (27) Spiegel modelled humans and baboons as arrays of cubical blocks. A complex polarization is used to generalize Equations 14 and 38 to:

$$\bar{P}_i^* = (K^*-1)\,\varepsilon_o \bar{E}_i \tag{63}$$

and

$$\bar{J}_i = i\omega \bar{P}_i^*. \tag{64}$$

$V^s(\bar{r})$ can be expressed in terms of this polarization as:

$$V^s(\bar{r}) = 1/4\pi\varepsilon_o \int_V \bar{P}_i^*(\bar{r}') \cdot (\bar{r}-\bar{r}')/|\bar{r}-\bar{r}'|^3\, d^3r' \tag{65}$$

and $i\omega\bar{A}^s(\bar{r})$ is negligible. Substitution of Equations 63 and 65 into 37 yields an integral equation to be solved for $\bar{P}_i^*$. The induced current distribution is then obtained from Equation 64. Spiegel found that $\bar{J}_i$ is not strongly dependent on the tissue electrical properties.

Chen et al. (28) use an integral equation approach with Equation 62. In this case the surface is not cylindrical, but arbitrary in shape. The surface potential V_s is not zero, but depends on the impedance to ground Z_g. The total current to ground, given by $I_g = V_s/Z_g$, is also the rate of change of the total surface charge. Hence:

$$V_s = (i\,\omega\,\sigma_{total}/Zg) = i\omega/Z_g \int_S \sigma(\bar{r}')d^2r'. \qquad (66)$$

Equation 66 can be substituted into Equation 62 to obtain an integral equation to be solved for $\sigma(\bar{r})$ by a moment method. The total current through any section j of the body is:

$$I_j = i\omega \int_{S_j} \sigma(\bar{r}')d^2r'$$

where S_j is the area of that section. These sectional currents can be used to find approximate values for the internal current densities and thus the internal fields and by Equation 25 local SAR values.

APPLIED MAGNETIC FIELDS

1. Analytical Models

Biological materials are not appreciably magnetized when subjected to an applied magnetic field. The internal magnetic field $\bar{B}_i(\bar{r})$ is related very simply to the external field $\bar{B}_e(\bar{r})$; they are equal. The magnetic field lines are undisturbed by the presence of the biological material.

A changing magnetic field produces an induced electric field $\bar{E}^i(\bar{r})$ given by Equation 3. The induced field lines form closed loops about $\bar{B}_i(\bar{r})$. If the magnetic field lines are incident upon a sphere, the top of a cylinder or are parallel to the symmetry axis of a spheroid, the exposed cross-sections are circular. The induced electric field lines will be circular loops of radius r. The left side of Equation 3b becomes $E^i(\bar{r})2\pi r$. If the magnetic field is spatially uniform, the right side becomes $-i\omega B\pi r^2$. Thus:

$$E^i(\bar{r}) = -i\omega rB/2. \qquad (67)$$

If the medium has a conductivity g, an induced conduction current given by Equation 10 will flow. Equation 24 yields the dissipated power density.

The induced current will itself produce a magnetic field which is opposite in direction to the applied field. For sufficiently high g and ω, this induced magnetic field could be comparable to the applied magnetic field. At low frequencies, however, it is negligible.

If the applied magnetic field is perpendicular to the symmetry axis of a spheroid or parallel to any of the axes of an ellipsoid, the cross-sections are ellipses. If the magnetic field is spatially uniform and in the y direction, the induced field is found from Equation 3b to be (14,18):

$$\bar{E}^i(x,z) = -i\omega(a^2+b^2)^{-1}(b^2x\bar{1}_z - a^2z\bar{1}_x) \tag{68}$$

where a and b are the ellipse's axes in the x and z directions. If the applied field is not directed along a symmetry axis, it may be decomposed into components along or perpendicular to those axes. Equations 67 and 68 may then be applied separately to the components.

Induced electric fields depend on ω and the shape of the object. Moreover, their magnitude increases from zero at the center to a maximum value at the periphery. They are independent of the electrical properties of the object and any surrounding medium. In contrast, the induced current and dissipated power depend directly on the object's conductivity.

Elliott et al. (29) have modelled a limb as a three-layered cylindrical section. A short current-carrying sheet wrapped about the limb produces a nonuniform changing magnetic field parallel to the cylinder's axis. The cross-section normal to $\bar{B}$ is thus formed of three concentric circles. The field's nonuniformity means that Equation 67 cannot be used. Because of the circular symmetry, the left side of Equation 3b is still $2\pi rE^i(\bar{r})$. The right side may be evaluated numerically in terms of elliptical integrals. The electric field lines will again be circular loops. Some numerical evaluation is required here because the field is nonuniform. The next section will discuss more complicated situations which require a complete numerical approach.

2. Numerical Models

The preceding models do not apply if the object is irregular in shape, inhomogeneous, or if the frequency is so high that the electric fields produced by the induced currents can no longer be neglected. Such situations are, for example, encountered in electromagnetically induced bone healing and hyperthermia.

Complicated geometry can even arise in an apparently simple physical situation. Suppose that a cylindrical sample is placed in the uniform magnetic field near the center of Helmholtz coils with the field parallel to the cylinder's axis. The induced field is given simply by Equation 67 because the object presents a symmetric, circular cross-section to the field. However, the situation is considerably more complicated if the sample's axis is perpendicular to the field. If the field is in the x direction, the cross-section presented to it by the sample is a rectangle in the y-z plane.

Equation 4 must now be used in addition to Equation 3. In the quasistatic approximation for magnetic fields the $\partial\bar{E}/\partial t$ term in Equation 4 is negligible compared to the $\bar{J}$ term. For low frequencies then:

$$\nabla \times \bar{B} = \mu_o \bar{J}_c. \tag{69}$$

McLeod et al. (30) combined Equations 3a and 69 to obtain an equation for the magnetic field in the sample:

$$\partial^2 B_x/\partial y^2 + \partial^2 B_x/\partial z^2 + i\omega g B_x = 0.$$

The simplest solution that will yield a non-zero field at the center is $B_x = C(e^{-qy} - e^{q(y-2a)}) \cos(m\pi z/2h)$. q is a parameter to be determined by the boundary conditions; m is any integer. But B_x must equal the applied field B_o at the four boundaries: $y = \pm a$ and $Z = \pm h$. This simple solution cannot satisfy these conditions. A linear superposition of such solutions for odd m, however, can. With the coefficients of the superposition so chosen, B_x can be expressed as an infinite series. From Equation 69, $\bar{J} = \nabla x B_x \bar{1}_x/\mu_o$. This solution applies to a particular x-y cross-section of the sample. For a cylindrical object, a varies with distance from the center of the coils.

Gandhi et al. (31) developed a very general approach for calculating the current induced by a changing magnetic field in an irregularly shaped, inhomogeneous object with anisotropic electrical properties (g and K direction dependent). The object is divided into a three-dimensional grid of arbitrary size and shape. Equation 3b is applied to a planar loop in this grid. An impedance $Z_j^* = l_j/g_j^* A_j$ can be associated with each side j of the loop. l_j is the length of one side; A_j is an area normal to it. Inhomogeneities and anisotropies are incorporated into the model by allowing g_j^* to change from one side to the next. If the complex-valued current in each side is I_j, then from Equation 3b:

$$\oint_{loop} \bar{E} \cdot \overline{dl} = \sum_j I_j Z_j \quad - \partial/\partial t \int \bar{B} \cdot \bar{n} d^2 r$$

can be evaluated over the area of the loop using the local magnitude and direction of B. Thus anisotropic fields can also be treated by this technique. This procedure is carried out for all the loops in the grid. In practice the resulting system of equations is expressed in terms of individual loop currents rather than I_j. An initial estimate of these currents is made and continuously refined by the successive over-relaxation method. A detailed picture of the current and dissipated power distributions in the material can be obtained in this way. The quasi-static approximation is assumed here; hence, this method is not valid beyond about 30 MHz. Armitage et al. (32) use a similar approach on an admittance network to obtain a set of equations for complex-valued nodal potentials.

The formalism developed earlier can be used to obtain expressions for the internal electric field. In this case $\bar{E}_o(\bar{r}) = -i\omega \bar{A}_o(\bar{r})$ where $\bar{A}_o$ is the vector potential produced by external sources. At low frequencies, $-i\omega \bar{A}^s$ is small compared to ∇V^s and $e^{-ik|\bar{r}-\bar{r}'|} \sim 1$. Hence:

$$\bar{E}_i(\bar{r}) = -i\omega \bar{A}_o(\bar{r}) - \nabla V^s(\bar{r}). \tag{70}$$

V^s can be expressed in terms of the induced interfacial charge density σ_i to yield:

$$\bar{E}^s(\bar{r}) = 1/4\pi\varepsilon_0 \int_S (\sigma_i(\bar{r})(\bar{r}-\bar{r}')/|\bar{r}-\bar{r}'|^3)d^2r' \qquad (71)$$

where S includes all interfacial surfaces.

Lunt (33) has modelled a limb subjected to a low-frequency alternating field as an infinite, insulating slab in the y-z plane bounded by a conductor at its ends ($x = \pm d/2$). The slab represents the bone; the conductor, soft tissue. The magnetic field is applied in the z direction. The resulting electric field in the limb is given by Equation 70. Because the slab is insulating, no current can flow within it. Its two boundaries with the conductor acquire a surface charge density $\sigma(y,z)$. The surface charges, in turn, produce an electric field given by Equation 71 with a component normal to the surface which exactly cancels the normal component of the induced field there. Thus the combined field in the conductor is entirely tangential at the boundary, and no current flows into the slab.

If the applied field is produced by Helmholtz coils, $\bar{A}_0$ and thus the normal component of the induced field are readily calculated. An initial assumption for $\sigma(y,z)$ is made and then successively refined until E_n is smaller than some pre-assigned value at all points on both boundaries. This surface charge distribution can now be used along with the $i\omega\bar{A}_0$ term to calculate the electric field everywhere in the bone and soft tissue. The contribution of the surface charges on each boundary is complicated by the requirement that the boundary conditions (Equations 15 and 17) for the field they produce must also be satisfied at the other interface. An infinite number of image charges is required. Lunt shows that, for conductivities typical of bone and soft tissue, the assumption of an insulating slab introduces little error.

In a more recent paper (34) Lunt has modelled a limb as a layered infinite cylinder. A bone core is surrounded by soft tissue and then air. Surface charges are developed at the bone-soft tissue and the soft tissue-air interfaces. The assumption that bone and soft tissue have the same permittivity is not overly restrictive because $g \gg \omega K \varepsilon_0$.

Hill et al. (35) used a two-dimensional quasistatic (e term ~ 1) version of Equation 41 to model the field induced in an arbitrarily shaped, inhomogeneous, infinite cylinder.

Substitution into Equation 70 yields an integral equation to be solved by a moment method for $\bar{E}_i(\bar{r})$. The details of Hill's approach will be mentioned later in the section on plane wave numerical analysis. One application of this model was to a three-layered cylinder subjected to a low-frequency solenoidal magnetic field.

DECOUPLED ELECTRIC AND MAGNETIC FIELDS

In some situations (for example, a transmission line array, or the near field of an antenna) an object is exposed to decoupled time-varying electric and magnetic fields. The net magnetic field in the object at low frequencies is simply the applied field. The net electric field is the vector sum of the direct field and the induced field discussed in the two previous sections. The direct field depends on the object's shape and conductivity, but not its absolute size. For an homogeneous object it is uniform. The induced field depends upon the object's size and shape, but not its conductivity. Its magnitude increases from the center to the periphery; its field lines are closed loops. On one side of the object the direct and induced fields are parallel; on the other they are opposite in direction. At the center the induced field is zero if the object is homogeneous; only the direct field is present there.

Spiegel (36) obtained an expression for the net electric field developed in a sphere embedded in a lossy medium. He applied it to the case in which the external medium was air and the applied $\bar{E}_0$ and $\bar{B}_0$ were perpendicular. For small spheres (mice) the direct field dominates whereas for larger spheres (men) the direct and induced fields are comparable over large portions of the body.

Hart and Marino (18) analyzed an ellipsoid embedded in a lossy medium. For $\bar{E}_0$ parallel to the longest axis and $\bar{B}_0$ perpendicular to $\bar{E}_0$ the induced and thus the net fields can vary widely from point to point within the body. The relative importance of the direct and induced fields depends upon the shape of the cross-section exposed to the magnetic field and the conductivity of the object.

The long wave-length approximation to be discussed in plane-wave analysis reduces to a separate consideration of the direct and induced electric fields as just described. For plane waves, however, $\bar{E}_o$ and $\bar{B}_o$ are not decoupled but are related by Equation 31.

PLANE WAVES

BASIC PRINCIPLES

The modeling of the interaction of plane waves with biological objects is reviewed in detail by Durney (37). The electric and magnetic fields associated with a plane wave are related by Equation 28. Unlike the low-frequency situation, it is not necessary to consider $\bar{E}$ and $\bar{B}$ separately. The internal electric field is usually determined. Then the absorbed power is found from Equation 23. Complications arise, however, in different ways. First, resonance phenomena occur for wavelengths comparable to the dimensions of the object. Second, at high frequencies the penetration depth may be comparable to the dimensions of the object; the wave amplitude may be significantly attenuated during passage through the object. Finally, waves entering a new medium may converge or diverge depending upon the electrical properties of the media and the curvature of the interface.

Consider first a plane wave incident normally on an infinite slab of thickness d. $\bar{k}$ is thus perpendicular to the interface; $\bar{E}$ is parallel to it. The slab has conductivity g and dielectric constant K. The medium on either side of the slab is free space. At the first boundary part of the incident wave is transmitted, and part is reflected. At the second boundary part of the transmitted wave is reflected back into the slab. The rest is transmitted into the free space beyond. The ratio of the reflected to the incident amplitudes is the reflection coefficient, r. According to Equation 33, the power depends on the square of the wave amplitude. The wave reflectance $R = r^2$. r can be found by application of the boundary condition Equations 17 and 18. Plane wave relationships are often expressed in terms of a complex index of refraction given by (1,2):

$$n^* = n + in_i = (k^*)^{1/2}. \tag{72}$$

The real and imaginary parts of k* and n* are related by $k = n^2 - n_i^2$ and $k_i = 2nn_i$. If c is the speed of the wave in free space, then $v_p = c/n$ is its speed in the medium. Because the frequency is independent of the medium, the wavelength must decrease from λ_o to λ_o/n upon entering the slab.

The wave reflected back from the second boundary interferes with the wave transmitted by the first. Moreover, if g is sufficiently high, both waves are being absorbed as they travel. The factor $e^{-d/\delta}$ is a measure of the transmitted amplitude. As $e^{-2d/\delta}$ becomes smaller, energy absorption increases. For frequencies on the order of 100 MHz and greater δ is comparable to the dimensions of human tissue (6). A significant fraction of the incident energy is absorbed.

The net wave amplitude is also determined by the interference of the individual waves. Assume for now that g is negligible. The wave reflected back from the second boundary suffers continual reflections at both interfaces. At each successive reflection its amplitude is decreased. The net wave amplitude inside the slab is the summation of the individual amplitudes with phase taken into consideration. One finds that (1):

$$R = 2r^2(1 - \cos B)/(1 + r^4 - 2r^2\cos B) \tag{73}$$

where $r = (n-1)/(n+1)$ and $B = 2kd = 4\pi nd/\lambda_o$. $R = 0$ for $B = 2m\pi$ where $m = 1,2,3....$ At these values there is no reflection. All the energy is coupled into the medium and then transmitted. This resonance condition is given by:

$$kd = 2\pi nd/\lambda_o = m\pi. \tag{74}$$

In general the incident wave is not normal to the surface. Its electric field will have components parallel to and perpendicular to the interface. Application of the boundary conditions yields more complicated expressions, the Fresnel equations, for the reflection coefficients of the two possible $\bar{E}$ polarizations. These expressions can be generalized to a conducting medium for which n is complex valued and absorption is important. The overall results are the same (6). The field in the medium is determined by an $e^{-d/\delta}$ factor and the presence of resonances related to kd.

At very high frequencies the wavelength may be small compared to the various radii of curvature associated with the object, which can then be modelled as a series of infinite slabs. The existence of a 180° phase change upon reflection must be checked for each interface. At such high frequencies the penetration depth is much smaller than typical tissue dimensions. The wave is completely absorbed near the object's surface.

More realistic models for biological objects are the sphere, the cylinder and the spheroid, which also illustrate the third factor complicating the modelling -- the focusing of the incident plane waves. Consider a plane wave incident upon an insulating sphere. The normal to the surface is everywhere directed radially. The angle of incidence θ_i between $\bar{k}$ and the normal varies from 0° at the equator to 90° at the poles. Suppose the exterior medium is free space, for which n = 1. The angle of refraction θ_t between $\bar{k}$ for the wave transmitted into the sphere and the surface normal is given by Snell's law:

$$\sin\,\theta_t = \sin\,\theta_i/n = \sin\,\theta_i/K^{1/2} \leq 1/K^{1/2} \qquad (75)$$

where n and K are the index of refraction and dielectric constant of the sphere. For biological materials K is generally large so that sin θ_t is small. A small angle of refraction indicates that the transmitted wave propagates nearly parallel to the normal for any angle of incidence. The incident plane wave is focused toward the center of the object.

For lossy materials the situation is more complicated, but the tendency to propagate along the normal is reinforced. The conclusions reached here for a sphere can be generalized to any convex surface -- the incident plane waves will be focused upon entry into a biological object. The increased energy density due to this focusing can be far more important than the decrease due to absorption (6).

A detailed analysis of the interaction of a plane wave with a sphere is presented in Stratton (4). The incoming wave propagating in the z direction can be expressed in rectangular coordinates as:

$$\bar{E}_{in}(\bar{r}) = E_o e^{-ikz}\bar{1}_x \tag{76}$$

where an $e^{i\omega t}$ time dependence is again assumed. At the surface, part of the wave is reflected or scattered back into space, and part is transmitted into the sphere. Because of the spherical symmetry their associated fields are most conveniently expressed in terms of spherical wave functions. The transmitted field, for example, may be written as:

$$\bar{E}_t(\bar{r}) = E_o \sum_{j=1}^{\infty} i^j \,(2j+1)/j(j+1)\; (a_j{}^t \bar{m}_j(r, \theta, \phi) - i b_j{}^t \bar{n}_j(r, \theta, \phi)). \tag{77}$$

$\bar{m}_j$ and $\bar{n}_j$ are complicated combinations of spherical Bessel functions or spherical Hankel functions for the radial r dependence, Legendre polynomials for the co-latitude θ dependence and trigonometric functions for the longitudinal ϕ dependence. The b_j coefficients generate fields corresponding to the oscillation of surface electric charges. The resulting waves are said to be of electric type or E waves. The a_j coefficients generate B waves or oscillations of magnetic type. Hankel functions diverge at $r = 0$ so that $\bar{m}_j$ and $\bar{n}_j$ contain only Bessel functions for $\bar{E}_t$. The corresponding magnetic field $\bar{B}_t$ is found using Equation 28.

A similar expression can be written for the scattered field $\bar{E}_r$ with coefficients $a_j{}^r$ and $b_j{}^r$. $\bar{E}_r$ is best represented by Hankel functions which behave as outgoing spherical waves for large r. An introductory treatment of these spherical wave functions is found in reference (1).

Continuity of tangential $\bar{B}$ and $\bar{E}$ for the θ and ϕ components at the surface yields four equations to find $a_j{}^r$, $b_j{}^r$, $a_j{}^t$ and $b_j{}^t$ for each 1. The analysis is facilitated by rewriting $\bar{E}_{in}$ in terms of the spherical wave functions with the a_j and $b_j = 1$. W_t, the total energy scattered and absorbed by the sphere, and W_s, the total energy scattered, can be found in terms of $a_j{}^r$ and $b_j{}^r$.

$$W_s = D\sum_{n=1}^{\infty} (2n+1)(\,|a_n{}^r|^2 + |b_n{}^r|^2) \tag{78}$$

and

$$W_t = D\sum_{n=1}^{\infty} (2n+1)(a_n{}^r + b_n{}^r) \tag{79}$$

where $D = \pi E_o^2 (K \varepsilon_o / \mu_o)^{1/2} / k^2$. The absorbed energy is then:

$$W_a = W_t - W_s \tag{80}$$

ANALYTICAL MODELS

This approach was applied to biological systems by Anne et al. (38) who modeled a skull as an homogeneous sphere of radius a. They found maxima and minima in the absorption of energy for ka between 0.2 and 2.0. For larger and smaller ka the absorption steadily decreased. The absorption of energy from plane waves is a maximum for wavelengths on the order of bodily dimensions. In this range complicated resonances are found. Kritikos and Schwan (39) extended this analysis to identify the location of "hot spots" within the sphere. According to Equation 32 the penetration depth δ decreases with increasing g and w. The conductivity g itself increases rapidly with frequency w in the microwave region for biological tissue. For $\lambda \ll a$ the energy is absorbed at the front surface of the sphere. At longer wavelengths hot spots are found within the sphere due to a combination of resonance and focusing. In addition to the ka dependence their locations are related to a/δ and thus the tissue conductivity.

Several groups (40-42) have developed a more detailed model of the skull by representing it as a multilayered sphere. The frequency dependencies of the dielectric constant and conductivity are specified separately for each layer. The additional layers produce significantly greater energy absorption at frequencies well above the homogeneous sphere resonance region. This increased absorption is due to resonances in the outer layers which have small thicknesses and thus resonate at higher frequencies. The locations of internal hot spots are also shifted by the added inhomogeneity. The details of these changes depend on the values assumed for the thickness, K and g of each layer.

Relatively simple analytical expressions can be obtained in the long wavelength limit ($\lambda \gg$ object size) for spheres, spheroids, and ellipsoids. Lin et al. (43) showed that in this limit Equation 77 reduces to the electric field obtained by the application of the quasistatic approximation separately to the incident electric and magnetic fields. $a_j^t = n^{*-1}$ and $b_j^t = (2j+1)\ n^{*-(j+1)}/j$. In this limit the $j = 1$ term dominates.

Durney et al. (44) used a perturbation approach to obtain a long wavelength approximation for the interaction of a plane wave with a conducting, prolate spheroid. Equation 76 can be rewritten as a power series expansion for the incident wave.

$$\bar{E}_{in} = \sum_{j=0}^{\infty} \bar{E}_j (-ik)^j. \tag{81}$$

The $\bar{E}_j$ do not have the dimensions of fields; they are merely expansion coefficients. Similar expressions are obtained for the scattered and internal fields. The three series are substituted into Maxwell's Equations 1-4 and the boundary condition Equations 16-19. The resulting set of equations can be solved for the $\bar{E}_j$ coefficients of the internal fields in each order. The long wavelength approximation uses only the $j = 0$ and $j = 1$ terms. For a good conductor $g \gg \omega K \varepsilon_0$, and the zeroth order internal field is negligible. The internal electric field is given by the $j = 1$ term. The orientation of the incident $\bar{E}$ and $\bar{B}$ fields relative to the major axis a of the spheroid is very important in determining the internal field and dissipated power (45). In magnetic polarization $\bar{B}$ is parallel to a, whereas in electric polarization $\bar{E}$ is. In cross polarization $\bar{k}$ is parallel to a. Energy deposition is greatest for E polarization, least for B.

Massoudi et al. (14,46) applied this approach to conducting ellipsoids. In this case the orientation of the incoming fields must be specified relative to three axes. Six possible polarizations result. It should be noted that resonance phenomena do not appear in the long wavelength approximation because $\lambda \gg a$. Massoudi et al. (47) further extended the approximation to dielectric ellipsoids for which it is no longer the case that $g \gg \omega K \varepsilon_0$ and the zeroth order field is negligible.

Different approximations are made at short wavelengths (high frequencies). It was noted previously that for very short wavelengths an object may be modeled as a series of infinite plane slabs. At wavelengths which are somewhat shorter than these but still beyond whole-body resonance, a cylindrical model is appropriate. The object is treated as an infinitely long cylinder in the z direction; hence, the solution contains no z dependence. In practice this means that the object's length $L >> \lambda$. However, the wavelength may be comparable to the object's radius R or the thicknesses of radial layers.

Ho (48) treated the interaction of a plane wave with a triple-layered lossy dielectric cylinder. The incident wave can be decomposed into a wave with the electric field parallel to the axis (E_z polarized or TM) and a wave with the magnetic field parallel to the axis (B_z polarized or TE). At the surface part of the wave is scattered back into space; part is transmitted into the layered cylinder. Because of the cylindrical symmetry the associated fields are most conveniently expressed in terms of cylindrical wave functions. Let ψ_m represent any one component of either the electric or magnetic field within a layer m. Then:

$$\psi_m = \sum_{n=-\infty}^{+\infty} (A_{nm}J_n(k_m r) + B_{nm}Y_n(k_m r))e^{in\phi} \quad . \qquad (82)$$

Here r and ϕ are radial and angular coordinates. J_n and Y_n are, respectively, cylindrical Bessel functions of the first and second kind. $k_m = 2\pi/\lambda_m$ where λ_m is the wavelength in layer m. The Y_n diverge as $r \longrightarrow 0$; hence, $B_{nm} = 0$ for the innermost layer. Although the incident wave could be expressed using Equation 76, it is more conveniently written as an infinite combination of $J_n(k_o r)$. The scattered wave is represented as an infinite combination of cylindrical Hankel functions $H_n(k_o r)$. Continuity of the tangential electric and magnetic fields at each interface yields for each n a set of equations that may be solved simultaneously to yield the A and B coefficients. The series is truncated after N terms. N is determined by requiring that the associated truncation error be less than some preassigned value. The higher the frequency of the wave, the greater N must be.

Massoudi et al. (49) have investigated the resonances associated with the layering. The locations of the maxima and minima appear to be independent of the polarization (TE or TM) and inversely related to the thickness of the layers. Energy absorption at the resonances can be doubled or halved relative to an homogeneous cylinder due to the layering.

At higher frequencies a geometrical optics approach may be used. A plane wave incident normally upon a cylinder will encounter angles of incidence ranging from 0° to 90°. Snell's Law (Equation 75) and the Fresnel equations may be used to compute the amplitude of the transmitted wave for any angle. At these high frequencies it may be assumed that all transmitted energy is absorbed. Massoudi et al. (50) have shown that homogeneous cylinders and prolate spheroids produce similar absorption characteristics at these high frequencies. At lower frequencies these geometrical optics results merge with those obtained with the cylindrical wave analysis just described.

Resonances due to layering may be described exactly for spherical and cylindrical models for which wave function expansion can be readily used. Layering resonances are not easily discussed for spheroids and ellipsoids which are treated by the long wavelength approximation or numerical methods. Barber et al. (51) have shown, however, that when the layer thicknesses are small compared to whole-body dimensions, layering resonances obtained from an infinite plane slab model may be superimposed upon the whole-body geometrical resonance response.

In summary (37,52) a complete analytical solution for all frequencies can be obtained for a layered sphere. At frequencies well below whole-body resonance the long wavelength approximation provides simple expressions for spheres, spheroids, and ellipsoids. Above whole-body resonance, complete analytical solutions can be obtained for infinite cylinder models. At very high frequencies the geometrical optics approximation is useful. The results are generally sensitive to the polarization of the incoming plane wave. Energy absorption is proportional to f^2 in the long wavelength region, to f^{-1} just above whole-body resonance, and independent of f at high frequencies. The calculation of internal fields and absorbed energy in the important region around whole-body resonance requires numerical methods for models more complicated than layered spheres.

NUMERICAL MODELS

Two methods have generally been used to numerically model the interaction of a plane wave with a biological object -- spherical wave expansion and moment methods. Other techniques involving finite element and finite difference methods are presently being introduced. For extensive discussions of these methods see the reviews by Durney (37), Spiegel (53), Chen (54), and Lin (55).

1. Spherical Wave Expansions

Ruppin (56) used the spherical wave expansion of Equation 77 to represent the incident, scattered, and transmitted waves for spheroids with very low eccentricity ($e = a/b < 1.5$). Continuity of the tangential electric and magnetic fields is imposed at a finite number of points on the spheroidal surface (point matching). The expansions are truncated after N terms. The resulting system of equations is solved for the desired coefficients.

Barber et al. (57) introduced the extended boundary condition method (EBCM) to the interaction of a biological object with an applied field. Outside the object the scattered field in Equation 34 can be regarded as being produced by currents induced only on the object's surface in a manner similar to Equation 48. The interior of the object in this first stage behaves as a perfect conductor; no fields or induced currents exist within. $\bar{E}^s(\bar{r})$ is expressed in terms of the current densities induced on the outer (+) surface, $\bar{n} \times \bar{E}_+$ and $\bar{n} \times \bar{B}_+$, and the dyadic Green's function $\overleftrightarrow{G}(\bar{r},\bar{r}')$. $\bar{n}$ is the outward directed normal to the surface S. Because there are no fields within the object, $\bar{E}_1(\bar{r}) = 0$ in Equation 34 which can then be written as:

$$-\bar{E}_0(\bar{r}) = \nabla \times \int_S (\bar{n} \times \bar{E}_+)\cdot\overleftrightarrow{G}(\bar{r},\bar{r}')d^2r' + \nabla \times \nabla \times (c^2/i\omega) \int_S (\bar{n} \times \bar{B}_+)\cdot\overleftrightarrow{G}(\bar{r},\bar{r}')d^2r'. \qquad (83)$$

In the second stage the interior of the object is reintroduced. It is not really a perfect conductor. There are, in fact, fields and currents within. Continuity of the tangential electric and magnetic fields at the surface means that $\bar{n} \times \bar{E}_+ = \bar{n} \times \bar{E}_-$ and $\bar{n} \times \bar{B}_+ = \bar{n} \times \bar{B}_-$ where (-) refers to the inner surface. These expressions are substituted into Equation 83. Use of Equation 28 then converts Equation 83 into an integral equation for $\bar{E}_-$. But $\bar{E}_-(\bar{r})$ and $\bar{E}_0(\bar{r})$ can be represented in terms of spherical wave functions as in Equation 77. A similar but more complicated spherical wave expansion can be made for $\overleftrightarrow{G}(\bar{r},\bar{r}')$. If these expressions are truncated after N terms, Equation 83 reduces to a set of 2N simultaneous equations to be solved for the expansion coefficients. Once these are known the electric field can be found everywhere within the object. Strictly speaking, spherical wave expansions may be used only beyond a sphere circumscribing the object and within a sphere inscribed by the object. The expressions for the fields, however, can be continued analytically by expansions about new origins to cover the entire region of interest (58).

Barber (57,59) has applied the EBCM to the absorption of plane waves by prolate spheroids near whole-body resonance. Various combinations of eccentricity, complex dielectric constant, wave polarization and angle of incidence are considered. Significant errors develop around resonance for large eccentricity, size, or dielectric constant of the spheroid because the spherical wave expansion inside the object converges poorly.

Iskander et al. (60) have developed an iterative EBCM (IEBCM) which overcomes these difficulties for objects of relatively high conductivity. The interior of the object is subdivided into overlapping spherical subregions. Individual spherical wave expansions are used in each. The EBCM solution can be used to obtain an initial estimate for the field in each subregion modeled as a perfect conductor. The resulting surface currents are used to recompute the subregional fields with the object's dielectric properties built in. The procedure is repeated until the surface currents converge within a preassigned limit.

2. The Method of Moments

The method of moments is a general technique used to convert an integral or differential operator equation into a matrix equation that can be solved by matrix inversion. As applied to integral equations for the internal field or current such as 40, 43, or 46, it is the most common numerical method. It is particularly useful for inhomogeneous objects of arbitrary shape. The primary reference is Harrington (61).

Given a function g and a linear operator L^{op} one seeks a function f such that:

$$L^{op}f = g. \tag{84}$$

This is accomplished by first expanding f in terms of expansion functions or basis functions f_i:

$$f = \sum_{i=1}^{N'} a_i f_i. \tag{85}$$

Next, the inner or scalar product of Equation 84 is taken with a set of weighing or testing functions w_j.

$$\sum_{i=1}^{N'} a_i \,(w_j, L^{op} f_i) = (w_j, g) \text{ for } j = 1 \ldots N'. \tag{86}$$

This set of N' equations can be written in matrix form as:

$$[l_{ji}][a_i] = [g_j]. \tag{87}$$

$[a_j]$ and $[g_j]$ are column matrices; $[l_{ji}]$ is a square matrix. If $[l_{ji}]$ has an inverse $[l_{ij}^{-1}]$, then:

$$[a_j] = [l_{ij}^{-1}][g_j]. \tag{88}$$

Equation 88 can be substituted into 85 to obtain the desired solution f. It is important that the testing functions are in the range of L^{op} and the basis functions in its domain.

Moment methods can be distinguished according to several factors: (1) the starting integral equation, (2) choice of basis and testing functions, (3) evaluation of l_{ji}, and (4) matrix inversion routine.

The integral equation can be solved for the internal current (40) or the field (43). For electric fields a surface integral equation (SIE) such as 47 or a volume integral equation (VIE) such as 43 or 46 may be the starting point.

The region of integration is decomposed into cells. The simplest basis functions are pulse functions: f_i = constant in cell i and 0 elsewhere. The simplest testing functions are Dirac delta functions. Use of pulse functions implies that the field is uniform in each cell. Delta testing functions evaluate the integral in l_{ji} at the center of each cell (point matching). A combination of pulse basis and delta testing functions thus places special emphasis on the field at the cell center. Satisfying the boundary conditions at each of a cell's boundaries may not be possible. Furthermore pulse functions are not really in the domain of $\nabla \cdot \bar{E}$ and may not strictly be usable in situations where this term is important. Linear functions, triangle functions, rooftop functions, and a combination of pulse and linear functions are more complicated alternatives for testing functions. Galerkin's method is a particularly accurate approach which uses the same choice for basis and testing functions.

Evaluation of the l_{ji} for a VIE involves an integration over the cell volume and thus depends on the cell shape. The cubical block is the simplest and most commonly used shape. Because problems may arise near the cube's edges, more complicated shapes such as cylinders and polyhedra have been used. The diagonal elements of $[l_{ji}]$ present a special problem. They represent the contribution to the field in a cell by that cell itself and may possess a singularity where $\bar{r}' \rightarrow \bar{r}$. Care must be taken in the evaluation -- hence the PV notation in Equation 46. This problem can be avoided by converting these volume integrals into surface integrals over the boundary of the cell. Evaluation of the off-diagonal elements is less difficult but still requires approximations.

Maximum cell size and thus the minimum number of cells N' required to represent adequately an object depend on the wavelength within the cell (62). N' increases with the frequency and cell dielectric constant. Because J_i and E_i are three-dimensional vectors, $[a_i]$ and $[g_j]$ contain 3N' elements and $[l_{ji}]$ contains $(3N')^2$. The increased memory space for storage and the computer time for the inversion of very large matrices place limitations upon the maximum frequency and economic demands upon the research budget. This paper will not discuss the special mathematical procedures developed to minimize this problem. Readers should consult recent papers in the area (63,64).

Moment methods were introduced into the bioelectromagnetics literature by Livesay and Chen (9) who used pulse basis functions, delta testing functions and cubical cells in Equation 46. Diagonal elements were obtained by approximating the cells as spheres of equal volume. The internal field was found to depend strongly on the orientation of the body relative to the polarization of the incident field. Chen and Guru (65-67) used such a block model to find the local internal field and, from Equations 23 and 25, SAR values inside humans facing plane waves having an electric field polarized either parallel or perpendicular to the long axis of the body. Computer requirements limited the frequencies considered to below 500 MHz but did include whole-body resonance. For homogeneous models of high water content tissue, part-body resonances ("hot-spots") are found, the location and magnitude of which depend upon the incoming wave's frequency and polarization. The introduction of inhomogeneities complicates the energy absorption pattern considerably though low-conductivity regions generally experience a higher SAR. Application was also made (67) to the SAR distribution within bodies exposed to a uniform high-frequency electric field for hyperthermia treatment.

Hagmann has modified this cubical block model approach in several ways (68-70). First, recall that the dyadic Green's function appearing in Equation 46 has a $\overleftrightarrow{I}$ factor originating in currents ($i\omega\bar{A}$) and a $\nabla\nabla$ factor produced by charge distributions ($-\nabla V$). This latter term is responsible for the principal value procedure required to evaluate the integral. But with a pulse basis, all inhomogeneities, and the resulting charge distributions, occur at the cell surface (71). Evaluation of the diagonal terms $[l_{ji}]$ can then be made using surface integrals. The boundary charges also serve to satisfy the boundary condition of discontinuous normal $\bar{E}$ for adjacent cells. Second, recall that pulse basis and delta testing functions emphasize the field at the cell center. Inhomogeneities can cause $\bar{E}$ to vary considerably from one cell to the next. A considerable amount of error can thus be introduced into the local SAR values which are assumed to be constant in each cell, equal to the value at the center. Hagmann et al. (68) interpolated the values at the centers of adjacent cells to obtain an estimate for the variation of the electric field within each cell. They also developed more complicated methods for representing the variation of the field within each cell.

These methods were applied (69) to a block model of man for the electric field parallel to the long axis of the body. Whole-body SAR levels over a wide frequency range up to 500 MHz and local SAR values near whole-body resonance (80 MHz) were found. The sizes and complex permittivities of the cells were chosen to provide a realistic model of the human interior. Hagmann et al. (70) used this model but with a greater number of cells in the head and neck region to investigate part-body resonances in that region. Much greater absorption was found for the head by this approach than was predicted by the analytical layered-sphere model. The polarization of the incident wave and the presence of the rest of the body are very important factors. The limitations of this cubical block model are still debated vigorously (72,73).

Well above whole-body resonance, computer limitations prevent the use of finely detailed, block-type numerical models. But as for analytical solutions, numerical methods can be applied to infinite cylinder models in this region. It should be noted that the term cylinder is not restricted to an object of circular cross-section, but can refer to elliptical or even irregular cross-sections that remain constant in the z direction. Infinite cylinders are also studied because with no z dependence they represent a two-dimensional boundary value problem. New modeling techniques can be tested in simple form on them. Recall that any incident wave can be decomposed into a wave with E parallel to the z axis (TM) or E perpendicular to the z axis (TE).

Ho (74) obtained solutions to Equation 40 for TM waves incident upon a three-layered, infinite, circular, dielectric cylinder. He expressed this equation in terms of the field by use of Equation 38. In this two-dimensional situation the volume elements reduce to area elements which he chose to be circles. To solve this volume integral equation (VIE) Ho used a moment method with pulse basis and delta testing functions. Ho also discussed the use of triangular basis functions with three-dimensional irregular dielectric bodies.

For a TM field Equations 3a and 4a can be combined to give Helmholtz's equation for E_z, the only scattered electric field component (61):

$$\nabla^2 E_z{}^s + k^2 E_z{}^s = i\omega\mu_0 J_{zi}. \tag{89}$$

J_{zi} is the induced current density in the cylinder. In two dimensions the solution to this scalar equation is:

$$E_z{}^s(\bar{r}) = -k(\mu_0/K^* \varepsilon_0) \int_S J_{zi}(\bar{r}') H_0{}^{(2)}(k|\bar{r}-\bar{r}'|)d^2r'. \tag{90}$$

This equation can be expressed in terms of two-dimensional Green's functions by $G(\bar{r},\bar{r}') = -iHo^{(2)}(k|\bar{r}-\bar{r}'|)/4$. For free space $k = k_o = 2\pi/\lambda_o$. Substitution of Equation 90 into 34 yields an integral equation for the internal field. For a TM wave $\bar{E}$ is purely tangential and thus continuous across any interface. Moment methods can be applied to solve this VIE. A similar approach can be used to obtain a solution for B_z with TE waves. $\bar{E}$ can then be found by application of Equation 4a.

Computer storage problems at high frequencies can still create difficulties even with two-dimensional problems. Use of a SIE approach reduces the computer requirements for homogeneous objects. Wu and Tsai (75) applied Equation 47 for TM waves to express the field inside one- or two-layered dielectric cylinders in terms of the electric fields and Green's functions evaluated on the boundary surfaces. The surface fields were found by noting that for TM waves E_z is continuous across each boundary. The fields at each boundary can be expressed in terms of Equation 34 with Equation 47 used to represent the field at a surface point produced by all the other surface elements. Divergences thus appear as $\bar{r}' \rightarrow \bar{r}$ in the Green's functions of Equation 47 and principal values must be taken. Two coupled integral equations are obtained for the surface field and its normal derivative. The solutions, obtained by a moment method with pulse basis and delta testing functions, are then used in Equation 47 to obtain the internal electric field. A parallel procedure is used to find the internal magnetic field for TE waves. Different resonance behavior is observed for TE fields.

Wu (76) extended the SIE approach to bodies of revolution, objects which possess rotational symmetry about the z axis (a spheroid, for example). The more complicated three-dimensional geometry requires a vector representation for the surface fields such as Equation 48. A modified Galerkin's method is used with triangular and trigonometric basis and testing functions to solve integral equations for the surface electric and magnetic current densities. These currents, in turn, can be used to obtain the electric field everywhere within the object. Wu found significantly different internal field distributions for E and B polarizations.

Massoudi et al. (77) used this SIE approach to obtain the average SAR for a hemispherically capped cylinder and K polarization. They provide interesting comparisons of the SAR dependence on polarization for people and for rats over a very wide frequency range. The SIE method can be used to higher frequencies and for more elongated objects than the EBCM. Computer requirements are not as severe for SIE as for VIE solutions (75,76); hence, the SIE method can be used to higher frequencies than the VIE for homogeneous objects. On the other hand, the VIE approach becomes more useful as the degree of inhomogeneity and thus the number of boundary surfaces increase.

TE fields are more difficult to treat for infinite cylinders than are TM fields because $\bar{E}$ is no longer purely tangential and continuous across interfaces. Discontinuities in the normal component induce charge distributions at interfaces. Although the problem can be formulated in terms of B_z, a direct solution for $\bar{E}$ is of interest. Hill et al. (35) applied a moment method to the two-dimensional, low-frequency limit of Equation 43 to obtain the TE solution for lossy inhomogeneous cylinders. In this limit $e^{-ik|\bar{r}-\bar{r}'|} \sim 1$ and the $i\omega\bar{A}^s$ term is negligible. A problem arises because the simple pulse basis functions are not in the domain of $\nabla'\cdot$ and cannot be used. The authors instead use Galerkin's method with a combination of pulse and linear functions. They also use arbitrarily shaped polygonal cells rather than blocks. Note that the $\nabla'\cdot$ term and thus the problem with pulse basis functions originates in the discontinuities of E-normal at interfaces. Such difficulties should not appear with TM fields.

The application of this approach (35) to model the field induced in a cylinder by the application of a uniform, low frequency, axial magnetic field was mentioned in an earlier section. At high frequencies, however, the simplifying assumptions in that section can no longer be made. In particular the $i\omega\bar{A}^s$ term in Equation 37 cannot be neglected. Lee and Chen (78) discussed the currents induced inside a body of revolution exposed to a uniform, high frequency, axial magnetic field. The object was divided into a set of coaxial circular rings placed top to bottom. Each ring had a square cross-section. Broader parts of the body require more rings. Because of the rotational symmetry induced currents do not cross rings and no charge distributions are produced at ring boundaries. The ∇V^s term in Equation 37 is absent.

Let $\bar{A}^{mn}$ be the magnetic vector potential at ring m due to the electric field in ring n. $\bar{A}^{mn}$ is given by Equation 42 where the field in ring n is $\bar{E}_i(\bar{r}')$. The total potential at ring m due to all the induced fields is $\bar{A}^m = \sum_n \bar{A}_{mn} = \sum_n C_{mn}\bar{E}_n$ where C_{mn} depends on the area, radius and electrical properties of ring n as well as the geometrical relationship between rings m and n. Substitution into Equation 37 yields an expression for the electric field at ring m, which is entirely circumferential:

$$E_m + iw \sum_n C_{mn} E_n = E_m^o$$

where E_m^o, the electric field directly induced in ring m by the applied magnetic field, is given by Equation 67. Such an equation can be written for each of the N rings. The N simultaneous equations which result can be solved by matrix inversion for E_m.

Borup and Gandhi (79,80) have modified the usual moment method to deal with the interaction of TM waves with inhomogeneous, infinite cylinders of irregular cross-section. They substitute Equation 90 for the scattered field into Equation 34. Pulse basis and delta testing functions convert this integral equation into an expression for the electric field at the center of a cell indexed by n and m in terms of the incident field E^o and the fields produced by the other cells indexed by i and j.

$$E_z^o(n,m) = E_z(n,m) + \sum_{i,j} (K^*_{ij}-1)E_z(i,j)M(n,m,i,j)$$

where M is an integral related to the separation of cell (i,j) and cell (n,m). With a rectangular grid one might, for example, regard n and m as x and y coordinate indices. Moreover, in that case the summation can be evaluated efficiently by a fast Fourier transform (FFT) procedure. The resulting reduction in computer storage and time demands allow this procedure to be used to higher frequencies and with more detailed objects.

3. Other Numerical Methods

Recently other numerical methods have been used to model the interaction of electromagnetic fields with biological systems. Spiegel (53) describes the finite-difference solution of Equations 3a and 4a. The object and a region of space surrounding it are represented by a three-dimensional grid. The time derivative of a component of $\bar{B}$ at a point $\bar{r}$ could be expressed as:

$$\partial B_z(\bar{r},t)/\partial t \sim (B_z(\bar{r},t+\Delta t/2) - B_z(\bar{r},t-\Delta t/2))/\Delta t.$$

The spatial derivatives of a component of B could be similarly represented as:

$$\partial B_y(r,t)/\partial z \sim (B_y(x,y,z+\Delta z,t) - B_y(x,y,z,t))/\Delta z.$$

Such expressions are substituted into Equations 3a and 4a to yield six coupled difference equations for E_x, E_y, E_z, B_x, B_y, and B_z. The time differences occur in half-interval steps. The new values of the fields at a grid element depend on their previous values at that element and the adjacent ones. These coupled equations are solved by "time-stepping" through the grid. This approach can be used to obtain the internal fields produced by transient, continuous or pulse waveforms. Although it has not yet been used extensively in bioelectromagnetics, this technique holds considerable promise for future applications.

A finite-element method has been used by Lynch et al. (81) to solve Equations 3a and 4a in their coupled form

$\nabla \times \nabla \times \bar{E} - k^2\bar{E} = 0$. $\nabla \times \nabla \times - k^2$ serves as the linear operator in Equation 84 with $f = \bar{E}$ and $g = 0$. This differential equation can be solved using a moment method. The authors use Galerkin's method applied on a finite element grid to convert the equation into matrix form. A basis and testing function is created for each node of the grid. The set of equations is completed by application of the boundary conditions. Use of this technique yields approximate solutions for the fields at each node but conserves energy exactly. Lynch et al. (81) obtain good agreement with analytical solutions for two-dimensional, high-frequency hyperthermia problems.

NEAR FIELDS

The exposure of biological objects to near field and induction zone radiation can be described using the principles of internal field calculations. In contrast to the plane wave case the unperturbed field $\bar{E}_0$ may vary significantly over the volume of the object in a manner dependent upon the details of the particular source. Thus Durney (37) and Spiegel (53) point out that few generalizations can be made for near-field problems, which must be considered on a case by case basis.

As a first approximation any reaction of the secondary fields produced by the object back on the source is neglected. Guy (82) calculated the fields produced inside a two-layer, infinite slab by a waveguide placed in direct contact with it (aperture source). The resulting fields are just those produced by a waveguide of a particular mode radiating into a layered medium of given electrical properties. The radii of curvature of the exposed tissues must be much greater than the dimensions of the waveguide aperture for the infinite slab model to apply. Ho et al. (83) discuss an aperture source in direct contact with a limb modelled as a three-layered, infinite cylinder. The field at the limb produced by such an aperture, considered by itself, can be written as a series of cylindrical waves. The fields inside each layer of the limb can be expressed similarly. Continuity of the tangential field components yields the coefficients of the series. This approach can even be used for a small spherical object placed within a waveguide if its radius is much smaller than the guide's dimensions (84). The resonant field can be regarded as a superposition of four plane wave fields. The interaction of the sphere with each wave can be treated using the methods described earlier. A small sphere evidently does not seriously perturb the field distribution within the guide because the resonant frequency is not shifted appreciably.

The field emitted by an aperture source can be regarded as being produced by equivalent electric and magnetic dipoles. Lakhtakia et al. (85) expand the field from such a source in terms of spherical wave functions as in Equation 77. The EBCM is then used to calculate the fields produced inside a nearby prolate spheroid.

Near field radiation patterns can be quite complicated. In particular longitudinal (parallel to $\bar{k}$) field components may be present. But the pattern can always be represented using a Fourier decomposition as a superposition of plane waves travelling in different directions. Chatterjee et al. (86) use this approach to describe the interaction of near field radiation from an RF sealer with a block model human. The field distribution for a particular source was measured directly in front of the human's position and the corresponding plane wave components found. These could be used to obtain the unperturbed source field in each block. The method of moments was then applied as in reference (69) to obtain the local net field and SAR distributions.

In many cases, however, the situation is complicated by the reaction of the fields scattered by the object back on the source. This "body-source coupling" must be considered in any complete near-field problem. To treat the interaction of a biological object with the near field of an antenna, Karimullah et al. (87) use Equation 34 with:

$$\bar{E}_0(\bar{r}) = \int_{\text{ant}} I(\bar{s}')\ \bar{s}' \cdot \overleftrightarrow{G}(\bar{r},\bar{s}')\ ds' \qquad (91)$$

for the unperturbed field. $\bar{s}'$ is an antenna coordinate. $I(\bar{s}')$ is the antenna current and $\overleftrightarrow{G}(\bar{r},\bar{s}')$ is the tensor Green's function. In an earlier paper Nyquist et al. (88) had regarded $I(\bar{s}')$ as a known function, simply substituted Equation 91 into 34, and solved the resulting integral equation by the method of moments. Karimullah et al. (87), however, regard $I(\bar{s}')$ as an unknown. They develop an additional integral equation relating the net (unperturbed plus scattered) tangential electric field at the antenna's surface to the impressed voltage. This set of coupled integral equations can be solved for the modified antenna current and the field inside the object.

Spiegel (89) used this approach to study the interaction of a $\lambda/4$ monopole antenna with a nearby ($d \ll \lambda$) block-model, full human, and the interaction of a $\lambda/2$ dipole antenna very close to a block-model head. The distribution of absorbed power differs considerably from plane-wave irradiation in both cases.

The importance of body-source coupling was investigated by Zhu et al. (90) who used a lossy dielectric sphere to model a human and a conducting sphere with a slot opening to model a nearby radiation source. This problem can be solved analytically by spherical-wave expansion methods either including or neglecting body-source coupling. The effects of successive scatterings are treated through a series expansion. If the coupling is omitted by considering only the first term, more than a fifty percent error in the computed SAR can result.

DISCUSSION

The basic principles underlying the mathematical modeling of the interaction of electromagnetic fields with biological systems are well understood. Reasonable order-of-magnitude agreement with experiment is obtained for whole-body currents or SAR with homogeneous numerical models and the more realistic analytical models. Choice of the model to be used in a particular situation depends on the shape of the object and the frequency of the applied field. Thus prolate spheroidal models are adequate for rodents whereas ellipsoidal models should be used for primates (46).

Given the shape, one may find that an approximate analytical solution is applicable. The quasistatic or long wavelength approximation in which the applied electric and magnetic fields are essentially decoupled is appropriate if $\lambda \gg$ object size. It should be noted that the smaller the object size, the higher the frequency range for which this model may be used. Thus it would be inappropriate for humans at 50 MHz but perfectly acceptable for rats (46). In fact, this approximation has been used out to 100 GHz for cells (91). Near whole-body resonance, numerical techniques such as the IEBCM and moments methods are required. Well above this region infinite cylinder models are useful.

The range of applicability of the numerical techniques also depends on the object's size and shape. Because the eccentricity of a rat is less than that of a human, the EBCM can be applied at higher frequencies for rodents than for people (59). Small animal size also leads to smaller cell size in moment methods and the use of these techniques out to higher frequencies for a particular set of computer requirements.

In summary, for a given animal one cannot usually apply the same model for all frequencies. Conversely, for a given frequency one cannot automatically use the same model for all objects. Their size must also be considered. At 1 GHz an infinite-cylinder model would be appropriate for humans, IEBCM for a rat, and the long wavelength approximation for a pea.

If local field, current density, or SAR values are desired, inhomogeneous models of realistic shape requiring numerical methods must be used at any frequency. Unfortunately, at present the results appear to be model dependent -- varying significantly with choice of cell geometry, weighing and testing functions.

The results of clinical experiments on laboratory animals must be extrapolated or scaled to humans. An important question is what physical parameter governs the scaling --- maximum surface field, average internal current density, or whole body SAR? Scaling cannot be carried out simultaneously for all these quantities, and a choice therefore must be made. Proper scaling is a difficult task. Even at low frequencies rats and humans exposed to the same free-space field experience different internal fields and current densities because of differences in shape.

At higher frequencies the difference in the internal fields produced by E, B, or K polarizations for the same incident field strength and frequency depends on the shape of the object. Polarization effects should be less important for rodents than for primates. Absolute size is important as well near resonance. Being smaller, rats experience whole-body resonance at a higher frequency than humans (37). Conversely, rats do not absorb 80 MHz radiation as well as do people. Effects observed in rats at a given field strength at this frequency should occur in humans at a much lower field strength.

At very high frequencies the ratio of penetration depth to animal size becomes significant. Radiation with $\delta \sim 1$ would be absorbed near the surface of humans, but deposit significant amounts of energy in the internal organs of mice. For all frequency ranges differences in the internal structure and detailed shape of animal systems must be considered in scaling the inhomogeneous numerical models used to calculate local fields and SAR. Part-body resonances should be more complicated for primates than for rodents.

Use of the SAR implies scaling according to energy deposition and a thermal origin of bioeffects. The observation of frequency and amplitude windows (92,93) should make researchers cautious about scaling all bioeffects according to SAR. Local field or current densities may be more appropriate in some situations.

The results obtained for the internal fields, current densities and SAR depend upon the values assumed for the electrical properties of the various tissues. Below 1 MHz $g \gg \omega K \varepsilon_0$ and tissue dielectric constants can usually be neglected. For accurate modeling the tissue electrical properties should be obtained from _in vivo_ rather than _in vitro_ measurements. Such information is available in the range 1-10,000 MHz (94). At lower frequencies where electrode effects become more important tissue conductivities may not be as reliable. In any case, tissue electrical properties are unlikely to be in error by an order of magnitude. The limiting factor at present in the accurate representation of the interaction of electromagnetic fields with biological objects is the difficulty in numerically modeling local fields and SAR.

Analytical models have probably been pushed to their limits. They provide a useful comparison for the results predicted by numerical models, a broad understanding of the basic phenomena and order of magnitude estimates for whole-body SAR and average internal fields. Most of the activity in mathematical modeling is now directed toward estimates of the local fields, current densities, and SAR for inhomogeneous objects of irregular shape.

As noted previously local field and SAR moment method results depend at present upon parameters such as the cell shape, choice of weighing and testing functions. All numerical methods are limited by computer storage and speed capabilities. Use of more elaborate cell shapes such as polyhedra and linear functions (35) may reduce the number of cells but at the expense of more complicated evaluations of the $[l_{ji}]$ elements. Larger and faster computers will permit greater refinements in the details of numerical models and their extension to higher frequencies. Newer techniques such as FFT (80), finite-element (81) and finite-difference (53) methods may provide useful alternatives to the more traditional numerical approaches in this area. But few groups have the computer facilities and expertise necessary to carry out such computations. Unlike analytical models which can be easily applied using a microcomputer, the complicated numerical methods are not at present readily transferable to a wide range of users. Calculation of local field and SAR for near field situations must be done on a case by case basis and requires numerical methods. Thus small-system researchers who wish to characterize near-field exposures for their particular apparatus face a serious problem. Their apparatus may not have been previously modeled in the literature, and their computer system is too small for adequate numerical modeling. Well-documented listings of successful numerical calculations should be more readily available to the bioelectromagnetics community. The development of software packages adaptable to a variety of problems for microcomputers should be investigated.

The mathematical modeling of the interaction of electromagnetic fields with biological systems has matured considerably since 1971. Whole-body exposure characteristics are now generally well understood. The next fifteen years may bring similar successes to the characterization of exposure at the local level.

ACKNOWLEDGEMENTS

I would like to express my thanks to Sue Armentrout and Jill Rawnsley for their reference work and to Barbara Hart for typing the manuscript.

REFERENCES

(1) Reitz, J.R, Milford, F.J. and Christy, R.W.: Foundations of Electromagnetic Theory, Addison-Wesley, Reading, MA, 1979.

(2) Lorrain, P. and Corson, D.R.: Electromagnetic Fields and Waves, 2nd Ed., Freeman, San Francisco, 1970.

(3) Jackson, J.D.: Classical Electrodynamics, 2nd Ed., Wiley, New York, 1975.

(4) Stratton, J.A.: Electromagnetic Theory, McGraw-Hill, New York, 1941.

(5) Pethig, R.: Dielectric and Electronic Properties of Biological Materials, Wiley, New York, 1979.

(6) Johnson, C.C. and Guy, A.W.: Nonionizing electromagnetic wave effects in biological materials and systems, Proc. IEEE 60:692-718, 1972.

(7) Tsai, C.T., Massoudi, H., Durney, C.H. and Iskander, M.F.: Numerical Evaluation of SAR Distributions in Biological Models: Comparison of Two Different Integral Equations and Basis Functions. Presented at the 7th Ann. Conf. Bioelectromagnetics Society, San Francisco, 1985.

(8) Van Bladel, J.: Some remarks on Green's dyadic for infinite space, IRE Trans. Antennas Propagat. AP-9:563-566, 1961.

(9) Livesay, D.E. and Chen, K.M.: Electromagnetic fields

induced inside arbitrarily shaped biological bodies, IEEE Trans. Microwave Theory Tech. MTT-22:1273-1280, 1974.

(10) Kaune, W.T. and Gillis, M.F.: General properties of the interaction between animals and ELF electric fields, Bioelectromagnetics 2:1-11, 1981.

(11) Schwan, H.P.: Biological Hazards from Exposure to ELF Electrical Fields and Potentials, NWL Technical Report TR-2713, 1972.

(12) Barnes, H.C., McElroy, A.J. and Charkow, J.H.: Rational analysis of electric fields in live line working, IEEE Trans. Power Appar. Syst. PAS86:482-492, 1967.

(13) Sheppard, A.R. and Eisenbud, M.: Biological Effects of Electric and Magnetic Fields of Extremely Low Frequency, New York Univ. Press, New York, 1977.

(14) Massoudi, H., Durney, C.H. and Johnson, C.C.: Long-wavelength analysis of plane wave irradiation of an ellipsoidal model of man, IEEE Trans. Microwave Theory Tech. MTT-25:41-46, 1977.

(15) Kaune, W.T.: Interactive effects in 60-Hz electric-field exposure systems, Bioelectromagnetics 2:33-50, 1981.

(16) Lattarulo, F. and Mastronardi, G.: Equivalence criteria among man and animals in experimental investigations of high voltage power frequency exposure hazards, Appl. Math. Modeling 5:92-96, 1981.

(17) Shiau, Y. and Valentino, A.R.: ELF electric field coupling to dielectric spheroidal models of biological objects, IEEE Trans. Biomed. Eng. BME-28:429-437, 1981.

(18) Hart, F.X. and Marino, A.A.: ELF dosage in ellipsoidal models of man due to high voltage transmission lines, J. Bioelectricity 1:129-154, 1982.

(19) Hart, F.X. and Marino, A.A.: Penetration of electric fields into a concentric-sphere model of biological tissue, Med. Biol. Eng. Comput. 24:105-108, 1986.

(20) Bernhardt, J. and Pauly, H.: On the generation of potential differences across the membranes of ellipsoidal cells in an alternating electric field, Biophysik 10:89-98, 1973.

(21) Marchesi, M. and Parodi, M.: Study of the electrical field inside biological structures, Med. Biol. Eng. Comput. 20:608-612, 1982.

(22) Kobayashi, T., Shimuzu, K. and Matsumoto, G.: Visualization of the electric field around a moving animal by numerical calculation, Trans. IECE Japan E65:565-571, 1982.

(23) Kaune, W.T. and McCreary, F.A.: Numerical calculation and measurement of 60-Hz current densities induced in an upright grounded cylinder, Bioelectromagnetics 6:209-220, 1985.

(24) DiPlacido, J., Shih, C.H. and Ware, B.J.: Analysis of the proximity effects in electric field measurements, IEEE Trans. Power Appar. Syst. PAS-97:2167-2177, 1978.

(25) Chiba, A., Isaka, K. and Kitagawa, M.: Application of finite element method to analysis of induced current densities inside human model exposed to 60-Hz electric field, IEEE Trans. Power Appar. Syst. PAS-103:1895-1901, 1984.

(26) Spiegel, R.J.: High-voltage electric field coupling to humans using moment method techniques, IEEE Trans. Biomed. Eng. BME-24:466-472, 1977.

(27) Spiegel, R.J.: Numerical determination of induced currents in humans and baboons exposed to 60-Hz electric fields, IEEE Trans. Electr. Compat. EMC-23:382-390, 1981.

(28) Chen, K.M. and Chuang, H.R.: Quantification of interaction between ELF-LF electric fields and human bodies, IEEE Trans. Biomed. Eng., In Press.

(29) Elliott, R.S., Harrison, W.H. and Storm, F.K.: Hyperthermia: Electromagnetic heating of deep-seated tumors, IEEE Trans. Biomed. Eng., BME-29:61-64, 1982.

(30) McLeod, B.R., Pilla, A.A. and Sampsel, M.W.: Electromagnetic fields induced by Helmholtz aiding coils inside saline-filled boundaries, Bioelectromagnetics 4:357-370, 1983.

(31) Gandhi, O.P., DeFord, J.F. and Kanai, H.: Impedance method for calculation of power deposition patterns in magnetically induced hyperthermia, IEEE Trans. Biomed. Eng. BME-31:644-651, 1984.

(32) Armitage, D.W., LeVeen, H.H. and Pethig, R.: Radiofrequency-induced hyperthermia: Computer simulation of specific absorption rate distributions using realistic anatomical models, Phys. Med. Biol. 28:31-42, 1983.

(33) Lunt, M.J.: Magnetic and electric fields produced during pulsed-magnetic-field therapy for non-union of the tibia, Med. Biol. Eng. Comput. 20:501-511, 1982.

(34) Lunt, M.J.: Theoretical model for investigating the magnetic and electric fields produced during pulsed magnetic field therapy for nonunion of the tibia, Med. Biol. Eng. Comput. 23:293-300, 1985.

(35) Hill, S.C., Durney, C.H. and Christensen, D.A.: Numerical calculations of low-frequency TE fields in arbitrarily shaped inhomogeneous lossy dielectric cylinders, Radio Sci. 18:328-336, 1983.

(36) Spiegel, R.J.: ELF coupling to spherical models of man and animals, IEEE Trans. Biomed. Eng. BME-23:387-391, 1976.

(37) Durney, C.H.: Electromagnetic dosimetry for models of humans and animals: A review of theoretical and numerical techniques, Proc. IEEE 68:33-40, 1980.

(38) Anne, A., Saito, M., Salati, O.M. and Schwan, H.P.: Relative microwave absorption cross sections of biological significance, in Biological Effects of Microwave Radiation, Vol. 1, M.F. Peyton, ed., Plenum, New York, 153-176, 1960.

(39) Kritikos, H.N. and Schwan, H.P.: The distribution of heating potential inside lossy spheres, IEEE Trans. Biomed. Eng. BME-22:457-463, 1975.

(40) Shapiro, A.R., Lutomirski, R. and Yura, H.T.: Induced fields and heating within a cranial structure irradiated by an electromagnetic plane wave, IEEE Trans. Microwave Theory Tech. MTT-19:187-196, 1971.

(41) Weil, C.M.: Absorption characteristics of multilayered sphere models exposed to UHF/microwave radiation, IEEE Trans. Biomed. Eng. BME-22:468-476, 1975.

(42) Joines, W.T. and Spiegel, R.J.: Resonance absorption of microwaves by the human skull, IEEE Trans. Biomed. Eng. BME-21:46-48, 1974.

(43) Lin, J.C., Guy, A.W. and Johnson, C.C.: Power deposition in a spherical model of man exposed to 1-20-MHz electromagnetic fields, IEEE Trans. Microwave Theory Tech. MTT-21:791-797, 1973.

(44) Durney, C.H., Johnson, C.C. and Massoudi, H.: Long-wavelength analysis of plane wave irradiation of a prolate spheroid model of man, IEEE Trans. Microwave Theory Tech. MTT-23:246-253, 1975.

(45) Johnson, C.C., Durney, C.H. and Massoudi, H.: Long-wavelength electromagnetic power absorption in prolate spheroidal models of man and animals, IEEE Trans. Microwave Theory Tech. MTT-23:739-747, 1975.

(46) Massoudi, H., Durney, C.H. and Johnson, C.C.: Long-wavelength power absorption in ellipsoidal models of man and animals, IEEE Trans. Microwave Theory Tech. MTT-25:47-52, 1977.

(47) Massoudi, H., Durney, C.H. and Johnson, C.C.: Comparison of the average specific absorption rate in the ellipsoidal conductor and dielectric models of humans and monkeys at radio frequencies, Radio Sci. 12(suppl):65-72, 1977.

(48) Ho, H.S.: Energy absorption patterns in circular triple-layered tissue cylinders exposed to plane wave sources, Health Phys. 31:97-108, 1976.

(49) Massoudi, H., Durney, C.H., Barber, P.W. and Iskander, M.F.: Electromagnetic absorption in multilayered cylindrical models of man, IEEE Trans. Microwave Theory Tech. MTT-27:825-830, 1979.

(50) Massoudi, H., Durney, C.H. and Johnson, C.C.: A geometrical-optics and an exact solution for internal fields in and energy absorption by a cylindrical model of man irradiated by an electromagnetic plane wave, Radio Sci. 14:35-42, 1979.

(51) Barber, P.W., Gandhi, O.P., Hagmann, M.J. and Chatterjee, I.: Electromagnetic absorption in a multilayered model of man, IEEE Trans. Biomed. Eng. BME-26:400-405, 1979.

(52) Durney, C.H., Iskander, M.F., Massoudi, H. and Johnson, C.C.: An empirical formula for broad-band SAR calculations of prolate spheroidal models of humans and animals, IEEE Trans. Microwave Theory Tech. MTT-27:758-763, 1979.

(53) Spiegel, R.J.: A review of numerical models for predicting the energy deposition and resultant thermal response of humans exposed to electromagnetic fields, IEEE Trans. Microwave Theory Tech. MTT-32:730-746, 1984.

(54) Chen, K.M.: Interaction of electromagnetic fields with biological bodies, in Research Topics in Electromagnetic Wave Theory, J.A. Kong, ed., Wiley (Interscience), New York, 290-347, 1981.

(55) Lin, J.C.: Digital computer methods for predicting field intensities, in CRC Handbook of Biological Effects of Electromagnetic Fields, C. Polk and E. Postow, eds., CRC, Boca Raton, FL, In Press.

(56) Ruppin, R.: Calculation of electromagnetic energy absorption in prolate spheroids by the point matching method, IEEE Trans. Microwave Theory Tech. MTT-26:87-90, 1978.

(57) Barber, P.W.: Resonance electromagnetic absorption by non-spherical dielectric objects, IEEE Trans. Microwave Theory Tech. MTT-25:373-381, 1977.

(58) Waterman, P.C.: Symmetry, unitarity and geometry in electromagnetic scattering, Phys. Rev. D. 3:825-839, 1971.

(59) Barber, P.W.: Electromagnetic power deposition in prolate spheroid models of man and animals at resonance, IEEE Trans. Biomed. Eng. BME-24:513-521, 1977.

(60) Iskander, M.F., Lakhtakia, A. and Durney, C.H.: A new iterative procedure to solve for scattering and absorption by dielectric objects, Proc. IEEE 70:1361-1362, 1982.

(61) Harrington, R.F.: Field Computation by Moment Methods, Macmillan, New York, 1968.

(62) Hagmann, M.J., Gandhi, O.P. and Durney, C.H.: Upper bound on cell size for moment-method solutions, IEEE Trans. Microwave Theory Tech, MTT-25:831-832, 1977.

(63) DeFord, J.F., Gandhi, O.P. and Hagmann, M.J.: Moment-method solutions and SAR calculations for inhomogeneous models of man with large number of cells, IEEE Trans. Microwave Theory Tech. MTT-31:848-851, 1983.

(64) Sultan, M.F. and Mittra, R.: An iterative method for analyzing the electromagnetic field distribution inside inhomogeneous lossy dielectric objects, IEEE Trans. Microwave Theory Tech. MTT-33:163-168, 1985.

(65) Chen, K.M. and Guru, B.S.: Internal EM field and absorbed power density in human torsos induced by 1-500-MHz EM waves, IEEE Trans. Microwave Theory Tech. MTT-25:746-756, 1977.

(66) Chen, K.M. and Guru, B.S.: Induced EM fields inside human bodies irradiated by EM waves of up to 500 MHz, J. Microwave Power 12:173-183, 1977.

(67) Chen, K.M. and Guru, B.S.: Focal hyperthermia as induced by RF radiation of simulacra with embedded tumors and as induced by EM fields in a model of a human body, Radio Sci 12 (suppl.):27-37, 1977.

(68) Hagmann, M.J., Gandhi, O.P. and Durney, C.H.: Procedures for improving convergence of moment-method solutions in electromagnetics, IEEE Trans. Antennas Propagat AP-26:743-748, 1978.

(69) Hagmann, M.J., Gandhi, O.P. and Durney, C.H.: Numerical calculation of electromagnetic energy deposition for a realistic model of man, IEEE Trans. Microwave Theory Tech. MTT-27:804-809, 1979.

(70) Hagmann, M.J., Gandhi, O.P., D'Andrea, J.A. and Chatterjee, I.: Head resonance: Numerical solutions and experimental results, IEEE Trans. Microwave Theory Tech. MTT-27:809-813, 1979.

(71) Hohmann, G.W.: Three-dimensional induced polarization and electromagnetic modeling, Geophysics 40:309-324, 1975.

(72) Massoudi, H., Durney, C.H. and Iskander, M.F.: Limitations of the cubical block model of man in calculating SAR distributions, IEEE Trans. Microwave Theory Tech. MTT-32:746-752, 1984.

(73) Hagmann, M.J.: Comments on "Limitations of the cubical block model of man in calculating SAR distributions," IEEE Trans. Microwave Theory Tech. MTT-33:347-350, 1985.

(74) Ho, H.S.: Dose rate distribution in triple-layered dielectric cylinder with irregular cross section irradiated by plane wave sources, J. Microwave Power 10:421-432, 1975.

(75) Wu, T.K. and Tsai, L.L.: Electromagnetic fields induced inside arbitrary cylinders of biological tissue, IEEE Trans. Microwave Theory Tech. MTT-25:61-65, 1977.

(76) Wu, T.K.: Electromagnetic fields and power deposition in body-of-revolution models of man, IEEE Trans. Microwave Theory Tech. MTT-27:279-283, 1979.

(77) Massoudi, H., Durney, C.H., Barber, P.W. and Iskander, M.F.: Postresonance electromagnetic absorption by man and animals, Bioelectromagnetics 3:333-339, 1982.

(78) Lee, J.H. and Chen, K.M.: Eddy currents induced by RF magnetic fields in biological bodies, Radio Sci. 17:61S-76S, 1982.

(79) Borup, D.T. and Gandhi, O.P.: Fast-fourier-transform method for calculation of SAR distributions in finely discretized inhomogeneous models of biological bodies, IEEE Trans. Microwave Theory Tech. MTT-32:355-360, 1984.

(80) Borup, D.T. and Gandhi, O.P.: Calculation of high-resolution SAR distributions in biological bodies using the FFT algorithm and conjugate gradient method, IEEE Trans. Microwave Theory Tech. MTT-33:417-419, 1985.

(81) Lynch, D.R., Paulsen, K.D. and Strohbehn, J.W.: Finite element solution of Maxwell's equations for hyperthermia treatment planning, J. Comput. Phys. 58:246-269, 1985.

(82) Guy, A.W.: Electromagnetic fields and relative heating patterns due to a rectangular aperture source in direct

contact with bilayered biological tissue, IEEE Trans. Microwave Theory Tech. MTT-19:214-223, 1971.

(83) Ho, H.S., Guy, A.W., Sigelmann, R.A. and Lehmann, J.F.: Microwave heating of simulated human limbs by aperture sources, IEEE Trans. Microwave Theory Tech. MTT-19:224-231, 1971.

(84) Yang, G.Y., Chou, C.K. and Guy, A.W.: Theoretical and experimental determination of SAR patterns for spherical tissue models in a rectangular resonant cavity, Bioelectromagnetics 5:89-99, 1984.

(85) Lakhtakia, A., Iskander, M.F., Durney, C.H. and Massoudi, H.: Absorption characteristics of prolate spheroidal models exposed to the near fields of electrically small apertures, IEEE Trans. Biomed. Eng. BME-29:569-576, 1982.

(86) Chatterjee, I., Hagmann, M.J. and Gandhi, O.P.: Electromagnetic energy deposition in an inhomogeneous block model of man for near-field irradiation conditions, IEEE Trans. Microwave Theory Tech. MTT-28:1452-1459, 1980.

(87) Karimullah, K., Chen, K.M. and Nyquist, D.P.: Electromagnetic coupling between a thin-wire antenna and a neighboring biological body: Theory and experiment, IEEE Trans. Microwave Theory Tech. MTT-28:1218-1225, 1980.

(88) Nyquist, D.P., Chen, K.M. and Guru, B.S.: Coupling between small thin-wire antennas and a biological body, IEEE Trans. Antennas Propagat. AP-25:863-866, 1977.

(89) Spiegel, R.J.: The thermal response of a human in the near-zone of a resonant thin-wire antenna, IEEE Trans. Microwave Theory Tech. MTT-30:177-185, 1982.

(90) Zhu, S.G., Chen, K.M. and Chuang, H.R.: Interaction of the near-zone fields of a slot on a conducting sphere with a spherical model of man, IEEE Trans. Microwave Theory Tech. MTT-32:784-795, 1984.

(91) Drago, G.P., Marchesi, M. and Ridella, S.: The frequency dependence of an analytical model of an electrically stimulated biological structure, Bioelectromagnetics 5:47-62, 1984.

(92) Adey, W.R.: Frequency and power windowing in tissue interactions with weak electromagnetic fields, Proc. IEEE 68:119-125, 1980.

(93) Blackman, C.F., Benane, S.G., House, D.E. and Joines, W.T.: Effects of ELF (1-120 Hz) and modulated (50 Hz) RF fields on the efflux of calcium ions from brain tissue in vitro, Bioelectromagnetics 6:1-11, 1985.

(94) Stuchly, M.A. and Stuchly, S.S.: Permittivity of mammalian tissues in vivo and in vitro: Advances in experimental techniques and recent results, Int. J. Electronics 56:443-456, 1984.

10

Recent Developments in the Theory of Ion Flow Across Membranes Under Imposed Electric Fields

MARTIN BLANK

Department of Physiology and Cellular Biophysics
Columbia University, New York, New York

INTRODUCTION

Electric and electromagnetic fields elicit a wide variety of responses in biological systems (1). The mechanisms that account for the responses are unknown, and each biological system has been considered essentially a black box with a particular transfer function. This has been justified by noting that the theory of ionic equilibrium across membranes cannot even explain all steady-state observations, so it is hard to extrapolate to transient processes in the more complex cells and tissues. However, in the last few years there has been significant progress in our understanding of the way in which electric fields interact with cell membranes, and this has increased our ability to deal with more complex effects.

In this chapter we shall show how a better understanding of ion transport across membranes can be obtained by focusing on surface processes, and how this development has the potential for explaining some of the unusual effects reported in different fields of bioelectricity. First, we shall consider ion flow across excitable membranes, show the inadequacy of the accepted theory and develop a model that can take into account: (1) ionic processes in the electrical double-layers at the surfaces of membranes; (2) transient as well as steady-state processes; and (3) the properties of voltage-gated channels in the membranes. Then we shall show how this approach, based upon electrodiffusion, leads to an understanding of selectivity in ion channels and ion movement during excitation. Finally, we shall consider the effects of alternating electric fields.

ION FLOW IN EXCITABLE MEMBRANES

The classic problem in bioelectricity, nerve excitation, has served as a model for understanding the interactions of electric fields with biological systems. Excitation has been studied by applying an electrical stimulus to a cell membrane and following the ensuing changes in membrane potential and ion fluxes. There are many similarities between the processes described in the different types of excitable membranes, but there is no accepted theoretical explanation for the ionic processes during excitation. The descriptions are phenomenological, since the observations of changes in electrical potential and ionic fluxes do not follow our expectations based on related processes in solution.

It is generally accepted that the electrical potential across a membrane in a steady state is due to ion movement. Ions diffuse as a result of gradients in the concentration, ΔC (or more accurately in the chemical potential $\Delta \log C$), and in the electric field, ΔE. The movement of ions (charges) generates electrical potentials, and the ions that have the greatest fluxes have the greatest effect on the membrane potential. A steady state arises when there is a balance between the different chemical forces for each ion and the single electrical potential across the membrane that results from their combined effects.

The permeability of the membrane to each ion, g_i, is different, and the resulting fluxes, J_i, can be written in terms of the differences in chemical and electrical potential across the membrane following the Nernst-Planck equation.

$$J_i = g_i C_i(58 \ \Delta \log C_i \pm \Delta E) \qquad (1)$$

where 58 is the numerical value in millivolts that includes the constants RT divided by the Faraday. Because of the greater permeability of the steady-state (resting) membrane to potassium ions than to the other cations and anions, the resting membrane potential is determined largely by the potassium ion distribution and, in the squid, has a value around -65 mV (Figure 1).

These ion fluxes would lead to the gradual dissipation of both chemical and electrical potentials, but active transport mechanisms utilize energy from ATP and pump ions against their chemical gradients to restore the original concentrations. To maintain the steady state, the pump fluxes, P_i are equal to and oppose the leak fluxes, J_i. If we add the pump and the leak fluxes to give zero net flux across the membrane, we obtain the Goldman equation, which defines the steady-state potential.

$$0 = g_i\ C_i(58\ \Delta\log C_i \pm \Delta E) \qquad (2)$$

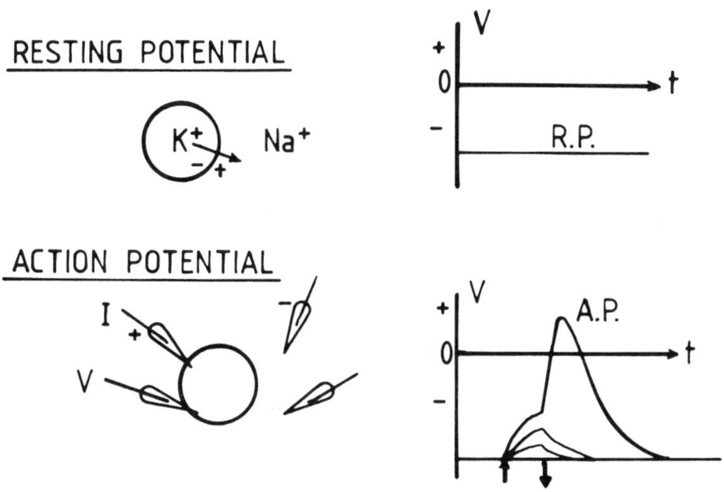

FIGURE 1. The resting potential (R.P.) is established by the greater diffusion of K^+ ions out of the cell than Na^+ ions in. When the membrane potential (V) is measured as a function of time (t), a steady value is seen. The lower diagram shows two sets of electrodes across the cell membrane, the upper pair (labeled I) to pass current with the cathode outside, and the lower pair (labeled V) to measure the membrane potential. When different magnitudes of current are turned on and off (as indicated by the arrows below the X-axis), the potential is displaced as shown. At a critical point, the potential (A.P.) rapidly reverses and then returns to resting level, the total period being about one millisecond. This is called the action potential. (Reproduced with permission from Plenum Publishing Corp.)

When current flows between electrodes placed across a membrane there will be a depolarization if the cathode is on the outside. For small depolarizations, there is a return to the resting value of the membrane potential. However, if the current exceeds a critical value, an action potential occurs (a rapid depolarization through zero, followed by a reverse polarization of about +40 mV and a return to the resting potential) (Figure 1). These changes cannot be explained, but are described in terms of permeabilities that are complicated functions of time and potential.

The ion flow during an action potential has been analyzed using the voltage-clamp method, a technique that imposes a fixed polarization across the membrane and maintains this value by sensing any departure due to the flow of current through the membrane and passing an equal and opposite current to compensate. The voltage-clamp method therefore measures the membrane current with time, at a particular potential. Typical curves show an early inward current followed by an outward current. From ion substitution experiments, the early inward current was found to be related to the external Na^+ ion concentration, and the later outward current was found to be associated with the internal K^+ ion concentration. The Hodgkin-Huxley model of the action potential is based on the results of the voltage clamp technique, and it assumes that depolarization of an excitable cell membrane leads to an initial inwardly directed current due to sodium ions followed by a later outwardly directed current, due to potassium ions. The entire current is then given as the sum of the two ionic currents plus a small leak current and also a capacitance current that is related to the rapid charging processes associated with changes of polarization.

The accepted explanation of the ion fluxes in excitable membranes today is essentially unchanged from that elaborated by Hodgkin and Huxley over 30 years ago (2). It is based on the existence of two types of voltage-gated channels with different kinetic mechanisms, each with absolute ionic specificity for either sodium or potassium. The quantitative kinetic description of the currents through the channels is based on three empirical differential equations for channel activation (i.e., opening) and inactivation (closing), each governed by three rate constants. There is also an empirical combination of the integrated forms of these equations for the total membrane current. This system of equations, with its many parameters, is very useful in that it precisely describes the total ionic current under a variety of conditions. The limitation of the formulation is the empirical nature of the equations and the absence of any theoretical basis for such difficult ideas as absolute ion specificity in an open channel.

The molecular picture of two ion-specific channels, one selecting sodium and the other potassium, that behave as complicated functions of membrane voltage and time is difficult to understand. The ion selectivity of a channel is usually considered in terms of a tight filter or bottleneck that discriminates on the basis of ion size, with factors such as ionic hydration and polarizability affecting the process. This approach leads to the paradox of a constricted pathway with a relatively large ionic conductance. The high rate of ion flow through the ion-selective channels of excitable membranes, which is not much lower than the rate in free solution, is incompatible with a bottleneck type of selective channel. Also, the selectivity is unusual in that lithium passes through the sodium channel almost as rapidly as sodium, and the potassium channel allows the larger thallium ion to pass through at almost twice the rate. It is apparent that ion selectivity is much more complex than filtration, and it may not even involve filtration in the usual sense.

Thus, the fundamental mechanisms of membrane responses to imposed electric fields are not well enough understood to permit an approach to the study of the responses of more complex biological systems. It is important to develop a better theoretical basis for understanding membrane processes. I shall describe an approach that may be helpful in this regard.

ELECTRICAL DOUBLE LAYERS - THE SURFACE COMPARTMENT MODEL (SCM)

One rather obvious limitation of the prevailing ideas on ion flow across membranes is the neglect of interfacial effects in the descriptions. Teorell (3) introduced and used the idea that surface potentials are not equal to bulk potentials, but this idea has not been applied generally to membrane-transport processes. Actually, many surface properties are different from bulk values. At charged surfaces, ionic concentrations and electrical potentials differ from values in the adjacent solutions. In these electrical double-layer regions, each ionic species is in equilibrium with the ions in solution, because the electrical potential difference is balanced by the chemical potential difference. If the charge on the surface is changed, by imposed electric fields or by the binding of charged species, the ions re-equilibrate. In membrane systems, two charged surfaces are involved and the re-equilibration of ions can also occur through ion channels connecting the two electrical double layers.

Membrane surfaces can also act like electrical capacitors. Interfacial capacitance is generally large, and varies with the potential and the ionic concentration. It is not easily detected because it is in series with the much lower dielectric capacitance. However, it can cause large changes in the partitioning of a potential imposed across a membrane.

In the last few years there has been renewed interest in the surface properties of membranes. As a result, much has been learned about electrostatic potentials and ion adsorption at membrane/solution interfaces (4). The numbers of surface charges on axon membranes have also been determined (5), and the increased surface concentrations due to ion binding and transference have also been considered (6). There have also been attempts to use surface properties in the analysis of passive ion-transport processes (7-12).

The surface compartment model (SCM) was introduced specifically to study the role of surface processes in ion transport, and to show that ionic concentration changes in the electrical double layer can contribute to unusual ion fluxes during excitation (13). First, a steady-state model of an axon membrane was presented (14); the model was then extended to transient conditions (15). Voltage-clamp solutions were obtained under constant high permeability (16). After developing a model for a voltage-sensitive channel (17), the transient SCM was studied with a gated channel (18), and under conditions where it was possible to show the kinetic basis for channel specificity (19).

The basic SCM equations used (13-19) included the effects of surface charge, surface capacitance, surface potential and ion-binding equilibria in the steady state. The highly coupled, independent differential equations are given below in terms of a membrane that consists of the discrete regions shown in Figure 2. Six equations describe the time variation of the ionic concentrations in the two surface compartments:

$$\dot{N}2 = (1/L2)(JN1 - JN - PN - \dot{N}22) \tag{3}$$

$$\dot{K}2 = (1/L2)(JK1 - JK - PK - \dot{K}22) \tag{4}$$

SURFACE COMPARTMENT MODEL

COMPARTMENT	OUTSIDE	OUTSIDE SURFACE	MEMBRANE	INSIDE SURFACE	INSIDE
Thickness	Reservoir	L2		L3	Reservoir
Potential	E1 = 0	E2		E3	E4
Capacitance		CAP 1	CAP 2	CAP 3	
Sodium	N1	N2		N3	N4
Passive Flux	JN1 →		JN →	JN3 →	
Pump Flux			PN →		
Potassium	K1	K2		K3	K4
Passive Flux	JK1 →		JK →	JK3 →	
Pump Flux			PK →		
Anion	A1	A2		A3	A4
Passive Flux	JA1 →		JA →	JA3 →	
Pump Flux			PA →		
Current	I →		I →	I →	
Surface Sites					
Charge		X2	JX →	X3	
Bound Sodium		N22		N33	
Bound Potassium		K22		K33	

FIGURE 2. Symbols used in the Surface Compartment Model. Fluxes between compartments are shown as arrows pointing in the positive direction. (Reproduced from Bioelectrochemistry and Bioenergetics, with permission).

$$\dot{A}2 = (1/L2)(JA1 - JA - PA) \tag{5}$$

$$\dot{N}3 = (1/L3)(JN + PN - JN3 - \dot{N}33) \tag{6}$$

$$\dot{K}3 = (1/L3)(JK + PK - JK3 - \dot{K}33) \tag{7}$$

$$\dot{A}3 = (1/L3)(JA + PA - JA3). \tag{8}$$

In the equations, the J's are fluxes driven by electrochemical potential differences and are given by Equation (1). The four equations that give the changes in the bound cations at the two surfaces are:

$$\dot{N}22 = BF(X2)(N2) - BR(N22) \tag{9}$$

$$\dot{K}22 = BF(X2)(K2) - BR(K22) \tag{10}$$

$$\dot{N}33 = BF(X3)(N3) - BR(N33) \tag{11}$$

$$\dot{K}33 = BF(X3)(K3) - BR(K33), \tag{12}$$

where BF and BR are the forward and reverse kinetic constants, respectively.

Changes in the negative surface charge on the sides of the membrane are given by:

$$\dot{X}2 = -(\dot{N}22 + \dot{K}22) - JX \tag{13}$$

$$\dot{X}3 = -(\dot{N}33 + \dot{K}33) + JX \tag{14}$$

where JX is the gating current associated with changes of polarization.

The currents in the surface compartments and the membrane during voltage clamp (E4 = 0) are given by:

$$I = FA(JN1 + JK1 - JA1) - (CAP1)\dot{E}2 \tag{15}$$

$$I = FA(JN3 + JK3 - JA3) + (CAP3)(\dot{E}3 - \dot{E}4) \tag{16}$$

$$I = FA(JN + JK - JA - JX + PN + PK - PA) + (CAP2)(\dot{E}2 - \dot{E}3) \tag{17}$$

where FA is the Faraday. (The $\dot{E}4$ term is necessary for the initial current when imposing the voltage clamp.) Equations (15)-(17) can be solved for the state variables E2 and E3 as well as the current, I.

To study the behavior of the SCM equations, we have used initial conditions based on published values for the squid axon. Although twelve parameters must be set, all but two of these values are fixed by experimental observations. The values that cannot be obtained from the literature can be estimated from their effects on the membrane current.

When the equations are solved for a voltage clamp, going from -65 mV to -20 mV or to 0 mV, a small fraction of the surface charge initially moves across the membrane, in line with the observed gating current. A voltage-gated channel is included as an equation for an increase in ion permeability upon depolarization. Under these conditions, we generally obtain an inward (positive) current, followed by an outward (negative) current, as observed in voltage-clamp experiments on squid axons (2)(Figure 3).

When the permeability to potassium is set equal to zero, the current resembles the curves obtained when introducing tetraethylammonium ions (TEA) in place of potassium. When the permeability to sodium is set equal to zero, the current resembles curves obtained when sodium ions are replaced by choline (or when the poison TTX is used). These results suggest that the voltage-clamp current can be resolved into an early inward component due to sodium ion and a late outward component due to potassium. However, the ion fluxes in both the inward and outward currents are due to both ions with the initial current dominated by the sodium ion, and the later current, by the potassium ion. The two ion currents interact and when removing either of the ions, we do not resolve the current into its original components. If the SCM results reflect processes that occur during excitation, the voltage-clamp currents can not be resolved into separate ionic currents for sodium (early) and potassium (late).

In general, when we use the SCM equations together with an equation for a channel whose cation permeability increases upon depolarization, we obtain membrane currents with an initial inward (positive) ionic current, followed by an outward (nega-

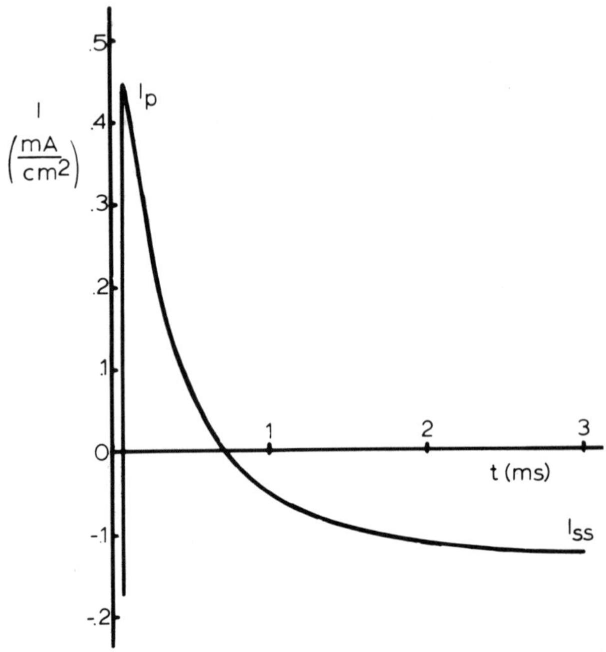

FIGURE 3. The SCM membrane current (I) versus the time (t) at a clamp voltage of 0 mV, under standard conditions. The initial inward current leading to a peak (I_p), is followed by an outward current (I_{ss}), as observed in squid axon. (Reproduced from Bioelectrochemistry and Bioenergetics, with permission).

tive) ionic current. However, when the gating current is increased or decreased, the result is an ionic current with greater apparent specificity. It appears that a sodium channel could be characterized by rapid-gating currents, and a potassium channel by slower-gating currents, with about an order of magnitude difference in the gating currents between the rate constants (13). A difference in the gating currents between the sodium and potassium channels is in agreement with measurements on ionic channels in excitable membranes. Figure 4 shows the ionic currents for the SCM under identical conditions except for gating charge conductance. The currents are due to all the cations in the system, but appear to be specific to one cation. A membrane composed of fast and slow channels gives rise to

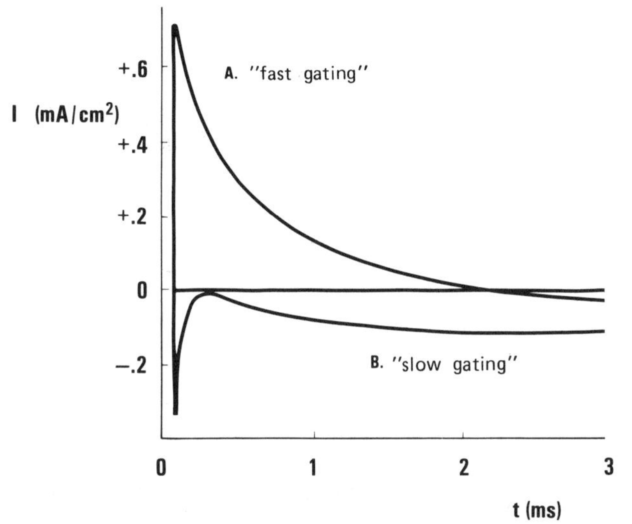

FIGURE 4. The ionic currents across a voltage-dependent channel as functions of time. Curve A is for a sodium channel, and curve B for a potassium channel. The physical properties of the channels are identical except for the one that controls the speed of the gating current. In curve A, the gating charge conductance is 15 times greater than in curve B. The values of the other parameters and the equations themselves can be found in references 14-19. Similar curves can be obtained under many different sets of conditions, including when the surface capacitances are asymmetric, with the outer surface having the greater capacitance. (Reproduced in modified form, with permission, from Bioelectrochemistry and Bioenergetics.)

currents similar to those normally observed during voltage clamp of squid axons, and the currents can be described by the electrodiffusion equations discussed earlier.

A KINETIC BASIS FOR ION SELECTIVITY IN CHANNELS

Using the results described in the last section, it is possible to account for the ionic fluxes in excitable membranes in terms of electrodiffusion. According to the SCM, the fluxes in membranes containing all types of channels would be given by the same equations; only physical constants such as gating current conductance, ion binding, and mobility parameters would differ. Since many ions would have access to the channel, the apparent selectivity of the current would arise largely from the kinetics of channel opening, and would be influenced by various electrical double-layer properties.

It should be remembered that the electrical potential that drives ions across the membrane is not the measured bulk value, but rather the difference between the two surface potentials. The magnitude of this potential difference depends upon the three capacitances in the system, which determine how an imposed change in polarization will be distributed across the membrane. The surface capacitances are especially important, since relatively small changes do not greatly affect the total capacitance, but can cause large changes in the potential difference. However, the dielectric capacitance has a much larger effect. Surface capacitances can be altered significantly by the adsorption of surface-active materials, and various physiologic and pharmacologic agents should be considered in these terms. Also, different types of channels can have different surface capacitances and can respond differently to various agents.

ALTERNATING ELECTRIC FIELDS IN THE SCM

The sodium channel is a unique structure with special properties, but its responses provide some insight into the general effects of electrical stimuli on both electrical double-layers and channels. We have therefore extended our studies with the SCM beyond the voltage-clamp conditions, and applied sinusoidal electric fields to the same system of equations (21). As expected, oscillating stimuli lead to periodic changes in the properties of the electrical double-layers. The currents oscillate out of phase with the voltage, due to the capacitance of the system, and a steady-state amplitude is reached, usually after a few cycles. The changes in the ionic concentrations in the surface compartments reach steady amplitudes, also after a few cycles, that are functions of the frequency of the oscillating electric field. Figures 5 and 6 show the dependence of the maxima of sodium concentration on the inner surface of the sodium channel, Na_i, and of potassium on the outer surface, K_o.

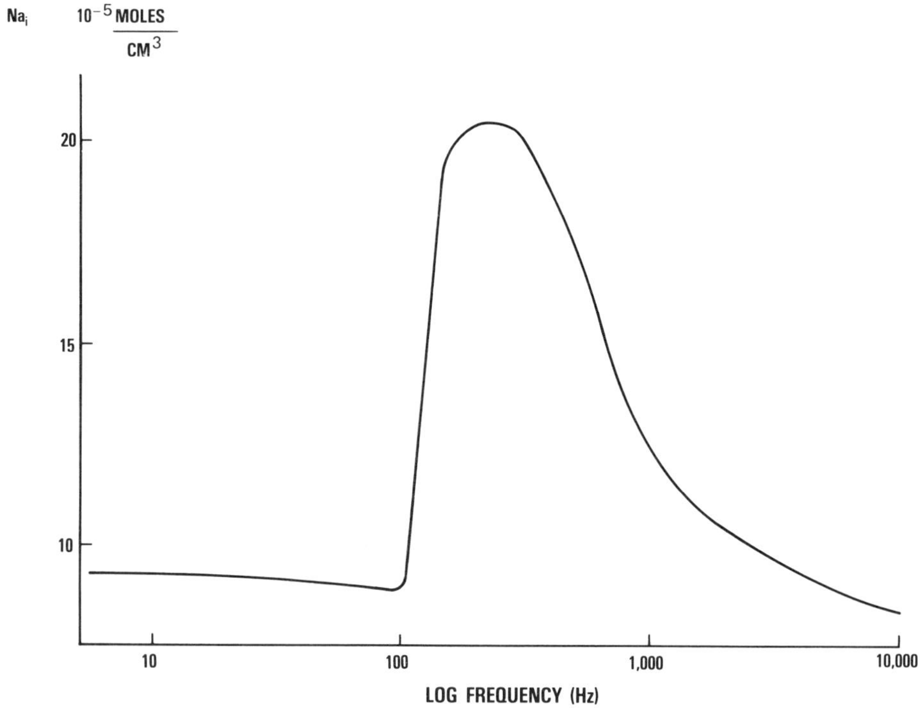

FIGURE 5. The maximum concentration of sodium ions in the electrical double-layer at the inner surface of the sodium-channel protein as a function of the frequency of an imposed sinusoidal voltage of 10 mV across the membrane. (Reproduced with permission from the J. Electrochem. Soc.)

It appears that all of the ionic concentrations, including bound ions, change with the application of oscillating electric fields, but that the percentage change is greatest in those concentrations with the lowest steady-state values. Na_i shows a three-fold increase in concentration for a 10 mV imposed AC signal at 200 Hz.

The SCM equations are frequency dependent and exhibit resonance phenomena. Under the conditions that apply to a sodium channel, the optimal effect occurs at about 200 Hz (Figures 5 and 6), but these maxima can be shifted if the magnitude of the stimulating voltage, the equilibrium binding constant or the ion binding rate constant are changed. The optimum

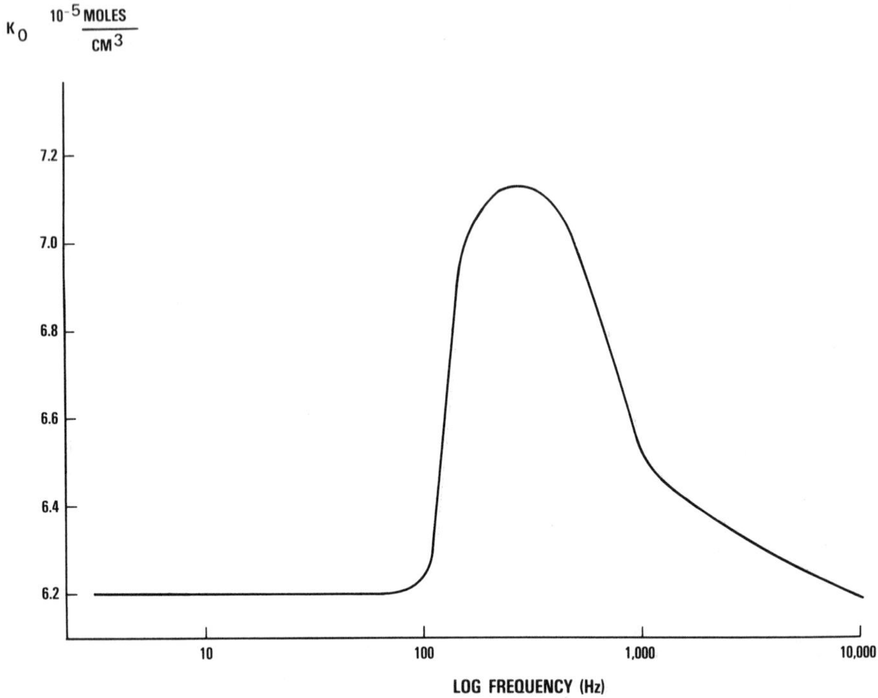

FIGURE 6. The maximum concentration of potassium ions in the electrical double-layer at the outer surface of the sodium-channel protein, as a function of the frequency of an imposed sinusoidal voltage of 10 mV across the membrane. (Reproduced with permission from the J. Electrochem. Soc.)

responses characteristic of a sodium channel can be obtained with several different combinations of parameters.

The steady-state ionic concentrations of Na_i and K_o normally control the activity of the ion-pump enzyme, the Na-K ATPase of cell membranes (22). Increases in these concentrations at particular frequencies should stimulate the enzyme and result in enhanced ion transport. This has been observed at comparable magnitudes of electric field in the ouabain-sensitive accumulation of rubidium by erythrocytes subjected to oscillating electric fields (23). The optimal frequency for the erythrocyte membrane was given as 1000 Hz, but the study included frequency values differing by factors of 10, and the maximum could not be defined more precisely. If the SCM studies included only the points used in the erythrocyte study, the two sets of results (ion accumulation at the surface, and cation accumulation in the erythrocyte) would be similar.

It is important to keep in mind that the SCM results reflect ionic processes at the two surfaces of a sodium channel. Parameters used in the computation have been chosen based on existing data for the sodium channel in the squid axon, and the effects of electric fields on this channel should depend quantitatively on the specific values. The qualitative aspects of the results could apply to the Na-K ATPase involved in the ion accumulation mechanism of the erythrocyte membrane, as well as to other voltage-gated channels. It has been proposed that a subunit of Na-K ATPase is structurally similar to the sodium-channel protein, and that they are encoded by genes that belong to the same family (24). The SCM results are therefore supporting evidence for a functional link between the two membrane protein structures.

Data showing an optimal release of Ca ions by brain slices stimulated at 16 Hz has been reported (1). These results may be due to similar processes, except that the channels responsible for transporting Ca ions have different physical properties. The observed resonance of calcium channels is close to the value calculated for a sodium channel.

A further outcome of our study of oscillating electric fields applied to the SCM is the prediction of unusual responses by the sodium channel that should influence excitability. The concentration changes that are calculated should be measurable, and should lead to changes in the measured electrical potential and ion fluxes across the membranes. The existence of optimal frequencies in these measurements would provide both a demonstration of the role of electrical double-layer processes in membrane function, and a test of the value of the SCM in predicting such effects.

ION-PUMPING PROCESSES

The SCM has shown that ionic processes in the electrical double-layer regions of membranes can lead to either inward or outward currents depending upon the magnitudes of the kinetic constants. The ability of an ionic flux to go in a direction opposite to that predicted on the basis of macroscopic electrochemical potentials, is the main attribute of an ion pump. Therefore, the SCM results provide a basis for considering the kinds of processes involved in ion pumping through channel structures in terms of electrodiffusion theory. In the SCM, a shift of charge from a high charge density region to a low charge density region, in response to a depolarizing stimulus, can cause a channel to open (Figure 7)(17). The same charge shift must free bound counter-ions in the high charge region and bind counter-ions in the low charge region. The two changes can result in a reversed local concentration gradient driving ions through the opened channel, and a diminished electrical gradient

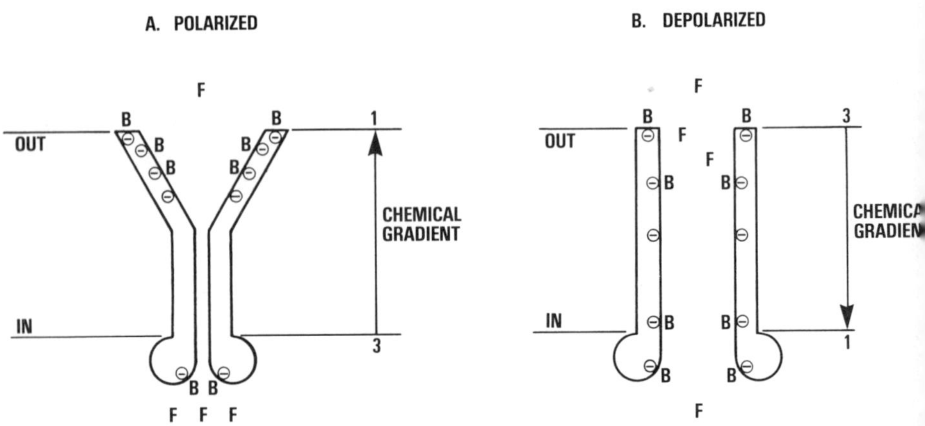

FIGURE 7. An outline of a hypothetical mechanism for ion pumping utilizing transient ion gradients. The arrows indicate the reversed directions of the chemical gradient for free (F) cations between the two electrical double-layers across a voltage-sensitive channel. A: In the resting polarized state, the distribution of bound (B) and free cations, which depends upon the distribution of negative charges, is shown with a closed channel. B: The electrical gradient is changed rapidly, the negative charges are redistributed, the channel opens and the ions re-equilibrate. A quick depolarization leads to a redistribution of B and F cations and a reversal of the F gradient as indicated. (Reproduced with permission from Studia Biophysica.)

across the channel. If specific adsorption is an additional factor, so that potassium or calcium ions are preferentially adsorbed, a gradient of that particular ion will be created. These kinds of processes, coupled with charge shifts brought about by an ATPase could create the local concentration gradients that lead to ion pumping.

SUMMARY

There is a great need for a theoretical model of ion transport in order to account for the observed effects of oscillating electric fields on biological membranes. The SCM appears to be an important contribution toward meeting this need. It is derived from first principles, and takes into account ionic processes in the electrical double-layers of membranes. A voltage-gated channel is included as a voltage-dependent permeability, and under voltage clamp, the SCM yields the currents shown by either sodium or potassium channels of squid axon. The SCM therefore accounts for the ionic fluxes during excitation in terms of electrodiffusion theory, and it explains the apparent selectivity of the two channels on the basis of the difference in the gating currents. It can also account for some of the observed effects of applied oscillating voltages, since the membrane-channel model leads to concentration changes that are frequency dependent. These results suggest explanations for the observed effects of oscillating electric fields on ion transport and provide insights into the functioning of ion pumps.

The SCM offers a physically based theory for considering the mechanisms of many membrane processes involving ion transport such as sensory and energy transduction. The ability of the model to explain the phenomena discussed here in terms of fundamental ionic processes suggests that it should be applied more frequently.

ACKNOWLEDGEMENT

The author acknowledges, with thanks, the support of the Office of Naval Research for the research summarized in this paper.

REFERENCES

(1) Adey, W.R.: Tissue interactions with nonionizing electromagnetic fields, Physiol. Rev. 61:435-514, 1981.

(2) Hodgkin, A.L. and Huxley, A.F.: A quantitative description of membrane current and its application to conduction and excitation in nerve, J. Physiol. (Lond.) 117:500-544, 1952.

(3) Teorell, T.: An attempt to formulate a quantitative theory of membrane permeability, Proc. Soc. Exp. Biol. Med. 33:282-285, 1935.

(4) McLaughlin, A.: Electrostatic potentials at membrane-solution interfaces, Curr. Top. Membr. Transp. 9:71-144, 1977.

(5) Gilbert, D.L.: Fixed surface charges, in Biophysics and Physiology of Excitable Membranes, W.J. Adelman, ed., Van Nostrand Reinhold, New York, 359-378, 1971.

(6) Blank, M. and Feig, S.: Electric fields across water-nitrobenzene interfaces, Science 141:1173-1174, 1963.

(7) Miller, I.R. and Blank, M.: Transport of ions across lipid monolayers: Reduction of polarographic currents of Cu^{++} by decylammonium monolayers, J. Colloid Interface Sci. 26:34-40, 1968.

(8) Britten, J.S. and Blank, M.: The effect of surface charge on interfacial ion transport, Bioelectrochem. Bioenergetics 4:209-216, 1977.

(9) Pilla, A.A. and Margules, G.: Dynamic interfacial electrochemical phenomena at living cell membranes: Applications to the toad urinary bladder membrane system, J. Electrochem. Soc. 124:1697-1706, 1977.

(10) Schmukler, R. and Pilla, A.A.: A transient impedance approach to nonfaradaic electrochemical kinetics at living

cell membranes, J. Electrochem. Soc. 129:526-528, 1982.

(11) Lorenz, W. and Schulze, K.D.: Macroscopic-dynamical model of membrane charge transfer, Z. Phys. Chem. (Liepzig) 262:1032-1040, 1981.

(12) Schulze, K.D. and Lorenz, W.: Surface controlled electrical instabilities at membranes, Z. Phys. Chem. (Liepzig) 263:27-32, 1982.

(13) Blank, M.: A physical interpretation of the ionic fluxes in excitable membranes, J. Colloid Sci. 20:933-949, 1965.

(14) Blank, M. and Britten, J.S.: The surface compartment model of the steady state excitable membrane, Bioelectrochem. Bioenergetics 5:528-540, 1978.

(15) Blank, M. and Kavanaugh, W.P.: The surface compartment model (SCM) during transients, Bioelectrochem. Bioenergetics 9:427-438, 1982.

(16) Blank, M., Kavanaugh, W.P. and Cerf, G.: The surface compartment model - Voltage clamp, Bioelectrochem. Bioenergetics 9:439-458, 1982.

(17) Blank, M.: The surface compartment model (SCM) - Role of surface charge in membrane permeability changes, Bioelectrochem. Bioenergetics 9:615-624, 1982.

(18) Blank, M.: The surface compartment model (SCM) with a voltage sensitive channel, Bioelectrochem. Bioenergetics 10:451-465, 1983.

(19) Blank, M.: Properties of ion channels inferred from the surface compartment model (SCM), Bioelectrochem. Bioenergetics 13:93-101, 1984.

(20) Hille, B.: Ionic Channels in Excitable Membranes, Sinauer Inc., Sunderland, MA, 1984.

(21) Blank, M. and Blank, J.N.: Concentration changes at ion channels due to oscillating electric fields, J. Electrochem. Soc. 133:237-238, 1986.

(22) Post, R.L., Merritt, C.R., Kinsolving, C.R. and Albright, C.D.: Membrane adenosine triphosphatase as a participant in the active transport of sodium and potassium in the human erythrocyte, J. Biol. Chem. 235:1796-1802, 1960.

(23) Serpersu, E.H. and Tsong, T.Y.: Stimulation of a ouabain-sensitive Rb^+ uptake in human erythrocytes with an external electric field, J. Membr. Biol. 74:191-201, 1983.

(24) Rossier, B.C.: Biosynthesis of (Na^+, K^+)-ATPase in amphibian epithelial cells, Curr. Top. Membr. Transp. 20:125-145, 1984.

11
Electrofusion of Cells

HERMANN BERG
Academy of Sciences of the GDR
Central Institute of Microbiology and Experimental Therapy
Department of Biophysical Chemistry
DDR-69 Jena, German Democratic Republic

INTRODUCTION

Since the first applications of electric pulses from a Volta pile by Johann Wilhelm Ritter (1776-1810) on tissues of his own body (1), many studies with nerves and muscles have been done. The first evidence for protoplast fusion by electric-field pulses was presented in 1979 and involved plant protoplasts of Rauwolfia and Hordeum (2), and diauxotrophic yeast mutants (3). In the first case, needle electrodes in glass capillaries controlled by a micromanipulator were used (Technique A), whereas for the second case a macrochamber with disk electrodes, an exponential pulse, and agglutinating polyethyleneglycol (PEG) was utilized for fusion and subsequent formation of prototrophic colonies (Technique B). Later, a valuable technique was introduced consisting of a microchamber with parallel wire electrodes; it used dielectrophoresis for cell collection and rectangular pulses for electrofusion under microscopic control (Technique C)(6,7). In all three cases, close contact between the cells to diminish adsorbed water layers and the repulsion of surface charges is the most important prerequisite for successful electrofusion. The initial applications of the three pulse techniques are shown in Table 1 (Figure 1).

Parallel to these pulse techniques, the influence of an alternating current on human red blood cells (dielectrophoresis) was studied, and the agglutination of cells, but not their actual electrofusion, was first described in 1976 (4,23,40).

For electroincorporation, the substances (DNA, proteins, drugs) for penetration are mostly adsorbed at the cell membrane prior to the electric field pulse application (12,13,60-62).

TABLE 1. Initial Applications of the Three Techniques for Electrofusion and Electrotransformation

TECHNIQUE	OBJECT	EVIDENCE	GROUP	REFERENCE	YEAR
Micromanipulator chamber	Plant protoplasts (*Rauwolfia* + *Hordeum*)	Morphological	Senda	2	1979
	Mouse blastomeres	Morphological	Berg	14	1982
	L 1210 Ascites cells	Morphological	Berg	14	1982
Macrochamber	Yeast protoplast mutants	Genetic	Berg	3	1979
	Dictyostelium discoid.	Morphological	Neumann	5	1980
	Mouse fibroblasts	Physiological	Tsong	8	1982
	H.s-virus plasmid into mouse lyoma cells (TK^-)	Genetic	Neumann	12	1982
	B. thuringiensis plasmid into *B. cereus* protoplasm	Genetic	Berg	13	1983
Microchamber (Dielectrophoresis)	Plant protoplasts (*A. sativa*)	Morphological	Zimmermann	6	1980 1981
	Sea urchin eggs	Physiological	Zimmermann	7	1981
	Yeast protoplast mutants	Genetic	Zimmermann	9	1982
	Hybridomas, human cells	Physiological	Zimmermann	10	1982
	Plant protoplasts (tobacco, barley)	Morphological	Berg, Senda	38	1981
	Hybridomas, mouse cells	Morphological	Berg	11	1982

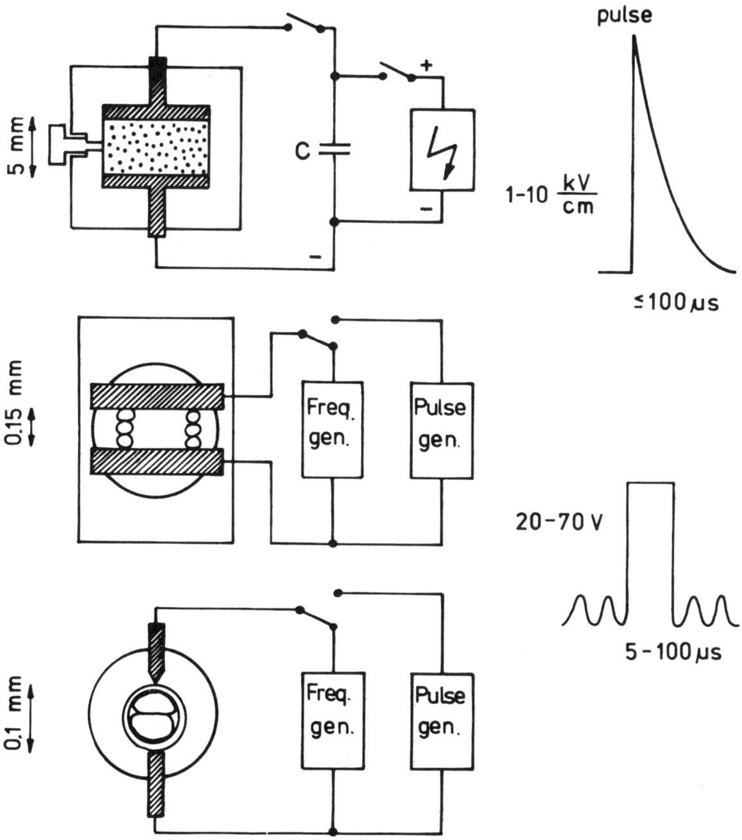

FIGURE 1. Three main techniques for electrofusion, described in the text as (from top to bottom) B, C, and A.

Such studies (electroporation) were performed in the early 1970's (15). Two remarkable examples should be mentioned for red blood cells: the incorporation of DNA from SV 40 or RNA (16), and of cancerostatic methotrexat (17).

In each case, enhancement of membrane pores and/or destabilization (reversible or irreversible breakdown) of the entire membrane structure occurs, usually followed by resealing, depending on the electrical pulse parameters, the preparation of the cells or protoplasts, and the composition of the solution. For each kind of cell a suitable pulse characteristic must be determined to get an optimum yield of fused cells. This brief review is devoted mainly to applications of electric-field pulses as a tool in modern genetics of prokaryotic and eukaryotic cells. In-depth treatments of the broad field are given elsewhere (21,36).

THEORETICAL AND EXPERIMENTAL PRINCIPLES

Successful electrofusion requires cell preparation, cell contact, appropriate equipment, and cultivation of the fused cells. Only some of these factors will be discussed in the following sections (22).

METHODS FOR CONTACTING CELLS

The yield of electrofusion depends strongly on the preparation of the membranes and their contact. For Technique A, a slight mechanical pressure against both cells by the electrode tips forming a contact is enough for starting fusion after the pulse or the dielectrophoretic alignment. For Technique B, agglutination by an agent such as PEG (molecular weight 4000-6000) or dextran (concentration greater than 10%) is most effective. The water layer is squeezed out from the membrane surface, bridges of dextran molecules are formed, and the electrostatic repulsion is decreased. For Technique C, dielectrophoresis (23) is a rather universal method for collecting particles or cells oriented in the direction of electric field lines as a "pearl-chain" formation (Figure 7). This process is brought about by an inhomogeneous alternating electric field which causes the cells to become polarized by dipole induction. The net force causes a motion towards the region of highest field intensity at the electrodes. The cells move along the field lines and approach one another. The dielectric force on a spherical cell of radius r in a field of strength E is proportional r^3 and to the divergence of E^2 (modified by the relative dielectric constants of the medium, ε_1, and the cell, ε_2) (23). The force increases with the square of the applied voltage U, and for cells having small r or for small differences in the dielectric constants ($\varepsilon_1 > \varepsilon_2$) the voltage must be stronger for effective pearl-chain formation. Electrostatic repulsion due to surface charges and water layers on membranes may be overcome by dielectrophoresis in the MHz range. On the other hand, voltage, heat production, and collection time must be controlled such that both breakdown and membrane stabilization are avoided. For cells having radii of 5 μm, the time constant for pearl-chain formation is 1-10 sec in an electric field of 10 kV/m. Along with pearl-chain formation, a remarkable effect occasionally has been observed since 1960 (18,23), namely the spinning of particles and cells at certain resonant frequencies of the electric field. For electrofusion, this rotation is a disturbing event that hinders tight contact, and it must be avoided by changing the frequency of dielectrophoresis. On the other hand, cell rotation and the influence on it by drugs and other substances in solution is now a valuable tool in cell research (18,25,26).

Besides dielectrophoresis, there are other physical possibilities for collection of cells, including centrifugal forces (5) and ultrasonics (24).

METHODS OF PULSE APPLICATIONS FOR ELECTROFUSION

The main aim is to generate a higher transmembrane potential. This induced potential difference ΔU depends on the cell radius, the angle θ between the electric field vector and the surface element considered, and the applied external field strength E_o, and is given (assuming the membrane conductivity is zero) for spheres by (25):

$$\Delta U = 1.5 \ r \ E_o \cos\theta \qquad (1)$$

A more general expression especially suitable for vesical membranes has been given (15).

From Equation (1) it can be seen that pulse effects are strongest at the membrane surface for $\theta = 0$, which occurs at the contact areas of the poles of the cells within the pearl chain. For larger cells the critical transmembrane breakdown voltage (on the order of 1 V) occurs at a lower external field strength. For example, for a cell of r = 5 μm, from Equation (1) we have:

$$E_o = 1 \text{ V}/(1.5 \times 0.5 \times 10^{-3} \text{ cm}) = 1.33 \text{ kV/cm}.$$

The three main experimental systems used to apply external electric fields to cells with r values between 1-50 μm are shown in Figure 1: the micromanipulator technique with two movable needle electrodes (Technique A)(2,14,27); the macrotechnique with disk electrodes and agglutinating polymers (Technique B) (3,5); and the microtechnique with parallel electrodes, in combination with dielectrophoresis (Technique C)(6,7,9-11,38).

Using Technique A for electrofusion of large cells under microscopic observation, the platinum needle electrodes must have a point diameter of less than 100 μm. In the special case of blastomere fusion within the zona pellucida of a mouse zygote (Figure 2) three electrode positions can be employed; 20 μm from the cell surface, a slight tangential contact at the cell surface, or a tight contact. These positions determine the optimum pulse parameters for viability of the fused cells.

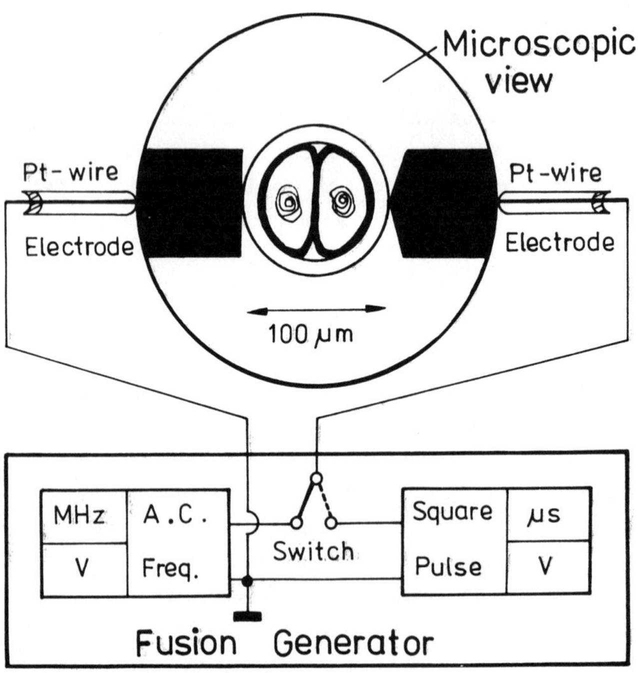

FIGURE 2. Technique A with a two-cell blastomere as the object between the needle electrodes.

Small cells can be aligned between the electrode tips using dielectrophoresis only in a low conducting medium.

Using Technique B, 1-5 ml of a cell suspension can be fused in the macrochambers with stainless-steel electrodes (Figure 3). Chambers having a variable electrode distance are also possible. The electric-field pulses are caused by the discharge of a high-voltage capacitor (0.075-2 μF). The exponential time course of the field has a time constant of τ = RC = 20 μsec and a heating time of RC/2 = 10 μs. The maximum field strength is of the order of 35 kV/cm. The same suspension may be subjected to several discharges. Agglutination of cells before pulse application is a prerequisite, and is sometimes accomplished by centrifugation.

With Techniques A and C, fusion can be observed microscopically and analyzed using an image analyzer. The microchamber, which is mounted on a slide, has two parallel platinum wires separated by > 0.1 mm (Figure 1). Sufficient membrane contact is achieved before the fusion pulse when a flattening in the membrane touching area is observed. Depending on cell diameter, the breakdown field strength is 0.3-12 kV/cm (compare

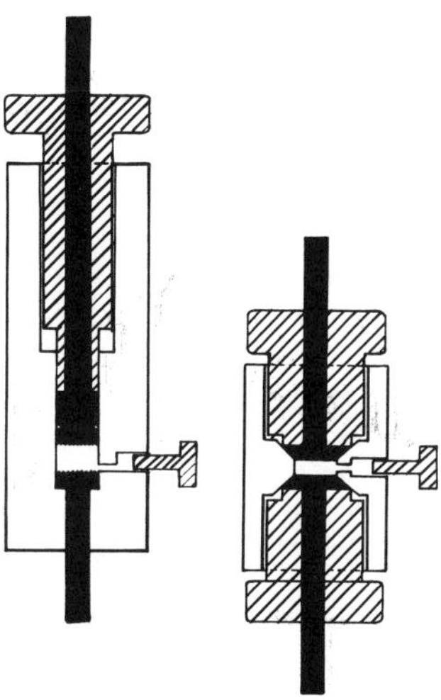

FIGURE 3. Macrochambers: left, for changing the volume continuously; right, for a fixed volume (black, stainless steel electrodes).

Equation (1)). Both the resealing time and the breakdown strength decrease with increasing temperature. Low conductivity of the solution is the prerequisite during this procedure.

RESULTS

THE MICROMANIPULATOR TECHNIQUE

1. Electrofusion of Blastomeres of Murine and Rat Zygotes

For electrofusion of blastomeres surrounded by the zona pellucida, both electrodes are in tight contact as shown in Figure 2. Several minutes after the pulse, the membranes between the blastomeres melt and only one cell is observed; this means that a kind of dedifferentiation has taken place (14,27, 28).

To preserve a high viability rate, as indicated by the ability of the fused cells to divide either within an incubator or after implantation (28), a low field strength ($\leq$ 1 kV/cm) should be applied in about 0.3 msec (Figure 4). Fusion by alternating current has also been detected (28).

Since the aim is to fuse different oocytes or zygotes in the early stages of embryonic development, the zona pellucida must be removed by pronase treatment. Consequently, lower pulse energies are needed and the naturally adherent membranes fuse first, followed by fusion of the outer membranes compressed by the tops of both needle electrodes (27).

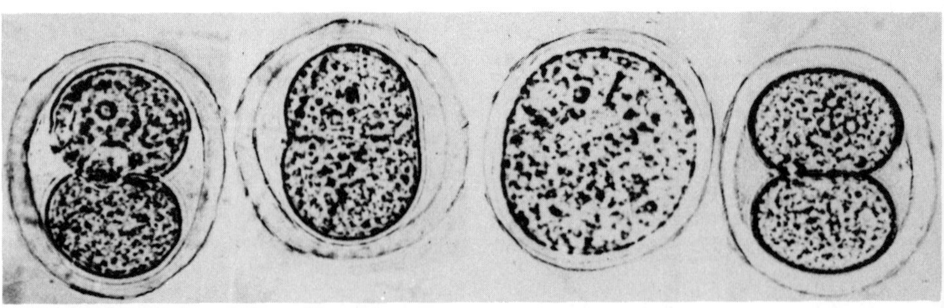

FIGURE 4. Electrofusion of rat blastomeres. Left to right, before the pulse (1 kV/300 μsec); after 2 min; fusion finished after 30 min; division after 24 hours incubation (28).

THE MACROCHAMBER TECHNIQUE

Prior to application of the fusion pulse, the cells must be agglutinated by an appropriate agent such as PEG, dextrane or polylysine. Although their mechanism of action is rather complicated, the basic effects are dehydration and destabilization of membranes, osmotic effects with shrinking of the cells, and finally tight contacts (3).

1. Electrofusion of Protoplasts of Yeast Mutants

Treatment of a suspension of mixed strains of auxotrophic yeast protoplasts in the presence of 30-40% PEG and 10^{-3} M Ca^{2+} ions by electric field pulses strongly enhances the fusion rate as determined from analysis of the number of prototrophic colonies formed on minimal nutrition medium (Figure 5 and Table 2). For optimum intergeneric fusion, higher pulse strengths are necessary (Table 2, *Saccharomycopsis lipolytica* + *Lodderomyces elongisporus*). Isolated prototrophic colonies,especially of the intraspecific type, are stable over more than 20 passages.

TABLE 2. Relative Number of Prototrophic Cell Colonies on Minimal Medium from Electrofusion of Auxotrophic Yeast Protoplast Mutants in Dependence of Field Strength

E/kV cm^{-1}	*S. cereus*	*S. lipolytica*	*S. lipolytica* + *L. elongisporus*
0 (control)	1	1	1
1.25	122	1	-
2.5	233	8	-
3.75	50	36	-
5.0	30	10	-
10.0	-	-	78
15.0	-	-	20
20.0	-	-	8

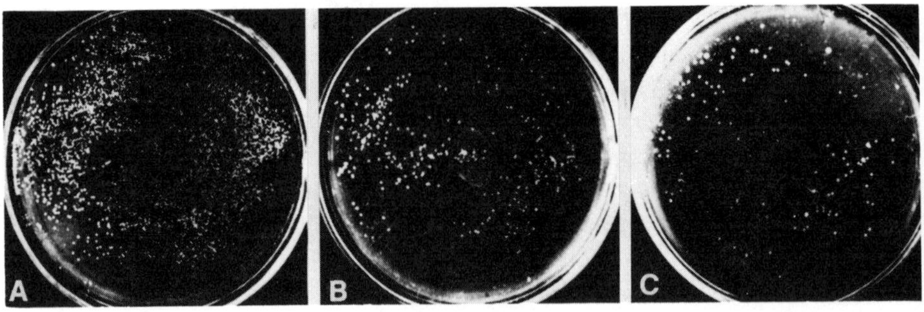

FIGURE 5. Prototrophic yeast colonies formation as a function of pulse intensity (see also the last column of Table 2). A, 10 kV/cm; B, 15 kV/cm; C, 20 kV/cm. Fusion products, 622, 160, 66 respectively (control, 8). (Courtesy H. Weber, Jena)

2. Electrofusion of Two Protoplast Strains of *Bacillus thuringiensis*

Conventional fusion by PEG did not result in any colony formation from the kanamycin-resistant strain and the kanamycin-sensitive strain, which produces a brown pigment on the selective kanamycin medium (39). Treatment with the electric pulse technique in the presence of PEG at 14 kV/cm was not sufficient, but three subsequent 5-μsec pulses at 20 kV/cm yielded recombinants that formed colonies which were kanamycin resistant and were also able to produce the brown pigment. The electrofusion frequency (the ratio of recombinant colonies to regenerated protoplasts) was found to be 10^{-3}. They were stable even after 30 passages on selective as well as on non-selective media. Moreover they were able to form the characteristic protein crystals that are toxic against insects.

3. Electrotransformation of *Bacillus cereus* by the Plasmid-DNAof *Bacillus thuringiensis*

B. cereus is suitable for biotechnological cultivation and therefore was combined with properties of B. thuringiensis by electrotransformation with the useful plasmid pUB 110. The stability of the protoplasts of B. cereus was tested, and no inactivation of cells up to 20 kV/cm was seen. Applying 14 kV/cm three times, a tenfold higher transformation frequency on selective medium occurred as compared to the control. These kanamycin-resistant colonies were stable after 30 passages, but the toxic protein crystals were not found microscopically.

THE MICROCHAMBER TECHNIQUE

1. Electrofusion of Plant Protoplasts

During the fusion process it can be seen that the membrane material from the breakdown area disappears into the cell body, resulting in a sphere from two protoplasts. Twenty-six percent of the original membranes are lost (Figure 6). Further examples and conditions are shown in Table 3 and are described in detail elsewhere (31-37).

2. Electrofusion of Animal Cells

The same basic events occur as in the case of protoplasts, but the cells have smaller diameters (Figure 7), and the necessary pulses are stronger (Table 4). Some peculiarities must be taken into account:

- Shortly after the pulse, the cells should be transferred to a nutrient medium to finish the rounding procedure.

- Red blood cells must be treated by neuraminidase to remove the glycocalix before the pulse application (49). Then giant cells consisting of about 1000 cells (100 μm in diameter) are formed by lateral fusion (48).

- Sea urchin eggs (unfertilized) must be treated with pronase to remove the vitellin layer (7).

- Electrofusion of myeloma cells with spleen cells to get hybridomas for the production of monoclonal antibodies causes difficulties with regard to breakdown because of the 3:1 difference in cell diameters. Addition of pronase (1 mg/ml) to the glucose solution containing the cell suspension may be helpful (10). The fusion products should be placed into wells for cultivation as quickly as possible (11). The ideal is 1:1 electrofusion, however one sometimes gets large multicellular bodies (Figure 8).

DISCUSSION

FUSION PRODUCTS

As can be seen from Tables 3 and 4, more kinds of plant and microbiological protoplasts have been electrofused than have animal cells. There are only a few examples where cultivation of fusion products was performed and unambiguous tests were made

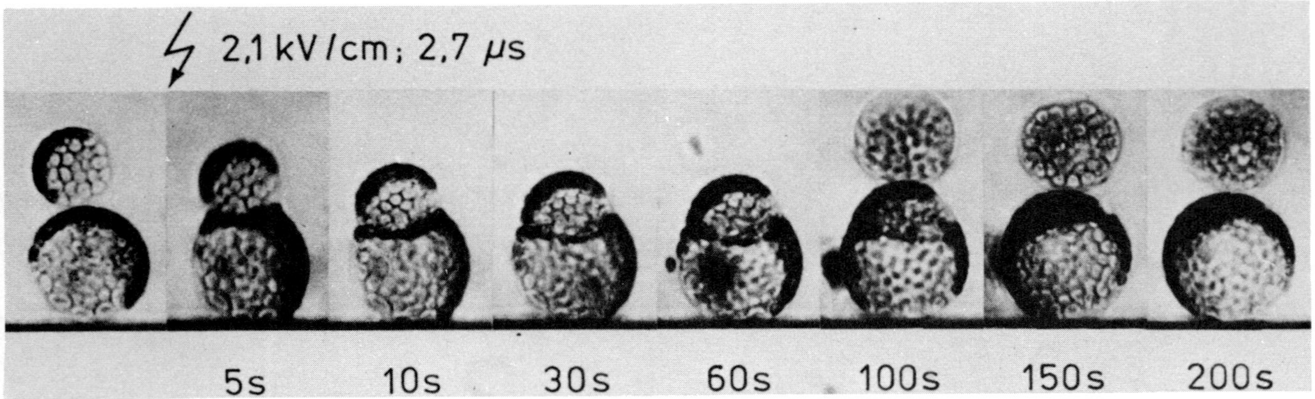

FIGURE 6. Electrofusion of two barley protoplasts (H. vulgare) by a pulse of 2.1 kV, 2.7 μsec using Technique C. (Courtesy H.-E. Jacob, Jena.)

TABLE 3. Electrofusion of Microorganisms and Plant Protoplasts (Different Cells)

OBJECT	ELECTRIC FIELD (kV/cm)	PULSE WIDTH (μsec)	AUTHOR	REFERENCE	YEAR
R. serpent. + H. vulgare	(12 μa)	5000	Senda	2	1979
Yeast mutants: Saccharomycopsis strains S113+26-10	3.75	100	Weber	3	1981
S. lipolitica + L. elongisporus	10	100			
B. thurintiengis + mutants	14	100	Shivarova	39	1983
V. faba*	1.65	50	Scheurich	6	1981
K. daigrem.* mesophyll + vacuole	0.5	50	Vienken	41	1981
A. sativa*	0.6	15	Zimmerman	42	1981
Sacch. c. respir. defect + sacch. c.	7.9	40	Halfmann	9	1982
N. tabacum 63 + N. tabacum 68	1.5	20	Kohn	43	1984
B. napus*	0.95	15	Zachrisson	44	1984
B. campestris + P. acaulis	0.48	15	Zachrisson	44	1984
L. esc. + L. peruvianum	2.2	20	Siegemund	45	1984
A. sativa + Z. mays	0.7	50	Bates	46	1983
N. plumbaginifolia + D. carota	0.8	50	Bates	47	1984
N. Glauca + N. Langsdorffi	1.0	200	Morikawa	57	1986

*different protoplasts of the same plant

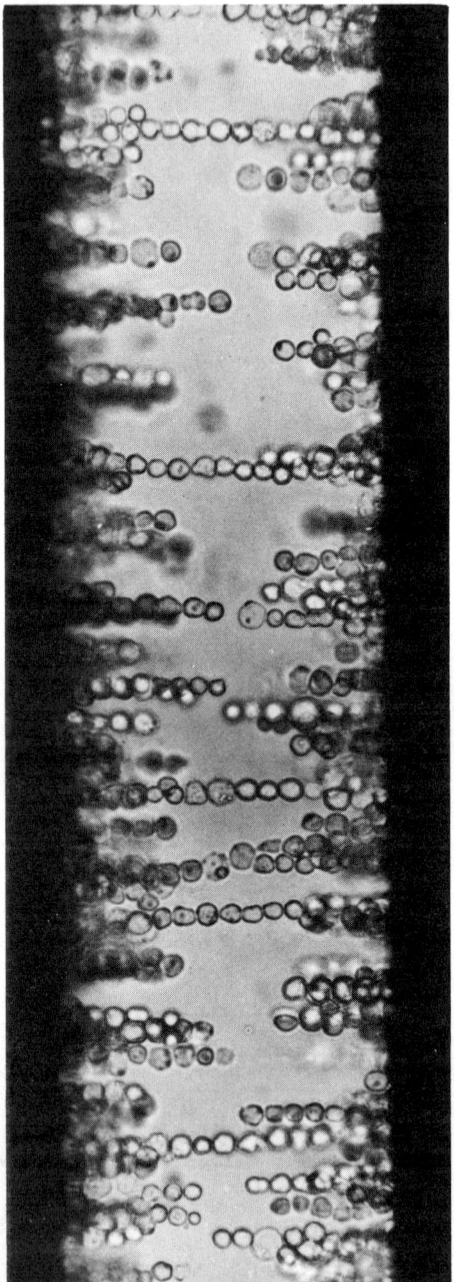

FIGURE 7. Pearl-chain formation of cancer cells L1210 by dielectrophoresis. (Courtesy H.-E. Jacob, Jena.)

TABLE 4. Electrofusion of Animal Cells

OBJECT	ELECTRIC FIELD (kV/cm)	PULSE WIDTH (μsec)	AUTHOR	REFERENCE	YEAR
Mouse blastomeres, rat blastomeres, mouse blastocytes (without and with zona pellucida)	0.75-5	100-300	Berg	14	1982
				27,28	1983
Mouse fibroblasts 3T3	1-2	100	Teissie	8	1982
Macrophages	5	5	Berg	22	1984
Human erythrocytes	2	3 (3x)	Scheurich	48	1981
Mouse L 1210	3.6	25	Berg	38	1981
Friend cells	2.7	20	Pilwat	49	1981
Sea urchin eggs	< 1	≤ 50	Richter	7	1981
Human myeloma + B lymphocytes	3.5	7	Bischoff	10	1982
Mouse fibroblasts 3T3	7	50	Zimmermann	50	1982
Mouse myeloma + spleen cells	3	50	Berg	11	1982
Mouse erythroleukaemic erythroblasts	2.7	20	Zimmermann	51	1984
Mouse myeloma + B lymphocytes	4(4x)	5	Lo	56	1984
Erythrocyte ghosts	5	200	Sowers	59	1986

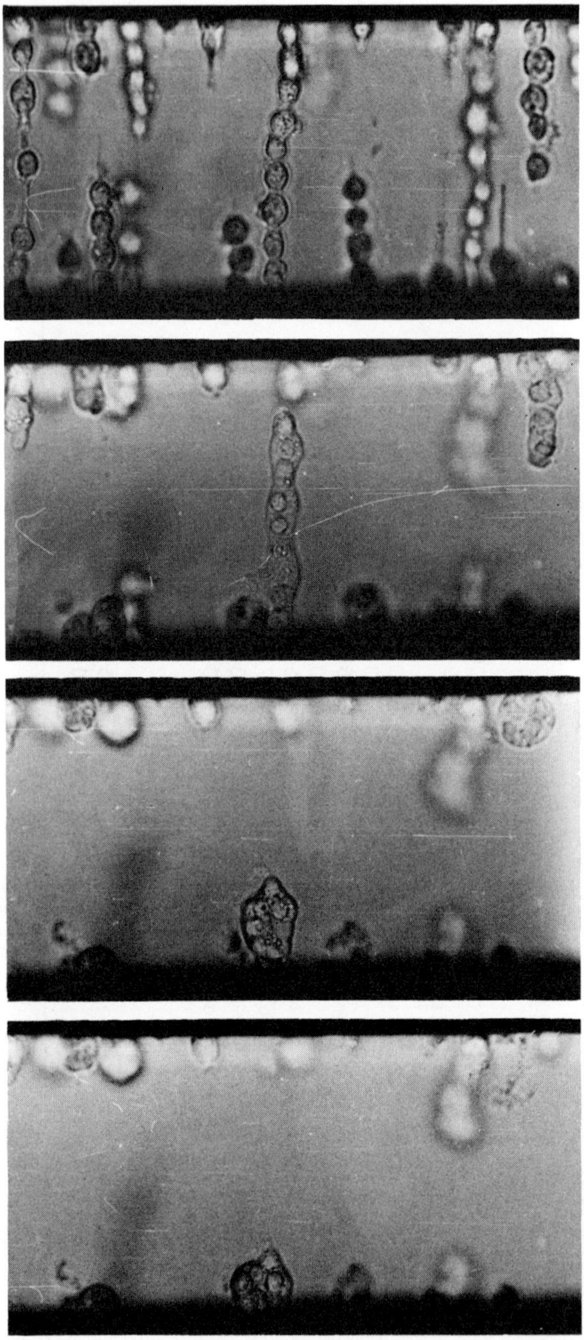

FIGURE 8. Pearl-chain formation and the stepwise fusion of myeloma cells. (Courtesy D. Berg, Jena.)

of the stability and viability of the fused structures; this was done for microorganisms (3,9,39), plants (57,58), zygotes (28), and oocytes (7). In the future, cultivation will be necessary for comparison with other fusion methods or for industrial use (63).

MODELS FOR PULSE EFFECTS ON MEMBRANES

The various theories are listed in Table 5 and explained elsewhere (37). The induced dipole repulsion model is schematically shown in Figure 9. The main idea is similar to that of electrodichroism. Membrane proteins can be oriented suddenly by the field pulse into the field direction, thereby causing destruction of the bilayer and intermingling of the adjacent parts of two cells thus starting the fusion.

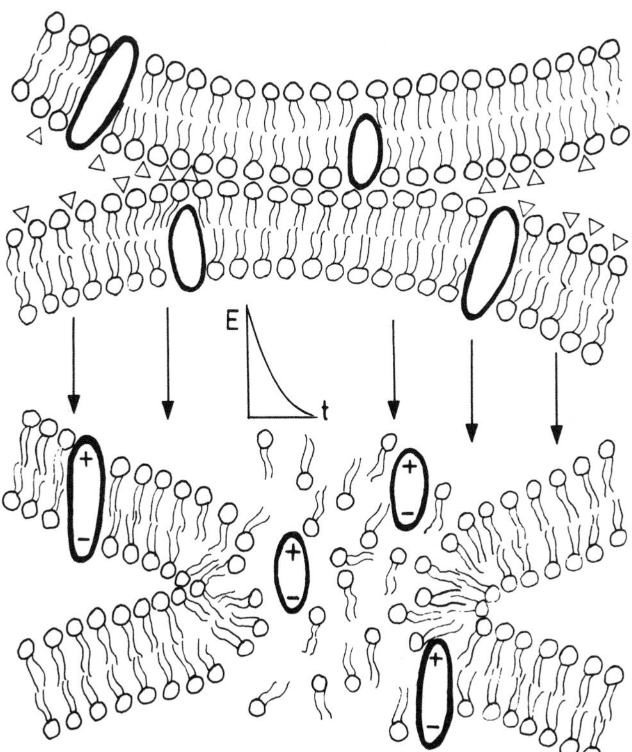

FIGURE 9. Scheme for breakdown fusion after a pulse caused by induced dipole formation and rectification.

TABLE 5. Models for Pore Formation and Breakdown

TYPE	AUTHOR	REFERENCE
Electromechanical	Crowley	52
Pore enhancement	Chernomordik	53
Pore enhancement	Weaver	54
Induced dipole repulsion	Berg	31
Periodic block-lipids	Sugar	55

SOME TRENDS

Using the three main techniques discussed (and some sophisticated modifications) one can assume that all fusion problems can be solved including inter-kingdom fusion between plant protoplasts and animal cells.

There are two main difficulties: (1) the validity of Equation (1) for cells of the same diameter and different membranes; and (2) the correct electrical window to guarantee the viability of the fusion products. Viability may be increased by using chemicals that stabilize or labilize the membranes. The combination of electrical and chemical techniques may yield a synergistic breakdown effect that decreases the applied electrical energy and increases viability. In the future, quasi-continuous electrofusion using pulsating electromagnetic induced currents may be possible. In any event, it is clear that electrofusion is a powerful tool in genetics and bioprocessing.

REFERENCES

(1) Ritter, J.W.: Beweiß, daß ein bestandiger Galvanismus den Lebensprozeß in dem Thierreich begleite, Weimar, 1798.

(2) Senda, M., Takeda, J., Abe, Sh. and Nakamura, T.:

Induction of cell fusion of plant protoplasts by electrical stimulation, Plant Cell Physiol. 20:1441, 1979.

(3) Jacob, H.-E., Förster, W., Weber, H. and Berg, H.: Electric field effects on single cells, Poster F7 at the 5th Symp. Bioelectrochemistry of BES, Weimar, Sept. 3-9, 1979. Reviewed by M. Blank, New York, in ESN 33-12, 495, 1980.

Enhancement of yeast protoplast fusion by electric field effects, Proc. 6th Int. Fermentation Symp., in 5th Int. Symp. on Yeasts, London (Canada), July 20-25, 1980.

Weber, H., Förster, W., Berg, H. and Jacob, H.-E.: Parasexual hybridization of yeasts by electric field stimulated fusion of protoplasts, Curr. Genetics 4:165, 1981.

Weber, H., Förster, W., Jacob, H.-E. and Berg, H.: Stimulation of yeast protoplast fusion by electric field pulses, Z. Allg. Microbiol. 21:555, 1981.

(4) Scheurich, P., Zimmerman, U., Mischel, M. and Lamprecht, I.: Membrane fusion and deformation of red blood cells by electric fields, Z. Naturforsch 35c:1081, 1980.

(5) Neumann, E., Gerisch, G. and Opatz, K.: Cell fusion induced by high electric impulses applied to Dictyostelium, Naturwissenschaften 67:414, 1981.

(6) Scheurich, P., Zimmermann, U. and Schnabl, H.: Electrically stimulated fusion of different plant cell protoplasts, Plant Physiol. 67:849, 1981.

(7) Richter, H.-P., Scheurich, P. and Zimmermann, U.: Electric field-induced fusion of sea urchin eggs, Develop. Growth and Differ. 23:479, 1981.

(8) Teissie, J., Knutson, V., Tsong, T. and Lane, H.: Electric-pulse induced fusion of 3T3 cells in monolayer culture, Science 216:537, 1982.

(9) Halfmann, H., Emeis, C. and Zimmermann, U.: Electrofusion of haploid Saccharomyces yeast cell of identical mating type, Arch. Microbiol. 134:1, 1983.

(10) Bischoff, R., Eisert, R., Schedel, I., Vienken, J. and Zimmermann, U.: Human hybridoma cells produced by electro-fusion, FEBS Lett. 147:64, 1982.

(11) Berg, D., Schumann, I. and Stelzner, A.: Electrically stimulated fusion between myeloma cells on spleen cells, Poster at the IX Jena Symp.: Photodynamic and Electric Effects on Biopolymers and Membranes, Sept. 20-25, 1982, Weimar, Proceedings: Studia Biophysica 94:101, 1983.

(12) Neumann, E., Schafer-Ridder, M., Wang, Y. and Hofschneider, P.: Gene transfer into mouse lyoma cells by electroporation in high electric field, Embo. J. 1:841, 1982.

(13) Shivarova, N., Forster, W., Jacob, H.-E. and Grigorova, R.: Stimulation of plasmid transformation of Bacillus cereus protoplasts by electric field pulses, Z. Allg. Mikrobiol. 23:595, 1983.

(14) Berg, H.: Fusion of blastomeres and blastocysts of mouse embryos, Bioelectrochem. Bioenergetics 9:223, 1982.

(15) Neumann, E. and Rosenheck, K.: Permeability changes induced by electric pulses, J. Membr. Biol. 10:279, 1972.

(16) Auer, D., Brandner, G. and Bodemer, W.: Dielectric breakdown of the red blood cell membrane and uptake of SV 40 DNA and mammalian cell RNA, Naturwissenschaften 63:391, 1976.

(17) Zimmermann, U., Pilwat, G. and Esser, B.: The effect of encapsulation in red blood cells on the distribution of methotrexate in mice, J. Clin. Chem. Clin. Biochem. 16:135, 1978.

(18) Chapman, D., ed.: Biological Membranes, Volume 5, Academic

Press, New York, 389, 1984.

(19) Berg, H. and Jacob, H.-E., eds.: Photodynamic and Electrical Effects on Biopolymers and Membranes, Studia Biophysica 94, 1983.

(20) Chayen, J. and Bitensky, L., eds.: Investigative Microtechniques in Medicine and Biology, Marcel Dekker, New York, 1984.

(21) Zimmermann, U.: Electric field-mediated fusion and related electrical phenomena, Biochem. Biophys. Acta 694:227, 1982.

(22) Berg, H., Augsten, K., Bauer, E., Förster, W., Jacob, H.-E., Kurischko, A., Mühlig, P. and Weber, H.: Cell fusion and transformation by electrostimulation, Bioelectrochem. Bioenergetics 12:119, 1984.

(23) Pohl, H.: Dielectrophoresis, Cambridge Univ. Press, 1978.

(24) Vienken, J., Zimmermann, U., Zenner, H., Coakley, W. and Gould, R.: Electro-acoustic fusion of cells, Naturwissenschaften 72:441, 1985.

(25) Schwan, H.P.: Field force effects, Ann. N.Y. Acad. Sci. 303:198, 1977.

(26) Glaser, R., Fuhr, G. and Gimsa, J.: Rotation of erythrocytes, plant cells and protoplasts in an outside rotating electric field, Studia Biophysica 96:11, 1983.

(27) Berg, H., Kurischko, A. and Freund, R.: Fusion of mouse blastomeres without and within zona pellucida, Studia Biophysica 94:103, 1983.

(28) Kurischko, A. and Berg, H.: Electric field effects on blastomeres of mice and rats, Bioelectrochem. Bioenergetics, In Press, 1986.

(29) Berg, H., Förster, W., Jacob, H.-E., Jungstand, W. and

Mühlig, P.: Electric field effects on single cells, Studia Biophysica 74:31, 1978.

(30) Jacob, H.-E., Förster, W. and Berg, H.: Microbiological implications of electric field effects. II. Inactivation of yeast cells and repair of their cell envelope, Z. Allg. Mikrobiol. 21:225, 1981.

(31) Berg, H.: Molecular biological implications of electric field effects, Studia Biophysica 90:169, 1982.

(32) Jacob, H.-E., Förster, W. and Hamann, M.: Photodynamic inactivation of yeast cells by thiopyronine after electric field pulse treatment, Studia Biophysica 94:71, 1983.

(33) Jacob, H.-E., Siegemund, F., Bauer, E., Mühlig, P. and Berg, H.: Fusion of plant protoplasts by dielectrophoresis and electric field pulse technique, Studia Biophysica 94:99, 1983.

(34) Jacob, H.-E., Siegemund, F. and Bauer, E.: Fusion von pflanzlichen Protoplasten durch elektrischen Feldimpuls nach Dielektrophorese, Biolog. Zentralblatt 103:77, 1984.

(35) Berg, H.: Zellfusion, Transformation und Pharmakainkorporation d. Elektrostimulation, Sitzungsber. d. Sächs. Akad. Wiss. (Leipzig) Akademie Verlag (Berlin) 177, 1986.

(36) Berg, H., Förster, W. and Jacob, H.-E.: Electrical field effects on biological membranes: Electroporation and electrofusion, in Bioelectrochemistry II, G. Milazzo, ed., Plenum Press, New York, In press, 1986.

(37) Berg, H.: Electromagnetic field effects on cell membranes and cell metabolism, in Electrochemistry in Research and Development, R. Kalvoda and R. Parsons, eds., Plenum Press, New York, 227, 1985.

(38) Berg, H., Weber, H., Forster, W., Jacob, H.-E. and Muhlig, P.: Poster at VIII Biophys. Congr., August 1981 (Mexico

City), Abstract of Posters TH-j-5, p. 237. Included unpublished results of M. Senda, H. Morikawa and H. Berg, Kyoto Univ., May, 1981.

(39) Shivarova, N., Förster, W., Jacob, H.-E., Berg, H. and Grigorova, R.: Fusion of Bacillus thuringiensis protoplasts by high electric field pulses, Bioelectrochem. Bioenergetics 11:181, 1983.

(40) Pohl, H., Pollock, K. and Rivera, H.: The electrofusion of cells, Int. J. Quantum Chem. 11:327, 1984.

(41) Vienken, J., Hampp, R. and Zimmermann, U.: Electric field-induced fusion of isolated vacuoles and protoplasts of different developmental and metabolic provenance, Physiol. Plant 53:64, 1981.

(42) Zimmermann, U. and Scheurich, P.: High frequency fusion of plant protoplasts by electrical fields, Planta 151:26, 1981.

(43) Kohn, H. and Schieder, O.: Somatic hybridization of two auxotrophic tobacco lines by electrofusion, International Symposium Plant Tissue Cell Cult. Appl. to Crop Improvement, Olomone (CSSR), 1984.

(44) Zachrisson, A. and Bornman, C.: Application of electric field fusion in plant tissue culture, Physiol. Plant 61:314, 1984.

(45) Siegemund, F. and Tschuch, G.: Elektrisch induzierte Fusion von Lycopersicon-Protoplasten unterschiedlicher Herkunft, Arch. Zuchtungsf. 14:163, 1984.

(46) Bates, G., Gayner, J. and Shekhawat, N.: Electrical fusion for optimal formation of protoplast heterokaryons in Nicotiana, Plant Physiol. 165:217, 1983.

(47) Bates, G. and Nea, L.: Successful culture of protoplast heterokaryons following electrical fusion, Plant Physiol.

Suppl. 75:133, 1984.

(48) Scheurich, P. and Zimmermann, U.: Giant human erythrocytes by electric field-induced cell-to cell fusion, Naturwissenschaften 68:45, 1981.

(49) Pilwat, G., Richter, H.-P. and Zimmermann, U.: Giant culture cells by electric field induced fusion, FEBS Lett. 133:169, 1981.

(50) Zimmermann, U., Pilwat, G. and Pohl, H.A.: Electric field mediated cell fusion, J. Biol. Phys. 10:43, 1982.

(51) Zimmermann, U. and Vienken, J.: in Cell Fusion: Gene Transfer and Transformation, R. Beers and E. Bassett, eds., Raven Press, New York, 171, 1984.

(52) Crowley, J.: Electrical breakdown of bimolecular lipid membranes as an electromechanical instability, Biophys. J. 13:711, 1973.

(53) Chernomordik, L. and Abidor, I.: The voltage induced local defects in unmodified BLM, Bioelectrochem. Bioenergetics 7:617, 1980.

(54) Weaver, J., Hao Ling and Sloan, S.: Conduction onset criteria for transient aqueous pores and reversible electrical breakdown in bilayer membranes, Bioelectrochem. Bioenergetics 15:229, 1986.

(55) Sugar, I. and Neumann, E.: Stochastic model for electric field-induced membrane pores electroporation, Biophys. Chem. 19:211, 1984.

(56) Lo, M., Tsong, T., Conard, M., Strittmatter, S., Hester, L. and Snyder, S.: Monoclonal antibody production by receptor-mediated electrically induced cell function, Nature 310:792, 1984.

(57) Morikawa, H., Sugino, K., Hayashi, Y., Takeda, J.,

Senda, M., Hirai, A. and Yamada, Y.: Interspecific plant hybridization by electrofusion in Nicotiana, Bio/Technology 4:57, 1986.

(58) Koop, H.-U. and Schweiger, H.-G.: Regeneration of plants after electrofusion of selected pairs of protoplasts, Eur. J. Cell. Biol. 39:46, 1985.

(59) Sowers, A.: A long-lived fusogenic state is induced in erythrocyte ghost by electric pulses, J. Cell. Biol. 102:1358, 1986.

(60) Shillito, R., Saul, M., Paszkowksi, J., Müller, M. and Potrykus, I.: High efficiency direct gene transfer to plants, Bio/Technology 3:1099, 1985.

(61) Hashimoto, H., Morikawa, H., Yamada, Y. and Kimura, A.: A novel method for transformation of intact yeast cells by electroinjection of plasmid DNA, Appl. Microbiol. Biotechnol. 21:336, 1985.

(62) Morikawa, H., Iida, A., Matsui, Ch., Ikegami, M. and Yamada, Y.: Gene transfer into intact plant cells by electroinjection through cell walls and membranes, Gene 41:121, 1986.

(63) Sowers, A., ed.: Cell Fusion, Plenum Press, New York, 1987.

IV

BIOLOGICAL EFFECTS OF ELECTROMAGNETIC ENERGY

12
Electromagnetic Energy and *Physarum*

EUGENE M. GOODMAN, MICHAEL T. MARRON*, BEN GREENEBAUM
Biomedical Research Institute
University of Wisconsin-Parkside, Kenosha, Wisconsin
*Office of Naval Research, Arlington, Virginia

INTRODUCTION

Fourteen years ago we initiated a program of research aimed at understanding the nature of the interaction between weak, low-frequency electromagnetic fields (EMFs) and biological organisms. We selected for study the slime mold Physarum polycephalum because it was an extremely well-characterized organism with a well-defined, easy-to-measure cell cycle (1-3). The field intensities and frequencies selected for examination were to a large extent dictated by a need in the early 1970's for developing a basis for assessing the hazards of exposure to weak fields. In recent years we have moved to emphasize mechanisms of interaction between the fields and organisms. This reflects both a change in attitude and in research strategy, since today there is no question that fields can affect organisms.

Our early work was funded by the US Navy as part of their research program related to potential bioeffects of EMFs associated with what was then called Project Sanguine. Project Sanguine was the name given to a large communication antenna that would be buried underground across hundreds of square miles in Northern Wisconsin and which would emit extremely-low-frequency electromagnetic radiation at 45-80 Hz. The project concept has changed design and names and its scope has been reduced several times since 1970, and is now referred to as Project ELF.

The antenna is designed for communication with submerged submarines and takes advantage of the inverse relationship between absorption of EMFs by water and the frequency of the fields. A dual antenna system located in Northern Wisconsin and Upper Michigan can communicate with submarines throughout the world. The field strengths in the ground near the Sanguine antenna as originally designed were expected to be a maximum of 0.07 V/m and 20 μT (the present antenna design does not involve a buried antenna, so lower ground fields than these will occur). Our experiments were designed to examine effects of fields up to ten times stronger than the proposed maximum fields. The initial values of the field for experimentation were thus chosen to be 0.7 V/m and 0.2 mT, rms, measured in the growth medium. The applied sinusoidal fields were in phase, as one would expect near the antenna.

Because of the low intensity of the electromagnetic fields involved, we reasoned that any observed bioeffect would probably be subtle. This thinking has guided much of our research; its application has resulted in the expenditure of a great deal of time and effort to obtain optimum resolution from every technique applied to the study of EMF bioeffects.

EXPOSURE TO ELECTROMAGNETIC FIELDS

Our initial experiments were designed to examine the effects of EMFs on the timing of the nuclear division cycle. Two incubators equivalent in all respects were used. One incubator served as a control environment and was equipped with an exposure system that was not energized; the second incubator was used to subject Physarum to the EMFs being examined. The incubators were designed so that the control and experimental chambers could be randomly switched by simply energizing or disconnecting a few circuits. Routine switching of control and experimental incubators was done to control for subtle incubator differences that might produce artifacts. A master-slave circuit was installed between the growth chambers that minimized temperature differences to less than 0.3°C (4). The fan motors supplied with the incubators were replaced with induction motors to reduce ambient magnetic fields.

An exposure vessel was designed and constructed that would facilitate optimal growth of the organism while providing for easy application of the fields. Electric fields were applied by using stainless steel electrodes placed in direct contact with the medium (Figure 1). Placement of the electrodes in direct contact with the medium assured us of being able to specify the internal fields far more precisely than if the electrodes were separated from the growth medium by an insulator. This design introduced the possibility that electrolysis products could be generated. To test for the possibility that electrolysis products might affect cells grown in these vessels we exposed batches of sterile growth medium to fields of 0.7 V/m and 0.2 mT for 48 hours without cells present. Unexposed cells were then added to this medium for 24 hours and the timing of mitosis was compared with control cultures grown in medium that had been placed in flasks 48 hours earlier but not previously exposed to fields. No differences were detected (5). Magnetic fields were

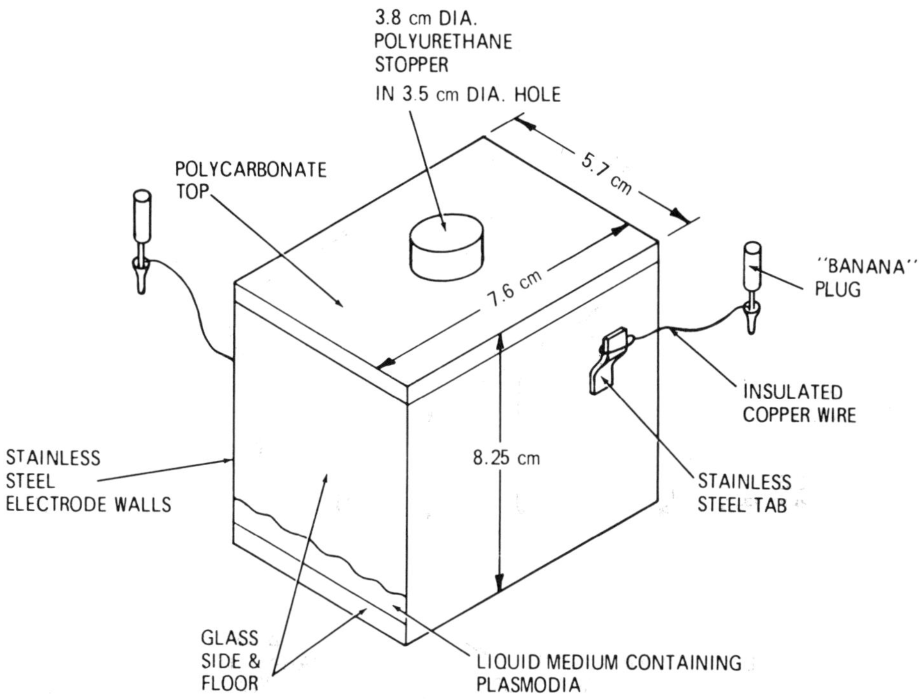

FIGURE 1. Rectangular flask used to maintain submerged shaker flask microplasmodia and amoebae. Parallel stainless steel electrodes comprise the sides of the vessel. The bottom and ends are made of glass, and the top is constructed of polycarbonate. A foam plug maintains sterility.

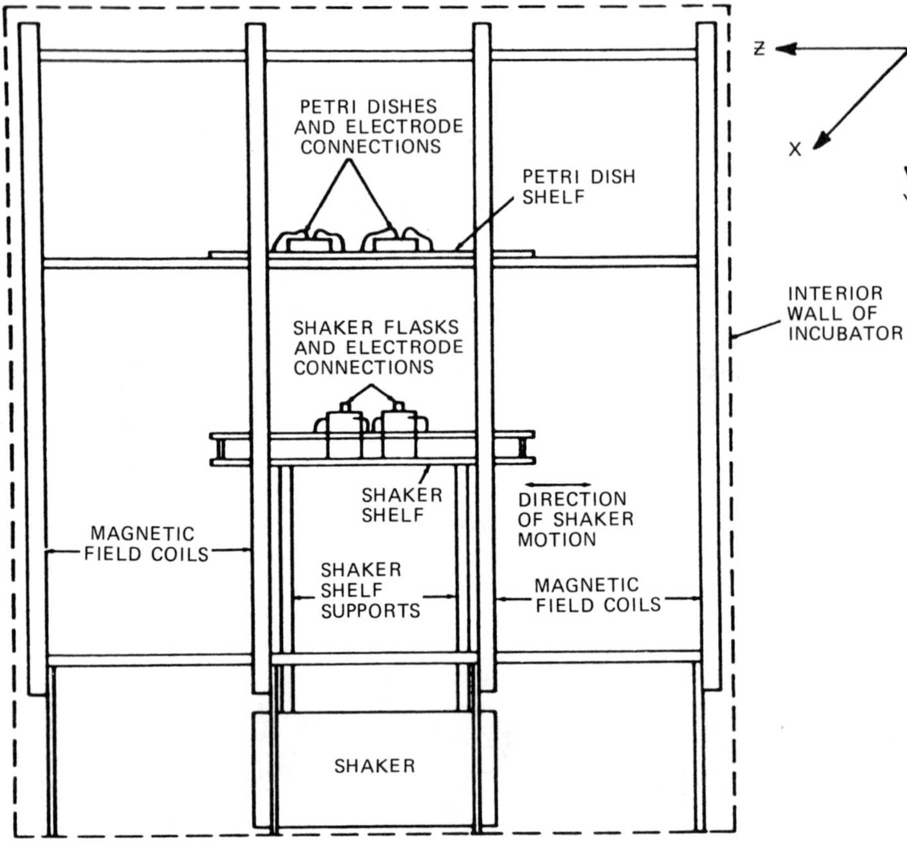

FIGURE 2. General view of the interior of an incubator. Coils surrounding the cultures generate the magnetic field. The electric fields are individually applied to each culture.

generated by coils placed inside the incubator surrounding a shaker and a platform for Petri dishes (Figure 2).

NUCLEAR DIVISION CYCLE

Experiments to assess the effects of weak fields were initiated by subculturing non-exposed, submerged, shake flask microplasmodia into eight exposure flasks: four cultures were maintained as controls and the remaining four were placed in the EMF-exposure incubator. Our first set of experiments sought to ascertain the effects of simultaneous 75-Hz electric and magnetic fields of 0.7 V/m and 0.2 mT on the length of the mitotic cycle. Control and EMF-exposed cultures were maintained in the appropriate incubators as submerged microplasmodia, and the length of their mitotic cycles was routinely measured. Cell-cycle measurements were performed using techniques that had been previously developed and used in earlier research (1,2). Aliquots of submerged microplasmodia in log-phase growth were placed on filter paper in Petri dishes and allowed to coalesce for 30 minutes. Following coalescence, nutrient medium was added to each dish, the time noted as the start of the experiment, and the cultures returned to the appropriate incubators. The Petri dish lids incorporated stainless-steel electrodes which were maintained in contact with the filter paper (5). Petri dishes in the EMF incubator were connected to field-generation equipment to continue field exposure over the next 24 hours during which the cell-cycle measurements were conducted; dishes in the control incubator were connected to unenergized circuits. The coalesced macroplasmodia were discarded upon completion of an experiment.

The time required for each culture to reach the metaphase configuration of the second post-fusion mitotic division was determined using ethanol-fixed smears that were scored with phase-contrast optics. In our first experiments we found that after about 120 days of constant EMF exposure, the mitotic cycle was lengthened by 30-60 min relative to control cultures (Figure 3). The exact length of exposure required to induce this effect could not be precisely determined; we used the criterion that a statistically significant difference in the mitotic cycle must occur for three consecutive experiments before we accepted the possibility that an effect had occurred. The experiments were continued for various time periods to statistically establish the occurrence of an EMF effect. After a significant effect was established, a new set of four cultures was introduced into the EMF incubator, and the experiment was repeated. In all experiments, a conclusion that an EMF effect had occurred was accepted only if the alteration could be reproduced with a new set of cultures at least once (5).

After we had repeated the observation of a longer cell cycle in cells exposed to 75-Hz fields, we changed the applied

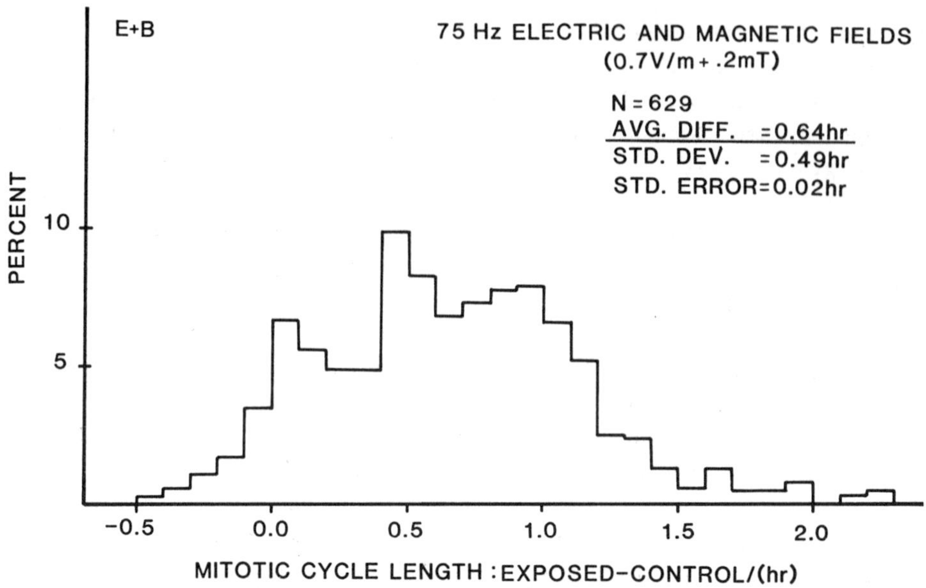

FIGURE 3. Distribution of differences in the length of the mitotic cycle for exposed cultures relative to the average daily control cell cycle (7). The average control value for each day has been subtracted from each observation made on exposed cultures. N is the total number of observations on exposed cultures represented in the histogram. Field-exposure conditions and histogram statistics are shown.

frequency first to 45 Hz and later to 60 Hz, while maintaining field intensities at 0.7 V/m and 0.2 mT. In these experiments the mitotic cycle was lengthened to a degree almost identical to that observed in cells exposed to 75-Hz fields. The only apparent difference appeared to be that the lower frequency fields required slightly less time to produce an effect. Unfortunately, this observation could not be more accurately quantified because we lack an unambiguous method for assessing the initial responses of Physarum to EMF exposure.

One of our more interesting cell-cycle experiments involved a series of cell fusions (6). In these experiments microplasmodia from control and EMF-exposed cultures were mixed in varying proportions and allowed to coalesce to form a mixed-nuclear macroplasmodium before timing their cell cycle (Table 1). Even

TABLE 1. Results of Mixing Experiments. The standard error of the mean is given in parentheses.

WEIGHT RATIO (control/exposed)	NO. DAYS DATA	TOTAL NUMBER CULTURES	AVERAGE DAILY FRACTIONAL SHIFT*	AVERAGE TIME TO SECOND METAPHASE (hr)		
				CONTROL	MIXTURE	EXPOSED
1:3	4	60	69(22%)	14.99(.09)	15.38(.12)	15.60(.08)
1:2	22	320	48(6%)	14.62(.03)	14.99(.04)	15.47(.04)
1:1.5	4	57	40(14%)	15.21(.10)	15.53(.08)	16.04(.05)
1:1	6	89	13(9%)	15.05(.11)	15.15(.12)	15.69(.14)
3:1	4	60	3(14%)	14.65(.10)	14.68(.12)	15.28(.09)

*[(M-C)/(E-C)] x 100

though cells with different cell-cycle lengths were mixed to form a single syncytium, all nuclei within the syncytium underwent mitosis synchronously in the manner characteristic of Physarum. When control and EMF-exposed microplasmodia were mixed in a 1:1 ratio, mitosis of the mixed culture occurred somewhat earlier than the time predicted by computing a simple average of the cycle lengths of the two unmixed cultures. Mixing control and experimental cultures in a ratio of 3 to 1 respectively resulted in a cell-cycle length that did not differ significantly from that of the control. In contrast, when the ratio of the mix was reversed (1 part control, 3 parts EMF-exposed), the length of the cell-cycle was significantly shorter compared to exposed plasmodia. We concluded from these and other mixing experiments that the control contribution to the combined plasmodium exerts a stronger influence on the mixtures than does the contribution from exposed plasmodia. One possible explanation is that control cultures have an excess of some factor(s) involved in mitosis that is in lower concentration in the experimental cultures. Whatever the reason, these experiments provide unequivocal evidence that cultures exposed to these weak electromagnetic fields exhibit significantly changed cell-cycle characteristics.

INDIVIDUAL ELECTRIC AND MAGNETIC FIELDS

Having convinced ourselves that lengthening of cell cycle was not an artifact, we proceeded to examine the role of the individual electric (E) and magnetic (B) fields, and to determine whether both were required to produce an effect (7). To address this question, cultures were exposed to either a 75-Hz E-field of 0.7 V/m, or a 75-Hz B-field of 0.2 mT, or to combined fields of the same strengths. The E+B experiment was a replication of the earlier work. When both fields were simultaneously applied the exposed cultures had a cell cycle that was lengthened by 0.64 hr. The cell cycle in cultures exposed only to B was lengthened by 0.46 hr; cultures exposed only to E were also lengthened by 0.39 hr (Figures 4 and 5). The smaller delay for either single-field exposure is significantly different from the combined-field exposures.

These data demonstrate individual E and B fields can slow the mitotic cycle, but that the effects of the individual fields

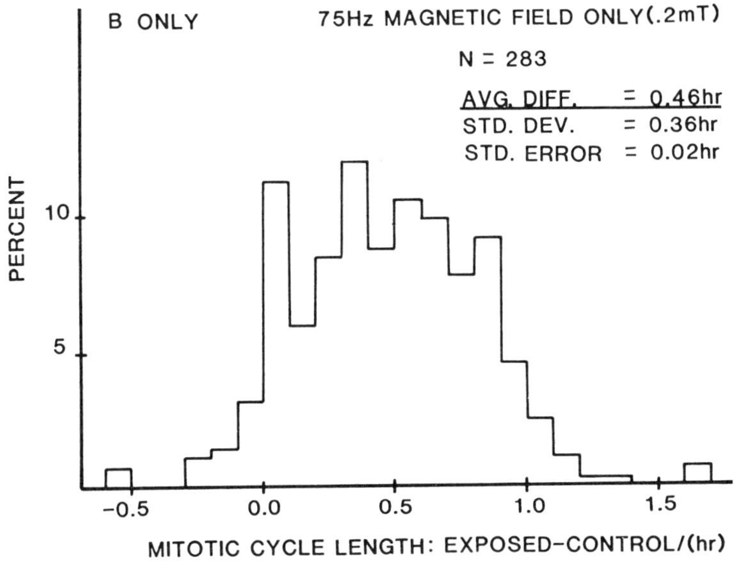

FIGURE 4. Distribution of differences in the length of the mitotic cycle for cultures exposed to 0.2 mT (7).

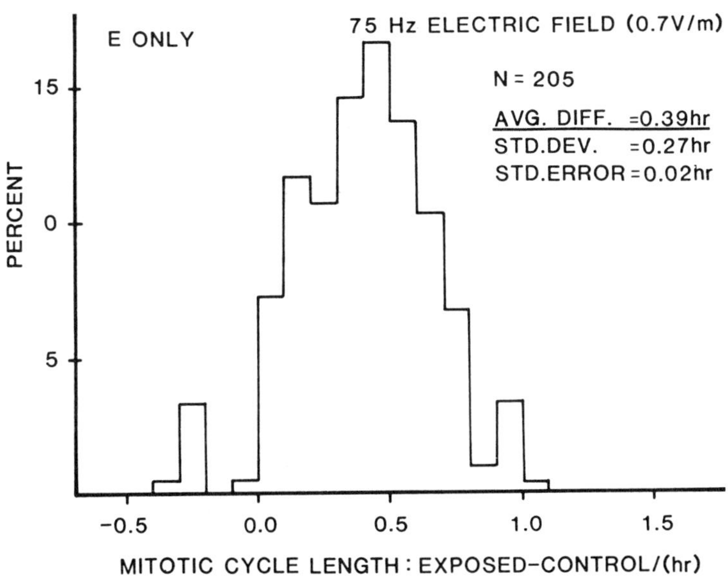

FIGURE 5. Distribution of differences in the length of the mitotic cycle for cultures exposed to 0.7 V/m (7).

do not appear to be additive. We regard the observation that electric and magnetic fields can each alter the cell cycle in nearly the same way as one of our most important observations. It presents a severe test to any mechanism proposed to account for these findings. There may well be two separate mechanisms that work either in concert or individually, to produce a lengthening of the cell cycle. Our work with pulsed electric and magnetic fields discussed below supports these observations.

NO LOWER THRESHOLD OBSERVED

We attempted to determine if a lower threshold existed by decreasing the field intensities by a factor of five to 0.14 V/m and 40 μT. The data (Figure 6) showed that the mitotic cycle was lengthened by 0.42 hr. Thus application of fields five-fold weaker induced a significantly smaller delay that is about the same magnitude as a more intense E field alone (0.39 hr) or an B field alone (0.46 hr).

In a related experiment using a modulated waveform quite similar to the sinusoidal 75-Hz fields, the applied field intensities were decreased by an additional factor of four to 0.035 V/m and 10 μT; a delay of 0.43 hr was obtained (Figure 9). Because we only performed experiments applying both the E and B fields, we are unable to determine whether a threshold for either the E field or B field was reached. Based on these data a dose-response relationship was not observed. In particular, a 20-fold decrease in the applied field intensities did not result in a proportional shortening of the mitotic delay.

MODULATED WAVEFORM

We also performed several experiments in which the waveform was modulated by shifting the frequency at the waveform peaks so that no discontinuities were introduced into the wave (7). This type of frequency modulation is known as minimum-shift-keying modulation and was selected because it had been proposed for use with the Sanguine/ELF communication system. In these experiments, the oscillator driving the field-generation equipment was set at a nominal center frequency of 76 Hz. It made random frequency shifts between 72 and 80 Hz on the average of eight times a second. Three pairs of field intensities were examined: (a) 0.7 V/m and 0.2 mT, (b) 0.14 V/m and 40 μT, and

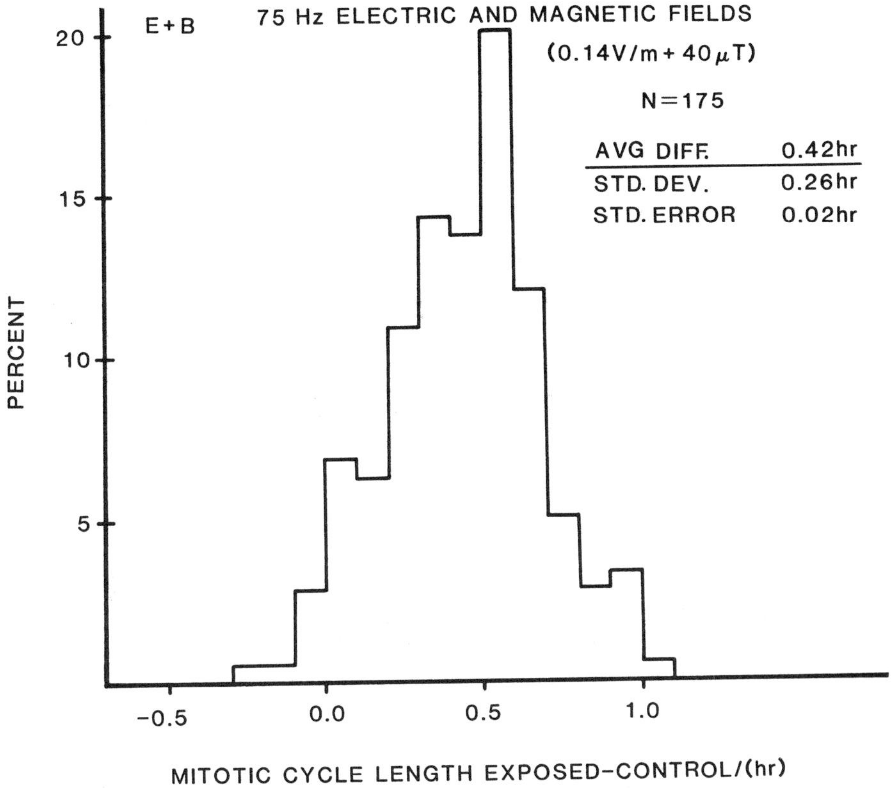

FIGURE 6. Distribution of differences in the length of the mitotic cycle for cultures exposed to 0.14 V/m and 40 μT (7).

(c) 0.035 V/m and 10 μT. The results are shown in Figures 7-9. We conclude from these data that this type of modulation adds nothing new to the fixed-frequency sinusoidal fields used in earlier experiments.

CONSISTENCY AND REPRODUCIBILITY

To insure internal consistency and reproducibility of the results, all data sets were analyzed by both a paired t test and a Wilcoxon signed-rank test (9). Both tests were used to compare two distributions of data to see if the means differed, or if the mean of a distribution differed from zero. The results of these comparisons are presented in Figure 10 (7).

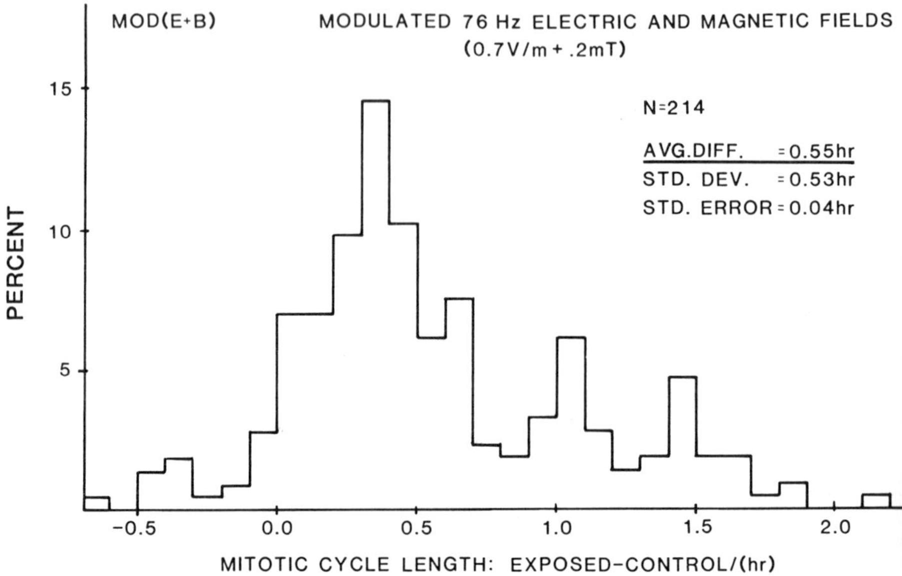

FIGURE 7. Distribution of differences in the length of the mitotic cycle for cultures exposed to modulated fields of 0.7 V/m and 0.2 mT (7).

Reproducibility was tested by constructing distributions for each individual set of cultures and comparing them with one another. Internal consistency of the data was examined by comparing a culture set with its replicate or by comparing the results of equivalent tests such as B1 vs E1 and B2 vs E1. The results of all such combinations are shown in Figure 11 (7). Based on these analyses, we conclude that the data are both reproducible and internally consistent.

RESPIRATION

EMF effects on respiration were also examined. In these experiments the same EMF-exposed and control microplasmodia that were used for the mitosis experiments were used to determine the respiratory quotient QO_2 (in units of microliters of O_2 consumed/min/mg protein). In our initial experiments we used a Warburg apparatus to measure respiration (7); in later experiments a Clark-type oxygen electrode was employed. Microplasmodia exposed to 75-Hz fields of 0.7 V/m and 0.2 mT exhibited a

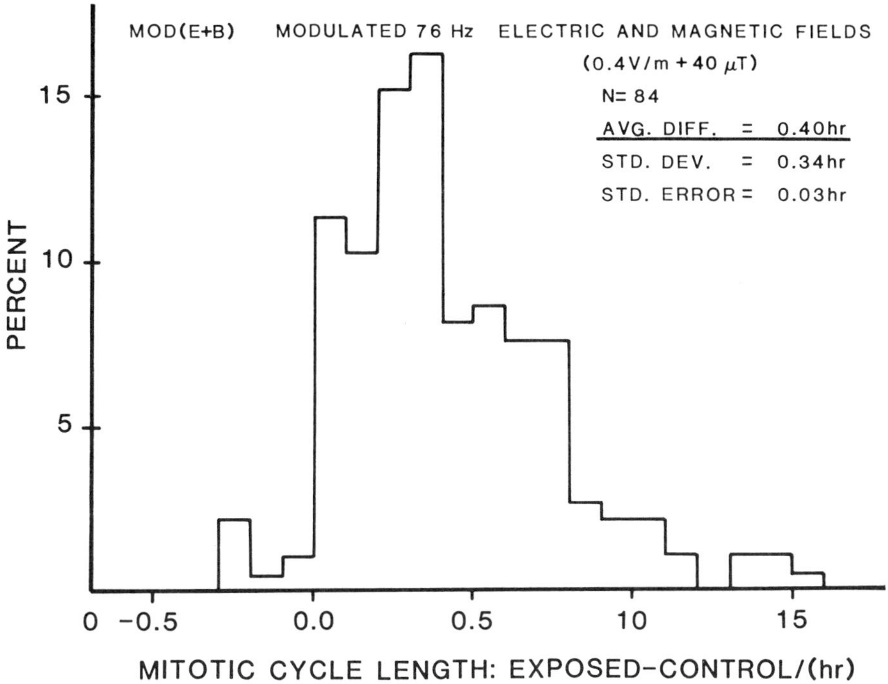

FIGURE 8. Distribution of differences in the length of the mitotic cycle for cultures exposed to modulated fields of 0.14 V/m and 40 μT (7).

16% decrease in QO_2 relative to controls. Exposure to either E or B fields alone decreased respiration by 8% and 9%, respectively. Application of modulated 76-Hz fields of 0.7 V/m and 0.2 mT produced a 7% decrease in QO_2, while weaker modulated fields of 0.14 V/m and 40 μT or .035 V/m and 10 μT diminished QO_2 by 3% and 4%, respectively (Figure 12).

These data show that EMF exposure resulted in a physiological depression in the respiration rate of the cell and, in general, they are consistent with our earlier cell-cycle experiments. However, whereas the individually-applied fixed-frequency E and B fields were not additive in their effects on the cell cycle, they appear to be additive in their effects on respiration. It is noteworthy that the magnitude of the depression was less in cells exposed to fields of modulated frequencies than in those exposed to a fixed-frequency field. These conclusions are

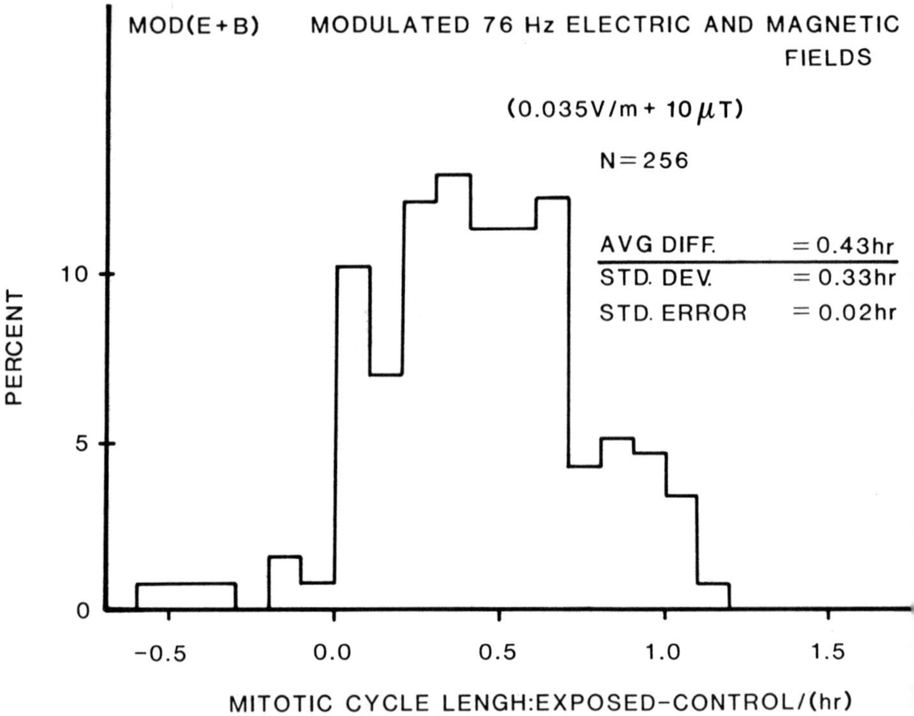

FIGURE 9. Distribution of differences in the length of the mitotic cycle for cultures exposed to modulated fields of 0.035 V/m and 10 μT (7).

weak, and may be biased by the broad distribution seen in the data obtained using modulated fields.

EFFECTS WITH TIME

Once an EMF effect has been induced in microplasmodia, its magnitude neither increases nor decreases with time. This fact has been substantiated by observing cultures that have been continuously exposed to EMF for periods as long as 5 years. To assess the ability of affected cultures to return to baseline levels, cultures exhibiting a lengthened mitotic cycle were removed from the field and placed in a control environment. The persistence of the lengthened mitotic cycle was periodically assessed; the results are shown in Figure 13. Within about 30 days, the mitotic cycle slowly shortened to the length of the

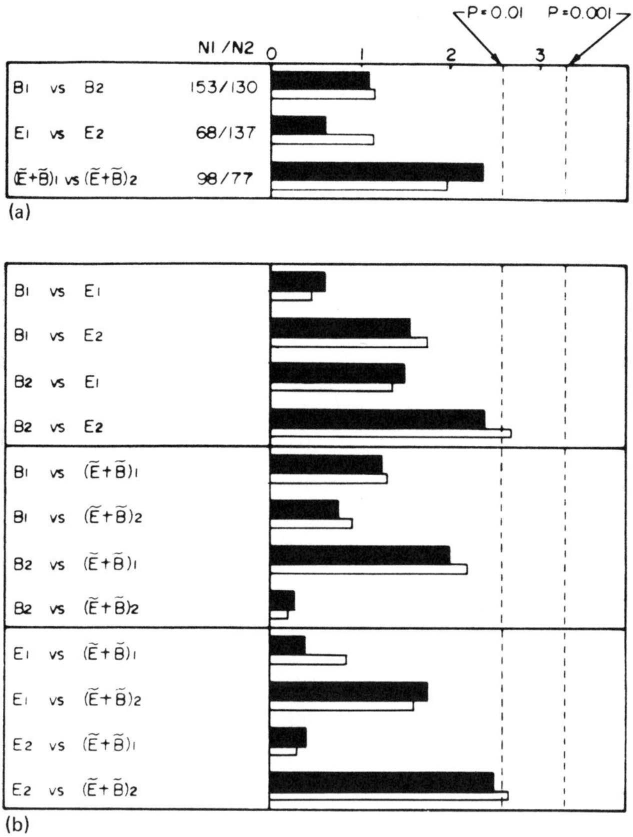

FIGURE 10. Statistical comparison of distributions to test reproducibility and internal consistency of the data. The solid bar gives the results for the Wilcoxon signed ranks test; the open bar gives the t-test statistic. The P = 0.01 and P = 0.001 levels for large samples are shown. (a) Test for data reproducibility by comparing distributions of cycle length data for two sets of cultures each exposed to the same field conditions but at different times. For example, cultures in set B_1 were exposed to a 75-Hz magnetic field of 0.1 mT; cultures in set B_2 were also exposed to a 75-Hz magnetic field of 0.1 mT but these experiments began almost 1 year later. N1/N2 is the number of data in the first/second set. (b) Test of internal consistency by comparison of equivalent distributions for different field conditions. There are three sets of four comparisons, all of which display internal consistency. Note also that the distributions are all statistically equivalent (7). Here, and in Figures 11 and 12, E = 0.7 V/m, B = 0.2 mT, $\tilde{E}$ = 0.14 V/m, $\tilde{B}$ = 40 μT, $\tilde{\tilde{E}}$ = 0.035 V/m, $\tilde{\tilde{B}}$ = 10 μT. The subscripts refer to replicate experiments.

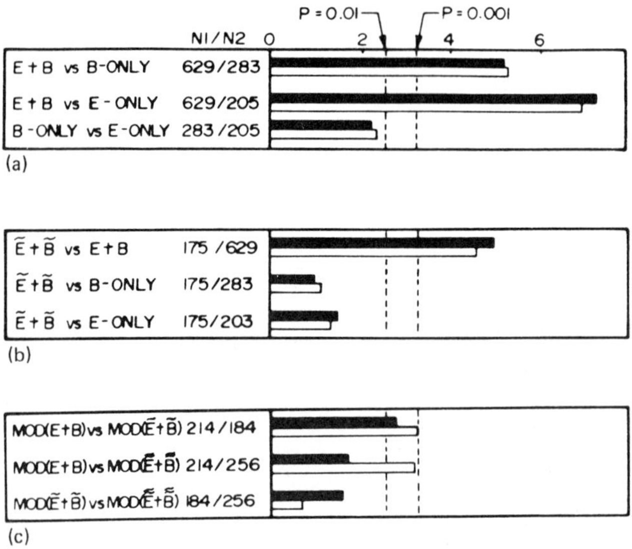

FIGURE 11. Comparison of mitotic cycle-length distributions for different field conditions. Data are pooled so that all data taken at a particular set of field conditions are treated as a single group. The solid bar gives the results for the Wilcoxon signed-rank test; the open bar gives the t-test statistic. The $P = 0.01$ and $P = 0.001$ levels for large samples are shown on the graph as vertical lines. Entries in the column headed by N1/N1 are the number of data in the first/second distribution. (a) Conclusions: $E + B \neq$ E-only or B-only. Simultaneous application of electric and magnetic fields produces a different mitotic cycle lengthening distribution than that observed when fields are applied individually. At these levels (0.2 mT and 0.7 V/m) the individual fields produce statistically identical results. (b) Conclusions: $\tilde{E} + \tilde{B}$ = E-only and B-only; $\tilde{E} + \tilde{B} \neq E + B$. Electric and magnetic fields applied at a level five times weaker cause a different result to occur. These weaker fields produce the same effect as either more intense E fields alone or more intense B fields alone. This finding agrees with those of Figure 10. (c) Conclusions: $\mathrm{MOD}(E + B) \neq \mathrm{MOD}(\tilde{E} + \tilde{B})$;

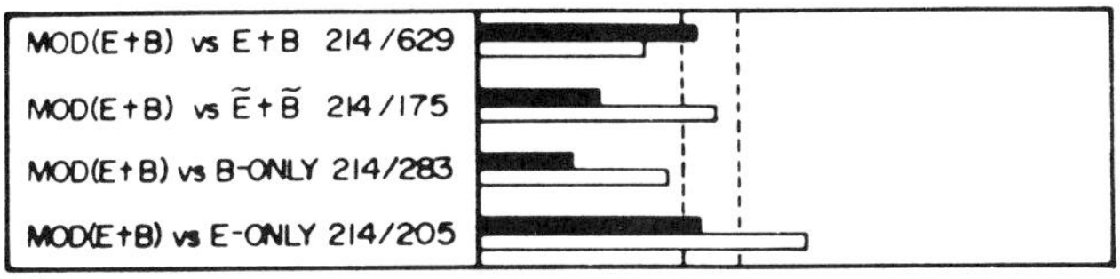

(d)

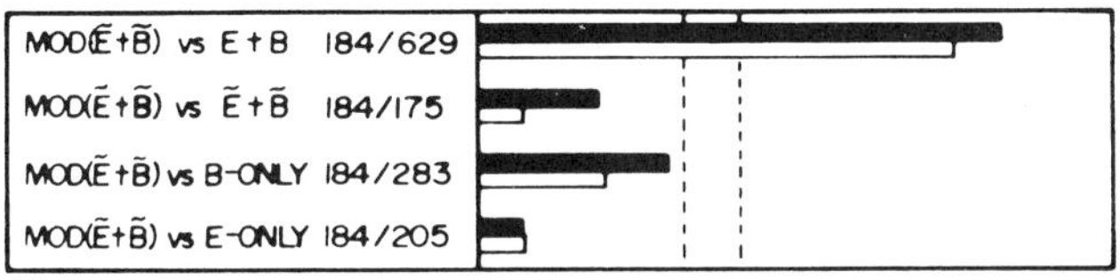

(e)

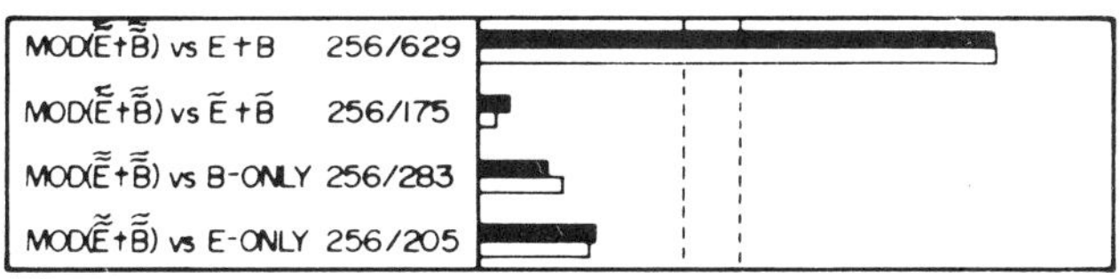

(f).

MOD(E + B) = MOD($\tilde{\tilde{E}}$ + $\tilde{\tilde{B}}$); MOD($\tilde{E}$ + $\tilde{B}$) = MOD($\tilde{\tilde{E}}$ + $\tilde{\tilde{B}}$). Reduction in field intensity by five times causes the mitotic cycle lengthening distribution to change, a further decrease in intensity by four times has no effect. (d) Conclusions: MOD(E + B) = E + B = $\tilde{E}$ + $\tilde{B}$ = B-only ≠ E-only. Modulation of the fields results in a mitotic cycle lengthening distribution similar to unmodulated fields of weaker intensity. Unlike the unmodulated fields, the MOD(E + B) results are similar to the B-only exposures. (e) Conclusions: MOD($\tilde{E}$ + $\tilde{B}$) ≠ E + B; MOD($\tilde{E}$ + $\tilde{B}$) = $\tilde{E}$ + $\tilde{B}$; MOD($\tilde{E}$ + $\tilde{B}$) = B-only or E-only. Modulated fields at 0.04 V/m and 40 μT produce results statistically identical to unmodulated fields of the same intensity. Like the unmodulated fields, MOD($\tilde{E}$ + $\tilde{B}$) data are indistinguishable from those produced by more intense unmodulated E-only and B-only fields. (f) Conclusions: MOD($\tilde{\tilde{E}}$ + $\tilde{\tilde{B}}$) ≠ E + B; MOD($\tilde{\tilde{E}}$ + $\tilde{\tilde{B}}$) = $\tilde{E}$ + $\tilde{B}$; MOD($\tilde{\tilde{E}}$ + $\tilde{\tilde{B}}$) = B-only or E-only. Decreasing field intensities further to 0.035 V/m and 10 μT has no apparent effect on the mitotic cycle lengthening distribution (7). MOD, modulated by minimum-shift-keying.

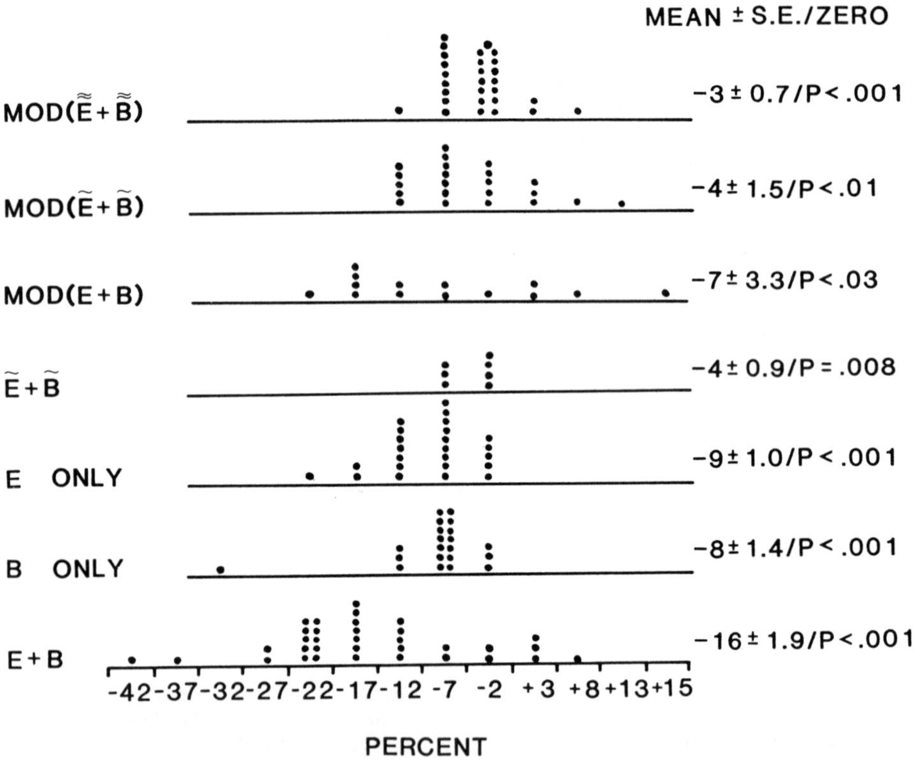

FIGURE 12. Respiration rates for exposed microplasmodial cultures relative to average daily control-culture respiration rates. Histogram bins are 5% wide; the center value of the bin is given below on the abscissa. Each point represents a set of rate measurements performed upon a single, exposed culture taken relative to the average of rate measurements in two control cultures determined at the same time. Data given on the right are average values ± standard errors computed from the raw data. The probability that the distribution mean differs from zero only by chance is also listed; these probabilities are derived using the Wilcoxon signed-rank statistic.

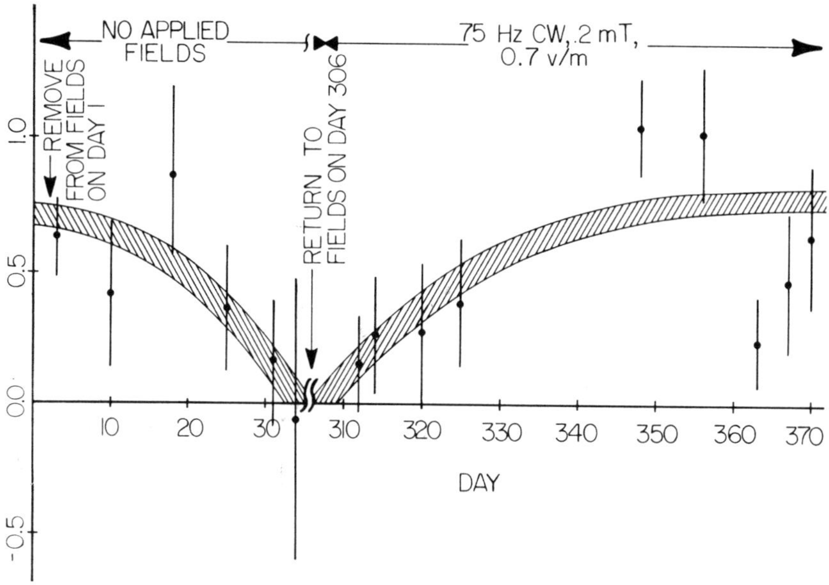

FIGURE 13. Microplasmodia exposed to 75 Hz, 0.2 mT, 0.7 V/m for 270 days and showing a lengthening of the mitotic cycle relative to control cultures. Ordinate shows the average difference of the nuclear division cycle from the time nutrient medium is added to a fused microplasmodium to the second metaphase configuration of the nuclei. Each point represents the average of 10 EMF-exposed cultures from which the average of 10 controls has been subtracted. Placing the EMF-exposed cultures into a control environment on day 1 (day 270 of exposure) resulted in the slow dissipation of the mitotic delay over a 30-day period. If the cells were reintroduced into the field (day 306), a mitotic delay became significant within 30 days.

control cycle. If the relaxed cultures were returned to the EMF environment, a lengthened mitotic cycle was re-established in about 30 days instead of about the 120 day period required when the cultures were initially exposed to EMFs.

CHROMOSOME EFFECTS

We also examined whether chronic exposure could induce a measurable drift in chromosome number. Since Physarum is a polyploid organism, distribution of chromosomes must be measured. Although the small size of the chromosome makes it extremely difficult to perform direct measurements on chromosome spreads, a distribution in the nuclear sizes may be examined for any gross change in nuclear material. The methods employed were based on the work of Mohberg et al. (8), who established that the distribution of nuclear sizes in Physarum is closely related to the distribution of chromosome numbers. Isolated nuclei were photographed under a phase-contrast microscope and the distribution of nuclear diameters were measured.

Measurements of exposed nuclei (75 Hz, 0.7 V/m, 0.2 mT, N = 104) yielded a mean diameter of 3.16 ± 0.05 μm (±95% confidence limits); the mean diameter of control nuclei (N = 213) was 3.14 ± 0.03 μm. These two means are not significantly different from each other. An estimate of how large the difference between the means would have to be before it could be detected can be gotten using the power function of the t test (9). This function estimates the sensitivity of a test from the scatter in the data at a predetermined confidence level. Variability in our data for nuclear diameters is such that we should be able to detect a difference as small as 0.09 μm between the means with 95% confidence; this difference corresponds to 3% of the mean values.

An interesting figure for comparison is the maximum difference in chromosome number. Using the measurements of Mohberg et al. (8) we found an empirical relationship between diameter, d, in micrometers, and chromosome number, c. Their data were fitted to first-, second-, and third-order equations, and the best fit was found for the linear equation $c = 41.2(d - 1.78)$. The fit for different powers of the mean diameter was nearly the same as measured by the correlation coefficient; 0.983 (1st power), 0.980 (2nd power), and 0.971 (3rd power). Based on these data we conclude that if a difference in chromosome number existed between exposed and control nuclei, it was less than 4 out of a total of approximately 60 chromosomes ($P < 0.05$).

The result of the chromosome measurements does not, however, rule out subtle changes at the level of the genes. An argument against the observed effects being attributable to slow, spontaneous genetic drift can be made based on the data concerning reproducibility of our mitotic cycle and respiration experiments described above. We have repeatedly placed sets of new stock cultures derived from our main stock control cultures into EMF environments. In all cases the effects induced were consistent with those induced in cultures introduced at other times. Any genetic drift affecting the measured parameters that does not occur reproducibly under the influence of EMF should create discernible differences among the various sets of cultures exposed to the same EMF environments.

SEARCH FOR A MECHANISM

As these data accumulated, we began to ponder the type of mechanism that might explain how such weak fields could interact with cells. A magnetic field of 0.2 mT produces a maximum current density of less than 10^{-5} A/m^2 in the exposure flasks, while an electric field of 0.7 V/m produces a current density of 0.55 A/m^2. Regardless of the mechanism invoked, it is difficult to imagine a scenario in which such weak fields or low current densities could produce a significant impact on a biological system. From a purely theoretical standpoint, it is improbable that both field components are interacting with a cell via the same mechanism. Based on a consideration of the current densities of the applied electric field and the impedance of the membrane, one may conclude that little or no current crosses the plasma membrane. In contrast, the membrane presents no barrier to the applied magnetic field. Since the individually applied fields can each induce a bioeffect, it seems probable that more than one mechanism of interaction is involved in producing the EMF effects we have observed.

One of our first experimental attempts at elucidating the mechanism of EMF interaction was aimed at determining whether

DNA replication in the diploid plasmodium of Physarum was altered. Our failure to detect significant differences in DNA levels in exposed cells set the tone for numerous future experiments. Whenever we attempted to examine a specific biochemical phenomenon we were unable to demonstrate an EMF response. In our laboratory, using our experimental system, EMF effects are more easily demonstrated in plasmodia if we examine parameters that represent the sum of numerous physiological processes. But a word of caution needs to be added to this observation by restating the obvious. Failure to demonstrate a significant effect may also be attributed to the resolution of the technique and cannot be construed as evidence that the parameter in question is not affected by EMFs.

AMOEBAL CELL PHYSIOLOGY

We realized the necessity of attempting to understand the mechanism of interaction between weak fields and cells, and chose to concentrate on the most likely initial site, the cell membrane. Because the plasmodium has a thick layer of slime (glycocalyx) surrounding its plasma membrane, we decided that the haploid amoeba stage of Physarum which lacks a readily discernible glycoalyx would be a more suitable model system for these studies. Amoebae (RSD-4) were grown in axenic medium in suspension culture and exposed to EMFs using the same exposure system described above. Physarum amoebae are 10-15 μm in diameter, have a doubling time in suspension culture of about 27 hours and, unlike the plasmodium, they undergo both karyokinesis and cytokinesis. Thus, in many respects they are more representative of a mammalian cell than is the plasmodium.

In our initial studies with amoebae, they were exposed to 1.0 V/m and 0.1 mT at 60 Hz (10). Subsequent experiments involved exposure to the individual fields. The results of QO_2 measurements made using an oxygen electrode are shown in Table 2. Respiration was depressed 20% in cells exposed to E+B fields, 7% in cells exposed to E fields alone, and no significant effects were observed in cells exposed to B fields only. Since mitosis cannot be readily followed in amoebae, we chose to examine the ATP content of exposed cells as an additional probe of the cells' physiology. ATP levels for amoebae exposed to identical field conditions are given in Table 3; 60-Hz E+B

TABLE 2. QO_2 ($\mu l\ O_2$ consumed/cell/min) in Amoebae Exposed to 60-Hz Sinusoidal Electromagnetic Fields

	E & B		E ONLY		B ONLY	
	CONTROL	1.0 V/m, 0.1 mT	CONTROL	1.0 V/m	CONTROL	0.1 mT
Overall Average	.620	.496	.645	.595	.660	.666
Average Difference	.124		.049		.006	
P (paired t-test)	<.001		.003		.765	
P (randomization)	.001		.016		.768	

TABLE 3. Concentration of ATP in Amoebae Exposed to 60-Hz Sinusoidal Electromagnetic Fields

	E & B (fmole ATP/cell)		E ONLY (fmole ATP/cell)		B ONLY (fmole ATP/cell)	
	CONTROL	1.0 V/m, 0.1 mT	CONTROL	1.0 V/m	CONTROL	0.1 mT
Overall Average	.496	.459	.509	.452	.492	.459
Average Difference	.037		.057		.033	
P (paired t-test)	<.007		<.004		<.006	
P (randomization)	<.007		.002		.016	

fields depressed the cells' ATP content by 8%, whereas an E field alone depressed ATP levels by 11%. A somewhat surprising finding, given our earlier observation of no change in QO_2, is the fact that the B field alone resulted in a 7% decrease in ATP. It should be noted once again that the absence of a significant depression in the QO_2 when a B field alone is applied, means that if an effect exists, it is below the resolution of our technique.

AMOEBAL CELL MEMBRANE

Once we convinced ourselves that amoebae could be affected by exposure to weak EMFs, we decided to examine their cell surface using the technique of cell partitioning in a two-phase aqueous polymer system. Dextran [poly(1,6-glucose)] and PEG [poly(ethylene glycol)] are water-soluble polymers that form a three-component solution of two immiscible phases when placed

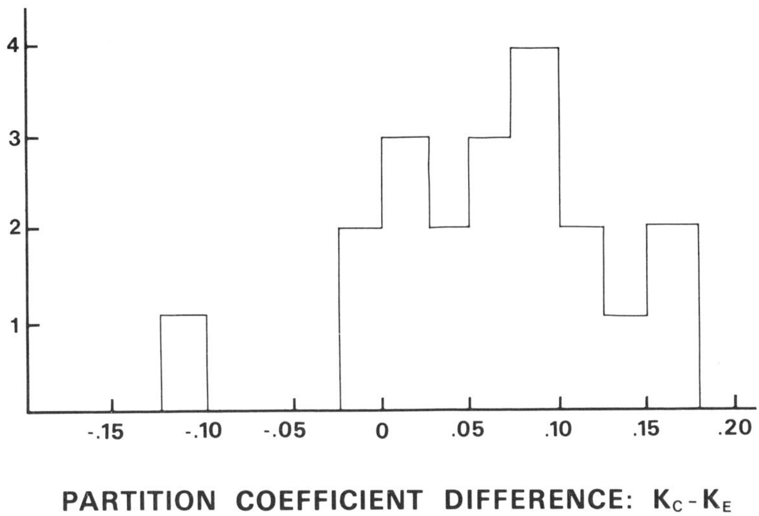

FIGURE 14. Distribution of differences in partition coefficients for amoebae exposed to EM fields (K_e and control amoebae K_c) note that partition coefficients must assume a value between 0 and 1. The histogram was constructed by computing a difference for each of 20 samples as the average value for the control sample less the average for an exposed sample; the ordinate indicates the number of observations. K_c has an overall average value 0.07 larger than K_e (12).

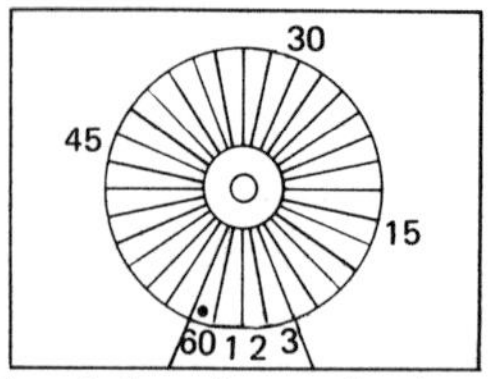

NOTE: Enlarged view of rotor chambers 60, 1, 2, 3.

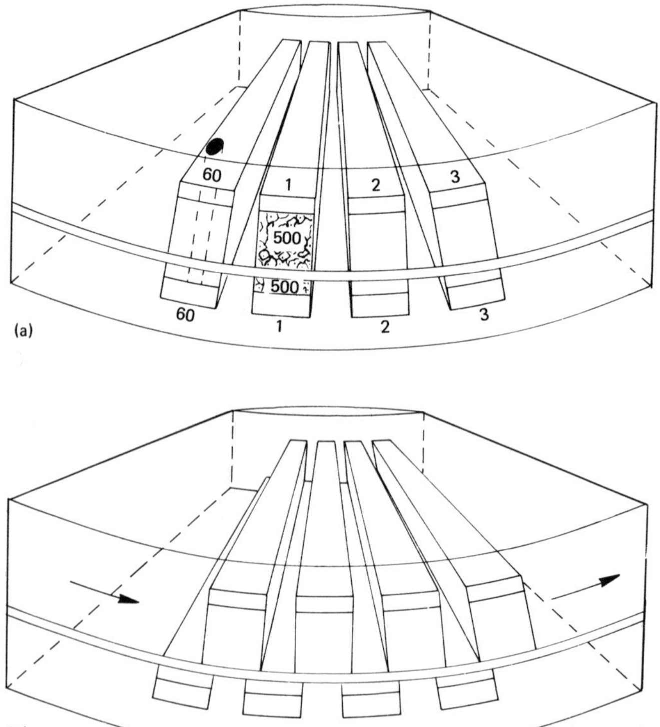

FIGURE 15. Schematic representation of three steps of a thin-layer countercurrent distribution procedure. Inset at the top is an overview of a 60 chamber rotor. (a) An expanded view of four thin-layer chambers at the beginning of an experiment: chamber 1 contains the original cell load, chambers 60, 2, and 3 contain only upper and lower phase solutions. (b) Following a

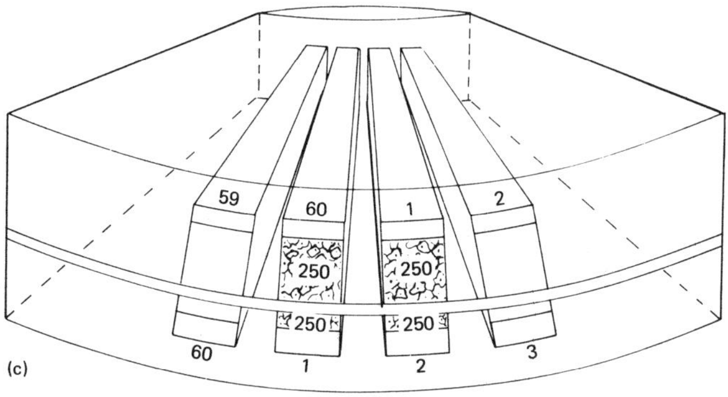

(c)

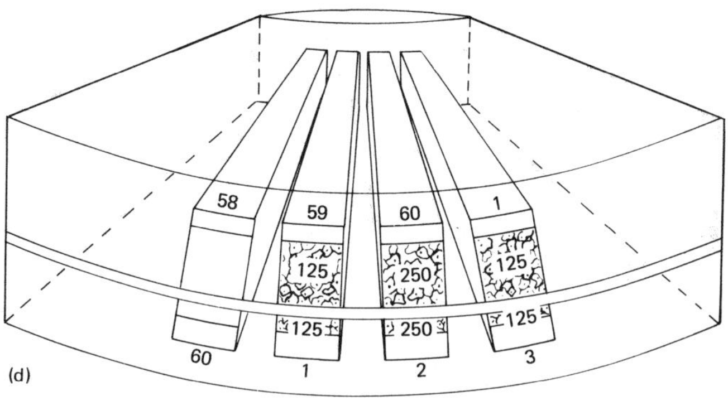

(d)

30 sec shaking period and a 10 min separation period, the top rotor is rotated in a counterclockwise direction by one chamber producing the situation depicted in (c): the 500 cells in the upper phase have been moved to the second chamber and new top phase has been brought into chamber 1. The rotor is shaken again and the phases are allowed to repartition before the top plate is moved counterclockwise by one chamber resulting in the redistribution of cells shown in (d).

together at concentrations above a critical point. The upper phase is rich in PEG and the lower phase is Dextran-rich. After cells are introduced into the PEG-Dextran system, the test tube is shaken and allowed to stand; after a short period the cells partition themselves between the interface and the upper phase based on their surface properties.

A partition coefficient is obtained by determining the fraction of the total number of cells located in the top phase at a fixed interval after mixing. Cell partitioning is dependent on the relative affinity of the membrane surface for the PEG-rich solution. The main surface components involved in a cell's affinity for the upper phase are the surface charge and lipid composition.

Data from these experiments showed that exposing amoebae to 60-Hz fields of 1.0 V/m and 0.1 mT resulted in a partition difference of about 7% (Figure 14)(12). These data permit us to conclude that either the amoebae cell surface charge or the cell surface composition or both, have been altered by EMF exposure.

Having shown that the cell surface is altered as a result of EMF exposure, we began to examine the partition technique as a probe for approaching and understanding the mechanism of action between EMFs and cells. To amplify the partition differences detected in the single-tube experiments, we applied a technique known as thin-layer countercurrent distribution (TLCCD), the principle of which is shown in Figure 15. Separation is achieved by a stepwise movement of the top phase while the interface and bottom phase remain stationary. In essence, it is a type of cell chromatography where the bottom phase is the stationary phase, and the top phase is the mobile phase. The partition steps are performed automatically in a 60-chamber circular plate (Bioshef, Sheffield, UK).

Two different phase systems were employed; a high-potential system similar to the one used in the single-tube partitions (5.5% Dextran T-500, 5.5% PEG-8000, 0.05 M/kg phosphate buffer, pH 7.0) and a low-potential system (5.0% Dextran, 4.0% PEG, 0.05 M/kg NaCl, and 0.01 M/kg potassium phosphate). The high-potential system creates a 1-3 mV electrostatic potential difference between the upper and lower phases such that the upper phase is positive relative to the lower phase. The low-potential system has little measurable potential difference between the upper and lower phases so that the cells partition principally on non-charge associated surface properties. Amoebae exposed to 60-Hz fields of 1.0 V/m and 0.1 mT and analyzed simultaneously in both high-potential and low-potential phase systems produced the TLCCD profiles shown in Figures 16a and b. Examination of the profiles shows that the EMF-exposed cells shifted to the right in the high-potential system when compared to control cells, and to the left in the low-potential system. A shift to the right reflects greater attraction for the upper phase.

To examine the effect of the individual electric and magnetic fields, cells were placed in 60-Hz fields of either 1.0 V/m or 100 μT for 24 hours. When the exposed amoebae were analyzed by TLCCD, cells exposed only to an E field were dis-

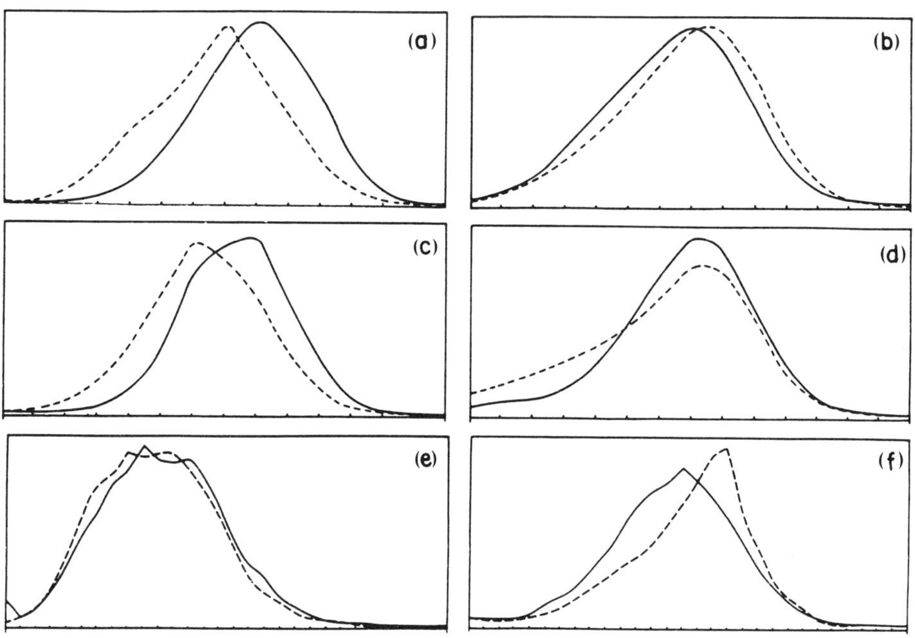

FIGURE 16. TLCCD profile of Physarum amoebae exposed for 24 hours to 60-Hz fields. The distributions occurred in a phase system in which the upper phase was positively charged relative to the bottom phase. (a) 1 V/m and 100 μT. (c) 1 V/m. (e) 100 μT. (b), (d), (f), respective are distributions in an uncharged phase system. -----control, ——— EMF-exposed.

placed to the right of the controls in the high-potential system but were coincident with the controls in the low-potential system (Figures 16c and d). To our surprise, when amoebae were exposed only to a B field the situation was reversed in two ways: First, the TLCCD profile of magnetic-field-exposed amoebae was coincident with that of the controls in the high-potential system and was displaced to the left in the low-potential system (Figures 16e and f). Thus, an increase in the net surface charge was observed in amoebae exposed to an E field, whereas cells exposed to a B field exhibited a change in non-charge associated surface properties. That is, they showed a decrease in cell surface hydrophobicity. Further, these data show that the individual E and B fields induced different cell-surface alterations.

PULSED EMFs

Pulsed EMFs are widely used in the medical community to facilitate bone repair in clinical cases of non-union. Although this therapeutic regimen appears to be reasonably effective, the mechanism of interaction between pulsed fields and cells remains elusive. We decided to examine the TLCCD profiles of cells exposed to a pulsed waveform (13).

Physarum amoebae were again used as the test organism, and the pulsed fields were applied using a generator and coils (Electrobiology, Fairfield, NJ). Erlenmeyer flasks containing amoebae were placed on a reciprocal shaker containing two coils. One coil was attached to the generator and the other was attached to a dummy load and served as the control. Fields were applied continuously for 24 hours as a burst of 22 sawtooth pulses of approximately 2 mT amplitude. Each pulse had a 200 sec rise and a 20 sec fall time and was separated by 5 sec; the bursts were repeated at a rate of 25 Hz. The charged and uncharged phase systems described above were used for the TLCCD experiment. The data were in agreement with our findings for cells simultaneously exposed to sinusoidal E+B fields (Figures 17a and 17b), that is, the exposed cells move to the right in the charged system and to the left in the uncharged system.

In an experiment analogous to the sinusoidal E-only experiment described above, amoebae were exposed to a pulsed E field

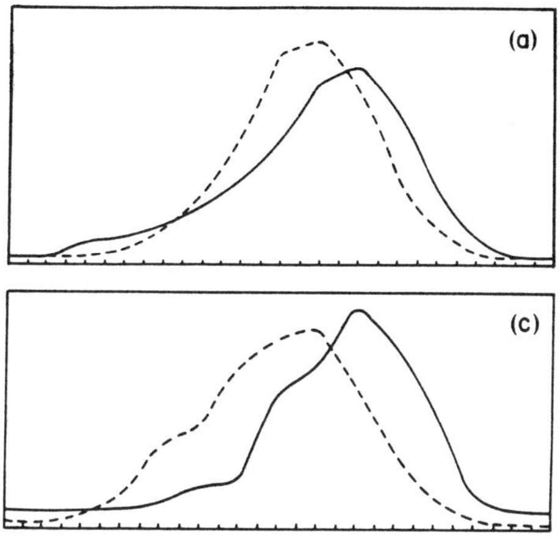

FIGURE 17. TLCCD profile of Physarum amoebae exposed for 24 hours to a pulsed electromagnetic field. Normalized cell count is plotted against the plate number. (a) and (c), cell distribution in a phase system in which the upper phase is positively charged relative to the bottom phase; (b) and (d), cell distribution in an uncharged phase system. -----control, ——— pulsed EMF exposed.

resembling the electric field induced by the coil system. Cells exposed to this field regimen for 24 hours were harvested and their TLCCD profiles determined. The data (Figures 17c and 17d) showed that the cells moved to the right in the charged system, but were coincident with the controls in the uncharged system. These data are consistent with the sinusoidal experiments and lend further support to the suggestion that E and B fields affect the membrane in distinctly different ways.

SUMMARY

The experiments discussed in this review admit the following conclusions:

1. Exposure of Physarum to sinusoidal fields of 45, 60, and 75 Hz, 0.35-1.0 V/m, and 10-200 μT lengthens the cell cycle, depresses the cell's rate of respiration and ATP content, and alters the cell membrane.

2. The time required to observe EMF effects varies from 24 hours to greater than 120 days depending on the particular parameter being examined and the mode of fields being applied.

3. The individual fields (E field and B field) appear to be interacting by different mechanisms.

ACKNOWLEDGEMENTS

The authors wish to acknowledge and express their appreciation for the excellent technical assistance of Mary Blanchard, Jean Fu, Judith Swanson, Pratima Tipnis, Kris Korlewitz, Kathy Carrick, James Lubniewski, and Todd Ratner. We would also like to acknowledge Drs. Tim Treffry and Paul Sharpe for their help with the TLCCD technique. Finally, we wish to acknowledge the support of the U.S. Naval Electronic Systems Command and NIEHS (ESO 2319).

REFERENCES

(1) Rusch, H.P.: Some biochemical events in the life cycle of Physarum polycephalum, in Advances in Cell Biology, D.M. Prescott, L. Goldstein and E. McConkey, eds., Vol. 1, Appleton-Century-Crofts, New York, 297-327, 1970.

(2) Goodman, E.M.: Physarum polycephalum: a review of a model system using a structure-function approach, in International Review of Cytology, G.H. Bourne and J.F. Danielli, eds., Vol. 63, Academic Press, New York, 1-58, 1980.

(3) Tyson, J.J.: Periodic phenomena in Physarum, in Cell Biology of Physarum and Didymium, H.C. Aldrich and J.W. Daniel, eds., Vol. 1, Academic Press, New York, 1982.

(4) Greenebaum, B., Marron, M.T. and Goodman, E.M.: Method for insuring comparable temperature in biological experiments using multiple incubators, Rev. Sci. Instrum. 48:937, 1977.

(5) Goodman, E.M., Greenebaum, B. and Marron, M.T.: Effects of extremely low frequency electromagnetic fields on Physarum polycephalum, Radiat. Res. 66:531, 1976.

(6) Greenebaum, B., Goodman, E.M. and Marron, M.T.: Magnetic field effects on mitotic cycle length in Physarum, Eur. J. Cell Biol. 27:256, 1982.

(7) Goodman, E.M., Greenebaum, B. and Marron, M.T.: Bioeffects of extremely low frequency electromagnetic fields: variation with intensity, waveform, and individual or combined electric and magnetic fields, Radiat. Res. 78:485, 1979.

(8) Mohberg, J., Babcock, K.L., Haugli, F.B. and Rusch, H.P.: Nuclear DNA content and chromosome number in the myxomycete Physarum polycephalum, Develop. Biol. 34:228, 1973.

(9) Pearson, E.S. and Hartley, H.O.: Biometrika Tables for Statisticians, Vol. II, Biometrika Trust, London, Chapter III and p. 46, 1976.

(10) Greenebaum, B., Goodman, E.M. and Marron, M.T.: Effects of extremely low frequency fields on slime mold: studies of electric, magnetic, and combined fields, chromosome numbers, and other tests, in Biological Effects of Extremely Low Frequency Electromagnetic Fields, R.D. Phillips, M.R. Gillis, W.T. Kaune and D.D. Mahlum, eds., Proceedings of the 18th Annual Hanford Life Sciences Symposium, 117, 1979.

(11) Marron, M.T., Goodman, E.M., Greenebaum, B. and Tipnis, P.: Effects of sinusoidal 60-Hz electric and magnetic fields on ATP and oxygen levels in the slime mold Physarum polycephalum, Bioelectromagnetics 7:307, 1986.

(12) Marron, M.T., Greenebaum, B., Swanson, J.E. and Goodman, E.M.: Cell surface effect of 60 Hz electromagnetic fields, Radiat. Res. 94:217, 1983.

(13) Goodman, E.M., Sharpe, P.T., Greenebaum, B. and Marron, M.T.: Pulsed magnetic fields alter the cell surface, F.E.B.S. Lett. 199:275, 1986.

13
An Electrochemical Consideration of Electromagnetic Bioeffects

ARTHUR A. PILLA
Bioelectrochemistry Laboratory, Department of Orthopedics
Mount Sinai School of Medicine, New York, New York

INTRODUCTION

Pulsed electromagnetically induced current (PEMIC) has been shown to stimulate the healing of delayed and non-union fractures (1-8). In addition, many cell and tissue systems have been affected by PEMIC having specific waveform parameters (9-18). It is important to consider the origins of the choice of these waveform parameters in order to relate them to a study of the mechanism of PEMIC bioeffects. This author was profoundly influenced by the early work of Becker (19) who proposed that electric fields play a substantial role in regeneration. Yasuda, Brighton, and Bassett (20-22) suggested that the pathway through which bone adaptively responds to mechanical input may be electrical. Pilla took the findings of these authors and used an electrochemical approach to predict a set of bioeffective electrical waveform parameters based on electrochemical kinetic interactions at the cell's surfaces (23-30). This approach ultimately led to the creation of PEMIC waveforms now in widespread clinical use for orthopaedic applications. It is the purpose of this chapter to review how electrochemistry has played a role in the electromagnetic modulation of cell and tissue behavior.

BASIC ELECTROCHEMICAL KINETICS APPLIED TO THE CELL SURFACE

Why consider electrochemical processes at the cell surface? The answer is that it provides a quantitative look at the cell's real-time responses to electromagnetic fields. Here it is important to note that the role of ions as transducers of information in the regulation of cell structure and function has gained widespread acceptance. Examples of ionic control mechanisms include: growth-factor activation of Na-K ATPase in fibroblasts (31,32), nerve growth-factor effects regulated by Na-K ATPase (33,34), Ca^{2+} regulation of the cell cycle via calmodulin (35,36), differential Ca^{2+} requirements of neoplastic vs. non-neoplastic cells (37,38), and Ca^{2+} dependent adenylate cyclase activation in macrophages (39). Ionic control mechanisms therefore represent a coupling mechanism for electromagnetic fields which can be quantitatively analyzed. The interaction of ions at the electrically charged interfaces of a cell is an example of a potential or voltage dependent process. The following is a review of the basic electrochemical kinetic approach to quantitate these ionic and/or dipolar interactions.

The working concept of electrochemical information transfer in vivo (23) uses the analogy between the electrified interfaces at the electrode/electrolyte and the membrane/fluid junctions. A change in the electrochemical microenvironment of the cell can cause the structure of its electrified surface regions to be modified by, for example, changing the concentration of a specifically bound ion or dipole which may be accompanied by a modification in the conformation of moleculer entities (such as enzymes) in the membrane structure. Basically, therefore, the regulatory interactions at a cell's surface are considered to have both potential and kinetic functions associated with the specific biochemical events to which these processes may be coupled.

If, as proposed, membrane structural changes that lead to a modification in membrane function involve or are caused by the basic electrochemical event of a specific adsorption (binding), it is easy to see how this process can be a key step in cellular response to the variety of external inputs which have been utilized. For skeletal tissue, the functional response to mechanical input can be envisaged to occur via at least two modalities involving electrochemical surface steps. If the cell membrane is not mechanically deformable (or does not experience significant force when in the collagen matrix), but is in direct membrane-collagen (matrix) contact via, for example, ligand bridging, then the known piezoelectric properties of the collagen bundles can modify the charge-charge interaction at the cell/collagen interface. On the other hand, certain membrane structures may be modified by mechanical input and result in new or increased specific adsorption and/or membrane transport. The latter may be more likely since it has been shown that the functional response of proliferative cartilage cells does not depend on whether they are bound to the matrix. Whatever structure responds directly to external force, it is clear that the cell membrane can exhibit new or modified charged-species interactions as required by the electrochemical information approach.

Under these conditions it would be expected that the cell membrane's electrical impedance would be modified, but in a different manner, dependent upon whether the cell exhibited direct or indirect response to mechanical input. In this context a variety of studies have shown that an electrical relaxation exists in bone tissue (alive or dead) when it is stressed (21,22). A portion of the slowly varying voltage function which is observable in bulk tissue at each stress input albeit very small, may be due to a transient change in the cell/collagen electrified interface structure. The time constant (kinetics) of this effect would be expected, as will be shown below, to be orders of magnitude smaller than the observed response function. Evidence that what the cell "sees" as a direct electrical input signal (in the absence of electrode effects) is indeed a much shorter-lived transient current, comes from attempts to mimic the stress-related relaxation signal. This has consistently failed to generate a biological response unless electrolysis and/or ion migration effects were obviously present.

When current is injected via electrodes, cellular response can again involve electrochemical surface steps if the electrochemical microenvironment is modified. Without considering the gross effects of electrolysis at this point, the ionic distribution changes that occur over a period of time during direct current (DC) flow can indeed couple to the concentration dependence (isotherm) of a specifically bound entity. It has already been observed that ionic changes in the extracellular fluid can cause a modification in cell state function (15,23-25). In fact, simple changes in extracellular ionic microenvironment can influence the rate of cell differentiation and even redirect its developmental pathway (40-42). If electrochemical information transfer is operative in cellular control, then there should be a direct functional response to the (pure) injection of current, provided it reaches the relevant cell surface and that the waveform parameters are chosen to modulate the kinetics of the directed electrochemical surface step. This chapter will consider the details of the type of charge interactions that could be involved in the cell's detection of, and reaction to, its immediate environment. In particular, the possible unifying nature of the electrochemical information transfer concept in cell regulation will be discussed in detail.

It is appropriate at this point to briefly consider the relation of membrane structure to function, and how interfacial electrochemical (non-faradaic) effects can be an integral part of this structure/function relationship. Among the various models of membrane structure (43,44) the dynamic fluid mosaic approach (45) appears to be the most consistent with observed behavior. The basic molecular components of cell membranes are lipids, proteins, and carbohydrates. Their molecular movements, conformations and interactions are without doubt influenced by the environment and can form part of a molecular feedback loop for cell regulation (46). Lipids are responsible for the structural integrity of the membrane. Membrane proteins are within this lipid fluid (integral proteins), or on the surface (peripheral proteins). They are highly mobile, and probably provide the structural modifications related to functional regulation.

Both the lipid and protein portions of the membrane contain hydrophobic and hydrophilic segments. The hydrophilic segments form the membrane side of the electrified interfaces at which specific adsorption (binding) can take place. Two distinct types of electrochemical interactions can occur at cell surfaces. The first involves all of the non-specific electrostatic interactions involving water dipoles and hydrated (or partially hydrated) ions. This structure is analogous to the electrode/electrolyte interface and can be contributory, along with lipid and protein asymmetry (47-49), to the observed dielectric response of the lipid and lipoprotein membrane structures (26,29). For small amounts of charge input, only minor modifications of this portion of a cell's surface structure would be expected. This is so for two reasons. The first relates to the fact that these non-specific electrostatic interactions are physically in series with the membrane dielectric structure, primarily due to the lipid bilayer fluid. Under these conditions any charge perturbation that could satisfy the kinetic requirements of these interfacial structures will primarily be experienced by the lipid dielectric. In addition, these non-specific interactions are, to a good first approximation, governed by a Boltzmann distribution with respect to the aqueous layer. Thus, over this portion of the interface, water dipoles would be expected to provide the first layer of charge interaction with the membrane surface as opposed to the more specific ion interactions discussed below. Water dipoles followed by a rigid layer of partially hydrated ions form an equilibrium structure which would be perturbed only to a negligible degree by low-level charge injection.

A second type of charge interaction at a cell surface involves potential-dependent specific adsorption (or binding). Here an ion or organic dipole can effectively compete with water dipoles and hydrated ions for specific membrane sites. This type of interaction involves, for the aqueous phase, the steps of dehydration, displacement and binding (50,51). If this is to engender a membrane function change, then the structure of the molecular entity within the membrane at which the binding occurs can undergo modification. For example, the allosteric nature of certain enzyme surely allows this to occur (52). In addition to enzyme activity it is known that biochemical reactions on the cell surface involve charged reactants (53), and the surface potential (and therefore structure) is experienced by an ionic species involved in membrane transport (54,55).

The most straightforward method to quantitate the above approach, which also provides unambiguous parameters capable of being experimentally tested, is to generate the electrical impedance of each relevant electrochemical pathway. All variables in this study will be given in terms of the complex frequency plane using the Laplace transformation (56). This frequency variable, s, has a real part, σ, and an imaginary, $j\omega$ part which define the axes of the Laplace plane. Utilization of the Laplace transformation allows a time domain function (such as a pulsating current) to be expressed in terms of its frequency content. Utilization of this transformation along the imaginary ($j\omega$) axis results in the familiar Fourier transformation by which the frequency spectrum of time domain signals is often expressed. Determination of the impedance, Z(s), of a cellular system will ultimately require, as will be shown in a later section, knowledge of the input pulsating current waveform and the pulsating voltage response of the membrane.

Here it will be considered that the physical passage of current into the membrane causes a change in its surface charge (transmembrane transport is neglected). The total current $i_T(s)$, can be considered, in light of the above discussion, to be the sum of a dielectric and double-layer charging portion $i_D(s)$ and a specific adsorption portion, $i_A(s)$. In other words, the total current can be written:

$$i_T(s) = i_D(s) + i_A(s) \quad (1)$$

Each of the above contributions to the total current can be related to the physical situation likely to exist at a living membrane.

Under linear conditions $i_D(s)$ represents the charging (discharging) of the surfaces of the lipid bilayer fluid which, because of structural asymmetry has different charge density on the intra- and extracellular surfaces, respectively. This structure behaves like a capacitor over the frequency range of interest for both observation and excitation (DC to 25 MHz). This is the well-known membrane dielectric capacitance, defined here as C_D. At each membrane/solution interface there exists, in addition, the electrostatic double-layer capacitance defined as C_i on the intracellular side and as C_e on the extracellular side. All charge separations corresponding to these membrane properties can be considered, to a first approximation, to lie on planes separated, for the dielectric capacitance, by 50-100 Å, and for the compact electrical double-layer by only a few Angstroms. Note that the latter is most certainly perturbed to some extent by the presence of carbohydrate on the extracellular surface of the membrane. However, this would be expected to be equivalent to surface roughness which would not change the basic capacitor analogy, but merely add a two-dimensional aspect. In view of all of the above, $i_D(s)$ may be related to the transient voltage response, E(s), by:

$$Z_D(s) = E(s)/i_D(s) = [1/s][1/C_i + 1/C_D + 1/C_e] \quad (2)$$

which describes the frequency behavior of the non-specific electrostatic portion of the membrane structure. Examination of Equation (2) shows that $Z_D(s)$ represents pure capacitive behavior for three capacitors in series.

Experimentally it is often observed that the majority of new charge is associated with C_D because it is much smaller (approximately 0.5 $\mu F/cm^2$) than either C_e or C_i which, by analogy with the electrode/electrolyte interface (23), are in the range of 10 $\mu F/cm^2$. This difference in interfacial and dielectric capacitance arises mainly because of the large difference in distance between the "planes" of charge separation associated with each capacitor (50-100 Å for C_D, and 1-6 Å for C_i and C_e). Because of this physical situation, it is important to realize that the majority of the voltage change, E(s), in response to $i_D(s)$ will appear across C_D, indicating that, in the absence of specific adsorption, most of the membrane charge acceptance will be associated with its dielectric structure. In most cases, therefore, $Z_D(s)$ can be represented by a single capacitor, C_D, the charging of which in response to low-level pulsating current would not be expected to alter membrane structure in a functional (regulatory) pathway since its equilibrium (or resting) structure would remain unaltered.

To quantitate the specific adsorption process, it is necessary to consider both its potential and concentration dependencies. The surface concentration, Γ, of the specifically adsorbed species can be equal to the number of ions (or dipoles) that penetrate the oriented water dipole layer. The specific adsorption current, $i_A(s)$, of Equation (1) will be utilized by the membrane to create a net change in the surface concentration, $\Delta\Gamma(s)$, of the bound species. In addition, the rate of specific adsorption can be considered, under linear conditions, to be adequately represented by first-order kinetics. Specific adsorption current $i_A(s)$ may be represented by:

$$i_A(s) = q_e s \Gamma_e \Delta\Gamma_e(s) \quad (3)$$

wherein only binding of a single species at the extracellular interface (subscript e) is considered (realizing that the adsorption of several species can occur, and at each interface), and q_e is a coefficient representing the dependence of interfacial charge upon the surface concentration of the bound species.

Equation (3) can lead to an expression for the impedance of specific adsorption, $Z_A(s)$, if the quantity $\Delta\Gamma_e(s)$ can be related to experimentally accessible parameters. This can be done, for the linearized conditions of this study, if a specific kinetic expression relating binding to potential changes is used. For this it is convenient to write:

$$\Gamma_e(s) = [v_e/\Gamma_e s]\ [\Delta\Gamma_e(s) + aE(s)] \qquad (4)$$

which states that the rate of change of surface concentration of the binding species is a function of the change in potential, E(s), via its potential dependence, a, and the exchange rate constant, v_e, taking into account that two species may not occupy the same site.

The adsorption impedance $Z_A(s)$ can now be written using Equations (3) and (4), as:

$$Z_A(s) = 1/q_e a[(1 + \Gamma_e s/v_e)/\Gamma_e s]. \qquad (5)$$

Inspection of Equation (5) shows that the specific adsorption process is functionally equivalent to a series R_A - C_A equivalent electric circuit. The heterogeneous adsorption process thus behaves as a lumped parameter system wherein the kinetic term is given by:

$$R_A = 1/q_e a v_e. \qquad (6)$$

As expected, R_A is inversely proportional to the exchange rate constant. C_A,which represents the accumulation of charged species at the kinetic site in question, is given by:

$$C_A = q_e a \Gamma_e. \qquad (7)$$

C_A is directly proportional to the resting concentration of adsorbed species about which its perturbation exists. It is now possible to construct the equivalent electrical model of a cell membrane at which the two considered processes exist. Inspection of Equations (1), (2) and (5) shows that the two current pathways consist of C_D, in parallel with R_A and C_A, which are themselves in series.

The above discussion illustrates (for a very simple case) the manner by which membrane charging can be utilized by the cell as a real-time event in its regulatory process. The relative rates of these non-specific and specific interfacial electrochemical steps is expected, from physical considerations alone, to be significantly different. In the context of electrochemical information transfer, this kinetic separation allows the concept of rate modulation selectivity to be considered. For example, if the specific adsorption process involves a regulatory enzyme, its average activity could be increased by affecting a net change in the surface concentration of bound ions. The first requirement for this is to satisfy the kinetics of, for example, the process described above during each current perturbation. As expected and observed (10,11,14,16,18), waveform duration is one of the most important parameters to achieve this result. The second requirement is to achieve sufficient charge injection to satisfy the potential dependence of the adsorption process. This would be relatively easy if there were no other adsorption processes with possible overlapping kinetics and potential dependences. This is certainly not the case, although for a given tissue in a given developmental, repair or maintenance phase it is possible that the overriding regulatory process involves a single family of membrane-bound entities. In view of this, it is reasonable to assume that a relationship will exist between waveform amplitude and width over the available selectivity range.

The real situation is not, however, as simple as just described. Because low-level perturbations are employed, linear or very near linear conditions can be expected in terms of the real-time direct response to the pulsating current input. It is therefore necessary to add the important variable of pulse repetition rate to the selectivity requirements. Of significance to cell function is the degree of kinetic coupling between biochemical follow-up reactions and the triggered electrochemical surface events. It thus becomes a question of the kinetics of the molecular control loop within the cell.

These considerations mean that the new boundary conditions for the control loop must be maintained each time the surface process is expected to be involved in the loop kinetics. The number of times the new loop conditions must be present for a functional consequence is, of course, unknown. In view of the physical nature of the surface processes involved in electrochemical information transfer, it is not expected that the repetition rate window will be as narrow as that in, for example, the alpha rhythm brain activity.

The above discussions provide a simplified quantitative picture of electrochemical information transfer in terms of cell surface regulation. It is obvious that the myriad surfaces and junctions which a cell exhibits may play a key role in its response to functional modification. However, recent evidence (57,59) tends to suggest that the cell surface is one of the major factors in cell regulation, and it may be the target area for current injection. It is obvious that the two membrane charging pathways given above allow an approach to selectivity in the choice of input current waveform by virtue of the different relaxation times associated with each process. Thus, for a given amplitude, pulse duration could be employed to achieve a selective response. Selectivity on the basis of pulse duration alone is, of course, not sufficient since, at the very least, the microenvironment of the cell, its state of function, position in cycle, etc., will contribute to the ultimate functional response.

The above discussion presents a working model of electrochemical information transfer by which the injection of low level current (in the $\mu A/cm^2$ range) can provide functional selectivity in the kinetic modulation of cell regulation. This method of predicting current waveform parameters is valid only when current is applied without concomitant electrolysis effects. When the latter are present the chemical microenvironment is under continual modification usually in an uncontrolled manner (with one notable exception (60)). Under these conditions the basic step of electrochemical information transfer may be present in the regulatory events occurring as the cell adapts to the modified environment. Interestingly, it is then possible to kinetically modulate the response to this new environment if the cell's surfaces and junctions are involved, by the superimposed injection of pure current. Indeed, if the kinetics of cell response to a modification in its chemical environment can be modulated, it then becomes possible to speak of synergistically enhancing the action of pharmacological agents, either by allowing a significantly reduced concentration to be employed, or through a basic enhancement of effectiveness. Preliminary experiments involving the lectin activation of peripheral human lymphocytes has shown this to be possible (62). A notable increase in the ability of concanavalin A to activate DNA synthesis was observed in the presence of injected current configured for specific adsorption selectivity. The implication of this in the possible enhancement of pharmacological agents is evident. In the context of bone repair, it is conceivable that the implanted electrode (used under controlled electrolysis conditions) and induced current techniques may well be synergistically employed for certain clinical situations.

GENERATION OF PEMIC WAVEFORMS - RELATION TO CELL IMPEDANCE STUDIES

To generate a voltage and current in tissue it was first decided that the induced waveform should have basically rectangular characterstics. This resulted from the kinetic analyses given above, which showed that excitation of real-time charge interactions can be more selectively accomplished if the driving voltage is relatively constant during the perturbation. In addition, the fact that inductive coupling results in a bipolar waveform has to be taken into account. In other words, if the model given in the previous section is valid, then the potential-dependent specific adsorption process can be more selectively perturbed if the driving waveform at the cellular level looks potentiostatic to the cell surface. This does not imply that waveforms other than rectangular will not provide a bioeffect. However, it does imply that rectangular waveforms may provide a greater degree of selectivity.

PEMIC waveform parameters in tissue are directly related to the electrical characteristics of the coil. For any coil the induced electromotive force (EMF) is proportional to the rate of change of current in the coil (dI_c/dt). The evaluation of this quantity for a given coil perturbation results in a description of the shape of the induced waveform in vacuum, air, and all nonmagnetic homogeneous conducting media in which the resulting current flow is not high enough to produce a sufficient back EMF for phase relationships to cause a waveform modification. The electrical characteristics of the coil relevant to PEMIC generation are its resistance, R_c, and inductance, L. The coil impedance is:

$$Z_c = R_c + Ls \tag{8}$$

from which dI_c/dt may be evaluated if the coil driving function is known. The coil-driving function V(t) may be arbitrary and constructed such that the coil current $I_c(t)$ rises from and returns to zero in an arbitrary manner. From the electrochemical information transfer concept it was decided that cell surface electrochemistry should be modulated using voltage control. This means that the electric field at the cell interface should have a rectangular shape. To generate this field, the coil-driving function, V(t), should be such that coil current, $I_c(t)$, rises and falls linearly from zero. The relative slopes may be equal or very asymmetrical, giving rectangles of differing amplitude and widths for dI_c/dt during coil charge and discharge. The Electro-Biology, Inc. coils are driven with a voltage step which is shut off in a time much shorter than the coil time-constant ($\tau_c = L/R_c$).

The coil current for an air-core inductor for a voltage step V_o is:

$$I_c(t) = V_o/R_c[1 - \exp(tR_c/L)] \tag{9}$$

Equation (9) shows that $I_c(t)$ rises exponentially to the short-circuit value (V_o/R_c) at a rate determined by the coil time constant, $\tau_c = L/R_c$. The waveform of the induced voltage is a direct function of dI_c/dt, which is

$$dI_c/dt = V_o/L\ [\exp(-tR_c/L)]. \tag{10}$$

Equation (10) clearly shows that, to achieve a rectangular-type induced waveform when V_o is applied, τ_c should be greater (by 10 times) than the desired pulse width. This can be achieved by proper choice of L and R_c. One modality is to keep L relatively small so that safe driving voltages (< 25 V) can be employed. Note that, as given by Equation (8), the maximum induced voltage (as $t \rightarrow 0$) is inversely proportional to coil inductance for a given V_o. Effective coil resistance can be kept small by utilizing heavy magnetic wire and connecting cable (14 to 16 B&S gauge). With the above taken into account, it is easy to see that, for a given τ_c, V_o can be applied to the coil for as long a time as the following relation is approximately valid:

$$dI_c/dt = V_o/L\ (1 - tR_c/L) \tag{11}$$

Over the time during which Equation (11) holds, an induced voltage waveform in the form of a step having some negative slope will be achieved. To maintain the relative rectangular nature of the induced voltage, pulse-shaping circuitry allows $I_c(t)$ to be kept linear during magnetic-field collapse.

The waveform described in Figure 1 has been utilized with pulse widths varying from 20 to 400 μsec in the main polarity T1, and from 2 μsec to 6 msec T2 in the opposite polarity. The pulses are repetitive, usually between 1-100 Hz and are sometimes set up as pulse bursts. The signal in primary use for recalcitrant bone-fracture repair is a 5 msec burst of pulses having 200 μsec main and 20 μsec opposite polarity. The repetition rate within the burst is approximately 4.4 KHz and the signal repetition rate is 15 Hz. To understand the rationale for the creation of a burst-type waveform it is useful to consider real-time cell-waveform interactions. For this it is useful to recall that the geometric dependence of the dosage of inductively-coupled current is predictable (63). Further, it has been shown that both Laplace and Fourier transforms are useful to describe various frequency characterizations of induced waveforms (30,64,65). Here this analysis is reviewed in the context of optimal coupling to cell surface electrochemical kinetics.

To quantitate waveform-cell interactions it must be pointed out that PEMIC appears to act as a trigger (30,65). Even the most powerful signal reported does not input sufficient energy to account for changes in processes as simple as ion fluxes

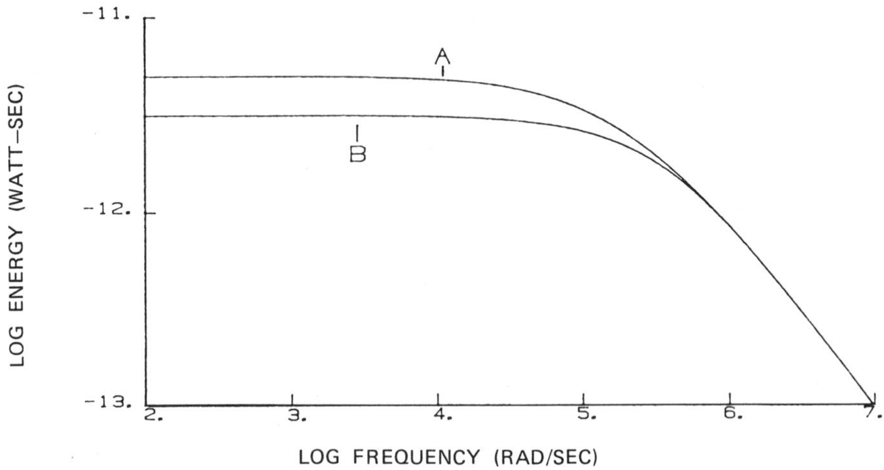

FIGURE 1. Schematic of the basic waveform of the (PEMIC) induced electric field vector considered in this study. The amplitudes of the main polarity (A) and opposite polarity (B) portions are controlled by the coil driving voltage. Note that A x T1 = B x T2 (T1 and T2 are the respective polarity pulse durations) so that for a larger T2, the opposite polarity amplitude (B) is lower when A is constant. The waveform exists in tissue and represents that voltage which can drive current into the specific adsorption-like pathway at the cell membrane. In a typical application the induced electric field is generated using two identical (10 x 10 cm) air-gap coils wound with 50 turns of No. 14 copper magnet wire set for approximately Helmholtz (magnetically-aiding) behavior (i.e., the intercoil distance is 6.5 cm). The effective amplitude reported for many in vitro and in vivo studies appeared to be that for which dB/dt ranged from 1-3.5 Gauss/μsec.

(65). Also, it is clear that the overall biological response is not related in a simple linear manner to the average total power in the signal. For example, several systems exhibit similar responses to a single pulse and a pulse burst containing 21 of these pulses, both at identical amplitude and repetition rate (16,18,66). These two waveforms differ by a factor of 21 in total average power. Furthermore, there is overwhelming evidence that, for bipolar PEMIC type waveforms, the parameters of the narrow high-amplitude portion of the signal are most correlatable with the observed bioeffect (16,18,65,66). Finally it has recently been reported (66) that nearly interchangeable repetition rate and amplitude dose curves can be obtained for repetitive single pulses having constant pulse width. Of most interest is the observation that the shape of the dose-response curve is greatly dependent upon pulse duration, being generally narrower as pulse width decrease (66). The above suggests that the dosage correlation for waveform-cell interactions should take into account response kinetics.

Impedance measurements at living cell membranes provide data for a time-constant due to the specific adsorption-like pathways at the cell membrane. The values obtained strongly correlate with the frequency range within which the single pulse and pulse burst waveforms have similar power levels. It is therefore of importance to examine the real-axis transformation of the power in the adsorption pathway, $P_A(\sigma)$. This is evaluated starting with $P_A = VI_A$ where I_A is obtained via the model for which Equation (4) is written. Thus, for a bipolar waveform having a main polarity duration, T1, and amplitude, A, and an opposite polarity duration, T2:

$$P_A(\sigma) = [A^2/(\sigma+1/TA)][1-e^{-(\sigma+1/TA)T1}+ (T1/T2)^2e^{-\sigma T1}(1-e^{-(\sigma+1/TA)T2})] \quad (12)$$

where TA is the specific adsorption (ion binding) time constant [see Equation (4)]. For T2 << T1, that portion of the waveform having duration T2 will have predominant power levels over the higher-frequency regions of this spectrum of the power. The main advantage of Equation (12) is that it clearly shows the effect of pulse duration on the power levels present over the frequency ranges of most effective coupling to TA. For example, Figure 2 shows the effect of a unipolar pulse having T1 equal to (B, Figure 2) and longer than (A, Figure 2). Clearly, pulses significantly longer than TA do not provide more usable power in the specific adsorption pathway over the relevant frequency regions. Use of the model-dependent spectrum of the power [Equation (12)] for various signal configurations allows the effect of differing ratios of T1/T2 to be predicted. This is performed by adjusting amplitude, A, at constant repetition rate

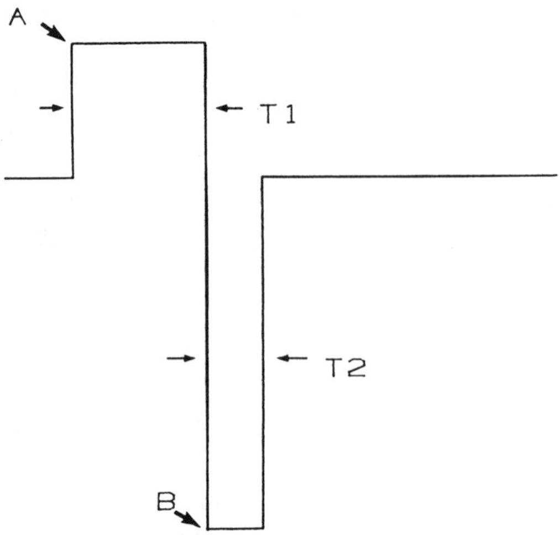

FIGURE 2. Spectra of the power via Equation (12) in an ionic membrane pathway functionally equivalent to specific adsorption for repetitive unipolar pulses having T1 = 5 (B) and 200 (A) μsec. For both pulses TA = 5 μsec. Shows that the power in the target pathway is governed by the kinetic response [TA, Equation (2)].

to obtain levels over the frequency range shown in Figure 2 that fall within the effective dose range observed for the system under study. This has been successfully tested on the cell adhesion assay system as described elsewhere (66).

The above shows that there is a real feasibility of using a model-dependent spectral analysis of the power in an ion binding pathway as a means to correlate the bioefficacy of electrical waveforms having widely different time parameters. The implication is that waveforms exhibiting power levels within given amplitude and frequency bands can effectively modulate cell and tissue function. Use of this analysis on results from a variety of different systems shows that Equation (12) offers a remarkable degree of correlation for the configuration of bioeffective waveforms. Of greatest mechanistic importance is the predictable effect of pulse width. As reported elsewhere, the cell adhesion assay allowed an effective assessment of the combined effect of TA and pulse width. These studies appear to show that the most effective predictions and correlations exist for TA of 1-5 μsec. Referring to the cell impedance data (26), it is clear the specific adsorption pathway appears the most likely to contain the ionic transduction process through which PEMIC couples to modulate cell function.

It may be noted that repetition rate was not specifically included in Equation (12). This was intentional since systematic results from several systems (66) appear to show a relative independence of effect on repetition rate over the range 2-60 Hz. This finding reflects the non-linear portions of the cellular reponse to PEMIC. The cause for this behavior may be due to the refractory time of the trigger site (65).

Notwithstanding the above analysis, the creation of the pulse-burst waveform in present clinical use was based on a consideration that the electrochemical processes to which PEMIC should couple had relaxation times in the millisecond range (23,26). This was prior to much of the _in vitro_ evidence now available. Given the desire to create a safe portable PEMIC delivery unit for clinical use it was reasoned that a burst waveform would enable an effectively wider pulse to be seen by the cell. The parameters were chosen such that the opposite polarity width was significantly shorter than that of the main polarity. In this way it was hoped that the narrow portion of the waveform would be too fast for the proposed slow kinetics. The cell process would respond to the envelope of the burst portion of the waveform. As subsequent studies have shown, this reasoning was erroneous because the cell/waveform interaction appears most effective when kinetics in the microsecond range are considered. Thus a single repetitive pulse having the same characteristics as the 21 pulses in the clinical pulse burst is essentially as effective as the burst waveform when applied with the same amplitude and repetition rate. This has been known since 1981 and it is surprising to this author that this waveform as well as others having much narrower main and opposite polarity pulse widths, which can offer significantly more efficient coupling to cell surface electrochemistry, have not yet seen clinical use.

REFERENCES

(1) Bassett, C.A.L., Pilla, A.A. and Pawluk, R.J.: A non-operative salvage of surgically-resistant pseudarthrosis and

non-unions by pulsing electromagnetic fields, Clin. Orthop. 124:128-143, 1977.

(2) Bassett, C.A.L., Mitchell, S., Norton, L. and Pilla, A.A.: A repair of non-unions by pulsing electromagnetic fields, Acta Orthop. Belg. 40:706-724, 1978.

(3) Bassett, C.A.L., Mitchell, S.N., Norton, L., Caulo, N. and Gaston, S.R.: Electromagnetic repairs of nonunions, in Electrical Properties of Bone and Cartilage, C.T. Brighton, J. Black and S.R. Pollack, eds., Grune and Stratton, New York, 605-630, 1979.

(4) Bassett, C.A.L., Mitchell, S.N. and Gaston, S.R.: Treatment of ununited tibial diaphyseal fractures with pulsing electromagnetic fields, J. Bone Joint Surg. 63A:511-523, 1981.

(5) Bassett, C.A.L., Caulo, N. and Kort, J.: Congenital pseudarthrosis of the tibia: Treatment with pulsing electromagnetic fields, Clin. Orthop. 154:136-148, 1981.

(6) Kort, J.S., Schink, M.M., Mitchell, S.N. and Bassett, C.A.L.: Congenital pseudarthrosis of the tibia: Treatment with pulsing electromagnetic fields, Clin. Orthop. 165:124-137, 1982.

(7) Bassett, C.A.L., Mitchell, S.N. and Schink, M.M.: Treatment of therapeutically resistant non-unions with bone grafts and pulsing electromagnetic fields, J. Bone Joint Surg. 64:1214-1220, 1982.

(8) Bassett, C.A.L., Mitchell, S.N. and Gaston, S.R.: Pulsing electromagnetic field treatment in ununited fractures and failed arthrodeses, J. Am. Med. Assoc. 247:623-628, 1982.

(9) Liboff, A.R., William, T. Jr., Strong, D.M. and Wistar, R.Jr.: Time-varying magnetic fields: Effect on DNA synthesis, Science 223:818-819, 1984.

(10) Norton, L.: Effects of a pulsed electromagnetic field on a mixed chondroblastic tissue culture, Clin. Orthop. 167:280-290, 1982.

(11) Chiabrera, A., Grattarola, M., Viviani, R. and Braccini, C.: Modeling of the perturbation induced by low-frequency electromagnetic fields on the membrane receptors of stimulated human lymphocytes, Studia Biophysica 91:125-131, 1982.

(12) Dixey, R. and Rein, G.: ^{3}H-noradrenaline release potentiated in a clonal nerve cell line by low-intensity pulsed magnetic fields, Nature 296:253-256, 1982.

(13) Goodman, R., Bassett, C.A.L. and Henderson, A.S.: Pulsing electromagnetic fields induce cellular transcription, Science 220:1283-1285, 1983.

(14) Luben, R.A., Cain, C.D., Chen, M.C-Y., Rosen, D.M. and Adey, W.R.: Effects of electromagnetic stimuli on bone and bone cells in vitro: Inhibition of responses to parathyroid hormone by low-energy low-frequency fields, Proc. Natl. Acad. Sci. USA 79:4180-4184, 1982.

(15) Chiabrera, A., Hinsenkamp, M., Pilla, A.A., J., Ponta, D., Belmont, A., Beltrame, F. Grattarola, M. and Nicolini, C.: Cytofluorometry of electromagnetically controlled cell dedifferentiation, J. Histochem. Cytochem. 27:375-384, 1979.

(16) Christel, P., Cerf, G. and Pilla, A.A.: Modulation of rat radial osteotomy repair using electromagnetic current induction, in Mechanisms of Growth Control, R.O. Becker, ed., Charles C. Thomas, Springfield, 237-250, 1981.

(17) Bassett, C.A.L., Valdes, M.G. and Hernandez, E.: Modification of fracture repair with selected pulsing electromagnetic fields, J. Bone Joint Surg. 64A:888-895, 1982.

(18) Smith, S.D. and Pilla, A.A.: Modulation of newt limb regeneration by electromagnetically induced low level pulsating current, in Mechanisms of Growth Control, R.O. Becker, ed., Charles C. Thomas, Springfield, 137-152, 1981.

(19) Becker, R.O.: The bioelectric factors in amphibian limb regeneration, J. Bone Joint Surg. 43A:643-656, 1961.

(20) Brighton, C.T., Friedenberg, Z.D., Zemsky, L.M. and Polis, R.R.: Direct current stimulation of non-union and congenital pseudarthrosis, J. Bone Joint Surg. 58A:368-377, 1975.

(21) Fukada, E. and Yasuda, I.: On the piezoelectric effect of bone, J. Phys. Soc. Japan 12:1158-1162, 1957.

(22) Bassett, C.A.L. and Becker, R.O.: Generation of electric potentials by bone in response to mechanical stress, Science 137:1063-1064, 1962.

(23) Pilla, A.A.: Electrochemical information transfer at living cell membranes, Ann. N.Y. Acad. Sci. 238:149-169, 1974.

(24) Pilla, A.A.: Mechanisms of electrochemical phenomena in tissue growth and repair, Bioelectrochem. Bioenergetics 1:227, 1974.

(25) Becker, R.O. and Pilla, A.A.: Electrochemical mechanisms and the control of biological growth processes, in Modern Aspects of Electrochemistry, J. O'M. Bockris, ed., Plenum Press, New York, Vol. 10, 289-338, 1975.

(26) Pilla, A.A. and Margules, G.S.: Dynamic interfacial electrochemical phenomena at living cell membranes: Application to the toad urinary bladder membrane system, J. Electrochem. Soc. 124:1697-1706, 1977.

(27) Pilla, A.A.: Membrane impedance as a probe for interfacial electrochemical control of living cell function, Adv. in Chem., 188:339-359, 1980.

(28) Pilla, A.A.: Electrochemical information transfer and its possible role in the control of cell function, in Electrical Properties of Bone and Cartilage, C.T. Brighton, J. Black and S.R. Pollack, eds., Grune and Stratton, New York, 455-490, 1979.

(29) Margules, G.S., Doty, S.B. and Pilla, A.A.: Impedance of living cell membranes in the presence of chemical tissue fixative, Adv. in Chem. 188:461-484, 1980.

(30) Pilla, A.A.: Electrochemical information transfer at cell surfaces and junctions -- application to the study and manipulation of cell regulation, in Bioelectrochemistry, H. Keyzer and F. Gutman, eds., Plenum, New York, 353-396, 1980.

(31) Mendoza, J., Wigglesworth, N.M., Pohjanpelto, P. and Rozengurt, E.: Na Entry and Na-K pump activity in murine, hamster, and human cells -- Effect of monensin, serum, platelet extract, and viral transformation, J. Cell Physiol. 103:17-27, 1980.

(32) Lopez-Rivas, A., Adelburg, E. and Rozengurt, E.: Intracellular K^+ and the mitogenic response of 3T3 cells to peptide factors in serum-free medium, Proc. Natl. Acad. Sci. USA 79:6275-6279, 1982.

(33) Boonstra, J., Van der Sagg, P.T., Moolenarr, W.H. and DeLaat, S.W.: Rapid effects of nerve growth factor on the Na^+,K^+-pump in rat pheochromocytoma cells, Exp. Cell Res. 131:452-455, 1981.

(34) Boonstra, J., Skaper, S.D. and Varon, S.J.: Regulation of Na^+,K^+ pump activity by nerve growth factor in chick embryo dorsal root ganglion cells, J. Cell Physiol. 113:28-34, 1982.

(35) Chafoules, J.G., Bolton, W.E., Hidaka, H., Boyd, A.E. and Means, A.R.: Calmodulin and the cell cycle: Involvement in regulation of cell-cycle progression, Cell 28:41-50, 1982.

(36) Whitfield, J.F., Boynton, A L., MacManus, J.P. Rixon, R.H., Sikorska, M., Tsong, B., Walker, P.R. and Swierenga, S.H.: The roles of calcium and cyclic AMP in cell proliferation, Ann. N.Y. Acad. Sci. 339:216-240, 1981.

(37) Boynton, A.L., Whitfield, J.F., Isaacs, R.J. and Trembley, R.G.: Different extracellular calcium requirements for proliferation of nonneoplastic, preneoplastic, and neoplastic mouse cells, Cancer Res. 37:2657-2661, 1977.

(38) Hazleton, B. and Tupper, J.: Calcium transport and exchange in mouse 3T3 and SV40-3T3 cells, J. Cell. Biol. 81:538-542, 1979.

(39) Gearsa, D., Seitz, M., Kramer, W., Grimm, W., Till, G. and Resch, K.: Ionophore A23187 raises cyclic AMP levels in macrophages by stimulating prostaglandin E formation, Exp. Cell Res. 118:55-62, 1979.

(40) Barth, L.G. and Barth, L.J.: The sodium dependence of embryonic induction, Develop. Biol. 20:236-262, 1969.

(41) Barth, L.G. and Barth, L.J.: ^{22}Sodium and ^{45}calcium uptake during embryonic induction in Rana pipiens, Develop. Biol. 28:18-34, 1972.

(42) Barth, L.G. and Barth, L.J.: Biol. Bull. 146:313, 1974.

(43) Davson, H. and Danielli, J.F.: The Permeability of Natural Membranes, 2nd Ed., Cambridge University Press, London, 1952.

(44) Robertson, J.D.: in Cellular Membranes in Development, M. Locke, ed., Academic, New York, 1964.

(45) Singer, S.J. and Nicholson, G.L.: The fluid mosaic model of the structure of cell membranes, Science 175:720-731, 1972.

(46) Mueller, G.C.: in The Cell Cycle and Cancer, R. Baserga, ed., Marcel Dekker, New York, 269, 1971.

(47) Bretscher, M.S.: Membrane structure: Some general principles, Science 181:622-629, 1972.

(48) Zwaal, R.F.A., Roelofsen, B. and Colley, C.M.: Localization of red cell membrane constituents, Biochem. Biophysica Acta 300:159-182, 1973.

(49) Tsai, K. and Lenard, J.: Asymmetry of influenza virus membrane bilayer demonstrated with phospholipase C, Nature 253:554-555, 1975.

(50) Parsegian, V.A.: Ion-membrane interactions as structural forces, Ann. N.Y. Acad. Sci. 264:161-174, 1975.

(51) Urry, D.W.: Basic aspects of calcium chemistry and membrane interaction: On the messenger role of calcium, Ann. N.Y. Acad. Sci. 307:3-27, 1978.

(52) Singer, S.J.: in Structure and Function of Biological Membranes, L.I. Rothfeld, ed., Academic, New York, 1971.

(53) Goldstein, L., Levine, Y. and Katchalsky, E.: A water-insoluble polyanionic derivative of trypsin. II. Effect of the polyelectrolyte carrier on the kinetic behavior of the bound trypsin, Biochemistry 3:1913-1919, 1964.

(54) McLaughlin, S., Szabo, G., Eisenman, G. and Ciani, S.M.: Surface charge and the conductance of phospholipid membranes, Proc. Natl. Acad. Sci. USA 67:1268-1275, 1970.

(55) McLaughlin, S.: The mechanism of action of DNP on phospholipid bilayer membranes, J. Memb. Biol. 9:361-372, 1972.

(56) Cheng, D.K.: Analysis of Linear Systems, Addison Wesley, London, 1959.

(57) Edelman, G.M.: Surface modulation in cell recognition and cell growth, Science 192:218-226, 1976.

(58) Rasmussen, H.: Cell communication, calcium ion, and cyclic adenosine monophosphate, Science 170:404-412, 1970.

(59) Harrison, R. and Lunt, G.G.: Biological Membranes: Their Structure and Function, Wiley, New York, 1975.

(60) Becker, R.O. and Spadaro, J.A.: Treatment of orthopedic infections with electrically generated silver ions, J. Bone Joint Surg. 60A:871-881, 1978.

(61) Conti, P., Gigante, G.E., Difone, M.G., Alesse, E., Ianni, G., Reale, M. and Angeletti, P.U.: Reduced mitogenic stimulation of human lymphocytes by extremely low frequency electromagnetic fields, FEBS Lett. 162:156-160, 1983.

(62) Conti, P., Gigante, G.E., Alesse, E., Cifone, M.G., Fieschi, C., Reale, M. and Angelette, P.U.: A role for Ca^{2+} in the effect of very low frequency electromagnetic field on the blastogenesis of human lymphocytes, FEBS Lett. 181:28-32, 1985.

(63) McLeod, B. and Parker, R.A.: Calculated induced currents in a model non-union fracture stimulated by Helmholtz coils or resistively coupled parallel plates, J. Bioelectricity 3:41-56, 1984.

(64) Pilla, A.A. and Kaufman, J.J.: Electromagnetic modulation of cell function: Frequency characterizations of input waveforms, J. Bioelectricity 3:3-18, 1983.

(65) Pilla, A.A., Schmukler, R.E., Kaufman, J.J. and Rein, G.: in Interactions between Electromagnetic Fields and Cells, A. Chiabrera, C. Nicolini and H.P. Schwan, eds., Plenum, New York, 423, 1985.

(66) Fox, K., Pilla, A.A., Hellman, K. and Ryaby, J.: Pulsating electromagnetic fields and rosette formation at macrophage membranes, Trans. 4th Ann. Bioelectrical Repair and Growth Society Meeting, Kyoto, Japan, P19, 1984.

14
Lymphocytes and Pulsing Magnetic Fields

RUGGERO CADOSSI, GIOVANNI EMILIA
GIOVANNI CECCHERELLI, GIUSEPPE TORELLI
Center for Experimental Haematology,
Department of Internal Medicine II, University of Modena
Modena, Italy

INTRODUCTION

Pulsing electromagnetic fields (PEMFs) are presently used in centers around the world to promote the healing of congenital and acquired pseudarthroses (1-6). Despite the large number of clinical studies, very little is known about the mechanism of action of PEMFs at the cellular level.

Several biological models have been used to investigate PEMF effects at the cellular level. Luben et al. (7) showed that PEMF exposure inhibited parathyroid hormone (PTH) action on cultured bone cells, but did not interfere with the action of vitamin D. It is interesting to observe that, while PTH binds to a receptor on the cell membrane, vitamin D acts directly within the cell in the cytoplasm. Dixey and Rein (8) observed that noradrenaline release by cultured nerve cells was significantly increased after 30 minutes of PEMF exposure. In an attempt to explain this result, they pointed out that cytoplasmic secretory phenomena are dependent on Ca^{++} influx, and they suggested that the observed phenomenon might be due to a PEMF effect on Ca^{++} influx across the cell membrane. Rodan et al. (9) found that cartilage cells exposed to 5 Hz electric fields exhibited increased DNA synthesis. They were able to inhibit the electric field effect by means of verapamil, a calcium antagonist.

Liboff et al. (10) reported enhanced DNA synthesis when cultured fibroblasts were exposed to an electromagnetic field during that phase of the cell cycle in which the transition from G_1 to S occurred. Goodman et al. (11) showed increased RNA transcription in dipterian salivary gland cells in cell cultures exposed to PEMFs. Ottani et al. (12) demonstrated that PEMF exposure increased the regeneration rate in rat liver after partial haepatectomy.

Grattarola et al. (13) and Conti et al. (14) employed different signals, but both reported that PEMF exposure decreased the response of normal human lymphocytes to lectin stimulation. Conversely, Hellman et al. (15) and Emilia et al. (16) reported that lectin-stimulated lymphocytes exposed to electromagnetic radiation exhibited increased DNA synthesis.

THE CELLULAR TARGETS OF PEMFs

The review of the published data (7-47) allows us to draw some general conclusions about PEMF mechanisms of action. PEMFs may act by interfering with those membrane processes that govern the interaction between an active substance and its receptor. The cell membrane is probably one example of a PEMF target. PEMFs also appear to produce an effect on Ca^{++} flux across the cell membrane, and they seem capable of triggering an increase in the pool of free intracellular Ca^{++}. In the light of these considerations, the present interest in the lymphocyte system is understandable, because this biological model is particularly helpful in the study of all the possible sites of action of PEMFs.

Chiabrera et al. used the lymphocyte system to study the PEMF effect on the kinetics of the reaction between the lectin and the membrane receptor (48). Cadossi et al. (49) and Conti et al. (42) estimated the effects produced by PEMF exposure on $^{45}Ca^{++}$ influx into the lymphocyte. Hellman et al. (50), Emilia et al. (16), and Conti et al. (14) used this system to evaluate the effect on DNA synthesis.

Before describing the details of lymphocyte cultures, we will briefly summarize the physiology of the lymphocytes and the changes that occur when lymphocytes are cultured in the presence of mitogens; we will focus on lectin stimulation.

THE LYMPHOCYTE-LECTIN MODEL

LYMPHOCYTE STIMULATION BY LECTINS

When cultured in the presence of a lectin such as phytohemagglutinin (PHA) (a glycoprotein extracted from

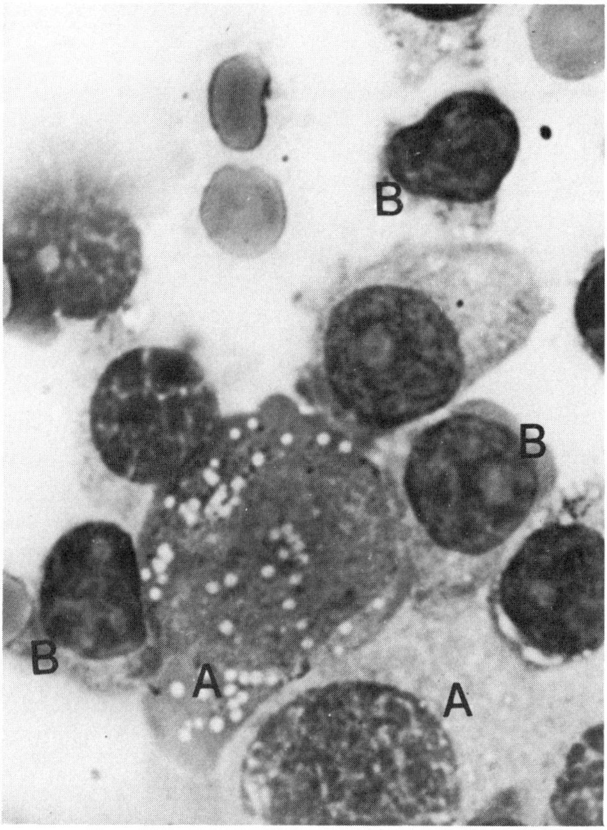

FIGURE 1. Normal human lymphocytes after culturing for 72 hours in the presence of PHA. A, activated blast lymphocyte; B, non-activated lymphocyte. May-Grunwald stain.

Phaseolus vulgaris), human normal lymphocytes obtained from peripheral blood undergo a series of morphological changes. The small, relatively inactive lymphocyte becomes transformed into a lymphoblast, a larger, metabolically active cell (51-54)(Figure 1). This morphological change is coupled with a cell transition from the resting phase, G_0, to the G_1 phase of the cell cycle.

A lymphocyte in G_1 does not necessarily enter the S phase. Some cells never begin DNA synthesis (55,56) even when they are morphologically transformed into blasts. In late G_1, there is a

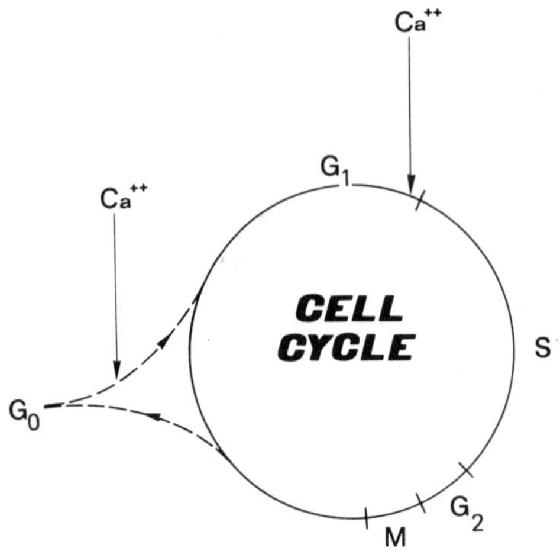

FIGURE 2. The cell cycle. Ca^{++} influx into the lymphocyte is required for the cell to move from G_0 to G_1 and to move from the G_1 to the S phase. DNA synthesis takes place during the S phase. In G_2, the DNA content of the cell is twice normal. During M phase mitosis is visible.

point that the cell must cross to enter the S phase of DNA synthesis. The crossing point is dependent on Ca^{++} influx. Once the cell has entered the S phase, it usually completes the cycle and doubles. It can then either proceed in the cycle or come out of it (Figure 2).

The reaction between the lectin and its receptor on the lymphocyte membrane is an essential phenomenon in lymphocyte activation. Once the bond is achieved, clustering of the receptor-lectin complexes occurs (57,58)(Figure 3). The impaired response of chronic lymphocytic leukemia (CLL) lymphocytes to PHA activation is partially explained by the small number of receptors present on the cell membrane (59).

Once the lectin-receptor bond occurs, an increase in intracellular Ca^{++} can be observed within the first 30-60 minutes (Figure 4). It results partially from an increased Ca^{++} influx and partially from a transfer of bound Ca^{++} present in the cytoplasm (60-66). If the culture medium contains a chelating agent, such as ethylene-diamine-tetracetic acid, that is capable of binding all the Ca^{++} present in the

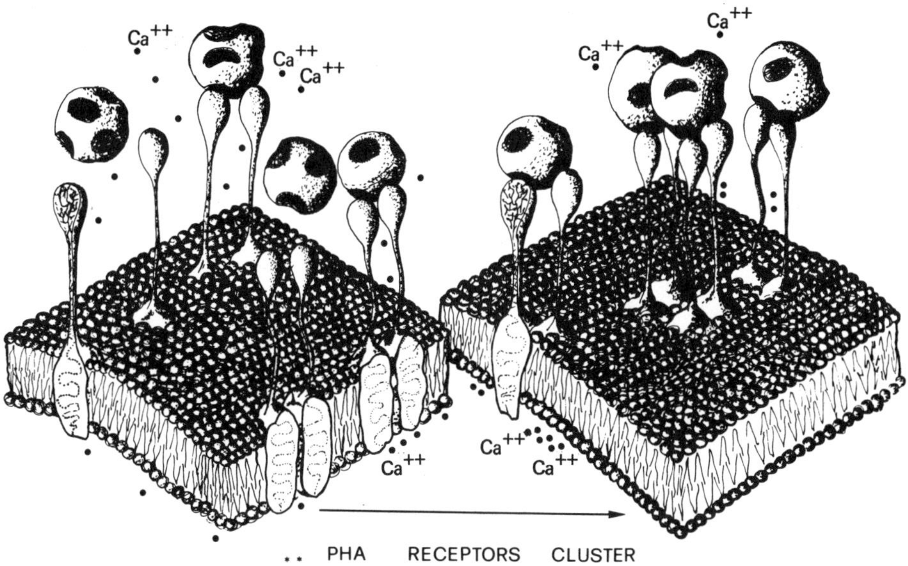

FIGURE 3. Left: PHA links its membrane receptors. Right: the PHA-receptor complexes cluster. This is accompanied by an increased Ca^{++} influx. (Modified from (48)).

medium, lymphocytic activation does not occur. Activation also does not occur in a culture medium that contains no Ca^{++} (66-68). On the other hand, if the culture is enriched with an ionophore such as A23187 which makes the membrane permeable, and hence increases the free intracellular Ca^{++} pool, lymphocyte stimulation takes place even in the absence of lectins (69-73). An increase in intracellular Ca^{++} seems to be a metabolic signal for cellular activation and progression in the cell cycle that is common to all eukaryotic cells (74,75). Ca^{++} influx is also required for transition from the G_1 to the S phase (55,75).

Since the various phases of lymphocyte activation are at least partially understood, the usefulness of this biological model can be recognized. The model applies to the study of several phenomena and especially to the investigation of PEMF-induced biological effects. But in addition to its advantages, the lymphocyte model presents some drawbacks that must be considered.

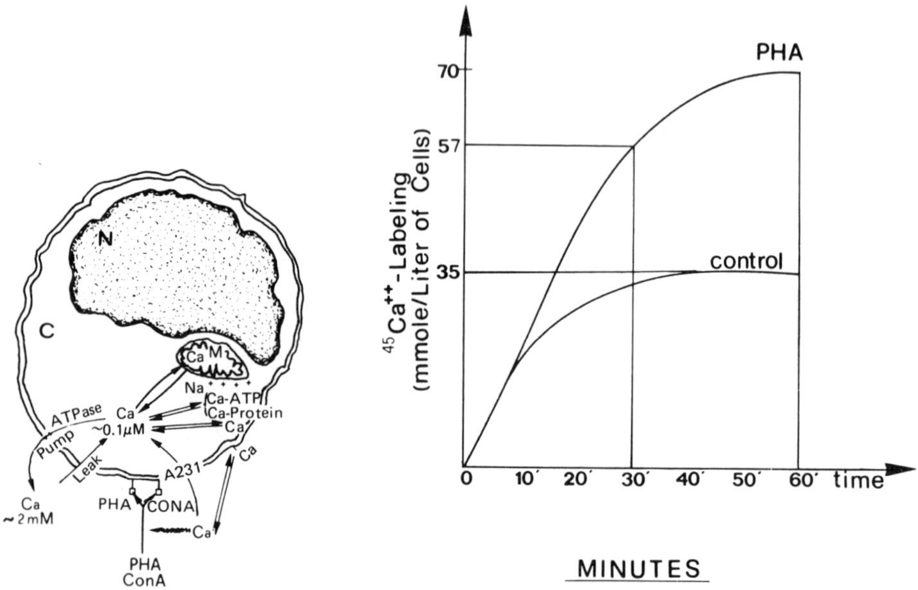

FIGURE 4. Right: immediately after the link between the lectin and its membrane receptor, Ca^{++} influx is observed. The increased free cytoplasmic Ca^{++} concentration is due both to Ca^{++} influx and to its mobilization from the mitochondrial stores. This increase starts a series of Ca^{++} dependent events. Left: kinetic of Ca^{++} influx into the lymphocyte following the lectin-receptor linkage.

LYMPHOCYTE-LECTIN MODEL ADVANTAGES

The lymphocyte is a normal, human eukaryotic cell. It is easily obtained by simple venous puncture. Lymphocytes from healthy donors are normal cells that have not undergone the changes that occur in cells passed in culture. Such cell lines often are obtained from malignant cells.

The lymphocyte is present in the peripheral blood, and is normally inactive. In suitable conditions however, its activation and duplication can be induced, thus permitting study of the phenomena that govern its entry and progression in the cell cycle. The preparation of the cultures is not particularly difficult if they are to be maintained no longer than 72-96 hours.

The lymphocyte is probably the most studied and best understood normal human cell.

LYMPHOCYTE-LECTIN MODEL DRAWBACKS

The amount of stimulation by lectins varies among different donors, and it may vary within the same donor depending on immunological status. The lymphocyte cell population is not synchronized, and cannot be made synchronous.

The obtainable quantity of lymphocytes is a limiting factor. The lymphocyte population lacks homogeneity in that several subpopulations can be delineated. Table 1 lists some of the most common surface markers used to identify B- and T-lymphocytes. The main categories recognized are B- and T-lymphocytes, whose responses to mitogenic stimuli induced by lectins are different as shown in Table 2. The lymphocytic

TABLE 1. Lymphocyte Identification by Surface Markers

CELL TYPE	SURFACE MARKERS					
	S Ig	C3	Fc	ME	SE	IAg
T	-	+	+	-	+	+
B	+	+	+	+	-	+

TABLE 2. Specificity of Commonly Used Lectins for B- and T-Lymphocytes

MITOGEN	LYMPHOCYTES			
	Human		Mouse	
	T	B	T	B
Phytohemagglutinin (PHA)	+	?	+	-
Concanavalin A (ConA)	+	-	+	-
Pokeweed (PWM)	+	+	+	+
Insoluble PHA, ConA, PWM	+	+	+	+

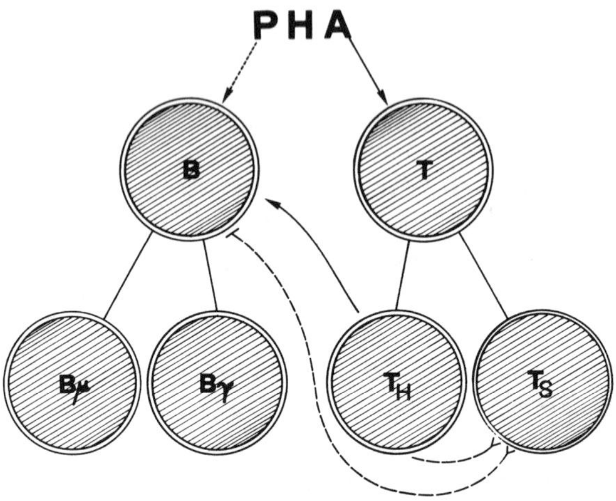

FIGURE 5. Schematic representation of the interaction between T- and B-lymphocytes responding to PHA stimulus.

response to lectins is also regulated by the release of growth factors that modulate the response of the different lymphocyte subpopulations (Figure 5).

Before examining the techniques used to prepare and grow lymphocyte cultures, we would like to review some basic information relating to lymphocyte physiology and to some of the metabolic changes that occur as a result of mitogenic stimulation. The wide variety of available data testifies to the interest in this biological model that has arisen because of its extreme sensitivity.

THE LYMPHOCYTE

The peripheral blood contains erythrocytes, platelets, monocytes, granulocytes and lymphocytes. Lymphocytes, which are involved in cellular immunity and antibody production, arise from a stem cell present in the bone marrow. The stem cell gives rise to the lymphoblast which differentiates into the lymphocyte. During the differentiation process, lymphocytes are primed as T-lymphocytes by thymus, or as B-lymphocytes by the lymphatic bursa equivalent organs (Figure 6).

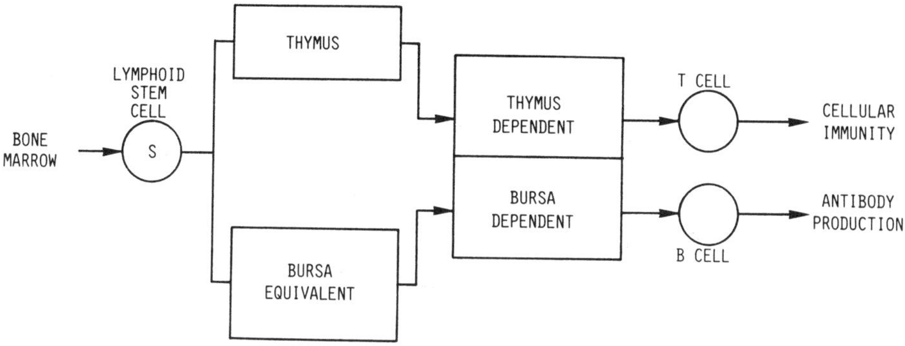

FIGURE 6. The lymphocytes are primed by the thymus or by the bursa equivalent organs. T-lymphocytes are responsible for the cell-mediated immunity (i.e., graft rejection). B-lymphocytes are primed for antibody production (i.e., defense from viral infection).

As we have already pointed out, lymphocytes cultured in the presence of mitogenic substances undergo a series of important morphological and metabolic changes (51). In addition to lectins, there are a great number of substances capable of eliciting lymphocytic proliferation. Table 3 shows a detailed list of the agents that are reported to induce lymphocyte proliferation in vitro.

The blast transformation of the small peripheral lymphocyte is also a property of the in vivo lymphocyte. When it contacts an antigen such as a viral particle or a structure that it recognizes as foreign, it becomes activated and transforms into a blast cell.

Within the first two hours after the lymphocyte-lectin interaction, an enhanced incorporation of precursors occurs in proteins (77), RNA (78), and in lipids (79). The formation of lactate and pyruvate also increases (80). DNA synthesis begins 24 hours after PHA addition (81). The significant change in RNA metabolism associated with blast formation is shown in Figure 7. If a culture is maintained for 48-72 hours, mitosis can be observed.

Some recent techniques of in situ hybridization, making use of messenger RNAs whose transcription is specifically related to

TABLE 3. Agents Reported to Induce Lymphocyte Proliferation

NON-SPECIFIC MITOGENS	SPECIFIC ANTIGENS
Chymotripsin	Ag-Ab complexes
Yeast zymosan	Dilantin
Lectins	Diphtheria toxoid
Microwave irradiation	Histoplasm
Papain	Penicillin
Staphylococcal filtrate	Polio vaccine
Streptolysin	Purified protein derivate
Trypsin	Ragweed pollen
	Some endotoxins
TISSUE ANTIGENS	Typhoid-paratyphoid vaccine
	Vaccinia vaccine
Cerebrospinal fluid	Varidase
Extract of white blood cells	
Fetal calf serum	
Homologous macrophages	ANTISERA
Leukocyte supernatant	
Lymphocyte homologous cultures	Monkey antihuman Ig
Platelets	Rabbit antihuman leukocyte
Sera	Some antisera

the phases of the cell cycle, allow us to show the cell transition through the different phases of the cell cycle from a molecular standpoint (82)(Figure 8).

CHRONIC LYMPHOCYTIC LEUKEMIA LYMPHOCYTES

Chronic lymphocytic leukemia (CLL) is a pathological condition characterized by the presence in the patients' peripheral blood of a large number of small lymphocytes (very often over 30,000/mm^3), whose metabolic activity is poor. These are generally extremely homogeneous populations (90%). CLL lymphocytes often share the membrane antigens that are specific for B-lymphocytes and, much more rarely, for T-lymphocytes. It is therefore the B-lymphocyte populations that show an impaired response to lectin stimulation (83,84). These characteristics make CLL lymphocytes very useful for lectin studies.

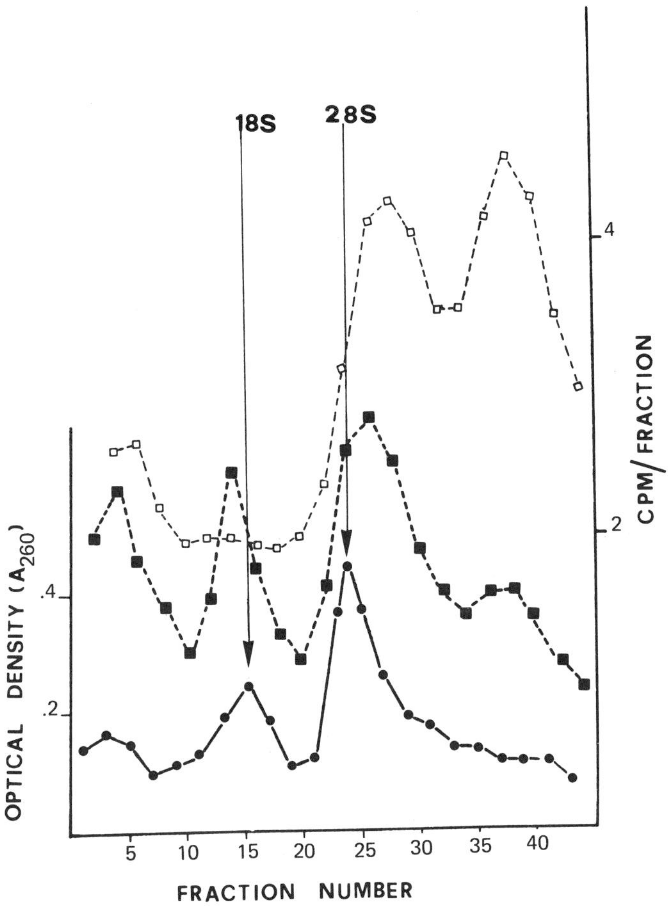

FIGURE 7. Sucrose (5-20%) sedimentation profile of total RNA extracted from human lymphocytes. Solid line, optical density of the RNA: 18 S and 28 S ribosomal RNA. Four hours ^{3}H-uridine pulse labeling of unstimulated lymphocytes (cpm x 10^{-2})(dashed line bounded by open squares) and of lymphocytes stimulated 72 hours with PHA (cpm x 10^{-3})(dashed line bounded by solid squares). In the last case the ribosomal RNA processing is evident.

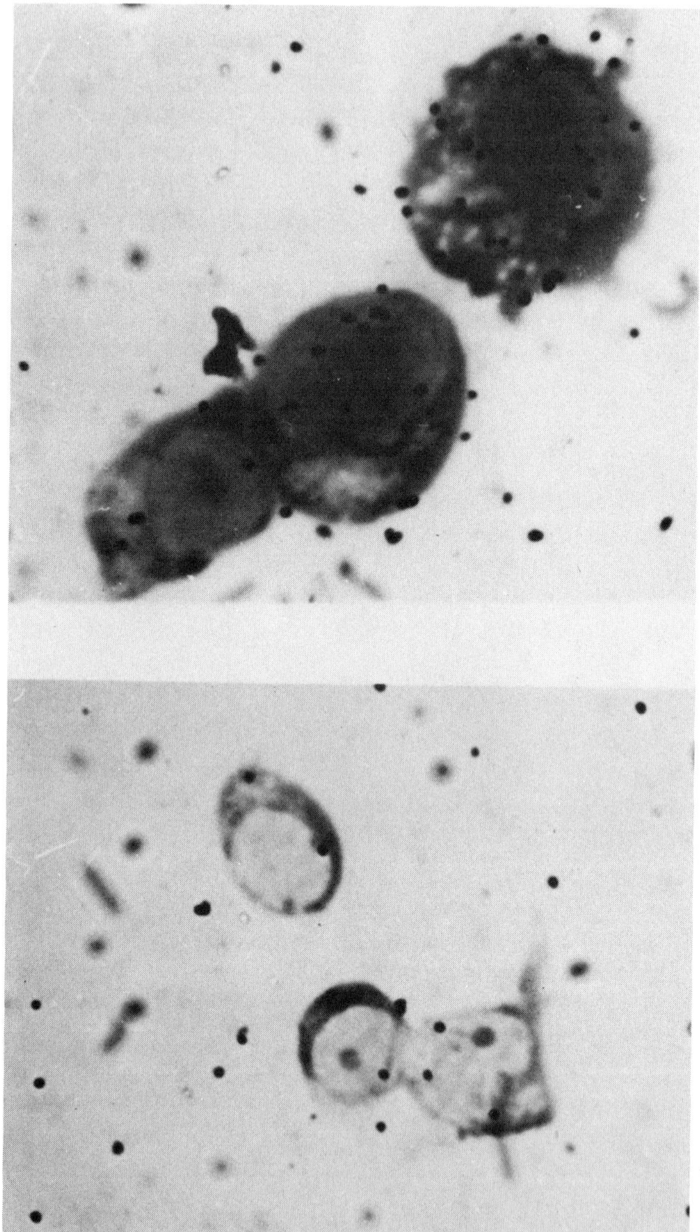

FIGURE 8. c-myb proto-oncogene expression in unstimulated (left) and stimulated (right) human normal lymphocyted, detected by in situ hybridization.

PREPARATION OF LYMPHOCYTE CULTURES

LYMPHOCYTE SEPARATION

Lymphocytes are obtained from the peripheral blood of donors. The venous blood is treated with heparin (heparin 5000 U/ml; 1 ml in 60 ml of blood). Macrodex 10-20% (dextran 6% in physiologic solution) is added to increase the sedimentation rate.

The blood sample is allowed to sediment at 37°C to separate the red cells. The plasma so obtained is diluted 1:2 with RPMI 1640 (Gibco) in a sterile plastic conic tube.

Mononuclear cells (monocytes and lymphocytes) can be separated from the other cells present by centrifugation on Ficoll-Hypaque 1077 (Sigma) (used 1:1 to the total volume of the diluted plasma). In this way the mononuclear cells are separated on the basis of their density.

CULTURE PREPARATION

A sterile culture is prepared containing 10% fetal calf serum (Gibco), RPMI 1640 90%, antibiotics (gentamycin 1 μg/ml) and lymphocytes to a final concentration of $0.5-2 \times 10^6$ lymphocytes/ml. The lymphocyte preparation is placed in either Falcon flask cultures (20 ml/flask) or microtiter plates (0.5 ml/well).

The culture is maintained at 37°C for 3 hours before the addition of the mitogen: in case of PHA (Wellcome), 20 μl/ml of culture medium are added. Control cultures to which the mitogen is not added are prepared to check spontaneous lymphocyte activation due to the presence of growth factors in the fetal calf serum. The culture is maintained at 37°C up to a maximum incubation time of 90-100 hours. Generally the culture is maintained 72 hours. For cultures lasting more than 100 hours, the culture medium is changed after 72 hours of incubation.

THE EFFECT OF PEMFs ON LYMPHOCYTES

The lymphocyte-lectin experimental model has been used by many authors, both to study the metabolic effect of PEMFs used clinically, and to evaluate possible side-effects due to chronic exposure to the electric fields of the high-voltage power-lines. All the studies that we will consider deal with non-thermal effects.

GENOVA GROUP

The investigators (13,38,40,48) studied the effect of PEMFs developed at Columbia University on normal human lymphocytes. Their PEMF generator supplied a pair of air-core coils with either a single-pulse signal or a burst signal. Both are used clinically for different indications. The signal employed is shown in Figure 9.

Lymphocyte cultures containing sub-optimal concentrations of PHA (2.5-10 μl/ml) were prepared from a sample of peripheral blood and incubated at 37°C for 72 hours. Four different samples were prepared: (a) control cultures; (b) PHA-containing cultures; (c) PEMF-exposed cultures; and (d) PHA-containing plus PEMF-exposed cultures. The samples were exposed to PEMFs for 72 hours. A careful and detailed analysis of the exposure system was given.

Both the change in lymphocyte volume corresponding to lymphocyte activation, and the number of cells stained by acridine orange (a measure of the DNA content of the cell) were evaluated. The results showed that PEMF exposure decreased the response of the lymphocytes to sub-optimum PHA concentrations.

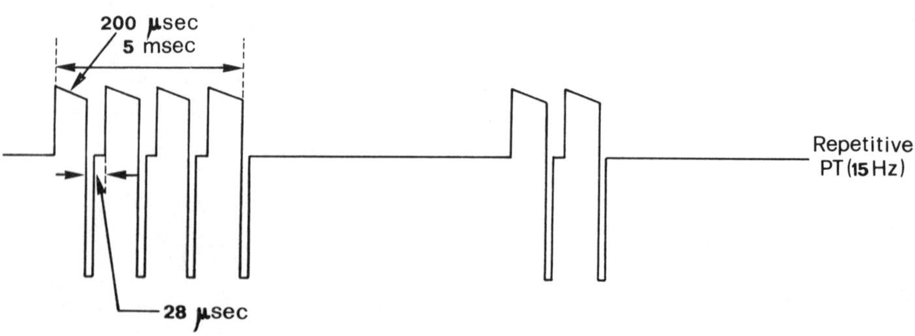

FIGURE 9. Waveform of the electrical signal used by the Genova group, as it is induced in a calibrated coil probe.

On the bases of these and other observations, the authors developed a mathematical model to fit the effect of PEMFs on the interaction between PHA and its membrane receptor. They concluded that PEMF exposure decreased the time spent by PHA in the proximity of its cellular receptor, thereby decreasing the response to the mitogenic stimulus.

The computerized model developed allowed a prediction that PEMF exposure increased the availability of free Ca^{++} near the membrane and the Ca^{++} influx via pre-existing channels.

L'AQUILA GROUP

A wide series of studies were performed on the effect of PEMFs on mitogen-stimulated lymphocytes (14,42,43). Normal lymphocyte cultures were prepared and DNA synthesis was evaluated by ^{3}H-thymidine incorporation, using the microtiter-plate technique. The lymphocytes were incubated for 72 hours at 37°C, and at the 66th hour the labeled precursor was added to the cultures. DNA synthesis was evaluated by tricloracetic-acid precipitation and counting.

The electromagnetic-field generator supplied a pair of coils with a "train of pulses of variable form and intensity;" the field value was 23-65 Gauss. The signal waveform and the actual field intensity are not given.

The effect of different frequencies on the response of lymphocytes to PHA, pokeweed mitogen (PWM) and concanavalin A (ConA) was tested. A window effect was observed, and the most inhibitory response occurred at 3 Hz. Table 4 shows the results obtained at different frequencies with the three lectins.

TABLE 4. Inhibitory Effect of PEMFs on the Lymphocyte Response to Lectins

MITOGEN INHIBITORY EFFECT	FREQUENCY			
	1 Hz	3 Hz	50 Hz	200 Hz
PHA	Yes	Yes	Yes	Yes
ConA	No	Yes	Yes	No
PWM	No	Yes	No	No

To test the hypothesis that PEMFs could interfere with ^{3}H-thymidine transport across the cell membrane, the cultures were exposed to PEMFs in the last 6 hours of incubation: no effect was observed as a consequence of PEMF exposure. The authors suggested that PEMFs could prevent the link between the receptor and the mitogen, and could decrease Ca^{++} influx into the lymphocyte. The authors have recently shown stimulating effect of PEMFs on the lymphocyte response to the mitogens when suboptimal concentration of lectins are used.

WASHINGTON DC GROUP

Spleen lymphocytes obtained from adult female BALB/c mice were exposed to the same signal used by Chiabrera et al. The lymphocytes were incubated at 37°C in the presence of PHA (50). The cultures were exposed for 24-48 hours to the electromagnetic field, and DNA synthesis was evaluated by ^{3}H-thymidine incorporation. A stimulatory effect of PEMFs on the response to the lectin was observed. A 50-70% inhibitory effect on the macrophage chemotaxis as a consequence of PEMF exposure was also observed.

MODULATED-FIELD EFFECTS

In 1983, using a 450 MHz field sinusoidally amplitude-modulated at 60 Hz (1.5 mW/cm^2), Lyle et al. reported up to a 20% inhibition of T-lymphocyte cytotoxic response (17). The response was evaluated by assaying the ability of T-lymphocytes to produce cytolysis of tumor target cell (H-2 B-lymphoma MPC-11). The murine lymphocyte cultures were exposed to the field 4 hours. When the field was modulated at other frequencies (3, 16, 40, 80, 100 Hz) the inhibitory effect progressively decreased. The results suggested a window effect of the field in which the response of the biological system was non-linear with respect to frequency. The cytotoxic activity was completely restored 12.5 hours after the end of the exposure. The unmodulated carrier wave had no effect on cytotoxicity.

In 1984 (85), human tonsil lymphocytes were exposed for 60 minutes to the same field (1 mW/cm^2) to test the cAMP-dependent protein kinase activity. It was not affected by exposure, but non-cAMP dependent activity was decreased (50%)

after 15 and 30 minutes exposure, and it returned to normal values after 45 and 60 minutes exposure. The non-modulated wave carrier had no effect.

OUR EXPERIENCE

We have studied the effect of exposure of normal and CLL lymphocytes to PEMFs (16,45-47,49). In most of the experiments, both a morphological analysis of the cultures, and a study of labeled precursor incorporation into DNA were performed. All the experiments were done in triplicate. For morphological analysis, 1000 lymphocytes were evaluated by three hematologists who were unaware of the nature of the experiments.

1. Material and Methods

(a) Morphological analysis: An aliquot of the lymphocyte culture is centrifuged at 600-800 g for 15 minutes. The recovered pellet is resuspended in RPMI 1640 and the cells are counted; slides are then prepared and stained with May-Grunwald-Giemsa stain.

The lymphocytes are classified as activated lymphocytes (which include blast and stimulated lymphocytes) and non-stimulated lymphocytes (Figure 1). The mitotic index (MI)(number of observed mitosis divided by the number of counted cells) is also determined.

(b) DNA synthesis assay: ^{3}H-thymidine, 1 μCi/ml of culture medium (New England Nuclear; 21.5 Ci/mM), is added (usually after the initial 48 hours of incubation). At the 72nd hour of incubation the cells are collected, and aliquots are used for assaying the labeled precursor incorporation into the DNA. Ten millileters of 10% tricloracetic acid (TCA) is added to the lymphocyte sample and, after 10 minutes at 4°C, the sample is filtered on Whatmann fiber-glass filters. The filters are washed twice with 10 ml of 10% TCA, dried, and then counted in Toluene-Permafluor (24:1) in a liquid scintillator (Packard).

(c) Karyotype preparation: Chromosomes are prepared according to the Summer ASG-technique for G band analysis (86,87). The technique yields evidence of translocations, deletions, and other damage that occurs as a consequence of exposure.

(d) Ca^{++} influx evaluation: In some experiments the cultured lymphocytes were used to test the effect of PEMFs on $^{45}Ca^{++}$ influx. One milliliter cultures containing 2 x 10^6 lymphocytes are prepared and maintained at 37°C for 3 hours before the addition of 10 μCi/ml of $^{45}Ca^{++}$ (Amersham Int., $^{45}CaCl_2$ 10 mCi/Mg-Ca). When the experiments are performed with PHA-stimulated cultures, $^{45}Ca^{++}$ is added immediately after PHA addition. Half of the cultures are used as controls, and the others are exposed to PEMFs.

After 1 hour incubation at 37°C in the presence of the $^{45}Ca^{++}$, 5 ml of cold NaCl (0.15 M) is added and the cultures are centrifuged at 800 g for 5 minutes. The lymphocytes are washed twice and then 1 ml of Soluene 350 (Packard) is added; the lymphocyte pellet is incubated overnight, and then 10 ml of Toluene-Permafluor is added and the sample is counted in a liquid scintillation counter.

(e) T-lymphocyte removal: T-lymphocytes can be removed from the culture medium because of the presence on their surface of specific receptors (88,89). Once the lymphocytes have been separated, 0.5 ml of culture medium containing 5 x 10^6 lymphocytes are added to 0.25 ml of fetal calf serum and to an equal volume of 10% sheep red blood cells sensitized with AET (Sigma), the mixture is centrifuged at 100 g per 10 minutes at 4°C and incubated overnight at 4°C. The sheep erythrocytes link to the T-lymphocytes and form rosettes (Figure 10).

By centrifugation on Ficoll-Hypaque 1077 at 4°C, rosetted T-lymphocytes can be removed (found on the bottom of the gradient). With this technique about 95-98% of the T-lymphocytes can be removed.

(f) The exposure system: For PEMF exposure, the cultures are placed between two coils as shown in Figure 11. The control and PEMF-exposed cultures are maintained in two separate incubators placed in different rooms. Control cultures are also maintained in the same incubator as the exposed cultures, but in a position where no current is induced in a coil probe. There is no significant difference between these control cultures and those maintained in the separate incubators. No difference is found in the temperature of the medium between the control and exposed cultures.

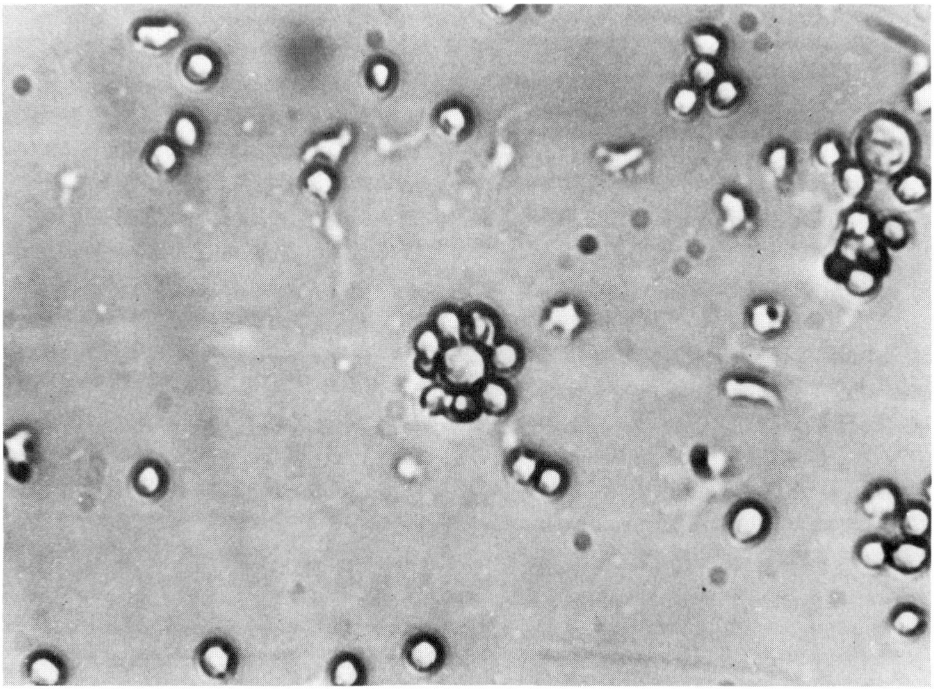

FIGURE 10. T-lymphocyte rosette. The lymphocyte in the center is surrounded by sensitized sheep erythrocytes.

The cultures were exposed to the PEMFs for 72 hours in all experiments unless stated otherwise. An electromagnetic-field generator (IGEA, Carpi, Italy) supplies a pair of air-core coils with 75-Hz pulses, each lasting 1.3 msec. Each coil (15 cm in diameter) was made of 1400 turns of copper wire (0.25 mm in diameter) with a resistance of 300 ohms. The peak value of the magnetic field was 2-2.8 mTesla. The induced electric field was measured with a coil probe (internal diameter 0.5 cm) made of 50 turns of 0.2-mm copper wire. A detailed analysis of the coil is

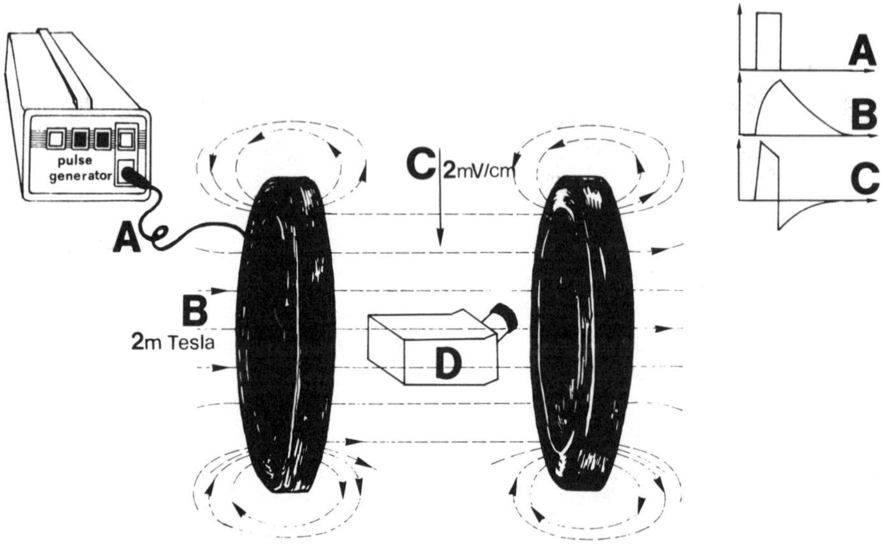

FIGURE 11. Schematic representation of the exposure system. A, waveform of the electrical current supplying the coils; B, magnetic field waveform flux lines; C, waveform of the electrical current induced in the culture medium.

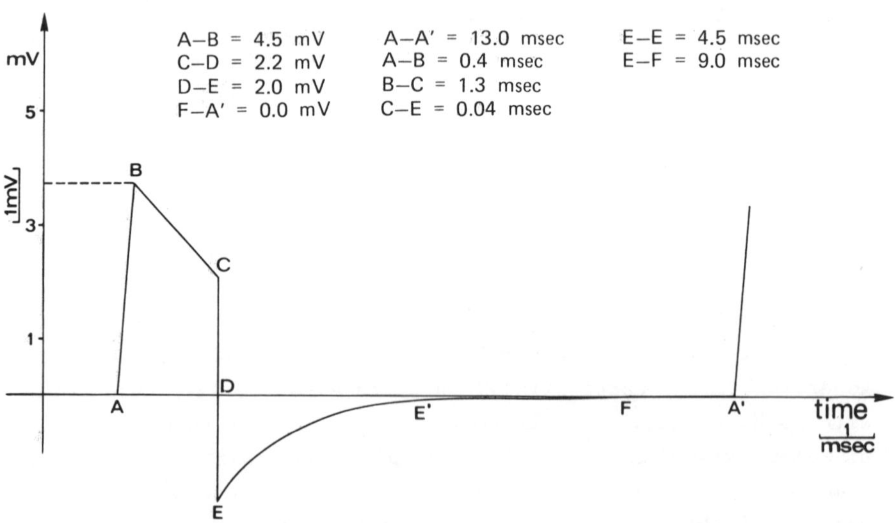

FIGURE 12. Detailed analysis of the characteristics of the waveform of the electrical current induced in the coil probe.

shown in Figure 12. If not specifically stated, the induced electrical voltage in the coil probe was 2 mV.

The relatively high impedance of the coils allows us to induce a voltage of more than 1 mV for more than 1 msec in the coil probe. We expect that the induced electromotive force moved ions present in the medium in one direction for more than 1 msec. Figure 13 shows the waveform of the induced electrical current recorded by connecting the coil probe to an oscilloscope.

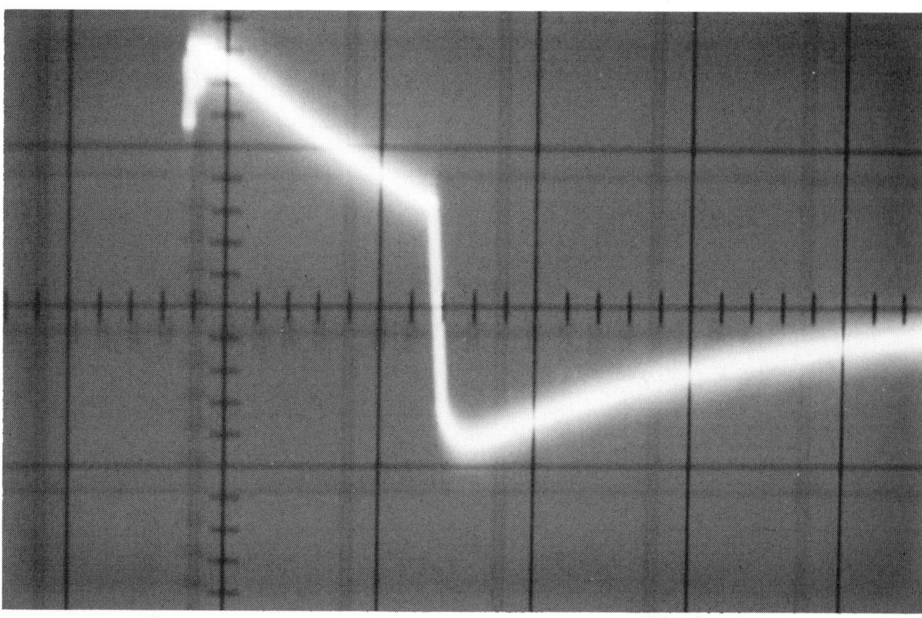

FIGURE 13. Actual oscilloscope recording of the voltage induced in the coil probe. 1 msec/cm time division and 1 mV/cm amplitude division.

2. Results

The cultured human lymphocytes exhibited some spontaneous stimulation because of growth factors contained in fetal calf serum. The amount of such stimulation was 4-8% in the case of normal lymphocytes, and even lower for CLL lymphocytes. This spontaneous lymphocyte stimulation was not modified by exposure to PEMFs.

When PHA or PWM were present in the culture medium, the percentage of activated normal and CLL lymphocytes increased significantly in the PEMF-exposed cultures compared to the controls. The same increase was observed when evaluating both DNA synthesis and mitotic index. The PEMF stimulatory effect was particularly evident in the CLL cultures. The increase in mitotic index suggested that PEMFs raised not only the number of lymphocytes activated by lectins, but also favored the entry into the S phase of the cell cycle of both lymphocytes stimulated by PHA alone, and of PHA + PEMFs stimulated lymphocytes.

TABLE 5. Effect of PEMFs on the Response of Normal (N = 10) and CLL (N = 14) Lymphocytes to PHA Stimulation.

PARAMETER EVALUATED	PHA	PHA + PEMFs
Normal Lymphocytes:		
% Activated	81%	98%*
^{3}H-thymidine (cpm x 10^{-3})	140	203*
Mitotic Index	0.065	0.085*
CLL Lymphocytes:		
% Activated	32%	48%*
^{3}H-thymidine (cpm x 10^{-3})	20	38*
Mitotic Index	0.01	0.03*

* $P < 0.01$, Student's T test

Table 5 shows the average values obtained after stimulating 14 CLL and 10 normal cultures with PHA. The statistical analysis showed a stimulatory effect of PEMFs over all the tested parameters. In all the experiments performed, the mitotic index was always increased in the exposed cultures. PEMFs in association wth PHA induced almost all the lymphocytes to enter the cell cycle.

Table 6 shows the results obtained using PWM. The same stimulatory effect of PEMFs was observed, indicating that it was not dependent on the lectin used.

Because of the importance of Ca^{++} fluxes for lymphocyte activation by lectins, we evaluated the effects of PEMF exposure on calcium. After 1 hour of exposure, the amount of $^{45}Ca^{++}$ present in the lymphocyte cytoplasm was increased both when PHA was present in the culture medium, as well as when it was not added. However the increased Ca^{++} influx in the case of cultures to which PHA had not been added was not enough to induce lymphocyte proliferation. Further experiments were conducted in the presence in the culture medium of a Ca^{++} antagonist, 10^{-6} M verapamil. Verapamil inhibited the lymphocyte response to PHA, but when the cultures were exposed to PEMFs, the verapamil block was antagonized so that the number of

TABLE 6. Effect of PEMFs on the Response of Normal (N = 3) and CLL (N = 3) Lymphocytes to PWM Stimulation.

PARAMETER EVALUATED	PWM	PWM + PEMFs
Normal Lymphocytes		
% Activated	65%	97%
Mitotic Index	0.05	0.07
CLL Lymphocytes		
% Activated	16%	31%
Mitotic Index	0.01	0.02

responsive cells increased to the value obtained with PHA alone (Figure 14). These data suggested that the PEMF effect on the cell could be explained on the basis of an effect on Ca^{++} fluxes across the membrane. If this was the case, short exposure times (1-2 hours) should be sufficient because Ca^{++} influx is an early event in lymphocyte activation (the G_0-G_1 transition phase).

PHA cultures were incubated in the presence of PEMFs for different periods of time and then removed from the field and maintained in a separate incubator up to the 72nd hour of incubation, when they were evaluated. It was found that a minimum exposure time to PEMFs of 10 hours was required to obtain the stimulatory effect.

The effect of PEMFs could not be explained solely as a consequence of the increased Ca^{++} influx, so some other cellular targets had to be involved. After 24 hours of incubation in the presence of PHA, T-lymphocytes are known to release B-cell growth factors that mediate the response of B-cells to PHA (90-92). Experiments were performed to test the possibility that PEMFs favored the release of B-cell growth factors. The effect of PEMFs on cell secretion properties has been described (8).

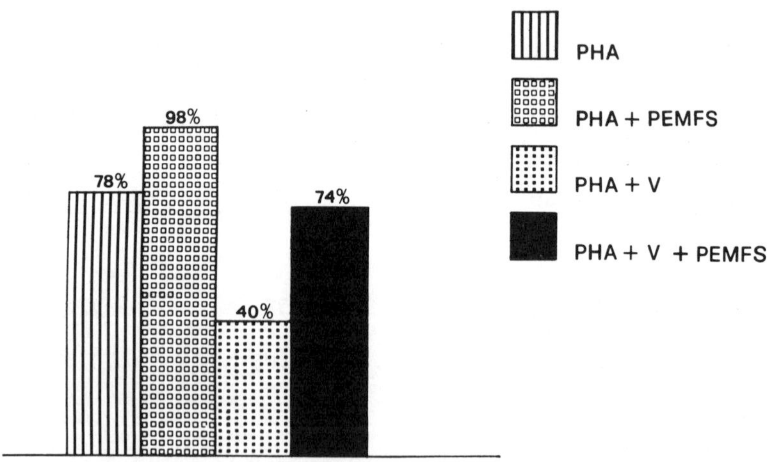

FIGURE 14. Percent stimulated normal human lymphocytes cultured 72 hours in the presence of PHA (20 μl/ml). PEMF exposure increased the number of activated lymphocytes and antagonized the verapamil (V) block.

T-lymphocytes were removed from CLL cultures by rosetting, after which PHA was added. The lymphocyte response in control cultures dropped to 0.5-2%. However, in the PEMF-exposed cultures, the response was always 3-4 times higher than in controls. These results suggested that PEMF-increased response was actually mediated by T-lymphocytes, and was not a direct effect of PEMFs on the reaction between PHA and B cells.

To investigate whether the cultures exposed to PEMFs contained more B-cell growth factors than control cultures, the following experiment was performed. CLL cultures were prepared, PHA was added, and half of the cultures were exposed to PEMFs and the remaining cultures were used as controls. The cultures were maintained at 37°C for 24 hours. At the end of the scheduled incubation time the cultures were centrifuged, the lymphocytes discarded, and the culture medium was used to prepare new cultures with fresh CLL lymphocytes obtained from the same donor. The new cultures were maintained at 37°C and none were exposed to PEMFs. After 24 hours of incubation, ^{3}H-thymidine was added and 24 hours later the cultures were counted and lymphocyte transformation was evaluated (Figure 15). In the cultures prepared with the medium obtained from PEMF-exposed cultures, a significantly higher number of activated lymphocytes and an increased DNA synthesis were observed. This is evidence of the presence in the culture medium of PEMF-exposed cultures of B-cell growth factors able to increase lymphocyte response to PHA. We do not know whether the increased B-cell growth-factor content in PEMF-exposed cultures can be attributed to an increased synthesis or to an augmented release from the T-lymphocytes. The latter effect could be dependent on increased cytoskeleton contraction, which requires Ca^{++}.

When lectins bind to membrane receptors, they cluster on the cell membrane. The lectin-receptor clustering event is necessary for Ca^{++} influx and lymphocyte activation (48,93,94). We evaluated whether PEMF exposure could increase lymphocyte response in the presence of sub-optimal PHA concentrations as a consequence of an effect on Ca^{++} influx. Cultures containing different concentrations of PHA were prepared, and half were exposed to PEMFs. Exposure to PEMFs increased the lymphocyte

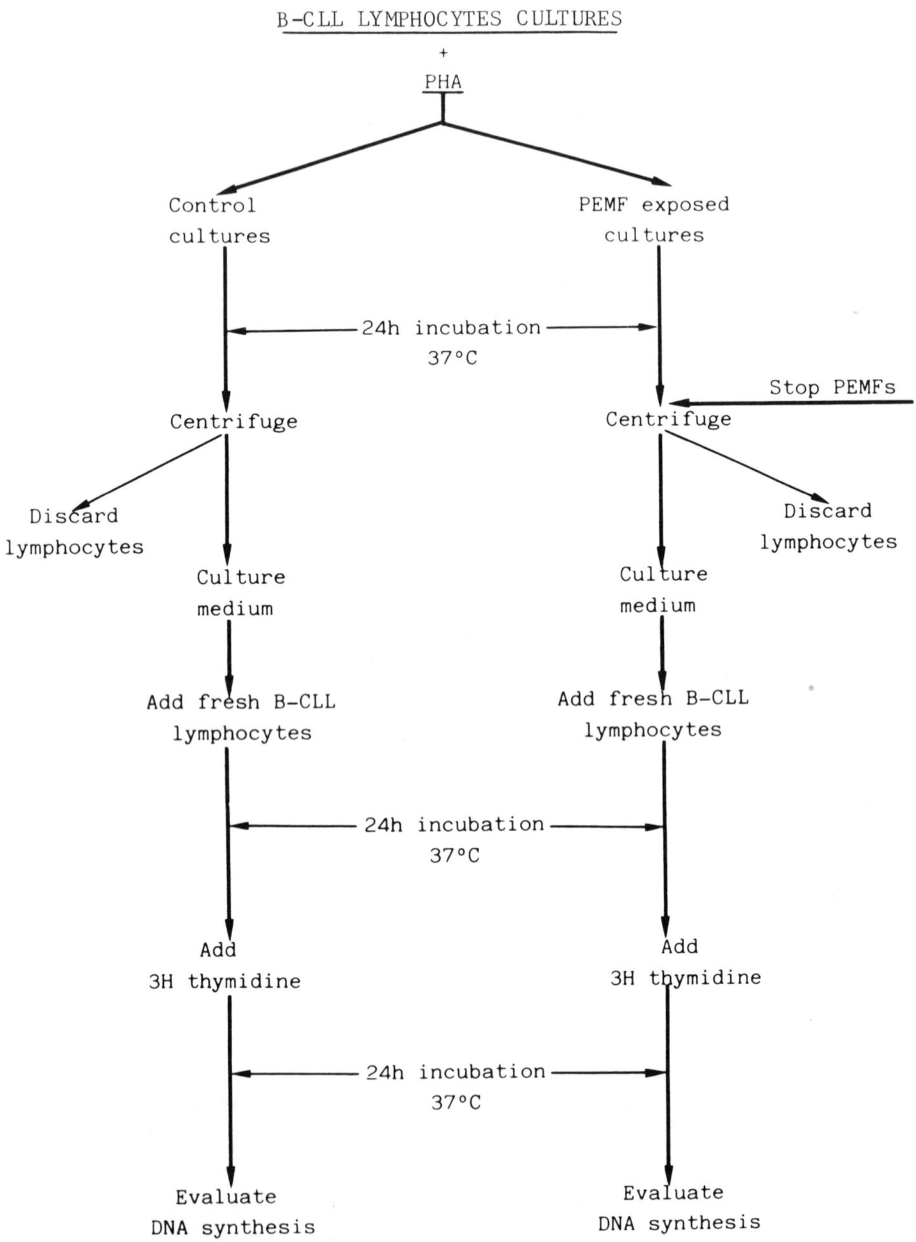

FIGURE 15. Experimental design to evaluate the amount of B-cell growth factors released in the culture medium.

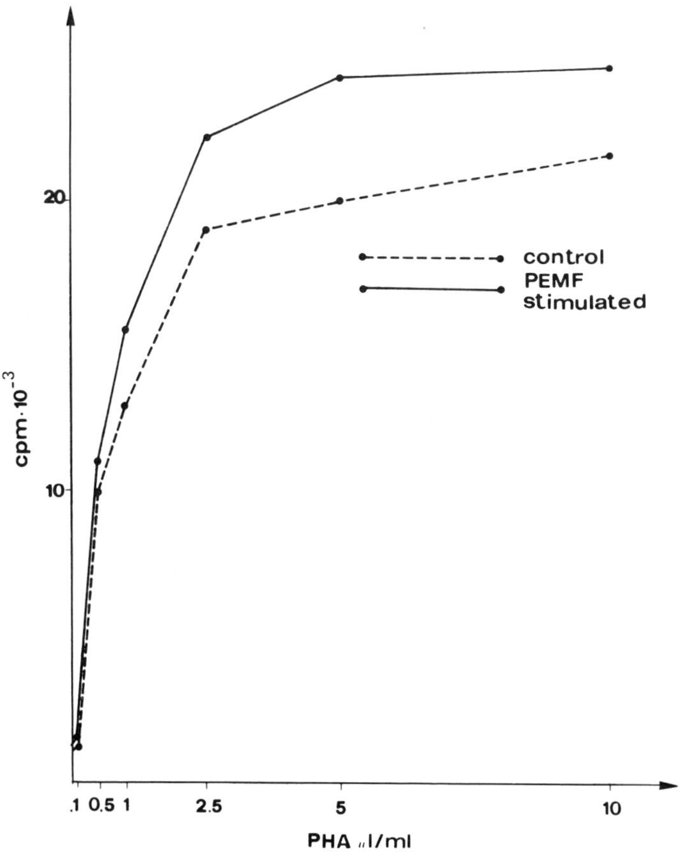

FIGURE 16. Stimulatory effect of PEMFs in the presence of sub-optimal concentration of PHA in the culture medium. The experiment was performed with normal lymphocytes.

response in the presence of low PHA concentrations, and a greater effect was observed at intermediate PHA concentrations (Figure 16). The same effect has recently been described by Conti (110) and Cantini (111).

We have already discussed the different effects obtained by other authors who exposed lectin-stimulated lymphocytes to PEMFs. One possible explanation is that the differences observed were due to the physical characteristics of PEMFs. PHA stimulated lymphocytes were exposed to different values of induced electrical current (1-10 mV) as measured with the coil probe described previously. Figure 17 shows that the response of the lymphocytes to PHA was increased at low voltages and inhibited at high voltages. We have no satisfactory explanation for the reversal effect observed at high voltages, but it suggests the importance of the physical parameters of the applied field.

3. Discussion

Our results demonstrate that PEMFs alone cannot induce lymphocytes to enter the cell cycle. When a lectin, which

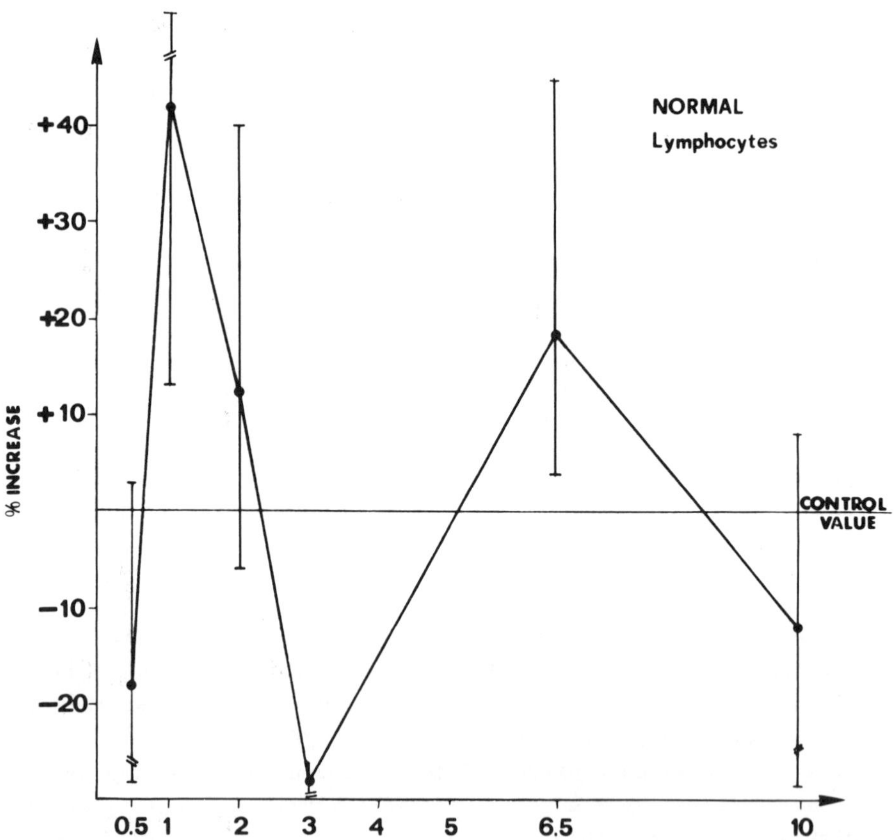

FIGURE 17. Effect of different values of electrical current induced in the coil probe on the response of lymphocytes to PHA.

itself can induce lymphocyte activation is present in the culture medium, PEMF exposure can significantly increase the effect. The exposure of cultured lymphocytes to PEMFs also increases Ca^{++} influx, but this effect cannot completely explain all the observed phenomena.

Further experiments allowed us to demonstrate that PEMFs favored the release by lymphocytes in the culture medium of B-cell growth factors. This effect emphasizes the need for long exposure times to observe the described effects.

The effect on Ca^{++} influx can probably account for the increased number of lymphocytes entering the S phase of the cell cycle because, as noted before, a Ca^{++} influx dependent restriction point is present in late G_1 that must be crossed to start DNA synthesis. The effect on Ca^{++} influx is also probably important when the lymphocytes are cultivated in the presence of suboptimal concentrations of PHA.

Finally, we have demonstrated that the lymphocyte response to lectins can be modulated by the amplitude of the signal induced. This observation helps to explain the different effects described by various authors (Table 7).

TABLE 7. Reported Effects of PEMFs on the Lymphocyte Response to Lectins.

AUTHOR	LYMPHOCYTES USED	Ca^{++} INFLUX	DNA SYNTHESIS	PEMF PEAK VALUE (mTesla)	Hz
Conti	Human	Decreased	Decreased	2-5	3-200
Chiabrera	Human	Not Tested*	Decreased	1.9	15
Hellmann	Mice	Not Tested	Increased	1.9	15
Emilia	Human	Increased	Increased	2	75

*Biphasic effect foreseen as a function of lectin concentration

ASSESSMENT OF HEALTH RISK

The lymphocyte model has been used for evaluation of possible damage induced by high-voltage powerlines. We will briefly mention some recent results obtained by exposing lymphocyte cultures to high electric fields.

Winters et al. (95) showed that the exposure to 60-Hz electric fields (300 mA/m^2) or magnetic fields (1.0 Gauss) had no effect on human lymphocyte response to a lectin mitogenic stimulus. However, they reported that a significantly increased response was observed in some exposed cultures. They concluded that there was no consistent or significant pattern in the observed responses.

Smith et al. (96) reported that 2-μsec electric-field pulses at up to 3.5 kV/cm did not affect DNA synthesis in mouse spleen lymphocytes incubated in the presence of ConA, PHA or LPS. One 10-μsec pulse, 2.4-3.5 kV/cm, produced a significant reduction in the response of lymphocytes to LPS, which was attributed to cell death.

Nordenson et al. (97), in an _in vivo_ investigation, found that 17 of 20 switchyard workers exhibited chromosomal anomalies. The rate of chromatid and chromosomes breaks was significantly higher than in controls. They also showed that 1 mA/cm^2 at 50 Hz did not induce any chromosome damage, but that 3-μsec spark discharges (peak strength of 3.5 kV/cm) caused chromosome breaks with a frequency similar to that induced by ionizing radiation.

CONCLUSIONS

The biological model considered here is of great interest, and will probably be used in future studies. For this reason, some points must be further discussed from a physical and a biological point of view. There are several techniques for preparing lymphocyte cultures, and we have shown that small differences in the amplitude of the induced electrical current can significantly change the results. Because of this, the shape of the container in which the lymphocytes are cultivated will play an important role, and should be taken into consideration. One of the most used techniques is the microtiter plate because it is convenient and allows testing of several small samples at the same time. Attention should be payed to magnetic-field homogeneity. If it is not uniform, as often happens in the peripheral wells, different amounts of current will flow in the wells giving rise to misleading results. The exposure conditions must be carefully controlled, and a clear and exhaustive description of the electromagnetic field used should always be given, to allow different authors to exactly reproduce the experimental conditions.

As we pointed out at the beginning, among the drawbacks of the lymphocyte-lectin biological model is the extremely wide range of variability which makes it necessary to repeat the experiments several times. The lymphocyte response to the mitogenic stimulus of the lectins can change at different times even within the same donor. The response among different donors varies significantly.

Finally, the response depends on immunological functional conditions that cannot be foreseen. The immediate consequence is that the amount of the response not only to lectins, but also to PEMFs can vary significantly.

EXPERIMENTS SUGGESTED BY *IN VITRO* RESULTS

In spite of the extreme variability of the biological response, the lectin-lymphocyte system has allowed us to obtain important information. However we are far from a satisfactory explanation for the mechanisms of action of PEMFs. The available data allow us to propose further experiments and suggest fascinating hypotheses, both for the interpretation of the obtained results in other areas of research (1,3,98-100), and for possible future clinical applications.

Since the initial results, interest has focused on the effect of PEMFs on ion-transport across the cell membrane. It was shown that PEMFs could interfere with Ca^{++} transport across the membrane, and most authors described an increased Ca^{++} influx with low-frequency fields (8,9,16,18,32). This PEMF effect is important because Ca^{++} is involved in nearly all the metabolic events of the cell (101,102). It has been observed that Ca^{++} increase in the cytoplasm can be considered a non-specific signal for proliferation that is common to all eukaryotic cells (74). For these reasons, the study of the effect of PEMFs on the transport of the ions across the cell membrane in the presence, in the culture medium, of specific inhibitors of ion transport will probably be one of the most interesting areas of investigation, and will probably give an enormous amount of information both on the mechanism of action of PEMFs, and on the mechanisms of action of ion-transport inhibitors.

Another interesting area of research will be the study of the relative effect of the different parameters characterizing the electromagnetic field on the same biological target. We have already shown that a change in the amplitude of induced electrical current can change a stimulatory effect to an inhibitory one. Thus, depending on the exposure condition chosen, we can modulate the same cellular function by inhibiting or stimulating it.

IN VITRO AND *IN VIVO* RESULTS

We have seen that the final result of exposing lymphocytes cultured in the presence of lectins to PEMFs is an increased number of lymphocytes that are activated by lectin: 100% in the case of normal lymphocytes, and 40% in the case of CLL lymphocytes. We have also shown that PEMF exposure increases lymphocyte duplication. A similar effect of PEMFs was described by Ottani et al. (12) studying the regenerating liver in rats. She observed that the number of cells involved in the repair process in the exposed animals was higher than in controls. The healing end-point remains the same for both control and experimental animals.

The same effect probably accounts for the results obtained by Law et al. (103) who treated sheep osteotomies with electromagnetic stimulation, and by Wahlstrom (104) who treated fresh rib fractures with PEMFs. Both the authors found that, while the final healing time was not affected by stimulation, at intermediate times the healing process was more advanced in PEMF-exposed bones. They observed a higher metabolic activity in the PEMF-exposed group. See also the chapter by Dal Monte.

PROSPECTIVE CLINICAL APPLICATIONS OF THE RESULTS OBTAINED *IN VITRO*

The data suggest that PEMFs can actually induce cells to enter both the cell cycle and the S phase of the cycle, but that they do not change the rate that the cells move through the cycle. It is interesting to observe that recently Barbiroli et al. (105), studying regenerating liver in the animals exposed to PEMF, described an increased expression of the c-myc proto-oncogene, which is transcribed by cells preparing to enter the cell cycle.

Zecca et al. (106) described a protective effect of PEMFs on the mortality of mice exposed to lethal doses of X-rays. Bone-marrow aplasia has been treated with moderate success using PHA; these experiments might be reconsidered in the light of our in vitro experiments (107-109). It should be interesting to determine whether PEMF exposure could shorten the recovery time of bone marrow after a therapeutically induced aplasia, by favoring a quicker cellular trigger. The possibility of having in the same period of time more cells entering into the S phase of the cell cycle can be used when trying to synchronize the cells. We are now planning to expose leukemic cells obtained from the peripheral blood of patients to PEMFs and to alkalating antitumoral agents to evaluate their combined effect.

Finally, we wish to stress that the study of the mechanism of action of PEMFs on lymphocytes gives us an opportunity to examine lymphocyte physiology from a completely original point of view. The increased response we obtained with CLL lymphocytes can allow an easier study of the karyotype of these malignant cells. Our results show that the impaired response of 20% of the lymphocyte population cannot be simply attributed to the lack or insufficient presence of PHA receptors. Otherwise they would have responded when exposed to PEMFs in the absence of the lectins.

Even though these studies do not allow a complete understanding of the mechanism of action of PEMFs, they do show the importance of the model with respect to a wide spectrum of possible clinical applications.

REFERENCES

(1) Bassett, C.A.L.: The development of pulsed electromagnetic fields (PEMFs) for ununited fractures and arthrodeses, Orthop. Clin. North Am. 15:61-88, 1984.

(2) Sedel, L., Christel, P., Duriez, J., Duriez, R., Evard, J., Ficat, C., Cauchoix, J. and Witvoet, J.: Fesultats de la Stimulation par Champ Electromagnetique de la Consolidation des Pseudarthroses, Rev. Chir. Orthop. 67:11-23, 1981.

(3) Fontanesi, G., Dal Monte, A., Rinaldi, E., Traina, G.C., Mammi, G.I., Giancecchi, F., Rotini, R., Poli, G., Negri, V., Virgili, B. and Cadossi, R.: The effect of low frequency pulsing electromagnetic fields for the treatment of congenital and acquired pseudarthroses, J. Bioelectricity 3:155-175, 1984.

(4) Sharrard, W.J.W.: Treatment of congenital and infantile pseudarthrosis of the tibia with pulsing electromagnetic fields, Orthop. Clin. North Am. 15:143-161, 1984.

(5) DeHaas, W.G., Watson, J. and Morrison, D.M.: Non-invasive treatment of ununited fracture of the tibia using electrical stimulation, J. Bone Joint Surg. 62:465-470, 1980.

(6) Dal Monte, A. and Poli, G.: Treatment of congenital pseudarthrosis with endomedullary nail and low frequency pulsing electromagnetic fields: A controlled study, J. Bioelectricity 4:195-209, 1985.

(7) Luben, R.A., Cain, C.D., Chi-Yun Chen, M., Rosen, D.M. and Adey, W.R.: Effects of electromagnetic stimuli on bone and bone cells _in vitro_, Proc. Natl. Acad. Sci. U.S.A. 79:4180-4184, 1982.

(8) Dixey, R. and Rein, G.: 3H-noradrenaline release potentiated in a clonal nerve cell line by low-intensity pulsed magnetic fields, Nature 296:253-256, 1982.

(9) Rodan, G.A., Bourret, L.A. and Norton, L.A.: DNA synthesis in cartilage cells is stimulated by oscillating electric fields, Science 199:690-692, 1978.

(10) Liboff, A.R., Williams, T. Jr., Strong, D.M. and Wistar, R. Jr.: Time-varying magnetic fields: Effect on DNA synthesis, Science 223:818-820, 1984.

(11) Goodman, R., Bassett, C.A.L. and Henderson, A.S.: Pulsing electromagnetic fields induce cellular transcription, Science 220:1283-1285, 1983.

(12) Ottani, V., Monti, M.G., Piccinini, G., Pernecco, L., Zaniol, P., Ruggeri, A. and Barbiroli, B.: Pulsed electromagnetic fields increase the rate of rat liver regeneration after partial hepatectomy, Proc. Soc. Exp. Biol. Med. 176:371-377, 1984.

(13) Grattarola, M., Viviani, R. and Chiabrera, A.: Modelling of the perturbation induced by low-frequency electromagnetic fields on the membrane receptors of stimulated human lymphocytes. I. Influence of the fields on the system's free energy, Studia Biophysica 91:117, 1982.

(14) Conti, P., Gigante, G.E., Cifone, M.G., Alesse, E., Ianni, G., Reale, M. and Angeletti, P.U.: Reduced mitogenic stimulation of human lymphocytes by ELF electromagnetic fields, FEBS Lett. 162:156-160, 1983.

(15) Hellman, K.B., Hellman, A., Strickland, A.G. and Fowler, A.K.: Effects of pulsed electromagnetic fields on the cells of the immune system, 3rd Ann. BRAGS, San Francisco, 9, 1983.

(16) Emilia, G., Torelli, G., Ceccherelli, G., Donelli, A., Ferrari, S., Zucchini, P. and Cadossi, R.: Effect of low-frequency low-energy pulsing electromagnetic fields on the response to lectine stimulation of human normal and chronic lymphocytic leukemia lymphocytes, J. Bioelectricity 4:145-161, 1985.

(17) Lyle, D.B., Schechter, P. Adey, W.R. and Lundak, R.L.: Suppression of T-lymphocyte cytotoxicity following exposure to sinusoidally amplitude-modulated fields, Bioelectromagnetics 4:281-292, 1983.

(18) Dihel, L.E., Sonneborn, J.S. and Middaugh, C.R.: Effect of an extremely low-frequency electromagnetic field on the cell division rate and plasma membrane of Paramecium tetraurelia, Bioelectromagnetics 6:61-71, 1985.

(19) Murray, J.C. and Farndale, R.W.: Modulation of collagen production in cultured fibroblasts by PEMFs, Biochem. Biophys. Acta 838:98-106, 1985.

(20) Chiabrera, A., Hinsenkamp, M., Pilla, A., Ryaby, J., Ponta, D., Belmont, A., Beltrame, F., Grattarola, M. and Nicolini, C.: Cytofluorometry of electromagnetically controlled cell differentiation, J. Histochem. Cytochem. 27:375-381, 1979.

(21) Pino, A., Ricci, R. and Piombo, G.: Absence of DNA damage in liver, spleen and kidney of rats after exposure to therapeutic magnetic fields, IRCS Med. Sci. 13:257-258, 1985.

(22) Rein, G. and Pilla, A.A.: Cell surface electrochemistry and the electromagnetic modulation of basic cell functions: Application of cell adhesion, 8th Symp. on Bioelectrochemistry & Bioenergetics, Bologna, 92, 1985.

(23) Norton, L., Pilla, A.A., Geller, S. and Tansman, L.: Pulsating electromagnetically induced current (PEMIC) augment chemoimmunotherapy in experimental cancer, 4th Ann. BRAGS, Kyoto, 36, 1984.

(24) Fitton-Jackson, S., Hickman, J. and Murray, C.: Effects of pulsed magnetic fields on acute tendon injuries, BEMS, San Francisco, 10, 1985.

(25) Fitton-Jackson, S., Marsland, T.P., Farndale, R.W. and

Boutle, A.R.: Physical factors involved in biological responses to pulsed magnetic fields, 2nd Ann. BRAGS, Oxford, 31, 1982.

(26) Norton, L.A.: Effects of a pulsed electromagnetic field on a mixed chondroblastic tissue culture, Clin. Orthop. 167:280-290, 1982.

(27) Korenstein, R., Somjen, D., Fischler, H. and Binderman, I.: Capacitative pulsed electric stimulation of bone cells. Induction of cyclic-AMP changes and DNA synthesis, Biochem. Biophys. Acta 803:302-307, 1984.

(28) Laub, F. and Korenstein, R.: Actin polymerization induced by pulsed electric stimulation on bone cells in vitro, Biochem. Biophys. Acta 803:308-313, 1984.

(29) Colacicco, G. and Pilla, A.A.: 669-transduction of electromagnetic signals into biological effects. Accounts of kinetics and energetics, Bioelectrochem. Bioenergetics 12:259-265, 1984.

(30) Colacicco, G. and Pilla, A.A.: Electromagnetic modulation of biological processes: Influence of culture media and significance of methodology in the Ca-uptake by embryonal chick tibia in vitro, Calcif. Tissue Int. 36:167-174, 1984.

(31) Lerner, E.J.: Biological effects of electromagnetic fields, IEEE Spectrum, 57-69, May, 1984.

(32) Bawin, S.M., Adey, W.R. and Sabbot, I.M.: Ionic factors in release of $^{45}Ca^{2+}$ from chick cerebral tissue by electromagnetic fields, Proc. Natl. Acad. Sci. U.S.A. 75:6314-6318, 1978.

(33) Blackman, C.F., Benane, S.G., House, D.E. and Joines, W.T.: Effects of ELF (1-120 Hz) and modulated (50 Hz) RF fields on the efflux of calcium ions from brain tissue in vitro, Bioelectromagnetics 6:1-11, 1985.

(34) Stevenson, A.P. and Tobey, R.: Potassium ion influx measurements on cultured Chinese hamster cells exposed to 60-Hertz electromagnetic fields, Bioelectromagnetics 6:189-198, 1985.

(35) Takashima, S. and Asakura, T.: Desickling of sickled erythrocytes by pulsed radio-frequency field, Science 220:411-413, 1983.

(36) Markov, M.S. and Todorov, N.G.: Electromagnetic field stimulation of some physiological processes, Studia Biophysica 99:151-156, 1984.

(37) Joley, W.B., Hinshaw, D.AB., Knierim, K. and Hinshaw, D.B.: Magnetic field effects on calcium efflux and insulin secretion in isolated rabbit islets of Langerhans, Bioelectromagnetics 4:103-106, 1983.

(38) Beltrame, F., Chiabrera, A., Gliozzi, A., Grattarola, M., Parodi, G., Vecchio, D., Ponta, D., Vernazza, G. and Viviani, R.: Electromagnetic control of cell reactivation, Symp. Electromagnetic Waves and Biology, 33-39, 1980.

(39) Chiabrera, A., Grattarola, M., Viviani, R. and Braccini, C.: Modelling of the perturbation induced by low-frequency electromagnetic fields on the membrane receptors of stimulated human lymphocytes. II. Influence of the fields of the mean lifetimes of the aggregation process, Studia Biophysica 91:125-131, 1982.

(40) Grattarola, M., Chiabrera, A., Bonanno, G., Viviani, R. and Raveane, A.: Electromagnetic field effects on phytohemagglutinin (PHA) induced lymphocyte reactivation, 5th ISPAB Course "Interactions between Electromagnetic Fields and Cells," Erice, 1984.

(41) Grattarola, M., Chiabrera, A., Viviani, R. and Parodi, G.: Interactions between weak electromagnetic fields and biosystems: A summary of nine years of research, J. Bioelectricity 4:211-225, 1985.

(42) Conti, P., Gigante, G.E., Alesse, E., Cifone, M.G., Fieschi, C., Reale, M. and Angeletti, P.U.: A role for Ca^{2+} in the effect of very low-frequency electromagnetic field on the blastogenesis of human lymphocytes, FEBS Lett. 181:28, 1985.

(43) Conti, P., Gigante, G.E., Cifone, M.G., Alesse, E., Fieschi, C. and Angeletti, P.U.: Effect of electromagnetic fields on two calcium dependent biological systems, J. Bioelectricity 4:227-236, 1985.

(44) Cifone, M.G., Boidi, E., Alesse, E., Reale, M., Gigante, G.E. and Conti, P.: Electromagnetic fields modify thromboxane B2 release by ionophore-activated neutrophils, IRCS Med. Sci. 12:719-720, 1984.

(45) Emilia, G., Cadossi, R., Torelli, G., Ferrari, S., Donelli, A. and Zucchini, P.: Low-frequency, low-energy pulsing electromagnetic fields: Stimulating effect on human lymphocytes of chronic lymphocytic leukemia (CLL), 3rd Ann. BRAGS, San Francisco, 69, 1983.

(46) Cadossi, R., Emilia, G., Torelli, G., Ferrari, S., Donelli, A., Zucchini, P. and Ceccherelli, G.: Effect of low-frequency low-energy pulsing electromagnetic fields on the blastic transformation of human normal and chronic lymphocytic leukemia lymphocytes by lectins, Proceedings of Third Southern Biomedical Engineering Conference, Birmingham, 230, 1984.

(47) Cadossi, R., Emilia, G., Torelli, G., Ceccherelli, G., Ferrari, S. and Ruggieri, P.: The effect of low-frequency pulsing electromagnetic fields on the response of human normal lymphocytes to phytohaemagglutinin, Bioelectrochem. Bioenergetics 3:115-119, 1985.

(48) Chiabrera, A., Grattarola, M. and Viviani, R.: Interaction between electromagnetic fields and cells: Microelectrophoretic effects of ligands and surface receptors, Bioelectromagnetics 5:173-191, 1984.

(49) Cadossi, R., Emilia, G. and Torelli, G.: Effect of low-frequency pulsing electromagnetic fields on the response to phytohemagglutinin of human normal and chronic lymphocytic leukemia lymphocytes, in Biological Effects and Dosimetry of Non-Ionizing Radiation: Static and ELF Electromagnetic Fields, M. Grandolfo and S.M. Michaelson, eds., Plenum Press, New York, 1985.

(50) Hellman, K.B., Brewer, P.P., Fowler, A.K., Hellman, A. and Swicord, M.L.: The effect of electromagnetic fields on lymphocyte function: Enhancement of mitogenic stimulation, BEMS, San Francisco, 10, 1985.

(51) Henry, P.H.: Composition and biochemistry of lymphocytes and plasma cells, in Hematology, W.J. Williams, ed., McGraw Hill, New York, 895-899, 1977.

(52) Nowell, P.C.: Phytohemagglutinin: An initiator of mitosis in cultures of normal human leucocytes, Cancer Res. 20:462, 1960.

(53) Inman, D.R. and Cooper, E.H.: Electron microscopy of human lymphocytes stimulated by phytohemagglutinin, J. Cell Biol. 19:441, 1963.

(54) Tanaka, Y., Epstein, L.B., Brecher, G. and Stohlman, F. Jr.: Transformation of lymphocytes in cultures of human peripheral blood, Blood 22:614, 1963.

(55) Maizel, A.L., Metha, S.R., Hauft, S., Franzini, D., Lachman, L.B. and Ford, R.J.: Human T lymphocyte/monocyte interaction in response to lectin: Kinetics of entry into the S-phase, J. Immunol. 127:1058-1064, 1981.

(56) Baserga, R.: The Biology of Cell Reproduction, Harvard University Press, Cambridge, 1985.

(57) Purrello, F., Burnham, D.B. and Goldfine, I.D.: Insulin receptor antiserum and plant lectins mimic the direct effects of insulin on nuclear envelope phosphorylation, Science 221:462-464, 1983.

(58) Cone, R.B.: The Lymphocytes: Structure and Functions, J.J. Marchalonis, ed., Marcel Dekker, New York, 565, 1977.

(59) Speckart, S.F., Boldt, D.H. and MacDermott, R.P.: Chronic lymphatic leukemia (CLL): Cell surface changes detected by lectin binding and their relation to altered glycosyl-transferase activity, Blood 52:681, 1978.

(60) Freedman, M.H., Raff, M.C. and Gomperts, B.: Induction of increased calcium uptake in mouse T lymphocytes by concanavalin A and its modulation by cyclic nucleotides, Nature 255:378-382, 1975.

(61) Maino, V.C., Green, N.M. and Crumpton, M.J.: The role of calcium ions in initiating transformation of lymphocytes, Nature 251:324-327, 1974.

(62) Lichtman, A.H., Segel, G.B. and Lichtman, M.A.: Effect of trifluoperazine and mitogenic lectins on calcium ATPase activity and calcium transport by human lymphocyte plasma membrane vesicles, J. Cell. Physiol. 111:213-217, 1982.

(63) Freedman, M.H. and Khan, N.R.: A rapid technique for measuring calcium uptake in mitogen-induced T and B lymphocytes, Can. J. Biochem. 57:1344-1350, 1979.

(64) Tsien, R.Y., Pozzan, T. and Rink, T.J.: T-cell mitogens cause early changes in cytoplasmic free Ca^{++} and membrane potential in lymphocytes, Nature 295:68-70, 1982.

(65) Hesketh, T.R., Smith, G.A., Houslay, M.D., Warren, G.B. and Metcalfe, J.C.: Is an early calcium flux necessary to stimulate lymphocytes?, Nature 267:490-494, 1977.

(66) Lichtman, A.H., Segel, G.B. and Lichtman, M.A.: The role of calcium in lymphocyte proliferation, Blood 61:413-422, 1983.

(67) Lee, K.S. and Tsien, R.W.: Mechanism of calcium channel blockade by verapamil, D600, diltiazem and nitrendipine in single dialysed heart cells, Nature 302:790-794, 1983.

(68) Whitney, R.B. and Sutherland, R.M.: Requirement for calcium ions by phytohemagglutinin, J. Cell Physiol. 80:329-338, 1972.

(69) Jensen, P. and Rasmussen, H.: The effect of A23187 upon calcium metabolism in the human lymphocyte, Biochem. Biophys. Acta 468:146-156, 1977.

(70) Truneh, A., Albert, F., Golstein, P. and Scmitt-Verhulst, A.M.: Early steps of lymphocyte activation bypassed by synergy between calcium ionophores and phorbol ester, Nature 313:318-320, 1985.

(71) Lichtman, A.H., Segel, G.B. and Lichtman, M.A.: Total and exchangeable calcium in lymphocytes: Effects of PHA and A23187, J. Supramol. Struct. 14:65, 1980.

(72) Hovi, T., Allison, A.C. and Williams, S.C.: Proliferation of human peripheral blood lymphocytes induced by A23187, a streptomyces antibiotic, Exp. Cell Res. 96:92, 1976.

(73) Resch, K., Bouillon, D. and Gemsa, D.: The activation of lymphocytes by the ionophore A23187, J. Immunol. 120:1514, 1978.

(74) Hesketh, T.R., Moore, J.P., Moore, J.D.H., Taylor, M.V., Rogers, J., Smith, J.A. and Metcalfe, J.C.: A common sequence of calcium and pH signals in the mitogenic stimulation of eukaryotic cells, Nature 313:481-484, 1985.

(75) Poenie, M., Alderton, J., Tsien, J.Y. and Steinhardt, R.A.: Changes of free calcium levels with stages of the cell division cycle, Nature 315:147-149, 1985.

(76) Wintrobe, M.M.: The lymphatic system, in Clinical Hematology, Lea & Febiger, Philadelphia, 271-293, 1981.

(77) Sell, S., Rowe, D.S. and Gell, P.G.H.: Studies on rabbit lymphocytes in vitro. III. Protein, RNA, and DNA synthesis by lymphocyte culture after stimulation with

phytohemagglutinin, with staphylococcal filtrate, with antiallotypte serum, and with heterologous antiserum to rabbit whole serum, J. Exp. Med. 122:823, 1965.

(78) Torelli, U.L., Henry, P.H. and Weissman, S.M.: Characteristics of the RNA synthesized in vitro by the normal human small lymphocyte and the changes induced by phytohemagglutinin stimulation, J. Clin. Invest. 47:1083, 1968.

(79) Hubert, H., Strieder, N., Winnler, H., Reiser, G. and Koppelstaetter, K.: Studies on the incorporation of 14C-sodium acetate into the phospholipids of phytohemagglutinin-stimulated and unstimulated lymphocytes, Brit. J. Haemat. 15:203, 1968.

(80) Packman, L.M.: The carbohydrate metabolism and respiration of isolated small lymphocytes: In vitro studies on normal and PHA stimulated cells, Blood 30:691, 1967.

(81) Loeb, L.A., Agarwal, S.S. and Woodside, A.M.: Induction of DNA polymerase in human lymphocytes by phytohemagglutinin, Proc. Natl. Acad. Sci. USA 61:827, 1968.

(82) Emilia, G., Donelli, A., Torelli, G., Ceccherelli, G., Zucchini, P. and Torelli, U.: Cellular levels of mRNA from C-Myc, C-Myb and C-Fes Onc-genes in normal myeloid and erythroid precursors of human bone marrow: An in-situ hybridization study, Br. J. Haemat. 62:287-292, 1986.

(83) Craddock, C.G.: The response of lymphatic tissue to antigens and mitogens, in Hematology, W.J. Williams, ed., McGraw-Hill, New York, 820-827, 1972.

(84) Rundles, R.W.: Chronic lymphocytic leukemia, in Hematology, W.J. Williams, ed., McGraw-Hill, New York, 880-896, 1972.

(85) Bius, C.V., Lundak, R.L., Fletcher, R.M. and Adey, W.R.: Alterations in protein kinase activity following exposure of cultured human lymphocytes to modulated microwave fields, Bioelectromagnetics 5:341-345, 1984.

(86) Moorhead, P.S., Nowell, P.C., Mellman, W.J., Battips, D.M. and Hungerford, D.A.: Chromosome preparations of leucocytes from human peripheral blood, Exp. Cell Res. 20:613-618, 1960.

(87) Summer, A.T., Evans, H.J. and Buckland, R.A.: New technique for distinguishing between human chromosomes, Nature New Biol. (London) 232:31-32, 1971.

(88) Bach, J.F.: Evaluation of T-cells and thymic serum factors in man using the rosette technique, Transplant Rev. 16:196, 1973.

(89) Froland, S.S. and Natvig, J.B.: Identification of three different human lymphocyte populations by surface markers, Transplant Rev. 16:114, 1973.

(90) Mingari, M.C., Gerosa, F., Carra, G., Accolla, R.S., Moretta, A., Zubler, R.H., Waldmann, T.A. and Moretta, L.: Human interleukin-2 promotes proliferation of activated B cells via surface receptors similar to those of activated T cells, Nature 312:641-643, 1984.

(91) Craddock, C.G., Longmire, R.L. and McMillan, R.: Function of lymphocytes and plasma cells in immunity, in Hematology, W.J. Williams, ed., McGraw-Hill, New York, 920-936, 1977.

(92) Greaves, M., Janossy, J. and Doenhoff, M.: Selective triggering of human T and B lymphocytes in vitro by polyclonal mitogens, J. Exp. Med. 140:1, 1974.

(93) McLaughlin, S. and Poo, M.M.: The role of electro-osmosis in the electric field-induced movement of charged macromolecules on the surfaces of cells, Biophys. J. 34:85-93, 1981.

(94) Lin-Liu, S., Adey, W.R. and Poo, M.M.: Migration of cell surface concanavalin A. Receptors in pulsed electric fields, Biophys. J. 45:1211-1218, 1984.

(95) Winters, W.D., Guest, G.F., Winters, B.T. and Phillips, J.R.: Human leukocyte responses after exposure to 60 Hz electromagnetic fields in vitro, BEMS, San Francisco, 55, 1985.

(96) Smith, G.K. and Cleary, S.F.: Effects of pulsed electric fields on mouse spleen lymphocytes in vitro, Biochem. Biophys. Acta 763:325-331, 1983.

(97) Nordenson, I., Hansson Mild, K., Nordstrom, S., Sweins, A. and Birke, E.: Clastogenic effects in human lymphocytes of power frequency electric fields: In vivo and in vitro studies, Radiat. Environ. Biophys. 23:191-201, 1984.

(98) Orgel, M.G., O'Brien, W.J. and Murray, H.M.: Pulsing electromagnetic field therapy in nerve regeneration: An experimental study in the cat, Plast. Reconstr. Surg. 73:173-182, 1984.

(99) Cacciatore, E., Cadossi, R. and Caselli, G.: Campos Electromagneticos Pulsantes de Baja Frequencia: su Empleo en las Enfermedades Vasculares Perifericas, Angiologia 32:212-216, 1980.

(100) Zecca, L., Costi, P. and Dal Conte, G.: Immune response and antiinflammatory activity of pulsed MHz waves, 2nd Ann. BRAGS, Oxford, 93, 1982.

(101) Robinson, K.R. and McCaig, C.: Electrical fields, calcium gradients, and cell growth, Ann. N.Y. Acad. Sci. 339:132-138, 1980.

(102) Whitfield, J.F., Boynton, A.L., McManus, J.P., Rixon, R.H., Sikorska, M., Tsang, B. and Walker, P.R.: The roles of calcium and cyclic AMP in cell proliferation, Ann. N.Y. Acad. Sci. 339:216, 1980.

(103) Law, H.T., Annan, I., McCarthy, I.D., Hughes, S.P.F., Stead, A.C., Camburn, M.A. and Montgomery, H.: The effect of induced electric currents on bone after experimental

osteotomy in the sheep, J. Bone Joint Surg. 67:463-469, 1985.

(104) Wahlstrom, O.: Electromagnetic fields used in the treatment of fresh fractures of the radius, 2nd Ann. BRAGS, Oxford, 26, 1982.

(105) Barbiroli, B., Ferrari, S., Battini, R., Monti, M.G. and Zaniol, P.: Pulsed ELF magnetic fields influence the expression of oncogenes during liver regeneration following partial hepatectomy, BEMS, San Francisco, 54, 1985.

(106) Zecca, L., Cal Conte, G., Furia, G. and Ferrario, P.: Effects of ELF magnetic fields on immune response and inflammation, Assoc. for Biomedical Applications of Electromagnetism, Venezia, 7, 1984.

(107) Aksoy, M., et al.: The results obtained from a triple treatment with androgens, corticosteroids and phytohaemagglutinins in aplastic anaemia of various aetiology, Med. Bull. Istanbul 3-4:56, 1970-71.

(108) Hayes, D.M. and Spurr, C.L.: Use of phytohemagglutinin to stimulate hematopoiesis in humans, Blood 27:78, 1966.

(109) Humble, J.G.: The treatment of plastic anaemia with phytohemagglutinin, Lancet 1:1345, 1964.

(110) Conti, P., Gigante, G.E., Cifone, M.G., Alesse, E., Fieschi, C., Bologna, M. and Angeletti, P.U.: Mitogen dose-dependent effect of weak pulsed electromagnetic field on lymphocyte blastogenesis, FEBS Lett. 199:130-134, 1986.

(111) Cantini, M., Cossarizza, A., Bersani, F., Cadossi, R., Ceccherelli, G., Tenconi, R., Gatti, G. and Franceschi, C.: Enhancing effect of low-frequency pulsed electromagnetic fields on lectin-induced human lymphocyte proliferation, J. Bioelectricity 5:91, 1986.

15
Effects of Electromagnetic Fields on Nerve Regeneration

BETTY F. SISKEN
Wenner-Gren Biomedical Engineering Laboratory and
Department of Anatomy, University of Kentucky
Lexington, Kentucky

INTRODUCTION

The problem of regeneration in the central and peripheral nervous system has been the topic of numerous investigations and a variety of techniques have been used to effect complete restoration (discussed in a later section)(1,2). In the past 10 years a resurgence of interest has developed in employing electric fields to stimulate regrowth. The purpose of this Chapter is to present results obtained with these fields and to explore future avenues of research.

IN VITRO STUDIES

The pioneering work of Ross Harrison (3) demonstrated that isolated nerve cells are capable of forming nerve processes (neurites) when isolated in a dish containing proper nutrient medium (tissue culture). This tissue-culture technique has been used to study the effects of various electric fields on neuronal growth. Since the tissues are denervated at the time of culture, the regrowth response is termed "neuronal regeneration" and it is assessed by determining outgrowth of neurites from the neuronal cell bodies, and survival of neurons with reestablishment of neurotransmitter synthesis and neurophysiological activity.

The first report of electric field effects on neuronal regeneration in vitro was by Ingvar (4), who found that regenerating cell processes oriented along direct current (DC) field lines (0.1-0.2 $\mu A/cm^2$). In 1934 Karssen and Sager (5) confirmed these findings using currents less than 1 $\mu A/cm^2$. In the same year Paul Weiss (6), a noted embryologist, maintained that nerves grow along the electrically-induced stress lines of the plasma clot so that growth was only secondarily affected. In 1946 Marsh and Beams (7) presented definitive proof that the neurons were directly influenced by DC (120 $\mu A/mm^2$). Not only did the newly formed axons grow preferentially to the cathode, but growth was reoriented in response to cathode placement.

Since that time, other studies using DC have demonstrated stimulation of growth and directional response to the cathode (8-14). More recently, noninvasively induced current via pulsed electromagnetic fields (PEMF) has been applied to cultured neurons resulting in both stimulatory and inhibitory effects (see below).

LEVELS OF APPLIED DIRECT CURRENT

Two schools of thought appear to exist for applying optimum levels of DC to neuronal systems in vitro. One concept stems from work with the application of DC current on the order of nanoamperes using in-dwelling metal (platinum or tantalum) electrodes for long time periods (15). A second view comes from the application of DC currents on the order of milliamperes using agar bridge electrodes for relatively short time periods, using a technique introduced by Marsh and Beams (7). Cathodally-oriented neurite growth has been observed in both systems.

The original justification for using fields of 100 $\mu V/cm$ or less arose from the work of Becker and Murray (16) and Pilla (17), who found that fields of 10^{-8}-10^{-4} V/cm provoked morphological changes in amphibian red blood cells (RBCs) to a "dedifferentiated" state. Pilla observed the greatest change at the electrodes where the calcium ions moved more slowly. Addition of bulk calcium ions to a solution of RBCs produced the same dedifferentiated response in 10% of the cells. Electrochemical tests of different electrodes indicated the superior quality of titanium or tantalum as electrodes; both demonstrated little faradaic current flow over a 2-volt range. In 1975, Sisken and Smith (8) reported on the stimulation of neuronal regeneration in vitro. DC (.001-11.5 nA/mm^2) applied via point platinum electrodes to trigeminal ganglion cultures enhanced neurite growth that was oriented to the cathode; the rate of growth was 0.1 mm/hr. Neuronal survival was also enhanced.

These studies have been extended using dorsal root sensory ganglia to include and compare PEMF effects with those produced by applied DC in a nonuniform field (18-21). Our aims have been to assess electric field effects on neuronal regeneration, including studies on the nerve cell body itself. We exposed our neuronal population to minute levels of DC for up to 3 days and observed effects at that time, or after an additional 3 days of culture. The results of these experiments indicated that not only was neurite growth enhanced, but protein content was significantly increased over control cultures (22). These observations were not dependent upon the number of non-neuronal cells (assessed by ^{3}H-thymidine incorporation, unpublished observations). This line of study differs from the studies of those who study electrophoresis-dependent neurite growth and movement of surface receptors.

More recently we tested different levels of DC on the peripheral-ganglia model using agar or metal (platinum, tantalum) electrodes (23), or tantalum electrodes driven by constant current sources (24). Maximal stimulation of regeneration was obtained with a constant current of 10 nA total current (agar electrodes) or 60 nA total current (tantalum electrodes) in chick sensory ganglia. Similar experiments using neuroblastoma cell lines (25) indicated that greater numbers of cells formed processes after application of 10 nA total current. Increasing the constant current above 10 nA did not increase transformation.

Milliampere levels of DC were employed first by Marsh and Beams (7) and then by Jaffe and Poo (9). The model developed by Jaffe and Poo used agar electrodes to expose dorsal root ganglia (cultured in the presence of nerve growth factor) to a uniform electric field created by applying up to 140 mV/mm across the dish for up to 20 hours. Field strengths of 70-140 mV/mm induced faster outgrowth of neurites oriented toward the cathode. The effective average current density in these experiments was 14.3 mA/cm^2, which was in the same range as that used by Marsh and Beams (Table 1). They postulated that the applied electric fields could cause electrophoresis of nerve growth factor receptors to the growing neurite. Hinkle et al. (10) used a similar system to expose single neurons from Xenopus neural tube. Preferential growth to the cathode was obtained

TABLE 1. Values of Electric Field, E, Current Density, J, and Time, T, (in hours) for *In Vitro* Experiments on Neuronal Tissue

	E	J	T
Ingvar (1920)	--	.0015 μA/mm^2	--
Marsh and Beams (1946)	65 mV/cm	12 mA/cm^2	29
Sisken and Smith (1975)	80 μV/cm	11.5 nA/mm^2	96
Jaffe and Poo (1979)	1000 mV/cm	14.3 mA/cm^2	4-8 (20 max)
Hinkle et al. (1981)	70-1900 mV/cm	1-27 mA/cm^2	18-20
Patel and Poo (1982)	100-10,000 mV/cm	1.4-143 mA/cm^2	6
Patel and Poo (focal) (1984)	3-30 mV/cm	0.2-2 pA/μm^2	.25
Sisken (1984)	0.6 μV/cm	9 nA/cm^2	72
Freeman (1985)	70-350 mV/cm	1 mA/cm^2	--

with exposure times of 18-20 hours; threshold values for this response were 6-8 mV/mm. Patel and Poo (13) applied steady electric fields of 0.1-10 V/cm to single Xenopus neurons and also found stimulated growth of neurites facing the cathode. The number of neurite-containing neurons was increased in these treated cultures as was neurite length. Addition of concanavalin A (ConA) abolished the electrically induced effects while fluorescently labeled ConA receptor accumulated at the cathode. This cathodal accumulation of growth-controlling surface glycoproteins was implicated as the mechanism by which electric fields exerted their effect.

The effects of electric fields on central nervous system regeneration in vitro have also been investigated. Khan and Gaik (26) placed rat embryo spinal cord explants on carbon fibers (8-10 μm diameter) in a culture dish and applied 20 nA DC to the fibers. Cathodally oriented growth (length and number of neurites) was significantly enhanced in this system. In our laboratory, spinal-cord explants of 8-day chick embryos or 16-day rat fetus were treated with 10 nA DC or single-pulse 72 Hz PEMF in the presence of varying concentrations of cytosine arabinoside, ara C (15). The area of the explant, total area of outgrowth, and area occupied by neurites were assessed on cultures stained with Bodian silver or on radioautographs of explants showing incorporation of ^{3}H-proline. Both DC (3-day, continuous) and PEMF (12 hr/day for 3 days) significantly stimulated neurite outgrowth relative to controls.

Freeman et al. (12) described the development and use of a circularly vibrating probe on central (retinal) neurons of the goldfish. This probe was capable of discriminating current densities of 5 nA/cm^2. They provided evidence for the existence of currents of 10-100 nA/cm^2 flowing into the filopodia of growth cones of cultured retinal ganglion cells and back out from near the junction of growth cone and filopodia. The currents were produced only during active growth and were believed to be carried primarily by calcium ions. To determine whether such endogenous fields were capable of directing filopodial growth toward or away from a point source of current, retinal ganglion cells were exposed to point sources. Threshold current for orientation was 40 nA (70 mV/cm). Endogenous currents generated an axial current within filopodia of

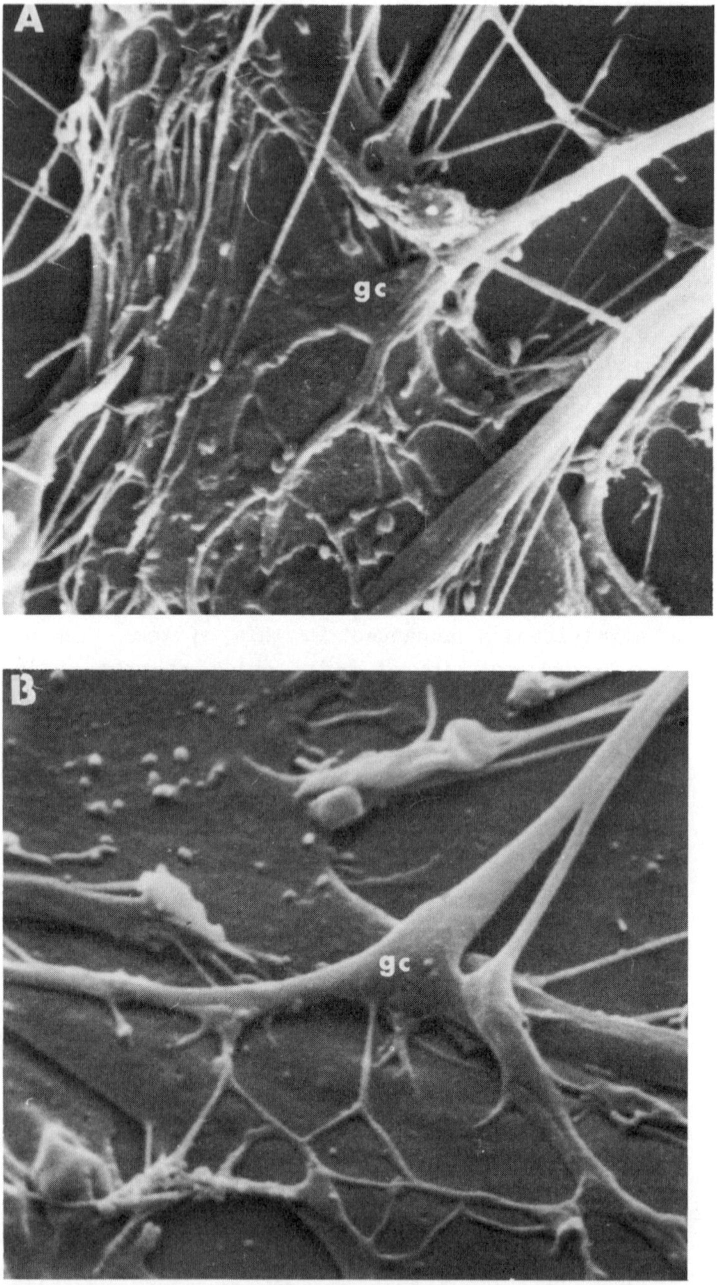

FIGURE 1. Scanning electron microscopy of neurite growth cones of 8-day chick embryo trigeminal ganglia after 4 days _in vitro_. The growth cones (gc) were found at the edge of the explant lying on top of the fibroblast mat. Note the expanded growth

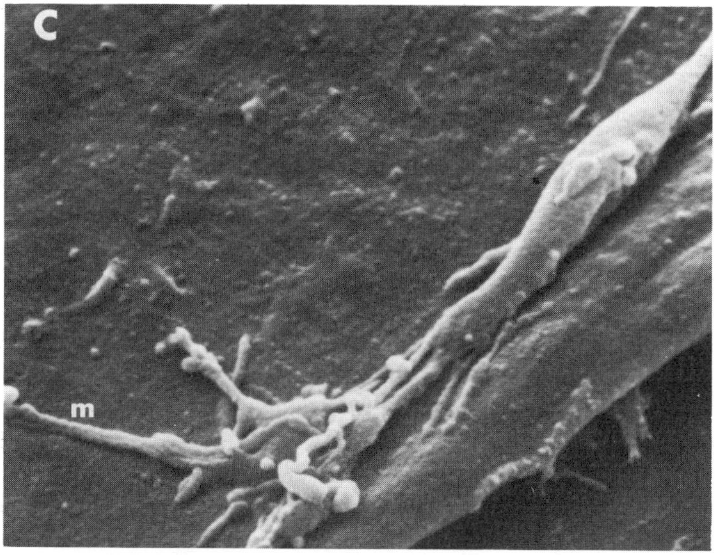

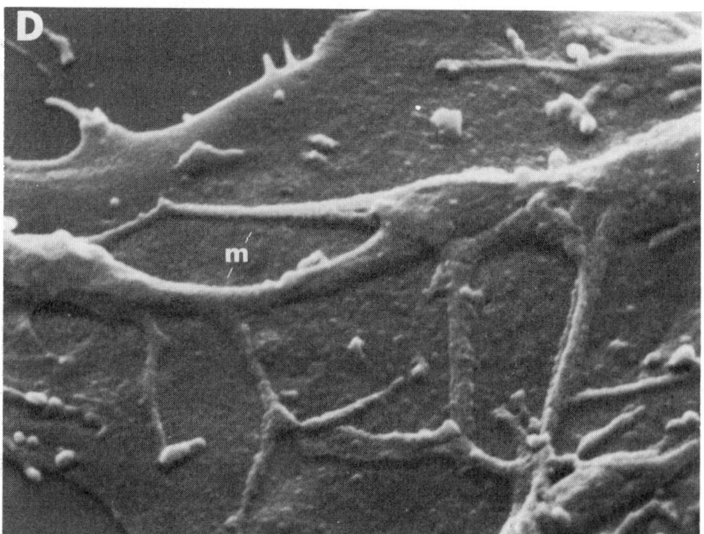

cones of the DC-treated ganglia and increased numbers of micro-spikes (m). A, (control culture) and B (10 nA DC-treated culture) x 2400. C, (control culture) and D (10 nA DC-treated culture) x 9750.

4 $\mu A/cm^2$, with an extracellular electric field of 0.3 mV/cm. These endogenous fields were 2 orders of magnitude lower than those used exogenously to cause lateral electrophoresis of surface macromolecules, but the authors suggested that they might be involved in polarization and lateral electrophoresis of molecules within the filopodia.

PULSED ELECTROMAGNETIC FIELDS

The first studies reporting the effects of pulsed electromagnetic fields on nerve tissue *in vitro* were those of Bawin et al. (27). Using 147 MHz, 0.8 mW/cm^2, amplitude modulated by slow sinusoidal signals, the release of bound $^{45}Ca^{+2}$ was increased maximally at a modulation frequency of 16 Hz. Such applied electrical gradients mimic the intrinsic extracellular electric field of the EEG of 50-100 mV/cm. Continuation of these studies (28) demonstrated that this response was unaffected by changing the calcium concentration of the bathing solution, but was inhibited by addition of H^+. These results indicated that these weak fields affected calcium bound to extracellular negative binding sites, and that H^+ competed for the same sites. There are many other studies describing PEMF effects on isolated pieces of normal tissue such as those by Wachtel (29), Wheeler (30), Bawin et al. (31), and Gundersen and Greenbaum (32).

Regeneration induced by PEMF applied to goldfish retina cultures for 24 hours was reported by Schwartz et al. (33). PEMF (perpendicular orientation of coils) exposure for 0.5 minute at 100 V, 1,000 Hz, 5 μsec and 100 V, 200 Hz, 25 μsec stimulated exaggerated process regrowth. PEMF-induced regrowth resembled that induced by brain-derived growth factors or glial cell conditioned medium. This is the only short exposure time study (0.5 min) found in the literature.

Studies in our own laboratory (34) on the effects of 72 Hz, single pulse PEMF on chick sensory ganglia indicated that threshold levels of 400 nA/cm^2 of induced current significantly enhanced neurite outgrowth. This stimulation was comparable to that obtained with DC application of 9 nA/cm^2. Using capacitor plates, Yoshioka et al. (35) reported that 27 MHz electric fields had no effect on rat sensory neurons at 1, 20, 50 and 610 mV/cm; significant stimulation was observed at 10 mV/cm.

Addition of nerve growth factor to cultures treated with 10 mV/cm did not change the growth response. In 1985, continuing results from Albert's (36) laboratory indicate that 1 mV/cm electric field repeated at 7 Hz inhibited neurite outgrowth of 14-15 day rat ganglia in the presence or absence of nerve growth factor.

Field effects on non-neural cells (Schwann cells) produced by 60-KHz capacitively coupled electric fields using various voltages (2-4 days exposure) were studied (37). A myelin marker (gal C) was preferentially retained at 250 and 375 V, but no differences between groups were obtained in ^{3}H-thymidine uptake studies.

The effects of spatially uniform pulsed fields, focally applied DC fields, and focally applied pulsed fields on orienting neurite growth from Xenopus neurons was described by Patel and Poo (14). Uniform pulsed fields of an equivalent time-averaged field intensity as uniform DC fields produced the same extent of neurite orientation. Unipolar electric current pulses applied focally through a micropipette to neuritic growth cones modulated the rate of neurite growth. Negative (sink) current increased growth rates while positive (source) currents were inhibitory. The threshold current density to obtain a growth cone response within 15 minutes with focal DC was 0.2-2 pA/μm^2; a similar response was obtained with focally-applied pulsed current of 4 pA/μm^2 at 10 Hz.

IN VIVO STUDIES

Electrical recordings of transected nerve activity were first reported by DuBois-Reymond (38). A more detailed study was performed by Grenell and Burr (39) using Ag-AgCl electrodes to measure potential differences on the limb surface of rabbits before and after nerve section, and in man after ulnar nerve injury. In each case a marked positive shift in potential was obtained, suggesting that such recordings could be used clinically to determine the extent of nerve injury.

EFFECTS OF ELECTRIC FIELDS ON PERIPHERAL NERVE REGENERATION

The application of electric fields to injured nerve tissue has been a recent event. A few early studies (40,41) using DC, or AC, have been followed by a fairly large number of reports. Hoffman (40) stimulated the spinal cord of large nerve trunks (10-60 minutes with 1.5 mA at 50-100 Hz) after transection of the 5th lumbar nerve of rats. Significant acceleration in reinnervation of denervated muscle fibers was found. Bodemer (41) stimulated nerve fibers of the brachial plexus _in situ_ in an attempt to increase nerve activity in an amputated stump. The electrical stimulation resulted in partial regeneration of the limbs in adult frogs (42,43)(also see S.D. Smith, this volume). A brief report was presented by Romero-Sierra et al. (44) in 1971 on the application of 27 MHz for 5-30 minutes to non-transected, desheathed sciatic nerves _in situ_. Histological investigation of such exposed nerves revealed various degrees of demyelination, Schwann cell damage and distortion of collagen fiber pattern close to the nerve. In the same conference, Yorde et al. (45) reported on the ultrastructural effects of DC on cortical synapses in biopsies of the monkey brain. DC (2.5 mA) was applied through surface electrodes for 1-2 minutes to the monkey cortex. A biopsy of this cortex was examined and synaptic vesicles counted; the depletion of vesicles after short periods of DC application was correlated with stimulation of synaptic transmission.

In 1976 Wilson (46) used a non-invasive technique to study the effects of PEMF on nerve and spinal cord regeneration. He applied a radiofrequency signal (Diapulse, 5-120 mW/cm^2) to transected median-ulnar nerves of rats for 15 minutes/day for 30-60 days. By 30 days, PEMF-treated animals showed significant restoration of nerve conduction activity and the histological presence of large-diameter nerve fibers.

The effects of daily stimulation on reinnervation and acceleration of nerve growth after lesioning the sciatic nerve of the rat were reported by Sebille and Bondoux-Jahan (47). Using an intensity 10 times that needed to elicit muscle contraction (30 mV at 50 Hz, pulse duration of 1 msec) for 30 minutes/day via electrodes attached to the leg skin, the toe-spreading reflex was observed and evaluated. Muscle stimulation significantly increased the rate of recovery. The increased muscle contractions resulting from the applied electric signal probably hastened the functional recovery of the end plate zone when the neurites reach the muscle membrane. A similar paradigm was used by Nix (48) after crushing the common peroneal nerve of the rabbit. Stimulation of the external digitorum longus muscle with 10-12 Hz for 8 hours/day via implanted electrodes into the muscle increased the time course of contraction and relaxation thus preventing denervation-induced slowing of muscle activity seen in controls.

In 1981, Winter et al. (49) reported that intraluminal insertion of Pt/Ag bimetallic electrodes (100 nA) stimulated regeneration of transected sciatic nerves. Regeneration was determined by analyzing the compound action potential obtained; only when the cathode was present and distally located was the DC effective. Maehlen and Nja (50) investigated the effects of electrical stimulation of pre- and postsynaptic cells on sprouting after denervation in the guinea pig superior cervical ganglion. Preganglionic stimulation on the cervical sympathetic trunk for 1 hour only immediately after denervation (100 pulses at 20 Hz every 25 seconds) increased the number of axons innervating each ganglion cell. This effect was abolished with hemamethonium, which blocked ganglionic transmission. The findings support a mechanism whereby retrograde transsynaptic trophic effects are modulated by impulse activity. Part of the stimulus for sprouting after denervation may be enhanced by a brief period of hyperactivity induced by the electrical stimulation followed by a period of subnormal activity.

In contrast to direct stimulation of nerves via electrodes, Ito and Bassett (51) subjected the entire body of rats to pulsed electromagnetic fields after transection of the sciatic nerve. The PEMF consisted of Helmholtz aiding coils delivering a repetitive single pulse of 380 μsec positive-going, quasirectangular waveform repeating at 72 Hz. All rats were treated for 12 hours/day. Motor function was evaluated by plantar-flexion force produced by stimulation of the nerve proximally. Return of motor function occurred within 4 weeks after PEMF in contrast to controls at 8 weeks.

Raji and Bowden (52) tested the effects of PEMF delivered by a Diapulse machine (1 mW/cm^2) on regeneration of the transected common peroneal nerve in rats. PEMF was administered for 15 minutes daily for periods of 3 days to 8 weeks. PEMF caused significant increases in skin, deep-tissue and rectal temperatures which returned to normal after treatment. The size of the intraneural blood vessels was increased after treatment and the amount of collagenous tissue fibrosis was reduced with PEMF. Most importantly, these studies support those of Wilson (46) on the PEMF-acceleration of regeneration and maturation of myelinated axons.

Nix and Hope (53) crushed the motor innervation to the soleus muscle in rabbits and stimulated the nerve proximal to the lesion with stainless steel electrodes sewn into a cuff placed around the nerve. Stimulation was performed for 4 weeks with a Grass S88 stimulator using rectangular pulses of 0.2 msec duration, at 4 Hz. Twitch force, tetanic tension, and muscle action potential amplitude measurements were taken pre- and post-operatively for each animal. In each case, significant differences were obtained as a function of treatment, and reinnervation was enhanced. The authors concluded that either the nerves grew faster or they established functional connections to the muscle sooner than untreated animals. They implicated the electrical stimulatory effect with maintenance of large myelinated fibers; it is well-known that increased motor activity enhances motor nerve regeneration. The slow-frequency pattern used for stimulation was chosen because the soleus muscle is a slow muscle; it could be that fast muscles were reflexly stimulated by this slow pattern.

Preliminary studies were reported on transected rat sciatic nerves treated with a PEMF clinical pulse-burst signal (15 Hz, Electrobiology, Inc.) for five days following transection (54). Assessment of regeneration indicated a faster return of neurophysiologically recorded function and more myelinated axons/mm^2 in the nerve distal to the transection than in untreated control animals.

Singer and Mehler (55) questioned whether increased 2-deoxyglucose uptake in axotomized motor neurons of the hypoglossal was associated with increased electrical activity or protein and RNA synthesis. They recorded spike activity in normal and axotomized hypoglossal nuclei, and observed 2-deoxyglucose localization radioautographically in the same nuclei. Increased uptake was noted in the axotomized nucleus but no differences were observed in numbers of action potentials. The authors conclude that increased uptake associated with axotomy was not the result of increased action potential activity, but rather correlated with synthesis of protein, RNA and lipid during regeneration.

The effects of PEMF on regeneration of the common peroneal nerve of the cat was determined using a multidisciplinary approach (56). Five days after transection, the cats were exposed to PEMF for 10 hours/day, 6 days/week for 12 weeks. Two different signals were tested: a pulse-burst signal used clinically for bone repair (15 Hz, 380 μsec positive, 20 μsec negative), and a single repetitive pulse (200 μsec positive, 6 msec negative, 72 Hz). Electrophysiologic data was collected pre- and postoperatively. Muscle biopsies were taken for fiber typing, the nerves biopsied for fiber counts, and retrograde transport of horseradish peroxidase to the motorneurons in the spinal cord were used for assessing regenerative events. No significant differences were noted between controls and either PEMF signal in muscle-fiber diameter, numbers of fibers/mm^2, axon-fiber caliber, areas of nerve compound action potential, or muscle compound action potential. However, the numbers of motor neurons retrogradely labeled in spinal cords of cats treated with the pulse burst signal were significantly increased (96.8% more than the unoperated side). This study represents the most in-depth exploration of electrophysiological and morpho-logical/morphometric parameters of peripheral nerve regeneration.

Pomeranz et al. (57) reported on accelerated sprouting of intact saphenous nerves after sciatic nerve transection. Electric current was applied using 1 μA DC or AC (20 Hz, 1000 μA/pulse) delivered through stainless steel electrodes placed in the skin of the digit of the hindpaw. No description of how the signals were generated was given. Only distally placed negative electrodes were effective. Roman et al. (58) applied 10 μA DC to rat sciatic nerves transected and left with a 5-mm gap. Proximal and distal ends of the nerve were placed inside a Silastic tube; the cathode-stimulator (Pt wire) was inserted into the tube close to the distal stump. Direct current (1 μA/cm^2) was administered using a battery and variable resistor. The contents of the Silastic tube were fixed at 3 weeks and examined histologically. A portion of the DC-treated tubes was filled with blood vessels and axon bundles; the total contents were two times larger than controls. The number of myelinated axons per tube was increased by a factor of three. In another series, the tube contained the proximal stump and the cathode stimulator alone, and no distal stump was present. Only 10 μA stimulated growth of nerve bundles.

EFFECTS OF ELECTRIC FIELDS ON SPINAL CORD REGENERATION

In 1976 Wilson (46) provided preliminary evidence on the effects of PEMF via Diapulse signals (see above) on hemicordectomies of the upper lumbar segment in cats. The experimental protocol was to expose the cats for 30 minutes/day for 30 days using 50 mW/cm^2, 400 Hz. Three months after lesioning, the cords were fixed and sectioned for histology. PEMF-treated cords exhibited decreased scarring and regenerating neurites traversing the lesioned area.

Cohen and collaborators (59) investigated the effects of applied DC (10 μA) on the regeneration rate of lamprey spinal cord neurons. Wick electrodes were placed distally to the lesion and current was delivered for 5 days. Axonal die-back was significantly correlated with the direction of the applied current; die-back of axons was increased with the anode and decreased (promotion of regeneration) with the cathode. The authors proposed that die-back was associated with entry of cations (primarily calcium ions) into the transected stump, and that the DC interacted with endogenous currents so that increasing or decreasing cation flow resulted in enhancement or reduction of axonal die-back, respectively.

A contusion-injured cat model has been used extensively by W. Young and his collaborators in studies on spinal cord injury. Application of a Diapulse signal (65 μsec pulses, 27.12 MHz, average power of 100 mW/cm^2) for 1 hour daily beginning 4 hours postinjury demonstrated beneficial effect on preservation and maintenance of function (60). Short-term analyses of PEMF effects (3 hours post-injury for 2 hours) on ionic changes showed that it had no effect on sodium, potassium, or water content of the contused area, but that it significantly decreased calcium accumulation in the cord (61,62). These results support the findings of beneficial effects of PEMF on restoration of peripheral nerve function cited above, and they provide another example of electric field action on decreasing calcium entry into injured spinal cord.

Functional electrostimulation (FES) has come of age following the original design of the pacemaker. Different types of FES are now in use or in different stages of development for treating various clinical states. FES has been used to alleviate pain, induce artificial respiration (diaphragm pacing), cause contraction (continence restoration) or evacuation (micturation reflex) of the bladder, and inhibit muscle atrophy and improve motor function.

THEORIES ON THE MECHANISM OF ACTION OF ELECTRIC FIELDS ON NERVE REGENERATION

The following discussion on mechanisms of action of imposed electric fields strongly implicates the role of calcium ions, and models originally proposed by Pilla (63), Jaffe et al. (64), and Bawin et al. (27) have addressed this point. In 1974, Pilla proposed a theoretical model to explain electrical effects on tissues and cells based on charges at the interface of the cell membrane and intra- and extracellular fluids. Strong electric fields exist at this site and energies of the order of 1 mW are capable of perturbing the interfacial structure. He suggested that alterations of a few millivolts could result in gross alterations of specifically adsorbed or bound species resulting in electrochemical information transfer. He demonstrated that DC fields of 0.01-100 μV/cm produced morphological changes in RBC, and implicated the slower migrating calcium ions in this phenomenon. This hypothesis was reinforced when addition of calcium alone to these cells induced similar morphological changes. At the same conference, Jaffe et al. (64) proposed an electrical hypothesis for localized growth, suggesting that the plasma membrane of a growth region becomes relatively leaky to cations such as calcium, magnesium, sodium and hydrogen which exist extracellularly at higher concentrations. As the cations enter the growth point of the cell, an electric current and field are generated which pull more negative cytoplasmic substances and vesicles (which form new membrane) toward the leaky area, thus generating more leaky membrane. This "positive feed-back loop" would increase the probability of localized growth and membrane expansions. This hypothesis was tested by measuring electric currents around developing fucoid eggs, finding growth points of current entry which contain calcium and sodium components. Application of fields of 200 mV/egg to naive eggs induced blister "fertilization-like" vesicle formation at the positive side of the egg.

Bawin et al. (27) implicated a specific class of calcium sites at the extracellular neuronal membrane that were responsive to low-frequency electric fields, and postulated that they play a role in regulating cell excitability. They found that RF fields modulated at brain wave-like frequencies increased calcium efflux from Ca^{45}-loaded cerebral cortex, and that efflux was enhanced in the presence of additional H^+ ions.

Other interpretations of bioelectric effects of external electric fields have been reported, and are reviewed in the following section.

DC LEVELS OF 100 μV/cm AND BELOW

As a consequence of the work of Becker and Murray (16) and Pilla (63) who described morphological changes in red blood cells using nanoampere currents, we began our studies on assessing the effects of these minute fields on cultured neurons (8). The stimulation of neuritic growth, which was cathodally oriented, was postulated to be correlated with changes of calcium ions at the boundary of the cell membrane. Studies on the growth cones of cultures subjected to DC (65) indicated that they were enlarged. Application of DC ($\sim$10 nA/cm^2) to trigeminal neurons significantly increased calcium efflux of preloaded ganglia relative to control or nerve growth factor-treated ganglia. Our hypothesis was that the DC acted by decreasing calcium entry into cells by enhancing calcium binding to the external membranous pool.

We tested the influence of calcium ions on growth processes by blocking calcium entry with lanthanum chloride or the drug Verapamil, or increasing calcium entry with added calcium chloride or the calcium ionophore A23187 which opens calcium channels allowing intracellular calcium concentration to increase. These studies indicated that long-term application (6 days) of compounds which inhibited calcium influx (lanthanum, Verapamil) increased neurite formation, while those that stimulated calcium influx (added calcium or the ionophore A23187) inhibited neurite formation (24). These results are consistent with those of Bray et al. (66), who found that low extracellular calcium induced the formation of growth cones all along the length of cultured sensory neurons. It appears very probable that long exposures to minute levels of DC mimic the drug-induced reduction of calcium entry into the neurons; both result in stimulation of nerve process growth.

The role of cations in neuronal growth and differentiation is slowly becoming clear. The regulation of sodium and potassium levels, and of sodium and potassium pump activities by nerve growth factor (NGF) in early stages of neuronal regeneration in culture has been characterized (67). Micromolar levels of calcium are required for growth-cone formation (the growth end of neurites), and mobility (actin network), but they are not required for neurite formation and (microtubule) stability.

An increase in the number of calcium channels has been correlated with active neurite outgrowth as neurons differentiate (12,68). This calcium requirement is reduced as a function of time in culture. Although the precise role of calcium in neurite formation is not known, Hammerschlag (69) postulated that amino acid uptake and protein transport are dependent upon calcium entry through the cell membrane, and Fukada and Kameyama (68) suggested that increased calcium is needed for protein synthesis for neurite membranous growth. In an elegant study, Anglister et al. (70) detected calcium action potentials in neuroblastoma cells, and found that voltage-activated calcium channels were less abundant in neurite processes, and more abundant in growth cones. Upon depolarization with excess potassium or electrical stimulation, the area of growth cones increased by 20-120% associated with increased neurite outgrowth. This expansion in growth cone area was inhibited with cadmium (calcium blocker) or in low-calcium media. The authors suggested that calcium ion entry functions as a trigger for neurite elongation.

Freeman et al. (12) employed a novel circularly vibrating probe (capable of measuring current densities of 5 nA/cm^2) to measure the direction and magnitude of endogenous current at the growth cone. Steady or slowly varying (not pulsatile) currents in the range of 10-100 nA/cm^2 entered the filopodia tip of the growth cone at the end of a neurite, and flowed down it and back out toward the base. The current appeared to be carried primarily by calcium ions, and was not derived from growth-cone motion or flow of medium over the surface of the plasma membrane. Average applied fields of 70 mV/cm were necessary to orient the growth cones to the cathode; the magnitude of the applied fields was 2 orders larger than those found endogenously. The authors concluded that the endogenous currents were too small to cause lateral electrophoresis of surface molecules, but may be the result of localization of clusters of voltage-sensitive calcium channels at the tips of growth cones. These calcium channel proteins are brought to the growing nerve tips by active transport and are incorporated into newly formed, expanding membrane, and sodium channels arrive later due to a "slower rate of lateral diffusion in the plasmalemma." Expanded growth of the plasma membrane that occurs by fusion of intracytoplasmic vesicles to the membrane requires calcium, some of which is current-generated. Other functions assigned to calcium ions include alignment of actin and myosin in the growth cone, and subsequent calcium-dependent contractile movements. Small changes in the membrane potential resulting from external electric field application affect calcium-ion entry. Depolarization on the side of the cell facing the cathode would increase calcium entry, while that facing the anode would inhibit entry. Excess calcium entering the growth cone not used for membrane expansion would be extruded as exocytosis of intracytoplasmic vesicles occurs, thereby maintaining calcium homeostasis.

The paradigm presented by Freeman et al. (12) is provocative. The sequence of processes suggested to occur at the level of the neuronal growth cone could be applied to other tissues and cells, and should be considered by other workers in this area. In this context, a hypothesis to explain stimulation of regeneration as a function of minute levels of direct current (~10 nA/cm^2) as studied in our own laboratory incorporates some of this thinking. The fields applied to neuronal (sensory/motor) explants are so small that membrane depolarization is perhaps in the order of 1 mV or less. Attempts to measure small changes have yielded equivocal results (Sisken and Ringham, unpublished results). However, the levels of DC used in our system (~10 nA/cm^2) is equivalent to that measured by Freeman et al. (71) (10-100 nA/cm^2) flowing into the tips of growth cones of retinal neurons. Adding any additional current to that present endogenously would increase calcium entry in growth cones facing the cathode; the cascade of events postulated by Jaffe (64) and Freeman et al. (12) would ensue, resulting in growth cone expansion. Such expansions with microspikes have been observed (65) and are demonstrated in Figure 1. Orientation of the newly formed growth cones to the cathodes by motile (calcium-dependent actomyosin system) growth cones would continue with ensuing lengthening of the neurites; such activity is normally suppressed as the neurons mature (72). Calcium entry would also be instrumental in stimulating amino-acid uptake and protein synthesis (65,42) needed for membrane proteins and microtubule formation. Exposure to DC over hours would be manifest by increased neurite outgrowth. As the neurons differentiate in culture, calcium channels are replaced by sodium channels (68,70) thereby reducing calcium entry. Continued application of minute levels of DC may act to increase the adhesion of non-entering calcium to the external membrane (18). Reduction of excess calcium entry also serves to stabilize the microtubules in the neurites. The foregoing activities require that the parent neuronal cell body be synthesizing relatively large amounts of protein to enable the processes to form and elongate; neurites of 100-150 μm have been observed in our DC-treated cultures. We have found that the protein content in such explants are significantly larger than control explants, and that they approach levels of NGF-treated sister cultures (unpublished observations).

The paradoxical role of DC may be related to the presence of calcium channels (12,71). DC increases calcium intracellularly by adding to preexisting endogenous (calcium) currents. When sodium channels replace calcium channels, calcium entry is regulated precisely by transport processes and depolarization phenomena. Addition of nanoampere levels of DC to relatively more mature neurons does not increase calcium entry, rather it fosters binding of the calcium to the negative sites on the external membrane.

DC LEVELS OF 10 mV/cm TO 1.5 V/cm

The agar wedge system employed by Marsh and Beams (7) was modified by Jaffe and Poo (9) to apply fields of ~100 mV/mm.

This model has been used by a number of investigators (see earlier section) who have examined the role of fields up to 1.5 V/cm on neurite growth and orientation. Jaffe and Poo (9) suggested that such fields caused lateral electrophoresis of nerve growth factor receptors along the plasma membrane. Hinkle et al. (10), however, favored the hypothesis that the imposed electric fields acted on the interior of the growth cone since the voltage drop across the microspikes of the growth cone was of the same magnitude (~0.7 mV/cell) as that found to orient neurites. The mechanism of action of these fields on neurite growth orientation most likely does not involve field-induced mechanical effects or chemical gradients (13), since constant perfusion with fresh medium did not change the observed effects. They proposed three mechanisms to account for oriented neurite growth: (1) an electric field-induced potential change causing redistribution of cytoplasmic material; (2) alteration of membrane potential asymmetry inducing preferential growth; and (3) an electrophoretic redistribution of charged surface molecules in the plasma membrane. Data obtained from their experiments favored the third hypothesis since field strengths used for neurite growth caused fluorescently bound ConA receptors to accumulate on the cathodal side of the neuron, and inhibition of receptor migration abolished the orientation effect.

The electromigration model (73) proposes receptor (acetylcholine, ACh) migration to the synaptic region. Transient fields of 1 V/cm are generated normally by synaptic currents; such fields applied exogenously cause electrophoretic migration of ACh receptors to synaptic contact areas. The proposal offered by these investigators states that "transient electric fields associated with neuronal activities serve to develop, maintain and modulate the topography of the membrane components responsible for these activities."

Finally, the role played by calcium ions in altering locomotor activity after imposition of 0.5-15 V/cm electric fields action was addressed by Cooper and Schliwa (72). Migration of keratocytes to the cathode was observed in these fields, and it was inhibited with calcium channel antagonists. Their hypothesis for cathodal migration requires asymmetric calcium entry into the cell through specific calcium channels to trigger contractile activity and direct orientated cell movement. In addition, they discuss the contradictory role played by calcium, specifically in neuronal systems. Anodal neurite retraction, observed by Patel and Poo (13) in field strengths of 5 V/cm hyperpolarized the neuron resulting in calcium entry and its subsequent axoplasmic filament degradation. However, depolarization at the cathode also increased calcium entry, with consequent increased growth-cone formation and neuronal survival. They did not resolve this discrepancy; it may finally rest on the absolute concentration of intracellular calcium levels which is as dependent upon calcium extrusion as it is on calcium entry.

PULSED ELECTROMAGNETIC FIELDS

The action of various pulsed fields on nerve tissue has been explored and the association of these effects with calcium changes has been implicated. Efflux studies of calcium after application of low frequency extracellular fields (1-300 Hz) (28,74,75) demonstrated a correlation of enhanced efflux with specific intensity ranges. The amplitude range and frequency of these fields are comparable to those of extracellular brain waves. The hypothesis offered by Bawin et al. (28) proposed that specific negative sites on the plasma membrane were occupied by calcium, and were affected by applied ELF.

Comparative studies of DC and PEMF on promotion of neurite growth (20) correlated stimulatory effects with current density (0.7 $\mu A/cm^2$) and with orientation of the Helmholtz coils. Only coils oriented so that the magnetic field was parallel to the bottom of the dish produced a significant response. Effective levels of DC and PEMF were equivalent (10^{-3} coulombs) on a time exposure basis. Mechanisms associated with the PEMF fields may also involve calcium ion changes (as discussed previously for DC effects) since the effective dosage of both systems is in the same range.

Thresholds of 3-30 mV/cm, DC, or 4 $pA/\mu m^2$, pulsed fields, applied focally to influence the direction of neurite growth were found to be similar to those associated with action potentials, and synaptic activity. Patel and Poo (14) suggest that the action of these fields is not to generally guide nerve growth in developing systems, but to modulate localized neurite orientation to specific areas of intense activity. In another study, square-wave pulses comparable to local extracellular AC fields produced by nerve action potentials (100 Hz)(76) redistributed ConA receptors to the cathodal area of cultured myoblasts. These findings add support to the electromigration model proposed by Fraser and Poo (73) that electrical events associated with nerve activity can modulate neuronal topography.

FUTURE STUDIES

In view of the rapid technological advances and the increasing numbers of workers in the area of bioeffects of externally-applied electrical fields, it is anticipated that more clinically-useful procedures will be developed to address nerve regeneration problems. Hopefully these methods will encompass stimulation of regeneration in the spinal cord as well.

Basic studies to determine the mechanism of action of these fields and how such imposed fields relate to normal activity and to activity following degeneration and regeneration should be conducted in parallel with the clinical investigations. It is hoped that advances made in the area of soft tissue regeneration in general, and nerve tissue specifically, will duplicate or exceed that found in the electrical treatment of bone repair.

ACKNOWLEDGEMENTS

The work from the author's laboratory cited in this chapter was supported by NSF BNS7813737, The Veteran's Administration Research Program, and ONR N0014-82-K-0105. I would like to express my appreciation to Ms. E. Barr and R.S. Estes for technical assistance, and to Mrs. L. Cunningham for help in typing this manuscript.

REFERENCES

(1) Kao, C., Bunge, R. and Reier, P. (eds): Spinal Cord Reconstruction, Raven Press, New York, 1983.

(2) Seil, F.: Nerve, Organ and Tissue Regeneration: Research Perspectives, Academic Press, New York, 1983.

(3) Harrison, R.G.: Observations on the living developing nerve fiber, Anat. Rec. 5:116-118, 1907.

(4) Ingvar, S.: Reaction of cells to the galvanic current in tissue cultures, Proc. Soc. Exp. Biol. Med. 17:198, 1920.

(5) Karssen, A. and Sager, B.: Sur L'influence de courant electrique sur la croissance des neuroblasts _in vitro_, Arch. Exp. Zellorsch 16:255, 1934.

(6) Weiss, P.: _In vitro_ experiments on the factors determining the course of the outgrowing nerve fiber, J. Exp. Zool. 68:393, 1934.

(7) Marsh, G. and Beams, H.W.: _In vitro_ control of growing chick nerve fibers by applied electric currents, J. Cell Comp. Phys. 27:139-157, 1946.

(8) Sisken, B. and Smith, S.: The effects of minute direct electrical currents on cultured chick embryo trigeminal ganglia, J. Embryol. Exp. Morphol. 33:29, 1975.

(9) Jaffe, L. and Poo, M.M.: Neurites grow faster toward the cathode than the anode in a steady field, J. Exp. Zool. 209:115, 1979.

(10) Hinkle, L., McCaig, C.D. and Robinson, K.R.: The direction of growth of differentiating neurons and myoblasts from frog embryos in an applied electric field, J. Physiol. 314:121-135, 1981.

(11) Freeman, J.A., Weiss, J.M., Snipes, G.J., Mayes, B. and Norden, J.J.: Growth cones of goldfish retinal neurites generate DC currents and orient in an electric field, Soc. Neurosci. 7:550, 1981.

(12) Freeman, J.A., Manis, P.B., Snipes, G.J., Mayes, B.N., Samson, P.C., Wikswo, J.P. Jr. and Freeman, D.R.: Steady growth cone currents revealed by a novel circularly vibrating probe: A possible mechanism underlying neurite growth, J. Neurosci. 13:257-283, 1985.

(13) Patel, N. and Poo, M.: Orientation of neurite growth by extracellular electric fields, J. Neurosci. 2:483-396, 1982.

(14) Patel, N. and Poo, M.: Perturbation of the direction of neurite growth by pulsed and focal electric fields, J. Neurosci. 4:2939-2947, 1984.

(15) Sisken, B.F.: _In Vitro_ Stimulation of Central Nervous System Regeneration by Direct Current and PEMF, Bioel. Repair and Growth Society Annual Meeting, October 1985, Boston, MA, 1985.

(16) Becker, R.O. and Murray, D.G.: The electrical control system regulating fracture healing in amphibians, Clin. Orthoped. 73:169-198, 1970.

(17) Pilla, A.A.: Electrochemical information transfer at living cell membranes, Ann. N.Y. Acad. Sci. 238:149-169, 1974.

(18) Sisken, B.F., Lafferty, J.F. and Acree, D.: The effects of direct and inductively coupled current and nerve growth factor on nerve regeneration _in vitro_, in Mechanisms of Growth Control and Their Clinical Applications, R. Becker, ed., Charles Thomas, Springfield, 1981.

(19) Sisken, B.F., Sechaud, P. and Barr, E.: Correlation of current delivered versus potential differences imposed on trigeminal ganglia _in vitro_, Presented May 9, 1982 Annual Meeting, The Electrochemical Society, Montreal, Canada, J. Electrochem. Soc. 129:579, 1982.

(20) Sisken, B.F., McLeod, B. and Pilla, A.A.: PEMF, direct current and neuronal regeneration: Effect of field geometry and current density, J. Bioelectricity 3:81, 1984.

(21) Sisken, B.F., Roberts, E. and Goetz, I.: Triethanolamine, tris, hepes, and cytosine arabinoside show neuritogenic activity in cultured chick embryo ganglia, Exper. Neurol. 88:27-43, 1985.

(22) Sisken, B.F., Estes, R.S. and Kryscio, R.: The effects of cytosine arabinoside on nerve regeneration, protein content and protein synthesis in NGF or electricity-treated sensory ganglia in vitro, Bioelectrical Repair and Growth Society 3:17, 1983.

(23) Sisken, B.F., Barr, E. and Estes, R.S.: Electrical effects on nerve ganglia in vitro, in Soft and Hard Tissue Repair: Biological and Clinical Aspects, T.K. Hunt, R.B. Heppenstall, E. Pines and D. Rovey, eds., Praeger Scientific, New York, 1984.

(24) Sisken, B.F. and Sisken, J.E.: Ionic and direct current effects on neurite differentiation in primary cultures of chick sensory ganglia in vitro, Bioelectrical Repair and Growth Soc. 4:22, 1984.

(25) Sisken, B.F. and Estes, R.S.: The effects of direct current on neurite growth in N18 neuroblastoma cells treated with Ara C, Trans. Bioelectrical Repair & Growth Society 4:12, 1984.

(26) Khan, T. and Gaik, G.: Preliminary studies of spinal cord explants cultured on carbon filaments and stimulated with direct current, Bioelectrical Repair and Growth Soc. 4:23, 1984.

(27) Bawin, S.M., Kaczmarek, L.K. and Adey, W.R.: Effects of modulated VHF fields on the central nervous system, Ann. N.Y. Acad. Sci. 247:74-81, 1975.

(28) Bawin, S.M., Adey, W.R. and Sabbot, I.M.: Ionic factors in release of $^{45}Ca^{2+}$ from chicken cerebral tissue by electromagnetic fields, Proc. Natl. Acad. Sci. USA 75:6314-6318, 1978.

(29) Wachtel, H.: Firing-pattern changes and transmembrane currents produced by extremely low frequency fields in pacemaker neurons, in Biological Effects of Extremely Low Frequency Electromagnetic Fields, R. Phillips, M. Gullis, W. Kaune and D. Mahlum, eds., U.S. Dept. of Energy, 132-146, 1979.

(30) Wheeler, D.D.: Effects of polarizing currents and repetitive stimulation on the uptake of amino acids by peripheral nerve, J. Neurosci. Res. 4:123-131, 1979.

(31) Bawin, S.M., Sheppard, A.R. and Mahoney, M.D.: Effect of electric field orientation and strength on neuronal excitability in the rat hippocampal slice, 6th Ann. Meeting Bioelectromagnetics Soc. 5, 1984.

(32) Gundersen, R. and Greenbaum, B.: The effects of ELF magnetic fields on synaptic function, Presented at the 7th Ann. Meeting of the Bioelectromagnetics Soc. 53, 1985.

(33) Schwartz, M., Korenstein, R., Mizarchi, Y., Newman, D. and Kimhi, Y.: Neuritic outgrowth from regenerating goldfish retina can be modulated by external factors, in Birth Defects 19:451-456, 1983.

(34) Sisken, B.F., McLeod, B. and Pilla, A.A.: PEMF and neuronal regeneration: Effect of field geometry, in Electrochemistry, Membranes, Cells and the Electrochemical Modulation of Cell and Tissue Function, A.A. Pilla and A.L. Boynton, eds., Springer Verlag, New York, 1984.

(35) Yoshioka, A., Cao, H. and Albert, E.: Effects of 27 MHz electric fields on rat DRG neurite growth in vitro, 6th Ann. Meeting Bioelectromagnetics Soc., 83, 1984.

(36) Albert, E.N., Yoshioka, A., Cao, H. and Imam, F.: Modulation of fetal rat DRG neurite growth by pulsing electromagnetic fields (PEMFs), 7th Ann. Meeting Bioelectromagnetics Soc., 11, 1985.

(37) Culp, R., Osterman, A., Brighton, C., Bora, F., Kreider, B. and Pleasure, D.: In vitro response of the Schwann cell to various capacitively coupled electrical fields, Bioelectrical Repair and Growth Soc. 4:24, 1984.

(38) DuBois-Reymond, E.: Vorlaufiger abrifs einer untersuchung uber den sogenannten froschstrom und uber die electromotorischen fische, Ann. Phys. U. Chem. 58:1, 1843.

(39) Grenell, R.G. and Burr, H.S.: Surface potentials and peripheral nerve injury: A clinical test, Yale J. Biol. Med. 18:517-525, 1946.

(40) Hoffman, H.: Acceleration and retardation of the process of axon-sprouting in partially denervated muscles, Aust. J. Biol. Med. Sci. 30:541-566, 1952.

(41) Bodemer, C.W.: Evocation of regrowth phenomena in anuran limbs by electrical stimulation of the nerve supply, Anat. Rec. 148:441, 1964.

(42) Sisken, B.F., Estes, S., Barr, E. and Kryscio, R.: Morphological and biochemical effects of NGF or applied electric fields in the presence or absence of non-neuronal cells of chick dorsal root ganglia, Soc. Neurosci. 9:5, 1983.

(43) Borgens, R., Vanable, J. and Jaffe, L.: Bioelectricity and regeneration I, Initiation of frog limb regeneration by minute currents, J. Exp. Zool. 200:403, 1977.

(44) Romero-Sierra, C., Halter, S. and Tanner, J.A.: Effect of an electromagnetic field on the sciatic nerve of the rat, in The Nervous System and Electric Currents 2, N.L. Wulfsohn and A. Sances, Jr., eds., Plenum Press, New York, 81-84, 1971.

(45) Yorde, D.E., Siegesmund, K.A., Sances, A. and Larson, S.J.: The effects of DC current on synaptic endings, in The Nervous System and Electric Currents 2, N.L. Wulfson and A. Sances, eds., Plenum Press, New York, 69-73, 1971.

(46) Wilson, D.H. and Jagadeesh, P.: Experimental regeneration in peripheral nerves and the spinal cord in laboratory animals exposed to a pulsed electromagnetic field, Paraplegia 14:12-20, 1976.

(47) Sebille, A. and Bondoux-Jahan, M.: Effects of electric stimulation and previous nerve injury on motor function recovery in rats, Brain Res. 193:562-565, 1980.

(48) Nix, W.A.: The effect of low-frequency electrical stimulation on the denervated extensor digitorum longus muscle of the rabbit, Acta Neurol. Scand. 66:521-528, 1982.

(49) Winter, W., Schnutt, R., Sisken, B.F. and Smith, S.D.: Effects of Low Levels of Direct Current on Peripheral Nerve Regeneration, 27th Ann. Orthop. Res. Soc., Las Vegas, 1981.

(50) Maehlan, J. and Nja, A.: The effects of electrical stimulation on sprouting after partial denervation of guinea-pig sympathetic ganglion cells, J. Physiol. 322:151-166, 1982.

(51) Ito, H. and Bassett, C.A.L.: Effect of weak, pulsing electromagnetic fields on neural regeneration in rat, Clin. Orthoped. 181:283-290, 1983.

(52) Raji, A.R.M. and Bowden, R.E.M.: Effects of high-peak pulsed electromagnetic field on the degeneration and regeneration of the common peroneal nerve in rats, J. Bone Joint Surg. 65:478-492, 1983.

(53) Nix, W.A. and Hope, H.C.: Electrical stimulation of regenerating nerve and its effect on motor recovery, Brain Res. 272:21-25, 1983.

(54) Parker, B., Bryant, C., Apesos, J., Sisken, B.F. and Nickell, T.: The effects of pulsed electromagnetic fields (PEMF) on rat sciatic nerve regeneration, Bioelectrical Repair and Growth Soc. 3:19, 1983.

(55) Singer, P. and Mehler, S.: Role of electrical activity in axotomy-induced increased glucose use, Exp. Neurol. 80:601-612, 1983.

(56) Orgel, M.G., O'Brien, W.J. and Murray, H.M.: Pulsing electromagnetic field therapy in nerve regeneration: An experimental study in the cat, Plast. Reconstr. Surg. 73:173-183, 1984.

(57) Pomeranz, B., Mullen, M. and Markus, H.: Effect of applied electrical fields on sprouting of intact saphenous nerve in adult rat, Brain Res. 303:331-336, 1984.

(58) Roman, G.C., Rowley, B.A., Strahlendorf, H.K. and Coates, P.W.: Electric low-intensity direct currents stimulate sciatic nerve regeneration, Neurosci. Lett., 1985.

(59) Roederer, E., Goldberg, N.H. and Cohen, M.J.: Modification of retrograde degeneration in transected spinal axons of the lamprey by applied DC current, J. Neurosci. 3:153-160, 1983.

(60) Gruner, J.A. and Young, W.: An Objective Test of Descending Motor Function to Study the Efficacy of Pulsed Electromagnetic Fields and Ganglioside Treatments of Spinal Injury, Presented at the American Assoc. of Neurological Surgeons, Atlanta, GA, 1985.

(61) Young, W.: Pulsed electromagnetic fields alter calcium in spinal cord injury, Central Nerv. Sys. Trauma 1:100, 1984.

(62) Young, W., DeCrescito, V., Flamm, E.S. and Ransohoff, J.: Pulsed electromagnetic fields (PEMF) and methylprednisolone (MP) alter calcium and functional recovery in spinal cord injury, Presented at the 1984 Ann. Meeting of American Assoc. Neurological Surgeons, San Francisco, 127-128, 1985.

(63) Pilla, A.A.: Mechanisms of electrochemical phenomenon in tissue repair and growth, Bioelectrochem. Bioenergetics 1:227, 1974.

(64) Jaffe, L.F., Robinson, K.R. and Nuccitelli, R.: Local cation entry and self-electrophoresis as an intracellular localization mechanism, in Electrically Mediated Growth Mechanisms in Living Systems 238, A.R. Liboff and R.A. Rinaldi, eds., New York Acad. Sci., 372-389, 1974.

(65) Sisken, B.F. and Lafferty, J.F.: Comparison of the effects of direct current and nerve growth factor on the regeneration of neurites _in vitro_, Bioelectrochem. Bioenergetics 5:459, 1978.

(66) Bray, D., Thomas, C. and Shaw, G.: Growth cone formation in cultures of sensory neurons, Proc. Nat. Acad. Sci. 75:5226-5229, 1978.

(67) Skaper, S.D. and Varon, S.: Nerve growth factor influences potassium movements in chick embryo dorsal root ganglionic cells, Exp. Cell Res. 131:353-361, 1981.

(68) Fukada, J. and Kameyama, M.: Enhancement of Ca spikes in nerve cells of adult mammals during neurite growth in tissue culture, Nature 279:546-548, 1979.

(69) Hammerschlag, R., Bakhit, C. and Chu, A.: Role of calcium in the initiation of fast axonal transport of protein: Effects of divalent cations, J. Neurobiol. 8:439-451, 1977.

(70) Anglister, L., Farber, I.C., Sharar, A. and Grinvald, A.: Localization of voltage-sensitive calcium channels along developing neurites: Their possible role in regulating neurite elongation, Dev. Biol. 94:351-365, 1982.

(71) Freeman, J.A., Mayes, B., Snipes, G.J. and Wikswo, J.P.: Real-time measurements of minute neuronal currents with a circularly vibrating microprobe, Soc. Neurosci. 8:302, 1982.

(72) Cooper, M.S. and Schliwa, M.: Electrical and ionic controls of tissue cell locomotion in DC electric fields, J. Neurosci. 13:223-244, 1985.

(73) Fraser, S.E. and Poo, M.: Development, maintenance, and modulation of patterned membrane topography: Models based on the acetylcholine receptor, Curr. Top. Dev. Biol. 17:77-100, 1982.

(74) Bawin, S.M. and Adey, W.R.: Sensitivity of calcium binding in cerebral tissue to weak environmental electric fields oscillating at low frequency, Proc. Natl. Acad. Sci. 73:1999-2003, 1976.

(75) Blackman, C.F., Benane, S.G., Kinney, L.S., Joines, W.T. and House, D.E.: Effects of ELF fields on calcium ion efflux from brain tissue in vitro, Radiat. Res. 92:510-520, 1982.

(76) Lin-Lui, S. and Adey, W.R.: Migration of cell surface concanavalin A receptors in pulsed electric fields, Biophys. J. 45:1211-1218, 1984.

16
Limb Regeneration

STEPHEN D. SMITH
Department of Anatomy, University of Kentucky
Albert B. Chandler Medical Center, Lexington, Kentucky

INTRODUCTION

I will begin with a brief summary of the events that occur in normal limb development, and then progress to a similarly brief outline of the processes involved in normal limb regeneration as it occurs in some vertebrates, with an attempt to demonstrate the similarity between ontogeny and epimorphic regeneration.

After comparing regenerating and non-regenerating forms, I will describe the efforts to alter regeneration by electric and magnetic field stimulation.

NORMAL LIMB DEVELOPMENT

The events leading to the formation of a normal limb during development have been studied and reviewed extensively (1-3). At the outset, the limb in the embryo is induced to begin its development by the somites in the future limb-forming region (4). The appearance of the limb is manifested by the formation of a small bump which contains a core of mesenchyme (loose embryonic connective tissue) covered by epithelium. The epithelium rapidly becomes specialized at the tip of the bump into a thickened ridge, usually called the apical ectodermal ridge (5). An epithelial-mesenchymal interaction ensues between the underlying tissue and this ridge such that the ridge directs differentiation and outgrowth of the limb, while the underlying tissue produces a substance or substances that preserve the existence of the ridge. Much literature documents the details of this interaction (6-9). In addition, events occur which are not entirely understood, but which result in polarization of the outgrowing limb along the proximodistal, dorsoventral, and craniocaudal axes. The polarization of the craniocaudal axis has been ascribed to the activities of a caudal organizing region near the base of the limb (9). In any event, the limb continues to extend and the tip flattens into a paddle-shaped structure. Internally, some of the mesenchyme condenses into models for the future bones.

Next, the paddle becomes altered by the appearance of bumps at its free margin. These are the finger buds, which gradually elongate to form the fingers. Internally, the mesenchymal bone models become chondrified, and the tissue between the fingers degenerates to separate them into discrete digits. Development is finished when the cartilage-bone models ossify, and the details of the fingers are laid down.

EPIMORPHIC REGENERATION AND ITS CONTROL

In some vertebrates, notably the Salamandridae group of Urodeles, adult animals can grow new limbs if the old ones are cut off. All of these adult regenerators are aquatic forms. Limb regeneration also occurs in many other amphibian larvae, including Anurans (frogs and toads). Again, these larvae are aquatic.

Land-dwelling forms generally do not regenerate as adults, though their offspring, if young enough, often do show some evidence of regenerative ability. This is true especially for animals whose young are born in a very immature state, and Mizell has provided a prime example with the opossum (10). Baby rats can exhibit regeneration (11), as can sub-pubertal humans who regenerate fingertips if the wound is not surgically closed (12).

In the case of the salamanders which regenerate complete limbs, the process mimics normal development quite closely. First, the epithelium closes over the wounded surface. Then it develops into a thickened structure similar to the apical ectodermal ridge. Simultaneously, a collection of loose, undifferentiated mesenchyme-like cells appears beneath the epithelium. Taken together, this combined structure is called a blastema. The origin of the mesenchyme-like cells has been a subject of much controversy for years. Some investigators have felt that they come from preexisting stem cells, while others have insisted that the limb tissues dedifferentiate into cells which can then redifferentiate into the new tissues as required. Hay (13) appears to have shown that muscle cells can dedifferentiate, and Oberpriller (14) has demonstrated that cells from a blastema derived from a regenerating intestine can be incorporated into the regrowing limb. The former observation seems to settle the question of whether old cells can dedifferentiate, and the latter lays to rest any requirement for specificity. The regrowing limb simply makes use of whatever cells it has at hand to make whatever tissue needs to be made.

The regenerating limb then follows a course that mimics normal limb development: a paddle-shaped structure forms, which then develops finger buds and elongated fingers. Internally, bone models form, chondrify, and then ossify as in normal development. A good review of the process, including a discussion of the evidence for various control systems and cell origins, is given by Rose (15). Generally speaking, the adherents of various control schemes have strongly propounded their own ideas.

What concerns us is the system responsible for initiating regeneration and polarizing the regenerate. Obviously, the animal must somehow learn that the limb has been cut off, so that it also knows that regeneration is necessary. At the conscious level this seems obvious. However, such awareness at the tissue level is far from obvious, and we are actually unaware of the nature of the signal. It has long been known that an injury produces a wound potential, generated by leakage from damaged cells (16). This potential makes the wounded surface strongly negative with respect to the surrounding tissue. Since the body can respond to electrical signals, the wound potential is a possible signal. Another possibility is suggested by experiments which show that the size of each part of the body appears to be regulated by products that it produces which inhibit the further development of more of the same tissue. This concept of specific inhibition suggests that when a part of the body is removed, these inhibitory products are also removed, and the remaining part of the body is then free to develop more of the tissue until the original volume is replaced (17). This concept works well for relatively homogeneous tissue such as the liver, but it is less obvious how it would operate in the case of a complex array of tissues like a limb. As we shall see, there is strong evidence that some kind of specific inhibition does exist for the limb.

Since regeneration does occur, we may assume that some sort of initiation signal does exist, whether electrical, chemical, or both. We now must examine the phenomenon of organization of the regenerate. It is intuitively evident that the regenerate must be polarized. Otherwise, the outgrowth would be a disorganized mass of tissues, bearing no visible relationship to a limb. What is the nature of the polarization? How does the information travel about so that every cell knows where it is in relation to every other cell? I will first discuss the presence of polarization, and then its transmittal to all tissues.

Proximodistal polarization of the regenerating limb has long been known. By cutting through reversed bits of limb grafted onto stumps, Kurz (18) discovered that the regenerates were always replacements for the tissues distal to the level of the cut surface. In other words, the graft always regenerated a limb, never a new salamander. Myriad experiments since then confirmed the existence of this Rule of Distal Transformation. There must be information present that allows the limb to distinguish distal from proximal (15).

The rules for polarization of the dorsoventral and craniocaudal axes have been worked out more recently. An elegant exposition has been presented by Vernon French and Peter and Susan Bryant (19). In their model, there exists a system of polar coordinates. If one examines the cut end of a limb, the polarizing information exists at a series of discrete points around the circumference of the limb, and also as a series of points along radii extending outward from the center. Thus, any cell can be located by a distance from the center of the stump and a kind of compass bearing.

The validity of this model has been extensively tested by grafting experiments, and it seems to explain most of the results seen so far, even the rather bizarre duplication seen when bits of limbs are recombined in various positional combinations (20). The result of this experimentation has been the establishment of rules which show that each cell exists in a defined three-dimensional network which allows it to know exactly where it is and what it is to become. The nature of the information is unknown.

Having established the existence of polarization of the regenerate, we come to the question of how the information is relayed. Some of the answers to questions about control modalities seem to have been answered for some time. Harrison (21,22) established that information travels only along axes of polarity. Turning a piece of tissue so that the axes no longer align releases that bit of tissue from control by the surrounding tissues. In 1946, Monroy (23) demonstrated that lines of control exist in the tissue, and that if grafts are made in such a way that the lines cross at right angles, regeneration is totally blocked. These lines are evidently labile, since they can be destroyed by X-rays (24). Indeed, X-rays can block the transmission of the morphogenetic information necessary to build regenerates, yet leave the cells alive and able to remain apparently healthy for years. This peculiar phenomenon has been known since 1937 (25). The regenerative ability can be restored by grafting either internal tissues (26) or epidermis (27) back into irradiated limbs, suggesting that the control influences can come from either source, although skin grafts resulted in greater restoration of regeneration. These results, together with some very interesting experiments on coelenterates, led Rose in 1962 to postulate that there is a system of information transfer within the limb stump that he called a "tissue arc" (28). The experiments with Tubularia indicated that the morphogenetic influences could migrate for short distances through agar bridges, and that, most importantly, the transmission occurs in only one direction. Material from the distal hydranth will pass down the stem toward the base, but not in the other direction.

The stage is now set for a discussion of the nature of the polarity that causes this unidirectional flow. Matthews (29) discovered in 1903 that coelenterates are electrically polarized. This was reconfirmed for coelenterates and a number of other forms by Lund in the 1920's (30,31). At about the same time, Child discovered that gradients of oxygen consumption existed in the same organisms, as well as others (32). This suggests that the electrical gradients seen in organisms are the result of redox potential gradients. This polarity, and the demonstration by Lund (33) that growth and differentiation could be controlled by applied direct electrical currents, led Burr and Northrop (34) to postulate in 1935 an "electrodynamic theory of life," modifications and extensions of which have been the foundations of the experiments discussed later in this Chapter, and in other chapters of this volume.

Taking these findings together with the discovery of the phenomenon of electrophoresis, the notion naturally arose that morphogenetically important molecules might be charged, and might be moved about in the organism by its own naturally-occurring fields. It simply remained to be shown that this could occur. I will discuss the use of electric fields to control regeneration later. Here I will only cite some experiments to show that morphogenetically important molecules can be moved in electric fields.

In 1963, I showed that molecules responsible for specific inhibition of anal collars in marine worms could be moved (35). Rose did it for Tubularia in 1966 (36). For higher vertebrates, I demonstrated in 1965 that the specific inhibitor of newt lens regeneration could be isolated electrophoretically (37), and Shaw demonstrated mobility of similar information-bearing molecules in frog embryos in 1966 (38). This concept is now well established, and the only remaining question is whether naturally-occurring fields are sufficiently large to do the same within the organism. Rose's experiments certainly seem to suggest that they are (39).

In summary, we can assume that a limb which is removed is recognized as being lost, that in some animals a signal stimulates the replacement of the lost tissues, and that control of the regeneration is carried out along straight lines between cells, using some form of regionally specific information which appears to move in an electric field, and which somehow gives each cell its specific location and task.

DIFFERENCES BETWEEN REGENERATORS AND NON-REGENERATORS

HISTOLOGICAL DIFFERENCES

The earliest event in regeneration, the rapid covering of the wound by epithelium, occurs in both kinds of animals.

Thereafter, the responses diverge. In regenerators, the epithelium thickens into a wound cap, and tongues of cells penetrate the underlying tissue. The epithelial cells may either act as phagocytes to remove debris from injured deep tissue cells, or contribute to the blastema (40). In addition, regenerating nerves send many neurites into the interstices between the overlying epithelial cells, thereby establishing a neural-epithelial link. As we shall soon see, this may be vital to regeneration. Dedifferentiation and morphogenesis ensue, ultimately restoring the lost parts.

In non-regenerators, after the wound is covered by its epithelium, connective tissue cells begin to position themselves beneath the epithelium. These cells form a layer several cells thick beneath the epithelium, and begin to lay down large amounts of collagen, thus forming a kind of scar. The connective tissue seems to block the invasion of the underlying tissues by epithelial cells, and also to prevent the contact between epithelial cells and regenerating nerves. If one examines the junction between this kind of healed skin and the underlying tissues, one sees that the regrowing nerve fibers have been diverted, growing parallel to the newly formed collagen fibers rather than into the epidermis.

It is now unquestioned that prolonged contact between the epidermis and deep tissues is an absolute requirement for limb regeneration. Tornier (41) demonstrated it in 1906, as did Schaxel in 1921 (42). If an amputation wound is covered by a seal of whole skin, regeneration is blocked. Godlewski (43) showed that if the wound is even sewn shut, regeneration is blocked in salamanders. More recently, there have been indications (12) that even in humans, the common clinical practice of closing amputation wounds may inhibit a substantial potential for regeneration. When whole skin covers a wound, the dermis seems to prevent the interaction between epidermis and underlying tissues which is a necessary factor for regrowth of a lost part.

The role of nerves is apparently equally important. Singer (44) and Taban (45) first showed the intimate relationships which were established between nerves and epidermis in the early stages of regeneration. Hay (46) subsequently demonstrated that junctions between nerve processes and epidermal cells exhibit many of the characteristics of synapses. The requirement for this contact in regeneration was first shown by Thornton (47). The full role of nerves in regeneration has been studied in detail by Singer and his students (44). The evidence shows that regeneration cannot occur unless there is a sufficient proportion of nervous tissue in relation to the other tissues of the extremity. If nerves are severed or blocked from regenerating, morphogenesis does not ensue. Conversely, regeneration can be stimulated in a variety of forms, including frogs (48), lizards (49,50), chicken embryos (51), and even mammals (10,52) when the nerve supply to the limb is augmented. Regeneration is not always perfect in these instances, but the attempt is at least made if the ratio is raised to a threshold level.

So, the roles of epidermal contact with underlying tissue and of nerve contact with epidermal cells seem clear. What about animals that regenerate as larvae but not as adults? The shift between regenerating and non-regenerating stages occurs at maturation, when the skin becomes toughened and more resistant to drying, and when the response to an injury becomes rapid closure with a connective-tissue scar.

Why the preponderance of aquatic forms or stages among regenerators? The question is easy to answer with respect to physical characteristics, but more difficult in terms of selective advantage. Aquatic forms tend to have thinner, less heavily keratinized skin which does not heal rapidly by formation of a scar. Thus, there is sufficient time for the interactions between epidermis, underlying tissue, and nerve to ensue following amputation.

It would seem just as important for a land-dweller to have fingers as for an aquatic salamander -- perhaps even more so. The answer may lie not in the advantage of regeneration, but in the disadvantage of slow wound healing. A small animal whose metabolism is relatively slow in cool water, and whose skin allows for a considerable degree of direct oxygen exchange, may be relatively insensitive to blood loss. Also, for an aquatic form, dehydration is a negligible problem. If an animal can get along without much blood, and is in no danger of drying out, it can afford to repair a loss at a leisurely pace. A land-dweller has no such luxury. Skin thin enough to exchange gas would also lose water vapor. Keeping exchange through the skin is necessarily rather limited. If metabolism is very rapid, or the animal is very large, oxygen exchange potential via the blood becomes critical, and neither blood nor fluid loss can be borne with impunity. For both reasons, rapid wound healing is a decided advantage in a land-dweller. Land forms have apparently given up the luxury of slow healing and regeneration in favor of fast fluid-tight healing.

ELECTRICAL DIFFERENCES

The first systematic study of electrical events during regeneration (at least in amphibians) was undertaken by Alberto Monroy in 1941. He examined surface potentials in salamanders (*Triton*) during tail regeneration (16). Similar studies have been done on limb regeneration at intervals ever since. The seminal work was done by Becker in 1961 (53). For regenerators, the initial measurements of the tip of the limb stump revealed that the wound surface became strongly negatively charged with respect to the center of the back. This initial charge is attributable to the injury potential and the products of damaged cells. As soon as the epithelium covered the wound, the potential reversed, and became strongly positive. In regenerators, Becker found that the potential then reversed again after about 4 or 5 days, becoming negative at about 10 to 25 mV. During the course of morphogenesis, the potential gradually rose toward a very slightly positive baseline value. The picture was different for non-regenerators. They underwent the initial negative spike, and the reversal to positive, but never reversed again to the negative potential of the regenerators. These findings have been the source of much controversy. Becker concluded that potentials arose in nerves, and demonstrated that they were abolished by anesthesia and denervation. Later investigators like Rose (54), Borgens, et al. (55), and Lassalle (56) did not find the denervation effect. Borgens and Lassalle attribute the electrical events to transepithelial potentials, produced by pumped ionic flow across the skin, and modifiable by changing the external ionic composition or by altering ionic permeability by various means. Lassalle has shown the same basic pattern of potential differences between regenerators and non-regenerators as reported by Becker. Lassalle attributes all measured potentials to epithelial status, but it is possible that the mechanisms postulated by both him and Becker are operative. It is clearly not reasonable to attribute all limb potentials in amphibians to transepithelial ionic passage. On the other hand, perhaps it is reasonable to conclude from Lassalle's data that surface potentials do not normally control regeneration in limbs. The story becomes more confusing when the data of Borgens et al. (57) are considered, since they found that artificial enhancement of the surface potentials can enhance limb regeneration in *Xenopus*, and that lowering it can inhibit regeneration in newts (58).

It is certainly possible to measure significant potential differences between the center of the back and the tips of the limbs in humans (unpublished data). Since humans are not usually wet, it would seem difficult to attribute these differences to ion-pumped transepithelial potentials. There may be many sources for such potentials, including potentials along nerves, streaming potentials derived from blood flow in arteries, and muscular activity.

The development by Jaffe and Nucitelli (59) of a miniature vibrating-probe electrometer has materially aided the study of electrical events associated with regeneration. Using this device, Borgens et al. (54,60) found that the charge distribution differed between adult newts, which regenerate, and adult frogs, which do not regenerate. They found that currents at the center of the limb in frogs were much lower than in newts, and attributed the difference to the large lymphatic spaces beneath frog skin which can shunt currents (61). These findings reflect the histological observations that only regenerators retain tight contact between epidermis and underlying tissues. They found no relationship between the stump currents and innervation.

Perhaps, as Lassalle suggests, the positive potentials observed in the limbs of amphibians are principally the result of transepithelial ionic passage, and can be altered slightly without affecting limb regeneration. The later-appearing strong negative potentials seen in regenerators may reflect nerve penetration of the epidermis. The final potential occurring at the surface of a limb during regeneration is perhaps a sum of the transepithelial potential which can vary considerably, and the internally derived potential, which may not have the same phase relationship as the transepithelial surface potential. Such a synthesis might explain some of the seemingly contradictory results obtained by the various laboratories investigating the phenomena.

In any case, it seems clear that the electrical behavior of regenerating and non-regenerating forms is quite different, whatever the source of the potentials. It now remains to be seen whether those differences can be exploited to advantage.

DIRECT CURRENT STIMULATION

The initial experiments aimed at electrical stimulation of limb regeneration were carried out by Bodemer (62) in 1964. He used an indirect method, that of stimulating the brachial plexus of frogs after forelimb amputations, and evoked a regrowth of tissue and elongation of the stump. I think that perhaps my efforts came next (63). Becker had suggested in 1961 that imitation of the regenerating electrical pattern in a non-regenerator might produce regeneration. The problem lay in devising a convenient means to do so. At the time, implants seemed too large for the small animals being used, and electronic miniaturization had not progressed to its present remarkable state. My thought was rather simple: why not let the frog be its own battery? Accordingly, I decided to implant a simple bimetallic couple consisting of silver and platinum wires, insulated except at their tips. I chose the metals based on their relative biocompatibility and electrochemical activities, hoping for some current flow in the salty extracellular fluids without major toxic effects of the metals. As it happened the choice was serendipitous, and the pair of metals generated the proper current levels to induce regeneration. I reported the results in 1967, demonstrating a rather remarkable degree of regeneration, including all of the tissues normally present in a new limb, plus some attempts at organization into wrist-like structures. Becker tried the same method in postnatal rats (64), adding a resistor between the silver and platinum to control the amount of current generated, and reported remarkable results. His paper generated a good deal of controversy, based on the brief period required to produce the results. To my knowledge, however, nobody ever has exactly repeated his experiments, so the criticisms were and are entirely speculative, and can be ignored until an accurate reproduction of the experiments is undertaken.

In these early experiments, the electrodes were simply inserted into the ends of the bones (into the marrow cavity) distally, as a means of fixing them in place. In my experiments maximal regeneration ensued if the silver end was placed distally. The results of these experiments were disappointing in that the organization of the regenerates did not closely resemble that of a normal limb. As information developed regarding control of polarity in the normally forming limb, I decided that the problem might lie in how the implants were being made. I reasoned that better results might be had if the stimulating electrode were placed at the wound surface in the spot where the apical ectodermal ridge normally forms. To do this, I had to design a new type of implant, using a battery in the center of the frog's back, with a long cathodal lead wire coming down the limb which could then be fixed in place at the end. This work was reported in 1974 (65). It confirmed that placement of the electrode in any position other than the dorsal postaxial quadrant of the wound surface resulted in poor regeneration. If, however, the placement were appropriate, regeneration of remarkable completeness could be obtained. In one instance, a perfect hand formed. Movable digits regenerated in 23% of the cases.

Borgens et al. (66) were concerned that electrode products might be causing the results rather than the electric fields, so they modified the experiment by using conducting wicks instead of wires to deliver the current to the end of the limb. They also succeeded in inducing limb regeneration, but did not obtain particularly good organization of the regenerated tissues. They used a current of 0.2 μA compared to 0.103 μA used in my study. Whether this difference or whether the electrode products their method was designed to eliminate may have contributed in my experiments are unresolved issues.

Sisken et al. (52,67) reported the results of a long series of experiments in rats in 1979 and 1984, showing tissue regeneration and outgrowth. The response was augmented by the simultaneous injection of nerve growth factor. Libbin et al. (11) demonstrated electrical stimulation of rat limb regeneration. In 1981, I reported that implantation of electrodes into the dorsal postaxial position in subadult rats initiated regeneration at a current of about 0.1 $\mu A/mm^2$ (68). The regenerates exhibited joints, muscles, cartilage, bone, and some suggestion of organization into wrist-like elements, but no complete regenerates were obtained. Sisken et al. (52) reported that implanted bimetallic strips enhanced the response of regenerating rat limbs to implanted fetal nerve tissue.

Thus, the evidence shows that regeneration can be initiated in normally non-regenerating forms. The regeneration generally proceeds only partially, and includes the formation of all of the new tissues required to form a limb. The final organization of the new tissues is incomplete, except in extremely rare instances, and then only in Anurans.

Why do DC fields stimulate regeneration? I think the answer may lie in the considerations raised in the previous Section. Which of the requirements for regeneration are satisfied by DC stimulation? Firstly, there seems to be no doubt that a DC field of appropriate strength provides an adequate signal to initiate regeneration. Experience in my laboratory suggests that the thresholds for such stimulation are narrow. Currents of less than 10 nA/mm^2 are ineffective, as are currents of 1 $\mu A/mm^2$. I obtain the best results at a current of about 100 nA/mm^2. The experience of Borgens et al. (66) suggests that even 200 nA may be too much in *Rana pipens*, the same test subject that I used. If the current is too high, the regenerate begins to consist more and more of just disorganized connective tissue. Above 1 $\mu A/mm^2$, only scar is produced.

Secondly, a DC field seems able to act as a polarity-inducing stimulus, or at least one which can act in concert with naturally occurring polarity. Our experiences with varying electrode placement supports this idea, although more work is needed.

Thirdly, some sort of information transfer surely does occur in these regenerates. The degree of organization, though generally imperfect, suggests that the regenerates are not simply bits of disorganized flesh. In Anurans, the degree of perfection of the regenerates varies widely, but can approach perfection quite closely on rare occasions. Something must be organizing the tissues. One possibility is that the field acts directly to organize information transfer, which in turn organizes the epithelial-mesenchymal interactions necessary for the initiation and completion of regeneration. This option implies that all of the information-bearing molecules are present normally, and that regeneration in non-regenerators only requires the right stimulus. A second possibility is that the DC field simply induces the rapid regeneration of neurites, which then penetrate the wound epithelium before scar formation can prevent it. It is the completion of a neural-epithelial circuit which produces the regeneration. It is possible to encourage nerve growth with fields and direct neurites toward a cathode. Thus, it is conceivable that the electrodes simply speed up neurite formation and direct their growth into a particular point in the epithelium which then establishes the necessary polarity and interactions to engender regeneration. So far as I can determine, there has not been a systematic study of this possibility. I have made preliminary observations that indicate that neurites do indeed penetrate the epidermis in frogs, but not in the large numbers characteristic of salamanders. Treatments that delay scar formation (69-71) also induce partial regeneration, so this option may indeed be correct. The nerve augmentation studies also support this hypothesis. A third possibility is that the fields somehow stimulate the cells to begin formation of the information necessary to regeneration, and also to proliferate to provide the tissue raw materials. Our own experience (72) suggests that pulsed fields can enhance thymidine uptake by cells in the skin's basal layers, and other chapters in this volume fully explore other stimulative cellular effects. There is a wealth of evidence to suggest that electric fields can engender transcription and cell proliferation, so this option is also a viable one. Perhaps tissue organization occurs as a result of a combination of these mechanisms.

It is necesssary to consider why regeneration is so seldom carried to completion, or is so often poorly organized, especially in mammals. Given the obvious fact that regeneration can be stimulated to begin, and that even mammals have the obvious capacity to at least partially organize a regenerate, why don't we obtain perfect limbs in non-regenerators? There are two possibilities to consider. Regeneration may stop because scar formation cannot be delayed indefinitely. Scar prevents information transfer, thereby halting regeneration. There have apparently been no experiments combining suppression of scar formation (with cortisone, collagenase, or some such agent) with electrical stimulation. Another possibility arises from the work of Stocum with salamanders (73). In some combinations derived by grafting bits of limbs together, the regenerates appear to progressively lose more of their organizational maps as extension of the regenerate occurs. This finally results in a spike-like outgrowth, rather than a normal limb. Under most grafting conditions, gaps of missing positional information are intercalated to restore a perfect map. However, if like areas are approximated, the limb may not recognize that something is missing, and fail to regrow a perfect regenerate. One of the characteristic forms of regenerates seen in Anuran and mammalian experiments is a spike-shaped outgrowth. This suggests that Anurans and mammals do not have complete positional information maps, and that regeneration therefore cannot be expected to be complete. This is a discouraging possibility because it suggests that there is no hope of inducing perfect limb regenerates in non-regenerators. But even if the pattern is incomplete, hope looms on the horizon in the form of retinoids. These analogues of vitamin A are apparently capable of inducing pattern duplications in the transverse axes of limbs (74). Their action is apparently stage-specific (75), but it may be possible to use retinoic acid in conjunction with DC fields to produce complete regenerates.

The final chapters of DC field stimulation of limb regeneration have by no means been written. The area remains largely unexplored, and should provide the basis for much experimentation.

PULSED MAGNETIC FIELD STIMULATION

One of the objections often raised in experiments with DC fields is the matter of implantation of electrodes. Aside from the irritation and potential for infections introduced by implantation procedures, one must consider the myriad electrochemical reactions that occur at an electrode surface. Study of the metabolic consequences of these reactions is in its infancy.

It would therefore be highly advantageous to be able to produce the same effects as DC fields using a non-invasive method. The simplest way to achieve this effect is to subject the animal to pulsed magnetic fields (PMF) which induce currents in the tissue that are directly proportional to the conductivity of the medium, the strength and shape of the pulses, and the geometry of the system. Use of pulsed magnetic fields introduces a whole new set of complexities, but they are at least a little better understood than the complexities at an electrode.

There is apparently only one published study of the effects of PMF on limb regeneration (76). After mid-forearm amputations, newts were placed in small individual aquaria and subjected to three types of PMF. One was a single pulse repeating at 72 Hz, (waveform 1), and the other two were complex pulse trains operating at 15 Hz. The individual pulses in the pulse-trains had widths of 22 and 6 μsec (Waveform 3 and 4 respectively) (76). The trio of pulses produced different energy inputs to the regenerating system (Figure 1). Figure 2 illustrates a distillation of the results. The single pulse treatment induced a premature differentiation of fingers in the regenerate. The mass of tissue was about the same as that of the control blastema, which had only reached the late cone or early paddle stage. Generally, these animals regenerated only three fingers, instead of four. Waveform 4 stopped regeneration completely. The wound was covered by a thin layer of epithelial cells, indicating that cell migration could proceed, but none of the usual events in regeneration occurred. Even the protruding bone left behind as the stump tissues contracted was not eliminated, as is usually the case within a few days. Regeneration was markedly enhanced by waveform 3. The regenerates at 21 days looked like controls at about 45-50 days. The regenerates were large, with four long finger outgrowths. In effect, the whole process had simply been accelerated, with everything in proper phase, so that a normal regenerate appeared in a very short time.

In sum, it appears that PMF can have a variety of effects including altered phase relationships that disturb the normal course of events, inhibition, and acceleration. The inhibition and acceleration may be due to simple modifications in the overall rate-controlling processes of regeneration. The

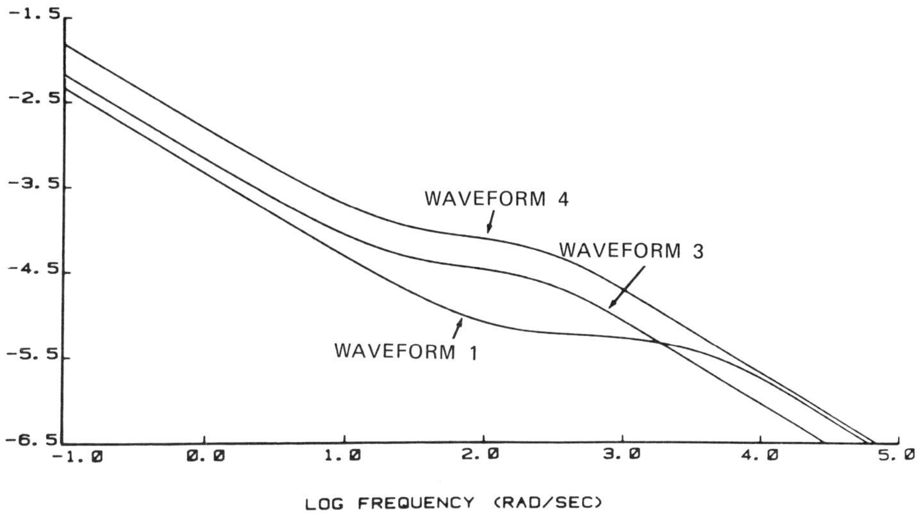

FIGURE 1. Real axis (LaPlace) frequency spectra for Waveforms 1,3, and 4. Waveform 1 is a simple asymmetrical pulse repeating at 72 Hz. Waveform 3 is a complex asymmetrical pulse burst with a 22 microsecond component which repeated at 15 Hz. Waveform 4 is similar to waveform 3, except that the 22 microsecond component is replaced by a 6 microsecond component. Note that waveforms 1 and 4 match at high frequency, 1 and 3 are close, but not matching at low frequency, and 3 and 4 are parallel at all frequencies, but not matching in amplitude. The Y axis is a dimensionless logarithmic relative amplitude scale, and is thus not labeled with specific units. (Reproduced from (76), with permission.)

premature differentiation is more difficult to explain. Since cell proliferation was stopped, and differentiation started well before their normally appointed times, something selective must have occurred. Since the field was active from the time of amputation, it suggests that there are multiple processes going on simultaneously in normal regeneration, and that the timing of the steps must be precise to obtain a normal regenerate. If one of the steps is accelerated, while others are not, the phase relationships become disturbed, and so does regeneration.

The fields induced by PMF are inherently less specifically polarized than DC fields. The induced currents are circular,

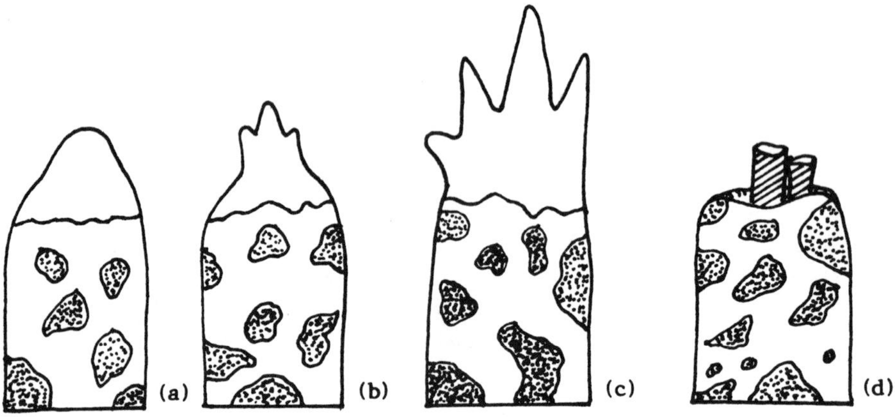

FIGURE 2. Drawings of typical results of PMF-treated regenerating newt limbs: (a) Control at 21 days. The regenerate is a late-cone early-paddle stage blastema; (b) This is a typical limb treated with waveform 1. The amount of tissue present is not much greater than controls, but differentiation is far advanced, indicating separation of the processes of tissue accumulation and differentiation; (c) A limb treated with waveform 3. Regeneration is far advanced, and apparently normal in morphology; (d) Waveform 4 has stopped regeneration completely. Ordinarily, the tips of the radius and ulna left exposed as the tissue retracts would have been eliminated by osteoclasts in the deep tissues. Here even that process has been interrupted.

and lie in a plane perpendicular to the magnetic field lines. However, an animal placed between two coils is not subjected to a field defined only at one particular point. Currents are simultaneously generated everywhere within the animal's body. It would seem, therefore, that unless one could find a pulse that produced a differential effect only upon the processes essential to regeneration, PMF would be of use only as a general technique to enhance and speed up the process. The results with the newt limbs suggest that both may be possible. Some selectivity is obviously possible, as well as generalized effects.

PMF have been used experimentally for at least ten years to stimulate bone healing in non-unions (77) and clinically-approved devices are now available. Dal Monte's chapter in this book reviews the evidence. I think it is safe to state that they are principally of use in stimulating generalized tissue regeneration such as bone healing. I cannot see how PMF can be used to satisfy the polarity requirements of limb regeneration. PMF apparently act primarily through alteration of small ion phenomena at the cell membrane, and unless the cells are regionally programmed to respond selectively, a territorially precise response to PMF seems unlikely. I am not suggesting that PMF are of no use in stimulating limb regeneration. They clearly are. If one of the problems in obtaining limb regeneration is providing enough tissue to make a regenerate, and doing so quickly, or perhaps in inhibiting scar tissue formation, PMFs may play an important role. One could certainly envision a regenerative scheme in which DC fields played a polarizing/initiating role, while simultaneously applied PMF stimulated blastema formation by cell proliferation, scar suppression, and ultimately a trigger to differentiation.

Such a scheme implies a great deal more information about the controls of regeneration, and about the details of action of PMFs than we now possess. We need accurate models which will predict the effects of an applied field, plus a detailed encyclopedia of the influences controlling normally-occurring regeneration. For example, to say that there is a map of positional information which controls regeneration is a step forward, but it does not allow us to recreate the map. We must know what the map is made of, and what determines how it is constructed. Without a knowledge of what the map constituents are, it is hard to devise precise means of stimulating whatever "map synthetases" there may be in an animal.

None of this is to suggest that experimentation with PMF and DC fields should stop until all of the details of regeneration are worked out in full. Nothing could be less useful. The history of science, biology in particular, is replete with case histories of the cart going before the horse. Most often, something is first found empirically, to be followed by experimentation and elucidation of the mechanisms. Limb regeneration need be no different. I only suggest that a confident prediction of success would be greatly enhanced by more perfect knowledge. We certainly know enough now about effective thresholds and useful waveforms to be actively engaged in experimentation. Sooner or later, I am quite certain that someone will, by rational progression of thought or by serendipity, discover the key to stimulated limb regeneration which progresses to complete perfection in most cases. I have attempted to suggest some avenues of approach which seem reasonable. I hope that someone will explore some of them -- the potential rewards both for science and the amputee are very considerable.

REFERENCES

(1) Warwick, R. and Williams, P.L.: Gray's Anatomy, 35th British Edition, W.B. Saunders Co., Philadelphia, PA, 122, 123, 1973.

(2) Sadler, T.W.: Langman's Medical Embryology, 5th Edition, Williams and Wilkins, Baltimore, MD, 139-141, 1985.

(3) Moore, K.L.: The Developing Human, 3rd Edition, W.B. Saunders, Co., Philadelphia, PA, 366, 367, 1982.

(4) Agnish, N.D. and Kochar, D.M.: The role of somites in the growth and early development of mouse limb buds, Dev. Biol. 56:174-183, 1977.

(5) Saunders, J.W. Jr., Cairns, J.M. and Gasseling, M.: The role of the apical ridge of ectoderm in the differentiation of the morphological structure and inductive specificity of limb parts in the chick, J. Morphol. 101:57-87, 1957.

(6) Saunders, J.W. Jr.: Control of growth patterns in limb development, in Normal and Abnormal Embryological Development, C.H. Frantz, ed., National Research Council, Washington, DC, 16-26, 1967.

(7) Zwilling, E.: Interaction between limb bud ectoderm and mesoderm in the chick embryo, J. Exp. Zool. 132:157-172, 1956.

(8) Amprino, R. and Camosso, M.: Analisi sperimentale dello sviluppo dell'ala nell'embrione di pollo, Roux Arch. Entwicklungsmech. d. Org. 150:509-541, 1958.

(9) Saunders, J.W. Jr. and Gasseling, M.T.: Ectodermal-mesodermal interactions in the origin of limb symmetry, in Epithelial-Mesenchymal Interactions in Development, R. Fleshmajer, ed., Williams and Wilkins, Baltimore, MD, 78-97, 1968.

(10) Mizell, M.: Limb regeneration: Induction in the newborn opossum, Science 161:283-285, 1968.

(11) Libbin, R.M., Person, S., Papierman, S., Shah, D., Nevid, D. and Grob, H.: Partial regeneration of the above-elbow amputated rat forelimb: II. Electrical and mechanical facilitation, J. Morphol. 159:439-452, 1979.

(12) Illingworth, C.M.: Trapped fingers and amputated fingertips in children, J. Pediatr. Surg. 9:853, 1974.

(13) Hay, E.D.: Electronmicroscopic observations of muscle dedifferentiation in regenerating Amblystoma limbs, Dev. Biol. 1:555-585, 1959.

(14) Oberpriller, J.: A radioautographic analysis of the potency of blastemal cells in the adult newt, Diemictylus viridescens, Growth 31:251-296, 1967.

(15) Rose, S.M.: Regeneration: Key to Understanding Normal and Abnormal Growth and Development, Appleton-Century-Crofts, New York, NY, 1970.

(16) Monroy, A.: Richerche sulle correnti elettriche dalla superficie del corpo di tritone adulti normale e durante la rigenerazione degli arti e della coda, Pubbl. Stn. Zool. Napoli 18:265-281, 1941.

(17) Rose, S.M.: A hierarchy of self-limiting reactions as the basis of cellular differentiation and growth control, Am. Nat. 86:337-354, 1952.

(18) Kurz, O.: Versuche über polaritätsumkehr am tritonenbein, Roux' Arch. Entwicklungsmech. d. Org. 50:186-191, 1922.

(19) French, V., Bryant, P.J. and Bryant, S.V.: Pattern regulation in epimorphic fields, Science 193:969-981, 1976.

(20) Stocum, D.: The urodele limb regeneration blastema: Determination and organization of the morphogenetic field, Differentiation 27:13-28, 1984.

(21) Harrison, R.G.: Experiments in the development of the forelimb of Amblystoma, a self-differentiating, equipotential system, J. Exp. Zool. 25:413-462, 1918.

(22) Harrison, R.G.: On relations of symmetry in transplanted limbs, J. Exp. Zool. 32:1-136, 1921.

(23) Monroy, A.: Ricerche sulla rigenerazione degle Anfibi urodeli. Nota III. Osservazioni su rigenerati formati si su doppie superfici de sezione e considerazione sui processi determinative della regenerazione, Arch. Zool. Ital. 31:151-172, 1946.

(24) Oberpriller, J.C.: The action of X-irradiation on the regeneration field of the adult newt, <u>Diemictylus viridescens</u>, J. Exp. Zool. 168:403-421, 1968.

(25) Brunst, V.V. and Scheremetjewa, E.A.: Ist es möglich, die Regenerations-fahigkeit einer <u>Triton</u> Extremität zu vernichten, ohne ihre Lebens-fähigkeit zu schädigen?, Byul. Eksperim. Biol. i Med. 3:397-399, 1937.

(26) Umanski, E.: A study of regeneration of the axolotl limb upon substitution of the inner tissues by dorsal musculature, Byul. Exsperim. Biol. i Med. 6:383-396, 1938.

(27) Umanski, E.: The regeneration potencies of axolotl skin studied by means of exclusion of the regeneration capacity of tissues through exposure to x-rays, Byul. Exsperim. Biol. i Med. 6:141-145, 1938.

(28) Rose, S.M.: Tissue arc control of regeneration in the amphibian limb, in Regeneration, D. Rudnick, ed., Ronald Press, New York, NY, 153-176, 1962.

(29) Matthews, A.P.: Electrical polarity in the hydroids, Am. J. Physiol. 8:294-299, 1903.

(30) Lund, E.J.: Experimental control of organic polarity by the electric current, I. Effects of electric current on regenerating internodes of Obelia commisuralis, J. Exp. Zool. 34:471-493, 1921.

(31) Lund, E.J.: Electrical polarity in the douglas fir, Publ. Puget Sound Biol. Sta. 7:1-28, 1929.

(32) Child, C.M.: Physiological dominance and physiological isolation in development and reconstitution, Roux' Arch. Entwicklungsmech. d. Org. 117:21-66, 1929.

(33) Lund, E.J.: Experimental control of organic polarity by the electric current, V. The nature of the control of organic polarity by the electric current, J. Exp. Zool, 41:155-190, 1923.

(34) Burr, H.S. and Northrop, F.C.S.: The electro-dynamic theory of life, Q. Rev. Biol. 10:322-333, 1935.

(35) Smith, S.D.: Specific inhibition of regeneration in Clymenella torquata, Biol. Bull. 125:542-555, 1963.

(36) Rose, S.M.: Polarized inhibitory control of regional differentiation during regeneration in Tubularia. II. Separation of active materials by electrophoresis, Growth 30:429-447, 1966.

(37) Smith, S.D.: The effects of electrophoretically separated lens proteins on lens regeneration in Diemictylus viridescens, J. Exp. Zool. 159:149-165, 1965.

(38) Shaw, C.: Ectodermal or presumptive skin disintegration in the frog embryo as a response to specific tissue factor(s), Dev. Biol. 13:1-14, 1966.

(39) Rose, S.M.: Polarized inhibitory effects during regenera-

tion in Tubularia, J. Morphol. 100:187-205, 1957.

(40) Rose, S.M.: Epidermal dedifferentiation during blastema formation in regenerating limbs of Triturus viridescens, J. Exp. Zool. 108:337-361, 1948.

(41) Tornier, G.: Kampf der Gewebe im Regenerat bei Begünstigung der Haut-regeneration, Roux' Arch. Entwicklungsmech. d. Org. 22:348-369, 1906.

(42) Schaxel, J.: Auffassunger und Erscheinigung der Regeneration, Arb. a.d. Geb. der Exp. Biol. 1:1-99, 1921.

(43) Godlewski, E.: Untersuchungen über Auslösung und Hemmung der Regeneration beim Axolotl, Roux' Arch. Entwicklungsmech. d. Org. 114:108-143, 1928.

(44) Singer, M.: The influence of nerves on regeneration, in Regeneration in Vertebrates, C.S. Thornton, ed., Univ. of Chicago Press, Chicago, IL, 59-80, 1959.

(45) Taban, C.: Les fibres nerveuses et l'epithelium dans l'edification des regenerats de pattes (in situ ou induites) chex le Triton, Arch. Sci. (Geneva) 2:553, 1949.

(46) Hay, E.D.: The fine structure of nerves in the epidermis of regenerating salamander limbs, Exp. Cell Res. 19:299-317, 1960.

(47) Thornton, C.S.: Epidermal modifications in regenerating and non-regenerating Amblystoma larvae, J. Exp. Zool. 131:373-394, 1956.

(48) Singer, M.: Induction of regeneration in the forelimb of the postmetamorphic frog by augmentation of the nerve supply, J. Exp. Zool. 126:419-471, 1954.

(49) Singer, M.: Induction of regeneration of body parts in the lizard, Anolis, Proc. Soc. Exp. Biol. Med. 107:106-108, 1961.

(50) Simpson, S.B. Jr.: Induction of limb regeneration in the lizard, Lygosoma laterale, by augmentation of the nerve supply, Proc. Soc. Exp. Biol. Med. 107:108-111, 1961.

(51) Fowler, I. and Sisken, B.F.: Effect of augmentation of nerve supply upon limb regeneration in the chick embryo, J. Exp. Zool. 221:49, 1982.

(52) Sisken, B.F., Fowler, I. and Romm, S.: Response of fetal rat limbs to fetal nerve tissue implants and direct current, J. Orthop. Res. 2:177-189, 1984.

(53) Becker, R.O.: The bioelectric factors in amphibian limb regeneration, J. Bone Joint Surg. 43A:643-656, 1961.

(54) Rose, S.M. and Rose, F.C.: Electrical studies on normally regenerating, x-rayed and on denervated limb stumps of Triturus, Growth 38:363-380, 1974.

(55) Borgens, R.B., Vanable, J.W. Jr. and Jaffe, L.F.: Bioelectricity and regeneration: Large currents leave the stumps of regenerating newt limbs, Proc. Nat. Acad. Sci. USA 74:4528-4532, 1977.

(56) Lassalle, B.: Are surface potentials necessary for amphibian limb regeneration?, Dev. Bio. 75:460-466, 1980.

(57) Borgens, R.B., Vanable, J.W. Jr. and Jaffe, L.F.: Small artificial currents enhance Xenopus limb regeneration, J. Exp. Zool. 207:217-225, 1979.

(58) Borgens, R.B., Vanable, J.W. Jr. and Jaffe, L.F.: Reduction of sodium dependent stump currents disturbs urodele limb regeneration, J. Exp. Zool. 209:377-386, 1979.

(59) Jaffe, L.F. and Nuccitelli, R.: An ultrasensitive vibrating probe for measuring steady extracellular currents, J. Cell Biol. 63:614-628, 1974.

(60) Borgens, R.B., Vanable, J.W. Jr. and Jaffe, L.F.: Bioelectricity and regeneration, BioScience 29:468-473, 1979.

(61) Borgens, R.B., Vanable, J.W. and Jaffe, L.F.: Role of subdermal current shunts in the failure of frogs to regenerate, J. Exp. Zool. 209:49-55, 1979.

(62) Bodemer, C.W.: Evocation of regrowth phenomena in anuran limbs by electrical stimulation of the nerve supply, Anat. Rec. 148:441-458, 1964.

(63) Smith, S.D.: Induction of partial limb regeneration in Rana pipiens by galvanic stimulation, Anat. Rec. 158:89-98, 1967.

(64) Becker, R.O.: Stimulation of partial limb regeneration in rats, Nature 235:109-111, 1972.

(65) Smith, S.D.: Effects of electrode placement on stimulation of adult frog limb regeneration, Ann. N.Y. Acad. Sci. 238:500-507, 1974.

(66) Borgens, R.B., Vanable, J.W. Jr. and Jaffe, L.F.: Bioelectricity and regeneration. I. Initiation of frog limb regeneration by minute currents, J. Exp. Zool. 200:403-416, 1977.

(67) Sisken, B.F., Smith, S.D. and Lafferty, J.F.: A comparison of the effects of direct current, nerve growth factor, and direct current plus nerve growth factor on amputated rat limbs, in Electrical Properties of Bone and Cartilage, C. Brighton, J. Black and S. Pollack, eds., Grune and Stratton, New York, NY, 267-287, 1979.

(68) Smith, S.D.: The role of electrode position in the electrical induction of limb regeneration in subadult rats, Bioelectrochem. Bioenergetics 8:661-670, 1981.

(69) Schotte, O.E.: Systemic factors in initiation of regenerative processes in limbs of larval and adult amphibians, in Regeneration, D. Rudnick, ed., Ronald Press, New York, NY, 161-192, 1961.

(70) Rose, S.M.: Methods of initiating limb regeneration in adult Anura, J. Exp. Zool. 95:149-170, 1944.

(71) Rose, S.M.: The effect of NaCl in stimulating regeneration of limbs of frogs, J. Morphol. 77:119-139, 1945.

(72) Smith, S.D. and Feola, J.F.: Effects of pulsed magnetic fields on thymidine uptake in mice, J. Electrochem. Soc. 129:132C, 1982.

(73) Stocum, D.L.: Distal transformation in regenerating double anterior axolotl limbs, J. Embryol. Exp. Morphol. 65[suppl]:3-18, 1981.

(74) Thoms, S.D. and Stocum, D.L.: Retinoic acid-induced pattern duplication in regenerating urodele limbs, Dev. Biol. 103:319-328, 1984.

(75) Niazi, I.A., Pescitelli, M.J. and Stocum, D.L.: Stage-dependent effects of retinoic acid on regenerating urodele limbs, Roux' Arch. Dev. Biol. 194:355-363, 1985.

(76) Smith, S.D. and Pilla, A.A.: Modulation of newt limb regeneration by electromagnetically induced low level pulsating current, in Mechanisms of Growth Control, R.O. Becker, ed., Charles C. Thomas, Springfield, IL, 137-152, 1981.

(77) Bassett, C.A.L., Pawluk, R.J. and Pilla, A.A.: Acceleration of fracture repair by electromagnetic fields. A surgically non-invasive method, Ann. N.Y. Acad. Sci. 28:242-262, 1974.

17

Behavioral Measures of Electromagnetic Field Effects

ROCHELLE MEDICI

Analytical Services
2220 El Molino Place
San Marino, California

WHY BEHAVIOR?

There are at least three major reasons for studying the behavioral effects of electromagnetic (EM) fields: First and most importantly, behavioral studies are a sensitive and reliable measure of the functioning of the central nervous system (CNS). Secondly, behavioral studies can validate or invalidate theories about CNS mechanisms of interaction. (Conversely, behavioral studies may lead to the formulation of such theories.) Thirdly, behavioral studies can provide us with sound ideas for practical applications of EM research. Such studies can define both the promise and the limitations of EM fields as a technique for changing or modifying human behavior.

Concern with practical issues of changing human behavior with EM energy has caught the attention of the press and the public in the last two decades. The "zapping" of the United States' Embassy in Moscow in the 1960s led to speculation that very weak pulsed EM fields might lead to dramatic thought disorders or physical illness in the Embassy staff. This focus has detracted from the sure, steady, and unglamorous results of behavioral studies, which have advanced our understanding of how EM fields affect living organisms, and our understanding of the role of the CNS in mediating these effects.

BEHAVIOR AS A SENSITIVE AND RELIABLE MEASURE OF CENTRAL NERVOUS SYSTEM EFFECTS

The major significance of behavioral studies is that they offer a sensitive measure of CNS function. It is possible to vary power, frequency, modulation and duration of exposure of EM fields, and to determine precise, dose-related changes in the behavior of an experimental animal, and thus assess effects on CNS function.

It is a common error to believe that behavioral measures of brain function are merely phenomenological, and that they are less precise, reliable, and real than physiological measures. This is not so. For example, electroencephalograms (EEGs) are in many ways a very gross measure of CNS function. Spectral analyses of EEGs from electrodes implanted deep in the brain of animals exposed to EM fields reveal great limitations in this technique. Questions of sampling adequacy and statistical inference are not easy to resolve. Medical imaging offers a limited picture of brain structure and pathology but, on occasion, behavior can tell us more. For example, the impact of low doses of drugs or EM fields may not permanently alter the brain. Behavioral studies can reveal important, but transient, changes. Studies of biochemical changes in vitro leave us with the problem of extrapolating the significance of any observed changes to the living animal.

A lack of knowledge about the science of behavior appears frequently in the nonionizing radiation area. Frequently, researchers regard all behavioral measures as equivalent. But actually there is no more equivalence between an activity measure of behavior and an inter-response time schedule of reinforcement than there is between measuring temperature by putting a hand on a child's forehead and measuring temperature with a precise gauge.

The appropriate model for behavioral studies of EM fields is that of research on low doses of drugs. Techniques for precisely assaying effects of minimal drug doses were developed in the United States in the 1930s. Essentially, these techniques involve training animals to perform simple, measurable behavioral tasks (schedule-controlled behavior, as originally described by B.F. Skinner). Following the training, one can then measure the degree to which defined doses of drugs -- or EM fields -- perturb that behavior. The perturbation of behavior, the measure of neural function, can be precisely measured.

With such techniques, it is possible to precisely study the effect of gradually increasing the power of the EM field, to measure effects of increasing the duration of exposures, to measure changes due to the introduction of modulation in a constant frequency field, and so on. The work of Thomas and his associates (1-7) is an excellent example of this approach. However, as we shall see in the review of the literature that follows, many other studies of EM fields were done with simple-minded and insensitive behavioral techniques.

BEHAVIOR AS A NECESSARY MEASURE OF THE VALIDITY OF THEORIES OF CENTRAL NERVOUS SYSTEM MECHANISMS OF INTERACTION

Behavioral studies provide the ultimate validation of hypotheses about CNS mechanisms. For example, Adey has proposed that since weak EM fields can affect calcium efflux (in *in vitro* chicken brains), present theories about CNS function must be radically changed. He argues that one should adopt a "nonequilibrium viewpoint" in which cells "whisper" to each other so that very low amounts of energy can affect vast arrays of neurons (8). Such theories have great intellectual appeal and fascination, but unless experiments are done to link them directly to changes in behavior, they remain an empty and trivial exercise. Earlier psychobiologists recognized the problem of theories that remain "locked in the mind." The most grandiose theory of neural functioning is useless if it is not anchored in relevant experimentation.

There are other requirements of a robust theory. It is essential that the theory do more than simply restate, in theoretical neurobiological terms, the empirical observations which led to its formulation (i.e., weak EM fields perturb behavior). Good theories should provide hypotheses that generate practical experiments. These new hypotheses must be testable. In the case of the nervous system and EM fields, this means testing at the behavioral level. The recent work of Thomas, et al. (5) is an example of behavioral validation of a neural theory.

Not only are behavioral studies necessary to the validation of neural theories, but they are also a rich source of ideas for the generation of such theories.

PRACTICAL APPLICATIONS OF BEHAVIORAL STUDIES OF ELECTROMAGNETIC FIELD EFFECTS

It has been said that man's egocentric concerns at the time

of Copernicus made it difficult for him to accept the notion that the sun rather than the earth is the center of the world. Our egocentric concerns may be drawn to rather sensational notions about the capability of EM fields to change human behavior. These concerns detract from the significance of solid, modest, laboratory studies. It is the implications of these laboratory studies for basic science which is ultimately significant.

It is conceivable that "...specific frequencies might affect different kinds of learning. One frequency might aid in memory retention; another might enhance performance in music, art, or mathematics since these are all very specific talents which involve different brain structures and different kinds of electrical activities" (9). However, the research to determine whether these applications are feasible, practical, or desirable has not been done. We need intensive, appropriate, research at the laboratory level. Epidemiological studies of EM effects reflect some of our concerns with hazard; such studies are often criticized for their lack of precision. However, if more laboratory studies were done, the guiding hypotheses for epidemiological studies could be stated far more definitively and better studies could be designed at the outset.

We need to understand and measure the dose-effect parameters of EM radiation including the effects of power, frequency, duration of exposure, and modulation. Then we may foresee both the limitations and benefits of this research for human application.

HISTORICAL PERSPECTIVES

Despite the fact that pioneering studies in nonionizing radiation were done in the 1960s (10), little relevant research occurred subsequently. Why?

Each new behavioral study that appeared was scrutinized, analyzed, criticized, and challenged by scientists who had been active in earlier EM research and in hazard standard-setting. As one of these scientists put it when he testified at the New York State Public Service Commission Hearings in 1976, "...whenever the claim was that no effect was observed, ... I was not further interested in digging into the material... I didn't see any motivation to dig very deeply into the statistics whenever the effect was reported null. I felt more motivated to dig into it if there was an effect reported... I think that adequately summarizes my approach to the evaluation..." (11).

Some scientists believed that it was simply impossible for low-level EM fields to affect behavior because the energy was small. Only a new awareness of neuroanatomy and the neural sciences eventually eroded this kind of objection.

There was also a problem from groups with vested interests. The military did not want to hear about possible hazards associated with radar installations. Microwave-oven manufacturers saw a threat to a new and booming business. Money for grants became very limited. At times it seemed that grant money was only available for investigators who were willing to do monolithic studies that used such insensitive biological measures that they were literally guaranteed to show that neither hazard -- nor effects of any kind -- occurred in the presence of weak EM fields. For example, Guy was awarded a multi-million dollar grant to do a long-term study at the University of Washington which used a variant of the open-field test as its only measure of the nervous system and behavior (12). This simplistic and insensitive behavioral measure would be guaranteed to show no effect to almost any kind of weak environmental stimuli. The politics of funding for EM research are discussed elsewhere (13).

In my opinion, another factor that slowed EM behavioral research was a lack of understanding of the science of behavior. Perhaps this is due in part to the inter-disciplinary nature of EM research. Behavioral studies were undertaken by physicists, engineers, veterinarians, physicians, and only occasionally by psychologists or psychobiologists. Elegant, sensitive, schedule-controlled tests of behavior were developed in the United States in the 1930s and have been widely used in toxicology and pharmacology to assess the effects of low doses of drugs. Yet much EM research has focused on insensitive, simplistic tests of behavior (open-field tests, activity tests). Or, ironically, they have focused on replication of Soviet techniques from conditioning studies that date back to about 1910 (e.g., foot withdrawal to shock).

In the 1970s, many EM researchers tended to equate one behavioral test with another. If sensitive, schedule-controlled tests showed effects, it was argued that these were negated by the lack of effects from an experiment in which an insensitive measure such as activity was used. A major thrust of this paper is to demonstrate that behavioral results can only be evaluated in the context of the adequacy or reliability of the specific behavioral measure which is used.

PRESENT STATUS OF ELECTROMAGNETIC BEHAVIORAL STUDIES

There are three major factors that affect experimental outcome (14): (1) the behavioral measure used; (2) whether the field is modulated; (3) whether the field is primarily magnetic or electric. Other variables such as the carrier frequency, intensity, and duration of exposure may also affect the result, but the primary importance of the three listed factors has been convincingly established.

The primary lesson to be learned from earlier reviews of the literature (15-17) holds true today. It may be stated rather simply: Behavioral techniques may be considered on a continuum proceeding from almost no external stimulus control (for example, open field tests and activity measures) to techniques in which the animal is required to respond to demanding elements of the task itself (as in escape or avoidance tasks). At one end of the continuum the behavior is too variable to adequately reflect the effect of a weak nonionizing field. At the other end of the continuum the animal is too preoccupied with the demands of the task to attend to the effect of the imposed fields. In between these two extremes are a variety of relevant schedule-controlled techniques, especially those which are time-based, that are both reliable and sensitive (15-17).

Among schedules of reinforcement (reward), there are two major categories: ratio schedules, in which food pellets are delivered to an animal depending on the number of responses the animal emits; and interval schedules, in which the animal must delay his response for a certain number of seconds before the reinforcer is available. Ratio schedules reinforce rapid responding; interval schedules reinforce precise time-based responding. Ratio schedules are relatively impervious to weak environmental stimuli or drugs; interval schedules are sensitive to even very low doses of drugs. When a behavioral task involves the imposition of strong external stimuli on an animal, the animal is likely to pay attention to those task stimuli rather than to the effect of a weak environmental EM field. The principle has been elegantly demonstrated in a study of pigeons working on a fixed consecutive number schedule of reinforcement. When the animals were injected with methyl mercury their performance became variable and unstable. However, if a light cue was added to the task, indicating when the animal should shift to the reinforcement key, the animal's behavior became stable and appeared normal. If the light was removed, the animal's behavior immediately deteriorated again. Depending on the precise conditions of the task, the effects of methyl mercury were either easily discernible or completely hidden. The study has obvious implications for EM experiments, as well as for epidemiological studies. Workers intent upon performing a task may show no immediate evidence of the effect of an EM field, just as a soldier in battle may be wounded and not realize he was injured until the action ends.

Keeton (18) demonstrated a similar point in his study of the homing of pigeons. He strapped tiny magnets to their backs and observed their homing behavior. If it were a sunny day, the pigeons paid attention to the sun as a guide to their behavior and ignored the magnets. If it were cloudy, their flight was disoriented by the presence of the artificial magnetic field.

Even now, expensive studies are being funded by the United States government which use archaic, insensitive 19th century behavioral endpoints (foot withdrawal to shock, swimming endurance, open field tests, etc.). This is done in spite of the fact that a critical scrutiny of earlier studies presents compelling evidence that effects of weak EM fields could be reliably demonstrated if time-based schedules of reinforcement are used (15,16). This was especially shown in the work of Thomas and his associates (1-4).

CATEGORIES OF BEHAVIORAL MEASURES

Behavioral studies prior to 1980 have already been reviewed (15-17). In the following paragraphs literature from 1980-1985 will be considered (14). This follows the general format of the earlier reviews. Experiments will be grouped according to the type of behavioral measure used: activity, escape and avoidance, thermoregulatory, Soviet techniques, schedule controlled behavior, etc.

1. Activity Studies

The Argonne laboratories (19,20) reported that 60-Hz fields showed little effect on activity or circadian rhythms -- as one would expect. D'Andrea et al. (21,22) reported a failure to replicate a study from the Soviet Union in which exploratory behavior and catalepsy were the behavioral endpoints in a 50-Hz modulated 40-MHz field. Variable results were seen when locomotion was measured during long-term exposure to 915 MHz (5 mW/cm^2). As has been pointed out many times, none of these results are at all surprising since the behavioral measures are too variable to detect subtle effects.

2. Escape and Avoidance Studies

Escape and avoidance studies continue to show only marginal or variable impact of exposure to microwaves (23,24). Only intense fields (16 mW/g or greater) produce reliable escape responding (24,25). Again, these results are to be expected, since escape measures of behavior make heavy demands on the animals and are relatively insensitive to weak environmental stimuli. It is interesting to learn that escape and avoidance measures are adequate to detect effects of relatively high strength 60-Hz fields. Creim et al. (26) reported effects on the avoidance behavior of rats in high intensity 60-Hz fields (75 kV/m or greater). Hjereson et al. (27) reported corroborative results in rats exposed to 60-Hz fields of 90 kV/m or more. Swine appear to respond similarly to weaker (30 kV/m) fields when long durations of exposure are used (28). One study of weak magnetic fields reported no effect when passive-avoidance techniques were used and activity was measured (29). A novel study by Beel et al. (30) indicated that post-trial exposure to high levels of pulsed microwaves (18-22 mW/cm^2) can affect active or passive avoidance learning.

3. Thermoregulatory Studies

Thermoregulatory studies continue to be done, and continue

to be largely unenlightening. These studies demonstrate only that if microwave levels are high enough, the animals will be heated and can learn to emit behavioral responses to lower their environmental temperature (31-36). The authors' interpretations of these studies often go beyond the data and suggest that the demonstration of thermoregulatory behavior implies that there can be no direct effects of nonionizing radiation.

4. Teratogenic Studies

Teratogenic studies of behavior generally present weak evidence of the effects of high-strength fields (30 mW/cm^2) (37). Mitchell et al. (38) presented some evidence that endurance tests (swimming) may be affected by pre-natal exposure. Frey (39) found a variety of teratogenic effects following exposure to weak 60-Hz fields (3.5 kV/m). These studies suggest that 60-Hz fields may have more impact on teratogenic behavior than microwaves.

5. Other Measures of Behavior

Other measures of behavior that cannot be easily categorized in the present scheme have also been used. Frey and Wesler (40,41) presented evidence that conditioned emotional responses (CERs) and Sidman avoidance may be affected by low-intensity 60-Hz fields at 3.5 kV/m. Cooper et al. (42) indicated that conditioned suppression was affected by high level 60-Hz fields (50 kV/m) in pigeons. Clarke and Justesen (43) reported that a paradigm using Pavlovian operant conditioning was sensitive to the effects of 60-Hz and DC magnetic fields in chickens.

Microwave exposure affects certain dopamine and opiate related behaviors according to Frey and Wesler (44-47). Seaman et al. (48) indicated that some sexual behavior in rats was responsive to pulsed microwave fields.

6. Techniques Used in the Soviet Union

Techniques used in the Soviet Union for studying behavior continue to be used in the United States. Monahan (49) reported failure to replicate a Soviet study in which exploratory behavior and avoidance behavior were the endpoints. D'Andrea et al. (22) looked at open-field behavior, avoidance, and some unspecified operant behavior in a replication of Soviet studies of weak microwave effects (500 $microwatts/cm^2$, 2450 MHz).

Swim-to-exhaustion tests are reportedly enhanced by exposure to 15-kHz fields at 1 kV/m but not at 2 kV/m (50). Lobanova et al. (51) reported effects on conditioned reflexes of 10 mW/cm^2 microwaves, and dose-related changes as duration of exposure was increased.

7. Schedule-Controlled Studies

Schedule-controlled studies of behavior occupy a significant place among behavior experiments. My early work on both ELF and modulated VHF fields used time-based schedules of reinforcements with monkeys, neonatal chicks, and wild mallard ducklings (52-55). These experiments offered considerable promise for the sensitive and reliable detection of EM effects on behavior.

The work of Thomas and his associates (1-4) is remarkable for both its subtlety and reliability. It is distinguished by the use of time-based schedules of reinforcement, by the exploration of the interaction of EM fields with low doses of drugs, and by the use of pulsed, rather than CW, EM fields (56). He found that pulsed fields did not affect the dose-effect function of chlorpromazine or diazepam; nor did CW fields affect behavior modified by diazepam or chlordiazepoxide. Earlier results had shown that pulsed fields, however, did affect the response to chlordiazepoxide. These results imply (1) "...that drug class alone does not adequately predict outcome" and (2) that field parameters (CW or pulsed) are an important variable. In another study, dextroamphetamine and pulsed microwaves were shown to affect time-based schedules of reinforcements in rats (3). At 10 and 15 mW/cm^2, Thomas and Banvard (4) found that pulsed microwaves selectively lowered response rates on a time-based schedule of reinforcement, and that CW fields did not affect the response rates. Attempts by Lovely et al. and Lundstrom et al. (57-59) to supposedly replicate some of Thomas' work met with failure, probably because they were not replications due to differences in field exposure conditions (e.g., the use of different pulse repetition frequencies).

Gage (60) reported that CW microwaves did not affect d-amphetamine/microwave interactions when a complex mixed schedule of reinforcement was used. He did report however, that length of exposure to 10 mW/cm^2 (2.0 W/kg) differentially affected a similar complex schedule (61).

Lebovitz (62-64) found that fixed-ratio responding in rats was not affected by microwaves more than was responding during time-out. He showed that externally-cued ratio-responding was less sensitive to microwaves than non-cued bar-pressing. Both findings corroborate our general understanding of schedule-controlled behavior and nonionizing radiation. Using a fixed-ratio/time-out schedule, Lebovitz could not detect any differences between pulsed and CW microwaves. However, some variation of a time-based schedule may have revealed such a difference. Lebovitz and Orr (65) found that the time-out portion of the fixed-ratio/time-out schedule was affected by CW microwaves, pulsed microwaves (3.5 mW/g), and low doses of phenobarbitol.

Extremely low frequency (ELF) modulation (3 Hz and 16 Hz) of EM fields (450 MHz) differentially affected fixed-time, schedule-controlled behavior of wild mallard ducklings (66). This study draws attention, again, to the significance of low-frequency modulation, and time-based schedules of reinforcement. It also suggests that species differences may be important and that migratory animals may be especially sensitive to EM effects, since neonatal chicks (54) did not show such a response.

Studies of the effect of ELF fields on schedule-controlled behavior by Feldstone et al. (67,68) have not yielded clear results. The research design appears to be overly complex. Stern et al. (69,70) reported that schedule-controlled behavior can be used to determine that the threshold for detection of 60-Hz fields generally lies between 4 and 10 kV/m for rats.

Finally, the study by Thomas, Schrot, and Liboff (5) is indeed one of the most dramatic of the 1980s. One can see that the significant variables in this study could be readily predicted from the existing data base (time-based schedules, low frequencies). In this study, rats were exposed to a 60-Hz field of 4×10^{-5} T rms, together with a static magnetic field of 2.61×10^{-5} T (half the geomagnetic field), and showed change in time-based schedules of behavior. The study has special interest because the 60-Hz frequency was chosen on the basis of the cyclotron resonance frequency of lithium ions.

PULSED OR MODULATED FIELDS vs CW FIELDS

Here one is looking not only for an effect, but for a differential effect. If the behavioral measure is not appropriate, a difference between pulsed and CW will not be observed. At present, the weight of evidence suggests that such a differential effect exists.

In behavioral studies of nonionizing radiation that were begun in 1966, Gavalas (Medici) examined the effect of low-frequency fields (7-75 Hz, 1-56 V/m)(71). Inter-response time schedules of reinforcement were performed by highly trained monkeys. These studies demonstrated that the animals' behavior was significantly modified (in the direction of shorter inter-response time). It was further shown that the animals were especially sensitive to the frequencies that were in the EEG range of the animals, that is 7 Hz, as contrasted with 45 Hz and 75 Hz. EEGs of the animals were analyzed and a change in the spectrum of the EEG was found when the animals were exposed to the nonionizing radiation.

In view of these results, Kaczmarcek, a young English neurochemist at UCLA, was asked to consider other ways to measure brain response to the fields. He initiated experiments with calcium efflux measurement following exposure to ELF fields. The studies on calcium efflux provided good concordance for the behavioral studies. Modulation was of key importance (72). Using modulated, 450 MHz fields, evidence was found for changes in calcium efflux from the in vitro brain of neonatal chicks. At the same time, a program of behavioral studies was begun, but not finished, in which effects with time-based schedules of reinforcement were to be compared using increasingly complex schedules.

Thus, there was evidence that in EM behavioral studies (1) the type of behavioral schedule used was very important; (2) the modulation frequency (the ELF frequency) of the field was very important; and (3) this frequency was relevant to what was going on neurophysiologically and neurochemically in the animal.

Unfortunately, those behavioral studies were not actively pursued. One of the major criticisms of the calcium efflux work, as it now stands, is that the observed neurochemical changes have not been linked experimentally to the behavior of the animal. The biological significance of the biochemical changes in the intact animal has not been adequately established.

The ELF modulation frequencies of the 450 MHz fields were selected on the basis of what was known about the EEG pattern of the monkey. This is a prime point that was lost on later researchers.

The importance of modulation can also be seen in the early work of Kalmijn (73) on detection of prey by sharks, which use passive electrosensing. He noted that it was important to simulate the ELF field produced by the breathing of the prey. The electrodes that he placed in the bottom of the shark's tank were not simply emitting DC fields but also contained a 4-Hz component to mimic the breathing of the prey. Again, the frequency was important and was particular to the organism and its ongoing activity.

In the years that followed, investigators were mindful of the possibly greater effect of pulsed vs CW fields. However, except for Frey and his experiments with brain-stem evoked responses (74), and heart responses (75), they looked at pulsed frequencies associated with common high-frequency field devices. None of the other investigators doing behavioral studies pursued the more precise idea of linking the modulation of the field to the exact ongoing physiological rhythms of the animal at the time of exposure.

Modulated vs CW fields in a variety of behavioral experiments will now be compared. Again, we will categorize these experiments according to the behavioral technique that was used.

In the 1970s some investigators, including Hunt et al. (76) found evidence for changes in activity in rats following exposure to pulsed microwaves. Servantie et al. (77) reported effects at intensities as low as 0.7 mW/cm^2. Other investigators such as Gage (78) and Roberti et al. (79) reported no effect on activity for CW fields. However, it is impossible to draw firm conclusions about the effect of pulsed vs CW fields in these studies because activity, as a measure, is so variable that real differences between the two field parameters may have been lost.

Studies of schedule-controlled behavior done in the 1970s revealed a mix of results. However, the studies of Thomas and his co-workers (1-4) are most noteworthy for their use of sensitive and reliable time-based schedules of reinforcement. Almost all these early experiments were done with pulsed EM fields. The pulse rate, however, was generally high (500 pps). Effects were found at low intensities (1 mW/cm^2).

In contrast, deLorge used less sensitive behavior measures, CW fields, and found largely negative results (except at very high intensities) (80,81).

Thus, there is an interaction between the kind of behavioral schedule used and the effects of modulation or pulsing. For example, when ratio schedules of reinforcement are used, even high intensity, pulsed fields may not affect behavior as may be seen in the work of McAfee et al. (82). Ratio schedules are relatively impervious to environmental change (or to low doses of drugs).

In tests of behavior which are less sensitive, and where there is a strong external stimulus controlling the task (for example, escape studies, avoidance studies, and taste aversion studies), the effects of pulsed vs CW fields are not clear. It is likely that the demands of the behavioral task override the impact of whether the field is pulsed. For example, Frey et al. (83) reported effects of pulsed microwaves on escape behavior at quite low intensities (0.2 mW/cm^2). On the other hand, Grove et al. (84) found that when relatively high intensity CW fields were tested, escape learning occurred only if the escape was cued by a light. Hjereson and Phillips (85) reported failure of avoidance with pulsed fields while Monahan and Henton (86) reported some success with CW fields. In escape and avoidance studies, results are mixed and it is not clear that modulated fields are more potent than CW fields.

Studies by Frey and his associates (87,88) have indicated that aggressive behavior produced by tail pressure may be affected by low-intensity fields (less than 0.5 mW/cm^2). In their experiments the field was always pulsed. These studies are distinguished by the fact that the behavioral measures were selected to evaluate the possible role of the dopamine system.

Lebovitz and his associates (62-65) directly addressed the question of pulsed vs CW radiation. However, the behavioral schedule chosen was a fixed-ratio/time-out task. As indicated earlier, a ratio schedule produces behavior which is unlikely to be perturbed by weak environmental stimuli. As one would expect, Lebovitz found that the time-out part of his schedule was more likely to show the effects of radiation than the fixed-ratio component (1.3 GHz at 1.5 mW/g or 2.7 mW/g). He reported, however, that there was no differential effect of pulsed vs CW fields on the time-out component of the schedule. Two major considerations are (1) the time-out component may not have been sensitive enough to detect a difference between the two conditions, and (2) the pulse repetition rate was 600 pulses per second with a pulse width of 1 microsecond. Such a repetition rate is well above the range of any biologically relevant frequencies for the rat. EEG patterns during such behavior would tend to have dominant frequencies of less than 25 Hz. Nevertheless these studies represent an important effort to examine directly the pulsed vs CW issue. It is interesting to note that Lebovitz and Orr (65) repeated this study with d-amphetamine and phenobarbitol and found effects on the time-out component of this schedule when phenobarbitol was used. That is, phenobarbitol affected behavior similarly to nonionizing radiation.

During the 1980s, Thomas and his associates directly compared CW and pulsed microwaves in two studies (6,90). In both studies, pulsed fields differentially affected the schedule-controlled behavior. In the first case, a lowering of response rate below a pre-conditioned level was observed at 10-15 mW/cm^2 for pulsed fields, but not for CW fields. Similar results were reported in the second study: the rate of appropriately timed responses declined in the presence of pulsed, but not CW fields. In a related study, Schrot et al. (56) reported that chlordiazepoxide effects on fixed-interval behavior were not affected by CW fields, whereas earlier results had indicated that pulsed fields did affect this interaction. Diazepam effects were not modified by pulsed fields.

The results of these studies suggest that differences between pulsed and CW fields will be consistently observed when the behavior schedule is appropriate (time-based). Thomas' studies used pulsed rates of 500 pps with a 2 microsecond pulse duration; it is disappointing that these investigators did not extend their research to much lower pulse rates or modulations, where even more dramatic results might have been observed.

D'Andrea et al. (22) reported a study in which 50-Hz modulation was used in a 40 MHz field. However, behavioral measures, which were modeled after a Soviet study, were very crude. Effects on exploratory behavior and catalepsy were recorded. Not surprisingly, no effects were observed.

Seaman et al. (48) reported that low-frequency pulsing of microwave fields (10 pps, 3100 MHz) affected selected aspects of mating behavior in rats.

Other studies done in the 1980s have used pulsed fields or low-frequency fields, but the results appear to be variable and isolated. Feldstone et al. (67,68) did some experiments on the effect of 60 Hz on a variety of behavioral measures in the baboon. Beel et al. (30) have done a suggestive study on the effects of rather high levels of pulsed microwave following passive and active avoidance training in mice. Lai et al. (91) have reported that a variety of drug-induced effects are differentially influenced by pulsed microwaves.

In general, it may be concluded that modulation of microwave fields is more likely to affect behavior than CW fields, and this will appear if the behavioral test used is appropriate.

Studies using ELF fields have also shown effects on behavior. Frey (39) reported that rats exposed _in utero_ to 3.5 kV/m, 60-Hz fields showed effects in a variety of typical teratogenic measures such as acoustic startle, and surface righting. In a Sidman avoidance task, rats exposed to a similar field showed a diminished avoidance to the field which "...may indicate a decrease in timing capacity or reduced sensory response" (40).

Stern et al. (69,70) looked at behavioral detection of 60-Hz fields in rats and concluded that the threshold for direct detection lies between 4 and 10 kV/m. Earlier, Stern expressed concern that the detection behavior in his studies was confounded by other variables. More recently he indicated that it was not the case. Hjereson and his colleagues (27,28) found evidence that both rats and swine will avoid 60-Hz fields in a shuttlebox experiment. Studies from the Argonne Laboratory (19,20) with 60-Hz fields are flawed by the use of very simplistic behavioral measures. Cooper et al. (42) used a conditioned suppression paradigm to demonstrate detection of 60-Hz fields (50 kV/m). Clarke and Justesen (43) found increased variability in simple operant responding for food following Pavlovian conditioning in chickens that were exposed to DC or AC magnetic fields. The authors pointed out that the effects of the DC field might have been due to modulation of the field by the movement of the animals.

Finally, and most dramatically, Thomas et al. in 1984 exposed rats on a time-based schedule of reinforcement to weak 60-Hz magnetic fields and found marked changes in their behavior (5). Liboff, earlier, had calculated cyclotron resonances for lithium ions at 60 Hz. This experiment brings together sensitive behavioral measures (time-based) with biologically relevant frequencies. The hypotheses suggested by the research of the 1960s have finally been tested.

In summary, the weight of evidence suggests that the pulsing of nonionizing radiation and the use of ELF nonionizing radiation are extremely important factors in studies of behavior. Effects will not be found unless appropriate tests of behavior are used, such as time-based schedules of reinforcement. It is disappointing that so few studies have followed the lead of the research of the 1960s which indicated that even more dramatic effects would be seen if pulsing or modulation were done at very low frequencies. None of the noted studies, except the Thomas et al. study with 60-Hz magnetic fields (5), have considered ongoing physiological or biological rhythms in the animal.

No studies have yet looked at the impact of gradually increasing the depth of modulation as Czerski (personal communication) suggested in the early 1970s. More studies need to be done at low modulation frequencies and more studies need to be done to directly compare, as Lebovitz, Frey, and Thomas have done, the effects of pulsed and CW fields. It may be especially interesting to compare ELF fields and microwave fields that are modulated at ELF frequencies; e.g., 60-Hz ELF fields and microwave fields that are modulated at 60 Hz.

ELECTRIC vs MAGNETIC FIELDS

The Thomas et al. study (5) brings us to a consideration of what must now be considered a third major variable of significance for the study of the effects of EM fields on behavior. It seems clear that magnetic fields may have evolutionary and biological significance, at least for some animals. In those cases, one may expect that magnetic fields will show more influence on behavior than will electric fields. Direct comparisons of electric and magnetic fields have not yet been made. The dramatic experiments of Delgado have been described (92), and the interested reader is referred to his article on magnetic fields, brain, and behavior.

CONCLUSIONS

This review of behavioral studies indicates that there is clear, solid evidence that (1) time-based schedules of reinforcement repeatedly reveal effects of nonionizing radiation even when power levels are very low; (2) pulsed fields have more impact than CW fields; and (3) magnetic fields are particularly influential in some, and perhaps all, species.

Many very interesting studies remain to be done. Studies need to be done with complex modulation of the EM fields. Studies need to be done to explore CNS mediators of the behavioral effects that are observed. Conversely, behavioral studies need to be done to validate the efficacy of CNS theories about mediators. Frequency-specific studies that are appropriate to a given species and a given kind of behavior need to be done. Long-term studies need to be done to determine if cumulative effects exist.

An exciting array of studies can be pursued with the sophisticated behavioral techniques that are available to us. Simplistic and inappropriate behavioral studies did little to enlighten the research of the past and offer no hope for the future.

REFERENCES

(1) Thomas, J.R, Yeandle, S.S. and Burch, L.S.: Modification of internal discriminative stimulus control of behavior by low levels of pulse and microwave radiation, in Biological Effects of Electromagnetic Waves, C.C. Johnson and M.L. Shore, eds., HEW(FDA) 77-8010, USNC-URSI, 201-214, 1975.

(2) Thomas, J.R., Finch, E., Fulk, D.W. and Burch, L.S.: Effects of low level radiation on behavioral baselines, in Biologic Effects of Nonionizing Radiation, P. Tyler, ed., Ann. N.Y. Acad. Sci. 247:425-432, 1975.

(3) Thomas, J.R. and Maitland, G.: Combined effects on behavior of low level microwave radiation and dextroamphetamine, International Symposium on the Biological Effects of Electromagnetic Waves, Arlie, VA, USNC-URSI, 121, 1977.

(4) Thomas, J.R. and Banvard, R.A.: Changes in temporal aspects of behavior by low levels of pulsed microwaves, National Radio Science Meeting, Bioelectromagnetics Symposium, Seattle, WA, 452, 1979.

(5) Thomas, J.R., Schrot, J. and Liboff, A.R.: Weak low frequency magnetic fields alter operant behavior in rats, Abstracts of Papers presented at the 7th Ann. Meeting Bioelectromagnetics Society, 1985.

(6) Schrot, J., Thomas, J.R. and Banvard, R.A.: Effects of 2.8 GHz microwave radiation in combination with tranquilizing drugs on fixed interval performance of rats, Bioelectromagnetics 1:203, 1980.

(7) Thomas, J.R. and Banvard, R.A.: Comparison of continuous wave and pulsed microwave exposures on conditioned temporal behavior in rats, Bioelectromagnetics 1:203, 1980.

(8) Adey, W.R.: Review of "The Body Electric" and "The Microwave Debate," N.Y. Acad. Sci. 52-58, 1986.

(9) Pines, M.: The Brain Changers, Harcourt Brace Jovanovich, Inc., New York, 1973.

(10) Frey, A.H.: Biological function as influenced by low power modulated RF energy, IEEE Trans. Microwave Theory and Tech. 19:153-164, 1971.

(11) Schwan, H.: Testimony under cross-examination, New York

State Public Service Commission Hearings, 1976.

(12) Guy, A.W., Chou, C-K., Johnson, R.B. and Kunz, L.L.: Study of effects of long-term low-level RF exposure on rats: A plan, Proc. IEEE 68:92-97, 1980.

(13) Steneck, N.H.: The Microwave Debate, MIT Press, Cambridge, MA, 1984.

(14) Medici, R.G.: Behavioral studies with electromagnetic fields: Implications for psychobiology, J. Bioelectricity 4:527-552, 1985.

(15) Medici, R.G.: Methods of assaying behavioral changes during exposure to weak electric fields, Proc. Conference XI: Abnormal Animal Behavior Prior to Earthquakes (II), U.S. Geological Survey Open File Report 80-453, Menlo Park, CA, 114-140, 1980.

(16) Medici, R.G.: Methods of assaying behavioral change as a function of exposure to weak electric fields, Abstracts of Papers presented at the 2nd Ann. Meeting Bioelectromagnetics Society, 200, 1980.

(17) Medici, R.G.: Where has all the science gone?, in Risk/Benefit Analysis: The Microwave Case, N.H. Steneck, ed., San Francisco Press, San Francisco, CA, 177-196, 1982.

(18) Keeton, W.: The orientational and navigational basis of homing in birds, Adv. Study Behav. 5:47-131, 1974.

(19) Ehret, C.F., Rosenberg, R.S., Sacher, G.A., Duffy, P.H. and Groh, K.R.: Biomedical Effects Associated with Energy-Transmission Systems: Effects of 60-Hz Electric Fields on Circadian and Ultraduab Physiological and Behavioral Functions in Small Rodents, Report No. DOE/TIC-1027653, Contract No. W-31-109-ENG-38, Document No. DE81-027655, available through NTIS, Springfield, VA, 1980.

(20) Rosenberg, R.S., Sacher, G.A. and Duffy, P.H.: Physio-

logical and behavioral effects of 60 Hz high voltage fields on mouse behavior, Bioelectromagnetics 1:201, 1980.

(21) D'Andrea, J.A., Gandhi, O.P., Lords, J.L., Durney, C.H., Astle, L., Stensaas, L.J. and Schoenberg, A.A.: Physiological and behavioral effects of prolonged exposure to 915 MHz microwaves, J. Microwave Power 15:123-135, 1980.

(22) D'Andrea, J.A., Bailey, C.M., Hagmann, M.J. and Gandhi, O.P.: Behavioral effects of exposure to 40 MHz modulated near-fields in rats, in Abstracts of 3rd Ann. Conference Bioelectromagnetics Society, 41, 1981.

(23) DeWitt, J.R., D'Andrea, J.A., Emmerson, R.Y. and Gandhi, O.P.: Behavioral effects of prolonged exposure to .5 mW/cm^2, 2450 MHz microwaves, in Abstracts of 5th Ann. Meeting Bioelectromagnetics Society, 116, 1983.

(24) Levinson, D.M., Riffle, D.W. and Justesen, D.R.: Influence of microwave dose rate and ambient temperature on escape behavior of rats, in Abstracts of 5th Ann. Meeting Bioelectromagnetics Society, 39, 1983.

(25) Riffle, D.W., Levinson, D.M. and Justesen, D.R.: The lever pressing operant as a cued or uncued escape response of rats motivated by intense microwave radiation, in Abstracts of 5th Ann. Meeting Bioelectromagnetics Society, 39, 1983.

(26) Creim, J.A., Hilton, D.I., Lovely, R.H. and Phillips, R.D.: Motivational aspect of electric-field avoidance in rats, in Abstracts of 5th Ann. Meeting Bioelectromagnetics Society, 113, 1983.

(27) Hjeresen, D.L., Kaune, W.T., Decker, J.R. and Phillips, R.D.: Effects of 60 Hz electric fields on avoidance behavior and activity in rats, Bioelectromagnetics 1:299-312, 1980.

(28) Hjeresen, D.L., Miller, M.C., Kaune, W.T. and Phillips, R.D.: A behavioral response of swine to a 60 Hz electric

field, Bioelectromagnetics 3:443-452, 1982.

(29) Tenforde, T.S., Davis, H.P., Mizumori, S.J., Allen, H., Bennett, E.L. and Rosenzweig, M.R.: Tests for behavioral alterations in mice exposed to a 1.5 Tesla DC magnetic field, in Abstracts of 3rd Ann. Conference Bioelectromagnetics Society, 93, 1981.

(30) Beel, J.A., Fisher, L.J., Gerren, R.E. and Luttges, M.W.: Posttrial microwave effects on learning and memory in mice, Soc. for Neurosci. Abstracts 9:644, 1983.

(31) Adair, E.R.: Microwaves alter thermoregulatory behavior, Fed. Prod. 38:1295, 1979.

(32) Adair, E.R. and Adams, B.W.: Microwaves modify thermoregulatory behavior in squirrel monkey, Bioelectromagnetics 1:1-20, 1980.

(33) Adair, E.R. and Adams, B.W.: Behavioral thermoregulation in the squirrel monkey: Adaptation processes during prolonged microwave exposure, Behav. Neurosci. 97:49-61, 1983.

(34) Adair, E.R., Adams, B.W. and Akel, G.M.: Minimal increases in hypothalamic temperature accompany microwave-induced changes in thermoregulatory behavior, in Abstracts of 5th Ann. Meeting Bioelectromagnetics Society, 105, 1983.

(35) Stern, S.: Microwaves: Effect on thermoregulatory behavior in rats, Science 206:1198-1201, 1979.

(36) Gordon, C.J. and Long, M.D.: Behavioral and autonomic thermoregulation in hamsters during microwave exposure, in Abstracts of 5th Ann. Meeting Bioelectromagnetics Society, 86, 1983.

(37) O'Connor, M.E., Bartsch, D.A., Chrobak, J., Proksa, J.C. and Indart, M.: Behavioral and developmental evaluation of rats exposed _in utero_ to microwave fields, in Abstracts of 5th Ann. Meeting Bioelectromagnetics Society, 60, 1983.

(38) Mitchell, C.L., Galvin, M.J., Tilson, H.A. and McRee, D.I.: Effects of microwave radiation pre- and perinatally on neurobehavioral functioning in rats, in Abstracts of 5th Ann. Meeting Bioelectromagnetics Society, 68, 1983.

(39) Frey, A.H.: Neural and behavioral consequences of prenatal exposure to 3.5 kV/m 60 Hz fields, in Abstracts of 4th Ann. Conference Bioelectromagnetics Society, 6, 1982.

(40) Frey, A.H. and Wesler, L.S.: Behavior modification in animals living in a relatively low intensity (3.5 kV/m) 60 Hz electric field, in Abstracts of 3rd Ann. Conference Bioelectromagnetics Society, 93, 1981.

(41) Frey, A.H. and Wesler, L.S.: Modification of the conditioned emotional response in rats living in a 60 Hz electrical field, Bull. Psychonomic Soc. 22:477-479, 1984.

(42) Cooper, L.J., Graves, H.B., Smith, J.C., Poznaniak, D. and Madjid, A.H.: Behavioral responses of pigeons to high intensity 60 Hz electric fields, Behav. Neural Biol. 32:214-228, 1981.

(43) Clarke, R.L. and Justesen, D.R.: Behavioral sensitivity of a domestic bird to 60 Hz AC and to DC magnetic fields, Radio Sci. 14:209-216, 1979.

(44) Frey, A.H. and Wesler, L.S.: Tail pressure behaviors modification associated with microwave energy exposure, Bioelectromagnetics 1:202, 1980.

(45) Frey, A.H. and Wesler, L.S.: A test of the dopamine hypothesis of microwave energy effects, J. Bioelectricity 1:305-312, 1982.

(46) Frey, A.H. and Wesler, L.S.: Dopamine receptors and microwave energy exposure, J. Bioelectricity 2:145-157, 1983.

(47) Frey, A.H. and Wesler, L.S.: Morphine effects appear to be potentiated by microwave energy exposure, J. Bioelectricity 3:373-383, 1984.

(48) Seaman, R.L., Murdock, G.K. and Conradsen, L.: Pulsed microwaves may affect rat mating behavior, in Abstracts of 3rd Ann. Conference Bioelectromagnetics Society, 1, 1981.

(49) Monahan, J.C.: Behavioral Methods for Assessing CNS Function Following Microwave Exposure, in Abstracts 5th Ann. Meeting Bioelectromagnetics Society, 68, 1983.

(50) Antipov, V.V., Dobrov, N.N., Kozlov, V.A., Nikitin, M.D. and Semonova, L.A.: Biological effects of the electrical component of superlong-wave electromagnetic fields, Ser. Biol. 3:419-425, 1983.

(51) Lobanova, E.A., Sokolova, I.P., Kitsovskaia, I.A., Rubtsova, N.B. and Lebed, E.K.: Relationship between the biological effects of microwave radiation and the intensity and duration of exposure, Gig. Tr. Prof. Zabol. 1:30-35, 1983.

(52) Gavalas, R.J., Walter, D.O., Hamer, J. and Adey, W.R.: Effect of low level, low frequency electric fields on EEG and behavior in Macaca nemestrina, Brain Res. 18:491-501, 1970.

(53) Gavalas-Medici, R. and Day-Magdaleno, S.: Extremely low frequency weak electric fields affect schedule controlled behavior in monkeys, Nature 261:256-258, 1976.

(54) Sagan, P.M. and Medici, R.G.: Behavior of chicks exposed to low power 450 MHz fields sinusoidally modulated at EEG frequencies, Radio Sci. 14:239-245, 1979.

(55) Medici, R.G.: Frequency specific effects of weak amplitude modulated VHF fields on schedule controlled behavior, in Abstracts of 3rd Ann. Conference Bioelectromagnetics Society, 5, 1981.

(56) Schrot, J., Thomas, J.R. and Banvard, R.A.: Effects of 2.8 GHz microwave radiation in combination with tranquilizing drugs on fixed-interval performance of rats, Bioelectromagnetics 1:203, 1980.

(57) Lovely, R.H., Lundstrom, D.L. and Phillips, R.D.: Dosimetric and behavioral analysis of microwave drug synergistic effects on operant behavior in the rat, Contract No. N00014-79-C-0819, Document No. AD-A115115, Available through NTIS, Springfield, VA, 1981.

(58) Lundstrom, D.L., Lovely, R.H. and Phillips, R.D.: Analysis of dose-response synergy of microwaves and chlordiazepoxide's effects on fixed interval behavior in the rat, Bioelectromagnetics 1:202, 1980.

(59) Lundstrom, D.L., Lovely, R.H. and Phillips, R.D.: Failure to find synergistic effects of 2.8 GHz pulsed microwaves and chlordiazepoxide on fixed interval behavior in the rat, in Abstracts of 3rd Ann. Conference Bioelectromagnetics Society, 43, 1981.

(60) Gage, M.I.: Interaction of electromagnetic radiation and drugs on schedule controlled behavior in rats, in Abstracts of 5th Ann. Meeting Bioelectromagnetics Society, 113, 1983.

(61) Gage, M.I., Ziriax, J.M., Crawford, D.L. and Highfill, J.W.: Duration of microwave exposure as a parameter affecting behavior of rats, in Abstracts of 4th Ann. Conference Bioelectromagnetics Society, 47, 1982.

(62) Lebovitz, R.M.: Prolonged microwave irradiation of rats: Effects on concurrent operant behavior, Bioelectromagnetics 2:169-185, 1981.

(63) Lebovitz, R.M.: Operant behavior of rats is similarly modified by CW and by PM microwave radiation, in Abstracts of 5th Ann. Meeting Bioelectromagnetics Society, 112, 1983.

(64) Lebovitz, R.M.: Pulse modulated and continuous wave microwave radiation yield equivalent changes in operant behavior of rodents, Physiol. Behav. 30:891-898, 1983.

(65) Lebovitz, R.M. and Orr, J.: Microwave radiation-induced alterations in operant behavior of rats: A comparison with

d-amphetamine and barbiturate, Soc. Neurosci. Abstracts 9:122, 1983.

(66) Medici, R.G. and Lesser, G.: Effects of weak amplitude modulated VHF fields on schedule controlled activity of wild mallard ducklings, Proc. International URSI Conference, Helsinki, Finland, 1978.

(67) Feldstone, C.S., Polonis, J.J., Gibson, E.G., Smith, H.D. and Dean, E.E.: A preliminary study of possible effects of high intensity 60 Hz electric fields on non-human primate behavior, Bioelectromagnetics 1:200, 1980.

(68) Feldstone, C.S., Polonis, J.J., Craig, D., Gibson, E.G. and Smith, H.D.: Effects of high strength 60 Hz fields on baboon behavior, in Abstracts of 3rd Ann. Conference Bioelectromagnetics Society, 6, 1981.

(69) Stern, S., Laties, V.G., Stancampiano, C.V., Cox, C. and deLorge, J.O.: Behavioral detection of 60 Hz electric fields by rats, in Abstracts of 5th Ann. Meeting Bioelectromagnetics Society, 114, 1983.

(70) Stern, S., Laties, V.G., Stancampiano, C.V., Cox, C. and deLorge, J.O.: Behavioral detection of 60 Hz electric fields by rats, Bioelectromagnetics 4:215-247, 1983.

(71) Gavalas (Medici), R.J., Walter, D.O., Hamer, J. and Adey, W.R.: Effect of low level, low frequency electric fields on EEG and behavior in Macaca nemestrina, Brain Res. 18: 491-501, 1970.

(72) Bawin, S.M., Kaczmarek, L.K. and Adey, W.R.: Effects of modulated VHF fields on the central nervous system, in Biologic Effects of Nonionizing Radiation, P. Tyler, ed., Ann. N.Y. Acad. Sci. 247:74-81, 1975.

(73) Kalmijn, A.J.: The electric sense of sharks and rays, J. Exp. Biol. 55:371-381, 1971.

(74) Frey, A.H.: Brain stem evoked responses associated with low intensity pulsed UHF energy, J. Appl. Physiol. 23:984-988, 1967.

(75) Frey, A.H. and Seifert, E.: Pulse modulated UHF energy illumination of the heart associated with change in heart rate, Life Sci. 7:505-512, 1968.

(76) Hunt, E.L., King, N.W. and Phillips, R.D.: Behavioral effect of pulsed microwave radiation, in Biologic Effects of Nonionizing Radiation, P. Tyler, ed., Ann. N.Y. Acad. Sci. 247:440-453, 1975.

(77) Servantie, B., Gillard, J., Servantie, A.M., Obrenovitch, J., Bertharion, G., Perrin, J.C., Creton, E. and Plurien, G.: Comparative study of the action of three microwave fields upon the behavior of the white rat, Intl. Symp. on Biological Effects of Electromagnetic Waves, Arlie, VA, USNC-URSI, 122, 1977.

(78) Gage, M.I.: Operant behavior of rats following overnight exposure to microwaves, Intl. Symp. on Biological Effects of Electromagnetic Waves, Arlie, VA, USNC-URSI, 118, 1977.

(79) Roberti, B., Heebels, C.H., Hendricx, J.O., DeGreef, A.H.A.M. and Wolthius, O.L.: Preliminary investigations of the effects of low level microwave radiation on spontaneous motor activity in rats, in Biologic Effects of Nonionizing Radiation, P. Tyler, ed., Ann. N.Y. Acad. Sci. 247:417-425, 1975.

(80) deLorge, J.: The effects of microwave radiation on behavior and temperature in rhesus monkeys, in Biological Effects of Electromagnetic Waves, C.C. Johnson and M.L. Shore, eds., HEW(FDA) 77-8010, USNC-URSI, 158, 1975.

(81) deLorge, J.O. and Ezell, C.S.: Vigilance behavior in rats exposed to 1.28 GHz microwave irradiation, Natl. Radio Sci. Meeting, Bioelectromagnetics Symp., Seattle, WA, 449, 1979.

(82) McAfee, R., Gordon, A., Longacre, J.R., May, J. and Elder, S.T.: Facial irradiation of freely responding Macaca mulatta by 9.3 GHz pulsed microwaves: A long term investigation, Intl. Symp. on Biological Effects of Electromagnetic Waves, Arlie, VA, USNC-URSI, 110, 1977.

(83) Frey, A.H., Feld, S.R. and Frey, B.: Neural function and behavior: Defining the relationship, in Biologic Effects of Nonionizing Radiation, P. Tyler, ed., Ann. N.Y. Acad. Sci., 433-439, 1975.

(84) Grove, A.M., Levinson, D.M. and Justesen, D.R.: Attempts to cue successful escape from a highly intense microwave field by photic stimulation, Natl. Radio Sci. Meeting, Bioelectromagnetics Symp., Seattle, WA, 454, 1979.

(85) Hjeresen, D.L. and Phillips, R.D.: Perception and response to pulsed microwave radiation by rats, Abstracts of USNC/URSI Ann. Meeting, Amherst, MA, 4, 1976.

(86) Monahan, J.C. and Henton, W.W.: Free operant avoidance and escape from microwave radiation, in Biological Effects and Measurements of Radiofrequency/Microwaves, D.G. Hazzard, ed., Symposium Proc., Rockville, MD, HEW(FDA) 77-8016, 23-33, 1977.

(87) Frey, A.H. and Specter, J.: Irritability and aggression in mammals as affected by exposure to electromagnetic energy, Abstracts of 1976 USNC/URSI Ann. Meeting, Amherst, MA, 93, 1976.

(88) Frey, A.H. and Wesler, L.S.: Modification of tail pinch consummatory behavior by microwave exposure, Natl. Radio Sci. Meeting, Bioelectromagnetics Symposium, Seattle, WA, 456, 1979.

(89) Maitland, G. and Thomas, J.R.: Behavioral effects of daily and weekly 1 mW/cm^2 electromagnetic radiation (EMR) in rats, Bioelectromagnetics 1:203, 1980.

(90) Thomas, J.R., Schrot, J. and Banvard, R.A.: Comparative effects of pulsed and continuous wave 2.8 GHz microwaves on temporally defined behavior, Bioelectromagnetics 3:227-235, 1982.

(91) Lai, H., Chou, C-K. and Guy, A.W.: Psychoactive drug response is affected by acute low level microwave irradiation, Bioelectromagnetics 4:205-214, 1983.

(92) Delgado, J.M.R.: Biological effects of extremely low frequency electromagnetic fields, J. Bioelectricity 4:75-92, 1985.

V

THERAPEUTIC APPLICATIONS OF ELECTROMAGNETIC ENERGY

18
The Modern Magnetotherapies

MICHAEL A. PERSINGER
Behavioral Neuroscience Laboratory, Department of Psychology
Laurentian University, Sudbury, Ontario, Canada

INTRODUCTION

Magnetotherapy can be defined as the treatment of subjective and objective ailments by exposures to magnetic fields. Historically, the use of magnetic fields in therapy has been remarkably periodic. When the magnetic dream was prevalent, it was pursued with adolescent enthusiasm. Frequently, the magnetotherapies were also paratherapies that were practiced beside and sometimes in opposition to the contemporary medicine.

In many ways, the development of the magnetotherapies has been similar to the clinical history of cannabis. During the last century, hemp or its many variants were used to treat an extraordinary number of complaints that ranged from migraine headaches to "women's difficulties." However the effect upon the presenting problem was rarely consistent and consequently the use of the plant was gradually dropped from the therapeutic repertoire.

The problem lay not in the therapy *per se* but in the failure to isolate the key chemical properties and to discriminate the importance of psychological factors. Now it is clear that the active chemical ingredient is THC (tetrahydrocannabinol). Now it is clear that expectancy about the effects of the drug can produce some of the effects as the drug itself. Furthermore, the personality of the client, his psychological expectancy (demand characteristics) and the drug can interact to produce enhanced effects or experiences.

There is also a sociological aspect to the use of magnetotherapy. Magnets (or other readily accessible devices) have been often described (out of print) as a "poor man's medicine." These procedures allow personal access to the treatment of personal ailments. In this context, magnetotherapy removes the person's dependence upon a complex and sometimes unfathomable medical establishment within which the individual has little control. It is no accident that surges in magnetic and other borderline treatments are often commensurate with problems in access to critical institutions and in the general feeling of social isolation.

The persistence of an interest in the effects of magnetic fields can also be traced to psychological factors. Every researcher's personal environment is a product of language and the processes by which it is generated. Despite maturational (developmental) shifts in the cognitive schemes by which we assimilate information, there are concepts from previous stages that remain. One of them is the fascination with invisible forces (animism). This idea serves as a conceptual core around which cluster ideas of infantile mysticism, paranormal experiences and sometimes a modified form of omnipotence. It is so closely tied to the concept of self that if care is not taken, magnetotherapies become a personal quest. It acquires dynamics of a belief.

Unfortunately this factor has been ignored, often with arrogant sarcasm, by the very scientists who practice it. However the powerful inertia that accompanies this conceptual core is repeatedly evident. It is manifested by the gradual shift from the use of magnetic fields to treat a specific ailment, to the treatment of all ailments, and then towards mystical or paranormal involvements. Ultimately there is a revelation of transmagnetism where the "critical factor" is contained within the magnetism of the spirit, self, or other word for invisible force. The pattern was conspicuous in the life of Mesmer and has been reiterated many times since.

Scientists are often disquieted by the introduction of these personal variables into technical discussions. However the recognition of these factors is essential for the objective evaluation of magnetotherapy. A researcher may have isolated specific effects but as he or she moves along the Mesmeric progression and loses "scientific respectability" these effects may be indicted as well. They are ignored and a potential discovery is lost.

The problem of the reality and of the strength of

magnetotherapy is complex; it is complicated both by methodological and psychosocial variables. This chapter will emphasize a moderate stance whereby methodology rather than theory or feasible mechanism is the criterion of evaluation. The review is limited to those experiments that have been published in journals that are accessible to the author. Claims from Chinese and Russian researchers are not included.

FUNDAMENTAL STRATEGIES OF MAGNETOTHERAPIES

Traditional approaches for reviewing a given area of research have involved organization according to measurement, methodological or conceptual (operational) similarities. However, considering the numerous modes, measurements, and methods that have been employed in magnetotherapies, another approach will be used. It involves the intrinsic metaphor or philosophical strategy of the method. There are four discernible strategies; they will be called: coercive, compensatory, corrective, and concentrative.

COERCIVE STRATEGIES

The coercive strategies involve the application of relatively high intensity, static or time-varying magnetic fields to all or part of the body. Exposure durations are brief (typically 15 minutes with ranges between 5 minutes and an hour). The approach has had a rich and often humorous history. Its serious beginning in the modern scientific literature can be seen with the experiments of Karen Hansen (1). She applied one pole of a horseshoe magnet to human skin and observed the consequent reaction to microinjections of histamine. The strategy can be found today with electromagnetic devices such as the Magnetotron (Germany). The treating solenoid coil is 0.5 m in diameter and 0.45 m in length and generates 50-60 Hz pulsed electromagnetic fields with strengths between 30-90 gauss. The client rests within the coil.

The operating metaphor for this strategy is that the magnetic field influences key target organs (vasculature, brain, etc.). There is usually a presumption that different organs of the body may respond optimally to particular field frequencies, shapes (square-wave, complex wave, etc.) or presentation patterns (high frequency carriers pulsed at lower frequencies). For pulsating fields in particular the therapeutic effect is argued to be mediated directly to the disease process through the induction of "pulsating magnetic energy." Within this concept is buried an implicit metaphor that disease processes contain undesirable energies that are somehow set free by the coerced application of magnetic fields.

One derivative of the coercion strategy is found in electrosleep-related therapies. Very large intensity electric currents are applied constantly or pulsed through the brain as electronic broad band noise between 5 Hz and 30 KHz at intensities that would be dangerous if given within a pure narrow band such as 60 Hz. Classic examples of such equipment included the Elektrodorm 1 (Austria), the Elektroson 2 (Russian) and the Electrorel GJP (Czechoslovakia). When electrodes are applied over the eyes, the criteria for sufficient current is based upon the report of phosphenes. Often when the electrodes are applied to the forehead, the criterion current is less clear.

COMPENSATIVE THERAPIES

Compensative strategies emphasize the theme that ailments are due to the disruption of normal biological conditions due to modern technology. Fundamental to this approach is that there are natural zeitgebers that maintain the fine-tuning of complex biological clocks. The two most common zeitgebers are the Schumann resonances and geogenic stimuli. Schumann resonances (2) are primarily sharp pulses or sine-wave like extremely low frequency (ELF) electromagnetic signals that are generated naturally between the ionosphere and the Earth's surface. Geogenic stimuli involve the "essences" within the Earth that are propagated to the surface (and to living organisms) by geomagnetic flux lines. The most recent scientific metaphor for geogenic stimuli are the vibrations of rock constituents. These vibrations are mediated to the surface by "soliton waves," a conceptual cousin to Larmor gyrofrequencies and hydromagnetic emissions.

Essential to the therapeutic strategy is that modern civilization has masked these signals. The intensity of

Schumann resonances and their harmonics is correlated with local subsurface water levels. They have fallen (in many places) because of heavy cultural demands (3). Steel structures and modern buildings prevent full-time exposure to these natural fields by faradic-like shielding. Asphalt and cement presumably prevent the emergence of geogenic stimuli. Travel across time zones, another consequence of modern civilization, adds a dynamic component to the disruption of the natural signals.

The main idea of compensative therapies is to simulate the natural condition by portable, always-present magnetic field devices. They are usually small "black boxes" that can be carried in the pocket or placed under the pillow. The field frequencies vary with the sophistication of the manufacturer, but always include the 1-20 Hz range. The simpler devices such as the Relaxit (Canada) generate fixed quasisymmetrical pulses from solenoids. Different devices generate different, fixed frequencies. More engineered devices such as the Mecos (Germany) apparatus emit MHz or KHz carriers that are pulsed at fixed frequencies or at variable frequencies. The latter option accommodates individual sensitivity. Field strengths are usually in the order of 1-10 gauss within 10 mm of the source.

More sophisticated devices have adjustable series frequencies to accommodate personal needs. Presumably clients often speak of a particular frequency as being most suitable. The quasi-theoretical basis of this finding is that the frequency resonates with the client's alpha rhythm. Induced nerve pulses by exterior loops have the greatest effect when they are in phase with each other (4). This explanation is the germinal center of perhaps the most frequent metaphor of magnetotherapy: resonance interaction.

Compensative therapy presumes that magnetotherapy acts as an artificial zeitgeber to correct disrupted body functions. Support for this idea stems from common sense and popular scientific knowledge about shifts in circadian rhythms during travel (jet lag) and the experiments of Rutger Wever (5). He found that human volunteers who were deprived of diurnal cues while living in an underground bunker showed disrupted, drifting and dissociative changes in their complex circadian clocks (temperature, activity). However the system could be "retuned" by the application of artificial 10-Hz electric (presumably with a magnetic component) fields.

Classic ailments that are associated with disruptive circadian rhythms are similar to those associated with deprivation of natural EM fields. They include sleep difficulties, exacerbation of pain complaints, flu-like symptoms, fatigue, problems of concentration and lethargy or other diffuse symptoms. Unlike the coercive treatments, the compensatory strategies involve long-term exposures; the devices are often carried for days to months.

Another interesting assumption of the compensatory strategy is that magnetotherapy works only if the person is showing the symptoms. The magnetic field devices will not produce any observable effects in the normal, symptom-free person. This theme is often forgotten in many experiments (including those of the reviewer) that attempt to test these apparatus in the laboratory. If this idea is valid then magnetotherapy may work much like the operation of aspirin. If you do not have a fever, aspirin will not affect your body temperature. If you have a temperature, the antipyretic effects of this simple drug are succinctly powerful.

CORRECTIVE STRATEGIES

These strategies attempt to solve the presenting symptom by correcting excessive, adverse sources. The source may be external (from the environment) or internal (within the person). External adverse sources are of two types: natural and man-made. Natural adverse sources include intense solar-induced magnetic storms and electromagnetic fields generated by extreme weather conditions (like the foehn, Chinook, Santa Anna, or sharav winds). Man-made sources include references to the ubiquitous 60 Hz buzz -- from energetic overhead powerlines to the magnetic fields generated from can openers, blow dryers, typewriters, and terminals.

Corrective strategies for external adverse stimuli involve shielding when possible or masking when it is not. Shielding involves living in special buildings or sleeping in (or under) special copper-containing bags or sheets. One example is Ludwig's (6) report that nocturnal arthritic complaints could be reduced by having susceptible sufferers sleep in special sleeping bags. The masking approach is to drown the adversive stimuli by elevating a more harmonic signal. The most popular technique involves the use of special metal sheets (like a sleeping blanket) that generate a dominant frequency.

A similar idea is to generate static fields around the sleeper by a special sheet (Energon). The field strengths are about three times the magnitude of the earth's static field (about 0.5 gauss). The specific value is based upon the belief that the geomagnetic field has dropped to 30% of its stable value within very recent human history (7).

A variant of the masking approach is a type of "dilution by averaging" procedure. Few magnetotherapists seriously assume that 60 Hz (presumably a disharmonic or adverse frequency) can be totally eliminated as a significant peak in the modern frequency spectra. Consequently this therapy offsets the adverse frequency by artificially adding a large number of conducive frequencies or their harmonics. In this way, the biosystem is less influenced by the single adversive stimulus because of the simultaneous presence of many positive, normalizing fields (8).

Correction of endogenous adverse "energies" is most clearly highlighted in MORA therapy (8); it was developed by Franz Morell and Erich Rasche (MORA being an acronym). The original insight occurred when Morrell noted that injection of a substance into an in vitro system occurred before the lag time required for the chemical reaction. He concluded (as many others have) that the effects of medicines are due to their electromagnetic oscillations. If you can transmit the drug's electromagnetic vibration to the body, it can obtain the same effects without the ingestion of the drug.

The idea was applied clinically. Healthy tissue was assumed to have a different complex electromagnetic spectrum than diseased tissue. Presumably a naturally occurring substance was found that differentially filtered harmonic (H) frequencies from disharmonic (D) frequencies. The latter are associated with disease while the former are correlated with healthy states. By the use of a complicated device, the subject's own electromagnetic patterns are separated (H and D). The H is then amplified and reapplied to the person (usually through a local contact like a bracelet but sometimes whole body exposure) as a positive feedback of harmonic frequencies while the D frequencies are inverted (negative feedback) in order to attenuate the frequencies of endogenous toxins and to help maintain the body's own protecting forces.

CONCENTRATIVE STRATEGIES

The last strategy accommodates two principles: (1) the enhanced efficacy of a force by focusing the area of concentration (the martial arts approach), and (2) the importance of spatial gradients within the body, that is, the whole body is not spatially homogeneous in either its function or effect. The latter, of course, is the fundamental concept of acupuncture. Connected to this strategy is the supposition that fine, focused stimulation of hot spots, trigger points, or whatever term for unique body space, removes or ameliorates disease processes at a distance. Disease is implicitly seen as something that interferes with the homogeneous unblocked flow of life forces. This metaphor pervades the chi force beliefs of oriental medicines. Although many western researchers think that the idea was eliminated along with vital force theorists of the early twentieth century, the idea recurs frequently today, invariably masked by fancy physics.

Application of the strategy ranges from crude to incredulous. The simplest form involves Rutkowski's technique (9) for insertion of needles into muscle in order to treat local pain. Similar forms involve application of electrode plates over tissue where current intensities are raised until paraesthesia (the symptom being treated) is enhanced artificially (10). Still other techniques involved focused magnetic fields applied through electromagnets. There is even one device that can generate focused electromagnetic fields with intensities of several thousands of gauss. Because the skull is transparent to magnetic fields (unlike electric current), focused stimulation of cortical tissue from electrical induction occurs. Stimulation of the motor cortex and induced muscle movements by human volunteers have been reported (11).

A (predictable) developing approach within the corrective strategy taps heavily from the traditions of homeopathic remedies. These folk-based beliefs have been a persistent antagonist of rational medicine. Homeopathy borders between sympathetic magic (found in many cultures) and the structure resonance speculations of Sheldrake. Essentially, certain substances give off energies which are usually intensified by dilution of concentration or by mechanical stimulation (shaking) of the solutions. Because of the fundamental aqueous base to treatment solutions, modern protagonists claim that dilution of homeopathic materials changes the structural configuration of aggregates in water molecules; these different, dominating aggregates generate the varied curative (electromagnetic) properties.

The magnetotherapeutic applications of this approach have ranged from passive to active techniques. One passive technique is the combination of the Energon sheet and the Beena method (7). The latter employs special strips containing beeswax, royal jelly or other substances. Both devices are installed in feather beds as treatments for sleeplessness, rheumatic ailments, etc. Active methods involve placing special homeopathic substances within an iron lattice that is surrounded by a coil; the person's resonances (via MORA-type devices) are passed through the coils and the net reapplied field is a combination of the biogenic patterns plus their modifications by the homeopathic substances within the iron lattice. The operating metaphor contains two parts: (1) amplify the person's harmonic frequencies, and (2) add the electromagnetic oscillations from special (curative) substances.

The most elaborate and probationary concentrative strategy involves the Indumed tradition that has been developed by Ludwig (8,12). This combines MORA therapy (9) with acupuncture concepts. By using a special ferromagnetic core, an upper transient frequency of 10 MHz is obtained; it is pulsed at various frequencies called series frequencies of 0.1-300 Hz. Although there is no convincing documentation of the claims, the specific frequencies (in the tradition of matter-as-electromagnetic-patterns) presumably affect particular functions. The hippocampal frequency of 7.8 Hz (also a Shumann frequency) allegedly assists in concentration while 10 Hz has an analgesic effect and 33 Hz facilitates general vitality. Their application is not whole body but focused within the target space (usually the diseased organ).

Although difficult to accept (even conceptually), this magnetotherapy emphasizes an important factor, namely response (personality) profiles. Perhaps they are the moderator variables of all magnetotherapies. Until this decade, the pharmacologically dominated medical model underestimated the importance of personality in therapeutic treatments. The most common personality classification has been borrowed from traditional biometeorological literature. Presumably every person can be designated on two orthogonal dimensions: autonomic type and stability. A person occupies some position between extreme sympathetic dominance and parasympathetic (vagotonic) dominance and as well some position between extreme stability and lability (13). There are many compatible variants of this idea, such as Curry's K (cold) type and W (warm) type (in response to types of air masses); the idea has an uncanny resemblance to Hans Eysenck's (14) two-dimensional personality structure of extroversion-introversion and stable/unstable.

Recent developments of concentrative therapies also emphasize different types of people. Ludwig (8) argues there are at least twelve types, which he categorizes on the basis of their color preference and some other relatively unusual personality criteria. Presumably, the series frequencies that most optimally affect the ailments of people are related to color preferences. Those that prefer red are optimally affected by 8.2 Hz series frequencies while those that prefer purple are influenced by 15.4 Hz series frequencies. Preferences of two similar colors (red-orange/orange) may involve frequency differences of 0.5 Hz. According to the strategy, the optimal treatment frequency is determined by the personality of the client. If this assertion is valid, then the failures of magnetotherapy may have been a consequence of the panacea mentality that ignored the personality dimension.

SUPPORTIVE EVIDENCE: MAJOR EXPERIMENTAL AND CLINICAL STUDIES

Most of the claims of magnetotherapy include disparate and diverse categories of ailments. There are two general clusters of symptoms that appear responsive to several forms of magnetic field patterns. These two clusters involve complaints of pain and relaxation/irritability. Loaded on the latter factor are traditional categories of sleep, depression (in a nonclinical sense) and anxiety.

TREATMENT OF PAIN

Responses to painful stimuli, noiceptive experiences and the cognitive evaluations of them are difficult phenomena to study. The inseparable interaction between psychological and biological processes in the pain experience has been known for decades; anticipation of an aversive stimulus (whose psychological correlates include references to apprehension or anxiety) can elevate subjective pain estimates quite substantially. It is now clear, from both pharmacological and neurochemical criteria, that pain experiences are not a single factor process. There are probably a multitude of different pain mechanisms that are mediated by various chemical transmitters in response to input from different parts of the body.

No doubt electrical stimulation can influence pain levels. Rutkowski (9) used electroanalgesia for chronic shoulder pain. Over a period of 5 years, 125 patients (76 women, 49 men), aged 38 to 68 years, were given 15-20 minutes of field treatment three times weekly (initially) to once every two weeks (10 to 35 sessions). The treatment involved 2-Hz fields delivered through two needle electrodes that were inserted into the back of the forearms plus a biphasic asymmetric rectangular-shaped spike current with a frequency of 1-10 Hz and amplitudes up to 800 μA. The method employed a concentrative strategy for trigger points. Clear improvement occurred in 80% of the cases and only 4 subjects were discharged without clear improvement. There were no placebo runs but a post-treatment follow-up indicated only 8 relapses during the next 5 years.

Magnetotherapy has been a popular treatment for pain. The most enthusiastic promotion of this treatment has been by Ludwig and his colleagues; they have combined competent engineering with keen marketing skills. Over several years Ludwig (11,15) and his group distributed over 920 ELF generators to over 860 clients; 220 of the generators were used as placebos. Two devices were used, those with fixed 9 Hz pulses and those with selectable ranges of 4-12 Hz. The rise time of the spiked fields was about 10 nsec with field strengths in the order of

0.1 mT. The primary measurement of the efficacy of the treatment was the reduction in the consumption of analgesic drugs over weeks to months of treatment.

The major results of these studies indicate that either ELF device (9 Hz or 4-12 Hz range options) substantially reduced reliance on pharmacological agents to treat chronic headache, migraine, rheumatic pain, and scar pain. Relative success with the magnetotherapy averaged between 87-96%, according to Ehrmann et al. (4). Reduction in reliance on drugs for volunteers who used the placebo devices occurred only about 20% of the time.

Superficially, these results suggest that painful experiences are reduced by magnetotherapy. However, the method of measurement indicates that the clients relied less on analgesics for treatment of their pain or ailments; there is no direct measure on whether the pain was reduced. Effectively, the volunteers may have been substituting one psychological dependency for another one.

There are several other problems with field (clinical) studies of this type; the first is the selection of clients. Readers do not know if the subjects were solicited, volunteered or selected randomly; this would clearly influence the direction of the effect. In addition, the assumption that placebo devices control for placebo effects is not warranted. This reviewer found that clients suffering from pain soon learned which device was placebo or real whenever they talked on a telephone. The actual ELF device generated such intense fields that audible clicks could be heard in the receiver (so much for blind conditions).

Exposures within experimentally controlled contexts have also been associated with the reduction of headache pain. Grünner (16) exposed 59 neurotic patients to either sham-field or field conditions (each subject served as his own control) in a series of blind procedures. The subject's head was placed within three quadratic coils that generated a continuous steady magnetic field of about 0.07 mT. Application of the fields was associated with reports of headache improvement in 85% of the patients while application of the sham conditions generated positive reports in 36% of the patients. Interestingly, Grünner suggested that the field facilitated the placebo effect, an idea certainly worthy of further consideration.

Grünner (17) found other field configurations to be associated with reduction in pain. Hourly exposures to a 225 Hz pulsating field (3.6 msec impulses) of 0.88 mT were associated with a relatively immediate regression of complaints. Another field configuration, characterized by an induction strength of 0.5 mT and a 260 Hz pulse was also successful in treating headache.

Problems with external validity are obviously complicated; usually field associated alterations in electrodermal skin response or electroencephalographic (EEG) activity are considered inferences of the change in the subjective (pain) state. Ehrmann et al. (4) claim that older patients showed a 50% increase in skin resistance (baseline near 100 Kohm) within 10 to 30 minutes of continuous exposure to 5 Hz or 10 Hz devices; the effect lasted for about one hour posttreatment. However, there was no reported correlation with subjective complaints. Grünner found that increases in electrodermographic resistance on the forehead over time was correlated with a reduction in headache intensity. This elevation was noted in both magnetic field and sham treatments, although increases were more common in the field (85% of the subjects) than sham (35%) condition. Grünner (16,17) has also employed EEG profiles to support his claims, although his approach was primarily visual inspection of the records. He showed that reduction of headache was associated with increased relative proportions in alpha activity and greater subjective relaxation. Again, the effect, although present in both sham and field conditions, was much more frequent during treatment with actual fields.

Grünner (18) has studied symptom-specific populations. He exposed 26 patients who were suffering from vasomotor (tension) headaches to either a stable homogeneous magnetic field (96 gauss), a 12 Hz (66 msec pulses) magnetic field with strengths of about 8 gauss, or sham-field conditions. Each subject received each treatment and each treatment was one hour in duration. Electrical skin resistance from the forehead was monitored before, during and after the treatment. Compared to pre-field values, the electrical skin resistance was elevated for subjects during the real field conditions; there was no apparent difference between the pulsed field and the continuous one. The analgesic effects were coincident with the elevation in skin resistance (indicating a shift to parasympathetic dominance). Exposure to the fields produced reports of a reduction of pain for all subjects; 65% of the patients said the headache had disappeared totally. After the sham treatment only 15% of the patients reported a mild retreat of the headache while 35% of them reported a worsening of the headache pain.

Posttreatment follow-ups to determine the long-lasting consequences of magnetotherapy are rarely completed. Grünner noted that 9 of his 15 patients were symptom-free after a month although they had had chronic headaches for years (17). Grünner also recognized the importance of discriminant validity. He appears to have realized that because headaches have different etiologies they should not respond homogeneously to the same magnetotherapy. In fact, he reported that the 260 Hz pulsed fields were analgesic for tension headaches and for those associated with depressive neurosis (presumably dysthymic disorder). However this treatment was not effective for headaches associated with tumors, anxiety neurosis or migraine (17,19,20).

If there are individual differences in field sensitivities and frequency-specific effects in magnetotherapy, then there should be reports of enhanced pain or headache during treatment. Ludwig and his colleagues depended upon the classic sympathotonic (sympathetic dominance) and vagotonic (parasympathetic dominance) dichotomy to describe such effects. They argued the sleep of vagotonics is facilitated by 8-12 Hz frequencies whereas sympathotonics report sleeping difficulties. They could sleep when pulse frequencies were adjusted to 4-6 Hz. There was also individual optimization with respect to intensity of the fields.

Grünner (20) candidly reported the effects of the Magnetodiapulse device upon 47 neurotic and depressive patients (although symptoms are not specified, they again appear to be dysthymic disorders using DSM III criteria). Subjects were exposed for about one hour to either sham conditions or to 1 msec trains of 27 MHz fields that were presented at nine pulses per second through non-contacting wire coil parietal electrodes. Sixty percent of patients reported the feeling of excitation that accompanies headache; another 25% of the patients reported unpleasant exhaustion or tension; only 15% of the patients reported drowsiness (relaxation). The sham-field treatment was associated with drowsiness or relaxation in 72% of the patients.

Studies to determine construct validity, that is the existence of some process within the body that is both associated with pain and affected by magnetic field exposures, are presented in the literature. Warnke (21) has argued that pain is triggered by the lack of oxygen in tissues and the related acidosis. He found that 5-10 minute exposures to a 3 mT magnetic field within an impulse frequency of 50 Hz, an impulse package frequency of 10 Hz and a resonance frequency of 150 Hz resulted in a 2-3 factor increase in oxygen partial pressure compared to prefield values. The effect disappeared within 5 minutes after the field was removed.

Warnke (21) cleverly demonstrated that the pulsating magnetic field effect evoked dilation of peripheral blood vessels in humans and horses by measuring infrared emissions. He argued that the results indicated a reduction in sympathetic tone (or increase in vagotonic dominance), a conclusion that is similar to that of Grünner. These results were also reported by Berner et al. (7). These authors state that magnetic foils (Energon) also evoked significant increases in peripheral blood circulation (as measured by thermovision) and arterial oxygen pressure (about 15 mm Hg) but distinctly reduced rheumatic pains. Warnke speculates that there may also be a reduction in activity of pain conducting sympathetic nerves that contain fibers of the A (myelinated) and C (nonmyelinated) category.

The second series of findings that lend credence to the construct validity of magnetotherapy in pain treatment involves the work with reduction in analgesia. A series of studies by Ossenkopp and Kavaliers (22) on rodents, indicate that brief field exposures (30 minutes to an hour) during the night (dark cycle) reduces the analgesia that is normally afforded by morphine. Effectively, rodents that were exposed to time-varying magnetic fields (1 mT, 10 mT) and then exposed to a controlled hot plate behaved as if they were saline-injected controls (no morphine). The effect was produced by rotating magnetic fields and by NMI (Nuclear Magnetic Imaging); however the critical component in the NMI exposures was not the kgauss static field nor the radio frequency component. Apparently, the critical component was the weak, ELF magnetic ripples that are used to enhance the image.

Interestingly, Ossenkopp and Kavaliers (23) found the reduction (abolishment) of nocturnal analgesia in rodents also occurred during a relatively intense geomagnetic storm. They had been suspicious that such factors might be involved on the basis of the high interday variability in nociceptive thresholds. This information is important in light of Grünner's (24) observation that neurotic complaints were enhanced during periods of very quiet geomagnetic activity or very intense geomagnetic activity. Although certainly not conclusive, the Kavaliers and Ossenkopp series appear to demonstrate a conceptual bridge between the mechanisms of analgesia, experimental magnetic field exposures and natural geomagnetic activity. An interface between these three factors would support a central theme of magnetotherapy.

RELAXATION AND SLEEP: ANXIOLYTIC EFFECTS

Considering the intimate relationship between pain and anxiety, it is not surprising that therapies that promise reduction of nociceptive experiences are also associated with anxiolysis. Although anxiety is probably mediated by a neuropeptide within the limbic system, the most well known correlate is muscle tension. In fact, anxious experiences are highly unlikely when muscles are relaxed, a principle that forms the basis of many conventional therapies (e.g., Wolpe's systematic desensitization). Several claims have been made that certain forms of magnetotherapy are also relaxation therapies.

Mantle and Persinger (25) performed one of the first actual double-blind experimental studies with ELF pocket generators. The devices (Relaxit) were manufactured by Electromedia (Rexdale, Ontario, Canada). Four groups of university students (16/group) served as subjects; they were exposed to either control, sham-field, 5 Hz or 9 Hz conditions. The sham devices looked like the devices that generated the 5 Hz or 9 Hz fields; however the sham field devices did not generate any detectable magnetic field. To insure that any effect was not due to a single device, each Relaxit condition involved two apparatus. The devices were worn (held by a belt) over the area of solar plexus. In the control condition, no belt or device was attached.

The 64 subjects were administered a series of statements

before and after the fields were switched on; these statements were Likert scale (1 to 7) descriptions of various vegetative (autonomic) states. During a 10-minute exposure the subjects filled out an anxiety scale questionnaire. Instructions to the subjects were given by tape and the experimenter did not know which subject was receiving which Relaxit condition; the code was broken only after the experiment was finished.

The results of the study indicated that brief exposures did not influence scores on the anxiety questionnaire. However, there was a significant increase in reports of relaxation (or items that loaded on that factor) after 10 minutes of exposure compared to the pretreatment administration for the 5 Hz group. These changes were significantly different from the 9 Hz, sham-field or control groups that did not differ from each other. In addition, there was no difference in scores between members of the two 5 Hz groups that had received exposures from two different apparatus.

Although the Mantle study was one of the more controlled brief exposure protocols in the area, there were several limitations. First, there was no measure of external validity. Statements that "my heart rate is reduced" or "I feel more relaxed" do not demonstrate there actually was a reduction in heart rate or muscle tonus. Secondly, there may have been a pre-test sensitization effect, that is, the subjects figured out that the study was about relaxation (after all, the device was called a Relaxit).

That the 5 Hz groups demonstrated a treatment effect but the 9 Hz and sham-field groups did not (and did not differ from controls) effectively eliminates the possibility of a placebo effect. However, these results cannot exclude the possibility that the 5 Hz effect was due to nonspecific arousal rather than relaxation. In other words, the 5 Hz field may have produced a low-level adrenalin-like arousal but the actual attribution of the experience (relaxation) was determined by the cognitive aspects of the situation; they were determined by the content of the autonomic state questionnaire (pretest sensitization). This effect would be similar to the studies of Schacter and Singer (26). They injected subjects with similar amounts of adrenalin. The emotional experiences of the volunteers were determined by the cognitive aspects (social demands) of the situation. People who were exposed to happy settings experienced joy while people who were exposed to sad contexts became tearful or sad.

Later unpublished studies by Mantle involving approximately 20 male and female volunteers suggested that there was a small change (decrease) in heart rate during 5 Hz field exposures compared to 9 Hz, control or sham conditions. The effects were not statistically significant. However the experiments were designed differently in that the subjects were not required to "think" (answer questions) during the brief field treatments. In addition, electrocardiograph electrodes were attached.

An indirect confirmation that relaxation may accompany brief exposures to 5 Hz Relaxits can be found in a slightly different series of studies. Michaud and Persinger (27) and Persinger and Nolan (28) were interested in the effects of theta frequency fields upon the memory of a three-minute narrative (The Dinosaur Egg Hunt). The basic idea was that magnetic fields applied along the side of the cranium (nearest the temporal lobe) while the subject was listening to a taped story might influence memory consolidation. This would be revealed by consequent discourse analysis. Both studies indicated that there were changes in recall of the narrative, although the type of changes varied. In the first study, there were differences in the form of memory while in the second study there was a reduction in the total amount of detail recalled. The effects of the three-minute exposure to the paratemporal 5 Hz fields were very small but statistically significant.

Although the narrative studies suggest that theta frequency fields interfered with or modified consolidation of memory (because there were no fields applied during the recall) of the narrative, the relaxation hypothesis is equally viable. Decreased arousal and relaxation would also interfere with the university subjects' attention or motivation to remember; the latter was relatively small anyway (optional two-point bonus to the final mark). Grünner noted that the appearance of alpha activity and enhanced reports of relaxation was one of the major features of patients who were exposed to ELF fields, although the frequency and field configurations were markedly different from the Relaxits in the narrative studies.

Grünner (24) also found that when he exposed the heads of

his 59 patients to 63 gauss homogeneous fields for one hour the rise in electrodermal response was accompanied by growing parasympathetic tonus and a decline in vigilance, leading in many cases to sleep. The paradigm of the Relaxit studies involved setting the subject in an acoustic chamber during the field exposure; such quiet conditions may have interacted with the ELF fields to have decrease vigilance. However no electrophysiological measures were taken in either the Michaud and Persinger or Persinger and Nolan studies.

There is little doubt that induction of direct currents within the brain can produce sleep in many people. The phenomenon of electroanesthesia has a long history that has been reviewed by Herin (29). The more brutal form of this therapy is often seen today as electroconvulsive shock. Electroanesthesia is mentioned within this chapter because of the conceptual and perhaps even synergistic overlap with magnetotherapies. One example of the combined sleep-inducing magneto-inductive and transtemporal electric currents (TEC) was reported by Photidades et al. (30). Subjects were exposed for one hour to either transtemporal pulsed current (100 pps; 0.1 msec durations and amplitudes of 0.5-2 μA) only or to this current plus a 1000 gauss magnetic field that was pulsed two times per second. The latter field was presented proximal to the face.

A total of 90 sessions, under relatively controlled conditions were recorded for the same 5 subjects; they were not told whether or not the field was present although controls for equipment noise artifacts were not reported. Depth of sleep was determined subjectively according to a 0-3 scale where 0 meant "no effect" and 3 meant "prolonged sleep." There were 45 sessions of transtemporal pulsed current only and 45 sessions of this procedure plus the magnetic field exposures. Unfortunately, the authors did not analyze their results statistically but the raw data were presented. Chi-square analyses by this reviewer indicated a highly significant difference ($\chi^2 = 47.29$, df = 3, $p < .001$) between the two treatment conditions. Whereas 38% of the sessions with both conditions were associated with the deepest prolonged sleep, only 8% of the TPC conditions were associated with the deepest prolonged sleep. Close scrutiny of the data indicated that the magnetic field facilitated the shift from deep sleep to deep prolonged sleep only.

However, sleep and relaxation from magnetotherapy may not be dependent upon coercive strategies. Some recent experiments reported by Subrahmanyam et al. (31) indicate that relaxation and anxiolysis can be accomplished by a combination of appropriate geomagnetic orientation of the client and the superimposition of relatively weak ultra-low-frequency fields. Maximum effects were found with pulse frequencies of either 0.01 Hz or 0.1 Hz (rather than 1, 10, or 20 Hz) at intensities of 20,000 or 40,000 gamma. Presumably, there were also correlative changes in autonomic measures.

These studies, if replicated, implicate the role of geomagnetic pulsations and ultra low frequency variations. Exacerbations of irritability, psychiatric complaints and related aversive behaviors have been frequently associated with magnetic storms. The Subrahmanyam results suggest that ailments may be associated with a lack of geomagnetic stimuli. Interestingly, Grünner (20,24) noted that the symptoms of neurotic patients were exacerbated on days when the geomagnetic oscillations were very quiet. The applications of these observations to magnetotherapy remain to be established.

Perhaps the strongest support for the role of magnetotherapies in the treatment of anxiety/irritability is again reported by Grünner (16). He used electronic broad-band noise (20 mA) for the treatment of patients who were diagnosed with endogenous depression, reactive depression, neurotic depression (dysthymic disorder) and anxiety neurosis. The effects were differential according to psychiatric category. Dysthymic patients reported dizziness and nausea while endogenous depressives reported heightened psychic (attentional) activity. There were no reported effects on reactive depressive patients and the treatment was not suitable for anxiety neurosis (although a square-shaped pulsed current of 1 mA appeared more effective). The discriminative nature of this magnetotherapy within a heterogeneous diagnostic population suggests credibility; if the same treatment would have produced similar effects in all four groups, the mechanism would be difficult to comprehend and the validity would be questionable.

There have also been reports of enhanced anxiety during magnetic field exposures. Sweetland et al. (32) exposed 175 volunteers to either MRI (magnetic resonance imaging), sham MRI or control conditions for about one hour. Pre- and post-treatment measures included scales from the Wechsler Adult Intelligence Scale (Block Design; Digit Symbol and Digit Span), the Wechsler Memory Scale of Paired Associate Learning, the Benton Visual Retention Test (Spatial Memory), several other memory indicators and a state anxiety inventory. Compared to the sham-MRI and control conditions, the MR imaging procedure interfered with Digit Span and enhanced anxiety scores. These authors concluded that the combinations of rf, steady state and time-varying fields associated with MRI elevated anxiety but did not influence gross cognition.

POTPOURRI STUDIES

A claim for cures of a variety of diverse and etiologically unrelated ailments has been a singular and discrediting trait of magnetotherapies. They are difficult to assess objectively because the data are rarely published or the primary information is written in obscure or eccentric language; most of the time the proofs are single-case clinical examples. In appropriate contexts, they can be very useful (e.g., the single episode of affluent aphasia that allowed accurate location of the lesion in the frontal lobe -- Broca's area).

There are many small laboratories and institutes that claim magnetic miracles. To dismiss them as fraudulent simply because they do not have access to traditional forms of expression may be as irresponsible as accepting the claims without question. One such example are the experiments of Charles Turley in Puerto Rico. Turley and his colleagues exposed volunteers to a Japanese manufactured 50-100 Hz pulsating magnetic field with strengths in the order of 100 gauss; exposure durations varied between 15-20 minutes and an average treatment involved between 15-20 sessions over a 5- to 20-day period.

One of the unpublished studies by Turley et al. (clinical report of January 3, 1976) reported a dramatic pre- to post-treatment drop in blood cholesterol of 10 subjects; 10 controls did not show any change. The mean pretreatment cholesterol levels of the treated group was 196 ± 26 mg/100 ml while the control group value was 209 ± 14 mg; whereas this value did not change appreciably (208 ± 12 mg) in the control group, the treated group's mean dropped to 141 ± 13 mg/100 ml. The time of treatment varied between 5 and 25 days.

They reported the individual data, so this reviewer analyzed them statistically. A three-way analysis of variance with one level repeated (pre-/post-cholesterol measures) and two main factors (treatment and sex) with covariance for duration of treatment was completed. The results demonstrated an extraordinarily powerful treatment difference ($F = 30.71$, $df = 1,15$, $p < .001$) that was due exclusively to the treatment by repeated measure interaction ($F = 61.80$, $df = 1,16$, $p < .001$) because of the large drop in post-treatment cholesterol for the magnetic field exposed subjects. There was no sex difference or significant covariance for duration of treatment.

Unfortunately this kind of data rarely reaches the scientific literature. Indeed there may be incompetent experimenters, fraudulent practitioners or simply unscrupulous promoters who might generate similar results. On the other hand, there may be actual discoveries that have not been expressed, in the tradition of Mendel or even Copernicus, simply because they cannot be published or have been published locally and lost to the general scientific community.

PLACEBO: A PHENOMENON WITHOUT A MECHANISM

A cursory glance at the ailments affected by the Mecos magnetotherapy would include chronic headache, general body aches, motion sickness, nervousness, fatigue, poor circulation, foehn sickness, and sleeplessness. Claims for the Relaxit are similar. Indupoint and MORA (corrective) therapies would also include allergy-related ailments. Magnetotron claimants would mention the neuralgias, angina, functional gastrointestinal disorders (irritable colon), gall bladder irritation and various paralytic problems. All would claim anxiolysis. The seasoned investigator would recognize this symptom as loading on a single factor: the placebo.

NATURE OF PLACEBO PHENOMENON

Traditionally, a placebo effect was defined as the production of improved conditions by administration of a pharmacologically inert substance. An important part of the placebo effect involves a degree of ritual, expectation, and suggestion. It does not require the ingestion of a substance but may only involve a situation. The "hello doctor, goodbye doctor" effect, where the sick patient suddenly feels better while just sitting in the doctor's office or the person with appendicitis suddenly "feels better" while in the emergency ward, are possible examples.

Disorders which have been found to be most responsive to placebo treatment are pain, headache, anxiety, depression, and fatigue; these conditions require some cortical processing. Peterson (33) lists the most frequent signs or symptoms that have been affected by placebo or expectancy treatments. Percentage of patients (in parentheses) that show amelioration of symptoms have been found for analgesia (28%), headache (62%), sleep disturbances (7%), sea-sickness (58%), neurosis (46%), hay fever (22%), colds (45%), rheumatism (49%), gastrointestinal disorders (22%), and menstrual difficulties (24%). Different side effects following a placebo treatment include (percentage responding in parentheses), headache (25%), sleepiness (50%), feelings of warmth (8%), and relaxation (9%).

The placebo phenomenon is often impugned because it is presumed to be "in the head." Yet a psychological etiology does not alter the power of the phenomenon or the compelling fact that cognitive processes, personality patterns, and traditional medical (biological) ailments are intimately connected. One striking placebo effect was summarized by Rossi (34). A patient with severe lymphosarcoma was refractory to all known therapies; he heard about "krebiozen," a miracle cure for cancer. The physician was dubious but injected the patient on a Friday, expecting a dead patient on Monday. However, by that time, the tumor had shrunk to half its size and ultimately regressed. When the patient read in newspapers that the treatment was "fake," the tumors reappeared and again he entered a terminal state. The physician found "some new remedy" and again the lymphomas shrunk only to reoccur (and this time the patient died) when the patient read that this remedy too was worthless.

THE PLACEBO REACTOR

In general, the placebo potency decreases in proportion to the number of doses received; usually by the fourth treatment the effect is substantially attenuated. Continued placebo effectiveness (fixed regime of administration) is influenced by the personality characteristics of the patient or client. Placebo reactors have been found to be more religious, older, easier to care for and to get along with, more dependent on outside stimuli, more conscientious, and less mature.

There is a clear relationship between suggestibility (hypnotizability) and placebo responding. Highly suggestible people tend to have rich fantasy-lives (35). Their fantasies are so intense that they are sometimes difficult to distinguish from actual events. They show a marked psychosomatic plasticity. This means that many of their body organs have come under control of cognitive processes. Very suggestible people can respond with the triple response of skin injury to mild tactile stimulation; this response involves the release of a histamine-like substance, localized dilation and increased permeability of minute blood vessels.

THE REFLEXIVE USE OF PLACEBO EXPLANATIONS FOR MAGNETOTHERAPEUTIC CLAIMS

Recently Barker (36) concluded that the symptoms affected by current magnetotherapies were due to placebo effects because similar symptoms have been produced by placebo procedures. This approach is popular among debunkers and hinges upon the relatively loose thinking that "similar plus similar equals the same." Using the same paralogical assumptions, Barker would also be forced to conclude that pepper does not make you sneeze (nor does looking at a bright light) simply because having a cold (a more frequent phenomenon) also evokes sneezing.

The actual neuropsychological mechanism of the placebo effect is unclear. Consequently, to state the magnetic effects are due to placebo phenomenon is useless labelling that reveals nothing about mechanism. It is possible, that whatever neuropsychological mechanisms are associated with placebo effects, might also be influenced directly by the application of therapeutic magnetic fields. This possibility has not been pursued.

H.W. Ludwig's older studies, and O. Grünner's earlier work were either double blind or the effects of the treatment were opposite to those of placebo conditions. However, these procedures and results still do not exclude the role of suggestibility. This issue was recently addressed by Ross and Persinger (37). A total of 40 university students were assessed for suggestibility using the procedure of Spiegel (Hypnotic Induction Profile); the characteristics of the population were similar to the norms of Spiegel (38).

The subjects were exposed to 3 minutes of Relaxit therapy (5 Hz or 9 Hz) or to sham-field conditions. They were told that that treatment was associated with relaxation and that they would be asked to recall a narrative after it was completed. The subjects were familiar with the supposition that relaxation interferes with memory consolidation. However, there was no significant correlation between the degree of suggestibility and the amount of recall, although the latter was influenced by the presence of the fields.

CURRENT STATUS, CRITIQUE AND SUGGESTIONS

The existence of valid magnetotherapeutic effects is highly probable. Meta-analyses of published data and pattern evaluation of the types of results that have been presented by the more rigorous researchers strongly suggest that the claims of some magnetotherapies are not due to simple artifacts. Whether or not these effects are of any practical significance beyond the role of a type of adjunct psychotherapy remains to be established. The routine substitution of pharmacological treatments with magnetotherapy is much less probable. However, if Liboff and Thomas' (39) lithium simulation (by bucking the geomagnetic field to half its normal value and superimposing a 60-Hz powerline-intensity magnetic field) shows external validity, then the contingencies will be definitely changed. After all, lithium is one of the simplest and most effective therapies for major bipolar affective disorders (manic depression).

METHODOLOGICAL ISSUES

The single greatest limitation of magnetotherapies is a consequence of the same factor that maintains it: individual interest and innovation. There are so many different types of magnetic field devices that interlaboratory comparisons or even systematic replications are rarely attempted. Because of the lack of standardization in exposure equipment, the types of methodological controls that are essential for understanding the phenomenon cannot be implemented. Science works by systematic replication whereby each new experiment adds to but at the same time reiterates existing data patterns. At this time, it is better to use the same apparatus in different laboratories for several years (at the risk of missing big effects) than to use different apparatus only once.

The following methodological issues must be addressed in magnetotherapy research; most of them are elementary aspects of experimental design. Controls should be instituted for pretest and posttest sensitization. In most experimental proofs of magnetotherapies there is no reference to the format of the questionnaire by which the subjective reports were obtained. Pretest sensitization can alter the person's expectations of the treatment. Similarly post-test sensitization can influence the person's recollection of experiences or the assessment of current status by the recondite themes carried by test items. The effects of test formats may appear to be subtle but they are critical determinants in the response by highly motivated clients; most clients in magnetotherapy are highly motivated.

Subject selection is also a critical issue. Magnetotherapy works mostly for people who display symptoms; the effects on symptom-free volunteers are less impressive. This fact is not necessarily a critique. Drug treatment works on the same principle. However, symptoms do not occur in isolation and the client with complaints is also a person with a definite personality, values and belief structure. They are a constant potential source of moderation. A moderating variable is one that must be present before a stimulus (e.g., magnetotherapy) can evoke a response (amelioration of symptoms).

Double-blind studies are so important to the support of drug efficacy that the pharmaceutical industry rarely proceeds without them. Single-blind studies (placebo controls) are frequently reported in magnetotherapies; they are never adequate. Few placebo procedures (actually sham-field exposures where the client is exposed to the apparatus without current production) are actually controls. There is no evident attempt to reproduce temperature, vibration or noise artifacts from the equipment. Double-blind studies, where the experimenter is also not aware of the treatment condition, are effectively non-existent in the literature. Considering the importance of the researcher's belief structure in magnetic field research, double-blind procedures are particularly important control measures.

The most likely form of successful magnetotherapy will probably involve the fine focus or concentrative strategies that effectively imitate neuroelectric therapies. Until recently the technology was not available to produce focused, high intensity fields with multiple frequency and complex waveforms. Considering the apparent success of Patterson et al. (40) in treating substance abuse, fine-focus magnetic therapies are a rational extrapolation. They would have the extra advantage of not requiring subcutaneous intervention by needles.

THEORETICAL DEVELOPMENTS

Theoretical developments in magnetotherapy (and for magnetic field effects in general) have been little more than projective tests; they reflect the contemporary fad or scientific metaphor. Many of the mechanisms that are suggested for magnetotherapeutic consequences appear to be obligatory rituals for the Discussion section rather than a realistic appraisal. A naïve reader can be flabbergasted (and that is the appropriate word) to see relatively mundane effects in a Results section transformed into a universal field theory within the Discussion. Often there is not even a rational (let alone empirical) connection between the results and the explanation except for the fact that both are strange. For simplicity, mechanisms of magnetotherapy can be grouped into two camps: single process versus field process.

Single mechanisms highlight or emphasize specific units. They change along levels of scientific discourse, from hemoglobin molecules to major organs; the basic models have not changed since the 1973 review of Persinger et al. (41). Three new advancements deserve attention. The first involves the potential effect of magnetotherapy upon immunological processes (42). As neuroimmunology develops and the cellular mechanisms are elucidated, the effect of magnetic field exposures on these phenomena will very likely become evident.

The second contemporary mechanism is actually a reification of an old metaphor. The pineal gland, primarily because of its unusual geometry and topological position within the brain, was considered to have special mystical qualities. Experiments by Semm (43) and his colleagues show that this structure is electrically and chemically (melatonin) sensitive to changes in near-nature magnetic field intensities. Because the pineal organ primarily inhibits the activity of key endocrine structures such as the thyroids, adrenals and gonads, changes in its activity can produce substantial whole body alterations. The nocturnal analgesia studies of Kavaliers et al. (22) are compatible with this approach.

The third contemporary mechanism involves special tissues within the body; they have organometallic properties that presumably detect or amplify magnetic fields. Some birds are argued to have "magnets" within their brains (44). Magnetite-like granules embedded within neural tissue have been reported in dolphins (45); the material is easily degaussed and demonstrates marked individual variation. The differential existence of such substances within human beings might help accommodate the notable individual variance in responsiveness to magnetic fields.

Another kind of special tissue involves the periventricular structures and circumventricular organs; the latter include the pineal organ, subfornical organ, subcommissural organ, and the area postrema. These structures have poor blood brain barriers and are highly localized vascular, neuronal, and cerebrospinal interfaces. Electrical stimulation of the periaqueductal gray has been reported several times to be associated with pain relief; there is now corroborative evidence that such stimulation elevates beta-endorphins within the ventricular fluid (46). Whether or not fine-focused magnetic fields can accomplish this stimulation or if magnetoresponsive patients have a particular sensitivity of these brain structures remains to be established.

The second metaphor that is used to explain magnetotherapies (and magnetic field effects) involves the field theories (47-49). The basic theme is that magnetic fields do not act upon an element or a structural unit (e.g., blood cells, pineal organ) but upon the field associated with it. The smallest field perspective concerns the electromagnetic oscillations of the cell. Herbert Pohl's (50) review of the natural time-varying electric fields within and about cells summarizes the theoretical and empirical basis of this contention. Individual nerve cells, for example, have been reported to display 50-500 Hz oscillations. A favorite explanation is that EM fields couple with these oscillations and information (in the form of energy) is transferred (oddly enough only in one direction).

The supreme development of the field metaphors is that consciousness is a field that is generated within (or by) the brain's neuroelectromagnetic matrix (47). Presumably the appropriate magnetic field configuration can directly influence the brain fields and hence modify consciousness. At the metaphysical end of this approach is the assertion that changes in the way we perceive the world (enlightenment) is associated with a restructuring of neural networks (48). The idea has been reinforced by exotic mathematical models, the most well-known being those of Burkhart Heim, where complexity and information are considered more important than intensity of the field or strength of the source.

Wolkowski (49) recently reviewed the role of field theory in biology. As he aptly points out, a field theory is an aggregate that accommodates properties that are not simply the sum of the parts, but it is also a system possessing an infinite number of degrees of freedom. This property makes empirical verification a bit difficult. In addition, most modern instrumentation is based upon digit enumeration of units; few instruments measure field properties of the organism.

The favorite mechanism of the field explanation is usually some type of resonance interaction. Because of the conspicuous graphic similarity between natural ELF (Schumann resonance) signals and the human electroencephalogram, the resonance model recurs periodically (51,52). Few people apparently realize that those EEG patterns are determined by instrumentation and by the characteristics of skull impedence. As stated by Callaway and Harris (53), the wave-like qualities of the EEG are not likely to be effective carriers of information. A 10-Hz wave needs at least 100 msec to be established and that is too slow a rate to account for the complex information processing of the brain.

Interestingly, the appearance of approximately 7-14 Hz waves in the EEG usually reflects a relative absence of information processing.

Despite these limitations the attractive field metaphor may still be vindicated. Cortical activity is dominated by thalamic input and thalamic structures have long been suspected to have special sensitivity to static and time-varying magnetic fields (54), although the causal factors were never determined. Midline thalamic structures contain pacemaker cells that determine the pattern of alpha rhythms; they occur primarily when the cortex is not actively processing. In addition, two types of rhythmic activities are found in thalamic neurons (55). The first is the well-known 7-14 Hz rhythm that consists of repetitive hyperpolarizations interrupted by burst discharges. A slower newly discovered rhythm has a periodicity of 10 seconds (0.1-0.2 Hz). Interestingly, experimental fields of these frequency ranges have been persistently associated with biobehavioral consequences (2).

NEUROCOGNITIVE MODERATION OF MAGNETIC FIELD EFFECTS

Just one decade ago, the possibility that neurocognitive factors might be the primary moderating variables in the magnetotherapies would have seemed unlikely; the zeitgeist has changed. Modern noninvasive techniques for assessing the dynamics of brain activity show an extraordinary complexity and sophistication. Many of these changes, such as regional blood flow, have been shown to be correlated with the type of cognitive activity in process. Evaluation of verbal symbols emphasize asymmetric changes in the left hemisphere while spatial contexts shift vascular activity to the right hemisphere. There is now evidence that electrical coupling between cortical potential shifts according to appositional (right hemisphere) and propositional (left hemisphere) problem-solving (53).

Positron emission tomography (PET) has demonstrated that substantial shifts in glucose uptake (and electrical activity) in the brain occur even in apparent sedate conditions. Perception, voluntary thinking, remembering, calculating, reading, discriminating, and speaking are all conscious activities that are characterized by an activation of the prefrontal areas in conjunction with motor areas or more posteriorly located cortical areas (56). Activation of some areas depends upon what thought processes are involved. For example, the posterior frontal cortex is activated by tasks that involve several degrees of freedom but ultimately lead to a choice.

The metabolic patterns within the brain form three-dimensional and temporal mosaics that are complex but tractable. These patterns respond to the cognitive aspects of the situation; that would include the expectancies, social demands, implicit assumptions and antecedent interpersonal interactions. Other PET studies emphasize the importance of individual differences. This is particularly appropriate for patients with temporal lobe electrical lability (complex partial epilepsy) where there is no general pattern of intrinsic brain activation across subjects, only within subjects.

Appreciation of these neurocognitive factors in regional and local brain activity changes the types of questions that should be addressed. For applied magnetic fields to have real therapeutic effects must there be specific metabolic mosaics within the brain? Do certain cognitive factors such as expectation, the major component of the placebo phenomenon, generate specific brain mosaics that facilitate the effects from the applied magnetic fields? Do repeated magnetic field exposures alter the operational pattern of brain activity and hence change the cognitive structure of the person?

Are there certain portions of the brain where the metabolic pattern or electrical processes are particularly sensitive to the applied fields or magnetotherapy? PET studies reaffirm the suspected special lability of the temporal lobes, particularly their subcortical components: the amygdala and the hippocampus. Both of these structures are electrically labile and prone to kindling. This is a process whereby the tissue displays electrical impulses (seizures) over time as it is exposed to repeated daily, depth stimulations of minute electric currents. Effectively, because of the repeated exposure to electric currents, the tissue learns to microseizure.

Several studies (57-60) have strongly suggested that certain personality and cognitive characteristics are associated with electrical lability of the temporal lobes. These individuals are more prone to report experiences of floating sensations before they fall asleep at night, hearing one's name called by an inner voice and intense meaningful sensations during early morning hours. The more numerous and frequent these signs, the larger the percentage of alpha activity (that occurs during waking activity) from temporal lobe regions. This temporal lobe factor is weakly correlated but not identical with suggestibility or imaginings.

This temporal lobe factor is a continuum; all of us display some of the symptoms. People at the upper end of the continuum display normal personality profiles without clinical indicators. People who show frequent temporal lobe signs (an inference of temporal lobe lability) are more likely to be anxious, suspicious, aloof, stereotyped in their behavior and ruminative. They are likely to be religious (with mystical overtones) or to experience frequent episodes of "a sense of presence;" they also show remarkable psychosomatic plasticity.

Because magnetic fields penetrate into deep brain space, key temporal lobe structures would be influenced. If the temporal lobe factor is a valid one, then people who display these signs would also have an enhanced electrical lability that could be influenced by therapeutic magnetic fields. Deep temporal lobe structures, such as the amygdala, are associated with the same class of phenomena that are most commonly associated with magnetotherapy. The experiences involve affective dimensions (from anxiety or irritation to euphoric relaxation) and the attribution of pain and pleasure to perceptual processing. In addition, the amygdala massively innervates structures that primarily control the immunocompetence, endocrine status and vasomotor tone of the total organism.

CONCLUSION

The magnetotherapies are at the threshold of verification. However, the diverse designs of apparatus and lack of systematic replication have delayed progress. There are strong suggestive patterns in magnetotherapeutic effects that contraindicate simple placebo explanations, even though the phenomena affected by magnetotherapies are loaded by affective and psychosomatic factors. There is a strong possibility that the key moderating variables are personality and neurocognitive conditions at the time of magnetic field application. Despite their intrinsic possibilities, the future of magnetotherapies is still precarious. A major episode of fraudulence or irresponsible attribution of a placebo effect could easily discredit them once again.

ACKNOWLEDGEMENTS

Thanks to Claudette Larcher for typing the manuscript and to Katherine Makarec for constructive comments. Special thanks to Dr. Douglas Goldsack, Dean of Science and Dr. Michael Dewson, Dean of Social Sciences, for obtaining the space and equipment for the Behavioral Neuroscience Laboratory.

REFERENCES

(1) Hansen, K.M.: Studies made for the end of finding an objective expression of the influence of magnetism on man and of ascertaining whether this influence is carried by way of the autonomic nervous system, Acta Medica Scand. CXXXV:448-457, 1949.

(2) Persinger, M.A.: ELF and VLF Electromagnetic Field Effects, Plenum Press, New York, 1974.

(3) Ludwig, W., Mecke, R. and Seelewind, H.: H. Elektroklimatologie, Arch. Met. Geoph. Biokl. 16:237-250, 1968.

(4) Ehrmann, W., Ludwig, H.W. and Sodtke, W.: Influence of alternating magnetic fields (frequencies between 1 and 20 Hz) on psychosomatic ailments, Second Bioclimatological Colloqium, Munich, West Germany, September 16-18, 1976.

(5) Wever, R.: Influence of electric fields on some parameters of circadian rhythms in man, Biochronometry, AAAS, Washington, DC, 117-133, 1971.

(6) Ludwig, H.W.: Wirkung einter nachtlichen Abschirmung der

elektrischen Feldstarke bei Rheumatikern, Arch. Met. Geoph. Biokl. 21:305-311, 1973.

(7) Berner, H.G., Hauss, R. and Ludwig, W.: The energon, beena and indumed therapies, J. Bioelectricity, In Press.

(8) Ludwig, W.: Biological signals in the ELF and VLF region. Preliminary report, J. Bioelectricity, in press.

(9) Rutkowski, B.: Electroanalgesia for chronic shoulder pain syndromes, in Pain, R. Rizzi and M. Visenthin, eds., Piccin, Italy, 121-124, 1984.

(10) Procacci, P. and Maresca, M.: Nothnagel-Schultze algoparaesthesic syndrome, in Pain, R. Rizzi and M. Visenthin, eds., Piccin, Italy, 35-39, 1984.

(11) Barker, A.T. and Jalinous, R.: Non-invasive magnetic stimulation of human motor cortex, Lancet, 1106-1107, 1985.

(12) Ludwig, W.: What is indupoint acupuncture therapy?, Am. J. Acupuncture 12:153-156, 1984.

(13) Persinger, M.A.: Weather Matrix and Human Behavior, Praeger, New York, 1980.

(14) Eysenck, H.J.: Behavioral Therapy and Neurosis, Pergamon, London, 1960.

(15) Ehrmann, V.W., Leitner, H.V., Ludwig, W., Persinger, M.A., Sodtke, W. and Thomas, R.: Therapie mit ELF-magnetfeldern, Physikalische Medizin 5:161-170, 1976.

(16) Grünner, O.: The cerebral application of pulsed magnetic fields, J. Bioelectricity, in press.

(17) Grünner, O.: The cerebral application of the constant homogeneous magnetic fields, J. Biometeorology, In Press.

(18) Grünner, O.: The effects of stable magnetic field and

interrupted electromagnetic field on vigility, Activ. Nerv. Sup. 22:114-115, 1980.

(19) Grünner, O.: Application of electronic noise and pulsed magnetic field in the treatment of sleep disorders, Clin. E Terapeutica (Brazil) 4:15-16, 1975.

(20) Grünner, O.: Application of electronic noise and pulsed magnetic field in the treatment of sleep disorders, in Sleep: Physiology, Biochemistry, Psychology, Pharmacology, Clinical Implications, W.K. Koella and P. Levin, eds., S. Karger, Switzerland, 442-445, 1973.

(21) Warnke, U.: Possible physical-physiological pain reducing mechanism in a pulsating magnetic field (PMF), in Pain, R. Rizzi and M. Visentin, eds., Piccin, Italy, 135-141, 1984.

(22) Kavaliers, M., Ossenkopp, K.-P. and Hirst, M.: Magnetic fields abolish the enhanced nocturnal analgesic response to morphine in mice, Physiol. Behav. 32:261-264, 1984.

(23) Ossenkopp, K.-P., Kavaliers, M. and Hirst, M.: Reduced nocturnal morphine analgesia in mice following a geomagnetic disturbance, Neurosci. Lett. 40:321-325, 1983.

(24) Grünner, O.: Use of natural and artificial changes of magnetic fields in balneopsychiatry, Activ. Nerv. Sup. (Praha) 16:4, 1974.

(25) Mantle, E.R. and Persinger M.A.: Alterations in subjective evaluations during acute exposures to 5-Hz but not 9-Hz magnetic field devices, J. Bioelectricity 2:5-14, 1983.

(26) Schachter, S. and Singer, J.E.: Cognitive, social and physiological determinants of emotional stress, Psychol. Rev. 69:379-399, 1962.

(27) Michaud, L.Y. and Persinger, M.A.: Geophysical variables and behavior: XXV. Alterations in memory for a narrative

following application of theta frequency electromagnetic fields, Percept. Mot. Skills 60:416-418, 1985.

(28) Persinger, M.A. and Nolan, M.: Partial amnesia for a narrative following application of theta frequency electromagnetic fields, J. Bioelectricity 4:481-492, 1985.

(29) Herin, R.A.: Electroanesthesia: A review of the literature (1819-1965), Activ. Nerv. Sup. 10:439-454, 1968.

(30) Photidades, D.P., Riggs, R.J., Ayivorh, S.C. and Reynolds, J.O.: Electrosleep in man by a combination of magneto-inductive and transtemporal electric currents, in The Nervous System and Electric Currents, N.L. Wulfsohn and A. Sances, eds., Plenum Press, New York, 153-158, 1970.

(31) Subrahmaryam, S., Sanker Narayan, P.V. and Srinivasan, T.M.: Effect of magnetic micropulsations on the biological systems: A bioenvironmental study, Int. J. Biometeorology 29:293-305, 1985.

(32) Sweetland, J., Kertesz, A., Prato, F. and Nantau, K.: The effect of magnetic resonance imaging on human cognition, Magnetic Resonance Imaging, 5:129-135, 1987.

(33) Peterson, E.: Psychopharmacology, W.C. Brown, Iowa, 1966.

(34) Pefferbaum, E.: Psychology of Mind-Body Healing, W.W. Norton, New York, 4-8, 1986.

(35) Barber, T.X.: Changing "unchangeable" bodily processes by (hypnotic) suggestions: A new look at hypnosis, cognitions, imagining, and the mind-body problem, in Imagination and Healing, A.A. Sheikh, ed., Baywood, Farmingdale, 69-127, 1984.

(36) Barker, A.T.: Nonionising electromagnetic therapy: Fact or fantasy?, Phys. Bull. 36:251-253, 1985.

(37) Ross, J. and Persinger, M.A.: Suggestibility, temporal

lobe signs and ELF electromagnetic field effects, J. Bioelectricity, In Press.

(38) Spiegel, H. and Spiegel, D.: Trance and Treatment, Basic Books, New York, 1978.

(39) Thomas, J., Shrout, J. and Liboff, A.R.: Low intensity magnetic fields alter operant behavior in rats, Bioelectromagnetics 7:349-357, 1986.

(40) Patterson, M.A., Firth, J. and Gardiner, R.: Treatment of drug, alcohol and nicotine addiction by neuroelectric therapy: Analysis of results over 7 years, J. Bioelectricity 3:193-221, 1984.

(41) Persinger, M.A., Ludwig, H.W. and Ossenkopp, K.-P.: Psychophysiological effects of extremely low frequency electromagnetic fields: A review, Percept. Mot. Skills 36:1131-1159, 1973.

(42) Joseph, R. and Sikyta, B.: Effects of magnetic field on inflammation, Experentia 31:1411-1412, 1975.

(43) Welker, H.A., Semm, R.P., Willig, R.P., Commentz, J.C., Wiltschko, W. and Vollrath, L.: Effects of an artificial magnetic field on serotonin N-Acetyltransferase activity and melatonin content of the rat pineal gland, Exp. Brain Res. 50:426-432, 1983.

(44) Walcott, C., Gould, J.L. and Kirschvink, J.L.: Pigeons have magnets, Science 205:1027-1029, 1979.

(45) Zoeger, J., Dunn, J.R. and Fuller, M.: Magnetic material in the head of the common Pacific dolphin, Science 213:892-894, 1981.

(46) Hosobachi, Y., Rossier, J., Bloom, F.E. and Guillemin, R.: Stimulation of human periaqueductal gray for pain relief increases immunoreactive β-endorphin in ventricular fluid, Science 203:279-281, 1979.

(47) Ravitz, L.J.: Electrodynamic field theory in psychiatry, South. Med. J. 46:650-660, 1953.

(48) Rascalla, L.M.: An electromagnetic field theory of consciousness, J. Am. Soc. Psychosom. Dent. Med. 21:40-51, 1974.

(49) Wolkowski, Z.W.: The field concept in physics and in biology, Synergy and Coherence in Biological Systems, University of Paris, 1983.

(50) Pohl, H.A.: Natural AC electric fields in and about cells, in Nonlinear Electrodynamics in Biological Systems, W.R. Adey and A.F. Lawrence, eds., Plenum Press, New York, 87-103, 1984.

(51) König, H.L.: ELF and VLF signal properties. Physical characteristics, in ELF and VLF Electromagnetic Field Effects, M.A. Persinger, ed., Plenum Press, New York, 9-34, 1974.

(52) Direnfeld, L.K.: The genesis of the EEG and its relation to electromagnetic radiation, J. Bioelectricity 2:111-121, 1983.

(53) Callaway, E. and Harris, P.R.: Coupling between cortical potentials from different areas, Science 183:873-785, 1974.

(54) Barnothy, M.: Biological Effects of Magnetic Fields, Plenum Press, New York, 1964.

(55) Steriade, M. and Deschenes, M.: The thalamus as a neuronal oscillator, Brain Res. Rev. 8:1-63, 1984.

(56) Roland, P.E.: Metabolic measurements of the working frontal cortex in man, Trends in Neurosciences, 430-435, November, 1984.

(57) Bear, D.M. and Fedio, P.: Quantitative analyses of interictal behavior in temporal lobe epilepsy, Arch.

Neurol. 34:454-467, 1977.

(58) Persinger, M.A. and Valliant, P.M.: Temporal lobe signs and reports of subjective paranormal experiences in a normal population: A replication, Percept. Mot. Skills 60:903-909, 1985.

(59) Makarec, K. and Persinger, M.A.: Temporal lobe signs: electroencephalographic validity and enhanced scores in special populations, Percept. Mot. Skills 60:831-842, 1985.

(60) Persinger, M.A. and Makarec, K.: Temporal lobe epileptic signs and correlative behaviors displayed by normal populations, J. Gen. Psychol. 114:179-195, 1987.

19
Electrical Silver Antisepsis

J. A. SPADARO
Department of Orthopedic Surgery
State University of New York, Health Sciences Center
Syracuse, New York

INTRODUCTION

One remarkable and potentially useful effect of exogenous, low-level, electrical currents in biology is the production of a sustained antibacterial environment near metallic silver electrodes. Attempts to use other materials as electrodes have not met with much success except with high current, toxic corrosive reactions, or resistive heating (1-4). The special antibacterial properties of silver seem to be metal-specific and far more efficient (5). The purpose of this chapter is to review the present understanding of this effect, especially the electrochemical features, the results of animal experiments, and clinical applications.

BACKGROUND

Silver in both compound and elemental form has been exploited for its medicinal properties for centuries (6). For what now seems like frivolous reasons, silver was used to treat epilepsy and other nervous disorders. In the 18th and 19th centuries silver nitrate (lunar caustic) was used successfully in the treatment of skin ulcers, compound fractures, and suppurating wounds. Noteworthy in this regard was John Higgenbottom of Nottingham who employed various concentrations of silver nitrate in solution for over 40 years in the treatment of burns, lacerations and infected compound fractures (7). In 1852, J. Marion Sims of Montgomery, Alabama reported development of a remarkably successful surgical repair of vesicovaginal fistula using silver sutures, a treatment that had always been frustrated by infection (8). As one of the few implantable metals then known, silver was used in a number of surgical applications.

As chemical antisepsis and aseptic surgery emerged in the late 19th century, silver continued to enjoy wide use against bacterial infection. In 1881 Carl Crede pioneered the installation of dilute silver nitrate in the eyes of neonates to prevent gonorrheal opthalmia, a technique which has had widespread use ever since (9). Von Naegeli and others in 1893 realized that the antibacterial effects were primarily due to the silver ion itself even in extremely small concentrations (10). Beginning in 1895, William Halsted at Johns Hopkins used silver foil as a dressing for surgical wounds (11). Thus silver was part of the surgeons armamentarium until the discovery of sulfonimides and penicillin in the 1930's. Today, silver sulfadiazine is an important topical treatment for burn wounds and silver nitrate is still used for prophylaxis in neonatal opthalmia (12-14). Silver is also used in water purification (15).

The mechanism by which silver affects bacterial cells is not clearly understood, but there is evidence of silver-ion binding to DNA (16), membrane changes (17), inhibition of respiration (18) and nonspecific protein binding and precipitation (19). Cowlishaw demonstrated early inhibition of beta-galactosidase enzyme activity in *Escherichia coli* near a low-current silver anode (20). *Staphylococcus aureus* organisms exposed to silver anode electrodes *in vitro* for 4 hours showed erratic septum formation, enlarged, dark-staining mesosomes, and plasma membrane separation (21). When similarly exposed to silver electrodes, *Pseudomonas aeruginosa*, a rod-shaped gram-negative organism, showed abnormal morphology and bleb formation after two hours (22). The mechanism of action of silver may vary with the organism as well as the method of delivery.

The importance of silver in electrophysiology is also worth noting. The silver/silver-chloride electrode is used as a stable reference electrode for *in vitro* and *in vivo* measurements, and also as a low impedance sensing electrode for electrocardiography and encephalography. These applications are based primarily on the reversible electrode reaction:

$Ag^+ + Cl^- \rightleftharpoons AgCl$, which establishes a stable electrode potential distribution at the metal-solution interface (23).

BACTERIAL INHIBITION BY SILVER ELECTRODES *IN VITRO*

During studies of the effects of various electrodes on agar gel cultures of bacteria, it was observed that direct current applied to metallic silver inhibited microbial growth in a region localized near the anode (5,24). This effect is easily demonstrated using the method and configuration seen in Figure 1. If 2-cm lengths of silver wire are used, anodic currents of 0.4-40 μA are routinely inhibitory if applied for a few minutes, although currents as low as 0.05 μA also demonstrate inhibition (20). Currents of 40 μA and higher inhibit

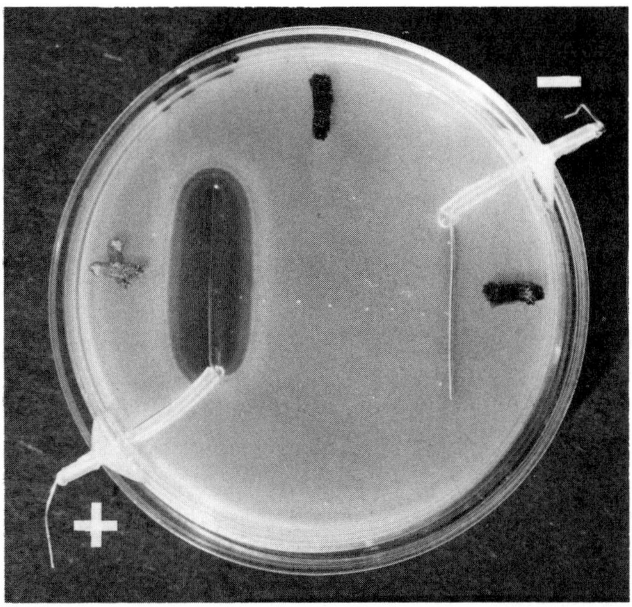

FIGURE 1. A bacterial culture plate of S. aureus in BHI agar, incubated for 20 hours. 0.4 μA DC was applied to the 0.38-mm diameter Ag wire during incubation. Central clear area at anode (+) is complete inhibition zone. Plate diameter is 5.5 cm. From (45), with permission.

bacteria when applied to other electrode metals (anode or cathode), but mainly in association with corrosion and hydrolysis. Of the metal electrodes tested, only silver anodes were inhibitory at low currents. If current is applied to silver electrodes after incubation of the seeded agar plate, a zone of inactive bacteria is created, which upon subculture are not viable (5). This demonstrates that for several organisms at least, the silver anode is bacteriocidal as well as inhibitory.

The dynamics of formation of zones of inhibition in semisolid media is not well understood. The generation of diffusable and/or ionized silver species at the electrode surface is one factor. The rate of diffusion of active species and the rate of bacterial cell division and growth are also clearly important factors. Since applied currents used to cause inhibitory zone formation at metallic silver are small (0.1-10.0 μA), electric fields in the bulk fluid are quite small (0.01-1 mV/cm) and electrophoretic movement of silver species is therefore only a minor factor in zone size. This conclusion is supported by the common observation that above a saturation level, inhibitory zone diameters are independent of current magnitude, and also by the fact that previously activated (anodized) silver surfaces can inhibit bacteria without simultaneous current application (5,45,47).

In work described below, inhibitory zone size <u>in vitro</u> is generally used as a measure of inhibition. Just as with conventional antibiotics, zone size is proportional to agent concentration, diffusion rate and sensitivity of the microorganism. The maximum extent of inhibitory zones with electrically activated silver is generally small (3-8 mm in radius) in contrast to soluble antibiotics. The reason for this is unclear but the mass of experimental evidence points to the formation of insoluble AgCl and protein complexes as the limiting factor. As noted below, this probably occurs <u>in vivo</u> as well. It is also noteworthy that the <u>in vitro</u> zone pattern is complex, typically consisting of an inner totally clear area surrounded by a 1-2 mm band of reduced colony density (partial inhibition) and then sometimes by a thin line of highly concentrated colonies. It is interesting that these various zones are rather sharply delineated, contrary to what one might expect from a continuous diffusion gradient from the electrode surface.

Initial experiments in vitro with silver and other electrodes dealt with four organisms commonly found in human wound infection (S. aureus, E. coli, Proteus vulgaris, and Pseudomonas aeruginosa)(5). All were found sensitive, although only a few strains were tested. Later studies showed that the spectrum of organisms susceptible to electrically generated silver ions was wide, and minimum inhibitory concentrations (MICs) compared quite favorably with silver sulfadiazine and other antibiotics (25-27). All gram-negative and gram-positive bacterial strains (including patient-derived mixed flora) tested so far by this author and collaborators have been found sensitive to anodically generated silver in vitro. Favorable results have also been found against the fungal microorganisms, Candida and Torulopsis (28).

Less information is available presently on the sensitivity of anaerobes to electrically activated silver. Taking advantage of the facultative nature of S. aureus, the ability of silver anodes to inhibit bacteria without oxygen was tested (29). The inhibition zones were measured in agar media around embedded silver anode wires after incubation with or without oxygen present during incubation and pre-incubation phases. The results, shown in Table 1, show that although clear zones from

TABLE 1. Silver Anode Inhibition of S. aureus under Aerobic and Anaerobic Conditions In Vitro

NUMBER OF TRIALS	OXYGEN PRESENT DURING ANODIZATION 24 HOURS BEFORE INOCULATION	OXYGEN PRESENT DURING INCUBATION 24 HOURS AFTER INOCULATION	MEAN INHIBITION ZONE DIAMETER (mm)
4	No	No	3.8
4	Yes	No	3.5
2	No	Yes	7.2
2	Yes	Yes	5.5
4	Simultaneous ⟶	No	5.0

aerobically cultured plates tend to be somewhat larger, all combinations showed significant inhibition. The presence of oxygen during electrical activation of the anode seemed to make no difference. These observations support the idea that the presence of oxygen is not required for the antibacterial effect to be obtained, and that the reaction with chloride in the medium is probably the important reaction during activation. Unreported observations with <u>Clostridia</u> species also showed inhibition under anaerobic conditions of electrical activation and incubation. This indicates that anodically activated silver may be useful in the treatment of localized anaerobic infection.

Other reports in the literature concur on the wide spectrum of bacterial organisms sensitive to silver in various chemical forms (12,30-33). It is also noteworthy that silver also counteracts protozoa (34,35) and viruses (36,37).

Resistance of clinically isolated organisms to silver seems to be uncommon but does occur, and is usually associated with multiple antibiotic resistance (38-41). The resistance is generally unstable during subculture and usually not transferable to other organisms (31,41,42). Some studies have indicated that silver resistance should be examined in light of the silver inactivation by certain growth media <u>in vitro</u> (43) and by serum and tissue proteins <u>in vivo</u> (44).

ELECTROCHEMICAL STUDIES OF SILVER INHIBITION OF BACTERIA

ELECTRODE POTENTIAL

Using elementary electrochemical techniques, it is possible to learn more about the process by which metallic silver becomes bactericidal under the activation of an anodic current. When a series of silver wire electrodes in agar cultures of bacteria are activated by direct currents so that the electrode potential is held to a predetermined value with respect to a reference electrode (calomel), the effect of electrode potential on inhibition can be determined. After application of the current for a given time (15-30 min is usually sufficient), the plate is cultured overnight and the resulting inhibitory zones measured. Typical results are shown in Figure 2, where the strikingly sharp dependence of inhibition on potential is evident (29,45).

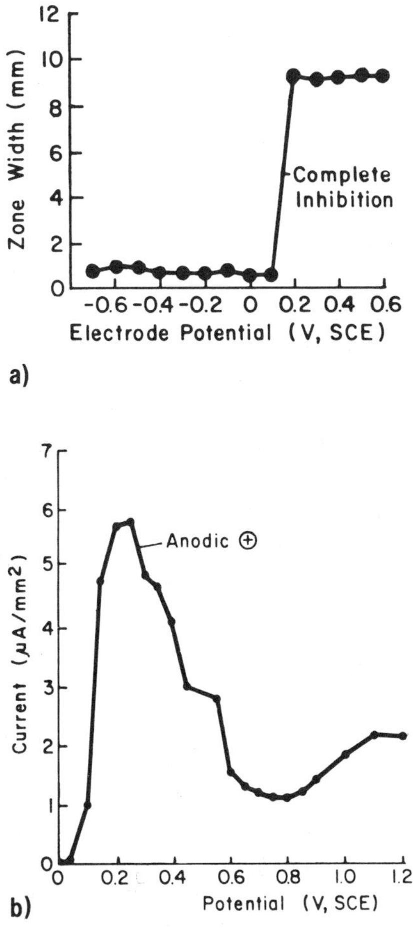

FIGURE 2. (a) Bacterial inhibition zone width (diameter) vs applied electrode potential (with respect to calomel reference) for a silver wire in BHI agar gel inoculated with S. aureus in same fashion as Figure 1. Current applied for 15 min. prior to incubation. (b) Anodic polarization curve for a silver electrode in cell culture medium with serum at 37°. The sharp increase in current (oxidation of silver) corresponds to the onset of bacterial inhibition.

Complete inhibition can be obtained under these conditions, in chloride containing media (BHI agar, GIBCO), with potentials of +0.2 volts or more with respect to the calomel reference. No particular enhancement is obtained above that level. Partial inhibition (reduced colony formation) begins between 0 and +0.1 volts. This suggests that the responsible reaction begins in that voltage range.

If one measures the DC current as a function of electrode potential applied to a clean silver wire in cell culture fluid, a typical polarization curve such as in Figure 2 is obtained (45). The sudden increase in current between 0.15 and 0.20 volts corresponds to the chloridization of the silver surface and coincides with the onset of the bacterial inhibition. Further increases in potential do not further increase current, but in fact decrease it; but this is predominantly due to the chloridization of the electrode and the building up of a resistive coating. This behavior is similar to a corrosion or oxidation curve, and confirms the relationship between the electrochemical oxidation at the surface and the inhibition of adjacent bacteria.

CURRENT MAGNITUDE

By measuring the size of the inhibition zone at a fixed time (in vitro in an agar gel medium as above) as a function of applied constant current, one finds a more gradual dependence on current than on electrode potential (29,45). A zone of partial inhibition appears at low currents. At higher currents, a zone of complete inhibition is found which reaches a size plateau at still higher values. Threshold and maximum values for the various parameters are shown in Table 2 for a typical case of a smooth silver wire in BHI agar with anodic constant direct current applied for 30 minutes. It is interesting to note that the current for a maximum inhibitory zone is comparable in magnitude to that for which cathode stimulation of osteogenesis has been demonstrated (46).

If one varies the duration of current application, an inverse relationship between current and duration seems to control maximum inhibition zone size. This suggests, as might be expected, a dependence on applied charge. A brief application of larger current can produce the same inhibition zone as a

TABLE 2. Effective Parameters for Silver Anode Inhibition of Bacteria *In Vitro*. For smooth drawn silver wire, 99.9% purity, 0.4 mm diam., 2 cm long. Tests based on constant direct current applied 0.5 hours. Inhibition determined from inhibition zone diameter for *S. aureus* in BHI agar gel medium at 38°. Ag oxidized calculated on the basis of Faraday's Law. (1 μAh = 0.0036 Coulomb)

	INHIBITION THRESHOLD	FULL INHIBITION
Total Current	0.5 μA	5 μA
Total Charge	0.25 μAh	2.5 μAh
Current Density	0.02 $\mu A/mm^2$	0.2 $\mu A/mm^2$
Charge Density	0.01 $\mu Ah/mm^2$	0.1 $\mu Ah/mm^2$
Ag Oxidized	4 nM/mm^2	40 nM/mm^2

weaker current flowing for a longer period. Therefore, we can introduce the concept of "activation charge," or that electrical charge (current x time) required to provide a given level of inhibition for a given configuration of silver electrode, organism, or medium.

The lack of zone size increase above the maximum potential or current confirms the concept noted earlier that diffusion processes, rather than electrophoresis is probably the limiting factor. Using Faraday's constant and Table 2, the amount of silver oxidized at the surface of a silver electrode wire in an agar culture, giving a maximum complete inhibition zone is about 40 nM/mm^2. As will be noted below, only a portion of this oxidized silver ever leaves the electrode surface and enters the bulk fluid.

PERSISTENCE AND REACTIVATION

Early experiments showed that metallic silver, once activated (anodized) electrochemically, would continue to inhibit bacteria after the current was stopped (5). But for how long? This question has relevance for both the mechanism of silver transport from the electrode as well as its clinical application. A partial answer to this was obtained from some simple experiments using coupons of electrochemically pre-activated silver wire incubated in cell culture fluid (RPMI-Gibco) which contained physiological salts and amino acids (44,47). The wires were anodized in normal saline solution for 1 hour at 1 $\mu A/mm^2$ and then added to vials of culture medium at 37°C. The fluid was changed 3 times per week. At intervals, coupons were removed and placed in bacterial culture plates inoculated with S. aureus, cultured for 24 hours and inhibition zones were measured. Experiments were repeated using various amounts of fetal calf serum added to the incubation fluid as well as to normal saline fluid.

The ability to inhibit bacteria persisted longest (at least 10 weeks) in plain normal saline solution, and was reduced to 4 weeks in culture fluid, and even less when 20% serum was added. Adding calf serum to normal saline showed a progressive reduction in the persistence of bacterial inhibition with

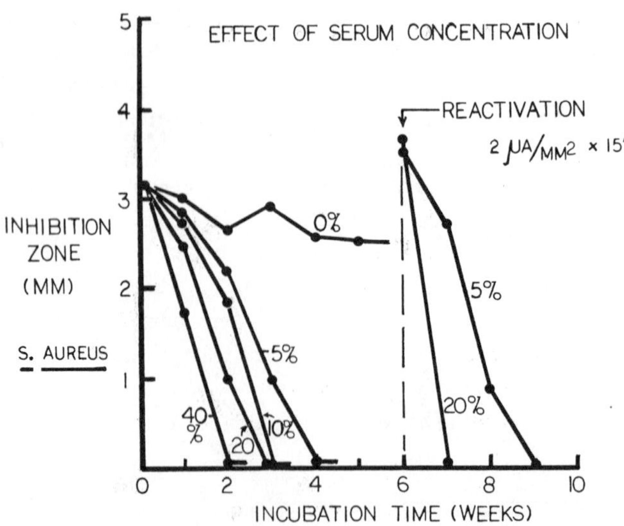

FIGURE 3. Bacterial inhibition zone (radius) from coupons of anodically activated Ag wire, after exposure to solutions of normal saline with various amounts of serum; S. aureus in BHI agar. Application of current a second time, after 6 weeks, revived the inhibition.

increasing amounts of serum (Figure 3). Natural extracellular fluids would correspond best to the 5% serum case, suggesting that this response is to be expected in vivo as well.

This reduction with time and protein content most likely results from protein binding to the electrode, thereby preventing diffusion from the zone of saturated AgCl near the electrode surface. Amino acids and other medium components can also contribute to this result (23,43,48). Some of the more soluble ionized silver may be lost by simple dissolution during the first hours of incubation, resulting in the initial reduction in inhibition observed, even in 0.1 M saline. By electrochemically reactivating additional coupons after 5% serum exposure, inhibition was restored (Figure 3) and again began to decrease. This result suggests the need to reactivate at intervals in order to maintain bacterial inhibition near a metallic silver anode.

SILVER CONCENTRATION IN SURROUNDING FLUID

Using constant anodic current activation, the amount of silver migrating into physiological fluid containing a silver electrode was measured in vitro. A current of 0.2 $\mu A/mm^2$ (60 μA total) was applied for 4 hours and the silver concentration in the 30 ml of Dulbecco's medium (20% serum) was measured by emission spectroscopy from aliquots removed at intervals (29). Results (Figure 4) showed that silver concentration increased steadily with time under these conditions. Centrifugation of the fluid did not remove any silver from the solution, whereas protein precipitation by trichloracetic acid removed 97% of it. This strongly suggests that the proteins in the fluid bind most of the silver leaving the electrode and that most of the complexed silver can remain in solution for a significant time. It is another indication of the important role played by proteins in the process of bacterial inhibition by silver electrodes. Other questions concerning the silver entering the fluid still remain to be answered. For example, what is the role of the fluid proteins in the rate of silver evolution? How does the binding to protein affect the inhibition of bacteria and the biocompatibility of the silver electrode?

It is also noteworthy that for smooth wire, the actual amount of silver in the surrounding fluid after a short time is much lower than the theoretical amount of silver oxidized by the

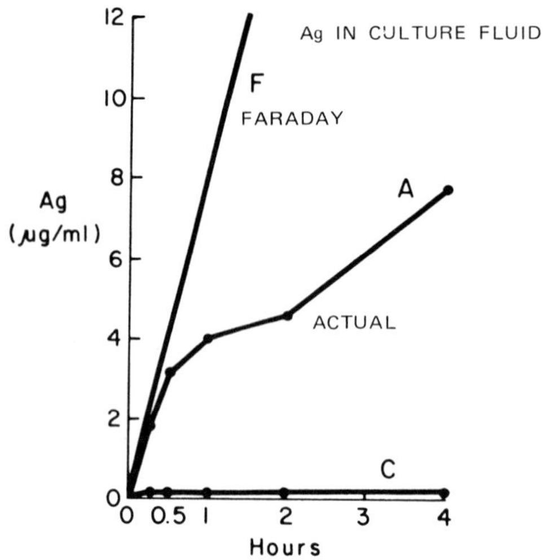

FIGURE 4. Ag concentration vs time in culture solution (nutrient broth) containing proteins and a silver anode carrying 2 μA/cm-length (0.083 μA/mm^2 area). A significant amount of the expected oxidized Ag (Faraday) does not appear in solution and remains attached to the electrode surface.

current (Faraday's constant = 4 μg/h/μA)(29,47,49). This is assumed to be the result of the formation of insoluble AgCl on the electrode surface. In this experiment approximately 75% of the oxidized silver remains behind on the electrode. The same behavior was observed in a brief study of the silver deposited in muscle tissue in vitro after insertion of a silver wire anode for up to 16 hours (47).

PENETRATION DEPTH

A key issue concerning the usefulness of anodically activated silver is the distance from the electrode that antibacterial doses can be formed. Many of the same phenomena that regulate activity at the interface, limit diffusion into the surrounding tissue or medium, chiefly protein binding and chloride precipitation. Tissue density and structure is another important factor. Penetration into bone, for example, will be limited because of its density and also because bone mineral, bone collagen, and tendon all have extremely strong binding affinities for silver ions (50,51). Silver, even in nitrate form, has been found to penetrate poorly into eschar and skin, presumably for the same reasons (6,52). As previously indicated, agar diffusion zones _in vitro_ for silver anodes are small in contrast to that of soluble antibiotics (and silver nitrate), usually measuring 2-5 mm from the electrode surface (Figure 1). Deitch et al. found similar values for inhibition distances above silver nylon fabric electrodes in agar (53). Electrophoretic enhancement of silver transport does not seem to occur at the low voltages used for activation (44,45). While the weak transport of silver ions through dense tissue appears to limit its antibacterial application, it may also be beneficial in preventing high silver concentrations from accumulating in body fluids and perhaps in allowing slow release of small quantities over a long period.

ANIMAL STUDIES OF BIOCOMPATIBILITY AND BACTERIAL INHIBITION

BIOCOMPATIBILITY

The literature contains a great deal of information on the toxicity of silver compounds (6,32,54,55). Investigators using _in vitro_ systems report some toxicity to animal cells (56,57). Animal and human data, on the other hand, suggest that silver compounds may be deposited in tissues and organs after prolonged exposure (argyria), but that toxicity is low, probably due in large part to protein binding and the formation of stable precipitates or reduction to metallic silver particles (6).

In the case of metallic silver and silver electrodes, a few studies are available. Silver-specific carcinogenicity probably does not occur (58). In implants into the peritoneal cavity of rats, silver did not elicit a foreign body reaction, although cellularity was reduced (59). Vascular smooth muscle lost its reactivity to norepinephrine with silver or silver chloride, except in the presence of serum protein (60). Johnson et al. found that intramuscular silver implants were less toxic in their reactivity than polyethylene and platinum, in contrast to _in vitro_ cell cultures which tend to show higher toxicity in general (57). Several reports have shown cell-specific changes in marrow cell populations exposed _in vitro_ to silver anodes

(1-2 μA) for 4 hours (25,61,62). Although some cell lysis was observed, neutrophils and eosinophils were increased in number, while immature blood cells were decreased, suggesting yet unexplained changes in staining characteristics and/or maturation. Silver cathodes did not show these effects and have furthermore been shown to stimulate osteogenesis in animals (46,61,62,63).

A series of percutaneous and totally implanted silver-plated transverse pins in the rabbit tibia for up to 6 weeks were studied microscopically for tissue reaction in bone, marrow, and overlying soft tissue following a single activation by anodic current at 20 μA for 1 hour (64,75). Upon comparison to plain stainless steel control pins in the contralateral tibia, no significant differences were observed in the number of inflammatory cells, necrosis, fibrosis, and bone sclerosis. Mechanical pull-out testing of the pins after sacrifice showed a slight increase in retention for silver-plated pins over the stainless controls. Microscopic particles, subcellular in size, were observed in tissue adjacent to the silver-coated pins and were assumed to be insoluble silver precipitate. No adverse reaction was associated with these particles. The results of this study are in agreement with others indicating a low toxicity of metallic silver implants in animals (6,46,59,60).

BACTERIAL INHIBITION

A few animal studies of the inhibitory or anti-infective properties of metallic silver anodes have been performed. Such experiments are difficult because of uncontrollable variables in dealing with implant infections in animal models, such as the natural resistance to bacterial infection, the difficulty in monitoring the bacterial concentrations, and in applying electrical current. Yuan et al. observed a reduction in the incidence of Enterobacter cloacae osteomyelitis in rabbits following inoculation and infection of the tibial medullary canal and insertion of a silver wire, activated with anodic direct current (65). Colmano et al., in a number of studies in rabbits, was able to inhibit S. aureus infection in the inoculated tibial canal using 9-12 μA applied to silver-coated stainless steel medullary pins for 1 hour after introducing the bacteria (66-68). The pins were coated with silver stearate monolayers (the best results), or were silver plated. Direct injection of AgNO3 was less effective than the coated pins.

Tamura treated infected rabbit tibiae with silver electrodes traversing the bone shaft using direct current, and compared the results to stainless steel electrodes (69). It was found that infections were reduced by 75% at silver in this model.

In a rat model of percutaneous implants, electrically activated silver was directly compared to conventional stainless steel in the ability to inhibit bacterial colonization (70,75). Percutaneous loops were inserted through the skin on either side

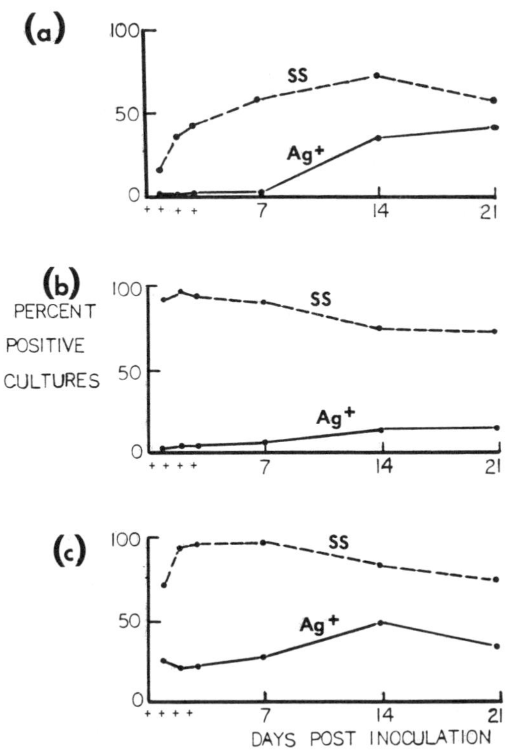

FIGURE 5. Bacterial culture tests of percutaneous tracts around electrically activated silver (Ag^+) and stainless steel control (SS) implants in rats (70). The ordinate is the fraction of streak tests positive for bacteria subsequent to inoculation at day 0 with S. aureus: (a) inoculum = 2×10^4 CFU/ml; (b) inoculum = 3×10^5 CFU/ml; (c) inoculum = 2×10^6 CFU/ml. The reduction in colonization at the Ag anode is statistically significant ($p < 0.05$). Anode activation (20 μAh) was applied once daily during the first 4 days.

of the spine, silver on one side and stainless steel on the other. Each loop was inoculated with various concentrations of *S. aureus* at the entry tracts. Anodic activation (5-50 μAh) was applied to the silver loops on days 0, 1, 2, and 3. Cultures taken from the skin tracts throughout the 3-week experiment showed that bacterial concentrations were significantly reduced at the silver implants (Figure 5). A surface-activation charge density of 1 $\mu Ah/mm^2$ gave maximal inhibition under these conditions. Starting DC activation 1 week after bacterial infection was established also resulted in clearing bacteria from the silver implants. Bacterial counts began to rise 7-10 days after activation, confirming the conclusion from previous *in vitro* experiments that some form of continued regular activation may be required to prevent recolonization (44,47). This is apparently the first study to demonstrate inhibition of bacterial colonization by electrical activation of silver at percutaneous implants *in vivo*.

CLINICAL APPLICATIONS

Several clinical applications have been suggested for electrically activated silver antisepsis, and at least 3 have been evaluated in preliminary patient studies.

WOUND MANAGEMENT IN CHRONIC OSTEOMYELITIS

Becker and Spadaro have used anodically activated silver-coated fabric as an antibacterial dressing for surgical wounds following debridement or resection of infected bone (71). In this method, conductive silver fabric is applied to exposed tissues of the entire wound bed immediately after debridement. The silver fabric is covered with moist gauze dressings and anodically driven with respect to a skin return electrode by means of a constant-current source (Figure 6). Approximately 1 $\mu A/cm^2$ of anode area was applied continuously. Silver fabric and dressings were changed daily with irrigation until the wound closed or skin grafting was performed.

The results of 25 patients treated with silver anode dressings for chronic osteomyelitis showed that wound closure was

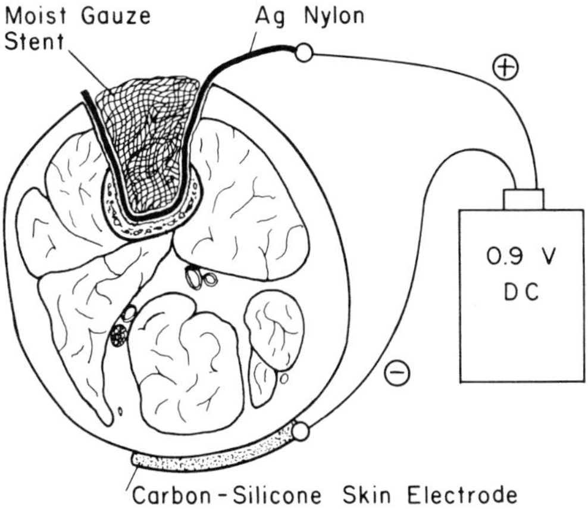

FIGURE 6. **Diagramatic cross-section illustrating the use of conductive, silver-coated fabric anodes in open wounds following surgical debridement in chronic osteomyelitis in the long bones (71,72).**

achieved in 64% (72). Bacterial counts taken from the open wounds decreased steadily during the first weeks of this treatment. The advantages of this application of electrically activated silver are that systemic antibiotics are avoided, and the conditions in the wound bed favor the formation of granulation tissue and promote healing.

Becker et al. introduced the idea of combining cathodic DC stimulation of fracture healing with anodic inhibition of bacterial infection around the implanted silver electrode (71,73).

TREATMENT OF CUTANEOUS ULCERS

A natural extension of the above is the application of silver-coated fabric anodes to cutaneous ulcers. The same principles and advantages apply as for the treatment of chronic osteomyelitis. Although no clinical trials have been reported, preliminary experience with this application has been positive (71). In vitro studies of inhibition by silver-coated nylon of bacteria commonly found in cutaneous wounds further suggests the usefulness of this approach (27). A study in pigs of dermal and epidermal wound healing under silver-coated fabric electrodes showed increased collagen production and epithelialization (74).

SILVER ANODE PERCUTANEOUS SKELETAL FIXATION PINS

This application has been suggested by in vitro and animal experiments which have demonstrated the biocompatibility and bacterial inhibition by percutaneous electrically activated silver pins and wires (64-66,69,70,75). Further testing in more advanced animal models is being pursued by several investigators.

DENTAL APPLICATIONS

Silver anode inhibition of oral bacteria has suggested its use in sterilizing pulp canals during endodontic procedures (26,76,77). Silver wires (without current) have been used extensively in the past for filling root canals following infection.

WATER PURIFICATION

The use of electrically activated silver to treat drinking water supplies has enjoyed some success and reduces the need for extensive chloridization (15,34).

FUTURE DIRECTIONS

There are perhaps more questions than answers raised by this review of electrical silver antisepsis. First, the biological interactions of silver ions and various microorganisms are not well understood. Although the spectrum of sensitivity seems broad, it is likely that different organisms are affected through different mechanisms. The question of resistance is part of this. Much of the past research on silver was not on electrode or metallic silver, but on soluble compounds and complexes.

The next area of future inquiry is a very practical one: silver transport from the electrode to the target cell or tissue. The limitation in inhibition zone size in vitro is symptomatic of the problem in intact tissue, where tissue planes and membranes place additional barriers to non-vascular ionic transport. Complexation to carrier molecules is one possible solution. Electrophoretic enhancement of transport may be made feasible. Protein binding followed by slow release may be proved beneficial.

At the electrode surface, the chloridization process microscopically demonstrates AgCl crystallite formation in dilute NaCl, but what is the effect of various current or potential levels, or various constituents in the body fluid? What is the advantage, if any, of low-level electrical current flow as compared to intermittent application at higher levels?

The biocompatibility of silver anode implants seems to be much less of a problem than might be anticipated on the basis of cell culture studies. But thorough studies with various materials and geometries have yet to be performed. The production of strong, adherent silver coatings on implantable base materials is a non-trivial technical question. Galvanic currents due to bimetallic composites may be a problem in some cases and beneficial in others. The discrepancy between cell culture indications of toxicity and those from animal studies and human applications, where toxicity is rarely observed, needs to be resolved.

In the areas of clinical interest, there is already evidence of antiviral and antitumor activity (78,79). Further research of these properties and their possible applications to localized lesions may be fruitful. Electrode-derived silver may be a useful alternative to topical pharmaceuticals for the treatment of burn wounds (where silver sulfadiazine is already the agent of choice) and other cutaneous lesions, as has already been noted. Lastly, observations of the possibility of stimulatory effects on wound healing and cell transformation by silver are intriguing and may herald a new direction for biomedical research (46,61,71,80).

REFERENCES

(1) Rowley, B.A.: Electrical current effects on _E. coli_ growth rates, Proc. Soc. Exp. Biol. Med. 139:929, 1972.

(2) Ebert, L. and Evtushenko, A.D.: Changes in bacterial

sensitivity to antibiotics under the action of constant current, Antibiotiki 16:641, 1971.

(3) Pareilleux, A. and Sicard, N.: Lethal effects of electric current on E. coli, Appl. Microbiol. 19:421, 1970.

(4) Anderson, A.K. and Finkelstein, R.: A study of electropure process of treating milk, J. Dairy Sci. 2:374, 1919.

(5) Spadaro, J.A., Berger, T.J., Barranco, S.D., Chapin, S.E. and Becker, R.O.: Antibacterial effects of silver electrodes with weak direct current, Antimicrob. Agents Chemother. 6:637, 1974.

(6) Hill, W.R. and Pillsbury, D.M.: Argyria - The Pharmacology of Silver, Williams and Wilkins, Baltimore, 1939.

(7) Higginbottom, J.: A Practical Essay on the Use of Nitrate of Silver in the Treatment of Inflammation, Wounds and Ulcers, Monograph, London, 1865.

(8) Sims, J.M.: On the treatment of vesicovaginal fistula, Arch. Gynaekol. 17:51, 1852.

(9) Crede, C.F.: Die Verfutung der Augenenzundung der Neugenborenen, Arch. Gynaekol. 17:50, 1881.

(10) von Naegeli, K.: Denkschr. Schweiz. Naturforsch. Ges. 33:174, 1893.

(11) Halsted, W.S.: Ligature and suture material, J. Am. Med. Assoc. 60:1119, 1913.

(12) Fox, C.: Silver sulfadiazine for control of burn wound infections, Int. Surg. 60:275, 1975.

(13) Hartford, C.E.: The bequests of Montcrief and Moyer - An appraisal of the topical therapy of burns, J. Trauma 21:827, 1981.

(14) Newell, F.: Centenary of crede prophylaxis, Am. J. Ophthalmol. 90:874, 1980.

(15) Chambers, C.W. and Clarke, N.A.: Control of bacteria in nondomestic water supplies, Adv. Appl. Microbiol. 8:208, 1966.

(16) Jensen, R.H. and Davidson, N.: Spectrographic studies of the binding of silver ions by DNA, Biopolymers 4:17, 1966.

(17) Coward, J.E., Carr, H.S. and Rosenkranz, H.S.: Silver sulfadiazine - Effect on the ultrastructure of Pseudomonas aeruginosa, Antimicrob. Agents Chemother. 3:621, 1973.

(18) Bragg, P.D. and Rainnie, D.J.: The effect of silver ions on the respiratory chain of Escherichia coli, Can. J. Microbiol. 20:883, 1974.

(19) Goodman, A. and Gilman, B.: The Pharmacological Basis of Therapeutics, 7th Edition, Macmillan Publ., New York, 966-967, 1985.

(20) Cowlishaw, J., Spadaro, J.A. and Becker, R.O.: Inhibition of enzyme induction in E. coli by anodic silver, J. Bioelectricity 1:295, 1982.

(21) Spadaro, J.A. and Becker, R.O.: Some specific cellular effects of electrically generated silver and gold ions, Bioelectrochem. Bioenergetics 3:49, 1976.

(22) Spadaro, J.A.: Unpublished observations using scanning electron microscopy of cells exposed in vitro to a silver anode.

(23) Janz, G. and Ives, D.: Silver-silver chloride electrodes, Ann. N.Y. Acad. Sci. 148:210, 1968.

(24) Barranco, S.D., Spadaro, J.A., Berger, T.J. and Becker, R.O.: In vitro effects of weak direct current on S. aureus, Clin. Orthop. 100:250, 1974.

(25) Berger, T.J., Spadaro, J.A., Chapin, S.E. and Becker, R.O.: Electrically generated silver ions - Quantitative effects on bacterial and mammalian cells, Antimicrob. Agents Chemother. 9:357, 1976.

(26) Thibideau, E.A., Handelman, S.L. and Marquis, R.E.: Inhibition and killing of oral bacteria by silver ions generated with low intensity direct current, J. Dent. Res. 57:922, 1978.

(27) Falcone, A. and Spadaro, J.A.: Inhibitory effects of electrically activated silver material on cutaneous wound bacteria, Plast. Reconstr. Surg. 77:455, 1986.

(28) Berger, T.J., Spadaro, J.A., Bierman, R., Chapin, S.E. and Becker, R.O.: Antifungal properties of electrically generated metallic ions, Antimicrob. Agents Chemother. 19:856, 1976.

(29) Spadaro, J.A.: Antibacterial effects of silver electrodes, Proc. 3rd Ann. Conf. Eng. Med. Biol. Soc. (IEEE), Houston, 215, 1981.

(30) Golubovich, V.N.: Toxic effect of ionic silver on different groups of microorganisms, Mikrobiologiya 43:783, 1974.

(31) Carr, H.S., Wlodkowski, T.J. and Rosenkranz, H.S.: Silver sulfadiazine - In vitro antibacterial activity, Antimicrob. Agents Chemother. 4:585, 1973.

(32) Petering, H.G.: Pharmacology and toxicology of heavy metals: Silver, Pharmacol. Ther. A. 1:127, 1976.

(33) AMA Drug Evaluations, 4th Edition, John Wiley and Sons, New York, 1980.

(34) Grier, N.: Silver and its compounds, in Disinfection, Sterilization and Preservation, S. Block, ed., Lea and Febiger, Philadelphia, 2nd Ed., 1977.

(35) Wlodkowski, T.J. and Rosenkranz, H.S.: Antifungal activity of silver sulfadiazine, Lancet 2:739, 1973.

(36) Chang, T. and Weinstein, L.: In vitro activity of silver sulfadiazine against herpesvirus hominis, J. Infec. Dis. 132:79, 1975.

(37) Tokumaru, T., Shimizu, T. and Fox, C.L.: Antiviral activities of silver sulfadiazine in ocular infection, Res. Com. Chem. Pathol. Pharmacol. 8:151, 1974.

(38) McHugh, G.L., Moellering, R.C., Hopkins, R.C. and Swartz, M.N.: Salmonella typhymurium resistant to silver nitrate, chloramphenicol and ampicillin, Lancet 2:235, 1975.

(39) Summers, A.O., Jacoby, G.A. and Swartz, M.N.: Metal cation and oxyanion resistances in plasmids of gram negative bacteria, in Microbiology, Schlessinger, ed., Am. Soc. Microbiol., Washington, DC, 1978.

(40) Joly, B. and Cluzel, R.: Las resistance de Pseudomonas aux antibiotiques et aux metraux lourds, Ann. Microbiol. Inst. Pasteur 127B:57, 1968.

(41) Bridges, K., Kidson, A., Lowbury, E.J. and Williams, M.D.: Gentamicin and silver resistant Pseudomonas in a burn unit, Br. Med. J. 1:446, 1979.

(42) Annear, D.I., Bee, B.J. and Bailey, M.: Instability and linkage of silver resistance, lactose fermentation, and colony structure in Enterobacter cloacae from burn wounds, J. Clin. Pathol. 29:441, 1976.

(43) Tilton, R.C. and Rosenberg, B.: Reversal of silver inhibition of microorganisms by agar, Appl. Environ. Microbiol. 35:1116, 1978.

(44) Spadaro, J.A.: Electrochemical activation of bacterial inhibition near silver implants, Trans. Electrochem. Soc. Mtg. Toronto, 681, May, 1985.

(45) Spadaro, J.A.: Bone formation and bacterial inhibition with silver and other electrodes, Reconstr. Surg. Traumatol. 19:40, 1985.

(46) Spadaro, J.A.: Electrically enhanced osteogenesis at various metal cathodes, J. Biomed. Mat. Res. 16:861, 1982.

(47) Spadaro, J.A., Webster, D.A. and Chase, S.E.: Direct current activation of bacteriostatic silver electrodes, Trans. Bioelec. Growth Repair Soc. 3:37, 1983.

(48) Gruen, L.C.: Interaction of amino acids with silver (I) ions, Biochim. Biophys. Acta 386:270, 1975.

(49) Marino, A.A., Deitch, E. and Albright, J.A.: Electric silver antisepsis, IEEE Trans. Biomed. Eng. BME-32:336, 1985.

(50) Spadaro, J.A., Becker, R.O. and Bachman, C.H.: Size-specific metal complexing sites in native collagen, Nature 225:1134, 1970.

(51) Spadaro, J.A. and Becker, R.O.: Effects of thermal denaturation on metal binding and ultrastructure in collagen fibrils, Biochim. Biophys. Acta 263:585, 1972.

(52) Stefanides, M.M., Copeland, C.E. and Kominos, S.D.: In vitro penetration of topical antiseptics through eschar of burn patients, Ann. Surg. 183:358, 1976.

(53) Deitch, E.A., Marino, A.A., Gillespie, T.E. and Albright, J.A.: Silver nylon: A new antimicrobial agent, Antimicrob. Agents Chemother. 23:356, 1983.

(54) Luckey, T.D., Venugopal, B. and Hutchison, D.: Heavy Metal Toxicity, Safety and Hormology, Academic Press, New York, 1975.

(55) Harrison, H.N.: Pharmacology of sufadiazine silver: Interaction with isolated DNA, Antimicrob. Agents Chemother.

114:281, 1979.

(56) Rangachari, P.K. and Matthews, J.: The effect of Ag^+ on isolated bullfrog gastric mucosa, Am. J. Physiol. 246:G443, 1985.

(57) Johnson, H.J. and Northrup, S.J.: Biocompatibility test procedures for material evaluation _in vitro_, J. Biomed. Mat. Res. 17:571, 1983.

(58) Furst, A. and Schlauder, M.C.: Inactivity of two noble metals as carcinogens, J. Environ. Pathol. Tox. 1:51, 1977.

(59) Keller, J.C., Marshall, G.W. and Kaminsky, E.J.: An _in vivo_ method for the biological evaluation of metal implants, J. Biomed. Mat. Res. 18:829, 1984.

(60) Jackson, W.F. and Duling, B.R.: Toxic effects of silver-silver chloride electrodes on vascular smooth muscle, Circ. Res. 53:105, 1983.

(61) Berger, T.J., Spadaro, J.A., Chapin, S.E. and Becker, R.O.: Orthopedic electrodes: Metal and polarity evaluation in short term marrow culture, Proc. 3rd Ann. Mtg. Soc. Biomat., 148, 1977.

(62) Marino, A.A., Berger, T.J., Becker, R.O. and Spadaro, J.A.: The effect of selected metals on marrow cells in culture, Chem. Biol. Interact. 9:217, 1974.

(63) Tarr, R., Luck, J. Jr., Snyder, S. and Mills, B.: Laboratory experiences with silver electrode bone stimulation, Trans. Orthop. Res. Soc. 29:253, 1983.

(64) Spadaro, J.A., Webster, D.A., Kovach, J. and Chase, S.E.: Biocompatibility studies of silver coated fixation pins, Proc. 10th Ann. Mtg. Soc. Biomat. 10:282, 1984.

(65) Yuan, H.A., Spadaro, J.A., Berger, T. and Becker, R.O.: Electrically generated silver ions as a bactericidal agent

in acute and chronic *Enterobacter cloacae* osteomyelitis in rabbits, Trans. Orthop. Res. Soc. 2:28, 1977.

(66) Colmano, G., Edwards, S. and Barranco, D.S.: Activated silver coatings for surgical implants, Virginia Med. J. 106:928, 1979.

(67) Colmano, G., Edwards, S. and Barranco, D.S.: Activation of antibacterial silver coatings on surgical implants by direct current: Preliminary studies in rabbits, Am. J. Vet. Res. 41:964, 1980.

(68) Colmano, G., Fainter, L.K., Edwards, S. and Barranco, D.S.: SEM of *S. aureus* on current-activated surgical pins coated with silver and silver stearate monolayers, Trans. Bioelect. Growth Repair Soc. 2:79, 1982.

(69) Tamura, K.: Some effects of weak direct current and silver ions on experimental osteomyelitis and their clinical application, J. Jpn. Med. Assn. 57:187, 1983.

(70) Spadaro, J.A., Chase, S.E. and Webster, D.A.: Bacterial inhibition by electrical activation of percutaneous silver implants, J. Biomed. Mater. Res., in press, 1986; also Trans. Bioelect. Repair Growth Soc. 4:30, 1984.

(71) Becker, R.O. and Spadaro, J.A.: Treatment of orthopedic infections with electrically generated silver ions, J. Bone Joint Surg. 60A:871, 1978.

(72) Webster, D.A., Spadaro, J.A., Becker, R.O. and Kramer, S.: Silver anode treatment of chronic osteomyelitis, Clin. Orthop. 161:105, 1981.

(73) Becker, R.O., Spadaro, J.A. and Marino, A.A.: Clinical experiences with low intensity direct current stimulation of bone growth, Clin. Orthop. 124:75, 1977.

(74) Alvarez, O.M., Mertz, P., Smerbeck, R. and Eaglestein, W.: The healing of partial thickness wounds is stimulated by

external electric current, J. Invest. Dermatol. 78:353, 1982.

(75) Spadaro, J.A., Webster, D.A. and Chase, S.E.: Antibacterial fixation pins with silver - Animal models, Trans. Orthop. Res. Soc. 9:335, 1984.

(76) Handelman, S.L., MacIntyre, B. and Ellis, F.H.: Electrophoretic effect of silver ions on oral bacteria, J. Endodontics 3:458, 1977.

(77) Ellis, F.H. and Thibodeau, E.: Material and Method for Introducing Ionic Silver to Dental Pulp, U.S. Patent #4,126,937, 1978.

(78) Shimizu, F.: Specific inactivation of herpes simplex virus by silver nitrate at low concentrations and biological activities of the inactivated virus, Antimicrob. Agents Chemother. 10:57, 1976.

(79) Spadaro, J.A., Berger, T., Chapin, S. and Becker, R.O.: Antitumor effects of silver electrodes _in vitro_, Trans. Soc. Biomaterials 3:149, 1977.

(80) Becker, R.O.: Processes and Products Involving Cell Modification, U.S. Patent #4,528,265, July 9, 1985.

20
Direct Current and Bone Growth

ANDREW A. MARINO
Department of Orthopaedic Surgery, Louisiana
State University School of Medicine in Shreveport
Shreveport, Louisiana

INTRODUCTION

To a significant degree, modern bioelectricity began with the study of the bioelectrical phenomena of bone. An introduction to the subject has recently been published (1), and much of the pertinent literature has been listed (2). My purpose here is to describe only what seem to me to be the most significant developments.

After a brief review of the composition, structure, and physiology of bone, the seminal experiments that led to the birth of bone bioelectricity are described. I paid particular attention to the rationale behind the early studies in the belief that it would be helpful to the reader to see how the threads that were the initiatives of different groups have come together to form the present fabric. It is the work of bold men who broke new ground and who were not deterred by the possibility of making some errors.

After the rush of new ideas in the 1960s, there ensued the present period of consolidation in which the major effort has been the sorting of the wheat and the chaff. A description of this work constitutes the principal part of the balance of the chapter. My view of the basis of electrical osteogenesis best warranted by the present evidence is presented in the last section. I think that there is a bright future for the clinical use of direct current for treating bone when the treatment is designed in a manner consistent with that basis.

BONE PROPERTIES

MATERIAL CHARACTERISTICS

Bone is mainly composed of collagen, a protein, and hydroxyapatite, an inorganic calcium salt. Collagen is secreted and then assembled extracellularly to form a matrix on which the hydroxyapatite is deposited, perhaps via epitaxial precipitation (3,4). The processes of collagen polymerization and hydroxyapatite precipitation occur in a spatial and temporal sequence that ultimately results in the formation of microscopically recognizable layers called lamellae. Lamellation is an end-stage structural pattern; sometimes it is preceded by woven bone, a less organized pattern in which the collagen fibers are randomly arranged like the fibers in a felt. Woven bone occurs in various pathological conditions, but in normal physiology it is an interim material, and is replaced by lamellar bone (5).

Bone occurs in two architectural forms. Compact (cortical) bone, typified by the shaft of the long bones, is relatively dense and its lamellae exhibit several patterns, the most common of which is concentricity around a vascular canal. Cancellous (trabecular) bone is a three-dimensional lacy network usually found inside bones, particularly at the ends of the long bones. Its lamellae, when present, usually run parallel to the trabeculae (5). Sometimes, such as in some instances of fracture healing, the cancellous architectural form appears on the outside of bones as a temporary means of stabilization called external callus.

CELLS OF BONE

Collagen is secreted by the osteoblast, a medium-sized cuboidal cell. Osteocytes are osteoblasts that have become enveloped by bone. They occupy melon-seed shaped lacunae within the bone and form a syncytium via cell processes that extend through channels called canaliculi. Osteocytes might have the ability to remove bone within about 1 micron of the lacunar surface (6), but the major bone-destroying cell is the osteoclast. Osteoclasts are large multinucleated cells that resemble foreign-body giant cells, but appear to lack the latter's phagocytic function (except with regard to the bone). Osteoclasts were previously believed to form via fusion of osteoblasts, but are now thought to arise from a specialized precursor cell that differs ultrastructurally from the cell that gives rise to the osteoblast (6,7).

Bone-lining cells are extremely flattened cells that cover the surfaces of cancellous and compact bone, as well as the surface of the osteonal canal (8). Although the bone-lining cell is the most common surface cell of bone (6,9), its function is not understood.

Behind the bone-lining cells are the osteoprogenitor cells, the immediate precursors of the differentiated cells of bone (10). Osteoprogenitor cells are capable of division, but the more differentiated cells (with the possible exception of the osteoblast (11)) do not divide.

The undifferentiated mesenchymal cell, which is present throughout the body, is also capable of producing bone in the postfetal organism. It forms heterotopic bone in the presence of an inducer and in certain pathological states, in contrast to the osteoprogenitor cell which forms orthotopic bone in the course of either normal development or repair. The undifferentiated mesenchymal cell and the osteoprogenitor cell are morphologically indistinguishable (12). Whether the former is a less differentiated precursor of the latter is an open question (13), but the distinction between them has been recognized since at least 1952 (14).

BONE GROWTH

The activity of bone is manifested on its various surfaces, each of which, at any given time, is either quiescent, forming, or resorbing bone (15). Excluding disease, there are four bone-forming activities that result in the production of orthotopic bone: development determines size, modeling determines shape, remodeling produces replacement bone, repair restores integrity. Although all bone-forming arises from the activity of osteoblasts, many aspects of the process differ among the listed activities. These include the histological environment in which the osteoblast functions, the pattern of the bone produced, the rate of bone production, the nature of the physiological stimulus that controls the process, and the nature of the artificial stimuli that can affect it. It is helpful to recognize the essential features of the known bone-forming activities so that they can be compared with the effects produced by electricity.

1. Development

In man, increase in length of bone occurs at the epiphyseal-metaphyseal complex (growth plate), where septa of calcified cartilage are covered with woven bone, thereby creating primary trabecular bone. The woven bone and calcified cartilage core is then removed and replaced by lamellar bone (6). Apositional growth in the developing bones leading to an increase in thickness occurs via periosteal formation of primary trabeculae, compaction to form cortical bone, and then replacement by lamellar bone (6). Periosteal bone growth does not involve the cartilage template step (called endochondral growth) and is termed intramembranous growth. Many factors including diet, drugs, disease, and exercise can affect bone growth in the sense of producing changes in cell kinetics or histomorphology. No factor, however, has been identified that actually controls the growth process. At some level of organization within the developing organism a genetic program is apparently read and executed. Embryonic tibiae when placed in tissue culture will continue to develop for a period in a manner resembling normal growth (16). Consequently, at least a portion of the control system resides in the developing bone itself.

During bone development, endochondral bone formation occurs at the growth plate, and intramembranous bone formation occurs along the periosteum. In both cases, woven bone is initially formed and subsequently replaced by lamellar bone. The intramembranous bone forms at different rates in different parts of a given bone, thereby producing an anatomic phenomenon known as modeling. Enlargement of the cranial vault and development of the ends of the long bones are typical examples (6).

Growing bone exhibits a plasticity principle (sometimes called modeling, sometimes remodeling) which permits it to adapt to mechanical forces by changing its geometry. The healed malaligned fracture is an example. Growth occurs on the concave side and resorption takes place on the opposite side, and the bone ultimately becomes straight. Another example of adaptive growth is the movement of teeth in the jaw caused by orthodontic appliances. In this case, resorption occurs in the portion of the bone in compression, thereby permitting the tooth to move in the direction of the applied force. The pattern of trabecular bone in the metaphysis is often said to reflect adaptive changes; if it does, that would be a third example of adaptive remodeling. The adaptive response of bone is usually manifest as a modification of ongoing development (as part of the maturation process of the bone), although it also occurs to a lesser extent in adult life (17).

2. Remodeling

By remodeling I mean the process by which existing bone is replaced by lamellar bone. Remodeling first occurs soon after birth when the primary osteons are replaced by the more highly structured secondary osteons that are characteristic of the adult. Remodeling continues throughout adult life as old secondary osteons are replaced by new ones. It is convenient to think of the remodeling of the primary and secondary osteons as a process of replacement of worn, fatigued, or weak bone by a stronger material, although there is little reason to believe that this is actually the reason that the process occurs. The initial event in remodeling is resorption of bone, followed directly by its replacement by lamellar bone on an equal volume basis (17). Its causes and control system are unknown (17). Remodeling also takes place during repair.

3. Repair

The histological appearance, tempo, and geometry of reparative bone growth depend on many factors including the particular bone and the nature of the injury. Consider for example a traverse fracture in a long bone produced by bending to failure. A blood clot forms in the gap between the bone fragments, and proliferation of the osteoprogenitor cells begins within about 24 hours. In a few days, the clot is replaced by fibrous tissue. Cartilage may appear, particularly if the bone fragments are unstable, but it is not an essential constituent of the healing process. When it does appear, endochondral ossification ensues, resulting in the production of woven bone in a trabecular architecture called callus. The callus may occur on the outside of the bone to form a bridging substance between the fragments, in which case it assumes a classic fusiform shape. The callus may also extend throughout the medullary canal in the region of the lesion. In the final stages of healing, the callus within the medullary canal and the external callus are resorbed, and the callus within the fracture gap along the path of the cortex becomes compacted and remodeled. Ultimately, the fracture gap becomes histologically indistinguishable from the uninjured portions of the bone.

If the bone fragments are rigidly approximated with a compression plate, neither cartilage nor callus are formed, and healing occurs by accelerated remodeling (18).

ELECTRICITY AND BONE: FOUNDATIONS

The beginning of modern bioelectrical research involving bone is generally traced to Iwao Yasuda, a Japanese orthopaedic surgeon (20-22). Yasuda was primarily concerned with the factor responsible for initiating callus formation. He began his work in the 1940's, a time when the realization was developing that bone callus formation was not inextricably linked with bone fracture. A variety of thermal, chemical, and mechanical stimuli were being identified that could produce bone callus in the absence of fracture (19). Yasuda applied mechanical force to long bones and observed callus formation in nontraumatized areas under compression. He also showed that viable and boiled bone yielded an electrical signal when subjected to mechanical deformation. Regions in apparent compression were negative and those in tension were positive with respect to unstressed areas. Yasuda's idea was that callus-producing cells in bone subjected to mechanical and other stimuli were not directly responding to the stimuli, but to a common-pathway signal. He theorized that electricity was the common-pathway signal for callus, and that application of electricity should produce callus. Some time between 1949-1953, he demonstrated that 1-100 μA of direct current (DC) produced callus in the medullary canal of rabbits (20). Thus one form of stimulus that produced callus — mechanical forces -- also produced electricity, and externally applied electricity produced callus.

Yasuda's work was distinctive because it emerged from his analysis of the developing knowledge of bone physiology, and his attempt to synthesize diverse observations. He was not the first to produce callus using DC current. In 1860, Garratt described the use of percutaneous insulated electrodes in the successful treatment of a nonunion of two years' duration, employing stimulation for 15 minutes a day, every third day, for 9 days (23). But Garratt's and other earlier reports were based on simple empiricism, and stemmed from a time at which electricity was a newly-discovered phenomenon. It was the novelty of electricity rather than its reasoned relevance, that formed the basis of the historical applications of electricity to living organisms.

In 1957, Fukada and Yasuda measured the numerical value of the piezoelectric constant of dry bone (24). Application of a shearing force along the bone axis caused a voltage to appear on bone surfaces parallel to the axis. Conversely, application of an external voltage to bone caused it to mechanically deform. The strength of the piezoelectric effect in bone (the polarization per unit stress or the strain per unit electric field) was about a tenth of that exhibited by quartz.

The first American investigator to report data regarding electricity and bone was the orthopaedic surgeon Robert O. Becker (25-28). Becker began studies of limb regeneration in the late 1950s. His particular interest was the factors that controlled or regulated the process, which he assumed were the same whenever regeneration occurred throughout the vertebrate phylum. The existence of surface electrical potentials (SEP) (slowly varying millivolt-strength electrical potentials that can be measured between any two points on a surface of a living organism) had been known for many years, although their origin and physiological significance remained unascertained. It had been suggested that SEPs were significant factors in embryonic development, and that they served to guide and perhaps determine organization, differentiation, and specialization in the developing organism (29,30). Since the time of Galvani, the existence of electrical changes at an injury site had also been known. Becker combined the two observations and hypothesized that the SEPs might have a role in the control of the regenerative process. He measured SEPs in salamanders, and observed a spatial pattern that roughly corresponded to the nervous system of the animal (25). He believed that the SEPs originated in neural or perineural tissue, and he concluded that the complexity of the pattern made it a candidate as a control mechanism for regeneration (26,27). Becker regarded the association with nerves as particularly important, because Singer had previously shown that the presence of a critical amount of nerve tissue within the amputation stump was required for regeneration to occur (116).

Becker amputated the forelimbs of salamanders (a regenerating species) and frogs (non-regenerators) and measured the SEP at the distal amputation stump relative to the proximal uninjured tissue in the limb (26). He found that the characteristic response following amputation was the occurrence of a positive spike in the SEP within about 1 day of the amputation, followed by a decrease to 20-30 mV at 5-10 days following amputation. Thereafter, the potential at the amputation stump approached the normal (negative) value from the negative side. The frogs also exhibited a positive peak within about a day of amputation, but thereafter the SEP decreased monotonically toward the normal (negative) value from the positive side. Thus, Becker identified a negative SEP, measured relative to adjacent uninjured tissue at about 5-10 days after amputation, with the phenomenon of regeneration.

Fracture-healing is also a regenerative phenomenon, and Becker found a similar temporal change in the SEP following a fracture of a long bone in the salamander (26). Again, the normal negative SEP at the fracture site became positive following fracture, and then became negative at 10-15 days thereafter.

Zachary B. Friedenberg, an orthopaedic surgeon at the University of Pennsylvania, conducted a detailed study of the SEP in fractured limbs of rabbits and patients (117). When Friedenberg's data is expressed similarly to that of Becker's (SEP at the injury site relative to adjacent uninjured tissue), it also shows that the injury site became positive (relative to uninjured bone) immediately after fracture, as described by Becker. In 16 patients with healing tibial fractures, the bone remained positive (relative to the proximal epiphysis), in apparent distinction to the negative values that were recorded by Becker in animals.

Becker presented his SEP data on regeneration and fracture healing at the 1961 meeting of the American Academy of Orthopaedic Surgery, and it led to a short but important period of collaborative research with C.A.L. Bassett, an orthopaedic surgeon at Columbia University. Bassett was particularly interested in the physiology of bone formation, and their mutual interest centered on the question of the control process for adaptive remodeling. They reasoned that electrical potentials had been associated with regeneration (citing Becker's work, and other references previously cited by Becker), and that electrical potentials might also be the basic link in the clinically observed adaptive response that occurs in children with healed malaligned fractures. In 1962, Bassett and Becker reported that moist human and frog bone yielded an electrical signal when subjected to cantilever bending (31). The signal decreased by only 5% when the tissue was dried, and was asymmetric in the sense that equal and opposite voltage pulses were not seen attendant loading and unloading the bone. Regions of the bone in apparent compression (concave side) were negative relative to regions in tension.

The interpretation of the stress-generated signal in terms of adaptive growth was subsequently described more fully (32). The endogenous electrical phenomenon was hypothesized to be capable of directing the activity of bone cells in such a way as to account for bone's adaptive behavior. Formation of insoluble bands of collagen was reported to occur when DC currents were passed through solutions containing collagen molecules. Based on this observation, the stress-generated signals in bone were hypothesized to also be responsible for orientation and extracellular aggregation of collagen (32).

I observed formation of bands in previously clear solutions of rat-tail tendon collagen within a few minutes of the initiation of current (1-75 μA) (33). The bands formed only in regions where the pH was raised above about 8.5 as a consequence of electrochemical changes at the cathode. Raising the pH of the collagen solution to 8.5 by the addition of sodium hydroxide quickly precipitated the collagen. Thus, if stress-generated electrical signals in bone do not alter local pH -- there is no good evidence to indicate that they do -- then it is unlikely that they can have the hypothesized non-cellular consequences.

What was the origin of the stress-generated electrical signals? It was argued that piezoelectricity could not account for the signals measured in bones that contained their normal water content (wet bone) because a piezoelectric material yields a symmetric signal upon application and release of the applied force, whereas asymmetric signals had been observed (32). In addition, both Becker and Bassett were conceptually dissatisfied with the piezoelectric mechanism as the source of the bone potentials because they believed that its inherent symmetry was inconsistent with its putative role in regulating adaptive bone growth. The idea was that a symmetric signal could not direct a long-term growth process because it was inherently incapable of sending a net signal to a target cell.

To explain the observed signal asymmetry, it was suggested that the molecular structure of the interface between collagen and hydroxyapatite actually formed a PN junction diode (32). Shamos and Lavine analyzed the measurement technique employed, and concluded that the electrical signal manifested by the wet bone was intrinsically symmetrical, and that the apparent asymmetry resulted from the choice of measuring circuit (34). Another factor that probably contributed to the observed asymmetry in voltage was viscoelastic flow (35).

Cochran made extensive measurements of stress-generated electrical signals from wet bone, which he assumed were due to the piezoelectric effect (36). He measured the signals produced in precisely machined strips of bone maintained under physiologic conditions and subjected to cantilever bending. The signal was a relatively insensitive property of the bone, and exhibited only minimal variation with thickness or physical treatment (boiling, autoclaving, formalin fixation, radiation with 100,000 r, heating to 200°C). On the other hand, the signal was obliterated when the porosity of the samples was changed by demineralization.

During 1962-1968 most investigators assumed that the stress-generated electrical signals from bone were piezoelectric in origin, irrespective of whether they were observed in wet or dry bone. This perception was altered by a dawning recognition of the existence of a class of electromechanical phenomenon in wet tissue called streaming potentials (37).

The nature of streaming potentials has been described by Pollack (38). At the interface of a solid and an ion-containing fluid, specific interactions occur resulting in surface-bound charges and in the creation of a region in which the ionic charge distribution differs significantly from that of the bulk fluid. The electrically altered region in the fluid phase is called the diffuse layer, and its boundary with the bulk fluid is the slip plane. The electrical potential at the slip plane is the zeta potential, and when it is zero the solid is said to be at its isoelectric point. If the diffuse layer is caused to move tangentially to the surface, the electrical potential of the slip plane is altered; this kinetic modification of the electrical potential at the slip plane is called the streaming potential. It is created by motion of the diffuse layer whenever the fluid pH is such that the surface is not at its isoelectric point.

Anderson and Eriksson reported the occurrence of a signal of the order of millivolts in tendon subjected to repetitive impulse loading; the signal vanished when the pH of the bathing fluid was such that the tendon was at its isoelectric point (39). From the absence of the voltage at the isoelectric point, they concluded that all electromechanical phenomena in wet tendon arose from streaming potentials and that piezoelectricity was absent from wet tendon. They performed similar experiments with bovine bone (40), and concluded from the observed change in the piezoelectric constant with pH, that streaming potentials were present in wet bone but could not completely account for its electrical signal.

It was not logical to conclude from the absence of an electrokinetic signal at the isoelectric point of human tendon that wet tendon was not piezoelectric. Another interpretation — and as it turned out, more probably the correct one -- was that the piezoelectric signal was simply not detected by the measuring system employed. Subsequent studies showed that wet biological tissue, including bone, are piezoelectric (41-43).

Streaming potentials are undoubtedly the physical basis of the electrical signals observed in wet bone (36), tendon (39), and cartilage (44,45) subjected to mechanical forces and measured with wick electrodes or metallic electrodes coated on the surface (46,47). The millivolt-strength signals disappear as the tissue is dried, and are replaced by microvolt-strength signals of piezoelectric origin (48) which were previously undetected because of rapid neutralization by the ions in the diffuse layer.

Whether the piezoelectric signal, which is not conveniently measured in wet bone, or streaming potentials (or possibly other mechanisms) should be identified with the electrical signal that hypothetically helps to mediate some bone adaptive responses has not been resolved. One reason that disposes me to the choice of piezoelectricity is the data obtained by McElhaney, who measured the piezoelectric charge distribution that appeared on the surface of an intact, embalmed, human femur (49). The bone was dried at 105°C for 2 weeks to remove adsorbed water, and more than 600 square electrodes (0.25 inches on a side) were attached. The ends of the bone were embedded in epoxy, and it was vertically mounted and cyclically loaded in compression (50-100 pounds, 1 Hz). Measurement of the charge appearing on each electrode yielded an apparent random distribution of positive and negative areas that did not correlate with stress distribution, wall thickness, curvature, or other topological features of the bone. Measurements made on sections cut from the bone and loaded in pure compression revealed a surface-charge distribution whose sign and magnitude varied strongly with circumferential position.

McElhaney's data included a posterior view of the outline of the right femur used in the study, showing the actual measured surface-charge densities. If the piezoelectric surface charge could act as a mediating factor in an adaptive osteogenic response, I reasoned that McElhaney's data ought to be interpretable as an adaptive signal according to a self-consistent scheme. I interpreted the medial and lateral charge distributions as signals to build or resorb bone in an amount directly proportional to the measured charge density (50). On the medial surface I identified a negative surface charge with bone building, and a positive surface charge with bone resorption. On the lateral surface the positive and negative surface-charge densities were identified with bone resorption and deposition respectively. An adaptive response (self-consistent change in femoral outline) was produced, and the integrity of the femur was preserved, as opposed to a random pattern that was expected if the measured charge distributions were unrelated to bone adaptability (50).

McElhaney demineralized some of the sections of the bone in

an attempt to ascertain the origin of the piezoelectric effect, but found that the bone matrix samples were too flexible, and hence unsuitable for piezoelectric measurements involving the application of stress. I employed a method developed by Fukada (24) in which an electric field is applied to the sample and a strain is measured (converse piezoelectric effect). Using this technique, I measured the piezoelectric effect in air-dried bone, and then chemically removed the bone mineral from half the samples and the bone collagen from the other half. The piezoelectric effect was manifested in the demineralized samples, but was absent in the decollagenated samples (51).

In addition to pursuing the role of stress-generated electrical signals in adaptive growth, Becker continued his study of the physiological significance of the other endogenous electrical property of bone -- the SEP. From a time-sequence study of fracture healing in the tibiofibularis of frogs, Becker and Murray concluded that the cells of the fracture callus originated from the erythrocyte, and not from mitotic activity of osteoprogenitor cells as occurs in the healing of mammalian fractures (52). They described a sequence involving the dedifferentiation of the mature amphibian erythrocyte into a stem cell, and its subsequent redifferentiation into the connective tissue cells capable of repairing the injury. They hypothesized that the SEPs at the injury site mediated the reparative response. The electrical events measured within several hours of the fracture were held to originate with the bone substance itself, and to have occurred as a result of the persistence of a residual stress in analogy with a phenomenon reported by Bonfield and Li (53). Electrical phenomena previously identified and associated with neural tissue (26-28) were felt to provide the subsequent control function of the healing response (52). To substantiate the portion of the theory dealing with the stimulus for dedifferentiation of the nucleated erythrocyte, Becker and Murray subjected amphibian erythrocytes to DC currents of 1-1000 μA, and directly observed a remarkable sequence of morphological changes in individual cells that was similar to the known maturation sequence of the amphibian erythrocyte, but which proceeded in the reverse direction -- from erythrocyte to stem cell.

The occurrence of a dedifferentiation response of amphibian

red blood cells was subsequently verified (54), but the factors responsible for its initiation when it occurs in vitro remain unclear. Dedifferentiation occurred in cultures through which DC current was not passed, and was attributed to the presence of static charges on the plastic chambers (52). The sequence of cellular changes was also observed at widely different current levels, leading the investigators to conclude that the sex and hormonal status of the donor were important factors in the response threshold.

APPLICATION OF ELECTRICAL ENERGY

INITIAL STUDIES

Bassett and Becker reasoned that if stress-generated electricity could directly affect bone cells thereby mediating adaptive bone growth, then externally applied electricity ought to affect bone growth in a polarity-dependent fashion. In particular, the negative electrode should be associated with growth, and the positive electrode with resorption. This view, which came from their considerations of the natural history of the healed angulated fracture (55), was tested in a controlled study in dogs reported in 1964* (56). Holes were drilled through one cortex of the femur, and platinum electrodes were inserted into the medullary canal. A simple series circuit was used to supply DC current to the electrodes, and inactive platinum electrodes were implanted in the contralateral femur. After 21 days, woven bone trabeculae were found at the active and control electrodes, but the greatest amount of grossly observable new bone occurred at the cathode. The current associated with the effect was about 3 μA.

The authors interpreted the data as supporting the hypothesized linkage of electrical negativity with bone growth and positivity with bone resorption, with the proviso that the failure to observe resorption at the anode was probably due to the

---

* It is incorrect to view the concave side of a healed malunited fracture as being in compression (57,58). It is incorrect to view bone in compression as being electrically negative, whether it is wet (59) or dry (49,60). Despite this, both ideas remain popular.

non-physiological nature of the electrical signal (continuous application for 21 days, rather than an intermittent electrical signal that would be associated with an adaptive osteogenic response). Another interpretation was that the observed osteogenic response was unrelated to the conceptualization that actually led to the study, but was basically an inflammatory response.

The experiment was repeated by O'Connor et al. with the incorporation of a numeric scale to quantitate the amount of bone formed at the electrodes (evaluation of X-rays by five naive observers) (61). In 7 of the 12 dogs studied the cathode had the most bone associated with it, but in 2 dogs it had the least. None of the average scores of the 4 electrodes (cathode, anode, proximal control, distal control) differed from the others by the paired t test.

Hambury et al. drilled holes through the cortex of both sides of the femur of rabbits and inserted platinum electrodes that were cut to end flush with the periosteum (62). An effective current of 3 μA was passed continuously for 21 days, and the extent of bone growth was then assessed quantitatively by measuring the strontium-85 uptake on the 21st day after surgery. In 17 animals, 7 had more bone growth near the active implant, 9 had more bone near the inactive implant, and in 1 animal there was no difference. Thus, the passage of 3 μA for 21 days could not be distinguished from the osteogenic response that occurred in response to the drilling of the holes through the cortex of the bone (as determined by strontium-85 uptake). Failure to observe electrical osteogenesis was also reported by Crelin and Dueker following implantation of electrical circuitry in mice for 2 weeks (63).

A major difficulty with DC studies of electrical osteogenesis in the late 1960s was the inability to control dose. The typical implanted circuit consisted of a battery in series with a resistor: such a circuit does not function at a predictable current level in the complex environment of animal tissue. This difficulty was overcome by Friedenberg et al. who employed current-controlled implantable circuitry, and thereby categorically established electrical osteogenesis as a real phenomenon (64). Two holes (1 cm apart) were drilled into the medullary canal in the rabbit femur. Stainless-steel electrodes were placed in the bone holes, and constant currents of 1, 5, 10, 20, 50, and 100 μA were administered in separate groups of animals. All the rabbits were sacrificed after 10 days, and both bone production and tissue destruction were analyzed at each electrode using semi-quantitative histology. It was concluded that 5-20 μA was optimum for bone formation at the cathode, and that tissue destruction occurred at the anode at currents as low as 1 μA (64).

ACCELERATED FRACTURE HEALING

The observation (56) and unequivocal verification (64) that electricity could make bone grow was largely a result of speculation about the nature of the control system for adaptive remodeling. About 1970, the thrust of the research in bone bioelectricity shifted rapidly to the essentially pragmatic consideration of whether and how electrical osteogenesis could best be used in the clinic.

In 1971, Friedenberg et al. reported data concerning the effects of DC current on fracture-healing in rabbits (65). Following bilateral fibular fractures, rabbits were divided into 5 groups depending on the position of the electrodes relative to the fracture site. All animals were recovered 18 days after fracture. There was significantly more callus (determined by X-ray evaluation) in the animals that were stimulated with the cathode in the fracture site (all animals received 10 μA). Mechanical testing revealed stiffer fibulae on the treated side.

Many subsequent studies in animals showed that DC currents could cause more callus formation at an injury site than would have otherwise occurred, and this condition was commonly called accelerated healing (66-69). In one such study (66), following bilateral osteotomies of the radius in rabbits, an electrode (it is unclear whether it was gold or platinum) was placed in the bone near (but not at) the osteotomy site. The other electrode was placed on the skin. Accelerated healing as determined by the roentgenologic appearance of periosteal callus was found only when the implanted electrode was a cathode; the maximum effect occurred at 15-20 μA. In the treated osteotomies, the periosteal reaction was seen at 1.5-2.5 weeks after surgery, which was about a week earlier than the bony callus that formed on the control side.

Connolly et al. (67) placed stainless steel pins above and below transverse osteotomies in dogs and passed 10-30 μA for 3-12 weeks. In a series of 70 dogs, a tendency for greater callus formation in the stimulated bone was noted based on mechanical testing, measurement of callus weight, and bone ash weight.

Rabbits were subjected to bilateral diaphyseal tibial osteotomies, and the bone fragments were fixed with compression plates (68). Holes for electrodes (material not specified) were drilled 5 mm above and below the osteotomy site, and 3-15 μA were passed for 21 days. Inactive electrodes were implanted in the contralateral tibia. The breaking strength of the tibias was 21% greater on the stimulated side ($P < 0.01$, paired t test). The DC current apparently elicited periosteal and endosteal callus which added to the mechanical strength of the osteotomy site (68). Accelerated healing of mandibular slot osteotomies (7x2 mm) was reported in dogs using 12 μA at 0.7 Hz (69). At 21-35 days more bone was present in the defect on the stimulated side as determined by gross examination, X-ray, and histological examination.

NON-UNION ANIMAL MODELS

Several attempts have been made to study the effect of electrical stimulation in a nonunion animal model. In one study, 88% of 87 radial osteotomies in dogs failed to heal in 12 weeks in the absence of internal fixation (70). The addition of 20 μA via platinum electrodes to the osteotomy site did not significantly alter the incidence of nonunions (85% of 13 osteotomies). The duration of the 12-week post-implant period during which the current actually flowed is unclear (70). Other animal nonunion studies reported more success (71-72).

A 1.5-cm section of the midshaft of the tibia in dogs was removed and replaced with a block of silicone (71). Eight weeks later, the silicone block was removed and the defect was externally stabilized. Four weeks later, a stainless-steel cathode was placed in the defect and platinum anodes were inserted into the medullary canal through bone holes located 1.5 cm above and below the defect. A current of 20 μA was passed for 4 weeks in 22 dogs, and an equal number of dogs served as sham-implanted controls. The existence of clinical union, the extent of technitium-99 activity, and the histological appearance of the defect were evaluated semi-quantitatively. Based on all three criteria combined, the data indicated superior healing on the treated side (Wilcoxon test, $P < 0.01$)(71). When the criteria were evaluated individually, clinical union, but not histological appearance, was statistically improved on the treated side ($P < 0.05$) (71).

In a similar study (72), the investigators stabilized the bone using an intramedullary rod through the silicone block. Eight weeks later, the rod and block were removed and a titanium cathode was placed in the defect. Rigid external fixation was achieved via transtibial pins connected by stainless-steel rods. The procedure was performed unilaterally on 30 dogs, which were recovered at 4, 8, and 12 weeks after implantation of the titanium. Half the animals in each group were stimulated using 20 μA (a platinum wire in the thigh was the anode), and the remaining animals served as controls. The response in the controls indicated progress toward a nonunion. An osteogenic response was observed at 4 weeks, both radiographically and histologically, but the extent of the response appeared to decrease at 8 and 12 weeks (72). An opposite trend was seen in the stimulated animals. Significantly less bone was present in the stimulated limbs compared to the controls at 4 weeks, but significantly more bone was present at 12 weeks after initiation of current flow. In contrast to the report of Friedenberg et al. (73) the bone that formed at the cathode did not occur in direct apposition to the wire.

The DC resistance (about 50,000 Ω) and the AC impedance (200-500 Ω) each remained essentially constant throughout the 12-week period (72). This data tends to devaluate any hypothesized importance of the role of time dependence of the electrical properties of bone in determining its biological response to an impressed voltage.

It is difficult to understand why the osteogenic response that occurred in the stimulated limbs following the second surgical procedure was reduced at 4 weeks, but increased at 12 weeks postoperatively compared to the control (72). It's as if the DC current sent two signals: one to build bone, and one to retard the process of bone building that had been triggered by another factor (removal of silicone block and intramedullary rod). Over time, their relative importance reversed, and the balance tended to favor the bone-building process.

THE MEDULLARY-CANAL MODEL

Friedenberg et al. (73) developed an animal model in which the role of the healing response to a defect in cortical bone was eliminated. A stainless-steel electrode entered the tibia of a rabbit at the level of the tubercle, and was passed down the shaft such that its uninsulated portion (1 cm) was located about 5 cm distal to the drill hole. The medullary electrode was operated as a cathode, and the electrical circuit was completed using an anode in the soft tissues of the thigh. DC current was applied for 21 days, and monitored regularly throughout the treatment period. The amount of intramedullary bone growth that occurred in a 2-cm section of the bone centered over the 1-cm exposed portion of the electrode was measured.

A technique was developed to quantitate the extent of bone formation in the medullary canal. A grid was superimposed on a cross-sectional image of the bone at the level to be measured, and the number of intersections on the grid that overlaid bone was determined and converted into an index that expressed the percentage of the medullary cross-sectional area occupied by bone. Merely placing the electrode in the medullary canal caused an increase of 1-2% of new bone growth in the canal. At 5-10 μA, the portion of the cross-sectional area of the canal occupied by new bone was increased to about 5%, indicating that bone callus formed around the cathode. Animals that received 20 μA exhibited about 20% new bone growth in the canal (more than 10 times the amount of bone produced by the inactive control electrode). Again, no histological changes occurred in the cortical bone adjacent to the callus. The highest current that could be accommodated by the walling-off response alone was 20 μA. At 30 μA, only 14% new bone was formed, and there was histomorphological evidence of tissue damage including destruction of marrow elements, enlarged Haversian canals in the cortical bone, and empty lacunae. These observations were made in 2 of the 6 animals that received 30 μA, in 4 of the 6 animals that received 40 μA, and in all animals that received 50 or 100 μA. Even at 100 μA, the extent of the reaction did not include the periosteum, which was unaffected in all animals.

Increased vascularity in the marrow cavity was also reported.

Increased vascularity, and the absence of mechanical instability at an injury site are factors that, separately, favor intramembranous bone formation (14), and this is what was observed (73) (cartilage was encountered only occasionally).

Histological examination of the DC-induced intramedullary bone revealed that it resembled a well-developed fracture callus (73). The implication, therefore, was that electrically induced bone was actually reparative bone and that the cells responsible for its formation were the same as those responsible for bone formation following a non-electrical stimulus. This idea was supported in a study in which the intramedullary model was modified to allow for recovery at 2-28 days after electrode insertion (bilateral implants, with 20 μA delivered to the right tibia) (74). Cells having essentially identical ultrastructural characteristics appeared on both sides (74). Even when initiation of electrical stimulation was delayed for 28 days following surgical insertion of the electrode (to allow for the trauma of insertion and the osteogenic effect of the presence of the wire in the medullary canal to become a minimum), distinctive ultrastructural characteristics on the stimulated side were not observed.

ROLE OF SIGNAL ELECTRICAL CHARACTERISTICS

Perhaps the most interesting aspect of the DC-induced callus formation reported by Friedenberg et al. (65,73) was its apparent localization to the cathode -- the observation fit well with the original rationale (56) for applying DC current to bone. But in 1972, Richez et al. (75) used platinum electrodes, and did not observe a polarity-dependent effect. Three holes (2 cm apart) were drilled into the medullary canal of the humerus of rabbits, and a platinum electrode (an anode, cathode, and control electrode) was placed in each hole. A current of 50 μA was passed for 1 second, and during the next second the stimulating electrodes were short-circuited. Stimulation was administered continuously for up to 3 weeks. A second treatment group received 250 μA for 1 second followed by a short-circuiting that lasted 9 seconds. A similar osteogenic response was seen at both active electrodes in the medullary canal, consisting of the formation of a trabecular network surrounding the electrode. The inactive electrode was simply covered with fibrous tissue. No differences in response were seen using the two stimulation signals (75).

Other reports also indicate that electrical polarity is not a fundamental factor in electrical osteogenesis (76-79). Two parallel osteotomies, 0.4 inches apart, were made normal to the sagittal suture and posterior to the coronal suture in the calvaria of rabbits (76). The defects were stimulated 15 hours/day, 6 days/week, for 3 weeks using platinum electrodes (anterior anode). The amount of bone present in the defects at sacrifice was determined by measuring the optical density of high-resolution radiographs of the excised calvaria. For reasons that were not explained, the control anode exhibited less healing than the more posteriorly located control cathode (4% vs. 35%). When the defects were stimulated using 10 μA DC, the amount of bone present at the electrodes was increased by roughly the same proportion at the anode and the cathode (4% increased to 8%, compared to 35% increased to 65%) (76).

A current of 7.5-30 μA administered via platinum electrodes inside the proximal metaphysis of rabbits produced alterations in the trabecular pattern (77,78). A reparative response occurred consisting of woven bone, and the histomorphological pattern was the same for both anodes and cathodes at the proximal metaphysis (in each case the other electrode was placed near the distal metaphysis).

In 20 dogs stimulated for 4 weeks via intramedullary electrodes, significantly more bone growth was seen at the anode as compared to the cathode (electrode material not specified) (79).

Current density can be an important factor in determining the magnitude of an osteogenic response. In a study by Chamoun et al. (80), two kinds of 1.4-mm-diameter stainless-steel cathodes were used; one was threaded (3 threads per mm), and the other was insulated except for 8 holes, 0.36 mm in diameter, that were drilled traversely through the electrode. The current density of the threaded electrode was smaller by more than a factor of 100, and it produced 20 times as much bone in the medullary canal (evaluated after 21 days' treatment) when both electrodes were powered with 20 μA (80).

Brighton et al., in 1981, presented evidence to indicate that pulsed DC currents were less effective than DC current in producing an osteogenic response in the rabbit medullary cavity (81). Stainless-steel electrodes were implanted bilaterally in rabbits, and one side was stimulated with DC while the other side was stimulated with 1-msec current pulses having an amplitude equal to that of the DC current (20 μA). The pulse repetition rate was varied from 10-750 Hz, but the pulsed current offered no advantage over direct current (Table 1).

IN VITRO STUDIES

Norton and Moore exposed pieces of 5-day-old rat calvaria in tissue culture to intense, low-frequency electric fields (100 kV/m, 5 Hz), in an experiment to determine whether bone development could be altered via the converse piezoelectric effect (82). Aberrant growth consisting of the formation of woven bone trabeculae was described. It is not possible to rule out a direct effect on the bone cells, or other phenomena such as an ozone-mediated effect (ozone is frequently associated with high-intensity electric fields). Nevertheless, the report provides some evidence that external electric fields can alter bone development in vitro.

An important experiment by Treharne et al. demonstrated that electrical energy could produce an osteogenic response in vitro (83). Fetal rat tibiae were grown in vitro for 8 days during which time they were subjected to 5-20 μA DC. The current was administered by passing a pointed stainless-steel cathode through the bone surface into the medullary canal. The circuit was completed by operating the stainless-steel raft on which the bone was placed as an anode. The thickness of the bone wall in the vicinity of the penetrating cathode was measured, and the tibiae that received 10 μA were more than 50% thicker than the controls. At 20 μA, the DC-induced increase in bone thickness was about 80%.

It follows from the Treharne et al. study that applied mechanical forces, the presence of neural tissue, and substances in autologous blood are all not required for electrical osteogenesis. In at least one instance electrical osteogenesis was observed in culture media that lacked the fetal calf serum that was usually present (110); this suggests that no blood-born substances are required for the effect. It would have been interesting to determine whether electrically-treated culture medium would have produced similar responses in bone growth, thereby directly implicating the electrochemical products produced at the electrodes.

Aro et al. (84) cultured callus fragments from 9-day rat tibial fractures, and cells that grew out of the explant were innoculated into 3.5-ml of culture media and electrically stimulated (100 μA pulses, 8 msec duration, 0.8 Hz repetition rate) using platinum/iridium electrodes. Cell confluence was reached about 80 hours after innoculation. The stimulated cell cultures showed a transient increase in tritiated thymidine uptake (but not in numbers of cells) at 33 hours after innoculation. The authors interpreted the study to indicate that cells from fracture callus were sensitive to electrical signals in vitro, but an equally valid interpretation is that the cells responded to the electrochemical byproducts produced at the electrodes.

CLINICAL STUDIES

The first systematic controlled study of the clinical efficacy of electrical osteogenesis was a report by Jorgensen involving accelerated fracture healing. He devised a clinical procedure for characterizing the degree of healing of tibial fractures (85-87), and applied it to the determination of the effect of electricity (88).

The procedure was built around an external fixation device commonly used to stabilize fractures of the long bones (Hoffmann). The stabilization device itself consisted of 2-3 pins drilled into the anteromedial face of the bone, 6 cm proximal and distal to the fracture site, and a metal bar that was attached to the pins to prevent movement at the fracture site. The bar could be replaced by a unit consisting of two separate metal bars and a micrometer, designed such that the micrometer would directly register the relative displacement of the bar ends along their axis when the tibia was loaded in bending. For the small deflections involved in evaluating the mechanical strength of the healing fracture, the angular deflection of the tibia when subjected to bending could be evaluated as the quotient of the micrometer reading and the distance of the measuring axis to the center of the bone (85). The leg was held by the examiner proximal to the proximal fixation screws, and loaded 6 cm distal to the distal fixation screws with a force of

5 kg (the bending plane was perpendicular to the anteromedial surface of the tibia). Working with autopsy specimens, Jorgensen found that the average deflection of the intact bone in women (average age 60) was 0.4°, and the average deflection in men (average age 65) was 0.2° (85).

In a group of patients having tibial fractures fixed with the Hoffmann apparatus, Jorgensen measured the tibial deflection at a time in the healing process at which the patients were clinically judged to be capable of full weight-bearing. Readings were made in the anteromedial and the posterolateral direction (which tended to open and close the micrometer gap, respectively) and then averaged. In 40 patients with fractures in the middle or distal third of the tibia, he found that 32 patients exhibited a deflection of 1° or less at the time they were judged to be clinically healed (86,87). Thus, a deflection of 1° was associated with full weight-bearing in a healing tibial fracture.

In another study, patients with fresh tibial fractures (2-10 days) were given electrical stimulation via the transtibial metallic pins (88). The applied current consisted of a constant component of 20 μA, and a 1-Hz component (intended to simulate signals that may be produced during walking) that had a peak value of about 500 μA. Following stabilization of the fracture, the patients were randomized into treatment and nontreatment groups, and the time required for healing to proceed to the point where the fracture was stable (exhibited less than 1° of bending) was determined. The polarity of the pins was reversed periodically throughout the course of the treatment.

The data showed that the clinical endpoint was reached more quickly in the stimulated group. In the series of 57 patients, 87% of the stimulated patients were healed (by the 1° endpoint) within 3 months, whereas only 45% of the control patients exhibited the endpoint ($P < 0.001$).

The patients apparently received continuous stimulation, but neither the extent of patient compliance nor the dependability of the stimulating device was discussed. It seems likely that there were significant times during the healing periods during which there was no stimulation.

Another study in which DC current was employed in an attempt to hasten the normal healing process in patients was reported by Mazureik and Eriksson (89). Forty patients with jaw fractures (anterior mental foramen) were treated with 10-20 μA via a platinum cathode percutaneously placed near the fracture site. The extent of the mobility of the fracture site was assessed clinically after 14 days. Of the 40 patients treated, 36 had a mobility ranging between excellent and good, whereas of 40 control patients, 35 had an estimated mobility between poor and fair. There was no difference in mobility between the groups after 6 weeks (the full period of immobilization of the jaw fracture in the study). The mobility of the jaw fracture in 5 stimulated patients and 5 control patients was quantified employing a device that measured the displacement of the bone fragments that occurred when a standard (1 kg) force was applied. The data parallelled the clinical observations (Figure 1).

The most frequent clinical use of electrical osteogenesis involving DC electrodes has involved the treatment of nonunions, which are fractures (usually of the long bones) that have failed to heal as expected. Several investigators have reported data from small groups (90-92), but the most complete studies have been performed by Carl T. Brighton, an orthopaedic surgeon at the University of Pennsylvania, and his colleagues (93-96).

In the Brighton procedure, electrodes were drilled into the nonunion site such that the bare tip (1 cm) of each stainless-steel cathode came to lie directly in the nonunion site (usually in the femur or tibia). The wires emerging from the leg were bent parallel to the leg and connectors were clamped to the exposed ends; the anode was a conducting pad placed on the skin. The power source was attached to the electrodes via the connectors and enclosed in a bandage, and a plaster non-weight-bearing cast was applied. Electricity flowed continuously for 12 weeks, and was monitored once every 4 weeks. Following 12 weeks' treatment, the cast and electrodes were removed and X-rays were made. Typically, the X-rays showed only little healing at this stage of treatment, and continued immobilization for another 12 weeks, without electricity, was provided. The initial report involved 57 patients who had an average duration of nonunion of more than 3 years (93). Among the first 18 cases, were 4 cases involving nonunion of the medial

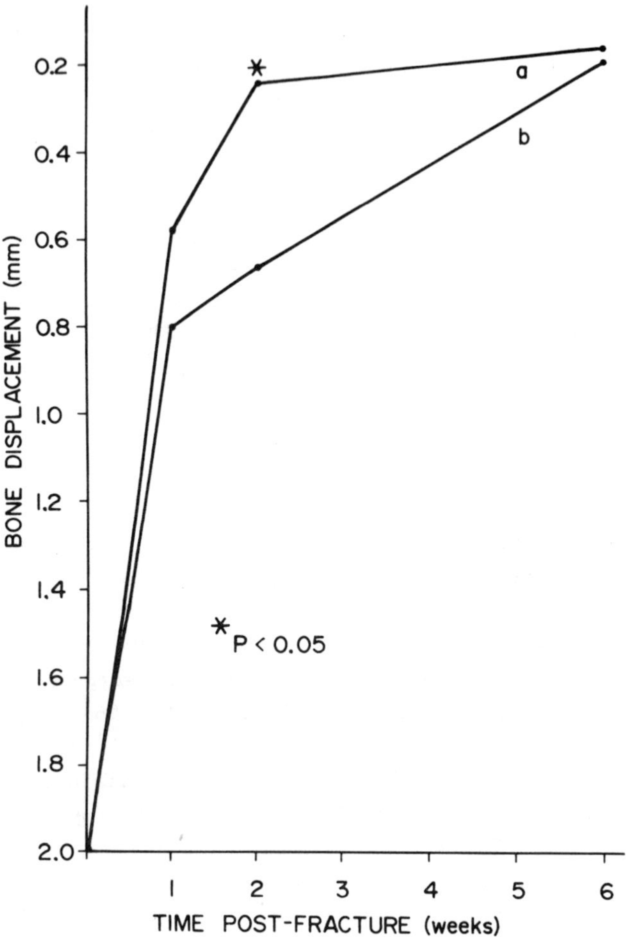

FIGURE 1. Average displacement of bone fragments for an applied force of 1 kg in patients with mandibular fractures (N = 5 in each group) (89). (a), (b) treated and control patients, respectively.

malleolus; each was treated using 10 μA, and each healed. One of 2 nonunions in the clavicle healed, but only 2 of 12 nonunions in a long bone healed. The data was interpreted to indicate that 10 μA administered via one percutaneous electrode was sufficient for a small bone like the medial malleolus, but was insufficient for a larger bone like the tibia. The use of 4 electrodes for the tibia or femur then became the standard procedure, with each electrode delivering 20 μA. Of the next 39 patients treated, 28 healed during the first attempt (72%).

The investigators concluded that the healing was due to electricity alone, and was not due to the cast immobilization. They reasoned that since the average duration of nonunion prior to treatment was 3.3 years, it would be unlikely for 12 weeks' immobilization alone to have brought about the therapeutic result.

The argument is obviously not conclusive, but it is plausible. To merely cast immobilize a 3-year-old nonunion is not a recognized therapeutic treatment. It therefore seems unjustified to impute a therapeutic consequence to this step when it is employed as part of a treatment regimen.

A factor that could falsify the assessment that the healing was due to electricity alone is the effect on the bone associated with drilling in the electrodes (which were actually Kirschner wires). In some cases the electrodes were passed into the nonunion gap through soft tissue, but in other cases they were drilled in through cortical bone. The investigators did not distinguish which technique was used in the various cases. A healing response will be initiated by the trauma associated with drilling through bone. Indeed, that was one of the reasons for developing the medullary-canal model (73). The possible role of mechanical stimuli delivered to the treatment site also does not seem to have been adequately considered. Although immobilization alone is not likely to have been a therapeutic factor, the chronic mechanical stimulation delivered to the treatment site by small mechanical motion of the ends of the K-wires that protruded through the cast remains undetermined. Recent evidence (97) indicates that mechanical motion of a percutaneous wire can produce an osteogenic response.

In 1979, a corrected success rate of 84% was reported in a

series of 168 patients, 72% of whom had had previous surgery (94). The use of technitium was developed to help identify patients that had developed synovial pseudarthroses, and who hence were not suitable candidates for the DC stimulation procedure (the synovial pseudarthrosis rate was about 6%). In 1981, a success rate of 79% in a series involving 189 nonunions was reported (95). Comparable results were found by 12 participating investigators (72% of 80 nonunions healed) (95).

In the final series involving 478 nonunions, the average uncorrected healing rate was 66% (96). This series included all patients treated with direct current, including those treated early in the study when the technique was initially being introduced into the clinic, with no patients having been lost to follow-up. The healing rate of the tibia, the most frequently represented bone in the series, was 72%, and the lowest rate of healing occurred in the humerus (33% of 45 bones treated). From an analysis of the failed cases, a series of practical clinical considerations was identified that, if followed, significantly improved the healing rate.

DISCUSSION

CHARACTERIZATION OF ELECTRICAL OSTEOGENESIS

Electrical osteogenesis (EO) is the production of orthotopic bone by use of electricity -- for purposes here, electrode-delivered electricity. In a typical EO study, a bone lesion is created, and electrodes are placed in the bone. The lesion itself, chemical activity of the electrodes, and mechanical motion of the electrodes each produce an osteogenic stimulus. An acceptable imputation to electricity alone of a causative role in the bone growth must involve suitable controls for these stimuli, and it is this additional amount of bone that is properly described as being caused by electricity.

The existence of electricity-produced bone can be established by quantitative morphometry (73), which is a technique that determines the amount of new bone in a standardized histological plane (98). Since EO promotes callus formation (66-69), mechanical testing in three-point or four-point bending is also a useful quantitative method for characterizing electrical osteogenesis. Additional callus often stiffens and strengthens healing bone (65). The technique of qualitative histology consists of subjectively characterizing the histological appearance of the treatment site in terms of an arbitrary numeric scale (99). The method is useful if the animal model is such that the DC current does not cause both osteogenic and osteolytic changes (71,73). Attempts have been made to assess EO using radio tracers such as strontium and technetium (62,71) but without significant success because radio tracers do not precisely delineate any particular stage in bone healing (100). Analysis of X-rays, either densitometrically (76) or more typically by subjective evaluation (61) is frequently incorporated into EO studies, but historically it has not proved useful for quantifying electrically produced bone.

Many parameters such as current density, frequency or repetition rate, polarity, and electrode metal can affect electrical osteogenesis, but the dominant influences are exerted by the magnitude and duration of the current. In actuality, for no discernible reason, most animal studies have involved 10-20 days duration of treatment, and most human treatments have involved 84 days. Consequently, current strength emerged as the experimentally important variable. There is broad agreement in the literature concerning the effects of magnitude of current on electrical osteogenesis in animal systems and patients: below 1-5 μA, either no response or only a minimal response is observed; at 5-20 μA, a maximum EO response occurs, accompanied by only a minimal osteolytic response. Above 20 μA, the relative importance of the stimulation and destruction effects of electricity are progressively reversed -- by 100 μA, destruction of bone is the completely dominant process.

Polarity is a relatively unimportant factor in EO produced via inert electrodes such as platinum (75-79). When active electrodes are used (such as stainless steel) the anode decomposes and liberates ions into the tissue. Stainless-steel anodes are not suitable for use in bone (73), and chemical toxicity of the electrode material is probably the underlying reason.

Alternating-current (AC) signals -- usually chosen to mimic an hypothesized endogenous signal -- of various types have been

TABLE 1. Comparison of the Effectiveness of Pulsed vs. Constant Current (20 μA) in Stimulating an Osteogenic Response in the Rabbit Medullary Cavity (81). N = 5-7 in each group. Total charge associated with control current (DC), 36.3 coulombs.

PULSE FREQUENCY (Hz)	CHARGE (coulombs)	PERCENT MEDULLARY CAVITY FILLED WITH NEW BONE	
		Pulsed Current	Control Current (DC)
10	0.36	2.8±0.8*	17.8±1.7*
50	1.8	3.5±2*	19.8±1.2*
100	3.6	5.2±1.2*	19.9±2.1*
200	7.3	6.8±1.5*	20.6±1.6*
500	18.1	12.1±2.1	18.8±3
750	27.2	13.3±0.7	16.9±1.8

*$P < 0.001$

employed to stimulate EO, but none has successfully demonstrated any advantage of AC over DC. The present evidence indicates that EO is related primarily to total charge passed through the electrodes, and consequently to signal repetition rate (Table 1).

Bone growth caused by electricity is reparative in nature, and consists initially of intramembranous woven-bone trabeculae that ultimately become remodeled and replaced by lamellar bone as occurs with any other injury. There exists no electrically specialized bone cell (74).

CLINICAL APPLICATION

The question of the clinical utility of EO involves three distinct issues, each of which requires its own methodological approach. One issue involves consideration of whether a therapeutic phenomenon occurred as a result of the treatment that employed EO. Every clinical course of treatment involves many factors such as immobilization, drugs, physical therapy, and patient motivation, and each factor obviously has some role in the overall result. Without attempting to apportion percent success among various factors in a treatment regimen, one may validly ask whether a particular regimen that employed EO produced a degree of success greater than that produced by a regimen not using EO. Ideally, this issue is addressed by a prospective clinical study in which the use of EO is controlled by the inclusion of a group of patients receiving standard therapy. A distinct issue, one that frequently cannot be addressed because of ethical or practical considerations, is whether a specific factor in a treatment regimen produces a therapeutic result -- that is, its inclusion is associated with a higher success rate than its exclusion, all other factors remaining constant. This issue can be addressed experimentally only by controlling for the putatively responsible factor. Thus, electrodes would be implanted in two groups of patients matched for all pertinent characteristics, but electricity would be applied in only one group and all other aspects of treatment would be identical in the two groups. Under these conditions, a higher success rate in the stimulated group could properly be attributed to the use of electricity.

The third, and most difficult issue, involves consideration of the clinical value of the contemplated treatment. A determination of value comes about as a result of the exercise of clinical judgment. As I have observed the process, ideally it proceeds in the following manner. The clinician considers data provided by a controlled clinical study and weighs the percent success associated with the new therapy against that provided by the standard therapy. A second, distinct, weighing involves the morbidity and convenience of the two courses of treatment. The clinical judgment of value comes about as a kind of overall weighing. If the new therapy is only as good as standard therapy, it will likely have only minimal value unless it is significantly more convenient or results in significantly less morbidity.

Figure 2 depicts an idealized time course of healing of a fracture. The pivotal point in the process is the time at which the injury is judged to be clinically healed. At that point immobilization devices can be removed, and the patient can return to a relatively normal lifestyle. A treatment that safe-

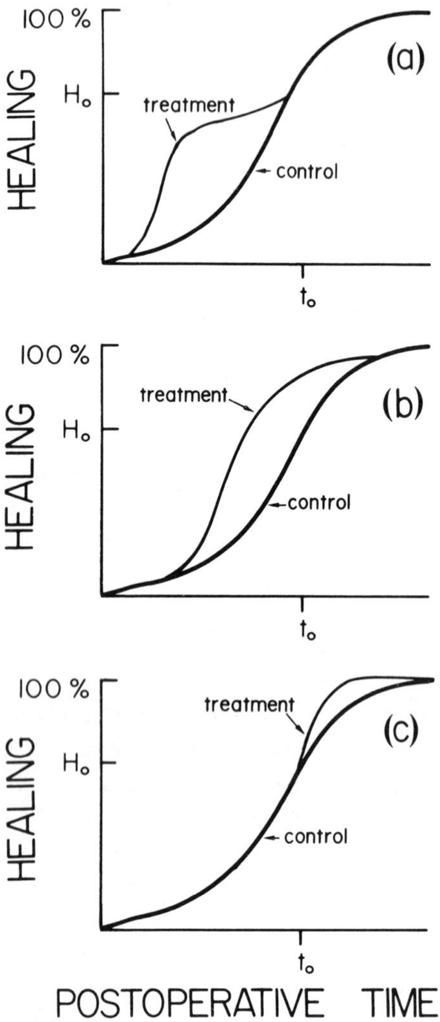

FIGURE 2. Hypothetical healing curve for a fracture. The treatment results depicted in (a) and (c) are not clinically useful. H_o, clinically significant degree of healing; t_o, postoperative day corresponding to H_o.

ly shortens time to clinical healing (Figure 2b) may have value. If the treatment produces an effect on healing only before (Figure 2a) or after (Figure 2c) the occurrence of clinical healing, it is unlikely to have clinical value -- irrespective of whether EO actually occurred.

Jorgensen showed that a particular combination of AC and DC current hastened the time to clinical healing in patients with tibial fractures (88). Since the control patients also had transtibial pins, but no electricity, the aspect of the therapy that produced the increased healing can properly be attributed to the EO. The clinical value of Jorgensen's treatment however is dubious for at least two reasons. A significant effort is required to maintain the integrity of the electrical circuitry, and the degree of effort may not be warranted by the degree of the effect produced. Also, Jorgensen's technique likely involved some morbidity as a result of corrosion at the transtibial (presumably stainless-steel) fixation pins which served as electrodes.

Masureik and Ericksson, in their controlled study of patients with jaw fractures, used platinum electrodes thereby eliminating the morbidity associated with corrosion (89). But it is not clear whether electrodes were also placed in the control patients. Thus, although a therapeutic effect was demonstrated, it cannot unambiguously be attributed to the EO. The existence of a therapeutic effect and the absence of morbidity due to electrode corrosion are factors supporting clinical use of the technique. The degree of patient compliance required to maintain the integrity of the electrical connections is a factor tending in the opposite direction.

Sustained efforts using EO in the treatment of nonunion have come from two groups (96,101). When EO via a titanium electrode was combined with surgical notching of the treated bone, and a bone graft, therapeutic results were demonstrated in an uncontrolled study (101). But there is no data by which to determine the element of the procedure that was responsible for the result -- perhaps it was the EO, perhaps not.

The nonunion studies of Brighton et al. were also uncontrolled against concurrent conventional therapy. But their patient population was sufficiently well characterized to permit a reasonable comparison with historical controls (96). On that basis, the conclusion that the treatment provided resulted in a success rate at least equal to that of bone grafting seems justified. A closer issue is that of the component of the therapy that was responsible for the success. Immobilization, drilling through bone, and mechanical stimulus to the treatment site via the relatively stiff electrodes may have, either singly or in combination, significantly contributed to the observed success rate.

Whatever the exact causative factors, the therapy produced a result comparable to that obtained with standard therapy (bone grafting), and with less morbidity than standard therapy (because the percutaneous insertion of the electrodes is less invasive than the full surgical procedure involving the harvesting and transplanting of autologous cancellous bone). The major shortcoming is the relatively high degree of patient compliance required for successful treatment. It seems to me that this factor has mitigated against the clinical value of the technique, thereby accounting for its relative lack of clinical popularity. If the duration of treatment could be shortened however, clinical judgment about the value of the technique might be vastly different.

MECHANISMS

The literature is essentially silent on the question of the specific factor at the cellular level that causes EO. For many years Brighton has intimated that EO was caused or in some manner associated with a relative lack of oxygen in the vicinity of the bone-forming cells (102-104), but the data is weak (103), and the counter-argument is persuasive (105).

There is probably no specific mechanism for the production of EO (106), any more than there is a specific cell by which it is brought about. When Küntscher placed a rusty iron wire in the medullary canal of femurs in dogs, florid callus formation ensued (19). Inflammatory agents such as croton oil also produced extensive callus formation. When large segments of the ulna in dogs were removed, thereby suddenly increasing the forces borne by the radius, it exhibited sudden and dramatic callus formation, thereby increasing its effective cross-sectional area (107). The application of heat also produced callus formation (108).

It seems likely that any somatic stimulus delivered to the bone-cell system results in a common-pathway signal that ignites osteogenesis, and that electricity produces essentially the same effect as do the more prosaic stimuli. The response consists of bone callus composed of woven bone, and demonstrates both an intensity threshold (below which no effect is produced) and a maximum reparative response (above which bone is destroyed, not produced). It proceeds via activation of resting osteoblasts and stimulation of osteoprogenitor cells, and occurs within 24 hours following delivery of the stimulus (11,109). The precise nature of the common-pathway signal actually present at the cell membrane is almost as obscure today as it was when Yasuda first considered the problem 30 years ago.

ENDOGENOUS ELECTRICAL SIGNALS FROM BONE

There are at least three endogenous electrical signals associated with bone. The piezoelectric signal is produced when mechanical forces are applied to bone in such a way as to cause shear along the collagen fibers (24). The piezoelectric signal is not ordinarily manifested in bone containing a normal moisture level because the signal is immediately neutralized by the motion of free charges in the fluid bathing the bone surface. In contrast, streaming potentials can only be measured in moist bone. They arise from the motion of charges in the diffuse layer near a bone-solution interface (38). SEPs are slowly varying electrical potentials that can be measured on the surface of bone, or on overlying tissue. The data presently available is insufficient to establish their origin or their significance in bone physiology.

One of the most fascinating, but unanswered, questions in bone bioelectricity is whether piezoelectricity or streaming potentials (or both or neither) should be identified with the feedback signal that regulates adaptive remodeling. Pollack has presented pertinent data regarding the importance of streaming potentials (59). He measured the potential manifested by moist bone across its cortical thickness when it was mechanically loaded at 1 Hz. The potential gradients were radially directed, and were correlated with osteonal structure: the potentials changed sign depending on whether the osteon was in tension or compression. But, to me, an even more impressive correlation

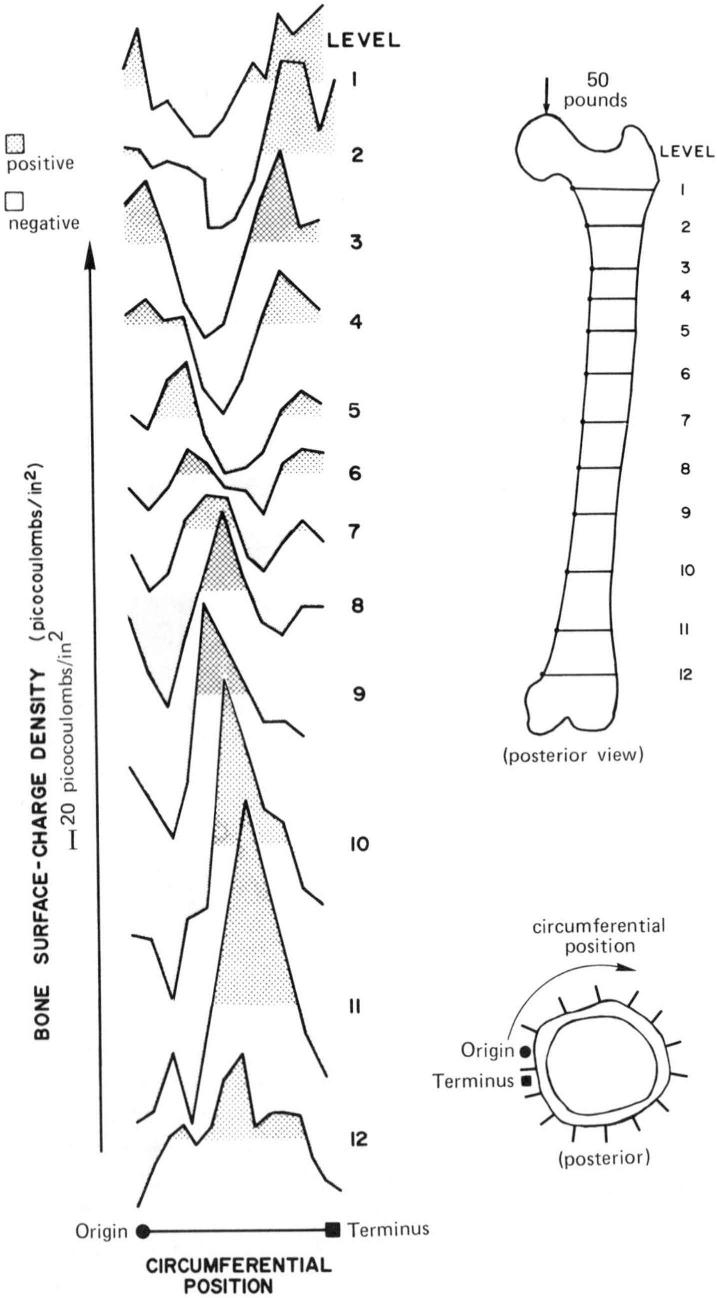

FIGURE 3. Piezoelectric surface-charge density as a function of circumferential position at the indicated levels (49).

between bone anatomy and mechanically-generated electrical signals is contained in McElhaney's data (49). He measured the surface charges that appeared on the surface of a human femur and, as described above, I previously reported a correlation between the measured charges and adaptive remodeling of the femoral outline (50). In his original publication, McElhaney also reported circumferential measurements of the charge density at specific levels of the bone, and this data (which I had not previously analyzed) is displayed in its entirety in Figure 3.

Assume that bone is deposited on areas exhibiting a positive charge, and resorbed on areas exhibiting a negative charge. Further, assume that the amount of bone deposited is directly proportional to the magnitude of the charge. Under these assumptions, McElhaney's data indicates the following adaptive response of the femur to the applied load. At levels 1-4 bone deposition occurs on the medial side, and resorption on the lateral side. At levels 5-7 the pattern of bone growth on one side of the bone and resorption on the other side continues, but the side of the bone exhibiting growth rotates clockwise (viewed from above), so that by level 7 growth occurs only on the anterolateral face. Thereafter, a consistent pattern of change occurs at levels 8-12 in which bone growth occurs on the lateral side of the bone, and resorption on the medial side. The amount of bone growth is greatest at levels 10-11.

Future studies will determine the limits of such an interpretation of electromechanical data from bone. Do long bones loaded non-physiologically generally yield a coherent remodeling response employing the listed assumptions linking surface charge, magnitude, and polarity with bone-cell activity? If 100 microns of the bone surface is removed, the piezoelectric surface charge is dramatically altered (60). All areas of such a free surface were never simultaneously in direct contact with bone cells, and therefore should lack the structural organization to be capable of eliciting an adaptive response.

A NEW BASIS FOR THE CLINICAL USE OF DC CURRENTS

The chronically administered factors associated with beneficial effects on bone growth characteristically function by triggering bone's plasticity principle -- orthodontic movement of teeth and straightening of malunited fractures are good examples. In contrast, electricity produces a reparative response which is a fundamentally different process from adaptive remodeling. There is no established scientific rationale for chronic administration of DC current to produce EO. Typically, chronic irritants are actually inimical to healing. Perhaps EO produced by chronically administered electricity is a net result of overlapping acute reparative responses (73) or is a response principally manifested only after cessation of stimulation (96). This idea leads to consideration of the initial cellular events following injury, and how DC current might be profitably used to enhance these events.

The initial cellular event after delivery of a stimulus to bone is the activation of osteoblasts lining the bone surface. The osteoblast pool existing prior to the injury is supplemented by mitotic activity from the osteoprogenitor cells, followed by differentiation of some of the daughter cells. Useful data regarding this process has been given by Tonna and Cronkite (11,111-113).

The femurs of 5-week-old mice were manually broken, and the extent of the periosteal cellular response was monitored for up to 14 days by flash labeling with tritiated thymidine one hour before killing (a technique that provides a measure of cells undergoing mitoses within one hour of the time at which the label is given) (11). The labeling index (percent of periosteal cells exhibiting the label) in the controls remained at about 2% throughout the post-fracture period. In the fractured animals, a response initially occurred at 8-16 hours after fracture, and peaked at 1-3 days (Figure 4a). The labeling index exhibited a sustained activity of about 10% (5 times more than the activity seen in the controls) during days 4-14. With 18-month-old mice, the maximum labeling index occurred at 4-5 days after fracture (111) (Figure 4b). Thereafter, the labeling index averaged about 6% over days 6-14 (about 30 times the level of the controls).

In another study (112), 5-week-old mice received femoral fractures on one side and 0.25-ml injections of either whole blood, serum, or saline into the soft tissue above the periosteum of the (intact) contralateral femur. The injected substances

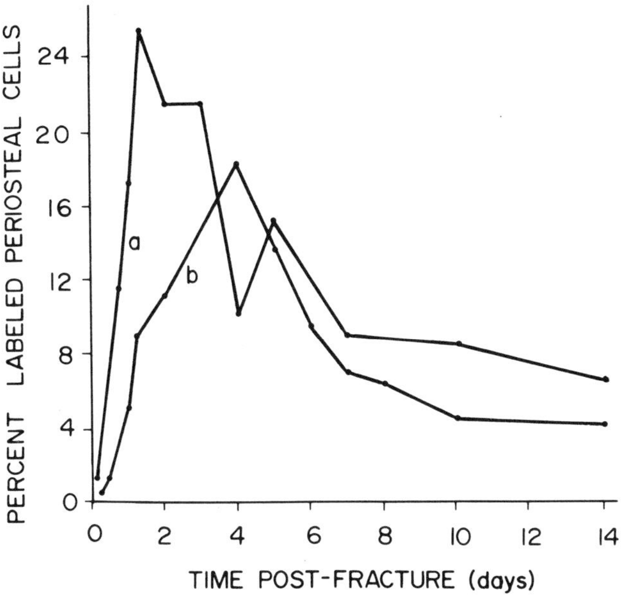

FIGURE 4. Uptake of tritiated thymidine by mice periosteal cells following fracture. Mice aged 5 weeks and 18 months in (a) and (b), respectively (11,111).

produced identical increases in labeling of about 9% (compared to 1% background); the labeling index in the fractured legs was about 16%.

It seems reasonable to assume that, all other things being equal, more bone will be built per unit time when more osteoblasts are present. The mitotic activity of the osteoprogenitor cells and the extent of differentiation of their daughter cells are the two interrelated factors responsible for the production of new osteoblasts. The number of new osteoblasts, B, on the nth day after an acute injury stimulus is

$$B = \sum_{n=1}^{N} a_n(1 - a_{n-1})!(L_n + 1)!P$$

where a_n is the fraction of osteoprogenitor cells that differentiate on the nth day ($a_o \equiv 0$), L_n is the labeling index on the nth day, and P is the number of osteoprogenitor cells present at the time of the injury. Employing Tonna's data for the first 4 days following fracture in the younger mice (Figure 4a), and assuming that $a_1 = a_2 = a_3 = a_4 = 0.2$, we find that the osteoblast population is increased by 23%, 23%, 22%, and 19% of the initial osteoprogenitor pool on successive days. If $a_1 = a_2 = a_3 = a_4 = 0.8$, the corresponding increases are 94%, 23%, 6%, and 1%. Thus, the number of new osteoblasts is a result of a complicated interplay between the rates of labeling and differentiation. Any factor that increases L_n and does not alter a_n produces an increase in osteoblasts on the $(n+1)^{st}$ day.

Because electricity is a reparative-eliciting stimulus, it too must result in a cell proliferative response similar to that reported by Tonna -- but in the absence of a bone lesion. In this view, some factor or combination of factors associated with the actual inflammatory reaction to the current -- exudation of leukocytes, activation of tissue enzymes, temporary tissue hypoxia are possibilities -- serves as the link between the stimulus and the activity of the osteoprogenitor cell. Furthermore, the osteoprogenitor response as determined by the labeling index is, in some sense, proportional to the extent of the injury (the response to a fracture was almost twice that of the response to 0.25-ml injections (112)).

Based on these considerations, I conclude that bone healing (or augmented bone healing) can be brought about by acute administration of DC current above some appropriate threshold to produce a pulse increase in the number of new osteoblasts, followed by a second (and possibly subsequent) doses of DC current at a time when the initial increase in mitotic activity has abated.

In a preliminary test of this idea, bilateral slot osteotomies (7 x 2 mm) were performed in rabbit mandibles and a stainless-steel electrode was attached to the bone and brought out through the skin in the area of the ramus (114). A current of 20 μA, negative polarity, was applied to one side, 4 hours/day for 2-4 days, and the other side served as the control. The extent of bone growth into the slot at 8 days after surgery was assessed qualitatively, and it was found that significantly more bone was present in the slot on the stimulated side in animals treated for 2-4 days. In animals stimulated on one side for the entire 8-day post-surgery time period, no bilateral differences in healing were observed (114).

These observations did not involve quantitative measurements of bone formation, and therefore must be regarded only as preliminary observations. In addition, an effect on the rate of fracture healing manifested as quickly as 8 days after surgery is unlikely to have clinical significance because it probably occurs at a time prior to clinical healing. Despite these limitations, the observation lends some support to the analysis of the existing literature given above.

The possibility that relatively brief stimulation delivered to an injury site during the immediate post-injury period (perhaps up to a week) actually enhances healing might significantly alter present evaluations of clinical value of EO. During an open reduction of a fracture an electrode could be placed directly over the periosteum or in the bony defect. The DC current could be administered with external equipment during the period of hospitalization, and the wire could be removed prior to discharge as would be the case with any other temporary indwelling device. If the DC current recruits additional osteoblasts in the absence of creating additional damage, and if the additional increment of bone produced adds stability more quickly, then the functional result would be accelerated healing. This process would not involve the basic step of direct communication with bone cells; instead, the effect of the DC current would be transduced to the common-pathway signal that triggers osteoprogenitor-cell mitosis. One advantage associated with providing a stimulus that does not function at the basic level of communicating directly with osteoprogenitor-cell membranes is that the danger of sending the wrong signal is proportionately reduced. That is, since the osteoprogenitor cell itself is not presented with a novel artificial environment, there seems little basis for concern about the possibility of triggering undesirable neoplastic growth (115).

REFERENCES

(1) Black, J.: Electrical Stimulation: Its Role in Growth, Repair, and Remodeling of the Musculoskeletal System, Praeger, New York, 1987.

(2) Spadaro, J.A.: Bioelectric stimulation of bone formation: Methods, models, and mechanisms, J. Bioelectricity 1:99-128, 1982.

(3) Neuman, W.F. and Neuman, M.W.: The Chemical Dynamics of Bone Mineral, University of Chicago Press, Chicago, 1958.

(4) Marino, A.A. and Becker, R.O.: Evidence for epitaxy in the formation of collagen and apatite, Nature 226:652-653, 1970.

(5) Albright, J.A. and Skinner, H.C.W.: Bone: Structural organization and remodeling dynamics, in The Scientific Basis of Orthopedics, 2nd Edition, J.A. Albright and R.A. Brand, eds., Apple and Lange Publishers, Norwalk, 161-212, 1987.

(6) Jee, W.S.S.: The skeletal tissues, in Histology; Cell and Tissue Biology, 5th Edition, L. Weiss, ed., Elsevier Biomedical, New York, 1983.

(7) Vaughan, J.: The Physiology of Bone, Clarendon Press, Oxford, 1975.

(8) Owen, M.: Cellular dynamics of bone, in The Biochemistry and Physiology of Bone, 2nd Edition, G.H. Bourne, ed., Vol. 3, Academic Press, New York, 271-298, 1971.

(9) Miller, S.C., Bowman, B.M., Smith, J.M. and Jee, W.S.S.: Characterization of endosteal bone-lining cells from fatty marrow bone sites in adult beagles, Anat. Rec. 198:163-173, 1980.

(10) Young, R.W.: Cell proliferation and specialization during endochondral osteogenesis in young rats, J. Cell Biol. 14:357-370, 1962.

(11) Tonna, E.A. and Cronkite, E.P.: Cellular response to fracture studied with tritiated thymidine, J. Bone Joint Surg. 43A:352-362, 1961.

(12) Owen, M.: The origin of bone cells, Int. Rev. Cytol., 28:213-238, 1970

(13) Owen, M.: Histogenesis of bone cells, Calc. Tiss. Res. 25:205-207, 1978.

(14) McLean, F.C. and Urist, M.R.: Bone: An Introduction to the Physiology of Skeletal Tissue, University of Chicago Press, Chicago, 1955.

(15) Parfitt, A.M.: The physiologic and clinical significance of bone histomorphometric data, in Bone Histomorphometry: Techniques and Interpretation, R.R. Recker, ed., CRC Press, Boca Raton, 143-223, 1983.

(16) Reynolds, J.J.: Skeletal tissue in culture, in The Biochemistry and Physiology of Bone, J.H. Bourne, ed., Academic Press, New York, pp. 69-126, 1972.

(17) Frost, H.M.: Mechanical determinants of skeletal architecture, in The Scientific Basis of Orthopedics, 2nd Edition, J.A. Albright and R.A. Brand, eds., Apple and Lange Publishers, Norwalk, 161-212, 1987.

(18) Hamm, A.W. and Harris, W.R.: Repair and transplantation of bone, in The Biochemistry and Physiology of Bone, Vol. III, G.H. Bourne, ed., Academic Press, New York, pp. 337-399, 1971.

(19) Kuntscher, G.: The Callus Problem, Warren H. Green, St. Louis, 1970.

(20) Yasuda, I., Noguchi, K. and Sata, T.: Dynamic callus and electric callus, J. Bone Joint Surg. 37A:1292-1293, 1955.

(21) Yasuda, I.: Mechanical and electrical callus, Ann. N.Y. Acad. Sci. 238:457-465, 1974.

(22) Yasuda, I.: Fundamental aspects of fracture treatment, J. Kyoto Med. Soc. 4:395-406, 1953; Transl.: N. Itada, Clin. Orthop. 124:5-8, 1977.

(23) Garratt, A.C.: Electro-physiology and Electrotherapeutics, 2nd Edition, Tickner and Fields, Boston, 1861.

(24) Fukada, E. and Yasuda, I.: On the piezoelectric effect of bone, J. Phys. Soc. Japan 12:1158-1162, 1957.

(25) Becker, R.O.: The bioelectric field pattern in the salamander and its simulation by an electronic analog, IRE Trans. Med. Elect. ME-7:202-208, 1960.

(26) Becker, R.O.: The bioelectric factors in amphibian limb regeneration, J. Bone Joint Surg. 43A:643-656, 1961.

(27) Becker, R.O., Bachman, C.H. and Slaughter, W.: The longitudinal direct current gradients of spinal nerves, Nature 196:675-676, 1962.

(28) Becker, R.O.: Search for evidence of axial current flow in peripheral nerves of the salamander, Science 134:101-102, 1961.

(29) Lund, E.J.: Bioelectric Fields and Growth, Univ. of Texas Press, Austin, 1947.

(30) Burr, H.S. and Northrop, F.S.C.: The Electrodynamic Theory of Life, Q. Rev. Biol. 10:322-333, 1935.

(31) Bassett, C.A.L. and Becker, R.O.: Generation of electric potentials by bone in response to mechanical stress, Science 137:1063-1064, 1962.

(32) Becker, R.O., Bassett, C.A.L. and Bachman, C.H.: The bioelectric factors controlling bone structure, in Bone Biodynamics, H.M. Frost, ed., Little Brown and Co., 209-232, 1964.

(33) Marino, A.A. and Becker, R.O.: The effect of electric current on rat-tail tendon collagen in solution, Calc. Tiss. Res. 4:330-338, 1970.

(34) Shamos, M.H. and Lavine, L.S.: Physical basis for bioelectric effects in mineralized tissue, 35:177-188, 1964.

(35) Steinberg, M.E., Bosch, A., Schwan, A. and Glazer, R.: Electrical potentials in stressed bone, Clin. Orthop. 61:294-299, 1968.

(36) Cochran, G.V.B., Pawluk, R.J. and Bassett, C.A.L.: Electromechanical characteristics of bone under physiologic moisture conditions, Clin. Orthop. 58:249-270, 1968.

(37) Cerquiglini, S., Cignitti, M., Marchetti, M. and Salleo, A.: On the origin of electrical effects produced by stress in the hard tissues of living organisms, Life Sci. 6:2651-2660, 1967.

(38) Pollack, S.R.: Bioelectrical properties of bone: Endogenous electrical signals, Orthop. Clin. N. Am. 15:3-14, 1984.

(39) Anderson, J.C. and Eriksson, C.: Electrical properties of wet collagen, Nature 218:166-168, 1968.

(40) Anderson, J.C. and Eriksson, C.: Piezoelectric properties of dry and wet bone, Nature 227:491-492, 1970.

(41) Marino, A.A. and Becker, R.O.: Piezoelectricity in hydrated frozen bone and tendon, Nature 253:627-628, 1975.

(42) Saha, S. and Lakes, R.S.: A new non-invasive device for monitoring the piezoelectric character of bone, in Electrical Properties of Bone and Cartilage, C.T. Brighton, J. Black and S.R. Pollack, eds., Grune and Stratton, New York, 57-68, 1979.

(43) Athenstaedt, H., Claussen, H. and Schaper, D.: Epidermis of human skin: pyroelectric and piezoelectric sensor layer, Science 216:1018-1020, 1982.

(44) Lotke, P.A., Black, J. and Richardson, S.J.: Electromechanical properties in human articular cartilage, J. Bone Joint Surg. 56:1040-1046, 1974.

(45) Lee, R.C., Frank, E.H. and Grodzinsky, A.J.: Oscillatory compressional behavior of articular cartilage and its associated electromechanical properties, J. Biomech. Eng. 103:280-292, 1981.

(46) Pollack, S.R., Salzstein, R. and Pienkowski, D.: Streaming potentials in fluid-filled bone, Ferroelectrics 60:297-309, 1984.

(47) Gross, D. and Williams, W.S.: Streaming potential and the electromechanical response of physiologically moist bone, J. Biomech. 15:277-295, 1982.

(48) Dwyer, N. and Matthews, B.: The electrical response to stress in dried, recently excised, and living bone, Injury 1:279-286, 1970.

(49) McElhaney, J.H.: The charge distribution on the human femur due to load, J. Bone Joint Surg. 49A:1561-1571, 1967.

(50) Marino, A.A. and Becker, R.O.: Piezoelectric effect and growth control in bone, Nature 228:473, 1970.

(51) Marino, A.A., Becker, R.O. and Soderholm, S.C.: Origin of the piezoelectric effect in bone, Calc. Tiss. Res. 8:177-180, 1971.

(52) Becker, R.O. and Murray, D.G.: The electrical control system regulating fracture healing in amphibians, Clin. Orthop. 73:169-198, 1970.

(53) Bonfield, W. and Li, C.H.: Deformation and fracture of bone, J. Appl. Phys. 37:869-875, 1966.

(54) Pilla, A.A.: Electrochemical information transfer at living cell membranes, Ann. N.Y. Acad. Sci. 238:149-170, 1974.

(55) Bassett, C.A.L.: Electrical effects in bone, Sci. Am. 213:18-26, 1965.

(56) Bassett, C.A.L., Pawluk, R.J. and Becker, R.O.: Effects of electric currents on bone in vivo, Nature 204:652-654, 1964.

(57) Currey, J.D.: The adaptation of bones to stress, J. Theoret. Biol. 20:91-106, 1968.

(58) Epker, B.N. and Frost, H.M.: Correlation of bone resorption and formation with the physical behavior of loaded bone, J. Dent. Res. 44:33-41, 1965.

(59) Pollack, S.R., Korostoff, E., Starkebaum, W. and Iannicone, W.: Microelectrode studies of stress generated potentials in bone, in Electrical Properties of Bone and Cartilage, C.T. Brighton, J. Black and S.R. Pollack, eds., Grune and Stratton, New York, 69-82, 1979.

(60) Martin, R.B.: Analysis of bone and other piezoelectric textures, in Electrical Properties of Bone and Cartilage, C.T. Brighton, J. Black and S.R. Pollack, eds., Grune and Stratton, New York, 141-154, 1979.

(61) O'Connor, B.T., Charlton, H.M., Currey, J.D., Kirby, D.R.S. and Woods, C.: Effect of electric current on bone in vivo, Nature 222:162-163, 1969.

(62) Hambury, H.J., Watson, J., Sivyer, A. and Ashley, D.J.B.: Effect of microamp electrical currents on bone in vivo and its measurement using strontium-85 uptake, Nature 231:190-191, 1971.

(63) Crelin, E.S. and Dueker, D.K.: The response of the femur to trauma, a foreign body, and a direct electrical current in mice, Yale J. Biol. Med. 43:71-75, 1970.

(64) Friedenberg, Z.B., Andrews, E.T., Smolenski, B.I., Pearl, B.W. and Brighton, C.T.: Bone reaction to varying amounts

of direct current, Surg. Gynecol. Obstet. 131:894-899, 1970.

(65) Friedenberg, Z.B., Roberts, P.G., Didizian, N.H. and Brighton, C.T.: Stimulation of fracture healing by direct current in the rabbit fibula, J. Bone Joint Surg. 53A:1400-1408, 1971.

(66) Anisimov, A.I.: Action of direct current on bone tissue, Byull. Eksper. Biol. Med. 78:100-102, 1974 (in Russian). English translation, Bull. Eksp. Biol. Med., Plenum, 1069-1071, 1975.

(67) Connolly, J.F., Ortiz, J., Price, R.R. and Bayuzick, R.J.: The effect of electrical stimulation on the biophysical properties of fracture healing, Ann. N.Y. Acad. Sci. 238:519-528, 1974.

(68) Weigert, M. and Werhahn, C.: The influence of electric potentials on plated bones, Clin. Orthop. 124:20-30, 1977.

(69) Shandler, H.S., Weinstein, S. and Nathan, L.E.: Facilitated healing of osseous lesions in the canine mandible after electrical stimulation, J. Oral Surg. 37:787-792, 1979.

(70) Harris, W.H., Moyen, B.J-L., Thrasher, E.L., Davis, L.A., Cobden, R.H., MacKenzie, D.A. and Cywinski, J.K.: Differential response to electrical stimulation: A distinction between induced osteogenesis in intact tibiae and the effect on fresh fracture defects in radii, Clin. Orthop. 124:31-40, 1977.

(71) Paterson, D.C., Carter, R.F., Maxwell, G.M., Hillier, T.M., Ludbrook, J. and Savage, J.P.: Electrical bone-growth stimulation in an experimental model of delayed union, Lancet 1278-1281, 1977.

(72) Collins, P.C., Paterson, D.C., Vernon-Roberts, B. and Pfeiffer, D.: Bone formation and impedance of electrical current flow, Clin. Orthop. 155:196-209, 1981.

(73) Friedenberg, Z.B., Zemsky, L.M., Pollis, R.P. and Brighton, C.T.: The response of non-traumatized bone to direct current, J. Bone Joint Surg. 56A:1023-1030, 1974.

(74) Brighton, C.T. and Hunt, R.M.: Ultrastructure of electrically induced osteogenesis in the rabbit medullary canal, J. Orthop. Res. 4:27-36, 1986.

(75) Richez, J., Chamay, A. and Bieler, L.: Bone changes due to pulses of direct electric microcurrent, Virchows Arch. Abt. A Path. Anat. 357:11-18, 1972.

(76) Hassler, C.R., Rybicki, E.F., Diegle, R.B. and Clark, L.C.: Studies of enhanced bone healing via electrical stimuli: Comparative effectiveness of various parameters, Clin. Orthop. 124:9-19, 1977.

(77) Ilfeld, F.W., Weinberg, C., Rosen, V. and August, W.: Direct current induced mosaic bone architecture, Clin. Orthop. 99:298-302, 1974.

(78) Weinberg, C., Ilfeld, F.W., Rosen, V., August, W., and Baddorf, R.L.: Electrical potentials in medullary bone, Clin. Orthop. 171:256-263, 1982.

(79) Janssen, L.W.M., Roelofs, J.M.M., Bisser, W.J. and Wittebol, P.: Hypothesis of bone remodeling and fracture healing by electrostimulation, in Electric Stimulation of Bone Growth and Repair, F. Burney, E. Herbst and M. Hinsenkamp, eds., Springer-Verlag, New York, 61-67, 1978.

(80) Chamoun, E., McCutcheon, M., Lemons, J., Henson, P. and Wilson, E.: A new cathode design for bone growth stimulation, in Biomedical Engineering III, L.C. Sheppard, ed., Pergamon Press, New York, 1984.

(81) Brighton, C.T., Friedenberg, Z.B., Black, J., et al.: Electrically induced osteogenesis: Relationship between charge, current density, and the amount of bone formed, Clin. Orthop. 161:122-132, 1981.

(82) Norton, L.A. and Moore, R.R.: Bone growth in organ culture modified by an electric field, J. Dent. Res. 51:1492-1499, 1972.

(83) Treharne, R.W., Brighton, C.T., Korostoff, E. and Pollack, S.R.: An in vitro study of electrical osteogenesis using direct and pulsating currents, Clin. Orthop. 145:300-306, 1979.

(84) Aro, H., Eerola, E. and Aho, A.J.: Determination of callus quantity in four-week-old fractures of the rat tibia, J. Orthop. Res. 3:101-108, 1985.

(85) Jorgensen, T.E.: Measurements of stability of crural fractures treated with Hoffmann osteotaxis: I. Method and measurements of deflection on autopsy crura, Acta Orthop. Scandinav. 43:188-206, 1972.

(86) Jorgensen, T.E.: Measurements of stability of crural fractures treated with Hoffman osteotaxis: II. Measurements on crural fractures, Acta Orthop. Scandinav. 43:207-218, 1972.

(87) Jorgensen, T.E.: Measurements of stability of crural fractures treated with Hoffman osteotaxis: III. The uncomplicated, terminal phase of healing of crural fractures, Acta Orthop. Scandinav. 43:264-279, 1972.

(88) Jorgensen, T.E.: The effect of electric current on the healing time of crural fractures, Acta Orthop. Scandinav. 43:421-437, 1972.

(89) Masureik, C. and Eriksson, C.: Preliminary clinical evaluation of the effect of small electrical currents on the healing of jaw fractures, Clin. Orthop. 124:84-91, 1977.

(90) Becker, R.O., Spadaro, J.A. and Marino, A.A.: Clinical experience with low intensity direct current stimulation of bone growth, Clin. Orthop. 124:75-83, 1977.

(91) Lavine, L.S., Lustrin, I. and Shamos, M.H.: Treatment of congenital pseudarthrosis of the tibia with constant direct current, Clin. Orthop. 124:69-74, 1977.

(92) Connolly, J.F.: Electrical treatment of nonunions: Its use and abuse in 100 consecutive fractures, Orthop. Clin. N. Am. 15:89-106, 1984.

(93) Brighton, C.T., Friedenberg, Z.B., Mitchell, E.I., et al.: Treatment of nonunion with constant direct current, Clin. Orthop. 124:106-123, 1977.

(94) Brighton, C.T., Friedenberg, Z.B. and Black, J.: Evaluation of the use of constant direct current in the treatment of nonunion, in Electrical Properties of Bone and Cartilage, C.T. Brighton, J. Black and S.R. Pollack, eds., Grune and Stratton, New York, 1979.

(95) Brighton, C.T., Black, J., Friedenberg, Z.B., et al.: A multicenter study of the treatment of nonunion with constant direct current, J. Bone Joint Surg. 62:2-13, 1981.

(96) Brighton, C.T.: The semi-invasive method of treating nonunion with direct current, Orthop. Clin. N. Am. 15:33-46, 1984.

(97) Spadaro, J.A., Mino, D.E., Chase, S.E., Werner, F.W. and Murray, D.G.: Mechanical factors in electrode-induced osteogenesis, J. Orthop. Res. 4:37-44, 1986.

(98) Jaworski, Z.F.G., ed.: Bone Morphometry, University of Ottawa Press, 1973.

(99) Marino, A.A., Cullen, J.M., Reichmanis, M. and Becker, R.O.: Fracture healing in rats exposed to extremely low frequency electric fields, Clin. Orthop. 145:239-244, 1979.

(100) Wood, M.J., Marino, A.A., Ashley, C. and Hackley, M.M.:

Uptake of Tc-99m MDP at fracture sites in rabbits following electrical stimulation, Proc. Ann. Meeting of Radiological Society of North America, 1985.

(101) Paterson, D.C., Lewis, G.N. and Cass, C.A.: Treatment of delayed union and nonunion with an implanted direct current stimulator, Clin. Orthop. 148:117-128, 1980.

(102) Brighton, C.T. and Friedenberg, Z.B.: Electrical stimulation and oxygen tension, Ann. N.Y. Acad. Sci. 238:314-320, 1974.

(103) Brighton, C.T., Adler, S., Black, J., Itada, N. and Friedenberg, Z.B.: Cathartic oxygen consumption and electrically induced osteogenesis, Clin. Orthop. 107:277-282, 1975.

(104) Brighton, C.T.: Biophysical studies of bone growth and repair (Shands Lecture), Combined Program of the Orthopaedic Research Society and the American Academy of Orthopaedic Surgeons, January 24, 1985, Las Vegas, Nevada.

(105) Spadaro, J.A.: Electrical osteogenesis - Role of the electrode material, in Electrical Properties of Bone and Cartilage, C.T. Brighton, J. Black and S.R. Pollack, eds., Grune and Stratton, New York, 189-197, 1979.

(106) Marino, A.A. and Becker, R.O.: Electrical osteogenesis: An analysis, Clin. Orthop. 123:280-282, 1977.

(107) Goodship, A.E., Lanyon, L.E. and McFie, H.: Functional adaptation of bone to increased stress, J. Bone Joint Surg. 61A:539-546, 1979.

(108) Richards, V. and Stofer, R.: The stimulation of bone growth by internal heating, Surgery 46:85-96, 1959.

(109) Ashihara, T., Kajawa, K., Kamachi, M., Inoue, S., Ohashi, T. and Takeoka, O.: H-thymidine autoradiographic studies of cell proliferation and differentiation in

electrically stimulated osteogenesis, in Electrical Properties of Bone and Cartilage, C.T. Brighton, J. Black and S.R. Pollack, eds., Grune and Stratton, New York, 401-426, 1979.

(110) Treharne, R.W., III: Application of Electric Currents to In Vitro Fetal Rat Tibiae, Dissertation, University of Pennsylvania, p. 87, 1976.

(111) Tonna, E.A. and Cronkite, E.P.: Changes in skeletal proliferative response to trauma concomitant with aging, J. Bone Joint Surg. 44A:1557-1568, 1962.

(112) Tonna, E.A. and Cronkite, E.P.: The effects of extra-periosteal injections of blood components on periosteal cell proliferation, J. Cell Biol. 23:79-87, 1964.

(113) Tonna, E.A. and Cronkite, E.P.: The periosteum: Autoradiographic studies on cellular proliferation and transformation utilizing tritiated thymidine, Clin. Orthop. 30:218-232, 1963.

(114) Marino, A.A., Gross, B. and Specian, R.D.: Electrical stimulation of mandibular osteotomies in rabbits, Oral Surg. Oral Med. Oral Pathol. 62:20-24, 1986.

(115) Becker, R.O.: Electrostimulation and undetected malignant tumors, Clin. Orthop. 161:336-339, 1981.

(116) Singer, M.: The influence of the nerve in regeneration of the amphibian extremity, Quart. Rev. Biol. 27:169-200, 1952.

(117) Friedenberg, Z.B.: Bioelectric potentials in bone, J. Bone Joint Surg. 48A:915-923, 1966.

21

Pulsed Electromagnetic Field Therapy in the Treatment of Congenital and Acquired Pseudarthrosis

A. DAL MONTE, G. FONTANESI,* R. CADOSSI, **
G. POLI, AND F. GIANCECCHI*

Department of Pediatric Orthopaedics
Instituto Ortopedico Rizzoli, Bologna, Italy
*Department of Orthopaedics and Traumatology
Arcispedale S. Maria Nuova, Reggio Emilia, Italy
**Center for Experimental Haematology
Department of Internal Medicine II
University of Modena, Modena, Italy

INTRODUCTION

The technique of pulsed electromagnetic fields (PEMFs) plays an important role within the larger area of the effects of electrical stimulation on osteogenesis. Its ease of application and high success rate have fostered a remarkable development and spread of this technique rather than other techniques such as those with implanted electrodes. As a result, the vast majority of electrically stimulated patients were treated with PEMFs (over 30,000 worldwide)(1).

Stimulation with implanted electrodes differs from that using PEMFs because in the former case the current is directly applied to the tissue by means of electrodes implanted in the pseudarthrosis site (2-4), whereas with PEMFs it is induced into the tissue by means of an electromagnetic field generated by one or two coils placed outside the skin over the area to be treated (5,6). Figure 1 shows how the treatment is performed, and Table 1 lists the parameters that characterize stimulation of osteogenesis with PEMFs and implanted electrodes.

The initial studies with PEMFs were intended to overcome problems associated with the use of implanted electrodes, such as the need for surgical intervention at the pseudarthrosis

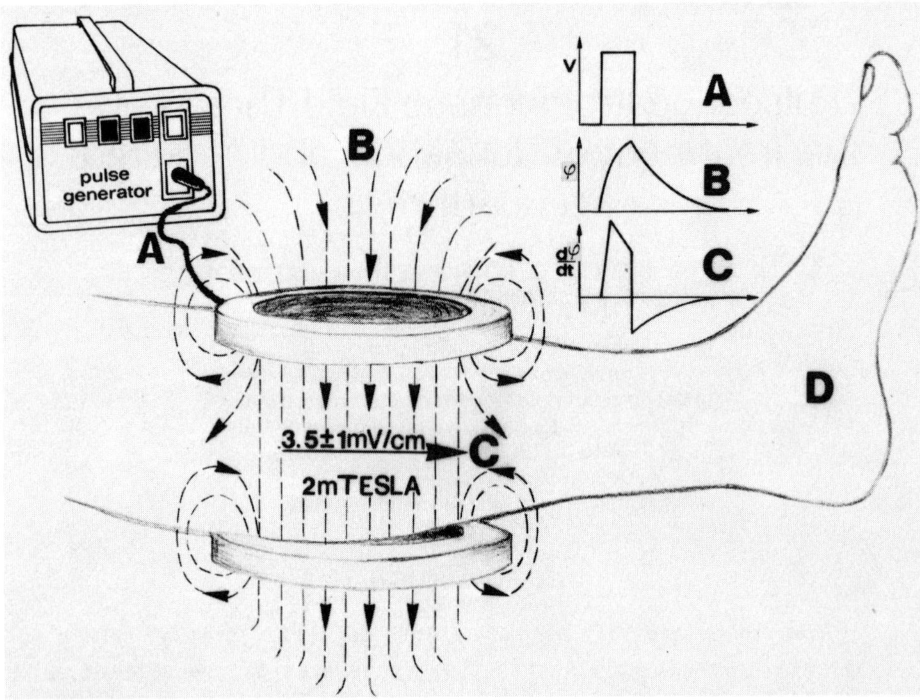

FIGURE 1. Schematic representation of the treatment conditions. Typical waveforms characterizing the stimulation. A, waveform of the electrical current supplying the coils; B, magnetic-field waveform; C, waveform of the electrical current induced in the body.

TABLE 1. Implanted Electrode and PEMF Techniques

CHARACTERISTICS	IMPLANTED ELECTRODES	PEMFs
Invasive	Yes	No
Electrical current	Directly applied	Induced
Intensity	20 μA (measured)	10–20 μA/cm^2 (calculated)
Waveform	Direct current	AC pulsed current
Anode	Distal from fracture	Indistinguishable
Cathode	In the fracture area	Indistinguishable

site, the risk of iatrogenic infections, and the breakage of electrodes (6). The proposed solution to these problems was an electrical current induced directly in the pseudarthrosis site from outside by PEMFs.

PEMFs induce an ionic current by means of an induced electromotive force. The waveform of the induced current (Figure 1C) depends on the characteristics of the coil current (Figure 1A), the electromagnetic field generated by the coils (Figure 1B), and on the inductance of the coils.

Different mechanisms are responsible for promoting osteogenesis in the cases of implanted electrodes and PEMFs (7-9). With PEMFs there exists no cathode, and the effect of the electrical current involves the entire pseudarthrosis site. Details of the process are being studied using computer models (10).

In 1974 Pilla (11) proposed that electrical currents induced by PEMFs, if properly set through control of the character-

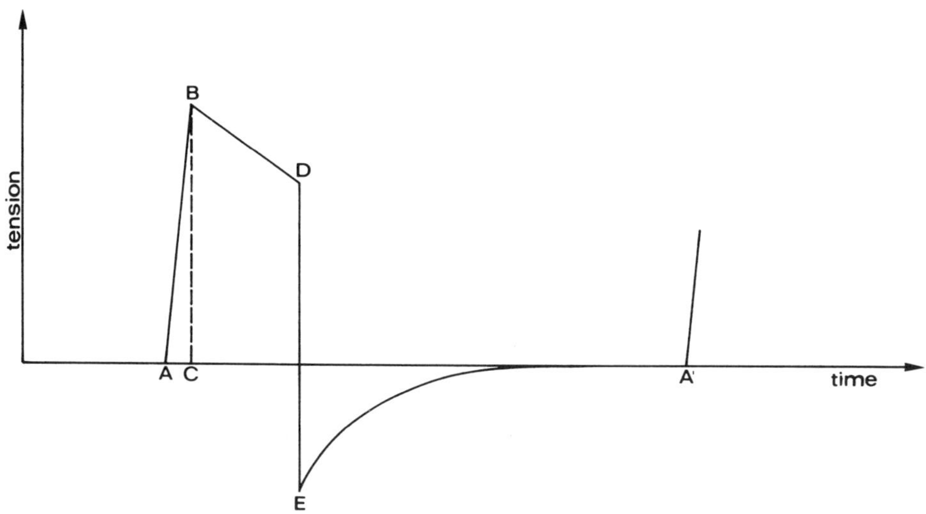

FIGURE 2. Voltage induced in a coil probe by the electromagnetic field. A-B, rising edge (the induced amplitude depends on how rapidly the magnetic field intensity varies); A-C, rise-time; B-D, pulse width; D-E collapse time of the electromagnetic field (depends on the coil inductance); E-A', period in which the electromagnetic field falls to zero; A-A', period (inverse of the frequency). From A to D the magnetic field strength increases, from D to A' it decreases.

istics of the current pulse (rise time, pulse width, amplitude and frequency)(12-15), could selectively interact with the cellular membrane and with ions located around cells, thus changing the cell's behavior. In particular, the need to use asymmetric pulses was emphasized by Pilla. Investigations are still in progress to determine the biological role played by each parameter characterizing the induced currents. There are data in the literature supporting the hypothesis that different signals produce different effects (16-22).

Even though PEMF studies were initially undertaken to study the link between electricity and osteogenesis (23-25), they have also developed into studies of the interaction between extremely-low-frequency, low-energy electromagnetic radiation and biological systems. Thus the studies offer prospects for new information that were originally unforeseen.

In our view, the data in the literature are not sufficient to conclude that each signal has its own information content, and therefore that the parameters characterizing such a signal produce specific effects. However, it is beyond doubt that different signals can modulate a specific cellular function in different ways (27). These observations have significantly affected the development of research in this field over the last ten years, and they are still the subject of a number of studies and debates.

CLINICAL EXPERIENCES

THE ELECTRODYNAMIC TECHNIQUE

In 1972 Kraus and Lechner (28) reported successful treatment of over 100 patients suffering from pseudarthrosis with a new technique to stimulate osteogenesis which they referred to as "electrodynamic." Their method was a link between invasive techniques with implanted electrodes that were already in use (2-4), and the completely noninvasive methods that were soon to develop. Their system involved a large coil, placed outside the lesion to be treated, supplied with a 12-22 Hz sinusoidal electrical current which generated a field of 30 Gauss. A pickup coil, 0.5 cm in diameter with a ferromagnetic core, was implanted in the pseudarthrosis area and connected to the bone via two platinium-iridium electrodes. The electromagnetic field generated by the external coil induced an electrical current (0.3-0.5 V/cm, 1-2 $\mu A/mm^2$) in the pickup coil which promoted osteogenesis (Figure 3). Because of the voltage (less than 1 volt) and frequency characteristics used, electrolysis of the steel alloys (such as Vitallium) did not occur. In some cases, the electrodes are connected with a suitably prepared Kuntscher nail.

In their studies, stimulation was always coupled with a surgical procedure at the pseudarthrosis site, namely stabilization by means of a plate and screws or by means of an intramedullary nail, and with a bone graft (29-35). Compared to the systems with implanted electrodes, their method offered the advantages of freedom from transcutaneous electrodes and from the need for a battery. The patients were treated for 6-7 weeks, initially for 4-8 hr/day and subsequently for 3 hr/day (36).

In 1981 results obtained from the treatment of 319 patients with pseudarthrosis were reported (37). A success rate of 93.6% was found despite the presence of infection in 107 patients. Along with electrical stimulation, the other factors involved in the pseudarthrosis management were: (a) internal fixation of the pseudarthrosis site by means of a nail or plate; (b) cancellous bone grafts; (c) resection of the pseudarthrosis site and any sequestrated bone fragments; (d) local and systemic treatment of any infection.

The major drawback in the Kraus and Lechner technique is the need for a surgical procedure. It is particularly difficult to evaluate their results because the effects produced by the cancellous bone grafts, by stabilization of the pseudarthrosis site, and by electrical stimulation cannot easily be distinguished.

The same technique was used for treatment of loose total-hip endoprostheses (38). Patients with loose cemented (426 cases) and noncemented prostheses (15 cases) were treated. The field generator was the same as that used in the treatment of pseudarthroses, except that the implanted pickup coil was not used. The results were assessed on the basis of clinical parameters including spontaneous pain, pain with motion, walking distance, and analgesic drug intake. The authors reported good

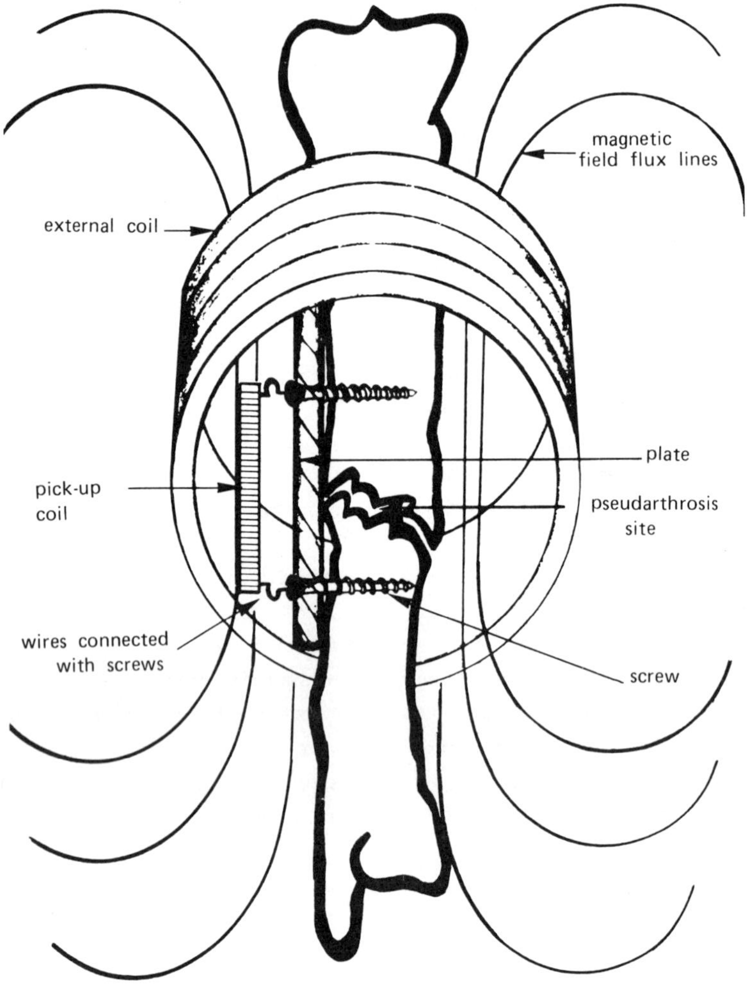

FIGURE 3. The electrodynamic method. The external coil generates an electromagnetic field which induces a voltage in an implanted pick-up coil. The voltage is applied by means of wires to A/O system screws (electrically insulated from the plate) placed on both sides of the pseudarthrosis.

results following 14 weeks' stimulation (performed 2-3 hr/day) in 67% of the patients. The system was also studied with regard to treatment of epicondylitis (success rate, 60-70%), first and second-stage Sudeck disease (success rate, 70%), juvenile osteochondritis (success rate, 80%), and avascular necrosis of the femoral head (36). In the treatment of avascular necrosis, a cancellous graft and implanted pickup coil were used and good results were seen in 21 of 22 patients in the experimental group, but only in 3 of 21 patients in the control group.

It is difficult to evaluate the specific effect of the PEMFs because the treatments were always coupled with complex surgical procedures that, alone, often yield high healing rates. It is also difficult to understand the usefulness of the implanted pickup coil because in some conditions (endoprosthesis loosenings, epicondylitis)(36,38) the technique seemed to work properly even in its absence. Furthermore, the technique does not avoid the need for surgery, and this is the chief advantage of the use of PEMFs for electrical stimulation of osteogenesis.

THE NEW YORK SYSTEM

The method of PEMF stimulation of osteogenesis most used world-wide is that studied and developed by Bassett in the 1970's. Over 20,000 patients have been treated (1,5,6,8, 39-57). In the original animal study, dogs were given fibular osteotomies and then stimulated with either low-frequency (1 msec, 1 Hz) or high-frequency (150 μsec, 65 Hz) pulses (50); the latter pulse was deemed to be more effective. Since that report, the system has evolved a great deal.

According to Bassett, proper stimulation involves use of a specific signal, and suitable orthopedic treatment including alignment of the pseudarthrosis stumps, absolute immobilization (in the majority of patients it is obtained by means of a plaster cast), and less than 1.5 cm of bone loss. The treatment lasts 10-12 hr/day and is maintained until consolidation is obtained. The reported healing stages are sclerosis of the fracture stumps, followed by progressive enhancement in the radiodensity of the pseudarthrosis gap (6). No periosteal bone callus formation is observed. On the contrary, when a "bulky elephant foot callus" was present, its resorption was often observed.

The healing process observed has been compared to that obtained with rigid internal fixation, and has been attributed to the organizational rather than the stimulatory properties of the signal, which reportedly favor calcification of the cartilage present in the pseudarthrosis site. This kind of consolidation was also observed by Sharrard et al. who reported that "no new external bone callus forms" (44).

When the pseudarthrosis site becomes radiodense, the patient begins a gradual program of axial compression of the pseudarthrosis to promote its consolidation (1). Stimulation has afforded many patients the opportunity to avoid surgery or even amputation.

The initial signal used in the experiments on beagles (50) has been improved, resulting in the selection of two presently used signals (Figure 4); one is a single pulse signal, and the

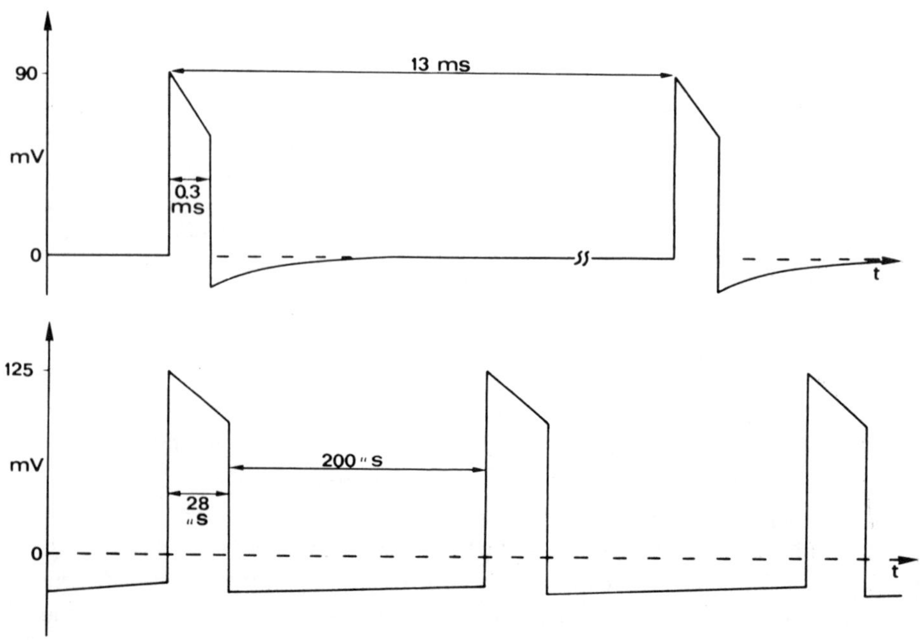

FIGURE 4. Waveform of the two signals used for pseudarthrosis treatment. Top, single pulse signal; bottom, burst signal. Each burst lasts 5 msec and is repeated at 15 Hz.

other is a pulse-burst. The two signals are said to have different properties: the former promotes vascularization, and the latter has calcificating properties and is mainly applied for treatment of nonunion. Several investigators have utilized this technique in different studies, with success rates over 75%. A significant concern in these studies has been avoiding starting stimulation immediately after surgical procedures or bone grafts, in order not to perform two treatments on the patients at the same time.

Most of the patients admitted to PEMF treatment were affected by nonunions that clearly could be healed only by means of surgery. X-rays made over the three months prior to initiation of stimulation were used to establish that the pseudarthrosis was not spontaneously evolving towards consolidation. Most patients were immobilized with a plaster cast before starting stimulation. As discussed below, this procedure inevitably caused much criticism.

In 1982 Bassett et al. (39) reported on the treatment of 1007 ununited fractures and 71 failed arthrodeses. Each patient had undergone an average of 2.2 surgical procedures before starting PEMF stimulation, and the average stimulation time was 5.5 months. The final success rate was 77%.

Others have used the same system. In 1981 Sedel et al. reported a success rate of 83% in 37 patients (43). In 1982 Sharrard et al. reported a success rate of 71.7% in 53 patients using an average stimulation time of 6 months (44). In 1984, Stein and Anzel reported (45) the results of stimulation performed on 17 patients affected by delayed union of the tibia who were treated with external fixators and PEMF stimulation. The treatment was 88% successful, with an average stimulation time of 5.6 months. A synergistic effect between bone grafts and stimulation when the latter was performed immediately after the operation was also reported (45). Bassett et al. reported the same observation in 1982 (49); when the patients were treated simultaneously with graft and stimulation, more than 90% of them healed (1).

PEMF stimulation is widely used for treatment of congenital pseudarthrosis which is an unsolved problem in modern orthopaedic surgery (6,48,51,52). In 1981 Bassett et al. (48) reported complete healing in 17 of 34 patients. In 1982, Sutcliffe and

Goldberg (51) reported positive results in 27 of 38 patients (71%). A discussion on the effectiveness of PEMF stimulation in the treatment of congenital pseudarthrosis has been presented (52). It was concluded that PEMF stimulation would improve prognosis by at least 20% with regard to the attainment of bony union.

Bassett has concluded that, in association with proper orthopaedic treatment, it is possible to obtain positive results in more than 90% of adult patients, and in 99% when the stimulation is combined with a bone graft.

New therapeutic applications have been proposed recently, including treatment of epicondylitis (58)(a double-blind study), avascular necrosis of the femoral head (59,60), failed spinal fusions (61), and persistent rotator cuff tendonitis (58).

In the tendonitis study (58), the investigators used the single-pulse signal because the lesion is often related to the vascular supply. The patients were treated 5-9 hr/day for 4 weeks. For all the parameters evaluated, the difference between patients treated with effective stimulators and those with placebo stimulators was statistically significant. This is a clear demonstration of the clinical effectiveness of PEMFs.

The available data (59,60) suggest that PEMF stimulation can be useful in treating avascular necrosis of the femoral head, as regards both functional recovery and relief from pain. Some patients showed improvement as judged by X-ray. A final assessment requires further investigation and an adequate follow-up.

Recently, a pilot study conducted in Great Britain (62) suggested that PEMF stimulation could significantly shorten the time required for recovery from Perthes disease.

PEMFs have been used to promote healing in cases of failed spinal arthrodesis with post-operative times of 40-52 months. In 10 of 13 patients body-to-body fusion was obtained (77%), indicating that stimulation could become a successful alternative to the treatment of failed lumbo-sacral spinal fusions (61).

THE SWANSEA TECHNIQUE

Another PEMF technique for the stimulation of osteogenesis is the Swansea method, chiefly studied by Watson (63-66). This system was developed simultaneously with that adopted by Bassett. The first result of the clinical application of this stimulator, the successful treatment of a nonunion of the tibia, was reported in 1975 (63). In Watson's unit, the electromagnetic-field generator was an iron-core electromagnet. The resulting magnetic field was 530 Gauss, which is significantly higher than the air-core electromagnetic systems. The frequency was less than 1 Hz. Daily treatment was performed for 20-22 hours, and the patients were immobilized in a plaster cast until X-ray healing was evident. This system has not yet been used on many patients.

In 1980, DeHaas et al. (65) reported on the treatment of 17 patients suffering from ununited fractures of the tibia. To assess the effectiveness of the stimulation, the authors compared its results with those of another series made up of 23 patients (67) with similar lesions, who had been treated with traditional techniques. In this series, 18 of the patients also underwent a surgical procedure and various kinds of bone graft. In the group of stimulated patients the success rate was slightly higher than that found in the surgically-treated patients (88% compared to 82%). The average treatment time was 5.9 months in the stimulated group, as against 10.7 months in the group undergoing surgery. Finally, whereas in the control group 4 failures were amputated, in the stimulated group the 2 failures were treated surgically and then healed. Formation of periosteal bone callus was not excluded even though a significant periosteal activation was not seen. "External callus formation is not conspicuous...some revascularization occurs at the fracture site, reducing the density of the bone ends. This is followed by very gradual interfragmentary callus formation...." The authors emphasized that further investigations were needed.

In 1983, Watson presented a new version of the stimulator which employed an asymmetric pulse (Figure 5)(68). The electromagnetic field was again generated by an iron-core electromagnet, but it was smaller than that used previously (20 Gauss at 11 Hz). The field induced a voltage about four times greater than that used in previous experiments (lower coil impedance). Preliminary data indicated a success rate lower than that found with the previous technique (65).

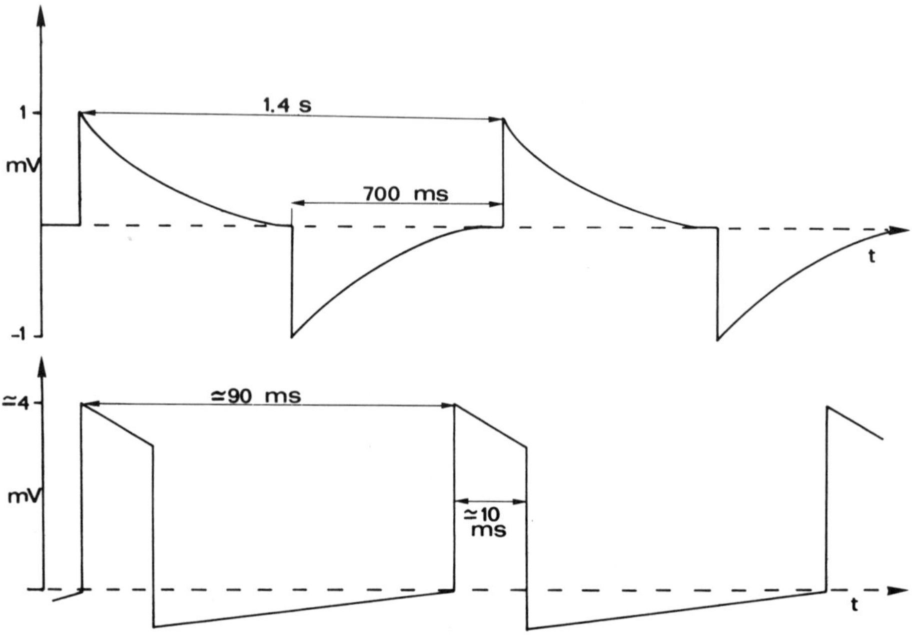

FIGURE 5. Waveform of the old (top) and new (bottom) signals of the Swansea method.

THE WHALSTROM AND THE HAIMOVICI TECHNIQUES

Thirty females with Colles fractures were randomly divided into two groups. All the patients received the same orthopaedic treatment, but only those in one of the two groups underwent PEMF stimulation (4 Gauss, 1-1000 Hz)(69). Healing was evaluated using technetium scintigraphy. Enhanced activity (about 30%) was observed in the stimulated fractures during the first 2 weeks, but not during the third or fourth week. Whalstrom concluded that PEMF stimulation "improves the early phase of fracture healing."

Bilateral osteotomies of the first metatarsal with plate and screw stabilization were performed by Haimocivi on 32 patients (70). One of the osteotomies in each patient was stimulated (1 hr/day) and the other was used as a control. The bilateral difference in healing time was significant. No delayed unions occurred in the stimulated bones, while in the control group two osteotomies developed a delayed union.

OUR CLINICAL EXPERIENCE

As early as 1979 we began to study the possibility of stimulating the healing of nonunions and congenital pseudarthrosis through PEMF stimulation (71-85). We used a pair of air-core coils placed over the area to be treated (Figure 1).

1. Stimulation Characteristics

The unit we used is the IGEA stimulator (Igea, Carpi, Italy). It produces single pulses of 3.5 ± 1 mV as measured using a standard coil probe (50 turns of copper wire (diameter 0.20 mm) internal diameter 0.5 cm (6))(Figure 6). The Fourier series of the induced signal is shown in Figure 7. Stimulation lasts 10-12 hr/day until complete healing.

2. Patient Selection

The patients admitted to the study suffered from either congenital pseudarthroses (classified according to Boyd) (86) or acquired nonunions. The latter group included 2 classes of patients. The first consisted of those with acquired pseudarthrosis, defined as the absence of any tendency to spontaneous healing. In these cases only a surgical procedure would be expected to bring about consolidation. Acquired pseudarthroses were of at least 10 months' duration. The second class consisted of patients having a delayed union; in this case the possibility of spontaneous healing still existed. For a delayed union, there was no evidence of healing for at least 5 months after the trauma.

The patients admitted to the study had undergone their last surgical procedure at least 3 months earlier; our approach is outlined in Figure 8. Unlike other studies, the immobilization technique was not changed just before the beginning of PEMF stimulation. Patients undergoing stimulation for longer than 12 months were considered treatment failures.

Since congenital pseudarthrosis is a relatively homogeneous illness, the patients could be divided into a control and an experimental group. All patients received the same surgical treatment consisting of excision of the pseudarthrosis site, reduction, and nailing (87)(Figure 9). The results were

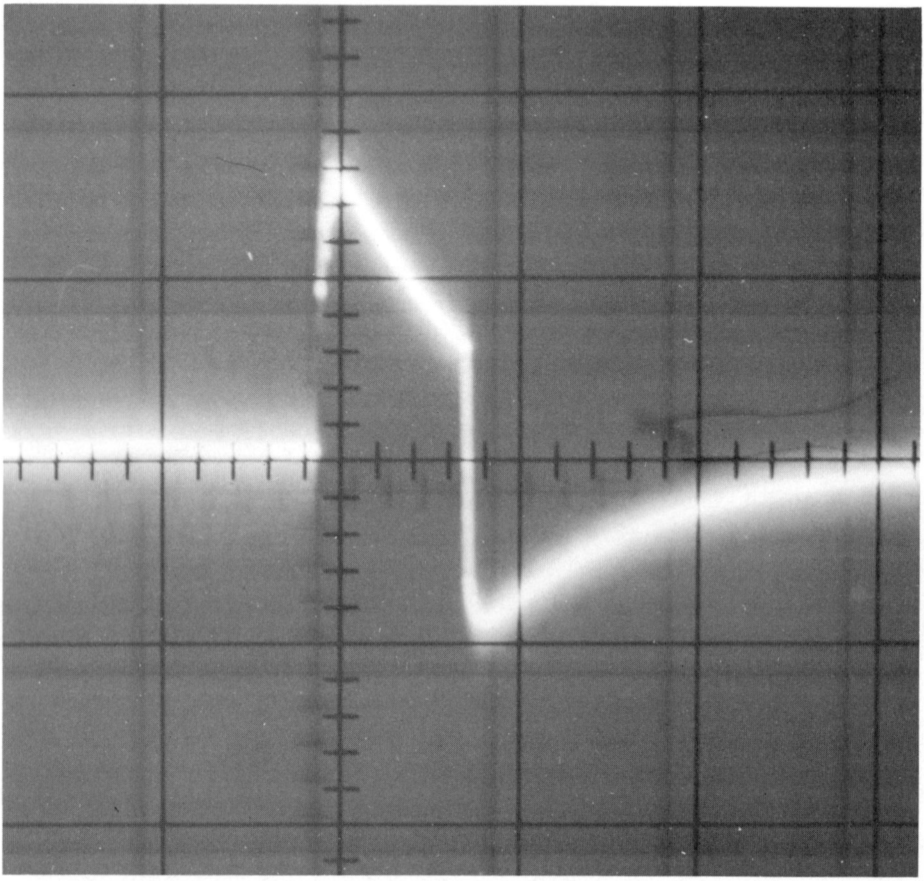

FIGURE 6. Waveform of the induced signal of the IGEA stimulator. Oscilloscope sensitivity, 2 mV/div, 2 msec/dv, for vertical and horizontal scales, respectively.

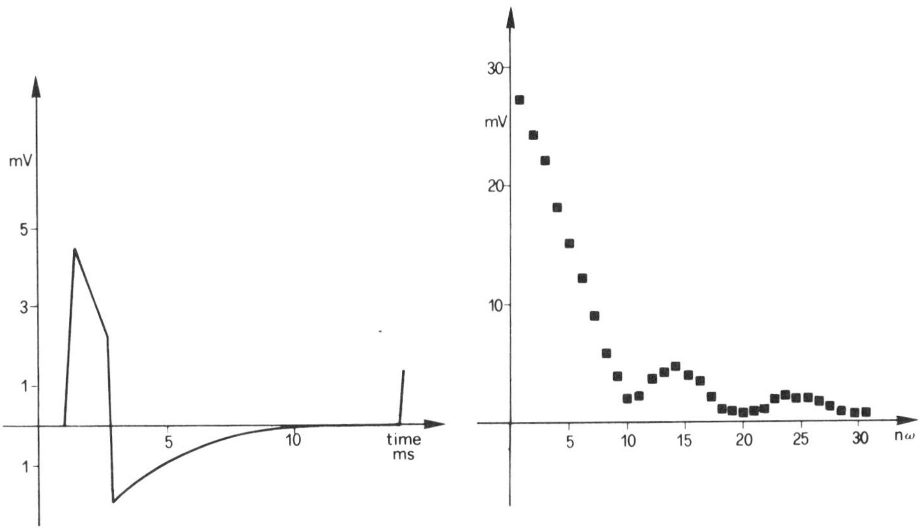

FIGURE 7. Fourier series (right) of the induced signal (left).

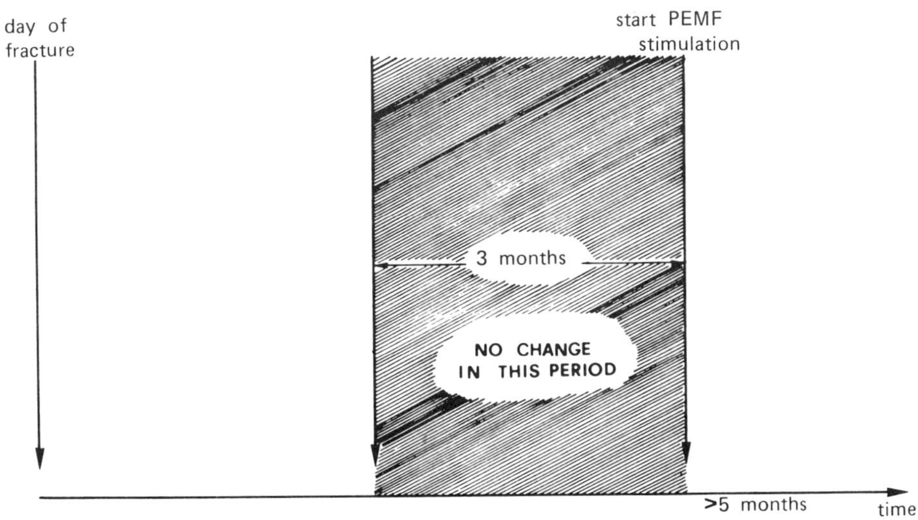

FIGURE 8. In the three months preceding the start of PEMF stimulation there was no surgery and no evidence of spontaneous healing.

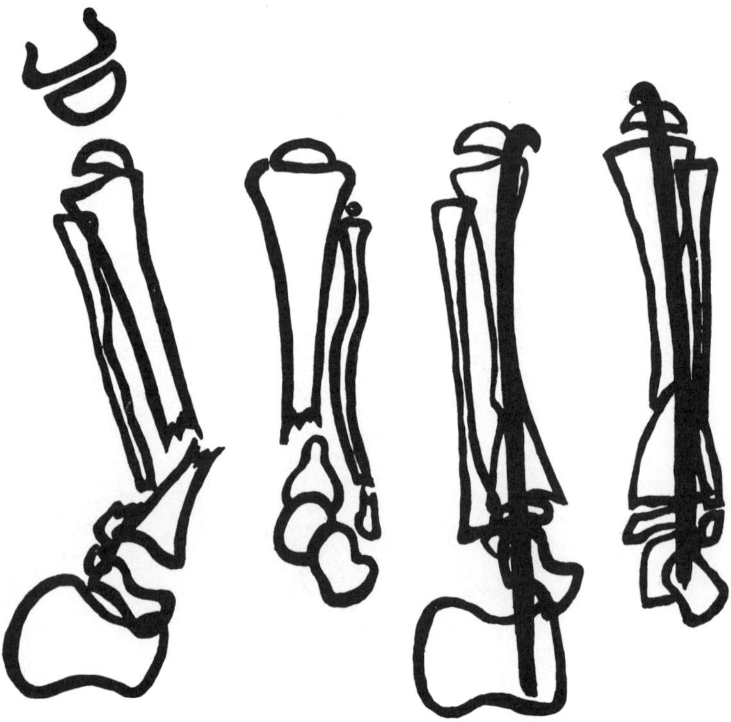

FIGURE 9. Schematic representation of the surgery all the patients suffering from congenital pseudarthrosis underwent: excision of the pseudarthrosis site, reduction and nailing with intramedullary nail.

assessed by determining (a) the need for additional surgical procedures, and (b) the increase in the imbalance between the healthy and the affected limb.

3. Results

Table 2 relates the data pertaining to the stimulation of patients suffering from congenital pseudarthrosis, and the statistical analysis of the results according to Fisher's test. Even though the number of patients is very small, a statistically significant difference is evident in the number of cases requiring reoperation.

Table 3 presents the general data for patients treated for acquired nonunion. The healing rate did not differ significantly among the different bones stimulated (Figure 10).

Table 4 lists the disability time prior to inclusion in the study of the patients with acquired nonunion; the success rate was not affected by the nonunion age.

Table 5 relates the results to the diagnosis; the success percentage was higher among the acquired pseudarthroses (91%) than among the delayed unions (80%). Table 6 shows the stimulation times: the group of failed patients showed a significantly lower average stimulation time than did the healed patients.

Table 7 shows the results obtained with the different immobilization techniques: external fixators had the lowest success rate (75%), and cast immobilization had a success rate of 85%. The success rate was not affected by the presence of

TABLE 2. Results of Congenital Pseudarthrosis Treatment

	CONTROL GROUP	STIMULATED GROUP
Number of cases	6	6
Cases reoperated	3*	0
Imbalance increase	3	1

*P < 0.001, Fisher's Test

TABLE 3. General Patient Data for Patients with Acquired Nonunion

CHARACTERISTIC	NUMBER
Patients treated:	248
Average age:	37
Males:	187
Females:	61
Healed:	208 (84%)
Failed:	40 (16%)

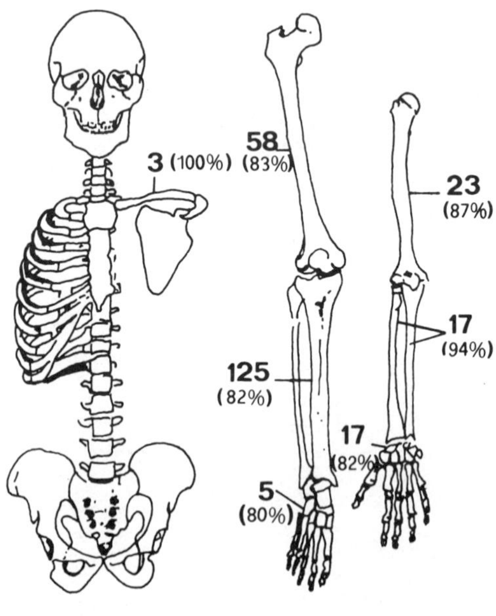

FIGURE 10. Bones treated with PEMF stimulation. The value in parentheses indicates the percentage success rate.

TABLE 4. Duration of the Acquired Nonunions (in Months)

DIAGNOSIS	HEALED	FAILED
Pseudarthrosis	17	13
Delayed union	6	5
Average time	10	7

TABLE 5. Result of PEMF Stimulation According to the Diagnosis

DIAGNOSIS	NUMBER OF CASES	HEALED	FAILED
Pseudarthrosis	100	91 (91%)	9 (9%)*
Delayed unions	148	117 (80%)	31 (20%)

*P < 0.05, Chi-square test

TABLE 6. Average PEMF Stimulation Time (in Months)

DIAGNOSIS	HEALED	FAILED
Pseudarthrosis	5	5.5
Delayed unions	2.8	3
Average time	4.3*	3.6

*P < 0.06, Student's T-test

TABLE 7. PEMF Stimulation with Respect to the Immobilization Technique Used

IMMOBILIZATION USED	NUMBER OF CASES	HEALED	FAILED
Cast	184	157 (85%)	27 (15%)
External Fixators	28	21 (75%)	7 (25%)
Orthopaedic Device	15	12 (80%)	3 (20%)
Others	21	18 (86%)	3 (14%)

infection (Table 8), which was treated with both local and systemic antibiotics. Table 9 lists the possible causes of the failed cases.

Figure 11 shows the result of the treatment of a typical congenital pseudarthrosis. Figures 12 and 13 show the results of the treatment of two acquired pseudarthroses.

4. Conclusions

PEMF stimulation seems capable of significantly improving the prognosis in cases of congenital pseudarthrosis when it is coupled with proper surgical treatment.

TABLE 8. Results of Acquired Nonunion Stimulation with Respect to the Presence of Infection

INFECTED	NUMBER OF CASES	HEALED	FAILED
Yes	90	76 (84%)	14 (16%)
No	158	132 (84%)	26 (16%)

TABLE 9. Possible Causes of PEMF Treatment Failure

NUMBER OF CASES	POSSIBLE CAUSE OF FAILURE
12	Inadequate immobilization
13	Short stimulation time
3	Excessive bone loss
12	Unknown

PEMF stimulation is particularly effective in the treatment of acquired nonunions. In the future, the success rate will probably be increased as a consequence of our experience. We think that many unsuccessful treatments can be avoided.

A successful stimulation seems to depend more on the stimulation time than on the seriousness of the lesion; indeed, 32% of failed patients discontinued the treatment within 60 days after the start of PEMF stimulation. This observation has also been made by others. Unlike what other investigators have observed, in these cases healing was always accompanied by the formation of periosteal bone callus. When a "bulky elephant foot callus" was present, we never observed its resorption.

Once bone repair was activated, it proceeded with a normal pattern and healing time. In our experience, the most significant changes in the X-ray picture occurred after the first 45-60 days of stimulation.

THE DOUBLE-BLIND PROBLEM

The average stimulation time of patients treated for acquired pseudarthroses was over 5 months. It is therefore difficult to evaluate the effectiveness of PEMFs as against the long period of time during which patients were immobilized in a plaster cast. Consequently, many have urged the performance of

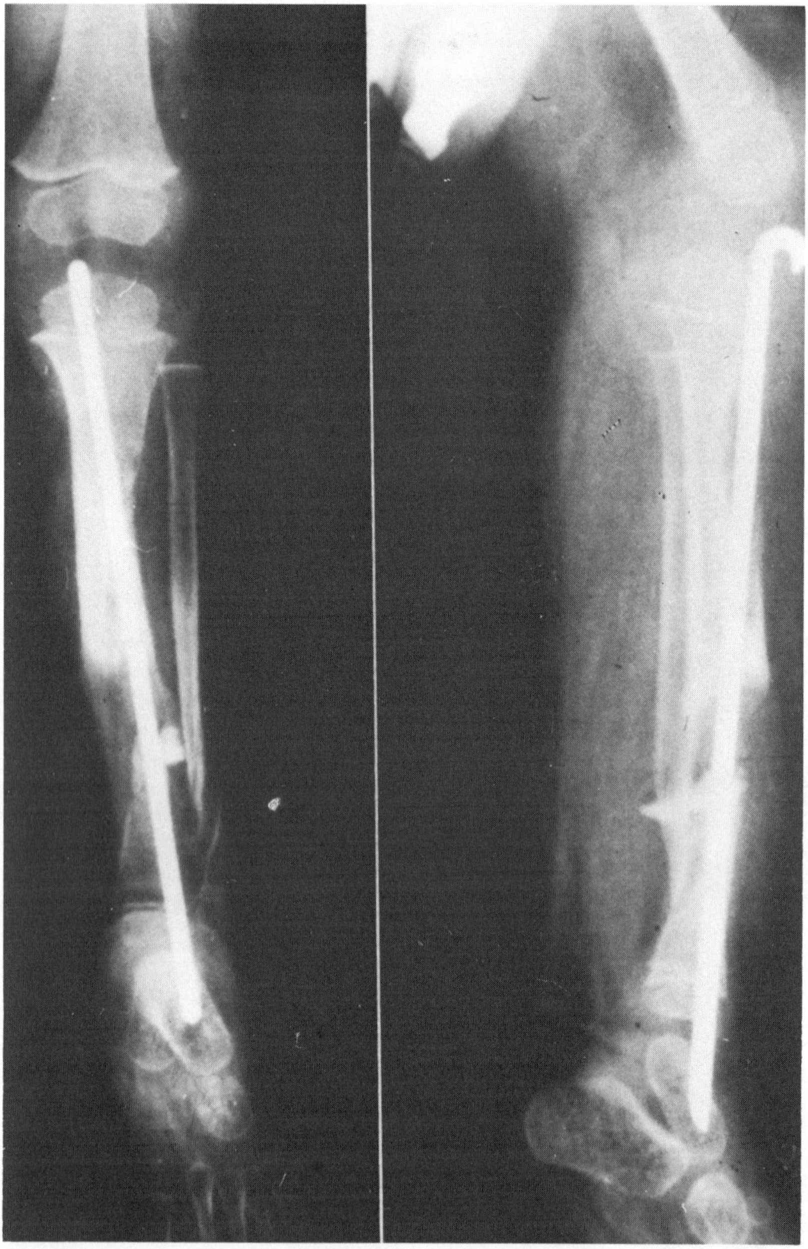

FIGURE 11. Male aged 4 years. Congenital pseudarthrosis of the tibia. Left: immediately after intramedullary nailing. PEMF stimulation began. Right: X-ray 12 months after the end of stimulation. Observe limb lengthening and the good bone callus quality.

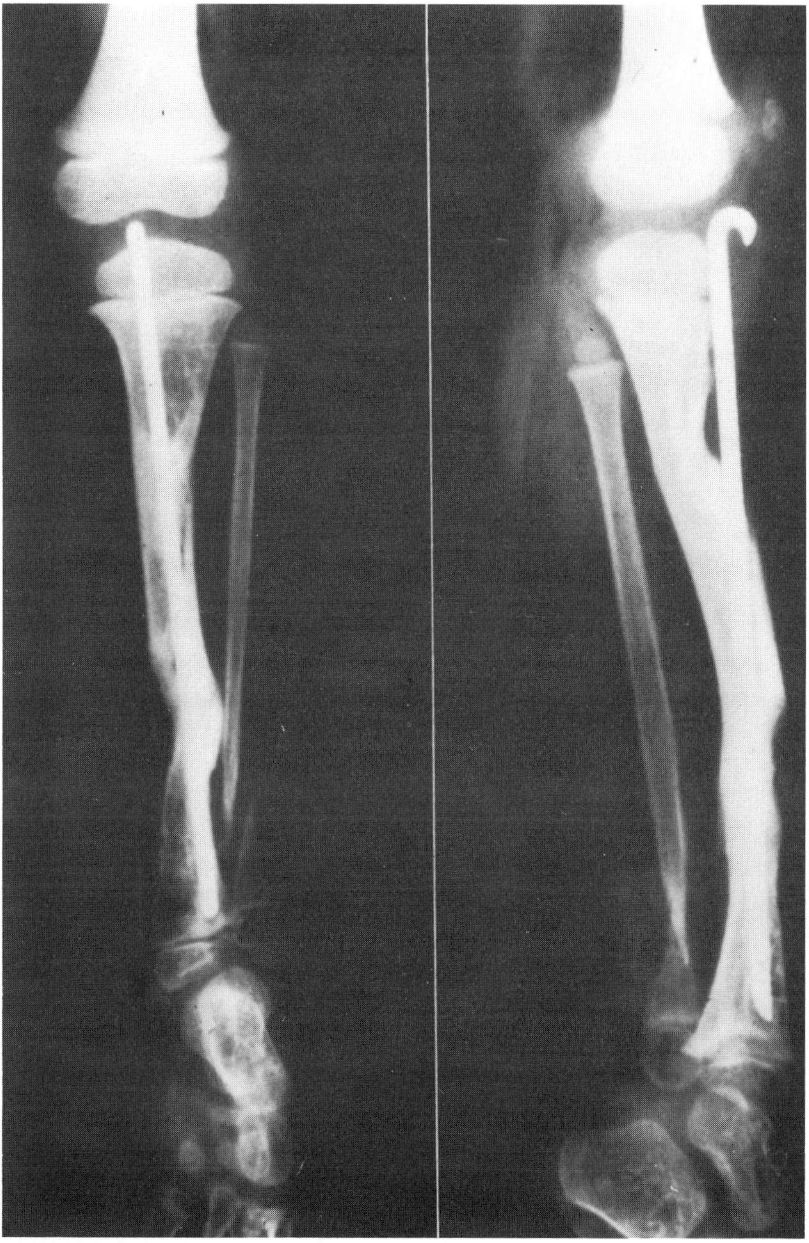

FIGURE 11. (Continued)

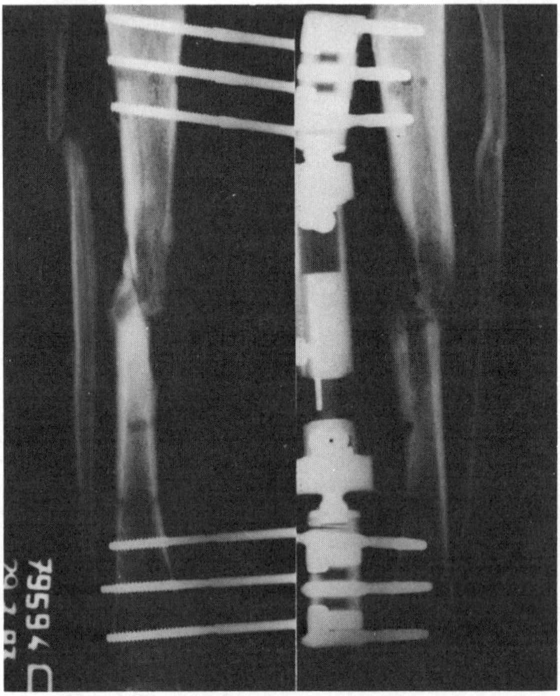

FIGURE 12. Male aged 18 years. Left: acquired nonunion of the tibia, 6 months after the trauma, chronic draining infection was present. At the time of the fracture serious vascular damage and soft tissue loss occurred. Top right: after 3 months stimulation complete recovery of bone and soft tissues. Infection healed. Observe the good periosteal bone callus formation. Bottom right: 7 months after the end of the stimulation (Courtesy of Prof. F. Pisano).

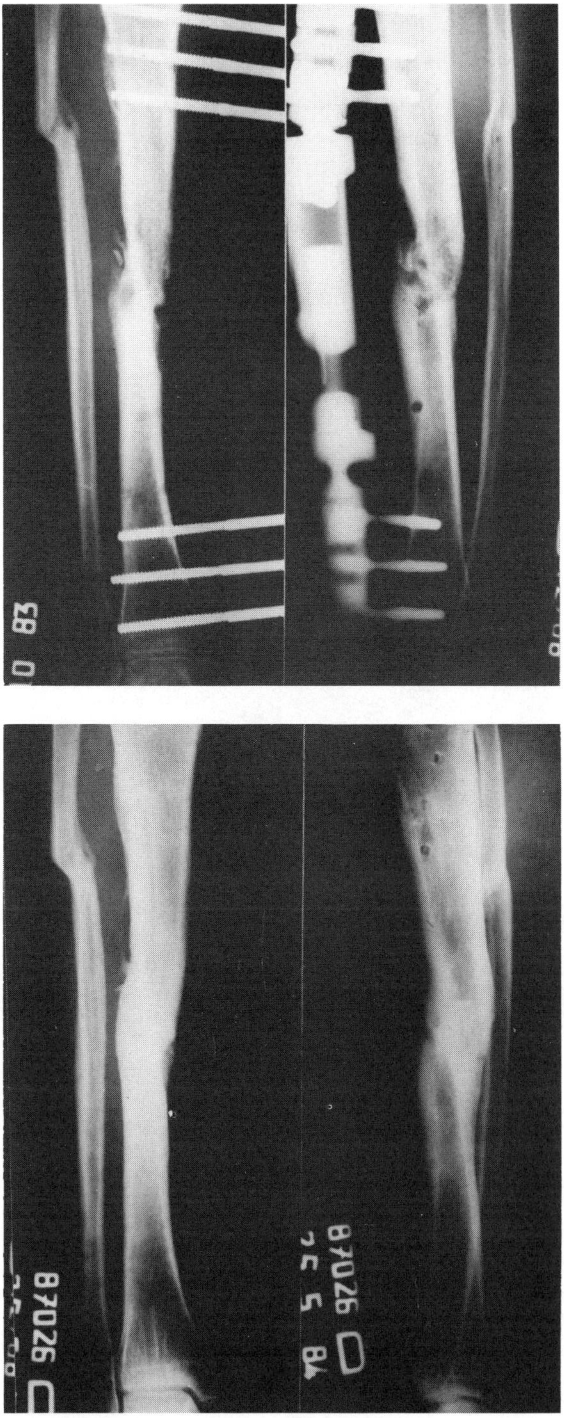

FIGURE 12. (Continued)

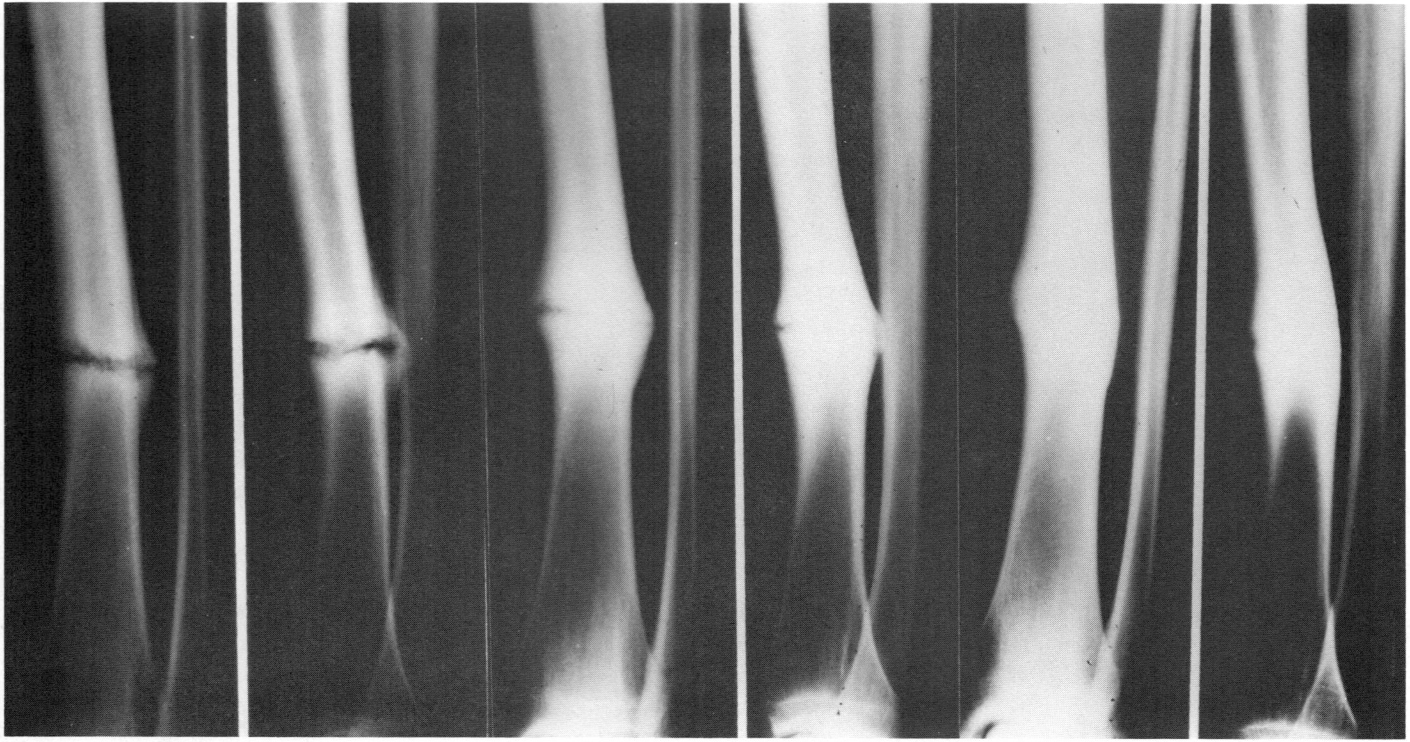

FIGURE 13. Male aged 24 years. Left: acquired nonunion of the tibia in hypertrophic evolution. Center: after 3 months stimulation. Right: X-ray at 3 months after the end of the stimulation. The bulky elephant foot callus has not been reabsorbed.

a double-blind study that would be capable of quantifying the effect of the PEMF itself. Sedel et al. (43) for example, believed that the stimulation effectiveness was unquestionable in 14 out of 35 patients, but that only a double-blind study could quantify it.

The plea for a double-blind study seems well-founded, but the arguments against it are also sensible. The extreme lack of homogeneity in the population of stimulated lesions, however, prevents investigators from having homogeneous groups unless they use a very large population. Also, the ethical aspect of requiring the patient to run the risk of a lesion worsening owing to protracted inactivity over several months should not be neglected. Many investigators have preferred to compare the treated patients with the patients previously examined, especially those who received bone grafts. Obviously, these considerations do not convince the double-blind study supporters (88,89).

In 1984 Barker (90) published preliminary results of a double-blind study conducted with stimulators making use of the burst signal used by Bassett. Even though statistical analysis of the results suggested that PEMFs were not effective in the treatment of nonunions, the research could not exclude a maximum positive effect of 33% due to stimulation. This research demonstrated that the plaster cast was by itself capable of healing 71% of pseudarthroses within 24 weeks, and that 81% of patients reached consolidation within 48 weeks. The soundness of the conclusions has been widely debated (91), both for the statistical analysis carried out and for the small number of patients examined (7 cases in the control group and 9 in the stimulated group). Besides, an analysis of the characteristics of the lesions admitted to the study revealed the relative lack of homogeneity between the two groups (92). The study has served to focus attention on conservative treatment.

COMPARISON BETWEEN TREATMENTS

Obviously, the most troublesome criticism facing the research carried out so far is attributable to the lack of a double-blind study (88). However, it is worth comparing the results obtained with PEMF stimulation with those that can be achieved by means of surgical techniques.

CONGENITAL PSEUDARTHROSIS

Sharrard (52) reported an effectiveness of at least 20%, and our series forecasts an improvement in the prognosis of at least 50%. A modern therapeutic proposal envisages the utilization of vascularized bone grafts (93). There are only a small number of such cases in literature, and failures have been reported. Only when the technique becomes more widespread, will we be able to compare it with the results achieved with stimulation. The possibility exists that PEMF stimulation could be used in conjunction with the surgery.

ACQUIRED NONUNIONS

Acquired nonunions are treated surgically to promote healing. In most cases mechanical stabilization of the nonunion is coupled with a bone graft so as to trigger osteogenic activity capable of leading to consolidation of the nonunion. PEMF treatment is performed with the aim of favoring activation of osteoprogenitor cells to trigger the osteogenetic activity in the pseudarthrosis site that will lead to consolidation.

While analyzing the technique to be used as an alternative to stimulation, it should be borne in mind that the patients we treated, and those stimulated by Bassett as well, came from a long, frustrating and ineffective orthopaedic experience. It should also be remembered that a comparison could be made only with those studies in which patients underwent surgery, since spontaneous healing could not be expected. This is the way the comparison between PEMFs and standard surgical techniques can be made, in that the two series can be considered sufficiently homogeneous.

Finally, the PEMF series cannot be compared with low-healing-rate fractures, since we are far beyond the 52 weeks analyzed by Watson and Jones (94). In our series, the average elapsed time from the trauma for the pseudarthrosis cases was 17 months.

1. Treatment of Nonunions with Surgery

In 1961, Boyd et al. (95) described the treatment of 842 patients affected by nonunion of the shaft of the long bones, and reported that 88% healed after one surgical procedure. In 1984, Lifeso and Faisal Al-Saati (96) obtained 93% healing after a surgical procedure in a series of 129 nonunions. In 1981, Weber and Brunner (97) described the results of the treatment of 249 nonunions of the tibia by means of a rigid internal fixation; only 6 failures were registered. Rechling and Walters (98) obtained a healing rate of over 90% using bone grafts. Berentey (99) described 122 cases treated with compression plates or intramedullary nails and recorded a success rate of 84.4%. Schellmann and Klemm (100) obtained a healing rate of 90% in tibia and femur pseudarthroses using interlocking nails.

2. PEMF Stimulation and Surgery Combined

PEMF treatment alone yields success rates that are very close to those obtainable from a surgical procedure; by our own experience 91 out of 100 patients suffering from pseudarthrosis were successfully treated. When a bone graft is used with stimulation, the success rate is very high: 100% in our series (81) and 99% in Bassett's series (1,49). Furthermore, it is worth pointing out that, whenever the PEMF treatment does not manage to bring about consolidation, it does not exclude the possibility of a subsequent surgical procedure.

Table 10 summarizes the success rate obtained by different authors with the first surgical operation as compared to PEMF treatment. Also, the second surgical operation success rate is compared to PEMF stimulation associated with surgery.

DISCUSSION

Figure 14 shows the characteristics of the various signals used, and Table 11 sums up the results obtained. Only a few authors failed to observe any significant effect of PEMF stimulation (101); none have described negative side-effects. Most authors, even those who are skeptical about PEMF stimulation (90,102,103), put more stress on the difficulties encountered in

TABLE 10. Nonunions Treatment with Surgery and/or PEMF Stimulation

AUTHOR	1st OPERATION SUCCESS RATE	2nd OPERATION TOTAL SUCCESS RATE	AMPUTATIONS NUMBER OF CASES
Boyd (95)	88%	94%	13 (1.5%)
Lifeso (96)	93%	—	?
Macnab (67)	83%	83%	4 (17%)
Weber (97)	88%	95%	?

AUTHOR	PEMFs SUCCESS RATE	PEMFs + SURGERY SUCCESS RATE	AMPUTATIONS NUMBER OF CASES
Bassett (39)	77%	99% (1)	?
De Haas (65)	88%	100%	0
Fontanesi (78)	84%	100% (81)	0
Heckman (64)	61%	90%	?

quantifying the PEMF effect than on their doubts about its soundness.

We think that the PEMF mechanism of action involves cellular recruitment: that is, that the number of activated cells over the same period of time is higher as a consequence of PEMF exposure. A similar hypothesis was proposed by Jacobs to explain the effects of stimulation with implanted electrodes (104). Several recent reports seem to support this view (69, 105). Some preliminary results we obtained from osteotomies of rat fibulae are a further confirmation: after 8 days, the callus looks more developed than in the controls, whereas after 23 days strength tests do not show any difference between stimulated and controls.

We have begun a study to determine the possible use of PEMFs in cases of closed fresh fractures of the tibia treated with cast immobilization. The aim is to shorten the healing

COMPARISON OF THE DIFFERENT SIGNALS USED FOR NON-UNION TREATMENT

UNIT	WAVEFORM	FREQUENCY	MAGNETIC FIELD PEAK VALUE	STIMULATION HOURS/DAY	INDUCED SIGNAL
E.B.I.	pulse	72	16 mTesla	10	
E.B.I.	burst	15	1.9 mTesla	10	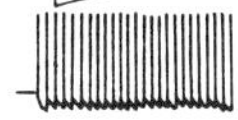
IGEA	pulse	75	2 mTesla	10	
MAGNETODYN	sine	10-22	2 mTesla	3-6	?
Swansea	pulse	1	53 mTesla	22	
Swansea	pulse	10	2 mTesla	10	
Wahlström	noise	1-1000	0.4 mTesla	8	

FIGURE 14. Comparison of the stimulation parameters produced by different apparatuses. The induced signal shown is the one recorded with the described coil probe.

TABLE 11. Results Obtained by Different Authors

AUTHOR	YEAR	NUMBER OF PATIENTS	STIMULATION TIME (months)	PERCENT SUCCESS	UNIT USED
Bassett	1982 (39)	1077	5.5	77	EBI
De Haas	1980 (65)	17	5.9	88	Swansea
Fontanesi	1984 (76)	248	4.3	83	IGEA
Heckman	1981 (54)	149	9	64	EBI
Lechner	1981 (37)	319	2	93	Magnetodyn
O'Connor	1985 (56)	32	7	80	EBI
Rinaldi	1985 (80)	16	5.6	75	IGEA
Sedel	1981 (43)	37	3.3	83	EBI
Sharrard	1982 (44)	53	6	72	EBI
Stein	1984 (45)	17	5.6	88	EBI

time and avoid complications such as the slow-healing fractures (94). Preliminary results indicate that both effects can be obtained with PEMF stimulation.

A slow healing response sometimes results from insufficient immobilization, but it can also result from an insufficient activation of the fundamental biological processes that lead to fracture-healing. It is important to recall that fracture healing is ultimately a biological process; this is often forgotten in favor of the mechanical aspect of the treatment of the fracture. It was the inhibition of some biological activity (the absolute lack of periosteal callus) that led to the onset of the well-known problems related to the utilization of A/O compression plates.

The effect of periosteal activation that we observed during the healing process in pseudarthroses reveals that after PEMF exposure the number of cells involved in the osteogenetic process has increased, and that the final healing comes from the triggering of a reparative process with suitable biological activity. We emphasize that, while the initial X-rays show significant changes, later on, once the healing process has been triggered, it proceeds with absolutely normal times and patterns (73). This interpretation, which is simplified, is different from that previously proposed (6). We think that stimulation triggers the activation, proliferation and differentiation of osteoprogenitor cells through mechanisms such as Ca^{++} influx, for example. We do not, however, want to exclude other effects.

If the proposed effect on cellular recruitment is accepted, the rationale for choice of PEMF stimulation resides in the need for promoting activation of osteogenetic biological processes. The proposal that PEMFs recruit a greater number of cells in a shorter period of time, as compared with the environmental factors only, could provide an interesting key to understand the large number of the described effects (106-110) and to interpret them in the light of a common mechanism of action. The orthopaedic surgeon should resort to PEMFs whenever he deems the biological response actually present to be not satisfactory.

COST/BENEFIT RATIO

The cost/benefit ratio for the patients and for the community is undoubtedly significant. The same is true of the risk/benefit ratio: no negative side effects have ever been described. However, on the bases of the data obtained in vitro (111), we feel it is important to emphasize that, even though an absolute specificity of signals does not exist, the possibility of inhibitory signals cannot be discarded. This suggests that a cautious approach to the use of electromagnetic fields should be adopted.

ACKNOWLEDGEMENTS

We thank Sabrina Casale for preparation of drawings.

REFERENCES

(1) Bassett, C.A.L.: The development of pulsed electromagnetic fields (PEMFs) for ununited fractures and arthrodeses, Orthop. Clin. North Am. 15:61-88, 1984.

(2) Brighton, C.T., Black, J., Friedenberg, Z.B., et al.: A multicenter study of the treatment of nonunion with constant direct current, J. Bone Joint Surg. 62:2-13, 1981.

(3) Zichner, L.: Repair of nonunions by electrically pulsed current stimulation, Clin. Orthop. 161:115-121, 1981.

(4) Paterson, D.C. and Simonis, R.B.: Electrical stimulation in the treatment of congenital pseudarthrosis of the tibia, J. Bone Joint Surg. 67:454-462, 1985.

(5) Bassett, C.A.L., Mitchell, S.N., Norton, L. and Pilla, A.A.: Repair of nonunions by pulsing electromagnetic fields, Acta Orthop. Belg. 44:706, 1978.

(6) Bassett, C.A.L., Mitchell, S.N., Norton, L. and Pilla, A.A.: A non-operative salvage of surgically-resistant pseudarthroses and non-unions by pulsing electromagnetic fields. A preliminary report, Clin. Orthop. 124:128-143, 1977.

(7) Bassett, C.A.L.: Pulsing electromagnetic fields: A new method to modify cell behavior in calcified and noncalcified tissues, Calcif. Tissue Int. 34:1-8, 1982.

(8) Bassett, C.A.L., Pawluk, R.J. and Pilla, A.A.: Acceleration of fracture repair by electromagnetic fields. A surgically noninvasive method, Ann. NY Acad. Sci. 238:242-262, 1974.

(9) Pilla, A.A.: Electrochemical information transfer and its possible role in the control of cell function, in Electrical Properties of Bone and Cartilage, C.T. Brighton, J. Black and S.R. Pollack, eds., Grune & Stratton, New York, 455, 1979.

(10) McLeod, B.R. and Parker, R.A.: Calculated induced currents in a model non-union fracture stimulated by Helmholtz coils or resistively coupled parallel plates, J. Bioelectricity 3:41-56, 1984.

(11) Pilla, A.A.: Electrochemical information transfer at living cell membranes, Ann. NY Acad. Sci. 238:149-169, 1974.

(12) Colacicco, G. and Pilla, A.A.: Bioelectrochem. Bioenergetics 10:119, 1983.

(13) Colacicco, G. and Pilla, A.A.: Electromagnetic modulation of biological processes: Influence of culture media and significance of methodology in the Ca-uptake by embryonal chick tibia in vitro, Calcif. Tissue Int. 36:167-174, 1984.

(14) Beltrame, F., Chiabrera, A., Gliozzi, A., Grattarola, M., Parodi, G., Vecchio, D., Ponta, D., Vernazza, G. and Viviani, R.: Electromagnetic control of cell reactivation, in International Symposium on Electromagnetic Waves and Biology, A.J. Berteaud and B. Servantie, eds., Paris, 33-39, 1980.

(15) Beltrame, F., Chiabrera, A., Grattarola, M., Guerrini, P.,

Parodi, G., Ponta, D., Vernazza, G. and Viviani, R.: Electromagnetic control of cell functions, Alta Frequenza 49:101-114, 1980.

(16) Bassett, C.A.L., Valdest, M.G. and Hernandez, E.: Modification of fracture repair with selected pulsing electromagnetic fields, J. Bone Joint Surg. 64:888-895, 1982.

(17) Conti, P., Gigante, G.E., Cifone, M.G., Alese, E., Ianni, G., Reale, M and Angeletti, P.U.: Reduced mitogenic stimulation of human lymphocytes by ELF electromagnetic fields, FEBS Lett. 162:156-160, 1983.

(18) Emilia, G., Torelli, G., Ceccherelli, G., Donelli, A., Ferrari, S., Zucchini, P. and Cadossi, R.: Effect of low-frequency low-energy pulsing electromagnetic fields on the response to lectine stimulation of human normal and chronic lymphocytic leukemia lymphocytes, J. Bioelectricity 4:145-161, 1985.

(19) Luben, R.A., Cain, C.D., Chi-Yun Chen, M., Rosen, D.M. and Adey, W.R.: Effects of electromagnetic stimuli on bone and bone cells _in vitro_, PNAS USA 79:4180-4184, 1982.

(20) Adey, W.R.: Frequency and power windowing in tissue interactions with weak electromagnetic fields, Proc. IEEE 68:119-125, 1980.

(21) Lyle, D.B., Schechter, P., Adey, W.R. and Lundak, R.L.: Suppression of T-lymphocyte cytotoxicity following exposure to sinusoidally amplitude-modulated fields, Bioelectromagnetics 4:281-292, 1983.

(22) Murray, J.C. and Farndale, R.W.: Modulation of collagen production in cultured fibroblasts by PEMFs, B.B.A. 838:98-106, 1985.

(23) Yasuda, I.: The classic fundamental aspects of fracture treatment, Clin. Orthop. 124:5-8, 1977.

(24) Fukada, E. and Yasuda, I.: On the piezoelectric effect of bone, J. Phys. Soc. Japan 12:121, 1957.

(25) Brighton, C.T., Friedenberg, Z.B., Black, J. et al.: Electrically induced osteogenesis: Relationship between charge, current density, and the amount of bone formed, Clin. Orthop. 161:122-132, 1981.

(26) Friedenberg, Z.B., Andrews, E.T., Smolenski, B.I. et al.: Bone reaction to varying amounts of direct current, Surg. Gynecol. Obstet. 131:894-899, 1970.

(27) Cadossi, R., Ceccherelli, G., Emilia, G., Torelli, G., Ruggeri, M.P., Monari, P., and Bersani, F.: Effect of low frequency pulsing electromagnetic fields (PEMFs) on the response of human cultured normal and leukemic lymphocytes to the lectins, IEE International Conference on Electric and Magnetic Fields in Medicine and Biology, Conference publication No. 257:4-7, 1985.

(28) Kraus, W. and Lechner, F.: Die Heilung von Pseudarthrosen und Spontanfrakturen durch strukturbildende elektrodynamische Potentiale, Munchen Mudizinische Woch 114:1814, 1972.

(29) Lechner, F.: Therapie resistenter Frakturen mit pulsierenden Magnetfeldern, Phys. Med. u. Reh. 11:504-516, 1977.

(30) Lechner, F. and Ascherl, R.: Grundlagen und Klinik der elektrodynamischen Feldtherapie bie Knochenheilungsstorungen. Basis and clinics of electrodynamic field therapy in disorders of bone healing, Med. Orthop. Tech. 2:43-49, 1978.

(31) Lechner, F., Oeller, G. and Ascherl, R.: Behandlung infizierter Pseudarthrosen mit der elektrodynamischen Feldtherapie, Fortschr. Med. 97:943-948, 1979.

(32) Lechner, F. and Ascherl, R.: Experiences and results of the electrodynamic field treatment in cases of pseudar-

throses and delayed bone repair, Acta Orthop. Belg. 44:699, 1978.

(33) Lechner, F., Ascherl, R., Schon, G. and Oeller, G.: Recent experimental results and clinical survey of the electrodynamic field therapy, Eur. Surg. Res. 12:25, 1980.

(34) Lechner, F.: Die Behandlung von Pseudarthrosen mit dem Magnetodyn-Verfahren nach Kraus und Lechner 1970-1980. Bundesforschungsbericht MT-267, Krankheitsforschung 1982, Aktualisierte Zahlen, August, 1982.

(35) Kraus, W.: The therapy with athermal electro- and magnetodynamic potentials, various results and an attempt to explain them. Vortrag, 2nd Meeting of the European Society of Biomechanics, Strasbourg, September 1979.

(36) Kraus, W.: Magnetfeldtherapie und magnetisch induzierte Elektrostimulation in der Orthopadie, Orthopade 13:78-92, 1984.

(37) Lechner, F., Ascherl, R. and Uraus, W.: Treatment of pseudarthroses with electrodynamic potentials of low frequency range, Clin. Orthop. 161:7181, 1981.

(38) Kraus, W., Lechner, F., Ascherl, R. and Blumel, G.: On the treatment of loosening of endoprostheses by means of pulsing electromagnetic fields, 2nd Annual BRAGS, U.K., 15, 1982.

(39) Bassett, C.A.L., Mitchell, S.N. and Sawnie, R.G.: Pulsing electromagnetic field treatments in ununited fractures and failed arthrodeses, JAMA 247:623-628, 1982.

(40) Bassett, C.A.L., Mitchell, S.N. and Gaston, S.R.: Treatment of ununited tibial diaphyseal fractures with pulsing electromagnetic fields, J. Bone Joint Surg. 63:511-523, 1981.

(41) Bassett, C.A.L.: Biomedizinische und biophysikalische Wirkung pulsierender elektromagnetischer Felder (PEMF), Orthopade 13:64-67, 1984.

(42) Marcer, M., Musatti, G. and Bassett, C.A.L.: Results of pulsed electromagnetic fields (PEMFs) in ununited fractures after external skeletal fixation, Clin. Orthop. 190:260-265, 1984.

(43) Sedel, L., Christel, P., Duriez, J., Duriez, R., Evard, J., Ficat, C., Cauchoix, J. and Witvoet, J.: Resultats de la Stimulation par Champ Electromagnetique de la Consolodation des Pseudarthroses, Rev. Chir. Orthop. 67:11-23, 1981.

(44) Sharrard, W.J.W., Sutcliffe, M.L., Robson, M.J. and Macheachern, A.G.: The treatment of fibrous non-union of fractures by pulsing electromagnetic stimulation, J. Bone Joint Surg. 64:189-193, 1982.

(45) Stein, G.A. and Anzel, S.H.: A review of delayed unions of open tibia fractures treated with external fixation and pulsing electromagnetic fields, Orthopedics 7:428-436, 1984.

(46) Hinsenkamp, M., Ryaby, J. and Burny, F.: Electromagnetic treatment of non-union. A European multicenter study, 2nd Annual BRAGS, U.K., 62, 1982.

(47) O'Connor, B.T.: Fractures and electrical stimulation: The treatment of difficult and long-standing ununited fractures with the Bio-osteogen system, SICOT, London, 49, 1984.

(48) Bassett, C.A.L., Caulo, N. and Kort, J.: Congenital "pseudarthroses" of the tibia: Treatment with pulsing electromagnetic fields, Clin. Orthop. 154:136-149, 1981.

(49) Bassett, C.A.L., Mitchell, S.N. and Schink, M.M.: Treatment of therapeutically resistant nonunion with bone grafts and pulsing electromagnetic fields, J. Bone Joint Surg. 64:1214-1222, 1982.

(50) Bassett, C.A.L., Pawluk, R.J. and Pilla, A.A.: Augmentation of bone repair by inductively coupled electromagnetic fields, Science 184:575-577, 1974.

(51) Sutcliffe, M.L. and Goldberg, A.A.J.: The treatment of congenital pseudarthrosis of the tibia with pulsing electromagnetic fields. A survey of 52 cases, Clin. Orthop. 166:45-57, 1982.

(52) Sharrard, W.J.W.: Treatment of congenital and infantile pseudarthrosis of the tibia with pulsing electromagnetic fields, Orthop. Clin. North Am. 15:143-161, 1984.

(53) Hinsenkamp, H., Ryaby, T. and Burny, F.: Treatment of nonunion by pulsing electromagnetic field: European multicenter study of 308 cases, Reconstr. Surg. Traumatol. (Th J.G. van Tens Edd.) 19:147-151, 1985.

(54) Heckman, J.D., Ingram, A.J., Loyd, R.D., Luck, J.U. and Mayer, P.W.: Nonunion treatment with pulsed electromagnetic fields, Clin. Orthop. 161:58-66, 1982.

(55) Kort, J. and Bassett, C.A.L.: Role of electricity in the treatment of congenital pseudarthrosis of the tibia, Reconstr. Surg. Traumatol. 19:140-146, Krager, Basel, 1985.

(56) O'Connor, B.T.: Treatment of surgically resistant nonunions with pulsed electromagnetic fields, Reconstr. Surg. Traumatol. 19:123-132, Krager, Basel, 1985.

(57) Cheng, N., Van Hoof, H., Delport, P.M., Hoogmartens, M.J. and Mulier, J.C.: Treatment of nonunion, congenital pseudarthroses and benign cystic lesions using pulsed electromagnetic fields, Reconstr. Surg. Traumatol. 19:118-122, Krager, Basel, 1985.

(58) Binder, A., Parr, G., Hazleman, B. and Fitton-Jackson, S.: Pulsed electromagnetic field therapy of persistent rotator cuff tendinitis. A double-blind controlled assessment, Lancet, March 31:695-698, 1984.

(59) Bassett, C.A.L., Schink, M. and Mitchell, S.N.: Pulsing electromagnetic field effects in avascular necrosis - A preliminary report, Trans. BRAGS 1:38, 1981.

(60) Bassett, C.A.L., Schink, M.M. and Mitchell, S.N.: Treatment of osteonecrosis of the hip with specific, pulsed electromagnetic fields (PEMFs): A preliminary report, in Proceedings of the Third International Symposium on Circulation in Bone, Hungerford, ed., Williams and Wilkins, Baltimore, 1983.

(61) Simmons, J.W.: Treatment of failed posterior lumbar interbody fusion (PLIF) of the spine with pulsing electromagnetic fields, Clin. Orthop. 193:127-132, 1985.

(62) Harrison, M.H.M. and Bassett, C.A.L.: Use of pulsed electromagnetic fields in Perthes disease: Report of a pilot study, J. Pediatr. Orthop. 4:579-584, 1984.

(63) DeHaas, W.G. and Watson, J.: Treatment of a non-union of the tibia using a pulsed magnetic field, Paper 1111 presented at the 1975 World Congress of the Societe Internationale de Chirurgie Orthopedique et Traumatologie, Copenhagen, Denmark, July 1975.

(64) Watson, J.: The electrical stimulation of bone healing, Proc. IEEE 67:1339-1352, 1979.

(65) DeHaas, W.G., Watson, J. and Morrison, D.M.: Non-invasive treatment of ununited fracture of the tibia using electrical stimulation, J. Bone Joint Surg. 62:465-470, 1980.

(66) Watson, J. and Downes, E.M.: Clinical aspects of the stimulation of bone healing using electrical phenomena, Med. Biol. Eng. Comput. 17:161-169, 1979.

(67) Macnab, I. and DeHaas, W.G.: The role of periosteal blood supply in the healing of fractures of the tibia, Clin. Orthop. 105:27-33, 1974.

(68) Watson, J.: A battery-operated portable orthopaedic stimulator, BRAGS 3, San Francisco, 55, 1983.

(69) Wahlstrom, O.: Stimulation of fracture healing with electromagnetic fields of extremely low frequency (EMF of ELF), Clin. Orthop. 186:293-301, 1984.

(70) Haimovici, N.: Utilizzazione Clinica dei Campi Magnetici Pulsati a Bassa Frequenza e Debole Intensita nell'Ambito dell'Ortopedia e Traumatologia. Esperienza di 6 Anni, in Campi Magnetici in Medicina, F. Bistolfi, ed., Edizioni Minerva Medica, Torino, 325-341, 1983.

(71) Fontanesi, G., Giancecchi, F., Rotini, R., Dal Monte, A., Poli, G., Capelli, A. and Cadossi, R.: L'Azione dei Campi Elettromagnetici Pulsati a Bassa Frequenza in Ortopedia e Traumatologia. Casistica e Presentazione di un Protocollo di Studio Pluricentrico, Aggiornamenti di Elettro-Magnetomedicina 1:2024, 1983.

(72) Cadossi, R., Giancecchi, F. and Fontanesi, G.: Low frequency pulsing electromagnetic fields in the treatment of delayed unions and acquired pseudo-arthroses, BRAGS 2, Oxford, 71, 1982.

(73) Fontanesi, G., Giancecchi, F., Rotini, R. and Cadossi, R.: Terapia dei Ritardi di Consolidazione e Pseudartrosi con Campi Elettromagnetici Pulsati a Bassa Frequenza, G.I. Ort. e Traum. IX:319-333, 1983.

(74) Poli, G., Capelli, A., Dal Monte, A. and Cadossi, R.: Il Trattamento della Pseudartrosi Congenita della Tibia con Campi Elettromagtici Pulsati. Studio Preliminare, C.O.M. 68:27-32, 1982-83.

(75) Fontanesi, G., Dal Monte, A., Rinaldi, E., Traina, G., Mammi, G.I., Pisano, F., Giancecchi, F., Poli, G., Rotini, R., Negri, V., Marenghi, P., Virgili, B. and Cadossi, R.: Non-invasive treatment of congenital and acquired pseudarthrosis with pulsing electromagnetic fields (PEMFs). A

multicenter study, Biomedical Engineering, L.C. Sheppard, ed., Birmingham, 3:226-229, 1984.

(76) Fontanesi, G., Dal Monte, A., Rinaldi, E., Traina, G., Mammi, G.I., Giancecchi, F., Rotini, R., Negri, V., Marenghi, P., Poli, G., Virgili, B. and Cadossi, R: Final results of a multicenter clinical trial performed to evaluate the osteogenic effect of low frequency pulsing electromagnetic fields, Trans. BRAGS 4:3, Kyoto, Japan, 1984.

(77) Fontanesi, G., Dal Monte, A., Giancecchi, F., Rotini, R., Poli, G. and Cadossi, R.: The first multicentre Italian clinical trial using low frequency pulsing electromagnetic fields (SFPEFs) for the treatment of ununited fractures and congenital pseudarthrosis, SICOT, 16th Congress, London 50, 1984.

(78) Dal Monte, A., Fontanesi, G., Giancecchi, F., Poli, G. and Cadossi, R.: Treatment of congenital and acquired pseudarthrosis with pulsing electromagnetic fields, Amer. Acad. Orthop. Surgeons, 52nd Annual Meeting, Las Vegas, USA, paper #143, 94, 1985.

(79) Fontanesi, G., Dal Monte, A., Rinaldi, E., Traina, G.C., Mammi, G.I., Giancecchi, F., Rotini, R., Poli, G., Negri, V., Virgili, B. and Cadossi, R.: The effect of low frequency pulsing electromagnetic fields for the treatment of congenital and acquired pseudarthroses, J. Bioelectricity 3:155-175, 1984.

(80) Rinaldi, E., Negri, V., Marenghi, P. and Braggion, M.: Treatment of infected pseudarthrosis of the inferior limb with low-frequency pulsing electromagnetic fields, J. Bioelectricity 4:251-264, 1985.

(81) Fontanesi, G. and Giancecchi, F.: Pseudarthrosis treatment with cancellous bone grafts and low-frequency pulsing electromagnetic fields, J. Bioelectricity 4:163, 1985.

(82) Dal Monte, A. and Poli, G.: Treatment of congenital pseud-

arthrosis with endomedullary nail and low frequency pulsing electromagnetic fields, J. Bioelectricity 4:195-209, 1985.

(83) Fontanesi, G., Dal Monte, A., Rinaldi, E., Traina, G.C., Mammi, G.I., Giancecchi, F., Rotini, R., Poli, G., Negri, V., Marenghi, P., Virgili, B. and Cadossi, R.: Multicenter clinical trial for evaluating the effects of low-frequency pulsing electromagnetic fields (PEMFs) in the treatment of congenital and acquired pseudarthrosis, Bioelectrochem. Bioenergetics 14:145-149, 1985.

(84) Poli, G., Dal Monte, A., Capelli, A., Fontanesi, G., Giancecchi, F., Rotini, R. and Cadossi, R.: A double approach to the treatment of congenital pseudarthrosis: Endomedullary nail fixation and stimulation with low frequency pulsing electromagnetic fields (PEMFs), Bioelectrochem. Bioenergetics 14:151-154, 1985.

(85) Giancecchi, F., Fontanesi, G., Dal Monte, A., Poli, G., Rotini, R. and Cadossi, R.: Use ELF electromagnetic fields in orthopaedics, IEEE Frontiers of Engineering and Computing in Health Care, 165-167, 1984.

(86) Boyd, H.B.: Pathology and natural history of congenital pseudarthrosis of the tibia, Clin. Orthop. 166:5-13, 1982.

(87) Sofield, H.A.: Congenital pseudarthrosis of the tibia, Clin. Orthop. 73:33-43, 1971.

(88) Barker, A.T. and Lunt, M.J.: The effects of pulsed magnetic fields of the type used in the stimulation of bone fracture healing, Clin. Phys. Physiol. Meas. 4:1-27, 1983.

(89) Editorial: Electromagnetism and bone, Lancet I:815-816, 1981.

(90) Barker, A.T., Dixon, R.A., Sharrard, W.J.W. and Sutcliffe, M.L.: Pulsed magnetic field therapy for tibial non-union, Lancet, May 5, 994-996, 1984.

(91) O'Connor, B.T.: Pulsed magnetic field therapy for tibial non-union, Lancet, 171-172, 1984.

(92) Sharrard, W.J.W.: Pulsed magnetic field therapy for tibial non-union, Lancet, 172, 1984.

(93) Weilland, A.J., Moore, J.R. and Daniel, R.K.: Vascularized bone autografts. Experience with 41 cases, Clin. Orthop. 174:87-95, 1983.

(94) Watson, J. and Jones: Slow union of fractures with a study of 804 fractures of the shafts of the tibia and femur, Br. J. Surg. 30:260-276, 1943.

(95) Boyd, H.B., Lipinski, S.W. and Wiley, J.H.: Observations on non-union of the shafts of the long bones, with a statistical analysis of 842 patients, J. Bone Joint Surg. 43:159-168, 1961.

(96) Lifeso, R.M. and Faisal Al-Saati: The treatment of infected and uninfected non-union, J. Bone Joint Surg. 66:573-579, 1984.

(97) Weber, B.G. and Brunner, C.: The treatment of nonunions without electrical stimulation, Clin. Orthop. 161:24-32, 1981.

(98) Reckling, W.R. and Waters, C.H.: Treatment of non-unions of fractures of the tibial diaphysis by posterolateral cortical cancellous bone-grafting, J. Bone Joint Surg. 62:936, 1980.

(99) Berentey, G.: in Pseudarthroses and their Treatment, G. Chapchal, ed., Thieme, Stuttgart, 63, 1979.

(100) Schellmann, W.D. and Klemm, K.: in Pseudarthroses and their Treatment, G. Chapchal, ed., Thieme, Stuttgart, 75, 1979.

(101) Enzler, M.A., Sumner-Smith, G., Waelchli-Suter, C. and

Perren, S.M.: Treatment of nonuniting osteotomies with pulsing electromagnetic fields. A controlled animal experiment, Clin. Orthop. 187:272-276, 1984.

(102) Hiss, E.: Technische Anmerkungen zur Magnetfeldtherapie. (Technical data of magnetic field therapy), Orthopade 13:93-96, 1984.

(103) Haupt, H.A., Electrical stimulation of osteogenesis, South. Med. J. 77:56-64, 1984.

(104) Jacobs, R.R., Luethi, U., Dueland, R.T. and Perren, S.M.: Electrical stimulation of experimental nonunions, Clin. Orthop. 161:146-153, 1981.

(105) Law, H.T., Annan, I., McCarthy, I.D., Hughes, S.P.F., Stead, A.C., Campburn, M.A. and Montgomery, H.: The effect of induced electric currents on bone after experimental osteotomy in the sheep, J. Bone Joint Surg. 67:463-469, 1985.

(106) Cacciatore, E., Cadossi, R. and Caselli, G.: Campos Electromagticos Pulsantes de Baja Frequencia: su Empleo en las Enfermedades Vasculares Perifericas, Angiologia 32:212-216, 1980.

(107) Ito, H. and Bassett, C.A.L.: Effect of weak, pulsing electromagnetic fields on neural regeneration in the rat, Clin. Orthop. 181:283-290, 1983.

(108) Sisken, B.F., McLeod, B. and Pilla, A.A.: PEMF, direct current and neuronal regeneration: Effect of field geometry and current density, J. Bioelectricity 3:81-101, 1984.

(109) Marchetti, P.G., Marchetti, N., Lissnti, M. and D'Elia, L.: La Magnetoterapia nelle Lesioni da Fucile da Caccia, Ital. J. Sports Traumatol. 3:213-224, 1981.

(110) Ottani, V., Monti, M.G., Piccinini, G., Pernecco, L., Zaniol, P., Ruggeri, A. and Barbiroli, B.: Pulsed elec-

tromagnetic fields increase the rate of rat liver regeneration after partial hepatectomy, Exp. Biol. Med. 176:371-377, 1984.

(111) Cadossi, R., Ceccherelli, G., Torelli, G., Emilia, G., Ruggeri, M.P. and Monari, P.: Modulatory effects of ELF electromagnetic fields on lectin-induced human lymphocyte proliferation, in Cell Biology and Clinical Management in Functional Electrostimulation of Neurons and Muscles, Padova, 37-40, 1985.

22
Electroacupuncture

MARIA REICHMANIS
1951 Bolin Road
North Augusta, South Carolina

INTRODUCTION

Traditional Chinese medicine, including acupuncture therapy, was developed empirically over the past several thousand years (1-7). Its philosophical basis is a holistic view of man in the universe, in which physical and mental health are maintained by an equilibrium between the twin forces yin and yang. This all-pervasive duality can be viewed as contrasting and coexistent aspects of nature: yin is the passive, dark, negative, feminine principle; while yang is the light, active, positive, masculine principle. Neither can exist without the other.

These forces control the circulation of the life energy ch'i through the meridians ching-lo of the human body. The meridians, or channels, are named after various organs: the six "solid" meridians, the liver, spleen, heart, pericardium, lungs, and kidneys; and the six "hollow" meridians, the stomach, small intestine, large intestine, gall bladder, urinary bladder, and triple warmer (for which we have no precise anatomical equivalent). In addition to the 12 pairs of meridians, there are 2 median meridians that do not form part of the organ meridian circulation. The anterior and posterior median meridians are called the conception vessel and the governor vessel, respectively. Any obstruction or imbalance of energy flow leads to illness. Good health can be restored by correcting the imbalance. This is done by stimulating one or more specific points on the meridians, points where the energy flow is most susceptible to alteration by outside means. There are several

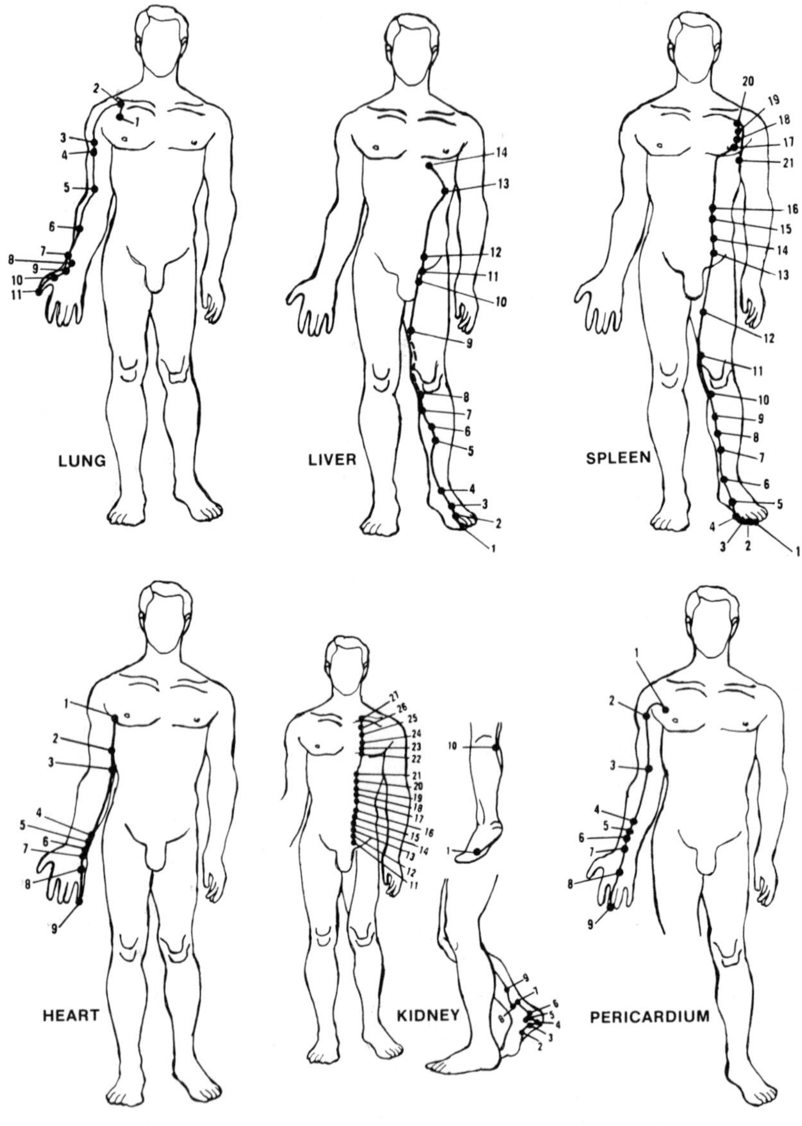

FIGURE 1A. The acupuncture points of the solid meridians of the human body (101).

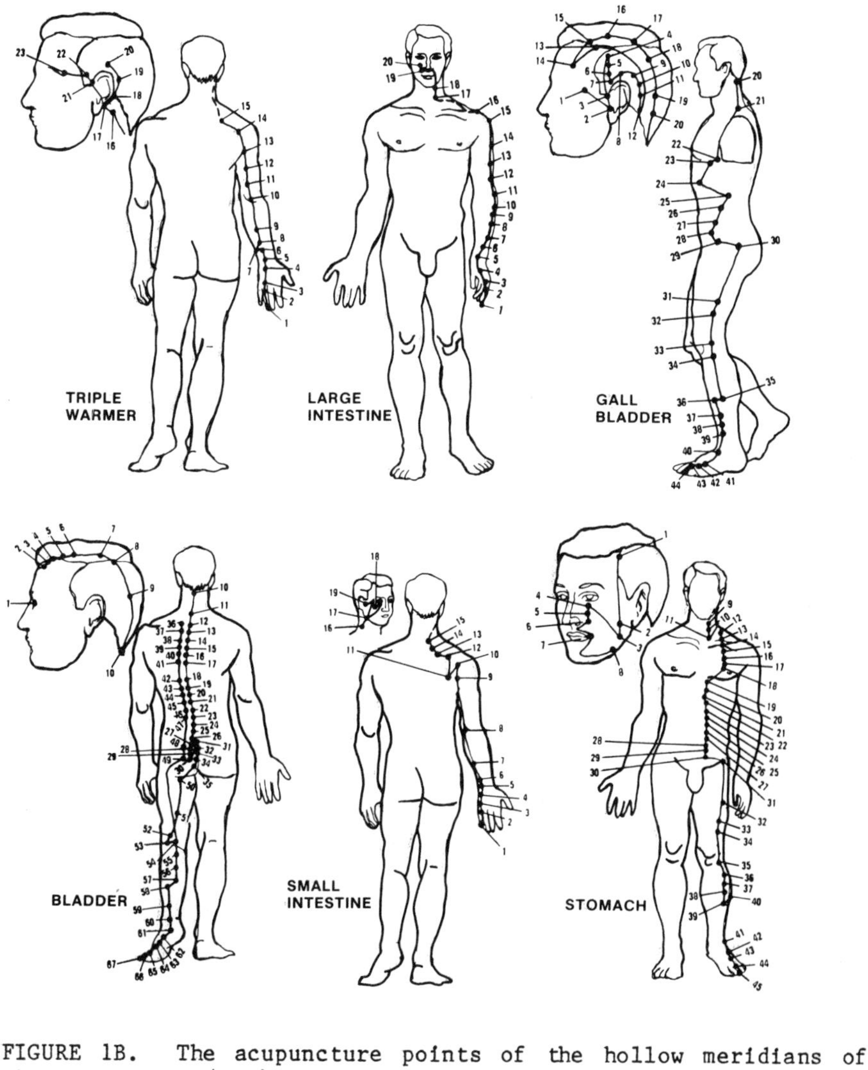

FIGURE 1B. The acupuncture points of the hollow meridians of the human body (101).

hundred such hsüeh, or acupuncture points, on the meridians (Figure 1)(4,6). Earlier, treatment was accomplished by manually twirling fine metal needles inserted into the chosen points. More recently, electrical currents have been used (3,4,6).

A traditional acupuncturist makes a diagnosis after consultation with the patient. He chooses the specific points to be treated after evaluating the patient's symptoms in the context of his past experience and knowledge of the acupuncture system. He may use different types of needles, of different metals, inserted to varying depths, and stimulated for varying durations, all depending on the nature of the illness. While there are many extant treatises on acupuncture, the earliest dating to perhaps 2500 B.C. (7), much of the traditional methodology has been passed down from teacher to student over the centuries (3,5). Many recent texts include compilations of recommended treatments for many combinations of symptoms, but still imply that successful treatment must be tailored to the individual patient (4,6,8).

The practice of acupuncture was first observed by Western travellers no later than the 17th century, and remained known throughout the 19th century (9-13). Some conventional therapies of that period were remarkably similar to traditional acupuncture; Sir William Osler described the insertion of needles several inches deep into the most painful areas, to remain *in situ* for several minutes, as a good treatment for lumbago and sciatica (14). However, interest in the applications of acupuncture, as with other folk remedies, waned with the development of modern medicine in the early part of this century. Much of the recent revival of interest in acupuncture is a result of the renewal of diplomatic, scientific, and cultural relations with the People's Republic of China in the early 1970's (1,15,16).

The Chinese themselves have undertaken to examine various traditional therapeutic modalities, including acupuncture, to determine which, if any, have a valid scientific basis and can be developed for general use. While renewed pride in their cultural heritage has played a major part in this development, economic factors are also important. Modern medicine, Western style, tends to ever more complex and expensive therapies, while the traditional methods may prove to be more cost-effective for Chinese society (17). Electroacupuncture and acupuncture analgesia resulted from this development.

THEORIES OF ACUPUNCTURE

The traditional theory of acupuncture is clearly inadequate to explain its effects. Many other explanations have been proposed, generally based on known or postulated neurophysiological mechanisms, or on the power of suggestion. The gate control theory is one example. First put forward by Melzack and Wall, it suggested that the perception of painful stimuli carried to the spinal cord can be blocked or masked by a different, non-painful stimulus (18). This is the basis for the current successful use of dorsal column stimulators for the relief of chronic pain (19). Similar phenomena include pain relief by counter-irritation, such as the application of hot and cold packs, or direct cutaneous stimulation. Man and Chen proposed a "two-gate" theory of acupuncture analgesia, with a second gate in the thalamus (20,21), while Melzack suggested a modified gate control theory (22,23). Current Chinese explanations involve as many as four gates, located in the spinal cord, brain stem, thalamus, and cerebral cortex (3,24).

Others have suggested that acupuncture is effective through hypnosis (25-27), or a combination of hypnosis, patient selection, and conditioning (19,28). Many observers have noted that not all patients are considered suitable candidates for acupuncture treatment (2,5,16). There is considerable evidence that pain perception in humans is influenced by ethnic and cultural factors. For example, Europeans, Asians, and other ethnic groups may exhibit measurably different pain thresholds (3,28). This, together with the total therapeutic environment, may partially explain the successful use of acupuncture in China. It would be difficult, however, to dismiss the reported effectiveness of acupuncture in children and animals as being due to hypnosis (24,29). In addition, although it can indeed be used to induce analgesia in carefully selected patients undergoing major surgery, at most 15-20% of the general population are good candidates (30,31). The Chinese experience indicates that up to 30% can successfully undergo acupuncture analgesia during major surgery. Furthermore, a higher proportion of Europeans experience a significant rise in pain threshold (albeit in a laboratory setting) with acupuncture (32-34).

ANATOMY OF THE ACUPUNCTURE SYSTEM

An important question is that of the physical reality of the acupuncture system. If any part of the system of points and meridians is, in fact, measurably different from the surrounding tissue, then there is a basis for a rational explanation of its effects and a further examination of its potential clinical applications.

Many reports have appeared relating the system to the human nervous system. Many acupuncture points appear to correspond to concentrations of sensory receptors (35-37); not all investigators see this relationship (38). Some acupuncture points are located at sites used for local or regional nerve blocks (37,39). The controversial studies of Bong Han suggest that previously unknown anatomical structures in the dermis and epidermis correspond to the classical system of meridians and points (40). However, his observations have not been substantiated (41). In general, histological studies have proved to be inconclusive, based as they are on extremely subtle distinctions between the experimental and control samples.

An area deserving of further investigation is that of the sensations reported by patients undergoing acupuncture. Insertion of a needle at a true acupuncture point results in the te chi phenomenon: a feeling of numbness or paresthesia, heaviness, and warmth at the site, which may in some cases travel slowly up or down the body (4,6). This is not felt at non-acupuncture points, so the patient's cooperation may be evident in the correct positioning of the needles (3-6). The therapist may notice some resistance to the needle, or muscle contraction opposing its withdrawal, again only at true acupuncture point sites (3).

There is some evidence that acupuncture points are distinguishable by means of their relatively low electrical resistance. This property is the basis of many commercially available point detection systems. The patient holds a large-area reference electrode, while the therapist lightly touches a small probe to the skin at various points. A small DC voltage, typically less than 10 volts, is applied between the two electrodes. A higher current flow corresponds to a lower DC resistance; lower current to higher resistance. The device is adjusted so that a light goes on or a buzzer sounds when the current is higher than a pre-determined threshold. These devices are subject to considerable error: it is possible to obtain a different reading by pressing the electrode more firmly against the skin, or by touching it more lightly (3,42,43). Reichmanis et al. described a series of controlled experiments in which this possibility was minimized or eliminated entirely. At least some acupuncture points were found to be real local resistance minima (Figure 2)(44-47). In addition, the resistance between two meridian, but non-acupuncture points tended to be lower than that measured between two non-meridian, non-acupuncture points (47,48). Electrical potential measurements are intrinsically more difficult because care must be taken to exclude perturbations due to electrode bias potential. Fleck and Spring recorded DC potentials from various acupuncture points, and found high-frequency variations only during electrode manipulation (49). In a preliminary study, Becker et al. found that some points exhibit a locally positive DC potential averaging a few millivolts (44). This correlates with their finding that the points tend to be resistance minima.

Buvet et al. recently studied the question of oxygen availability at acupuncture points in the rat by means of cyclic voltammetry. A standard calomel reference electrode was placed in contact with the intact skin, while a fine metal probe was inserted at the acupuncture point. The current between the electrodes was measured while the potential difference was varied continuously over the range -0.75 to 0.25 V. Their results indicated that oxygen availability was greater at acupuncture points than at nearby non-acupuncture points. The authors concluded that the acupuncture system may thus correspond to a network of greater energy flow (50).

The Japanese system of Ryodoraku is quite similar to the acupuncture system. It is based on the observation that some

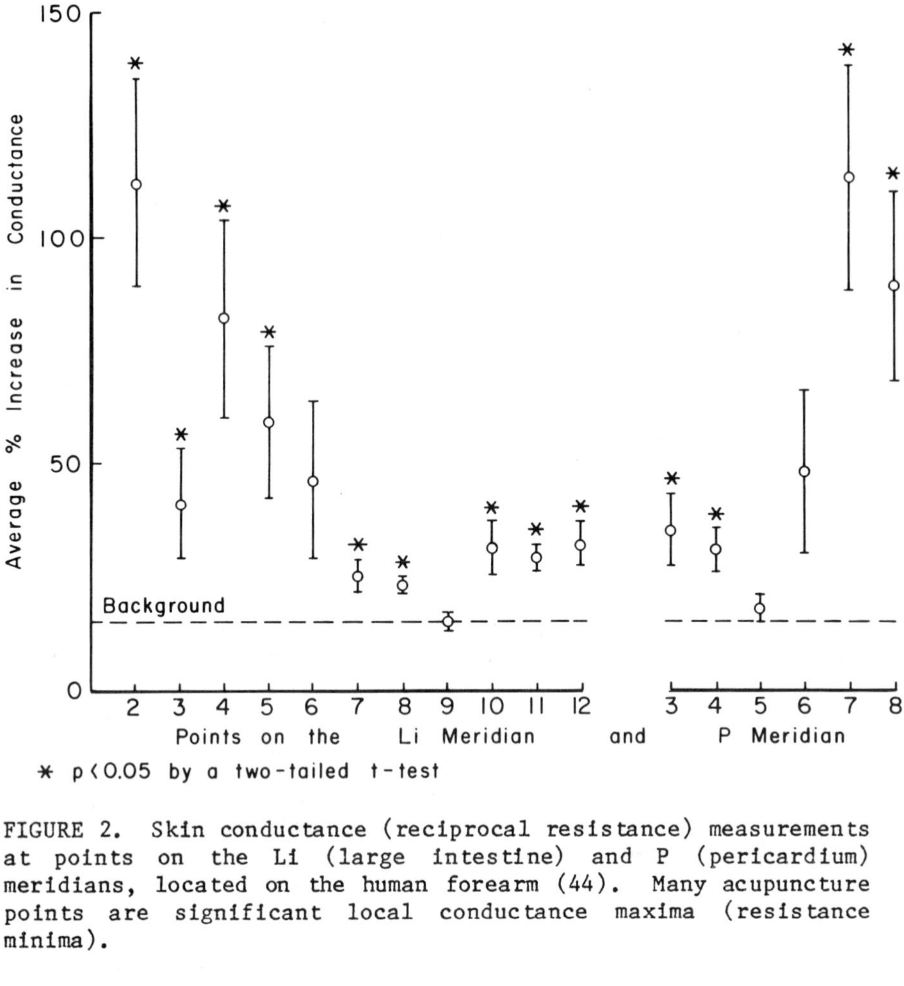

FIGURE 2. Skin conductance (reciprocal resistance) measurements at points on the Li (large intestine) and P (pericardium) meridians, located on the human forearm (44). Many acupuncture points are significant local conductance maxima (resistance minima).

points on the skin are local resistance minima. These points, when connected by lines, resemble the acupuncture system of points and meridians. Diagnosis and treatment are according to each patient's resistance readings (3,51).

It is unclear at present how these findings relate to the anatomy of the acupuncture system. Any anatomical distinctions must be extremely subtle, and relate to structures or properties not previously recognized to constitute a discrete system. We must conclude that no simple anatomical correlates to the acupuncture system have yet been observed, but that the experimental evidence strongly indicates that it constitutes some form of energy transmission network (15,44).

EFFECTS OF ACUPUNCTURE

ANALGESIA

Perhaps the most dramatic application of acupuncture is for analgesia during surgery. There have been many reports on its use in China and elsewhere (1,3,16,33). These effects may depend partly on subjective factors as noted above, but many carefully-controlled laboratory studies confirm that acupuncture treatment can indeed induce significant analgesia. A more accurate assessment of the effectiveness of various treatments is possible in a laboratory situation, which allows for more control of the subject's environment than does a clinical setting (52).

A number of studies have been made on acupuncture analgesia for noxious thermal stimuli. Both traditional manual acupuncture and electroacupuncture have been used. Both are more effective when used at known acupuncture points than at control points (53-56), although some investigators note that the former treatment is more effective (56). Stewart et al. reported that acupuncture was more effective than suggestion in increasing pain tolerance, but that treatment at non-acupuncture points on the same dermatomes was equally effective as treatment at acupuncture points (57). Experiments in animals also support the conclusion that acupuncture is more effective than placebo treatment in reducing pain thresholds (3,52).

Dental pain thresholds can be increased by electroacupuncture treatment (Figure 3)(32,58). Surface electrodes tended to be more effective than the use of needles, and low-frequency currents (on the order of 1 Hz) more effective than higher frequencies (10 Hz, 100 Hz)(52,59,60). Some reports indicate that acupuncture alone may provide sufficient analgesia for dental treatment (61), but others conclude that it is fully effective in only a small number of subjects (62). In a study of the effectiveness of local anesthetics such as novocaine compared to acupuncture, only the former were found to lead to significant analgesia (63).

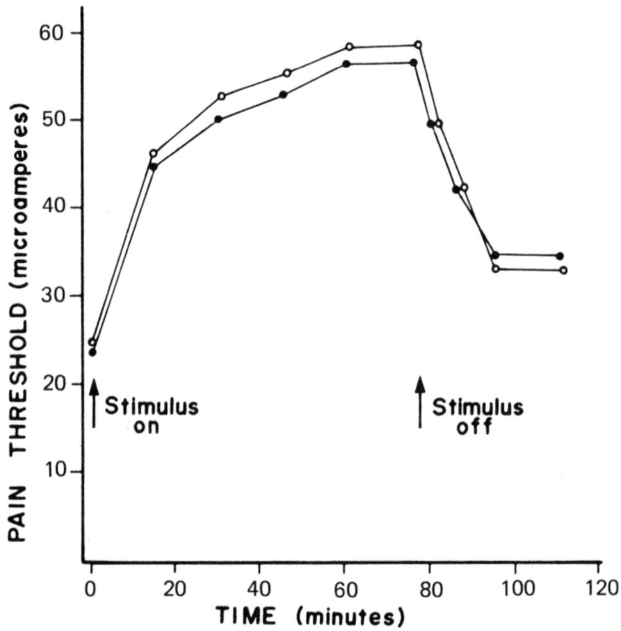

FIGURE 3. Increase in dental pain threshold during acupuncture treatment. Open circles, surface electrodes. Solid circles, needles (3,32).

Electroacupuncture can increase pain thresholds due to ice-water immersion (64) and ischemia (65,66). It has been reported that pain due to electrical stimulation is relieved by acupuncture (19), but another report shows no such effect (67).

Chapman et al. compared 33% nitrous oxide inhalation with low-frequency electrical stimulation at the ho-ku point (Li-12) by means of signal-detection theory. Both treatments produced increased response bias as well as reduced pain sensitivity in their subjects (58). Clark and Yang found that their subjects were less likely to admit to pain while undergoing acupuncture, but experienced the same sensations as with placebo treatment (68). Gwei-Djen and Needham have raised an interesting point concerning such conclusions: is it reasonable to insist that a person who states that he is not in pain, is merely refusing to admit to pain (3)?

Clinical studies of acupuncture have included surgical analgesia and the relief of chronic pain syndromes. Gwei-Djen and Needham reported on their personal observations of surgery conducted with acupuncture analgesia, and discussed its present uses in China and elsewhere (3). Acupuncture is used in 15-30% of cases overall in China, but in selected cases its use can be over 90%. It tends to be most effective in surgery of the head (including dental procedures), neck and thorax. It is less effective in abdominal surgery due to the lack of adequate muscle relaxation in conscious patients. Acupuncture is often supplemented by small amounts of medication or a local anesthetic (3).

Sufian et al. (5) reported a study on acupuncture treatment of chronic pain due to cervical osteoarthritis. Half of their experimental patients received traditional manual acupuncture, and half received electroacupuncture. The control group received placebo therapy. Only those patients who had true acupuncture had any improvement, as measured by objective and subjective criteria. In a later group of studies, they found that 83% of their patients experienced significant improvement, while none reported a negative result (5).

Hansen and Hansen evaluated acupuncture and placebo acupuncture for chronic facial pain. Patients were treated by traditional acupuncture and placebo acupuncture in a cross-over design, each receiving both modalities. The subjective effectiveness of treatment was reported to be significantly more effective with true acupuncture (69). They noted that some of their patients reported different sensations during placebo treatment (a lack of the te chi phenomenon; see above), which suggests the difficulty of setting up a truly controlled clinical trial. Several other reports indicate that acupuncture is effective in the relief of chronic pain (39,70-72), while others found no effects (73-76).

PHYSIOLOGICAL EFFECTS

A variety of physiological changes following acupuncture treatment have been reported. Electroacupuncture is known to suppress the jaw-opening reflex in rabbits (77), monkeys (78), and rats (79). Willer et al. reported that electroacupuncture induced a slow depression in the R2 component of the human blink reflex, in contrast to a rapid depression during high-frequency transcutaneous nerve stimulation. This effect was naloxone-reversible (80). The polysynaptic reflex in pre-collicular decerebrate rats was suppressed following electrical stimulation, while stimulation at points as little as 0.5 mm away had no effect (81).

The viscero-somatic reflex evoked by electrical stimulation of the splanchic nerve in cats was suppressed by electroacupuncture at several points on the hind limbs. This effect was lessened by spinal transsection at the cervical or upper thoracic levels, but not by decerebration (82).

The magnitude of evoked potentials in cat sciatic nerve decreased significantly after electroacupuncture at several points, while stimulation at nearby non-acupuncture points had no effect unless the magnitude of the stimulus was increased (83). Thalamic neurons that respond differentially to painful stimuli were affected by electroacupuncture (84,85).

Ikezono et al. measured evoked responses from the brain, spinal cord and muscle in a small number of human volunteers. After manual acupuncture at the tsu-san-li point (St-36), the amplitude of the early component of the scalp evoked response decreased. There were no changes in the evoked electrospinogram, but the amplitude of the H wave and the ratio of the H to the M waves of the evoked electromyogram decreased (86). Chen and Hung, however, reported that acupuncture had no significant effect on evoked cortical potentials (87). Pauser et al. found that thalamic evoked potentials in human subjects were significantly depressed after electroacupuncture at several different sites (88).

Bresler studied the effects of traditional manual acupuncture on a number of parameters in humans. Changes were observed in the electroencephalogram, electrocardiogram, electrooculogram, galvanic skin response, galvanic skin potential, respiration rate, and body temperature (61,89). Doenicke et al. measured serum albumin, triglycerides, cholinesterase, free fatty acids, hematocrit, and blood sugar levels in human volunteers before and after acupuncture treatment. Both manual stimulation and low frequency electrical stimulation were used. They found that albumin, hematocrit, and serum cholinesterase levels decreased and free fatty acid levels increased after treatment (90). Aso et al. found changes in the plasma levels of LH, FSH, progesterone and estradiol after low frequency electroacupuncture in female subjects (91).

In other animal studies, Lee reported that RBC velocity and carotid arterial pressure decreased significantly during electroacupuncture in the rat. The opposite effect occurred during control stimulation at non-acupuncture points (92). Cardiac output and stroke volume decreased during electroacupuncture treatment in dogs (93). Electroacupuncture has also been observed to affect immune responses in rats and guinea pigs (94).

Many investigators have reported that acupuncture analgesia can be reversed by administration of naloxone (15). Sjölund et al. (95) found that saline injection did not have this result, and concluded that acupuncture analgesia may be mediated by the release of endogenous opiates. Injection of dynorphin antibody blocks electroacupuncture analgesia in the rabbit, indicating that dynorphin may reduce nocifensive responses in the spinal cord and thereby mediate the observed analgesic effect (96). Chou reported that electroacupuncture analgesia in the rat is potentiated by intrathecal injection of captopril and bestatin (97). Taken as a whole, these reports suggest that electroacupuncture analgesia, and the other observed physiological effects of acupuncture treatment, are induced by release of endorphins, the body's naturally-produced analgesic agents.

AURICULAR ACUPUNCTURE

A new development in acupuncture is treatment of the external ear, based on the claim that its musculature has connections with all other parts of the body. The ear points, called fan ying tien or resonance points (Figure 4), can be used for diagnosis and treatment in the same fashion as the other acupuncture points (3,6).

ELECTROACUPUNCTURE

Electroacupuncture is a relatively recent development, dating to the late 1950's (3). Instead of manual twirling, an

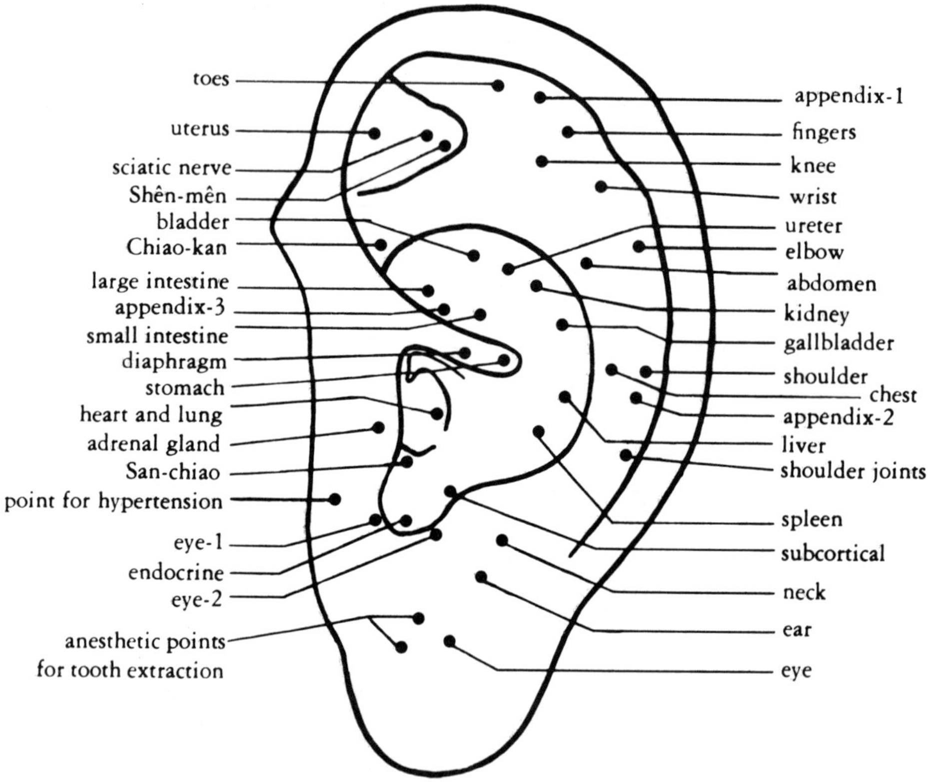

FIGURE 4. Ear acupuncture points (6).

electrical current is used to stimulate the needles. A power source, usually a low-voltage power supply or battery, is connected to two implanted needles and the voltage is increased slowly until the current reaches the desired level (generally on the order of several microamperes). The patient experiences a similar set of responses to those elicited by manual stimulation: a feeling of fullness or heaviness at the needle site, mild muscle spasm, paresthesia, etc. It has been noted that the patient's tolerance of electrical stimulation tends to increase during treatment, so that the voltage may have to be increased to maintain the desired therapeutic effects (6).

The advantages of electrical over manual stimulation include the possibility of more accurate adjustment of the stimulus, and the greater practicality of extended periods of treatment (whch may last up to several hours (1,6,16). The indications for treatment and choice of points are the same as in traditional manual acupuncture. Many pairs of points, or one reference point and several other points, may be treated simultaneously. Low frequency AC stimulation (on the order of 1-5 Hz) or DC stimulation has been found more effective than higher frequency AC stimulation (52).

COMPLICATIONS OF ACUPUNCTURE TREATMENT

Acupuncture is considered by the Chinese to be a generally safe, as well as simple and effective modality (1,6,16). Nevertheless, some complications may occur, particularly if the proper precautions are not observed and the practitioner is not suitably qualified.

Aseptic technique can minimize the possibility of infection at the needle sites. Minor bruising is occasionally observed, especially if the treatment is continued for extended periods of time (6). Damage to a nerve or artery is possible, as is pneumothorax if points on the thorax are used in treatment. Bonica has cited a number of other possible complications, including dizziness and fainting due to apprehension in some patients, breakage of the needles, penetration of the peritoneum or urinary bladder, and damage to various internal organs (1,16). Such complications are nevertheless rarely encountered in practice (3,6).

INDICATIONS FOR ACUPUNCTURE TREATMENT

Traditional acupuncture has been used for the treatment of essentially every known disorder or illness, either alone or in conjunction with other therapeutic modalities (1,3,4,6,16). More recently, attention has been focussed on the treatment of acute and chronic pain (although not pain due to malignant neoplasms), as well as acupuncture analgesia for surgery (3,16).

The Chinese use acupuncture for treatment of various neurologic disorders, such as Bell's palsy, nerve deafness, paraplegia and hemiplegia. It is their experience that 85% of patients are helped to some degree by this therapy (1,16). Gwei-Djen and Needham have observed that acupuncture is used more for conditions of malfunction (whether of internal or external cause) than for infectious diseases. Chronic conditions appear to respond better to treatment than acute ones (3).

Acupuncture is also used in conjunction with psychiatric treatment (98) and has replaced the use of electroconvulsive therapy in China (3). A new development is its use in the treatment of addiction, including tobacco and alcohol as well as narcotics. Patterson notes that electroacupuncture is particularly effective in alleviating withdrawal symptoms (99,100).

CONCLUSION

An examination of the history and current state of knowledge of acupuncture therapy leads to the conclusion that the classical system of points and meridians does indeed have some physiological basis. The distinct electrical and bioelectrochemical properties of some acupuncture points, and the large body of evidence concerning the measurable effects of treatment at these points, make it impossible to deny their reality. The precise nature of the system and its mode of action are not yet fully understood. The physiology of the nervous system appears to be an important factor, but the endocrinological and immunological systems may also be involved. While acupuncture therapy is clearly useful in some cases, particularly the treatment of chronic pain syndromes, further research is needed to clarify both its physiological basis and clinical applications.

REFERENCES

(1) Bonica, J.J.: Therapeutic acupuncture in the People's Republic of China, J. Am. Med. Assoc. 17:1544-1551, 1974.

(2) Chen, J.Y.P.: Acupuncture, in Medicine and Public Health in the People's Republic of China, J.R. Quinn, ed., CHEW Publication No. (NIH) 72-67, pp. 65-90, 1972.

(3) Gwei-Djen, L. and Needham, J.: Celestial Lancets: A History and Rationale of Acupuncture and Moxa, Cambridge University Press, Cambridge, 1980.

(4) Mann, F.: Acupuncture: The Ancient Chinese Art of Healing, Vintage Books, New York, 1972.

(5) Sufian, S., Paulides, C., Fischer, C.R., Matulewski, T. and Matsumoto, T.: Acupuncture for chronic pain and anesthesia, Surg. Rounds, 38-49, Jan., 1982.

(6) Tan, L.T., Tan, M.Y.C. and Veith, I.: Acupuncture Therapy: Current Chinese Practice, Temple University Press, Philadelphia, 1973.

(7) Veith, I.: Acupuncture therapy -- past and present, J. Am. Med. Assoc. 180:478-484, 1962.

(8) Anon.: An Outline of Chinese Acupuncture, Compiled by the Academy of Chinese Traditional Medicine, Foreign Languages Press, Peking, 1975.

(9) Carrubba, R. and Bowers, J.Z.: The Western world's first detailed treatise on acupuncture: Willem Ten Rhihne's de acupunctura, J. Hist. Med. 29:371-398, 1974.

(10) Cassedy, J.H.: Early uses of acupuncture in the United States, with an addendum (1826) by Franklin Bache, M.D., Bull. N.Y. Acad. Med. 50:892-906, 1974.

(11) Haller, J.S.: Acupuncture in nineteenth century Western medicine, N.Y. State J. Med. 73:1213-1221, 1973.

(12) Quen, J.M.: Acupuncture and Western medicine, Bull. Hist. Med. 49:196-205, 1975.

(13) Rosen, G.: Lorenz Heister on acupuncture: An eighteenth century view, J. Hist. Med. 30:386-388, 1975.

(14) Veith, I.: Sir William Osler -- acupuncturist, Bull. N.Y. Acad. Med. 51:393-400, 1975.

(15) Becker, R.O. and Marino, A.A.: Electromagnetism and Life, State University of New York Press, Albany, New York, 1982.

(16) Bonica, J.J.: Acupuncture anesthesia in the People's Republic of China, J. Am. Med. Assoc. 229:1317-1325, 1974.

(17) Russell, C.: Chinese medicine: Old and new, Science News 116:292-295, 1979.

(18) Melzack, R. and Wall, P.: Pain mechanism, a new theory, Science 150:971-979, 1965.

(19) Saletu, B.: Acupuncture and experimentally-induced ischemic pain, Proc. NIH Acupuncture Res. Conf., H.P. Jenerick, ed., DHEW Publication No. (NIH) 74-165, pp. 63-64, 1974.

(20) Man, P.L. and Chen, C.H.: Acupuncture "anesthesia" -- a new theory and clinical study, Curr. Ther. Res. 14:390-394, 1972.

(21) Man, P.L. and Chen, C.H.: Mechanism of acupunctural anesthesia, the two-gate control theory, Dis. Nerv. Syst. 33:730-735, 1972.

(22) Melzack, R.: Acupuncture and pain mechanisms, in Recent Progress in Anesthesiology and Resuscitation, A. Arias, et al., eds., Excerpta Medica, Amsterdam, 27-30, 1975.

(23) Melzack, R. and Jeans, M.E.: Acupuncture analgesia, a psychophysiological explanation, Minn. Med. 57:161-166, 1974.

(24) Geiger, H.T.: How acupuncture anesthetizes, the Chinese explanation, Med. World News 14:51, 1973.

(25) Kroger, W.S.: Hypnotism and acupuncture, J. Am. Med. Assoc. 220:1012-1013, 1972.

(26) Kroger, W.S.: Acupunctural analgesia, its explanation by conditioning theory, autogenic training, and hypnosis, Am. J. Psychiatry 130:855-860, 1973.

(27) Kroger, W.S.: A scientific explanation for the rationale of acupunctural analgesia, in Biologic and Clinical Effects of Low-Frequency Magnetic Fields, J.G. Llaurado et al., eds., Charles C. Thomas, Springfield, Ill., 220-226, 1974.

(28) Wall, P.D.: An eye on the needle, New Scientist 55:129-131, 1972.

(29) Katz, R.L.: Acupuncture studies in progress, in Proc. NIH Acupuncture Res. Conf., H.P. Jenerick, ed., DHEW Publication No. (NIH)74-165, pp. 73-75, 1974.

(30) Brown, P.E.: The use of acupuncture in major surgery, Lancet 1:1328-1330, 1972.

(31) Kroger, W.S. and Lee, S.T.: The use of hypno-anesthesia for caesarean section and hysterectomy, J. Am. Med. Assoc. 163:442-450, 1957.

(32) Andersson, S.A., Ericson, R., Holmgren, E. and Lindquist, G.: Electroacupuncture: Effect on pain threshold measured with electrical stimulation of teeth, Brain Res. 63:393-396, 1973.

(33) Dias, P.L.R. and Subramanian, S.: Minilaparotomy under acupuncture analgesia, J. Royal Soc. Med. 77:295-298, 1984.

(34) Mann, F., Bowsher, D., Mumford, J., Lipton, S. and Miles, J.: Treatment of intractable pain by acupuncture, Lancet 2:57-60, 1973.

(35) Dornette, W.H.L : The anatomy of acupuncture, Bull. N.Y. Acad. Med. 51:985-987, 1975.

(36) Fleck, H.: Acupuncture and neurophysiology, Bull. N.Y. Acad. Med. 51:903-913, 1975.

(37) Matsumoto, T.: Acupuncture and US medicine, J. Am. Med. Assoc. 220:1010, 1972.

(38) Bull, G.M.: Acupuncture anesthesia, Lancet 2:417-418, 1973.

(39) Matsumoto, T., Levy, B. and Ambruso, V.: Clinical evaluation of acupuncture, Am. Surg. 40:400-405, 1974.

(40) Bong Han, K.: Kyungrak system and theory of sanal, Proc. Acad. Kyungrak DPRK, Medical Science Press, Pyongyang, 1965.

(41) Kellner, G.: Bau und Funktion der Haut, Presented at the 13th International Acupuncture Congress, Vienna, 1975.

(42) Noordergraaf, A. and Silage, D.: Electroacupuncture, IEEE Trans. Biomed. Eng. BME-20:364-366, 1973.

(43) Reichmanis, M. and Becker, R.O.: Physiological effects of stimulation at acupuncture loci: A review, Comp. Med. East & West 6:67-73, 1978.

(44) Becker, R.O., Reichmanis, M., Marino, A.A. and Spadaro, J.A.: Electrophysiological correlates of acupuncture points and meridians, Psychoenergetic Systems 1:105-112, 1976.

(45) Reichmanis, M., Marino, A.A. and Becker, R.O.: Electrical correlates of acupuncture points, IEEE Trans. Biomed. Eng. BME-22:533-535, 1975.

(46) Reichmanis, M., Marino, A.A. and Becker, R.O.: Laplace plane analysis of transient impedance between acupuncture points Li-4 and Li-12, IEEE Trans. Biomed. Eng. BME-24:402-405, 1977.

(47) Reichmanis, M., Marino, A.A. and Becker, R.O.: Analysis of impedance between acupuncture points H-3 and H-4, Comp. Med. East & West 5:289-295, 1977.

(48) Reichmanis, M., Marino, A.A. and Becker, R.O.: Laplace plane analysis of impedance on the H meridian, Am. J. Chin. Med. 7:188-193, 1979.

(49) Fleck, H. and Spring, M.: The acupuncture point: Study of events, NY State J. Med. 74:1060-1064, 1974.

(50) Buvet, R., Sechaud, P., Darolles, J., Thirion, G. and Braslawsky, C.: Acupuncture et bioelectrochimie de l'oxygene cutane, Agressologie 25:1117-1121, 1984.

(51) Chan, S.H.H.: What is being stimulated in acupuncture: Evaluation of the existence of a specific substrate, Neurosci. Biobehav. Rev. 8:25-33, 1984.

(52) Reichmanis, M. and Becker, R.O.: Relief of experimentally-induced pain by stimulation at acupuncture loci: A review, Comp. Med. East & West 5:281-288, 1977.

(53) Berlin, F.S., Bartlett, R.L. and Black, J.D.: Acupuncture and placebo: Effects on delaying the terminating response to a painful stimulus, Anesthesiology 42:527-531, 1975.

(54) Croze, D., Antonietti, C. and Duclaux, R.: Changes in burning pain threshold induced by acupuncture in man, Brain Res. 104:335-340, 1976.

(55) Day, R.L., et al.: Evaluation of acupuncture anesthesia: A psychological study, Anesthesiology 43:507-517, 1975.

(56) Haider, M., Groll-Knapp, E. and Reichmann, C.: Experimental studies and theoretical considerations about acupuncture effects, in Recent Progress in Anesthesiology and Resuscitation, A. Arias, et al., eds., Excerpta Medica, Amsterdam, 700-701, 1975.

(57) Stewart, D., Thompson, J. and Oswald, I.: Acupuncture analgesia: An experimental investigation, Br. Med. J. 1:67-70, 1977.

(58) Chapman, C.R., Gehrig, J.D. and Wilson, M.E.: Acupuncture compared with 33% nitrous oxide for dental analgesia, Anesthesiology 42:532-537, 1975.

(59) Holmgren, E.: Increase of pain threshold as a function of conditioning electrical stimulation, Am. J. Chin. Med. 3:133-142, 1975.

(60) Omura, Y.: Electroacupuncture: Its electrophysiological basis and criteria for effectiveness and safety -- Part I, Acupunct. Electrother. Res. 1:157-181, 1975.

(61) Bresler, D.: Acupuncture analgesia in oral surgery, in Proc. NIH Acupuncture Res. Conf., H.P. Jenerick, ed., DHEW Publication No. (NIH) 74-154, pp. 68-69, 1974.

(62) Brennan, R.W. and Veldhuis, J.: Acupuncture anesthesia and dental pain: A controlled study, in Proc. NIH Acupuncture Res. Conf., H.P. Jenerick, ed., DHEW Publication No. (NIH) 74-154, pp. 46-49, 1974.

(63) Stich, E.: Electrical measurement of the stimulus threshold with special reference to the effect of acupuncture, in Recent Progress in Anesthesiology and Resuscitation, A. Arias, et al., eds., Excerpta Medica, Amsterdam, 704-707, 1975.

(64) Anderson, D.G., Jamieson, J.D. and Man, S.C.: Analgesic effects of acupuncture on the pain of ice water: A double blind study, Canad. J. Physiol. 28:239-244, 1974.

(65) Smith, G.M., Chiang, H.T., Kitz, R.J. and Antoon, A.: Acupuncture and experimentally-induced ischemic pain, Adv. Neurol. 4:827-832, 1974.

(66) Smith, G.M.: Acupuncture and experimentally-induced ischemic pain, Proc. NIH Acupuncture Res. Conf., H.P. Jenerick, ed., DHEW Publication No (NIH) 74-165, pp. 63-64, 1974.

(67) Baum, M.: Measurement of pain threshold (algesimetry) during acupuncture, in Recent Progress in Anesthesiology and Resuscitation, A. Arias, et al., eds., Excerpta Medica, Amsterdam, 702-703, 1975.

(68) Clark, W.C. and Yang, J.C.: Acupunctural analgesia? Evaluation by signal detection theory, Science 184:1096-1098, 1974.

(69) Hansen, P.E. and Hansen, J.H.: Acupuncture treatment of chronic facial pain -- A controlled cross-over trial, Headache, 66-69, March, 1969.

(70) Chen, G.S., Hwang, Y-C. and Song, S.J.: Long term effect of acupuncture therapy on headache, Am. J. Acupunct. 6:23-32, 1978.

(71) Chia, J.N., Mao, W., Toomey, T.C. and Gregg, J.M.: Acupuncture and chronic pain mechanisms, Pain 2:285-299, 1976.

(72) Man, P.L. and Chen, A.C.: Two-year acupuncture followup study on 182 chronic pain patients, Am. J. Acupunct. 3:143-146, 1975.

(73) Lee, P.K., Anderson, T.W., Modell, J.H. and Saga, S.A.: Treatment of chronic pain with acupuncture, Ann. Intern. Med. 84:381-384, 1976.

(74) Lewith, G.T., Field, J. and Machin, D.: Acupuncture compared with placebo in post-herpetic pain, Pain 17:361-368, 1983.

(75) Mendelson, G., Selwood, T.S., Kranz, H., Loh, T.S., Kidson, M.A. and Scott, D.S.: Acupuncture treatment of chronic back pain, Am. J. Med. 74:49-55, 1983.

(76) Moore, M.E. and Berk, S.N.: Acupuncture for chronic shoulder pain, Ann. Intern. Med. 84:381-384, 1976.

(77) Fung, D., Hwang, J.C.C., Chan, S.H.H. and Chin, Y.C.: Electroacupuncture suppression of jaw depression reflex elicited by dentalgia in rabbits, Exp. Neurol. 47:367-369, 1975.

(78) Ha, H., Tan, E.C., Fukunaga, H. and Aochi, O.: Naloxone reversal of acupuncture analgesia in the monkey, Exp. Neurol. 73:298-303, 1981.

(79) Kawahita, K. and Funokoshi, M.: Suppression of the jaw-opening reflex by conditioning a-delta fiber stimulation and electroacupuncture in the rat, Exp. Neurol. 78:461-465, 1982.

(80) Willer, J.-C., Roby, A., Boulu, P. and Bourcau, F.: Comparative effects of electroacupuncture and transcutaneous nerve stimulation on the human blink reflex, Pain 14:267-278, 1982.

(81) Chan, S.H.H. and Fung, S.J.: Suppression of the polysynaptic reflex by electroacupuncture and a possible underlying mechanism in the spinal cord of the cat, Exp. Neurol. 48:336-342, 1975.

(82) Shen, E., Ts'ai, T. and Lan, C.: Supraspinal participation in the inhibitory effect of acupuncture on viscero-somatic reflex descharges, Chin. Med. J. 1:431-440, 1975.

(83) Rohner, J.J. and Planche, D.: Modification des potentiels evoques associatifs par stimulation des points d'acupuncture chez le chat, C.R. Soc. Biol. (Paris) 176:1180-1185, 1973.

(84) Linzer, M. and Van Atta, L.: Effects of acupuncture stimulation on activity of single thalamic neurons in the cat, Acta Neurol. 4:799-811, 1974.

(85) Linzer, M. and Van Atta, L.: Electrophysiological assessment of acupuncture's effect on single thalamic neurons in the cat: Neural coding for pain on a thalamic level, Am. J. Chin. Med. 1:305-316, 1973.

(86) Ikezono, E., Ohama, K., Nagayama, K., Sawa, T., Yoshida, T., Yokogama, K. and Shimoji, K.: The effects of acupuncture needling on the evoked responses of brain, spinal cord and muscle in man, Am. J. Chin. Med. 4:53-59, 1976.

(87) Chen, R.C. and Hung, T.P.: Acupuncture: Its effect on somatosensory evoked potentials in man, J. Formosan Med. Assoc. 47:341-344, 1975.

(88) Pauser, G., Benza, H., Bischko, J., Ganglberger, J., Haider, M., Mayrhofer, O., Schmid, H., Semsroth, M. and Thoma, H.: Klinische und experimentelle Ergegnisse mit der Akupunktur-Analgesie, Anaesthetist 25:215-222, 1976.

(89) Bresler, D.: A comparison of the effects of acupuncture, local anesthetic and systemic analgesic drugs, and hypnosis on experimental pain thresholds, in Proc. NIH Acupuncture Res. Conf., H.P. Jenerick, ed., DHEW Publication No. (NIH) 74-154, pp. 50-51, 1974.

(90) Doenicke, A., Kamipik, G., Praetorius, B. and Schmidt, M. Veränderung blutchemischer Parameter bei gesunder Versuchspersonen durch Akupunktur, Anaesthetist 25:235-238, 1976.

(91) Aso, R., Motohashi, R., Murata, M., Nishimura, R., Kakizaki, K. and Kimura, S.: The influence of acupuncture stimulation on plasma levels of LH, FSH, progesterone, and estradiol in normally ovulating women, Am. J. Chin. Med. 4:391-401, 1976.

(92) Lee, G.T.C.: A study of electrical stimulation of acupuncture locus tsu-san-li (St-36) on mesenteric microcirculation, Am. J. Chin. Med. 2:53-66, 1974.

(93) Lee, M.O., Lee, D.C., Kim, S. and Clifford, D.H.: Cardiovascular effects of acupuncture at tsu-san-li (St-36) in dogs, J. Surg. Res. 18:51-63, 1975.

(94) Chu, Y.M. and Affronti, L.F.: Preliminary observations on the effect of acupuncture on immune responses in sensitized rabbits and guinea pigs, Am. J. Chin. Med. 3:151-163, 1975.

(95) Sjölund, B. and Eriksson, M.: Electroacupuncture and endogenous morphines, Lancet 2:1085, 1976.

(96) Han, J.S. and Xie, G.-X.: Dynorphin: Important mediator for electroacupuncture analgesia in the spinal cord of the rabbit, Pain 18:367-276, 1984.

(97) Chou, J., Tang, J., Del Rio, J., Yang, H.-Y.T. and Costa, E.: Action of peptidase inhibitors on methionine^5 - enkephalin - arginine^6 - phenylalanine^7 (YGGFMRF) and methionine^5 - enkephalin (YGGFM) metabolism and on electroacupuncture antinociception, J. Pharmacol. Exp. Ther. 220:349-352, 1984.

(98) Kao, J.J.: Acupuncture therapy in the treatment of schizophrenia, Am. J. Chin. Med. 5:181-189, 1977.

(99) Patterson, M.A.: Addictions Can Be Cured: The Treatment of Drug Addiction by Neuroelectric Stimulation (Especially Acupuncture) - An Interim Report, Lion, Berkhampstead, England, 1975.

(100) Patterson, M.A.: Electroacupuncture in alcohol and drug addictions, Clin. Med. 81:9-15, 1974.

(101) Duke, M.: Acupuncture, Pyramid House, New York, 1972.

VI

HEALTH HAZARDS OF ELECTROMAGNETIC ENERGY

23

Evolution and Results of Biological Research with Low-Intensity Nonionizing Radiation

ALLAN H. FREY

Randomline, Inc.
County Line and Mann Roads
Huntingdon Valley, Pennsylvania

INTRODUCTION

Until the late 1940's, the primary interest in the biological effects of radiofrequency (RF) radiation was in heating for medical applications (1). Thus, the dominant theme of the research was the use of high powers to heat tissue. With the development and increasing use of RF radiation as radar in the 1940's, however, questions were increasingly raised about the possibility that exposure to this energy would have adverse biological effects on military personnel and workers.

In the mid-1950's, the Department of Defense's (DoD) RF hazards assessment establishment contracted for research to determine if there were adverse biological effects of RF radiation: the Tri-Service Program. The primary thrust of the program was essentially determined by the implicit assumption upon which prior work was based. It was assumed that the only way the energy could affect an organism was through overloading its heat-dissipation mechanism. Thus, little effort was expended to determine the effect of low-intensity energy. This assumption also resulted in an acrimonious dispute between those who contended that only thermal effects could occur and those who thought that nonthermal effects could also occur. But the fruitless argument was really the result of a semantic problem. The participants were talking past each other, for there never was a common definition of the words thermal and nonthermal. It was also assumed in the Tri-Service Program that nervous-system function and behavior could not be affected, so the possibility that modulation would be of consequence was essentially ignored (2). The Tri-Service Program was terminated in 1961, after gathering some data on overloading the temperature-regulating system.

Through the 1960's and early 1970's, there was some research on the biological effects of low-intensity RF energy. This was a distinct departure from the pattern of prior research. It was possible because the very limited funds available for the research were not controlled by those whose interests were in hazards or medical applications. This funding for research on low-intensity RF bioeffects continued through most of the 1970's.

Beginning in the early 1970's, a new program of research on high-intensity radiation effects, again primarily sponsored by the DoD's RF hazards establishment, was superimposed on and overshadowed the low-intensity research. By the late 1970's, the low-intensity research was being squeezed out because of the concentration of control of the funding into the hands of those in the DoD RF hazard establishment. The details on the control of research funds and its effects can be found elsewhere (1,3,4).

In the early 1960's, I initiated some of the early research in this country on bioeffects of exposure to low-intensity RF radiation. I was the most active investigator in this area during the 1960's and on into the 1970's. Thus, I have been given the task to trace chronologically, through my own research and the research of others in the US, the development in this country of biological research with low-intensity RF energy. The objective is not to give a comprehensive in-depth review of all aspects of RF biological research during those two decades and into the 1980's. Rather, the objective is to indicate the development of the more significant patterns of research, and to indicate where the research would likely lead if pursued as science. Since I was almost alone in doing research on low-intensity RF radiation bioeffects in this country during the 1960's, the beginning of this narrative will be primarily about my research.

THE QUIET DECADE

In the late 1950's, neurophysiological theory on information transfer in the nervous system did not provide much understanding of neural function. It was about this time that I became curious about electric fields and the possibility of their interaction with the nervous system. In 1960, I was working at General Electric's Advanced Electronics Center at Cornell University doing biological research. One line of research I had initiated there was concerned with electrostatic fields and nervous-system function. I was also experimenting with air ionization and its biological effects. Late that year, while attending a small conference sponsored by the General Electric Company (GE), I happened to talk to a GE technician whose job was to measure RF radiation in the vicinity of radars. He mentioned that he "heard" radars. I found this to be interesting, since I, as well as everyone trained in the life sciences, had been taught that people hear acoustic energy and see, as light, electromagnetic energy. He was rather surprised when I asked if he would take me to a site and let me hear the radar. It seemed that I was the first person he had told about hearing radars who did not dismiss his statement out of hand.

A few weeks later, I went to the radar site and I heard the RF radiation. I performed a few simple tests to assure myself it was not an artifact. I then undertook a series of experiments which resulted in the publication of a brief article about the phenomenon in 1961 and a more detailed article in 1962 (5,6). I laid out the data from a variety of tests with humans. I suggested that there were probably multiple mechanisms for such an effect, but there was not sufficient data to specify mechanisms. Although the articles provoked interest in some members of the biological community and disbelief in others, there was little immediate activity by others to pursue the findings.

I searched the literature for information about the nature of RF field interaction with biological organisms and tissues. In essence, I found that there had been little effort in this country to consider sensory phenomena such as I was reporting, nor was there any significant research on neural effects of RF radiation. Virtually all of what little research existed had been done in the Soviet Union, but the translations were generally of poor quality and almost uninterpretable.

I expanded my literature search and published a

comprehensive paper in 1965 (7). I assessed biological interactions with a wide portion of the electromagnetic spectrum, from the infrared down to low frequency. I critiqued the literature available, offered suggestions as to the portions of the spectrum with which the more significant research could be done, and pointed out the possibility of micron-wavelength emission from active nerves. This analytical review evoked a considerable amount of interest, for I received almost 5000 reprint requests.

After the initial exploratory work with the hearing phenomenon and concurrent with the preparation of the analytical review, I initiated further research with RF radiation. The series of experiments I carried out in the 1960's centered about four major themes: (1) experimental controls and techniques; (2) brain function and behavior; (3) sensory function; and (4) heart function.

Although the results of my work on experimental controls and techniques are too extensive to review in detail here, they are critical for accurate data collection. As a sampling, comparative studies of biological data recording techniques were done, including assessment of recording electrode systems in RF experiments. It was found that certain conventionally used systems yielded artifacts as data, due to induced currents stimulating the tissue as well as feeding into the recording preamplifier. It was found that filtering had limited usefulness and that lead placement was of consequence. New types of recording electrodes were developed which showed excellent characteristics in the RF field. In fact, the sponsor of one of these studies had the electrode patented (8,9).

Experiments were also carried out to develop techniques to remotely monitor the activity of nerves in an RF field. A method to record neural activity with no recording devices in the field was developed (10). Studies were made of restraint devices to hold animals, and of the RF field distorting effects of these devices. Polystyrene head holders were developed for use with cats. Teflon and nylon chairs and restraints were developed in studies with monkeys, and wooden enclosures and restraints were developed for use with cats (11,12).

Experiments were carried out using three-dimensional field plots to investigate the effect of the biological object itself on the field within an RF anechoic enclosure. Similar studies were made on the perturbing effect of field measurement devices on the field. Standardized methods of measurement and reporting of measurements were developed. Experimentation was also carried out to determine the effect of body position and its orientation on results. Studies were made of shielding materials and their usefulness in experimentation (11). I found in my other experiments that carrier frequency and modulation had to be controlled because they were critical in the effect of low-power-density RF radiation on some functions of higher organisms (6,11,13-15).

As may be seen from this sampling, there are many variables that need to be controlled and special techniques that must be used in biological work with RF radiation. But the literature shows that many of these variables have not been controlled.

Turning now to the second theme, data specific to brain function, I shall summarize the information obtained. Cats were illuminated with pulse-modulated RF radiation and evoked activity in the brain was observed (11). The threshold average power density necessary to evoke activity was approximately 20 microwatts/cm^2. The controls used indicated that the activity was neural evoked activity rather than an artifact of the situation. Using an Echosorb shield to cover the entire cat, or head, or body, it was found that the head must be exposed to the radiation in order to have an effect occur. Within the carrier frequency range used, there appeared to be a reduction of effect at the highest frequency. Variation in power density had a distinct effect on the evoked activity. Polarization of the energy, whether perpendicular or parallel to the spine, did not seem to matter. As pulse repetition frequency (PRF) was changed, the evoked activity did not change significantly until the PRF was greater than approximately 50 pulses per second (pps). In general, recording from the rostral brain stem did not yield evoked activity as diffuse and persistent as recording from the caudal portion of the reticular formation of the brain.

In view of what I was seeing in using RF radiation to influence neural tissue, and because of ideas I had about neurophysiological theory, I expanded the brain function experimentation to assess the possibility that nerves, when active, would emit coherent electromagnetic energy. It seemed that the channel capacity indicated by neurophysiological theory was insufficient to encompass the results of many neurophysiological and behavioral experiments, and that there might be communication between nerves via the emission of electromagnetic energy at micron wavelengths. To assess this possibility, I set up an experiment using some of the equipment I developed for remote sensing. I used live nerves from the legs of blue crabs because of their characteristics. I sought to determine whether there was emission of micron wavelength energy when the nerves were active. I found that the emission was considerably greater than what would be expected from a black-body nerve model. I established that the emission was not an artifact, that the emission was from the surface of the nerve, and calculated the amount of emission and its spectral band (10). A number of subsequent papers by others used these findings in their development, at the molecular level, of new conceptualizations of neural function. These include Lee's concepts on the role of excitons and phonons on nerve permeability and propagation of impulses (16), Cope's micron-wavelength concepts on phonon coupling and IR involvement in nerve (17), and Maurel and Galzigna's (18) definition of the involvement of the dipole moment of acetylcholine in neural chemical transmission. There are other similar papers relevant to low-intensity RF radiation bioeffects (19-32).

The third theme was an extension of the RF hearing research and an exploration for other sensory effects. No visual effects were found at that time, but tactile stimulation in humans at very-low frequency (VLF) carrier frequencies was found (13,14). An attempt was made to determine the locus of the RF hearing mechanism. I searched for cochlear microphonics in guinea pigs and cats exposed to RF radiation, but found none(13,14). The in-air RF hearing thresholds for humans were determined for two carrier frequencies. Since they were quite different, a mathematical model of layers of head tissue was constructed. As RF energy passes through each layer of tissue, the absorption of the energy differs as a function of carrier frequency. Thus, I sought to determine mathematically where in the head the RF energy from the two frequencies became equal. Such an equality point, the crossing of signal strengths, would suggest where to look for the sensing mechanism. In constructing the model, all tissue electrical values were selected in advance, standard values for tissue thickness were used, and first reflections were taken into consideration. The calculations indicated that the RF energy crossing was in the fluid at the first bone/soft-tissue interface. This suggested a locus in the cochlea or at the surface of the cerebral cortex.

Experimentation was also carried out with cats, using the avoidance conditioning technique to determine if they could sense RF energy. Cats avoided the radiation and thresholds were established. In experiments with rhesus monkeys, avoidance behavior also appeared.

The last major theme of my 1960's experimentation concerned heart function. The isolated frog heart, stripped of its neural and hormonal buffer systems, was exposed to RF radiation (15). It was found that the heart was responsive to RF radiation when the pulses were synchronized with certain phases of the heart cycle. When the RF pulse occurred about the time the QRS complex occurred, the beat rate increased. In half the cases, arrhythmias occurred, and occasionally the heart ceased beating after a period of arrhythmia. No such effect appeared when the heart was illuminated at earlier points in the cycle.

During the 1960's, others also reported on experiments with low-intensity RF radiation. For example, Hearn (33) explored the effect of long continued low-intensity RF energy on visual acuity. He found significant differences in the flicker thresholds of irradiated as compared to nonirradiated subjects. Korbel and Thompson (34) exposed rats to what they believed to be low-intensity RF energy. They found that irradiated subjects were more active than nonirradiated subjects for a short period of time during the early part of the experiment, but they became less active than the nonirradiated subjects as the days of radiation exposure increased. In a follow-up study, Korbel and Fine (35) explored a possible relationship between RF frequency range and activity level, but they had equipment problems that left their results in doubt. Bourgeois (36) found that exposure to RF radiation resulted in a significant decrease in auditory thresholds in humans. The threshold change was found to be a function of the type of modulation used, since auditory thresholds were significantly lower upon exposure to 1000-Hz modulated RF radiation than upon exposure to 300-Hz modulated RF radiation.

The foregoing summarizes the primary lines of biological research with low-intensity RF radiation in this country during the 1960's. I had spent most of the decade laying a foundation in data for the study of RF radiation interaction with biological organisms and tissues. Although I would get reprint requests in the thousands for some of my reports on experiments, it was a rather quiet and lonely effort that was, however, quite interesting.

THE LIVELY DECADE OF THE 1970's AND INTO THE 1980's

INTRODUCTION

The period of my quietly doing research came to an end in 1969 with the passage of Public Law 90-602, the Radiation Control for Health and Safety Act. The purpose of the law was to protect the public health and safety "...from the dangers of electronic product radiation." The Bureau of Radiological Health, Department of Health, Education and Welfare, became active in the area because of the law. The hazards people of the DoD, who had been involved in the Tri-Service Program, again became active in the area.

The Bureau convened a symposium in September of 1969 in Richmond, Virginia, that I helped organize. The topic of the symposium was "Biological Effects and Health Implications of Microwave Radiation." I presented a paper entitled "Effects of Microwaves and Radio Frequency Energy on the Central Nervous System" (37). In it, I detailed why there was so much misunderstanding and confusion in the area, and summarized some of my research. I spelled out lines of research I considered to be worth pursuing, techniques that could be used, and the controls that had to be used in order to get valid data. During the next few years, I found myself spending a large proportion of my time answering phone calls and letters from scientists. DoD had started funding research in the area. The world was not so lonely any more.

MECHANISMS

The decade of the 1970's opened for me with the preparation of a paper in which I presented some of my thinking on possible mediators or mechanisms for biological effects of very low-intensity RF radiation (2). It is my nature to look at the broad picture and to integrate. Most of my experimentation is done because I have reached a choice point in my theorizing. In order to decide which way my thinking should go, I do an experiment to provide data for the choice. This is why I do such a diversity of experiments.

In the preparation of that paper, I made explicit some of my thinking (2). Much of what I said then is still relevant, for much of the research that was done during the 1970's was irrelevant to the questions about the biological effects of low-intensity RF radiation. The DoD sponsors who determined what would be done appear to have been primarily interested in research that used high power levels or used techniques relevant to thermoregulation questions.

In that paper, I identified the mistaken assumptions that formed the basis of Schwan's notions about nervous-system function. Those notions had inhibited research on low-intensity and nervous-system effects since the 1940's. He had set up a mathematical model of the axon membrane, and assumed that it was a reasonable representation of the nervous system (38). His calculations with the model indicated that at field strengths that are "not thermally significant," the induced potentials across the nerve membrane are many orders of magnitude smaller than the nerve resting potential. He stated that such induced fields applied to the resting potential of the axon cannot excite the nerves, and essentially, on the basis of this, he concluded that the nervous system could not be influenced by low-intensity RF radiation.

I pointed out that there were at least two faults in his reasoning. One was that his implied model of the nervous system was unrealistic. Nerves function, and the resting potential is only one extreme of a continuum of potentials on the axon. He ignored most of the nerve cell, including the most important part, when he considered only the axon in his model. Further, nerves interact, and the points of interaction on the cell bodies are the most sensitive to disturbance, not the axon.

Thus, his model, based upon the resting potential of the axon, did not correspond to reality, at least not to the reality of the nervous systems of man, monkey, cat, or frog.

Another fault was his assumption that we have a good understanding of nervous system function. Our understanding then, and even now, of how information is coded, transferred, and stored in the nervous system is negligible. We have, at best, only multiple hypotheses. None of these have much support, nor are generally accepted as truth. Thus, a conclusion such as Schwan's, based upon calculations using assumptions about information coding, transfer, and storage in the nervous system is hardly acceptable.

I also showed that by simply changing one of his model assumptions to a more realistic one, and then doing his calculations leads to the conclusion that the nervous system would have to be affected by RF radiation.

I went on in the paper to discuss some of the possible mechanisms for electromagnetic-field effects considering only the electrophysiology of the nervous system. I noted that Valentinuzzi (39) used a mathematical approach to explore the possibility that magnetic fields affect the nervous system. He used equations to evaluate the effects of magnetomotive force on ions, magnetic induction of electrical fields, and magnetic changes of inductance. The equations indicated that the effect of a static magnetic field upon an axon would be almost undetectable. But he recognized that living organisms are composed of more than a single axon. Thus, he extended his equations to consider the nervous system functioning as a whole. The results of this extension led him to suggest that an appropriate magnetic field may influence the activity of the nervous system. And these kinds of effects are now being observed (32).

I also discussed the analyses and hypothesis of MacGregor (40). I noted that he started with a reasonably realistic model of a functioning nervous system. Using it, he mathematically explored the idea that the electrical component of RF radiation might induce transmembrane potentials in nerve cells and thereby disturb neural function and behavior. He considered steady and modulated RF fields and estimated through a series of equations the transmembrane currents and potentials that could be expected. He concluded that the intracranial electric fields associated with low-intensity RF radiation "may induce transmembrane potentials of tenths of millivolts (or more) and that, therefore, such externally applied fields may disturb normal nervous function through this mechanism." His analyses indicated that the induced transmembrane potential would exhibit a maximum at frequencies in the UHF band. Further, he found that large cell components in regions of high cell density should be most influenced by extracellularly applied fields. Recent (41-43) and older (11) experiments indicate the usefulness of this line of thought.

I also pointed out some of the more recent conceptualizations of nervous-system functioning, and suggested how they could guide research to determine the biological interactions of low-intensity RF radiation. I noted that Szent-Gyorgyi (44) showed that molecules with low reactivity and with a major role as metabolites or hormones can give off an electron and form a free radical. This suggested to him that charge transfer might be one of the most common and fundamental biological reactions. On the basis of his experimental results, he proposed a quantum mechanical view of biology that had relevance to the mechanisms of RF radiation effects (45).

I wrote that a variant of Szent-Gyorgyi's view, a view specific to the nervous system, had been detailed by Wei (46-48). He suggested that the neuron has the structure and potential profile of a PNP transistor and may function like it. In part, Wei used as evidence for the accuracy of his model the discovery by Segal (49) of negative fixed surface charge on membranes, the birefringence phenomenon observed by Cohen et al. (50), and the report of micron wavelength emission from nerve by Fraser and Frey (10). Wei also suggested that what are considered to be transmitter chemicals in the central nervous system (e.g., acetylcholine) are electrical dipoles which when oriented and arranged in a large array could produce an electric field strong enough to drive positive ions over the junction barrier of the postsynaptic membrane. I noted that this hypothesis could provide an explanation for cleft size at synapses, synaptic delay, subthreshold integrations, facilitation with repetition, and the effects of calcium and magnesium. There was another line of thinking related to Wei's hypotheses that I described. Becker (51) had carried out an extensive series of experiments exploring what he considered to be a neural semiconduction control system in various living organisms. On the basis of the data he had gathered, he suggested as a conceptual framework the movement of mobile charge carriers within a solid-state neural system.

I also detailed how Cope (52), using nuclear magnetic resonance spectroscopy, reported that the water in brain tissue is bound in a highly ordered structure that can best be described as crystalline. Also mentioned was the Fritz and Swift (53) study of the high-resolution proton magnetic resonance spectrum of the sciatic nerve of the frog while active and resting.

I also pointed out that new concepts on the nature of neural function relevant to RF radiation effects have also come from Batteau (54) as a result of his study of the mechanisms of hearing. He suggested that sensation in the organism was due to the shifting of the probability of transition of electrons from the excited state to the ground state in organic molecules. He obtained data that he interpreted to indicate a transition energy gap of about 0.35 eV, corresponding to a 3 micron wavelength signal.

I went on to detail a number of other concepts and data that could have been used as a guide for research with RF radiation. I brought out the possibility of what could be called microthermal effects involving localized temperature gradients. I pointed out that some of my calculations indicated that this was a real possibility. Recent research indicates that such effects exist and are significant (55,56).

My conclusions in that paper were that if we break out of the thermoregulatory mind set, use realistic assumptions in our conceptualizations, and recognize that our knowledge of how the nervous system codes, transfers, and stores information is almost nil, a number of possible mechanisms through which RF radiation could affect higher organisms were apparent (2). The question, I said, is not whether there is a possible mechanism, but rather which of numerous possible mechanisms to use as a guide for experimentation. I also noted the seemingly forgotten possibility that more than one mechanism can be involved in RF radiation effects.

EXPERIMENTAL STUDIES

The above paper on possible biological mediators or mechanisms was published in January, 1971 (2). By that time, I had laid a foundation of data on low-intensity RF radiation bioeffects as I had intended to do. My thinking on possible biological mediators had jelled, and, as can be seen in the foregoing, I could spell out a number of these possible mediators as guides for research. Interest was rising in the area and funds were becoming available for other scientists to begin experiments which could provide data that would support a quest for biological mediators. Thus, in 1971 I moved on to such a quest. I anticipated finding multiple mechanisms and the quest was formulated with that expectation in mind. This led to my pursuing multiple interacting lines of biological research with RF radiation.

It appeared to me that the nervous system and its function would be one of the systems most sensitive to RF radiation. Thus, my quest for biological mediators in the 1970's was heavily weighted to experiments involving nervous system function.

One line I pursued was to look at a known electrosensing system in action to develop an understanding of how that system worked. Thus, I undertook experimentation on weakly electric fish. Another line was to identify the locus of the mechanism of the RF hearing phenomenon. Although I had found that one could induce skull vibration with very high power pulses, the data with very low-intensity energy suggested that the hearing effect was occurring in the cochlea. A third line stemmed from my expectation that brain chemistry would be involved in RF radiation effects. After considering data from the pharmacological literature, I developed the hypothesis that the dopamine systems of the brain could be influenced by exposure to this radiation. This led me to do a line of neuropharmacological experiments which, subsequently, led to a broadening of the hypothesis to include the opiate systems of the brain. The fourth line of experiments involved selectively permeable membrane function. I hypothesized that RF radiation exposure would disturb biological membranes, and this was explored in a series of studies involving the blood-brain barrier of the brain and the blood-vitreous barrier of the eye. This also led, in part, to my experiments with RF radiation and the placental barrier and also to experiments exploring developmental effects.

During this decade, I also carried out experiments that do not fall into the broad categories above. Some I did to clarify an aspect of my theorizing as it evolved, or I did an experiment to explore an idea. An example of the former are the additional experiments I did with the heart. An example of the latter is an experiment I did to explore the possibility of affecting the treatment of leukemia by modifying the blood-brain barrier with RF radiation to influence the passage of methotrexate. Chang and her associates have pursued this possibility in a somewhat different way (57,58). They have found an interaction between RF radiation exposure and methotrexate effect.

Of the four primary lines of research that I pursued, the first was to determine how a natural electrosensing system worked. There exists a number of weakly electric fish that generate a field and gather information about their environment apparently by analysis of perturbations in their field. My analysis of the literature and the experimental work I did with electric fields, sensors, and objects in various sized bodies of water led to the development of a mathematical model of the fishes' electrosensing system. This allowed me to interpret much of the data that had been previously published on the sensory system of the fish. I suggested a working hypothesis for an electrosensor mechanism. I showed the linkage among the various neural coding schemes that had been suggested for the fish and showed their essential identity (59).

RF HEARING

The earlier RF hearing research led to an extensive series of experiments by others, as well as several more by me. Most of the former were supported by DoD agencies concerned with hazards of RF radiation. The investigators controlling the work, such as Guy and Lin, were not biologists. Consequently, the work was hazards oriented. There was also inappropriate mathematical modelling and some biological experiments that had an unrecognized confounding variable. In papers published in 1979 and 1980 (60,61), I identified the confounding factor and noted the fault in the assumptions underlying their conclusions from the modelling. Besides the research leading to the paper on the confounding variable noted above, I carried out psychophysical experiments exploring modulation effects (62). I also explored the use of the periodicity pitch phenomenon (63,64). The results of these indicated that the locus of the effect is in the cochlea.

During this time and subsequently, RF hearing research was also done in the Soviet Union (65), and in other laboratories in the United States (66-73). Most notable was the experimentation of Wilson and Joines (55), which showed that at low power densities the locus of the mechanism of the hearing effect is in the cochlea. Thus, the efforts by multiple laboratories, using a variety of techniques, showed that the mechanism of the hearing of low-intensity RF radiation occurs in the cochlea. This finding opened up the possibility of new techniques for studies of ordinary hearing, and offered a new approach to understanding how hearing takes place.

The results of the experiments on the RF hearing mechanisms also clearly indicate that questions can be raised about whether RF radiation affects the labyrinthine system and implies a possible effect on the balance mechanism. I carried out an experiment to assess this possibility. I exposed animals on a rotating rod and determined how quickly they fell off. I found that they fell off more quickly when exposed to the radiation (74,75). But there is a confounding factor which will be noted later.

One other line of study that derived in part from the RF hearing work has significant medical implications. The pattern of clinical reports from the Soviet Union, as well as some observations by me and others in this country, suggest that a small percentage of people may incur what is called closed head injury as a result of exposure to RF radiation. This type of injury has sometimes been referred to as "post-concussive syndrome." Closed head injury may not show up in neurologic examinations such as CAT scans or EEGs because they are not appropriate tests of the functions affected. But such head injury is shown by clear-cut behavioral changes which include reduced attention span, impaired complex information processing, memory disturbance, and personality changes such as increased emotional lability, irritability, anxiety, and depression.

The neural damage and subsequent behavioral dysfunction in such head injuries is considered to be due to a shear-strain type of injury. It has been hypothesized that acceleration results in axonal tearing and neural degeneration in the brain stem as well as stretching and tearing of fibers within the cortex. This interpretation is supported by head injury research which found microscopic lesions within white matter and in particular brain stem structures. Behavioral measures seem to be the most sensitive indicators (76-78).

After publication of my 1968 paper giving further details on the RF hearing phenomenon, White (personal communication) wrote to me suggesting that the hearing effect was due to conversion of the RF energy into pressure waves via a thermoacoustic conversion mechanism, possibly in the cochlea. White, a physicist, had done the original work on defining the thermoacoustic conversion phenomenon of electromagnetic energy in material a few years earlier (79). It was immediately apparent to me that, if such conversion was taking place in the brain, damage could be caused that could account for the numerous reports of behavioral changes after long-term RF exposure such as reduced attention span, memory disturbance, irritability.

The reports of Olsen and Lin indicated the existence of the requisite pressure waves in the brain (80,81). Spiegel et al. (82) demonstrated mechanical motion in the brain as a consequence of field exposure. These data provide a reason to believe that pressure pulses occur in the brain.

It is ironic that it is such a shear-strain effect in the brain that the engineers concerned with hazards were implicitly assuming when they were trying to explain away the RF hearing effect as not being an indication of hazard (83,84). They never realized that shear-strain due to thermoacoustic expansion in brain tissue would itself damage the brain (61). I would expect that this effect would be noticeable particularly in those who have had long-term exposure to modulated RF radiation.

SENSORY AND CARDIAC EFFECTS

I also did incidental experiments to extend several earlier lines of research, in particular the sensory effects and cardiac effects lines. The possibility of other sensory effects of the energy was explored using several techniques. One was a shuttle box avoidance experiment using a box I developed for use in RF fields (85,86). One side of the shuttle box was shielded from exposure to the RF radiation. Each animal was placed in the box separately and the box was exposed to the radiation. It was found that the animals could detect and would avoid the pulse-modulated RF energy. The avoidance behavior occurred in response to exposure to power densities less than 1 mW/cm^2. It was found that varying the pulse modulation did not significantly influence this behavior. Unmodulated energy appeared to have little effect; the animals either did not perceive or did not care about exposure to unmodulated energy.

The numerous other behavioral experiments I did as measures of neural function are cited in various places in this chapter. In the late 1970's and early 1980's, behavioral experiments by others were reported. These latter are well described in a recent analytical review by Medici (87), and by her chapter in this volume. She also offers useful guidance for behavioral research with RF radiation. Particularly notable among the experiments she described are those of Thomas and his associates. In their most recent paper (88), they reported that the central nervous system is affected by a combination of a magnetostatic field and a 60-Hz magnetic field, as evidenced by changes in operant behavior.

The heart research was extended with an experiment involving intact frogs (89). Because of the low pulse repetition rate used, the average power used was 3 $microwatts/cm^2$. The heart rate was modified in the group exposed to pulses synchronized with the R wave of the ECG. The group with exposure synchronized with the T wave and the sham exposed group were significantly different from the R wave group. Others have carried the heart research much further.

Lords et al. (90) exposed isolated turtle hearts to RF fields using CW exposure. They found they could induce a bradycardia. They suggested the effect might be due to the stimulation of the parasympathetic and sympathetic nerve remnants in the turtle heart. In a follow-up experiment using an isolated turtle heart with the same exposure arrangement, Tinney et al. (91) found that over a narrow power range there was apparent stimulation of sympathetic and parasympathetic nerve remnants which could increase or decrease the heart rate, respectively. Galvin et al. (92) reported that RF radiation induced alterations in the cardiac cell membrane. Schwartz et al. (93) have shown RF fields with certain modulations will significantly increase calcium efflux from the heart. Other experiments along this line have also been reported (94-100).

BRAIN CHEMISTRY

My third line of research in this decade involved brain chemistry. In the early 1970's, I hypothesized that the dopamine systems of the brain could be influenced by exposure to RF energy (101). If this was the case, then certain behavioral measures of neural function should reflect such an effect. I initially set up two experiments to test this hypothesis. One used the intermittent tail pressure technique with rats, which provided a measure considered to be one of irritability and aggression. The other was a rotating rod experiment which would indicate if there was influence on motor coordination through the nigrostriatal tract or via a labyrinth effect.

In the aggression experiment, animals were placed in a box in pairs and intermittent pressure was applied to their tails. This induced them to emit aggressive actions, behaviors that are clear cut and which can be readily quantified. A clear-cut "docility" effect was seen when the animals were exposed to incident average power densities of 50 microwatts/cm^2 during the procedure. The RF radiation exposed group took significantly longer to begin aggressive behavior, the number of episodes and duration of the behavior was significantly less, and it ended significantly sooner than in the control group. This result was consistent with what would be expected with an RF radiation influence on the dopamine systems of the brain (101).

If there is such an effect on the dopamine system, it would also be expected that fine motor coordination would be influenced by exposure to the energy. This was tested, as was a possible labyrinthine effect as noted before, by placing animals on a horizontal rotating rod. The rod speed was slowly increased until the animal fell off. The animal was exposed to pulse-modulated RF radiation. Exposure power levels of less than 1 mW/cm^2, average power density, were effective in disrupting their balance capability (74,75).

In view of the outcome of the foregoing research suggested

by the dopamine hypothesis, additional experiments were undertaken. The influence of an electromagnetic field on emotionality was explored using the conditioned emotional response (CER) technique (102,103).

The CER has been used in drug testing as an animal analog of anxiety. The specific prediction was that a CER learned and tested outside the field would be affected in an animal living in a low-frequency field. The animals lived in a low-intensity 60-Hz field and were trained and tested outside of the field. It was found that living in such a field significantly increased emotionality as defined by the test. It was also found, as we had observed in several other experiments, that some animals were particularly sensitive to exposure to such fields. This finding is not particularly remarkable since individual hypersensitivity reports are common in the pharmacological literature. But such idiosyncrasies in sensitivity had not been recognized in the RF radiation literature of the United States. (It had been recognized in the Soviet Union.) The recognition of it is important though, since, as I suggested, it has to be considered in the data analysis of an RF radiation experiment. It can lead to an erroneous conclusion of no effect when in fact an effect is real, what is called in statistics a type II error.

The dopamine hypothesis also suggested that internal timing capability would be disturbed by exposure to RF radiation. To test this, a Sidman avoidance experiment was carried out using exposure to 60-Hz fields. With the Sidman technique, an animal must make sensitive use of timing behavior to minimize work, but, in doing so, work enough to avoid an aversive stimulus. It was found that there were significant effects on timing behavior with animals living in a low-intensity low-frequency field. The exposed animals incurred significantly more aversive stimulation than those sham exposed. After this was established, the exposure conditions for the exposed and sham exposed groups were reversed. There was no change in the pattern of Sidman behavior for the first 24 hours, but after 48 hours the behaviors of the two groups reversed (102).

Direct changes in dopamine were reported about that time by Stith and Erwin (104). They found that tyrosine hydroxylase in the hypothalamus and brain stem is significantly decreased by exposure to low-intensity RF radiation. Dopamine is synthesized from tyrosine and tyrosine hydroxylase acts on tyrosine in a rate-limiting enzyme step.

The dopamine hypothesis also suggested another experiment that would involve an animal analog of a human clinical state. An isolated rat subjected to a steady light tail pressure emits certain behaviors such as chewing, licking, and gnawing that are situationally peculiar. These types of behaviors are referred to as stereotypic behaviors and are considered to be analogs of clinical states in humans involving dopamine system defects. Experimentation was undertaken to determine if RF radiation exposure would influence stereotypic behaviors. It was found that it did (105,106). Average power densities on the order of 8 microwatts/cm^2 were sufficient to modify stereotypic behavior. In addition, interactions between pressures, certain odors, and pulse repetition rates influenced the nature of the effect. Certain odors, for example, heightened the RF effect. Further, the responses to RF radiation in the exposed animals were bimodally distributed. We had noted similar effects on distribution in other experiments (107). This again called attention to the importance of individual differences in sensitivities when low-intensity RF radiation is used. The bimodal distribution finding has important implications. RF radiation effects may well have appeared but been overlooked in some experiments, for commonly used statistical tests are not appropriate when data is distributed bimodally. Use of them in such circumstances would mislead the experimenter into believing there was no effect when, in fact, there was an effect of the radiation (a type II error). Thus, several lines of evidence indicated that dopamine and/or dopamine receptors were involved as a mediator in RF energy exposure effects.

By the time the above experiment was being done, the advances in knowledge in pharmacology reported in the literature provided a basis to expand the dopamine hypothesis to involve the opiate systems of the brain. The literature by the late 1970's indicated that there is a complex interaction between the dopamine and opiate systems. It also suggested that the dopamine system may be the final step through which opiate system function is expressed. Thus, I extended the theorizing about RF radiation and the dopamine systems to include certain interactions with the opiate systems (107,108).

It was relatively straightforward for me to predict that RF radiation exposure would disturb these interactions and to set up a simple experiment to test the hypothesis. I set up and carried out a direct test of the predictions using a classic test of opiate system function -- tail flick. In this test, the animals' tails are exposed to an aversive stimulus. The measure is the time it takes for them to flick their tails away from the aversive stimulus (106,109,110).

I hypothesized that apomorphine, a dopamine agonist, which at low doses was reported to inhibit presynaptic DA release by stimulating presynaptic receptors, would increase tail-flick latency, an effect similar to a high dose of morphine. Further, it had been found that stimulation of postsynaptic DA receptors with high doses of apomorphine antagonized morphine effects, thereby decreasing latency. I used high and low dose apomorphine groups with and without RF radiation exposure to assess the interactions that could be predicted. The group getting only a low dose of apomorphine had a significantly increased flick latency. The group getting only a high dose of apomorphine had a significantly decreased latency. The comparable groups that were, in addition, exposed to RF radiation showed that such exposure blocked the effects of apomorphine on tail flick latency. It was also found that exposure to RF radiation alone decreases latency. This suggested the possibility of increased presynaptic firing or direct stimulation of postsynaptic receptors. The results supported the dopamine-opiate system hypothesis.

I did additional experiments with apomorphine, as well as with librium, nalaxone, haloperidol, and morphine, with results consistent with the hypothesis. For example, in the morphine experiments with RF radiation, tail flick latency was again used as a measure (109,111). It was found that a high dose of morphine had a significant effect on flick latency but a low dose did not. But when the low dose of morphine was given with exposure to RF radiation, tail-flick latency increased significantly. Thus, RF radiation exposure potentiated morphine analgesia, an effect similar to classic DA inhibitors such as haloperidol. This effect occurred with average incident power densities of 200 microwatts/cm^2.

The dopamine-opiate system hypothesis suggested that one way these effects could occur was through an alteration of dopamine (DA) receptor sites by changing protein conformation at the neuronal membrane. Binding of DA would be inhibited and calcium metabolism would be altered. Frey and Wesler's data suggest the former is occurring and the latter is indicated by the results of Blackman et al. (112) and Bawin and Adey (113).

Other investigators pursued the line of study involving brain chemistry. Lai et al. (114) explored the effects of apomorphine, d-amphetamine, and morphine in animals exposed to RF radiation. They found irradiation enhanced apomorphine effects and attenuated amphetamine effects. Morphine effects were also enhanced. Schrot et al. (115) extended previous work with drugs using chlordiazepoxide, chlorpromazine, and diazepam. They found that the latter two drugs decreased response rate with animals on an operant conditioning schedule, while chlordiazepoxide increased response rate. They found some complex interactions between drugs and field, and suggested that the field parameters may be an important variable in the nature of the effect obtained with various psychoactive drugs. In a study by Ashani et al. (116), the combined effects of low-intensity RF radiation and anticholinesterase drugs were investigated. They found complex interactions and suggested that use of such drugs would provide significant information on the effects of RF radiation. Nakas et al. (117) and Jamakosmanovic et al. (118) considered the effect of repeated daily exposure to RF radiation on levels of acetylcholinesterase in the brain. They found that chronic exposure to the radiation induced significant depression of the acetylcholinesterase activity in the brain. They also found other effects which led them to suggest that the central nervous system is very sensitive to low-intensity RF radiation. In two other studies, Lai et al. (119,120) report RF radiation potentiated an effect of apomorphine and attenuated an effect of ethanol and also amphetamine. They also found that pulsed, but not continuous, radiation decreased hippocampal choline uptake, a measure of cholinergic nerve activity. This effect was blocked by the opiate antagonist naloxone as would be expected by the dopamine-opiate hypothesis. Miller et al. (121), using 60-Hz magnetic fields, found that exposure to the fields attenuated the behavioral response of mice to morphine.

One interesting implication of the dopamine-opiate hypothesis is that RF radiation could affect the hypothalamic set point for body temperature regulation. As I have pointed out before (107), the mechanism that sets the body's temperature is located in the hypothalamus. Dopamine is believed to play an important role in the adjustment of this mechanism (122). One can, of course, dump enough energy into a mammal to raise its temperature directly. But it seems likely that at low-intensity exposures an effect of the RF radiation on the hypothalamic set point via the dopamine system could result in a temperature shift. This is an ironic twist in view of the hazards establishment's efforts to press all findings of radiation effects into the "due to thermal effects" mold. It may be the other way around, with some of the "thermal effects" a result of an effect in the brain.

BLOOD BARRIERS

The remaining line of investigation that I pursued in my quest for an understanding of the mediators or mechanisms of RF radiation effects involved the blood barriers of the body, i.e., blood-brain barrier, blood-vitreous-humor barrier, and placental barrier.

The initial experimentation in this line of research was with the blood-brain barrier, which separates the brain and cerebral spinal fluid of the central nervous system from the blood. The locus of the barrier is considered to be at the interface between them, i.e., in the choroid plexus, the blood vessels of the brain and subarachnoid space, and the arachnoid membrane. The barrier consists of cells connected with tight junctions in an essentially continuous layer that regulates intercellular diffusion. Solutes that are lipid soluble readily penetrate the barrier. But lipid insoluble substances or proteins encounter a set of regulatory interfaces between the blood and the nervous system that control their transport.

Several fluorescent dyes bind to serum protein when injected into the blood stream. These have been used to study the nature of these regulatory interfaces and have been found to be quite useful. I used one of these, sodium fluorescein, to explore the effects of exposure of the animal to RF radiation. The procedure was to irradiate the animals with RF energy, inject them intravenously with the dye, and then several minutes later, exsanguinate, perfuse, section, and measure the fluorescence of the brain sections under ultraviolet light. It was found that there was penetration of the barrier as indicated by this classic technique in response to exposure to RF radiation. Fluorescence was seen in the diencephalon level of the brain as well as, to some extent, in the mes- and metencephalon. The differences in brain fluorescence between the sham-exposed and the exposed animals was statistically significant.

Oscar and Hawkins (123) extended the work by exposing rats to RF radiation to assess the uptake of several radioactive neutral polar substances in the brain. Barrier permeability increases were observed for mannitol and inulin but not for high molecular weight dextran. The apparent permeability change, which was reversible, was greatest in the medulla, followed, in decreasing order, by the cerebellum and hypothalamus. It was also found that RF radiation of the same average power, but with different pulse characteristics, produced different uptake levels.

Albert (124) exposed Chinese hamsters to RF radiation and injected them with various electron dense tracers. Specimens were then prepared for light and electron microscopic examination. The exposed and sham-exposed groups differed in that exposed animals showed tracer penetration of the barrier in the cerebral and cerebellar cortices, medulla, thalamus, and hypothalamus.

Merritt et al. (125) reported doing a replication of the first two studies and stated "No transfer of parenterally administered fluorescein across the BBB of rats after 30 minutes of 1.2 GHz radiation at power densities from 2-75 mW/cm^2 was noted." But a statistical analysis of the data presented in their paper by several scientists showed that, in fact, their data supported the opposite conclusion and provided a confirmation of the findings of Frey et al. (126). They may also confirm Oscar and Hawkins, but sufficient data were not provided in their published paper to allow an appropriate statistical analysis.

Preston et al. (127) reported on the permeability of the barrier to mannitol under RF radiation. They reported that they did not confirm Oscar et al.'s work. But they made a type II statistical error in their data analysis. Using binomial tests to correct for this, Frey (126) found that their data revealed significant RF radiation effects that were consistent with Oscar et al.'s results.

Thus, Merritt et al. and Preston et al. did, in fact, find a RF radiation effect in the brain. What appeared at first to be conflicting and controversial data actually are quite consistent and indicate that something involving the blood-brain barrier occurs in the brain subjected to low-intensity RF radiation. Other blood-brain barrier experiments of varying quality have been reported (128-135). In an interesting one, Quock et al. (136) found that RF radiation exposure facilitated a quaternary ammonium derivative antagonism of pilocarpine and oxotremorine. They suggest this may be due to an enhanced passage through the barrier. In another interesting study, Lange et al. (137) found that RF radiation facilitated depletion of brain catecholamines by drugs otherwise restricted to peripheral effects. But the DoD terminated funding for such research, so the nature and extent of such effects have not been determined (1).

But before the funding was terminated, I extended the work to other blood barriers. The eye was the first, since it is derived from the neuroectoderm, as is the brain. It has blood barriers much like the blood-brain barrier, i.e., the blood-aqueous barrier and the blood-vitreous barrier. The blood-vitreous barrier is important inasmuch as it regulates the composition of the vitreous humor and is involved in controlling the ionic and metabolic environment of the retina.

In a blood-vitreous barrier experiment, the animals were exposed to average incident power densities of 75 microwatts/cm^2 for 25 minutes then injected with sodium fluorescein. There were several variants of this basic procedure used. A small but significant increase in vitreous humor fluorescence occurred with the animals exposed to RF energy (138,139).

The analysis of Neelakantaswamy and Ramakrishnan (140,141) could be extended from the lens to provide an explanation for this finding. Their work indicated that RF radiation could induce bending moments and stresses in the tissue that could upset physiochemical processes in the eye. This might also provide an explanation for the finding that RF radiation apparently causes cataracts in humans (142).

The writer's work on barriers was extended to placental

barriers with exposure of rats in utero to 60-Hz fields. But it was not possible to complete this line of work and the question of permeability changes in the placental barrier was unresolved until recently. Vodicnik et al. (143) exposed pregnant mice to RF radiation in a single dose. They found a significant change in fetal drug uptake as well as placental uptake. The effect for mannitol and glucose occurred only if the RF exposure occurred prior to the administration of the drug. Other experiments involving prenatal exposure to RF radiation have now shown effects on various measures in the newborn (144-146).

EFFECTS ON THE BRAIN

There are a variety of other experiments that indicate an influence of low-intensity RF radiation on the brain. Sanders and his associates have studied the effects of RF radiation on the energy metabolism of the brain. They found that exposure to the energy modified the ATP levels in the brain. They also found that CP-kinase, a divalent metallo-enzyme, was susceptible to interaction, and concluded that there was a resonant energy inhibition of CP-kinase (147,148). They have also explored the effect of a field on ninicotinamide adenine dinucleotide (NADH), adenosine triphosphate (ATP), and creatine phosphate (CP) in the brain. They found changes in NADH levels and ATP and CP concentrations. They believed the results suggested that RF radiation inhibited electron transport chain function in brain mitochondria and caused a decreased energy level in the brain (149). Wilson et al. (150), in a series of studies, have shown changes in the energy metabolism of the brain using $[^{14}C]$-2-deoxy-D-glucose.

McKee et al. (151) found morphological changes in the central nervous system, particularly in the hippocampal, hypothalamic, and cortical regions of the brain. Albert et al. (152) also reported morphological change in the brain as a consequence of exposure to RF radiation.

In other work bearing on the central nervous system, Sheridan et al. (153) reported lipid chain disorder induced by low-intensity RF radiation. Stevens (154) concluded that the rate at which brain protein makes the transition from one conformational state to another depends on the particular relationship between the membrane potential and the equivalent dipole moment change of the protein.

Barnes and Hu (30) offered a useful model for RF radiation effects on biological membranes. They indicated that shifts in ion concentrations across the membrane and the orientation of the long-chain molecules are possible. Others, such as Schwartz (155), Lee (29), and Frohlich (156) offered mechanisms that might be implicated in RF effects in the brain. Liboff (32) and McLeod and Liboff (157) offered a particularly noteworthy conceptualization involving cyclotron resonance as an explanation of calcium efflux effects of RF radiation. A prediction from it involving lithium ions was recently supported by the results of a behavioral experiment by Thomas et al. (88).

Wachtel (158) and Wachtel et al. (159) reported experiments which indicate minimum RF threshold energy pulse widths for evoking fast neuronal effects as well as RF induced hypokinesis. Bearing on this is the report of Brown and Larsen (160). They used change in resting birefringence of crab nerves, which coincides with propagation of the action potential, as a measure of nerve response to RF radiation. They found that pulsed radiation influenced birefringence.

In another study, McRee, working with Wachtel, reported RF radiation effects on nerve vitality. They indicated that long-term regulatory processes were involved (161,162). Wachtel and his associates also offered a mathematical model for the selection of RF pulse parameters. They have done other modeling and suggest an influence on electrogenic pumps in nerves (41-43).

Another line of investigation with the central nervous system which has been fruitful has been concerned with calcium efflux. This line stems from the behavioral experiments with nonionizing radiation that Medici began in 1966 while she was at UCLA (163). She examined the effect of low-frequency fields using inter-response time schedules of reinforcement with monkeys. Her studies showed that the animals' behavior was significantly modified by the fields. She also showed that the animals were particularly sensitive to frequencies that were in the EEG range of the animals, that is 7 Hz, as contrasted with 45 Hz and 75 Hz. A change in the spectrum of the EEG was found when the animals were exposed to RF radiation.

Because of these results, Kaczmarek, a neurochemist at UCLA, was asked to consider other ways to measure brain response to the fields. He initiated experiments in which calcium efflux was measured following exposure to RF fields (164). A series of experiments followed from this that have provided useful information about the effects of RF radiation on the brain.

At first, Bawin and her associates (113,165) carried forward the work with calcium efflux. They found that isolated brain tissue responds to exposure to an RF field of a narrow frequency band and a narrow amplitude window. Skeletal muscles did not respond to the radiation in a similar fashion. Lin and Adey (166) extended the work to synaptosomes and concluded that their data indicated that whole cells or organized tissue are not required for the effect to occur.

In a series of studies Blackman and his associates verified and extended the findings of Bawin and her associates (112, 167-176). They also found frequency windows and power density windows in which calcium efflux was enhanced by exposure to modulated RF radiation. They also found that they could shift carrier frequencies and get effects. In another experiment they found that the enhanced calcium efflux phenomenon was more dependent on the electric field intensity in the brain than on the power density of the incident radiation. They hypothesized that brain tissue is electrically nonlinear at specific field densities, and that this nonlinearity demodulates the carrier and releases a particular modulated signal within the tissue. They suggested that the signal is selectively coupled to the calcium ions by some mechanism, perhaps a dipolar type (Maxwell-Wagner) relaxation which enhances the efflux of calcium ions. Tenforde (177) has suggested that the field interaction involving calcium may involve a resonance or cooperative energy transfer mechanism.

IMMUNE SYSTEM

There has been scattered work concerning RF radiation and the immune system. Some of it has used radiation that I would not expect to influence the immune system; other work has used more appropriate, though not optimal, frequencies and has revealed effects that are of consequence. Schlagel and Wiktor-Jedrzejczak (178) and Wiktor-Jedrzejczak et al. (179) studied complement receptor positive cells and exposure to RF radiation. They found an increase in complementary positive spleen cells in mice. They concluded that the response was not due to alterations of lymphocyte recirculation patterns, but rather was mediated by a soluble, humoral factor produced by cells within the spleen which induced an increase in complement receptor positive B cells. In a later study, Schlagel et al. (180) concluded that complement receptor positive cell response was not regulated by the T-cell population. They suggested that the data indicate that the RF radiation induced change in the population of cells with specific cell surface receptors. Further, they suggested that susceptibility to these changes was under genetic control; that endotoxin, corticosteroids, and regulatory T cells do not play a significant role in the mechanism regulating the increases that they found. Liddle et al. (181) explored the effect of RF radiation on the circulating antibody response of mice. They immunized the mice and then exposed them to the pulse-modulated RF radiation. The $\log_2$ hemagglutination titers among the RF-exposed mice were significantly higher than among those sham irradiated. They concluded that the increase in circulatory antibody response provided some protection to the mice. Schlagel and Yaffe (182) and Schlagel and Ahmed (183) report that the transitory increase in complement receptor bearing lymphocytes induced by a single exposure to RF radiation is under genetic control. In the report of their study with several strains of mice, they concluded that the essential regulatory gene was to the right of the Pgm-1 locus and to the left of the rd locus on chromosome 5. RamaRao et al. (184) reported that exposure of hamsters to RF energy resulted in viricidal activation of peritoneal macrophages and extended survival of hamsters injected with a lethal dose of vesicular stomatitis virus. They also reported highly significant increases in plaque production, measured using the hemolytic plaque assay, when animals were injected with sheep red blood cells.

In sum, the immune system data suggests a responsiveness to RF radiation. But the investigators have used radiation of marginal frequency and modulation for inducing such effects. The optimal RF parameters for exploring such effects have not been used.

DNA

There are some suggestions that DNA may be influenced by RF radiation. Swicord and Davis (185), using an optical heterodyne technique, found that the DNA of E. coli in aqueous solution absorbed RF radiation. Edwards et al. (186) carried out similar studies using dielectrometric methods. They report significantly increased absorption of RF radiation by a DNA solution relative to the solvent. They also stated that the data suggested a chain length dependence of the absorption. Liboff and Homer (187) studied the uptake of ^{3}H-thymidine in cell culture as a function of sinusoidal magnetic field intensity. Their data led them to infer that DNA synthesis in the exposed cells was significantly greater than in unexposed cells. The effect occurred with a wide range of modulation frequencies and was maximum during the mid-S phase of the cell cycle.

Kohli and VanZandt (188) calculated what the absorption of RF radiation would be by a DNA molecule. They presented curves of expected absorption vs frequency for various damping parameters and reflection conditions. In a somewhat similar approach, VanZandt et al. (189) calculated absorption of DNA double helix in aqueous solution and compared the theoretical results with experimental data. They report that the calculated absorption values agreed strikingly with experimental values for the relative absorption of water and DNA solution. Edwards et al. (186), concluded that their studies, noted above, partially supported the theoretical work of Kohli, Prohofsky, and VanZandt.

CONCLUSIONS

What has been done in most of the biological research with RF radiation in the United States has been determined by the thermoregulatory mindset of those involved in the research. That research has not been considered to any extent in this review since it contributes little to our knowledge and understanding of the biological effects of low-intensity electromagnetic fields.

As I stated in 1971 (2), the question is not whether there is a mediator or mechanism for biological effects of low intensity RF radiation. Rather, the question is how many are there and how are they related. There are some answers to these questions, but a discussion of them is beyond the scope of this chapter.

It is clear that there are a diversity of biological effects of RF radiation, effects such as those on neuropeptides. Understanding the nature of these various effects can lead to significant advances in biology. Neuropeptides, for example, have been shown to have behavioral and neuroendocrine effects. Data indicating that they exert effects on immune system function is accumulating. These ligands and their receptors are abundantly distributed in areas of the brain that mediate emotion and higher cognitive functions. Thus, as Ruff et al. (190), writing about other matters, put it, "...it seems plausible that the same neurohumoral mediators of various mood states in brain may also communicate to monocytes and other cells involved in healing and homeostatic processes. Short signal peptides (neuropeptides) and their surface receptors define a group of cells whose function may be to integrate information from the central nervous, immune, and endocrine systems through a psychoimmunoendocrine network, thereby altering the behavior of the whole organism." Angeletti and Hickey (191) also see such linkages. And it is in this direction that the low-intensity RF radiation bioeffects data is taking us.

As another example, consider memory and RF radiation. Recent findings in pharmacology have led to the hypothesis (192) that one mechanism of memory involves "Brief bursts of high frequency activity (that) cause a transient elevation of calcium in spines that activates a membrane-associated calpain. This enzyme then breaks up a localized portion of the fodrin network, producing structural and chemical changes in the region of the postsynaptic membrane. As a result, previously occluded glutamate receptors are exposed, thereby increasing the size of the postsynaptic response to the released transmitter. More prolonged or repetitive bursts of activity can be expected to produce a larger calcium disturbance and more widespread activation of the calcium-dependent proteinase, events that we propose will produce alterations in the ultrastructure of the dendritic spine." The implications of RF radiation exposure with such a memory mechanism are significant.

Appropriate research with low-intensity RF radiation, and I include magnetic fields, will likely lead to substantial advances in knowledge of biological processes. But that research will not be done in the United States in the foreseeable future. The more interesting and more significant research with electromagnetic fields will be and is now being done in other countries. An analytical review of that work, though, is beyond the scope of this chapter.

As I noted earlier, the significant research, that which does not use high intensities and is not thermoregulatory oriented, began to be tapered off about 1980. Though such research received only a very small fraction of the huge amount that has been spent (several hundred millions of dollars) on RF bioeffects research since 1970, it has been largely squeezed out for reasons unrelated to science. These reasons can be found in a recent book by a historian of science who has made a study of this field of research (1).

But this will only hinder and delay science, for the significant research with electromagnetic fields will continue to move forward in other countries. And this will contribute to the broad revolution in biology that is now taking place.

REFERENCES

(1) Steneck, N.H.: The Microwave Debate, Cambridge, MA, MIT Press, 1984.

(2) Frey, A.H.: Biological function as influenced by low power modulated RF energy, IEEE Trans. on Microwave Theory and Techniques, MTT-19:153-164, 1971.

(3) Frey, A.H.: From the laboratory to the courtroom: Science, scientists, and the regulatory process, in Risk/Benefit Analysis: The Microwave Case, N.H. Steneck, ed., San Francisco, San Francisco Press, 197-226, 1982.

(4) Medici, R.G.: Where has all the science gone?, in Risk/Benefit Analysis: The Microwave Case, N.H. Steneck, ed., San Francisco, San Francisco Press, 177-196, 1982.

(5) Frey, A.H.: Auditory system response to radio frequency energy, Aerospace Med. 32:1140-1142, 1961.

(6) Frey, A.H.: Human auditory system response to modulated electromagnetic energy, J. Appl. Physiol. 17:689-692, 1962.

(7) Frey, A.H.: Behavioral biophysics, Psychol. Bull. 63:322-337, 1965.

(8) Frey, A.H.: A transition from an electrode to coaxial cable, Behav. Res. Methods and Instrumentation 2:94, 1970.

(9) Frey, A.H., Fraser, A., Siefert, E. and Brish, T.: A coaxial pathway for recording from the cat brain stem during illumination with UHF energy, Physiol. Behav. 3:363-364, 1968.

(10) Fraser, A. and Frey, A.H.: Electromagnetic emission at micron wavelengths from active nerves, Biophys. J. 8:731-734, 1968.

(11) Frey, A.H.: Brain stem evoked responses associated with low intensity pulsed UHF energy, J. Appl. Physiol. 23:984-988, 1967.

(12) Frey, A.H. and Thornton, S.J.: A restraint device for cats in an ultrahigh frequency electromagnetic energy field, Psychophysiology 2:381-383, 1966.

(13) Frey, A.H.: Human response to very low frequency electromagnetic energy, Naval Res. Rev., 1-4, August, 1963.

(14) Frey, A.H.: Some effects on human subjects of ultra high frequency radiation, Am. J. Med. Electronics 2:28-32, 1963.

(15) Frey, A.H. and Siefert, E.: Pulse modulated UHF energy illumination of the heart associated with change in heart rate, Life Sci. 7:505-512, 1968.

(16) Lee, C.Y.: The role of excitons and optical phonons in the change of ionic permeability and the propagation of nerve impulse, Bull. Math. Biol., 1976.

(17) Cope, F.W.: Electron phonon (trapped photon) coupling and infrared coaxial transmission line theory of energy transport in mitochondria and nerve, Bull. Math. Biol. 35:627-644, 1973.

(18) Maurel, P. and Galzigna, L.: Dipole moment of acetylcholine and its relevance to the chemical synaptic transmission, Biophys. J. 11:550-557, 1971.

(19) Margineanu, D. and Moisescu, D.: Energy of nerve impulse, Bull. Math. Biophys. 32:151-153, 1970.

(20) Margineanu, D.G.: Activation parameters of the nerve impulse conduction (II), Experientia 28:1286-1288, 1972.

(21) Spiegel, R.J. and Joines, W.T.: A semiclassical theory for nerve excitation by a low intensity electromagnetic field, Bull. Math. Biol. 35:591-605, 1973.

(22) Hodson, C. and Wei, L.Y.: Comparative evaluation of quantum theory of nerve excitation, Bull. Math. Biol. 38:277-293, 1976.

(23) Almeida, S.P., Bond, J.D. and Ward, T.C.: The dipole model and phase transitions in biological membranes, Biophys. J. 11:995-1001, 1971.

(24) Sherebrin, M.H., MacClement, B.A.E., and Franko, A.J.: Electric field induced shifts in the infrared spectrum of conducting nerve axons, Biophys. J. 12:977-989, 1972.

(25) Wei, L.Y.: Quantum theory of nerve excitation, Bull. Math. Biophys. 33:187-194, 1971.

(26) Wei, L.Y.: Possible origin of action potential and birefringence change in nerve axon, Bull. Math. Biophys.

33:521-537, 1971.

(27) Wei, L.Y.: Dipole theory of heat production and absorption in nerve axon, Biophys. J. 12:1159-1170, 1972.

(28) Arndt, R.A. and Roper, L.D.: Quantitative comparison of dipole models for steady state currents in excitable membranes, Bull. Math. Biophys. 34:305-324, 1972.

(29) Lee, C.Y.: Molecular mechanism of sodium conductance changes in nerve: The role of electron transfer and energy migration, Bull. Math. Biol. 45:759-780, 1983.

(30) Barnes, F.S. and Hu, Chia-lun J.: Model for some nonthermal effects of radio and microwave fields on biological membranes, IEEE Trans. on Microwave Theory and Techniques, MTT-25:742-746, 1977.

(31) Taylor, L.S. and Cheung, A.Y.: The mechanisms of microwave biological effects, A report on a workshop held at the University of Maryland, College Park, MD, 1979.

(32) Liboff, A.R.: Cyclotron resonance mechanism for electromagnetic energy transfer to cells, 7th Ann. Meeting Bioelectromagnetics Society, San Francisco, CA, 39, 1985.

(33) Hearn, G.E.: Effects of UHF Radio Fields on Visual Acuity and Critical Flicker Fusion in the Albino Rat, Ph.D. dissertation, Baylor University, Waco, TX, University Microfilms 65-10:002, 1965.

(34) Korbel, S. and Thompson, W.D.: Behavior effects of stimulation by UHF radio fields, Psychol. Rep. 17:595-602, 1965.

(35) Korbel, S. and Fine, H.L.: Effects of low intensity UHF radio fields as a function of frequency, Psychonomic Sci. 9:527-528, 1967.

(36) Bourgeois, A.E.: The Effect of Microwave Exposure upon

the Auditory Threshold of Humans, NASA Science and Technical Information Service N68-23132, 1967.

(37) Frey, A.H.: Effects of microwave and radio frequency energy on the central nervous system, Proc. Symposium Biological Effects and Health Implications of Microwave Radiation, Richmond, VA, BRH/DBE 70-2, PB 193 898:134-139, 1969.

(38) Schwan, H.P.: Interaction of microwave and radio frequency radiation with biological systems, Proc. Symposium on Biological Effects and Health Implications of Microwave Radiation, Richmond, VA, 1969.

(39) Valentinuzzi, M.: Notes on magnetic actions upon the nervous system, Bull. Math. Biophys. 27:203-214, 1965.

(40) MacGregor, R.J.: A model for responses to activation by axodendritic synapses, Biophys. J. 8:305-318, 1968.

(41) Wachtel, H.: Microwave modulation of neural activity, Biomed. Sci. Instrum. 18:45-50, 1982.

(42) Wachtel, H., Adey, W.R., Chalker, R. and Barnes, F.S.: Temperature rise rate as a causal factor of rapid neural responses to microwave absorption, 4th Ann. Meeting Bioelectromagnetics Society, Los Angeles, CA, 34, 1982.

(43) Wachtel, H. and Barnes, F.S.: Field detection and thermal rate models for explaining neural effects of wide microwave pulses, 5th Ann. Meeting Bioelectromagnetics Society, Boulder, CO, 27, 1983.

(44) Szent-Gyorgyi, A.: Intermolecular electron transfer may play a major role in biological regulation, defense, and cancer, Science 161:988-990, 1968.

(45) Szent-Gyorgyi, A.: Molecules, electrons, and biology, Trans. N.Y. Acad. Sci. 31:334-340, 1969.

(46) Wei, L.Y.: A new theory of nerve conduction, IEEE Spectrum 3:123-127, 1966.

(47) Wei, L.Y.: Electric dipole theory of chemical synaptic transmission, Biophys. J. 8:396-414, 1968.

(48) Wei, L.Y.: Role of surface dipoles on axon membrane, Science 163:280-282, 1969.

(49) Segal, J.R.: Surface charge of giant axons of squid and lobster, Biophys. J. 8:470-489, 1968.

(50) Cohen, L.B., Keynes, R.D. and Hille, B.: Light scattering and birefringence change during nerve activity, Nature 218:438, 1968.

(51) Becker, R.O.: The neural semiconduction control system and its interaction with applied electrical current and magnetic fields, Presented at the 11th Int. Congress of Radiology, September, 1965.

(52) Cope, F.W.: Evidence for semiconduction in aplysia nerve membrane, Defense Documentation Center, Alexandria, VA, Rep. DDC 687343, 1969.

(53) Fritz, O.G. and Swift, T.J.: The state of water in polarized and depolarized frog nerves, Biophys. J. 7:675-678, 1967.

(54) Batteau, D.W.: Localization of Sound - Part 5: Auditory Perception, U.S. Naval Ordnance Test Station, China Lake, CA, NOTS tp-3109, 1963.

(55) Wilson, B.S. and Joines, W.T.: Mechanisms and physiologic significance of microwave action on the auditory system, J. Bioelectricity 4:495-525, 1985.

(56) Tigranyan, R.E.: Hypothesis on the acoustic nature of the mechanism behind the biological action of pulsed UHF

fields, Biofizika, In Press, 1985.

(57) Chang, B.K., Huang, A.T. and Joines, W.T.: Inhibition of DNA synthesis and enhancement of the uptake and action of methotrexate by low power density microwave radiation in L1210 leukemia cells, Cancer Res. 40:1002-1005, 1980.

(58) Chang, B.K., Huang, A.T. and Joines, W.T.: Microwave treatment of intracerebral L1210 leukemia: Schedule-dependent partial reversal of the effects of methotrexate, Bioelectromagnetics 2:77-80, 1981.

(59) Frey, A.H. and Eichert, E.S.: The nature of electrosensing in the fish, Biophys. J. 12:1326-1358, 1972.

(60) Frey, A.H. and Coren, E.: Holographic assessment of a hypothesized microwave hearing mechanism, Science 206:232-234, 1979.

(61) Frey, A.H. and Coren, E.: Holographic assessment of microwave hearing (the authors reply), Science 209:1144-1145, 1980.

(62) Frey, A.H. and Messenger, R.: Human perception of illumination with pulsed ultrahigh frequency electromagnetic energy, Science 181:356-358, 1973.

(63) Eichert, E.S. and Frey, A.H.: Human auditory system response to low power density, pulse modulated, electromagnetic energy: A search for mechanisms, Presented at the 11th Microwave Power Symposium, Belgium, 1976.

(64) Frey, A.H. and Eichert, E.S.: Psychophysical analysis of microwave sound perception, J. Bioelectricity 4:1-14, 1985.

(65) Khizhnyak, E.P., Shorokhov, V.V. and Tyazhelov, V.V. Two types of microwave auditory sensation and their possible mechanisms, URSI Symposium "Ondes Electromagnetiques et Biologie," Paris, France, 101-103, 1980.

(66) Cain, C.A. and Rissmann, W.J.: Microwave hearing in mammals at 3.0 GHz. Biological Effects of Electromagnetic Waves, URSI Annual Meeting, Boulder, CO, 79-88, 1975.

(67) Cain, C.A. and Rissmann, W.J.: Mammalian auditory responses to 3.0 GHz microwave pulses, IEEE Trans. Biomed. Eng. 25:288-293, 1978.

(68) Lebovitz, R.M. and Seaman, R.L.: Single auditory unit responses to weak, pulsed microwave radiation, Brain Res. 126:370-375, 1977.

(69) Joines, W.T. and Wilson, B.S.: Field induced forces at dielectric interfaces as a possible mechanism of RF hearing effects, Bull. Math. Biol. 43:401-413, 1981.

(70) Wilson, B.S., Joines, W.T. and Casseday, J.H.: Identification of sites in brain tissue affected by nonionizing radiation, Bioelectromagnetics 1:208, 1980.

(71) Wilson, B.S., Joines, W.T., Casseday, J.H. and Kobler, J.B.: Responses in the auditory nerve to pulsed, CW, and sinusoidally modulated microwave radiation, Bioelectromagnetics 1:237, 1980.

(72) Lebovitz, R.M.: Auditory single unit response to pulsed microwave radiation, Bioelectromagnetics 1:238, 1980.

(73) Seaman, R.L.: Auditory system model which accurately predicts microwave hearing thresholds, Bioelectromagnetics 1:251, 1980.

(74) Frey, A.H. and Gendleman, S.: Motor coordination or balance degradation during microwave energy exposure, Bull. Psychonomic Soc. 14:442-444, 1979.

(75) Frey, A.H. and Meles, D.: Motor coordination or balance differences between rats exposed or sham exposed to 60 Hz electrical fields, URSI Ann. Meeting, Seattle, WA, 455, 1979.

(76) Levin, H.S., Eisenberg, H.M., Wigg, N.R. and Kobayashi, K.: Memory and intellectual ability after head injury in children and adolescents, Neurosurgery 11:669-673, 1982.

(77) Newcombe, F.: The psychological consequences of closed head injury: Assessment and rehabilitation, Brit. J. Accident Surg. 14:111-136, 1982.

(78) Barth, J.T., Macciocchi, S.N., Giordani, B., Rimel, R., Jane, J.A. and Boll, T.J.: Neuropsychological sequelae of minor head injury, Neurosurgery 13:529-533, 1983.

(79) White, R.M.: Generalization of elastic waves by transient surface heating, J. Appl. Phys. 34:3559, 1963.

(80) Olsen, R.G.: Evidence for microwave induced acoustic resonances in biological material, Bioelectromagnetics 1:219, 1980.

(81) Olsen, R.G. and Lin, J.C.: Microwave pulse induced acoustic resonances in spherical head models, IEEE Trans. Microwave Theory and Techniques 29:1114-1117, 1981.

(82) Spiegel, R.J., Alie, J.S., Joines, W.T. and Peoples, J.F.: The measurement of mechanical motion of brain tissue exposed to low frequency electric fields, 7th Ann. Bioelectromagnetics Society Meeting, San Francisco, CA, 40, 1985.

(83) Lin, J.C.: Microwave auditory phenomenon, Proc. IEEE 68:67-73, 1980.

(84) Chou, C-K. and Guy, A.W.: Microwave induced auditory responses in guinea pigs: Relationship of threshold and microwave pulse duration, Radio Sci. 14:193-197, 1979.

(85) Frey, A.H. and Feld, S.R.: Avoidance by rats of illumination with low power nonionizing electromagnetic energy, J. Comp. Physiol. Psychol. 89:183-188, 1975.

(86) Frey, A.H., Feld, S.R. and Frey, B.: Neural function and behavior: Defining the relationship, in Biological Effects of Nonionizing Radiation, P. Tyler, ed., Ann. N.Y. Acad. Sci. 247:433-438, 1975.

(87) Medici, R.G.: Behavioral studies with electromagnetic fields: Implications for psychobiology, J. Bioelectricity 4:527-552, 1985.

(88) Thomas, J.R., Schrot, J. and Libboff, A.R.: Weak low frequency magnetic fields alter operant behavior in rats, Preprint, 1985.

(89) Frey, A.H.: Cardiac and neural effects of modulated RF energy, 23rd ACEMB, Washington, DC, 175, 1970.

(90) Lords, J.L., Durney, C.H., Borg, A.M. and Tinney, C.E.: Rate effects in isolated hearts induced by microwave irradiation, IEEE Trans. Microwave Theory and Techniques 21:834-836, 1973.

(91) Tinney, C.E., Lords, J.L. and Durney, C.H.: Rate effects in isolated turtle hearts induced by microwave irradiation, IEEE Trans. Microwave Theory and Techniques MTT-24:18-24, 1976.

(92) Galvin, M.J., Lieberman, M. and McRee, D.I.: Interaction of 2.45 GHz microwave radiation with embryonic quail hearts, Physiologist 22:42, 1979.

(93) Schwartz, J.L., Delorme, J. and Mealing, G.A.R.: Effects of low frequency amplitude modulated radiofrequency waves on the calcium efflux of the heart, Biophys. J. 41:295a, 1983.

(94) Chalker, R., Wachtel, H. and Barnes, F.: Non-contact measurement of microwave induced changes in frog heart beating rhythms, 3rd Ann. Meeting Bioelectromagnetics Society, Washington, DC, 80, 1981.

(95) Seaman, R.L., Burdette, E.C. and DeHaan, R.L.: RF radiation alteration of cardiac cell aggregate electrical parameters, Bioelectromagnetics 1:215, 1980.

(96) Reed, J.R., Lords, J.L. and Durney, C.H.: Microwave irradiation of isolated rat hearts treated with ANS blocking drugs, URSI Annual Meeting, Amherst, MA, 115-116, 1976.

(97) Olsen, R.G., Durney, C.H., Lords, J.L. and Johnson, C.C.: Low level microwave interaction with isolated mammalian hearts, in Microwave Power Symposium 1975 Proceedings, International Power Institute, Ontario, Canada, 76-78, 1975.

(98) Olsen, R.G., Lords, J.L. and Durney, C.H.: Microwave induced chronotropic effects in the isolated rat heart, Ann. Biomed. Eng. 5:395-409, 1977.

(99) Kloss, D.A. and Carstensen, E.L.: Effects of ELF electric fields on the isolated frog heart, IEEE Trans. Biomed. Eng. BME-30:347-348, 1983.

(100) Galvin, M.J., Hall, C.A. and McRee, D.I.: Microwave radiation effects on cardiac muscle cells _in vitro_, Radiat. Res. 86:358-367, 1981.

(101) Frey, A.H. and Spector, J.: Irritability and aggression in mammals as affected by exposure to electromagnetic energy, URSI Ann. Meeting, Amherst, MA, 1976.

(102) Frey, A.H. and Wesler, L.S.: Behavior modification in animals living in a relatively low intensity (3.5 kV/M) 60 Hz electric field, 3rd Ann. Meeting, Bioelectromagnetics Society, Washington, DC, 93, 1981.

(103) Frey, A.H. and Wesler, L.S.: Modification of the conditioned emotional response in animals living in a 60 Hz electrical field, Bull. Psychonomic Soc. 22:477-479, 1984.

(104) Stith, R.D. and Erwin, D.N.: Effect of exposure to microwaves on the neuroendocrine status of the rat, Bioelectromagnetics 1:236, 1980.

(105) Frey, A.H. and Wesler, L.S.: Tail pressure behaviors modification associated with microwave energy exposure, 2nd Ann. Meeting Bioelectromagnetics Society, San Antonio, TX, 202, 1980.

(106) Frey, A.H. and Wesler, L.S.: A test of the dopamine hypothesis of microwave energy effects, J. Bioelectricity 1:305-312, 1982.

(107) Frey, A.H. and Wesler, L.S.: Modification of tail pinch consummatory behavior by microwave energy exposure, URSI Ann. Meeting, Seattle, WA, 456, 1979.

(108) Frey, A.H.: Biological mediators and mechanisms for low intensity electromagnetic effects, Invited lecture at the 3rd Ann. Meeting Bioelectromagnetics Society, Washington, DC, 1981.

(109) Wesler, L.S. and Frey, A.H.: Microwave energy interactions with dopamine and opiate systems in the brain, 3rd Ann. Meeting Bioelectromagnetics Society, Washington, DC, 25, 1981.

(110) Frey, A.H. and Wesler, L.S.: Dopamine receptors and microwave energy exposure, J. Bioelectricity 2:145-157, 1983.

(111) Frey, A.H. and Wesler, L.S.: Morphine effects appear to be potentiated by microwave energy exposure, J. Bioelectricity 3:373-383, 1984.

(112) Blackman, C.F., Benane, S.G., Joines, W.T., Hollis, M.A. and House, D.E.: Calcium ion efflux from brain tissue: Power density versus internal field intensity dependencies at 50 MHz RF radiation, Bioelectromagnetics 1:277-283, 1980.

(113) Bawin, S.M. and Adey, W.R.: Calcium binding in cerebral tissue, Proc. Symposium on Biological Effects and Measurement of Radiofrequency Microwaves, HEW Publication (FDA) 77-8026, D.G. Hazzard, ed., 305-313, 1977.

(114) Lai, H., Chou, C.K. and Guy, A.W.: Psychoactive drug response is affected by acute low level microwave irradiation, Bioelectromagnetics 4:205-214, 1983.

(115) Schrot, J., Thomas, J.R. and Banvard, R.A.: Effects of 2.8 GHz microwave radiation in combination with tranquilizing drugs on fixed interval performance of rats, Bioelectromagnetics 1:203, 1980.

(116) Ashani, Y., Henry, F.H. and Catravas, G.N.: Combined effects of anticholinesterase drugs and low level microwave radiation, Rad. Res. 84:496-503, 1980.

(117) Nakas, M., Jamakosmanovic, A., Drecun, M. and Shore, M.L.: Effect of repeated 2450 MHz microwave radiations on the acetylcholinesterase activity in developing rat brain, Period Biol. 81:173-174, 1981.

(118) Jamakosmanovic, A., Nakas, M., Drecun, M. and Shore, M.L.: The levels of ATP, ADP, and AMP in rat brain during postnatal development following in utero 2450 MHz microwave radiation, Period Biol. 81:151-152, 1981.

(119) Lai, H., Horita, A., Chou, C.K. and Guy, A.W.: Acute low level microwave exposure affects drug actions, 4th Ann. Meeting Bioelectromagnetics Society, Los Angeles, CA, 100, 1982.

(120) Lai, H., Horita, A., Chou, C.K. and Guy, A.W.: Effects of low level microwave exposure on hippocampal cholinergic function, 7th Ann. Meeting Bioelectromagnetics Society, San Francisco, CA, 78, 1985.

(121) Miller, D.B., Blackman, C.F. and Ali, J.S.: Behavioral responses of morphine treated mice to ELF magnetic fields,

7th Ann. Meeting Bioelectromagnetics Society, San Francisco, CA, 54, 1985.

(122) Cox, B.: Dopamine, in Body Temperature, Drug Effects and Therapeutic Implications, P. Lomax and E. Schonbaum, eds, Marcel Dekker, New York, 213-255, 1979.

(123) Oscar, K.J. and Hawkins, T.D.: Microwave alteration of the blood brain barrier system of rats, Brain Res. 126:281-293, 1977.

(124) Albert, E.N.: Light and electron microscopic observations on the blood brain barrier after microwave irradiation, Symp. on Biological Effects and Measurement of Radio Frequency/Microwaves, Rockville, MD, 294-304, 1977.

(125) Merritt, J.H., Chamness, A.F. and Allen, S.J.: Studies on blood brain barrier permeability after microwave radiation, Radiat. Environ. Biophys. 15:367-377, 1978.

(126) Frey, A.H.: On microwave effects at the blood brain barrier, Bioelectromagnetics Society Newsletter 18:4-5, 1980.

(127) Preston, E., Vavasour, E.J. and Assenheim, H.M.: Permeability of the blood brain barrier to mannitol in the rat following 2450 MHz microwave irradiation, Brain Res. 174:109-117, 1979.

(128) Albert, E.N.: Reversibility of the blood brain barrier, Int. Symp. on Biological Effects of Electromagnetic Waves, URSI Annual Meeting, Airlie, VA, 166, 1977.

(129) Albert, E.N.: Reversibility of microwave-induced blood brain barrier permeability, Radio Sci. 14:323-327, 1979.

(130) Albert, E.N.: Current status of microwave effects on the blood brain barrier, J. Microwave Power 14:281-285, 1979.

(131) Albert, E.N. and Kerns, J.M.: Reversible microwave

effects on the blood brain barrier, Brain Res. 230:153-164, 1981.

(132) Oscar, K.J.: Determinants of brain uptake, URSI Annual Meeting, Airlie, VA, 167, 1977.

(133) Sutton, C.H.: Effects of microwave induced hyperthermia on the rat blood brain barrier, URSI Annual Meeting, Airlie, VA, 165, 1977.

(134) Ward, T.R., Elder, J.A. and Long, M.D.: A comparative study of microwave and high ambient temperature exposures on the blood brain barrier, Bioelectromagnetics 1:207, 1980.

(135) Lin, J.C. and Lin, M.F.: Power time relations of microwave induced blood brain barrier permeation, Bioelectromagnetics 1:207, 1980.

(136) Quock, R.M., Lange, D.G. and Fujimoto, J.: Microwave facilitation of methylatropine antagonism of central cholinomimetic (ACh) drug effects, Fed. Proc. 41:1576, 1982.

(137) Lange, D.G., Quock, R.M., Sadowski, J.A. and Fujimoto, J.M.: Microwave facilitation of central neurochemical effects on peripherally acting drugs, Fed. Proc. 42:1122, 1983.

(138) Coren, E. and Frey, A.H.: Modification of the blood vitreous humor barrier of the eye with microwave energy, 3rd Ann. Meeting Bioelectromagnetics Society, Washington, DC, 81, 1981.

(139) Frey, A.H.: Possible modification of the blood vitreous humor barrier of the eye with electromagnetic energy, J. Bioelectricity 3:281-284, 1984.

(140) Neelakantaswamy, P.S. and Ramakrishnan, K.P.: Microwave induced hazardous thermal stresses in the ocular lens of

the human eye, Biomed. Tech. (Berlin) 23:109-113, 1978.

(141) Neelakantaswamy, P.S. and Ramakrishnan, K.P.: Microwave induced hazardous nonlinear thermoelastic vibrations of the ocular lens in the human eye, Biomechanics 12:205-210, 1979.

(142) Frey, A.H.: Data analysis reveals significant microwave induced eye damage in humans, J. Microwave Power 20:53-55, 1985.

(143) Vodicnik, M.J., Fujimoto, J.M. and Lange, D.G.: Drug disposition changes in fetal and placental tissue following microwave exposure in mice, Fed. Proc. 41:1576, 1982.

(144) O'Connor, M.E. and Monahan, J.C.: Effects of 2.45 GHz microwaves at 1 mW/cm^2 on mouse fetuses exposed during three independent stages of development, 3rd Ann. Meeting Bioelectromagnetics Society, Washington, DC, 99, 1981.

(145) Berman, E. and Carter, H.B.: Evidence for stunting in mice offspring as a result of in utero exposure to 2450 MHz microwaves, 3rd Ann. Meeting Bioelectromagnetics Society, Washington, DC, 99, 1981.

(146) Frey, A.H.: Behavioral and Prenatal Effects of 60 Hz Fields, Final Report for the U.S. Department of Energy under Contract No. DE-AC02-80RA50293, 1983.

(147) Sanders, A.P., Rafal, M.D. and Joines, W.T.: Effect of 200 MHz and 591 MHz exposures on rat brain CP-kinase, Bioelectromagnetics 1:217, 1980.

(148) Sanders, A.P., Rafal, M.D. and Joines, W.T.: Effect of temperature on 591 MHz, CW exposures on brain energy levels, Bioelectromagnetics 1:217, 1980.

(149) Sanders, A.P., Schaefer, D.J. and Joines, W.T.: Microwave effects on energy metabolism of rat brain, Bioelectromagnetics 1:171-181, 1980.

(150) Wilson, B.S., Zook, J.M., Joines, W.T. and Casseday, J.H.: Alterations in activity at auditory nuclei of the rat induced by exposure to microwave radiation: Autoradiographic evidence using [^{14}C]2-deoxy-d-glucose, Brain Res. 187:291-306, 1980.

(151) McKee, A.D., Dorsey, C.H., Eisenbrandt, D.L. and Woollen, N.E.: Ultrastructural observations of microwave induced changes in the central nervous system of hamsters, Bioelectromagnetics 1:206, 1980.

(152) Albert, E.N., Sherif, M.F., Papadopoulos, N.J., Slaby, F.J. and Monahan, J.: Effects of RF (100 MHz and 2.45 GHz) on rat cerebellar purkinje cells, 2nd Ann. Meeting Bioelectromagnetics Society, San Antonio, TX, 206, 1980.

(153) Sheridan, J.P., Priest, R., Schoen, P. and Schur, J.M.: Techniques of raman spectroscopy applied to study the effects of microwaves upon synthetic and naturally occurring phospholipid membranes, in The Physical Basis of Electromagnetic Interactions with Biological Systems, Available through NTIS, Springfield, VA, Document No. AD A051218, 1978.

(154) Stevens, C.F.: Interactions between intrinsic membrane protein and electric field, Biophys. J. 17:264a, 1977.

(155) Schwartz, G.: Mechanisms of interaction: Basis of direct electrical effects on biological function, in Interactions of Electromagnetic Waves, URSI Ann. Meeting, Washington, DC, 28, 1981.

(156) Frohlich, H.: What are non-thermal electric biological effects? Bioelectromagnetics 3:45-46, 1982.

(157) McLeod, B.R. and Liboff, A.R.: Dynamics of biological ions in low frequency multi field configurations, 7th Ann. Meeting Bioelectromagnetics Society, San Francisco, CA, 38, 1985.

(158) Wachtel, H.: A model for predicting minimum RF threshold energy pulse widths for evoking fast neuronal effects, 3rd Ann. Meeting Bioelectromagnetics Society, Washington, DC, 27, 1981.

(159) Wachtel, H., Barnes, F.S., Stavinoha, W.B., Modak, A.T. and Deam, A.P.: Microwave induced hypokinesis -- effects of varying pulse intensity and width, 3rd Ann. Meeting Bioelectromagnetics Society, Washington, DC, 28, 1981.

(160) Brown, P.V.K. and Larsen, L.E.: Differing effects of pulsed and CW microwave energy upon nerve function as detected by birefringence measurement, IEEE Trans. Microwave Theory & Technique 28:1126-1133, 1980.

(161) McRee, D. and Wachtel, H.: Pulsed microwaves have phase independent effects on nerve vitality, Bioelectromagnetics 1:206, 1980.

(162) McRee, D.I. and Wachtel, H.: The effects of microwave radiation on the vitality of isolated frog sciatic nerves, Radiation Res. 82:536-546, 1980.

(163) Medici (Gavalas), R.J., Walter, D.O., Hamer, J. and Adey, W.R.: Effect of low level, low frequency electric fields on EEG and behavior in _Macaca nenestrina_, Brain Res. 18:491-501, 1970.

(164) Kaczmarek, L.K.: Cation binding models for the interaction of membranes with EM fields, Neurosci. Res. Program Bull. 15:54-60, 1977.

(165) Bawin, S.M., Sheppard, A.R. and Adey, W.R.: Models of long range order in cerebral macromolecules: Effects of ELF, VHF, and UHF fields in calcium binding, Int. Symp. on Biological Effects of Electromagnetic Waves, URSI Ann. Meeting, Airlie, VA, 100, 1977.

(166) Lin-Liu, S. and Adey, W.R.: Low frequency amplitude modulated microwave fields change calcium efflux rates

from synaptosomes, Bioelectromagnetics 3:309-322, 1982.

(167) Blackman, C.F., Elder, J.A., Weil, C.M., Benane, S.G., Eichinger, D.C. and House, D.E.: Induction of calcium ion efflux from brain tissue by radiofrequency radiation: Effects of modulation frequency and field strength, Radio Sci. 14:93-98, 1979.

(168) Blackman, C.F., Benane, S.G., Joines, W.T., Hollis, M.A. and House, D.E.: The influence of modulated 50 MHz radiation on calcium ion efflux from brain tissue, 2nd Ann. Meeting Bioelectromagnetics Society, San Antonio, TX, 212, 1980.

(169) Blackman, C.F., Benane, S.G., Elder, J.A., House, D.E., Lampe, J.A. and Faulk, J.M.: Induction of calcium ion efflux from brain tissue by radiofrequency radiation: Effect of sample number and modulation frequency on the power density window, Bioelectromagnetics 1:35-43, 1980.

(170) Blackman, C.F., Joines, W.T. and Elder, J.A.: Calcium ion efflux induction in brain tissue by radiofrequency radiation, ACS Symp. Ser. 157:299-314, 1981.

(171) Blackman, C.F., Benane, S.G., Joines, W.T. and House, D.E.: Effects of ELF fields between 1 and 120 Hz on the efflux of calcium ions from brain tissue, _in vitro_, 5th Ann. Meeting Bioelectromagnetics Society, Boulder, CO, 28, 1983.

(172) Joines, W.T. and Blackman, C.F.: RF-induced calcium ion efflux from brain tissue: Influence of the carrier frequency, Bioelectromagnetics 1:213, 1980.

(173) Joines, W.T. and Blackman, C.F.: Power density, field intensity, and carrier frequency determinants of RF energy induced calcium ion efflux from brain tissue, Bioelectromagnetics 1:271-275, 1980.

(174) Spiegel, R.J., Joines, W.T. and Blackman, C.F.: Calcium

efflux from isolated brain tissue: Is it caused by an electromagnetically induced pressure on the brain? 4th Ann. Meeting Bioelectromagnetics Society, Los Angeles, CA, 31, 1982.

(175) Blackman, C.F.: Effects of modulated radiofrequency electromagnetic radiation on the efflux of calcium ions with brain tissue in vitro, J. Electrochem. Soc. 129:132C, 1982.

(176) Blackman, C.F. and Wilson, B.S.: Distribution of label in studies on the effects of nonionizing radiation on the association of calcium ions with brain tissue, 5th Ann. Meeting Bioelectromagnetics Society, Boulder, CO, 73, 1983.

(177) Tenforde, T.S.: Thermal aspects of electromagnetic field interactions with bound calcium ions at the nerve cell surface, J. Theoret. Biol. 83:517-521, 1980.

(178) Schlagel, C.J. and Wiktor-Jedrzejczak, W.: Possible humoral mechanism of 2450 MHz microwave induced increase in complement receptor positive cells as studied using a modified diffusion chamber culture technique, Bioelectromagnetics 1:247, 1980.

(179) Wiktor-Jedrzejczak, W., Schlagel, C.J., Ahmed, A., Leach, W.M. and Woody, J.M.: Possible humoral mechanism of 2540 MHz microwave induced increase in complement receptor positive cells, Bioelectromagnetics 2:81-84, 1981.

(180) Schlagel, C.J., Sulek, K., Ho, H.S., Leach, W.M., Ahmed, A. and Woody, J.N.: Biological effects of microwave exposure II. Studies on the mechanisms controlling susceptibility to microwave induced increases in complement receptor positive spleen cells, Bioelectromagnetics 1:405-414, 1980.

(181) Liddle, C.G., Putnam, C.G., Ali, J.S., Lewis, J.Y., Bell, B., West, M.W. and Lewter, O.H.: Alteration of

circulating antibody response of mice to 9 GHz pulsed microwaves, Bioelectromagnetics 1:397-404, 1980.

(182) Schlagel, C.J. and Yaffe, L.J.: Genetic studies of 2450 MHz microwave induced augmentation of complement receptor bearing lymphocytes, 4th Ann. Meeting Bioelectromagnetics Society, Los Angeles, CA, 55, 1982.

(183) Schlagel, C.J. and Ahmed, A.: Evidence for genetic control of microwave induced augmentation of complement receptor bearing B lymphocytes, J. Immunol. 129:1530-1533, 1982.

(184) RamaRao, G.V., Tompkins, W.A.F. and Cain, C.A.: Effects of microwave exposure on the hamster immune system, 5th Ann. Meeting Bioelectromagnetics Society, Boulder, CO, 44, 1983.

(185) Swicord, M.L. and Davis, C.C.: Microwave absorption of DNA between 8 and 12 GHz, Biopolymers 21:2453-2460, 1982.

(186) Edwards, G.S., Swicord, M.L. and Davis, C.C.: Microwave absorption characteristics of highly purified E. coli DNA, 5th Ann. Meeting Bioelectromagnetics Society, Boulder, CO, 3, 1983.

(187) Liboff, A.R. and Homer, L.D.: Eddy current effects on DNA synthesis at geomagnetic intensities, 5th Ann. Meeting Bioelectromagnetics Society, Boulder, CO, 3, 1983.

(188) Kohli, M. and VanZandt, L.L.: Microwave absorption by folded DNA chains, Biopolymers 21:1399-1410, 1982.

(189) VanZandt, L.L., Kohli, M. and Prohofsky, E.E.: Absorption of microwave radiation by DNA double helix in aquo, Biopolymers 21:1465-1468, 1982.

(190) Ruff, M.R., Pert, C.D., Weber, R.J., Wahl, L.M., Wahl, S.M. and Paul, S.M.: Benzodiazepine receptor -- Mediated chemotaxis of human monocytes, Science 229:1281-1283,

1985.

(191) Angeletti, R.H. and Hickey, W.F.: A neuroendocrine marker in tissues of the immune system, Science 230:89-90, 1985.

(192) Lynch, G. and Baudry, M.: The biochemistry of memory: A new and specific hypothesis, Science 224:1057-1063, 1984.

24
Electromagnetic Energy and Cataracts

MILTON M. ZARET

Director of Research
Zaret Foundation
1230 Post Road
Scarsdale, New York
and
Clinical Professor of Ophthalmology
New York Medical College
Valhalla, New York

INTRODUCTION

Ophthalmology has frequently been at the leading edge of state-of-the-art technology by incorporating both contemporary scientific electromagnetic theory and advanced electronic research techniques into the pragmatic practice of clinical medicine. Concomitantly, eye injury can provide the first sign of hazards.

The biological effects of laser/maser radiation were first reported in 1961 (1). In 1965, Zaret (2) described how the special features of various laser/maser irradiation could be applied to ophthalmology and other specialties of medical practice for therapeutic, diagnostic, or research purposes, and most of these applications have now been realized. Also in 1965, Zaret (3) described potential health hazards for humans, including cataractogenesis, that could occur as a consequence of direct or indirect exposure either to laser/maser irradiation itself, or to the electronic smog created by either the laser/maser generating circuitry or other electronic equipment.

EMBRYOLOGICAL AND ANATOMICAL FACTORS

It is necessary to describe the embryology and anatomy of

the intraocular lens to permit an understanding of the cataractogenic process in humans. The lens is composed of a non-rigid colony of cells, known as the lens substance, enclosed within a tight fitting elastic-like membrane known as the lens capsule. The lens is located within the anterior segment of the eye immediately behind the iris and its pupillary opening, and is approximately 1 cm in diameter and 0.5 cm at its widest depth along its optical axis.

The cells of the lens substance are not totally transparent. Instead, like other embryonal cells, their protoplasm has a translucent appearance so that the lens substance is visible when viewed in a dark surround while being directly illuminated as by a slit-beam of light at an oblique angle of incidence to the viewing axis of a biomicroscope. In vivo, the aqueous in front of and the vitreous behind the lens are ordinarily optically empty, and thereby serve as the dark surround. The obliquely placed slit-beam of light provides an illuminated, thin cross-section of visible lens substance through the thickness of the lens. The various depths, from either the anterior or posterior surface under observation, can be determined by the focal plane of the biomicroscope. Relative distances from the anterior or posterior refringent edge of the optical beam can also be determined.

Under normal conditions the capsule is not visible by slit-lamp examination because it is thin (a few microns thick), and not relucent. However, its site is always recognizable by slit-lamp examination because it is demarcated by the edge of the lens where the light beam is entering or exiting. The normal capsule is elastic and it exerts a compressive force on the lens substance which can thereby be molded to permit focussing an image of regard onto the retina. The capsule also acts as a Donnan's membrane permitting selective permeation of dissolved metabolites into and waste products out of the lens.

The capsule elasticity compresses the lens cells against each other relatively compactly, and the sparse extra-cellular fluid is ordinarily not perceivable. Occasionally a small sphere of fluid, termed a vacuole, is formed and may be observable by slit-lamp biomicroscopy for a few weeks or months. Such a transient microscopic vacuole represents the largest collection of extra-cellular fluid ever found in the normal lens.

Also occasionally, dot-like opacities located in the lens substance can be observed by slit-lamp examination. These, too, are usually present only for weeks or months and are believed to represent salt precipitates temporarily present in the extracellular fluid. Although vacuoles and dot opacities can both be observed by slit-lamp examination as if they were lens substance defects, neither are, of themselves, relatable to cataractogenesis.

From its earliest recognizable embryonic stage of development as a discrete organ of accommodation, at about the one-month stage of fetal evolution, the lens has all of the elements that it will ever contain. It has become sequestered from the rest of the eye and separated from direct connections to the circulatory system so that, thereafter, no cells can enter or leave it. Although still microscopic in size, the lens is predestined to grow throughout life only by multiplication and differentiation of the cells contained within its capsule.

Soon after this stage of its evolution, the life-long organization of the lens becomes recognizable. Lining and in intimate direct contact with the entire anterior capsule is a unicellular sheet of cuboidal epithelial cells. Filling the remainder of its volume are the primitive or primary lens fiber cells (lens fibers). At the equator of the lens, new lens fiber cells, termed secondary lens fibers, are formed by differentiation of the peripherally located (around the equatorial circumference) epithelial cells. As the secondary lens fibers are forming, they become elongated with one end extending posteriorly into the potential space between the pre-existing lens fibers and the posterior capsule, and the other end extending anteriorly into the potential space between the pre-existing lens fibers and the posterior (inner) surface of the unicellular sheet of cuboidal epithelial cells that separate the lens fibers from contact with the anterior capsule. Thus, the oldest lens fiber cells are located in the center of the lens and the newest are at its periphery. When partial opacification of a small region of the lens substance occurs, it may be possible to determine the approximate time during life when the event took place by the depth of the opacity from the surface of the lens.

Secondary lens fibers, although retaining their original dimensions in cross-sectional diameter, nevertheless continue to grow in length until, near the anterior and posterior poles of the lens, they abut against the ends of similar lens fibers coming from the opposite direction. When that occurs, axial lens fiber growth stops, apparently because of contact inhibition.

When new lens fiber formation occurs the sheet of epithelial cells enlarges concomitantly by cell multiplication. The cells retain their embryonal size and cuboidal shape, and each cell remains in contact with the capsule anteriorly so that the areal enlargement of the unicellular sheet continuously extends to the equatorial region of the enlarging lens. There, the most peripheral epithelial cells continue in the process of forming new secondary lens fibers throughout life. In this fashion, the growth of the lens continues at a slow rate, by the addition of new lens fibers.

With continuing addition of lens fibers, suture lines occur at the anterior and posterior poles of the enlarging lens where contact inhibition takes place. Some sutures become visible by slit-lamp biomicroscopy because the tips of the elongated fibers do not make complete contact with other fibers, and the space becomes filled with extra-cellular fluid, which has a different index of refraction from lens fibers. Such visualizations of the sutures can appear to be defects in the lens substance.

Once fully formed, lens fibers can be compressed tightly against each other. The older, more centrally located lens fibers normally appear to be more transparent or less relucent than the younger, more peripherally located lens fibers. This results in a difference in luminosity of lens fibers so that at different depths, different regions can be identified. Generally, the lens can be divided into two regions according to luminosity, a nucleus at its center and a cortex at its periphery. At either site, side-to-side adhesions or axial adhesions of small groups of adjacent lens fibers can occur. They give rise to the slit-lamp appearance of banding and riders, respectively. When larger areas are affected, a localized haze can be seen.

All of the above-mentioned types of lens imperfections or defects were diligently searched for as part of the ophthalmic examinations I performed on several thousand scientists and technicians who worked in radiant-energy environments. For present purposes, the principal concern will be with alterations in the appearance of the central or nuclear portion of the lens, the posterior polar subcapsular sutural region of the lens, and the capsular surface or refringent edge of the lens. Other pathological features of the lens will also be discussed.

PHYSIOLOGICAL OPTICS

It is appropriate to define the manner in which the lens is involved with normal visual function. The lens is suspended by circumferentially located zonules attached to the peripherally located ciliary muscle which controls focussing by its state of tonus. The lens lies immediately behind and in the plane of the iris whose posterior surface rests against the suspended lens's anterior capsular surface. The center of the pupillary aperture in the iris is practically co-axial with the optical axis of the lens.

The posterior surface of the iris is lined with a layer of densely pigmented cells which also line the inner surface of the ciliary body and the outer portion of the retina. These pigmented tissues are comparable to the black, light-tight lining of a camera; they minimize the entry of stray light that could interfere with the quality of perceived images. Thus, all of the light useful for vision enters the eye through the pupil.

Under ordinary conditions of vision, the object of regard must be illuminated sufficiently so that its reflected light can be imaged onto the macular portion of the retina. Simultaneously, the remaining ambient light entering the eye from the remainder of the field of vision must be kept below intensity levels that would interfere with the discriminatory functions of the macula. A major method for adjusting the amount of light inside the eye is by the tonus of the iris musculature controlling pupil size. Thus, the pupil does not have a fixed diameter, but instead fluctuates, averaging approximately three millimeters in diameter.

The major refracting surface of the eye is the cornea, but the fine focussing is accomplished by alteration of the curvature of the lens via the process of accommodation. It is mediated by the elasticity of the lens capsule via the tension maintained on it.

Thus, two major factors that affect visual acuity are the ambient background light within the inner eye, and the refractive state of accommodation.

CATARACTS

The term "cataract" has different meanings in different contexts. Many scientists (including a few ophthalmologists) label as a cataract any lens defect acquired by a laboratory animal during the course or as a consequence of an experiment. On the other hand, there are a few ophthalmologists who maintain that a lens opacity in humans is a cataract only if it reduces visual acuity. Visual acuity is only one of many measures of visual function. It is inadequate and frequently misleading to rely on visual acuity alone as the measure of cataracts. For the reasons given below, a better, more clinically useful definition is that a cataract is any type of lens defect that interferes with visual function (4).

Light rays emanating from an illuminated object of regard enter the lens only through the pupillary aperture, and these occupy only the axial and paraxial projection of the pupillary area through the lens. In other words, only the central axial 20% of the lens volume is used for visual acuity determination. The remaining 80% of the lens substance lies in the relative penumbra and umbra of the iris while viewing an object of regard. Under these conditions, if there are any opacities of the lens substance located in the penumbral or umbral zone and such opacities are illuminated by a peripherally located light source, then those opacities become secondary luminaires within the eye thereby producing glare and reducing visual acuity. Thus, visual acuity may be excellent under the contrived darkroom conditions usually used for testing, but may be considerably impaired under the ordinary life situation of driving at night or into the direction of a low-lying sun.

Another optical factor to consider is the nodal point of the lens. This lies in the lens substance, along the optical axis, near the posterior pole in the region of the posterior suture. As will be discussed later, this is the site where posterior polar cataracts occur. Prior to opacification of the lens substance, extra-cellular fluid can collect at the nodal point. This can produce a prismatic optical effect at the interface between the lens fibers and the extracellular fluid, resulting in monocular diplopia or polyplopia with each of the multiple images having a visual acuity of 20/20.

Another factor to be considered is the state of refraction for distance vision. Ordinarily, corrective lenses are prescribed to provide a visual acuity of 20/20, which is the pragmatic equivalent of placing the retina in focus with infinity. Abnormal hydration of the lens causes it to become swollen and an infrequently encountered stage of this is known as intumescence. This results in an abnormal increase in convex curvature which then induces a relative myopia. Under that circumstance, visual acuity can still be corrected to 20/20, but a change in the basic eyeglass prescription is needed. Subjectively, if a patient were nearsighted he would think his vision got worse, if the patient were emmetropic (normal vision) he would become nearsighted, and if the patient were farsighted he would think his vision got better.

An additional factor, capsulopathy in its early stages, may also induce an aberration in visual function without materially affecting visual acuity. In this example, consider that the posterior refringent surface of the lens is defined by the capsule, and our immediate concern is limited to that portion of the capsule situated along the optical axis of the eye and lying within the areal projection of the pupil. Should this part of the capsule, which is only a few microns thick, exhibit the optical qualities of becoming recognizable by slit-lamp biomicroscopy as if it were roughened, thickened and faintly opacified, it would behave optically as if it were a microscopically glazed and thin diffraction filter. As such, it would not reduce visual acuity, but the patient nevertheless would describe vision as if it had a misty or hazy veiling even though visual acuity was still correctable to 20/20.

Although more factors could also be discussed concerning abnormalities in visual function that can occur without a reduction of visual acuity, a final item to be presented here concerns anomalous change in transmission of color through the lens. The chromatic content of an illuminated object being imaged on the macula is ordinarily the same for both eyes of an individual so that there is no perceptible difference in color value when comparing one eye to the other. Just as irradiation can cause all of the other examples listed above, it can also cause a change in the colorific value of some lens proteins, thereby causing that lens to act as a partial color filter and thus affect the perceived hues of the observed object. Should the lens of one eye receive more irradiation producing this effect than the other eye, then the patient's color sense can be different for the two eyes, a condition I termed anisometachromatopsia (5,6).

The net result of examining the factors of pathophysiological optics of the lens, as described above, is to recognize that there are many anomalies of visual function in addition to visual acuity. Any one or more of these factors could be due to an early stage of radiant-energy injury, and thereby be the harbinger of an evolving incipient cataract. Prevention of an advanced stage of radiant-energy cataract can frequently be achieved. However, that usually depends upon establishing the diagnosis of lens injury early, before visual acuity has degraded appreciably. This is especially true where, while visual acuity is still correctable to 20/20, there has been an interference with some other function, such as anisometachromatopsia, intermittent misty or hazy vision, temporary blindness from a light in the field of vision such as oncoming lights during night driving, or occasional diplopia or polyplopia. In such cases, and especially if the lens pathology is present only in one eye, the probability exists of either preventing a loss of visual acuity or significantly delaying its onset or the future need for cataract surgery by identifying the irradiation source and protecting the patient from additional exposures (7).

The difficulty of relying solely on visual acuity as a test for degradation of visual function was demonstrated clearly by Zaret and Snyder (8) in 1977 when they chronicled how an air traffic controller with 20/20 visual acuity in each eye repeatedly put aircraft into mid-air collision courses because of other disturbances in visual function. Also in 1977, Zaret and Snyder (9) presented evidence in an air traffic controller that the earliest symptom of a radiant-energy cataract was a transient visual problem, such as looking through a misty fog or wet glass. This finding of altered visual function antedated by several years the subsequent development of reduction in visual acuity. Moreover, in 1979 Zaret (10) described the case of an airline pilot who, despite having a visual acuity of 20/20 in each eye, nevertheless exhibited polyplopia during the landing of a commercial airplane.

ELECTROMAGNETIC ENERGY AND CATARACTOGENESIS

Duke-Elder discovered that exposure to radiant energy, including long-term direct exposure to sunlight, was either the causal or a contributory etiological factor for most cataracts acquired after birth (11). His data antedated the electronic revolution, and my data has been collected after its inception, so that there have been both qualitative and quantitative changes in the composition of electronic smog. Nevertheless, my data supports his original thesis, adds detailed refinement to some of his observations, and extends our knowledge into other regions of the electromagnetic spectrum that formerly were not, but now are, clinically important.

Much of what appears here is in large part the result of my unique personal experience, during the past 35 years, investigating some aspects of radiant-energy cataractogenesis under both laboratory and clinical research conditions. An important aspect of this work is the unprecedented opportunity it afforded me to investigate actual situations where cataract problems were occurring. This permitted discovery of differing types of cataractogenesis at varying stages of their evolutions, institution of preventive measures designed to reduce or eliminate additional exposures, and follow-up investigations by repeated ophthalmological surveillance of the subjects to determine the subsequent course of the cataractogenic process and the effectiveness of the preventive measures.

The following account represents an integration of some classical teaching that I found to be valid with some new information. The reader should bear in mind that there are a number of different mechanisms that can ultimately result in cataracts, and that several different pathways may be operant simultaneously or sequentially. Some specific mechanisms will be presented as being primarily an ionizing radiation effect, an ultraviolet radiation effect or a nonionizing radiation effect, while recognizing that none of these mechanisms occurs in the total absence of the others.

IONIZING RADIATION

One concept of ionizing radiation injury of the lens involves mutagenic changes that are usually expressed as aberrant lens fiber formation, resulting in some of the newly forming lens fibers become aborted in that process. The aborted cells either drift or are propelled by succeeding normally developing lens fibers into the posterior subcapsular space, and after 3-5 years start to collect at the posterior pole of the lens where they can be recognized as a posterior polar cataract.

ULTRAVIOLET RADIATION

One currently popular concept of how ultraviolet radiation may induce cataractogenesis is based upon partial, _in situ_ denaturation of cytoplasmic protein contained within lens fiber cells. The evidence is derived from _in vitro_ analysis of extracted lens substance protein where a chromophoric change to a brown color can occur following exposure to ultraviolet radiation. This was the theorized rationale whereby the nuclear sclerosis type of cataract, which is ordinarily gray, sometimes turns brown. Nuclear sclerosis and its occasional derivative, brunescent cataract, are considered by many ophthalmologists to be stages of the aging process and, as such, are ordinarily described as senile cataracts. However, senescence can occur without any evidence of nuclear sclerosis and I have examined many older patients who never exhibited nuclear sclerosis or other evidence of cataractogenesis. So, if nuclear sclerosis is a sign of senility, it is a sign of abnormal aging and abnormal aging, itself, can be an irradiation effect.

NONIONIZING RADIATION

Nonionizing radiation cataract _per se_ has evolved as a recognizable clinical entity only since World War II, which fueled the electronic revolution and led to the subsequent explosive proliferation of military, intelligence, industrial, communication, medical, educational, consumer product, and amusement-related electronic devices. This type of cataract is dependent upon exposure to radiant energy of any frequency in the nonionizing spectrum. Its distinctive feature is lens capsule injury.

Capsulopathy, at its inception, is recognizable by slit-lamp biomicroscopy as the appearance of roughening, thickening and opacification at the refringent edge of the lens. The fact that acquired capsular cataract could serve as a signature of nonionizing radiofrequency injury of the lens, especially when all other differential diagnostic criteria were also satisfied, was discovered and first reported in 1964 by Zaret (12), and independently verified by Bouchet and Marsol (13) in 1967.

IRRADIATION RELATIONSHIPS

When a cataract is produced by exposure to a specific spectral band of electromagnetic energy, it is common to find other changes in the eye. These include posterior polar cataract induced by ionizing radiation, brunescent cataract from ultraviolet radiation, and capsulopathy, an early stage of capsular cataract from nonionizing radiation.

These distinctions are based on clinical findings in humans who, during their lifetimes, have experienced variable exposures to a wide variety, both qualitatively and quantitatively, of irradiations. To some degree, prior irradiations are both additive and cumulative. A more detailed discussion of many of these factors as well as a classification generally suitable for any type of radiant energy cataract is given elsewhere (4).

CATARACTS INDUCED BY NONIONIZING RADIATION

It was noted earlier that lens fibers are generally packed tightly together, side by side with parallel axial alignment. This structural regularity accounts for not only the optical-quality transmission of images to the retina, but also for the normal appearance of the lens itself. However, following irradiation, because adjacent contiguous lens fibers are kept in a practically immobile contact with each other, the cell walls can adhere to each other as if some contact points between these cells have agglutinated. Then, should hydration occur in such a localized region of the lens, thereby converting the potential extracellular space between lens fibers into a real space, this could result in the tissue appearing as if it were a syncytium because the regular parallelism of the adjacent lens fibers had been altered.

Thereafter, whenever this syncytial region of the lens substance is illuminated, it causes internal scatter and reflection of the illuminating light, resulting in a misty or hazy opacified appearance in that region of the lens. Such hydration in the lens substance can be the initial stage of cataractous change, and it ordinarily preceeds protein coagulation. When this type of opacification results from hydration alone, the opacity will seem to disappear when the slit-lamp beam and viewing optics are aligned co-axially.

Ordinarily, the space between the posterior capsule and the posterior surface of the lens substance is optically clear. But it is an area where aberrant lens fibers can collect, frequently in a decaying or liquefying state. Their proteolytic products cause opacification of adjacent lens fibers, resulting in clinically recognizable posterior subcapsular opacities. When such opacification interferes with visual function, it is termed posterior subcapsular cataract (PSC). This form of PSC can frequently be observed in its early stages to be separated from and not in direct contiguous contact with the capsule when it arises from exposure to ionizing irradiation.

Another form of posterior subcapsular opacification occurs in which the cataractogenic process begins at the capsule itself. This process is a capsulopathy and it also can occur in the anterior capsule. Eventually it can result in reduced visual function, at which time it may be recognized as a capsular cataract.

Another feature of capsulopathy is that patchy areas of the capsule appear to lose their ability to function effectively as a selectively permeable membrane, thereby permitting a greater than usual local fluid transudation. This results in the accumulation of vesicles lying in contact with the capsule. Vesicles differ from vacuoles in many ways such as being macroscopic in dimension. They are usually present in large numbers, and rarely disappear completely. Vesiculation is another sign of abnormal hydration and it usually leads to cataractous changes in the lens substance. In contrast, vacuoles are seldom found in clusters, are usually resorbed, and generally do not result in cataracts.

Capsulopathy exhibits a variable slit-lamp appearance during its development. It first appears as if small, scattered regions of the capsule are roughened and thickened. Later, as the number and size of these areas increase, confluence of regions occurs and opacification of the capsule becomes evident and then denser. This process leads to a variety of localized forms that are variously described as a honey-comb (especially when accompanied by vesiculation), brush-mark, lace-cloth, spider web, breadcrumb, peppered-surface, depending upon the imaginative perception of the observer. Eventually, a coalescence of some of the areas occurs as the process becomes more wide-spread, resulting in sheet-like areas several millimeters long. However, the entire lens capsule never becomes uniformly or homogeneously dense white. Instead, even when almost the entire lens capsule appears to be involved, some areas of clear capsule can still be recognized. Most ophthalmologists would recognize this end stage as capsular cataract, formerly an extremely rare condition, but now appearing to increase in prevalence commensurate with the electronic revolution.

Acquired posterior polar cataracts usually originate as a small volume of fluid in the subcapsular space. Although it appears to be optically empty, like a vacuole or a vesicle, it can differ because it is not spherical. Moreover, it acquires an irregular outline and may become somewhat lobulated as it enlarges and becomes invaginated into the lens substance. Subsequently, as a consequence of new lens fibers encompassing it from behind, it becomes partially incorporated within the lens substance at the posterior suture in or near to the optical axis of the lens. (It is sometimes referred to as an "oil drop" cataract, but this is misleading because that conjures a false impression of the composition of the fluid.) It is primarily an extremely localized area of phakohydritis. Should it spread, as it usually does, more and more of the adjacent lens fibers will undergo localized lens protein coagulation as the cataractous process enlarges. Ultimately, a disturbance in visual function occurs and occasionally this may progress to a stage where surgery is required. Because it is located near the optical nodal point of the lens, it can give rise to monocular diplopia or polyplopia due to prismatic effects.

Still another type of lens substance opacity that can be acquired following irradiation is nuclear sclerosis. It is often erroneously referred to as a "senile" type of cataract because it has long been known to increase in prevalence with advanced chronological age. However, as discussed earlier, nuclear sclerosis especially in a pre-senile age group (i.e., under age 65-70) can itself be considered to be a sign of premature, abnormal aging and pathological aging, by itself, can be a sign of radiation injury.

Additional types of lens-substance defects can also be observed by slit-lamp biomicroscopy. These include opacification along the linear axis of adjacent lens fibers, and a change in relucency over a larger segment of the lens which is usually pie-shaped with the apex towards the optical axis. These defects depend upon the geometry of the lens fibers, and on their state of hydration and coagulation.

The role of radiation-induced phakohydritis (as well as the absence of lens inflammation, termed phakitis) was described in 1974 (4). Here, a few pertinent factors will be reviewed relating to whether the hydration was immediate or delayed, localized or wide-spread, and reversible or irreversible.

It is rare for the entire lens to be irradiated homogeneously, or for the effects to be present uniformly throughout the lens. Thus, intumescence of the entire lens (hydrops) is an extremely uncommon finding and, when present, is usually due to nonionizing radiation, much less frequently due to ionizing radiation and, in my experience, rarely, if ever, due to ultraviolet radiation.

The sudden appearance of intumescence ordinarily signifies that the lens has been irradiated recently. Usually, intumescence is localized to specific regions of the lens such as to a quadrant following beta radiation therapy for recurrent pterygium, to a partial posterior capsular vesiculation following radiofrequency irradiation, and to posterior polar hydration following X-ray irradiation. In these instances, partial intumescence ordinarily becomes apparent from months to years following exposure. Once established, however, it usually remains irreversible. Nuclear sclerosis is a form of partial intumescence, just sufficient to produce a syncytial-appearing arrangement of the lens fibers of the nucleus.

Generalized intumescence of the lens can be caused by many factors, such as radiation, trauma, toxic substances, and metabolic sources, and it can follow a variable course from being effervescent to being irreversible. However, in most types of phakohydritis, the etiological factors can usually be estimated if searched for diligently.

DIAGNOSTIC CRITERIA

It is an error to ascribe all of the perceived biological effects to only one physical factor simply because it may be easy to detect and measure that factor (14). For example, consider ionizing radiation where biological effects of concern such as mutagenesis or cataractogenesis are not apparent until years after the suspect exposure, and where the measured parameter and presumed sole etiological factor is ion-pair formation. Because the physical models and the physicists' mathematics were so elegantly constructed, it appears that an important physical-biological phenomenon escaped notice. Namely, in addition to stripping electrons, the irradiation also pumps other electrons into higher energy orbits without ejecting them from their atoms. When the higher-energy-level electrons revert to their original, stable energy state, emission of nonionizing radiation occurs. In other words, some of the incident ionizing radiation energy is converted, _in situ_, into nonionizing radiation. Nonionizing radiation can also be cataractogenic.

It has frequently been recognized that cataractogenesis itself can proceed via any one or more of several pathways, singly, sequentially, or in various simultaneous combinations. Catractogenic exchange energy relationships are not so simplistic as to be describable solely in terms of electron pairs created by ionizing radiation, or as biochemical degradations stimulated only by specific spectra such as that of ultraviolet radiation.

Even more misleading is the concept that, for humans, the cataractogenic potential of chronic exposure is relatable only to some contrived temperature quotient. Microwave cooking may indeed result in acute cataractogenesis due to a thermal injury. Fortunately, however, such instances are rare, and are usually so dramatic in occurrence as to be diagnosed readily; but one should never accept an overt burn as the threshold for radiation injury, *per se*. Instead, the ordinarily encountered human cases of nonionizing radiation cataractogenesis are delayed in onset, and are caused by repeated or chronic irradiation at field intensities either too low to be perceived directly, or too brief in duration to result in recognizable elevation in core or local temperature.

The diagnostic standard applied to human pathology is not scientific certitude, but rather the test of medical reasonableness. Nevertheless, the first cluster of human nonionizing radiation cataract cases that I discovered had an exposure commonality and a specific end-point pathological criterion response, lenticular capsulopathy, that fulfilled all the qualitative criteria of scientific certitude better than any past or proposed experimental protocol. The relevant cases involved individuals where the viewing (and exposure) was performed only with the dominant eye through a peep-hole, and where the contralateral non-exposed eye then served as a control for comparison. This provided the best possible investigative model, had it been done intentionally, for determining the specific cataractogenic effects of radiation in humans.

From other clusters of cases exhibiting similar pathological changes in the lens capsule, two important additional discoveries subsequently became apparent. The first was that nonionizing radiation was cataractogenic throughout its spectral distribution. This includes leakage from microwave ovens (7), from cathode-ray tube display units initially used by air traffic controllers (8-10), from those units now in widespread use by office workers and school children (5,6), and from radio transmissions such as citizens-band radios and hand-held walkie-talkies as used by policemen and firemen.

The other important observation was the total absence of any clinical relationship between delayed-appearing cataractogenic effects of chronic exposure to nonionizing radiation and intra-ocular temperature. Thus, the health-safety criteria for nonionizing radiation in the United States, all of which are based on presumptive thermal effects, are meaningless. The appropriate methodology for monitoring human injury is different from that which has been generally applied. When a lens is injured by irradiation, it will ultimately display a pathological response that may proceed to cataract formation despite the inappropriate assumptions that underly the present applicable regulations.

Cataracts are not born fully formed and, as mentioned above, may evolve via several different pathways. The PSC is an example. At its inception, minute gray opacities take origin in the lens substance at a few scattered loci near, but separated from, the capsule. These may remain stationary for long periods of time without interfering with vision, but they ordinarily increase in number, size and density and eventually interfere with visual function. They may be diagnosed as PSC by ophthalmologists when these are first observed or when they first interfere with vision. Since ophthalmologists ordinarily do not make such precise distinctions between opacities and cataracts located in the posterior subcapsular region of the lens substance, it is understandable that interpretation of the results of clinical investigations may be confusing.

An epidemiological study of lenticular imperfections in the eyes of a sample of microwave workers compared to a control population was undertaken because cataracts were known to be occurring in microwave workers, especially radar technicians and scientists working in both military and civilian occupational environments (15). Prior to this study, there were two important unknowns -- what was the earliest indicator of impending radiofrequency cataractogenesis, and what exposure time duration for occupational environments was required before that pathological process reached a stage where the diagnosis of cataract could be established with medical certainty? I found that posterior subcapsular cataract, as diagnosed by most ophthalmologists, was an indicator, and it would take about 20 years before that became evident. A recent report by Hollows and Douglas (16) confirmed both findings: posterior subcapsular cataract was found to be the earliest indicator of impending, occupationally-induced, nonionizing radiation cataractogenesis.

DISCUSSION

There is no question that chronic exposure to nonthermal levels of both ionizing and ultraviolet radiations can be cataractogenic, and that no data exist upon which meaningful human safety can be formulated for either form of radiation. The situation regarding nonionizing radiation is exactly the same.

The only published data that implies otherwise was provided by Colonel Appleton, then the chief ophthalmologist in the United States Army and staff ophthalmic consultant to the Army's Surgeon General, who together with McCrossan, an optometrist at Fort Monmouth, New Jersey, conducted an ophthalmic epidemiological study in which they claimed that the microwave environments at Fort Monmouth were not cataractogenic (17). However, Frey's statistical analysis of Appleton and McCrossan's data (18) revealed that there actually was a difference between their exposed and control groups, and that their conclusion should have been the opposite of what was stated (19).

CONCLUSION

Here we have touched on a few highlights. Although it is gratifying to find that a few major discoveries of mine have been substantiated, like the confirmation by Bouchet and Marsal that capsular cataract can be the result of injury by nonionizing radiation as well as the confirmation by Hollows and Douglas that a valid epidemiological study would demonstrate that chronic exposure to nonionizing radiation would result in an increased prevalence of posterior subcapsular opacities. Nevertheless, there is still much more information to be acquired.

It has become apparent that our societal need is not quantity of research, but instead, our need is for better quality and relevance. That can only be achieved by enlisting the aid of our best, independent scientists.

It behooves the scientific community interested in these matters to adopt a code of ethics in order to protect society from narrow parochial and partisan vested interests, without ignoring the relative importance for and contribution to society of those very same vested interests. Predictably, the next major test will evolve around the medical diagnostic use of nuclear magnetic resonance imaging where Smith (20), in preliminary experiments, has demonstrated a cataractogenic effect in bovine lenses.

REFERENCES AND ENDNOTES

(1) Zaret, M.M., Breinin, G.M., Schmidt, H., Ripps, H. and Siegel, I.M.: Ocular lesions produced by an optical maser (laser), Science 134:1525-1526, 1961.

(2) Zaret, M.M.: The laser application in the field of medicine, J. Appl. Math. Physics 16:1, 1965.

(3) Zaret, M.M.: Analysis of factors of laser radiation producing retinal damage, Fed. Prod., Suppl. 14, 24:S62-S64, 1965.

(4) Zaret, M.M.: Selected cases of microwave cataract in man associated with concomitant annotated pathologies, in Biologic Effects and Health Hazards of Microwave Radiation, Polish Medical Publishers, Warsaw, p. 294-301, 1974.

(5) Zaret, M.M.: Cataracts following use of cathode ray tube displays, Comite Nationale Francais de Radioelectricite Scientifique International Symposium on Electromagnetic Waves and Biology, Juoy-en-Josas, France, 1980.

(6) Zaret, M.M.: Cataracts and visual display units, Conference on Health Hazards of VDU's, Loughborough University of Technology, England, 1980.

(7) Zaret, M.M.: Cataracts following use of microwave oven, N.Y. State J. Med. 74:2032-2048, 1974.

(8) Zaret, M.M. and Snyder, W.Z.: Cataracts and avionic radiations, Brit. J. Ophthal. 61:380-384, 1977.

(9) Zaret, M.M. and Snyder, W.Z.: Cataracts in aviation environments, Lancet, Feb. 26, 1977.

(10) Zaret, M.M.: Air safety: Role of family physician, N.Y. State J. Med. 79:1694-1698, 1979.

(11) Duke-Elder: The pathological action of light upon the eye, Lancet, 1188, 1926.

(12) Zaret, M.M.: An experimental study of the cataractogenic effects of microwave radiation, U.S. Air Force, Technical Documentary Report No. RADC-TDR-64-273, 1964.

(13) Bouchet and Marsol: Cataracte capsulaire bilaterale et radar, Arch. Ophthal. (Paris) 27:593, 1967.

(14) Zaret, M.M.: Nonionizing Radiational Injury of Humans, Presented at the IXth International Congress of the SFRP (French Society of Radioprotection), Nainville-les-Roches, France, May 22-26, 1978.

(15) Zaret, M.M., et al.: A study of lenticular imperfections in the eyes of a sample of microwave workers and a control population, U.S. Air Force, Technical Report No. RADC-TR-63-125, 1963.

(16) Hollows and Douglas: Microwave cataract in radio linemen and controls, Lancet ii:406, 1984.

(17) Appleton and McCrossan: Microwave lens effects in humans, Arch. Ophthal. 88:259-262, 1972.

(18) Frey: Data analysis reveals significant microwave-induced eye damage in humans, J. Microwave Power 20:53-55, 1985.

(19) There were major flaws in the study by Appleton and McCrossan. They designated as "controls" workers at risk of eye injury due to radiations other than microwaves. The control group was composed <u>entirely</u> of individuals who worked with "...laser, xenon arcs, ultraviolet and welding equipment (to include plasma torches)" -- all of which are recognized to be potentially cataractogenic. Also, they used an unorthodox examination technique that renders some lens opacities invisible, and included a score for iridescence of the lens capsule, a finding that has no diagnostic significance nor comparative value and only served to dilute the significance of positive findings.

More incredible than the report itself, was the subsequent

failure of the editors of the publishing journal. After initially accepting for publication a Letter-to-the-Editor from me containing a detailed description of the defects in the Appleton and McCrossan article, the editor, for unstated reasons, subsequently refused to publish my letter. This raises some serious scientific and ethical questions for the Surgeon General of the Army (who did not affix his usual disclaimer) and for the ophthalmic journal editor, who appears to me to have relinquished his proper role.

(20) Smith: Cataracts related to nuclear magnetic resonance exposure, University of Salford Report, England, 1985.

25
Immunologic and Cancer-Related Aspects of Exposure to Low-Level Microwave and Radiofrequency Fields

STANISLAW SZMIGIELSKI, MARIAN BIELEC
SLAWOMIR LIPSKI, GRAZYNA SOKOLSKA
Department of Biological Effects of Nonionizing Radiation
Center for Radiobiology and Radiation Safety
Warsaw, Poland

INTRODUCTION

Microwave electromagnetic radiation (MW), 300-300,000 MHz (wavelength in air 1 m to 1 mm), and radiofrequency electromagnetic radiation (RF), 0.3-300 MHz (wavelength in air 1,000-1 m), are both relatively new but rapidly intensifying occupational and environmental factors. For about 40 years there has been a steadily increasing exposure of both occupational groups and the general population to various intensities of these radiations due to the use of MW/RFs in radar, navigation, communication, and television, as well as for multiple industrial and household purposes. Thus, biological effects and possible health hazards of MW/RFs have become an important problem to be solved in connection with the elaboration of valid safety standards. Despite numerous experimental studies and epidemiologic observations (1,2), and distinct philosophies of MW/RF safety standards in the U.S. and West European countries versus Eastern European countries, it is still not possible to prove the existence and character of specific molecular, cellular or system-related damages that may be evoked by exposure to low-level MW/RF fields. Most of the observed effects are inconsistent, transient and difficult to confirm and interpret. Absorption of a certain amount of electromagnetic energy in cells, biological tissues, and in living organisms results in a thermal load that cannot be dissipated to the environment.

After exceeding the capacity of the thermoregulatory and adaptation mechanisms, it leads to an increase of temperature with all the known consequences of local or whole-body hyperthermia. Physiologic and pathologic effects of short-lasting MW hyperthermia have often been misinterpreted as being directly due to the influence of the radiation, and only recently, after progress in the measurement of specific absorption rates (SARs), the two phenomena have been differentiated.

The situation occurring in animals irradiated with MW/RFs is additionally complicated by the concomitant stress reaction. In subjects exposed to thermogenic fields, the stress reaction and the resulting well-known general adaptation syndrome can occur. But stresses related to low-level MW/RFs exposures may also occur. There exist anecdotal reports (3) that rodents (the main subject in MW/RF exposure studies) can perceive low radiation levels in an unknown way, and that they seem to be aware of being irradiated in weak fields and try to escape from the irradiation area or at least move to areas with lower field intensity. Thus, possible behavioral effects due to discomfort caused by weak MW/RF fields may occur and in turn influence neurohormonal pathways. A typical stress reaction and adaptation syndrome with stimulation of the hypothalamo-hypophyseal-adrenal axis and release of catecholamines and adrenal steroids seem to be a reasonable consequence. These problems will be discussed.

The influence of non-thermal MW/RF fields on immune reactions and immune status, as well as on development and growth of neoplasms will be reviewed. Theoretically, these phenomena may be influenced both by stress and by possible specific effect of MW/RFs. There is general agreement that whole-body hyperthermia results in remarkable immunologic and cancer-related effects (4). Thus, when analyzing the relation of observed phenomena to MW/RF exposures, care must be taken to differentiate the three possibilities -- specific interaction of MW/RFs at the molecular/cellular level, non-specific stress with adaptation syndrome, and possible thermal effects. Biological effects of MW/RFs at the cellular and subcellular levels were recently reviewed (5) with the general conclusions that no consistent changes in molecular or subcellular systems exposed in vitro can be attributed to specific MW/RF interactions. No consistent effects of these radiations have been demonstrated on growth and colony-forming ability of single cells, although there is an indication that sodium and potassium ion transport across red blood cell membranes can be affected in a manner different from generalized heating by exposure in vitro.

The possibility that specific cellular interactions of MW/RFs are connected with the pulse modulation of the carrier wave should be also considered. In Russian and East European literature of the 1960's and early 1970's (6,7) there exist different opinions concerning whether pulse-modulated MW radiation exerts stronger neurologic, behavioral, and immunologic effects compared with continuous-wave radiation of the same frequency. There is no convincing evidence that millisecond pulses of MW radiation at mean power densities not leading to detectable thermal effects may influence the function of living organisms to a higher degree than continuous-wave radiation. Certain cellular disturbances however, may be attributed to specific interactions of MW/RFs modulated at low frequencies (1-10 Hz). Adey and his group (8-10), on the basis of 20 years of experience in searching for cellular effects related to low-level sinusoidally-modulated MWs and RFs of different frequencies and pulse modulations in the low-frequency range, have found much evidence that the amplitude-modulation characteristics appear to be a prime determinant of the nature of interactions at the cellular level. The authors have stressed the existence of windowing in many of the observed interactions in both the frequency and amplitude (of pulses) domains; most of these interactions were connected with cell-membrane function (8). More recently, a series of enzyme responses as intracellular markers of events that are sensed at the cell membrane as a result of interactions with weak modulated MW/RF fields have been investigated (10,11). A strong inhibition of cAMP-independent protein kinases (messenger enzymes important for protein synthesis in the cell) occurred in cultured human lymphocytes exposed to a 450-MHz field, sinusoidally amplitude-modulated at 16 Hz (10). The effect was strongly dependent on the modulation frequency, with diminishing responses at 40 and 60 Hz and no response at 80 or 100 Hz. The authors also observed a strong time-dependence for inhibition of protein kinases -- the effect occurred only in the first 15 to 30 minutes of exposure and disappeared thereafter, despite continued exposure.

These and other events occurring in cells exposed to low-level MW/RF fields are not fully understood or resolved. They indicate the complexity of the interactions that may be connected with the function of immunocompetent cells and the whole immune system, as well as with development and growth of neoplasms.

IMMUNOLOGIC RESPONSE TO LOW-LEVEL MICROWAVE AND RADIOFREQUENCY FIELDS

The complicated immune system provides a multifactoral non-specific and specific defense for the organism against various pathogens (bacteria and viruses), and protects against development of neoplasms. Besides the basic immunocompetent cells, originating from hemopoietic stem cells (lymphocytes, macrophage-monocyte system and granulocytes) many humoral factors (e.g., opsonins, complement system, interferons, interleukens) play an important role in non-specific and specific immunity. The immune system exhibits multistage internal cooperation and self-regulation based mostly on feedback mechanisms and extensive neurohormonal and endocrine control. The complexity of the immune system and its internal and external interrelations impede evaluation of immune functions under the influence of environmental factors. Some of these factors, including MWs and RFs, exert only inconsistent, non-specific and transient effects (12). Unfortunately, there exists no single test (or set of tests) that allows evaluation of the whole system, and thus most experiments are directed toward one or a few parameters of immunity. An integrated evaluation of immunity is possible on the basis of observations of course and/or final results (survival) of stabilized experimental bacterial or viral diseases, development of neoplastic tumors, or the widely used graft versus host reactions. Although such integrated evaluations do indicate immunosuppression or immunostimulation, they yield little information regarding the immunologic mechanisms underlying the observed changes in the host's resistance.

Another difficult problem involves the relation between the observed immunological responses and the investigated factor (Figure 1). Theoretically, MW and RF radiations may exert

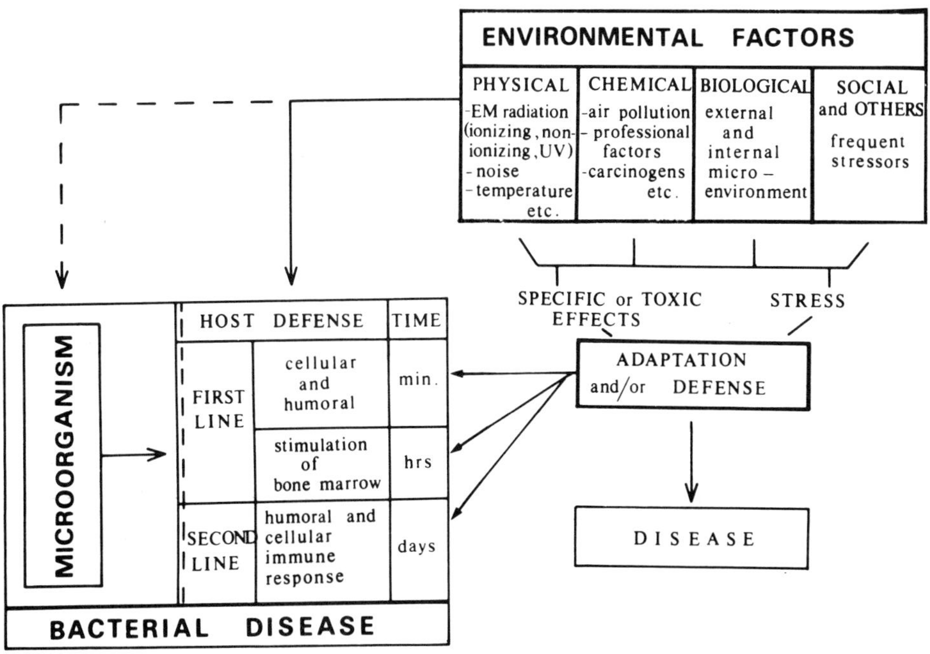

FIGURE 1. Possible interactions of environmental factors, including electromagnetic fields, with antibacterial resistance of the host. Environmental factors exert specific effects on various levels of biological organization (subcellular, cellular, organ, system), and a concomitant non-specific stress reaction. For defined factors the two elements of the biological reactions are expressed at certain levels (e.g., for electromagnetic radiation the stress reaction predominates). Specific effects and the stress reaction trigger adaptation and/or repair mechanisms, as a form of the host's defense against harmful effects. When the capacity of the adaptation mechanisms is exceeded, the factor-related disease (environmental disease, occupational disease) occurs. However, both the specific effect of the factor and certain elements of the triggered adaptation mechanisms (e.g., stimulation of hypophyseal-adrenal axis) may influence non-specific and/or specific immune reactions and result in increased susceptibility to bacterial infections (45).

various specific effects directly on immunocompetent cells and their cooperative/regulatory mechanisms, or they may influence neurohormornal regulation of the immune system. Both the possible specific interactions of MW/RFs and the concomitant stress reaction trigger various adaptation mechanisms that may or may not be beneficial for the host's immunity. Because of adaptability and redundancy in the immune system and its regulatory mechanisms, the host can generally survive subtle and transient perturbations in single elements of immunity. Thus, the subtle effects generated by MW/RFs and the concomitant stress reaction may not lead to clinically detectable immune dysfunctions.

In view of the above general comments, it is not surprising that the available literature on immunological responses to MW/RF radiations (4,12) does not offer convincing evidence for specific non-thermal interactions of the radiation with hemopoietic or immunocompetent cells. Animals exposed at different frequencies, modulations and power densities of MW/RFs using various facilities have shown inconsistent and transient changes in elements of the immune system (Tables 1-3) with the observed effects suggesting both immunostimulation and immunosuppression or inconsistent changes. In view of the large amount of conflicting and sometimes misleading information on the effects of MW/RF radiation on the immune system, we have grouped the data as follows: alterations in immunocompetent cells irradiated in vitro; responses to low-level short-term exposures in vivo; response to low-level long-term exposures in vivo; and integrated evaluation of immunity in MW-exposed animals.

ALTERATIONS IN IMMUNOCOMPETENT CELLS IRRADIATED *IN VITRO*

Several studies (Table 1) have attempted to determine whether in vitro exposure of lymphocytes and other immunocompetent cells to MW/RF radiations at various intensities leads to metabolic or functional alterations in these cells. In an early study, often cited in the literature, Stodolnik-Baranska (13) exposed unstimulated cultures of human lymphocytes to 3,000-MHz pulse-modulated (millisecond pulses) MWs for 3-5 days at 7 mW/cm^2 (4 hours daily) or 14 mW/cm^2 (15 minutes daily). The MW-exposed cells exhibited a fivefold increase in lymphoblastoid transformation, as measured morphologically by counting

TABLE 1. Alterations of Immunocompetent Cells Exposed to Microwave Fields *In Vitro*

OBSERVED EFFECTS	TYPE OF CELLS	CONDITIONS OF EXPOSURE	REFERENCE	REMARKS
Increased spontaneous blastic transformation of lymphocytes	Human lymphocytes	3000 MHz, PM, 4 hr daily, 3-5 days at 7 mW/cm^2	13	Effect not confirmed. SAR not specified, field measurement doubtful
No increase of spontaneous or mitogen-induced blastic transformation of lymphocytes exposed to non-thermal MW fields	Murine splenic rat lymphocytes	2450 MHz, CW, 1-4 hr 10 mW/cm^2, SAR 10 mW/g 2450 MHz CW, 1-4 hr 5-20 mW/cm^2	15	No increase of medium temperature
Increased spontaneous blastic transformation of lymphocytes at elevated temperatures	Human murine lymphocytes	water bath	17,18 87,88	Maximal transformation at 39°C; rapid inhibition of transformation above 41°C
No change in viability or growth	Human lymphoblasts (cultured)	2450 MHz CW, waveguide, 15 min 10-500 mW/cm^2	20	
Depressed phagocytosis	Murine peritoneal macrophages	2450 MHz, 50 mW/cm^2 30 min with cooling of medium	21	Effect not confirmed. Possible microthermal cell injury
Release of lysosomal enzymes, lowering of viability	Rabbit granulocytes (isolated)	3000 GHz PW, 15-60 min 1-5 mW/cm^2 (SAR not determined)	24	Field measurements doubtful, possible thermal effects
Suppression of T-lymphocyte cytotoxicity	Murine lymphocytes	450 MHz, modulated at 60 Hz, 1 mW/cm^2, 4 hr	9	Effect depends on modulation, windowing of the effect
Lowering of protein kinase activity	Human tonsil lymphocytes	450 MHz, modulated at 3-100 Hz, 1.5 mW/cm^2 15-60 min	10	As above, maximal effect at 16-60 Hz

TABLE 2. Immunological Response to Short-Term Exposures to Microwaves *In Vivo*

OBSERVED EFFECTS	SPECIES	CONDITIONS OF EXPOSURE	REFERENCE	REMARKS
Increased antibody levels to SRBC, reversible increase in proliferation of lymphocytes	Mice	3000 MHz PW, 0.5-5 mW/cm^2 3 hr/day, 6 weeks	14	SAR not specified, field measurement doubtful
Transient enhancement of cell mediated and humoral immunity	Mice	2450 MHz, 10 mW/cm^2 4 hr/day, 4-5 days	28	
Increased circulating antibody response to SRBC	Mice	9000 MHz PW, 10 mW/cm^2 2 hr/day, 5 days	29,30	At 1 mW/cm^2 negative findings, threshold below 10 mW/cm^2
	Mice	3000 MHz PW, 1-12 mW/cm^2 1 hr/day, 2 days	31	Maximal effect at 7 mW/cm^2; threshold between 1-5 mW/cm^2
No detectable effects after short-term exposures in non-thermal MW/RF fields	Mice	2450 MHz CW, 5-35 mW/cm^2 1-22 days, 15-30 min/day	48	
	Rats	100 MHz CW, 46 mW/cm^2, up to 57 days, 4 hr/day	90	
		2880 MHz PW,5-10 mW/cm^2 3-7.5 hr/day	46,47	
Responses to Brief Microwave Hyperthermia				
Increased number and faster maturation of B lymphocytes in spleen; Enhanced response to B lymphocyte mitogens	Mice	2450 MHz, wave-guide, SAR 14 mW/cm^2, 30 min. or 3 time	89,36	Threshold at about 12 mW/cm^2, effect strain-specific for mice

	Mice	2450 MHz CW, anechoic chamber 40 mW/cm^2, 30-min single session	37	
Increase in number of T lymphocytes without change in B cells	Mice	2450 MHz, 15 mW/cm^2 30 min/day, 9 days	35	
Increase of T and B lymphocytes in spleen, reduction in lymphocyte distribution	Mice	2600 MHz 25-800 mW/cm^2 15-60 min	38,39	Similar effects after administration of glucocorticoids
Suppression of non-specific cell mediated and humoral immunity	Mice	2450 MHz, CW, 30 mW/cm^2 90 min, 2-9 sessions	40	
	Hamsters	2450 MHz, 40 mW/cm^2 60 min/day, 1-5 days	42	Similar effect after glucocorticoids
Increase of macrophage phagocytosis and viricidal capacity	Mice Hamsters	As above	40,42	
Increase of macrophage cytotoxicity to virus-infected cells	Hamsters	2450 MHz, 25 mW/cm^2 60 min/day, 1-5 days	22,23	
Increased survival of herpes infections (encephalitis)	Mice	2450 MHz, 40 mW/cm^2, 2 hrs/day, 8 days	43	
Stimulation of cell-mediated immunity, followed by transient suppression	Mice	2450 MHz, 40-60 mW/cm^2 2 hr/day, 1-14 days	4,44	Stimulation after 1-4 days of hyperthermia, suppression after 10-14 days of exposure

TABLE 3. Immunological Response to Low-Level Long-Term Exposures to Microwave Fields *In Vivo*

OBSERVED EFFECTS	SPECIES	CONDITIONS OF EXPOSURE	REFERENCE	REMARKS
Lowered circulating antibody levels to *Salmonella*	Rabbits, Mice, Guinea pigs	10 GHz CW, 10 mW/cm^2 few months	25	Methods and conditions of exposure not described in detail
No detectable effects, including responsiveness of lymphocytes to mitogens	Rats	2450 MHz CW, 5mW/cm^2 15-30 min/day, 22 days	48	Relatively short period of exposure
Transient suppression of phagocytic capacity of granulocytes, suppressed cell-mediated and humoral immunity	Rats	2450 MHz, 0.5 mW/cm^2 1-3 months, daily exposures not specified	50	
Slight suppression of B-lymphocyte reactivity	Rabbits	2450 MHz 10 mW/cm^2 23 hr/day, 6 months	49	Results obtained in small number (4) of rabbits
No detectable changes in blood picture and reactivity of lymphocytes to mitogens	Rabbits	2450 MHz CW, 0.5-5 mW/cm^2 23 hr/day, 90 days	52	
	Rats	2450 MHz CW, 0.48 mW/cm^2 23 hr/day for life span	53	
Increased susceptibility to staphylococcal infections with lowered phagocytic capacity of macrophages	Mice	2450 MHz CW, 5-15 mW/cm^2 6 or 12 weeks, 2 hr/day	45	Transient changes, effects after 12 weeks exposure
Accelerated development of neoplasms, lowered natural antineoplastic resistance	Mice	2450 MHz CW, 5-15 mW/cm^2 1-6 months, 2 hr/day	61,62	

the number of blastoids (transformed small lymphocytes with active synthesis of nucleic acids and proteins), compared with unexposed controls showing weak spontaneous transformation. The intriguing observation was, however, never confirmed despite many attempts performed under better controlled conditions of irradiation, and with the use of more objective methods for evaluation of blastoid transformation of lymphocytes (14-16). On the other hand, numerous studies have shown an increased mitogenic response of human blood lymphocytes in culture after temperature elevation above 37°C (17-19). Human lymphocytes cultured at 38-40°C respond with faster and increased blastoid transformation both in unstimulated cultures (elevated temperature acts as a transforming factor), and in cultures stimulated with phytohemagglutinin (PHA) or concanavalin A (ConA), the two mitogens commonly used for studies of lymphoproliferative response _in vitro_. Roszkowski et al. (19) found that maximal transformation of human lymphocytes occurred at 39°C, and that above 41°C there was inhibition of lymphocyte function and an unresponsiveness to PHA and ConA. A similar blastogenic response was found when cultures were heated to 39°C in a water bath, or after exposure to 2450-MHz MW field (unpublished results). Inhibition of the transformation occurred above 41°C, independently of the source of thermal energy.

Smialowicz (15) exposed splenic murine lymphocytes to 2450-MHz continuous-wave radiation for 1-4 hours at 10 mW/cm^2 (SAR, 19 mW/g). Immediately after completing the irradiation session the temperature of the exposed cultures did not differ from that of the controls, and the viability of the cells was unchanged. Following irradiation, the exposed and control cultures were treated with non-specific T and B lymphocyte mitogens (PHA, ConA, lipopolysaccharides), and the proliferative response was measured after 72 hours by incorporation of ^{3}H-thymidine. The blastogenic response of MW-irradiated and sham-irradiated cultures did not differ in cultures stimulated with mitogens or in unstimulated cultures. In a similar experiment, Hammrick and Fox (16) exposed rat lymphocytes to 2450-MHz continuous-wave radiation for 4, 24, or 44 hours at 5, 10 and 20 mW/cm^2 (SARs of 0.7, 1.4 and 2.8 mW/g, respectively) and measured the transformation of unstimulated and PHA-stimulated cultures by incorporation of ^{3}H-thymidine. These investigators also did not observe an influence of MW irradiation on blastic transformation of lymphocytes. The effect of the same frequency of MW radiation on the growth and viability of cultured human lymphoblasts was studied by Lin and Peterson (20). Continuous cultures of human lymphoblasts (established cell lines Daudi and HSB_2) were irradiated in a wave-guide facility for 15 minutes at 10-500 mW/cm^2 (SARs of 25-1200 mW/g). Due to cooling of samples kept during irradiation in capillary tubes, no increase in temperature was observed, even at the highest power densities tested. No changes were observed in the viability and growth of MW-exposed lymphoblasts compared with unexposed controls.

The above studies, performed under well-controlled conditions of exposure, provide evidence that no detectable changes in lymphocyte activity occurs following MW exposure _in vitro_ when proper control of culture temperature is achieved and no elevation of the medium temperature occurs. Thus, the original observation of Stodolnik-Baranska (13) cannot be related to specific, non-thermal influence of MW radiation and must be due to other causes. In the original experiments, performed using facilities at the Institute of Aviation Medicine in Warsaw, Poland, the following conditions were present: 3000 MHz millisecond pulse MW generator; open field from horn antenna; absorbing screen opposite the antenna; no anechoic chamber. No measurement of field power density during irradiation was made; instead, an estimation of the power density based on earlier measurements was given. The lymphocyte cultures were irradiated in thick-walled, glass culture vessels. Under the above conditions, absorption of MW energy inside the vessels was undeterminable, and there also existed a possibility of large difference of power density compared with the listed levels of 7 or 14 mW/cm^2. It is reasonable to assume that the temperature of the medium was elevated during irradiation, and that the hyperthermia resulted in stimulation of blastic transformation of the irradiated lymphocytes.

Other immunocompetent cells have rarely been the subject of experiments with _in vitro_ exposure to non-thermal MW/RF fields. Information concerning the influence of this radiation on viability, function, and morphology of macrophages, monocytes and granulocytes is scarce and fragmentary. In an early study, Mayers and Habeshaw (21) found depressed phagocytosis of isolated peritoneal murine macrophages exposed to 2450 MHz at 50 mW/cm^2. During irradiation, a 2.5°C increase of medium temperature was noted, however the final temperature in the culture vessel did not exceed 36.2°C. The suppressed phagocytic activity returned to normal after discontinuation of MW irradiation. The mechanisms by which the observed effect occurred are unknown. Such a slight increase in temperature normally results in stimulation, and not suppression, of phagocytosis; this process is optimal for human macrophages at 38-39°C and for murine peritoneal macrophages at 39-40°C.

More recently, Rama Rao et al. undertook broad investigations on the effects of MW exposure on the hamster immune system (22,23). Since the animals were irradiated in vivo (2450 MHz, 25 mW/cm^2, 60 minutes) the results will be discussed in detail in the next section, however it is worthwhile to mention here that peritoneal macrophages isolated from the irradiated hamsters were found to be activated, as measured by their viricidal activity to vaccinia viruses (22).

Szmigielski (24) exposed isolated rabbit granulocytes in vitro to continuous 3000-MHz MW radiation at 1 and 5 mW/cm^2 for 30-60 minutes and observed an increased number of dead cells (increase in nigrosine staining), and an enhanced liberation of lysosomal hydrolases (a symptom of sublethal cell damage). Since the same facility, used by Stodolnik-Baranska (13) and Czerski (14) (described above) was used for this study, there is no certainty regarding the field power density, and significantly higher intensities than those reported were possible. Unfortunately, the experiments with granulocytes were never repeated under more controlled exposure conditions.

In summary, there is no convincing evidence for metabolic and/or functional alterations in immunocompetent cells irradiated in vitro in non-thermal MW/RF fields. Thus, the reported immunologic phenomena observed in animals exposed in vivo must be explained on the basis of alterations of humoral control and/or regulatory mechanisms of the immune system or nonspecific stress reaction of the irradiated subjects.

Our present knowledge of non-thermal effects of MW/RFs on the function of immunocompetent cells is still scarce and fragmentary. Only simple experimental systems have been tested, and specific functions of immunocompetent cells have only rarely been investigated. Virtually nothing is known about the influence of RFs (0.3-300 MHz) on these cells, and on effects related to pulse modulation of the carrier wave. We have mentioned earlier that Adey and his group (8) have evidence that pulse and amplitude modulations of the carrier wave appear to be a prime determinant of the nature of interaction at the cellular level. Recently this group reported suppression of T-lymphocyte cytotoxicity following exposure to sinusoidal, amplitude-modulated 450-MHz MW fields (9), and alterations in protein kinase activity following exposure of cultured human lymphocytes to MW fields of the same frequency and modulation (10). In both experiments, well-controlled conditions of irradiation and measurements were provided (anechoic chamber for 450-MHz MWs, a Crawford cell exposure) and described in detail (9). The incident power densities were relatively low (1.0-1.5 mW/cm^2) with sinusoidal amplitude modulation at 3-100 Hz. Murine cytotoxic T lymphocytes (line CTLL-1) exhibited a significant inhibition of allogeneic cytotoxicity against the target cells MPC-11, when a 4-hour assay of cytotoxicity was conducted during irradiation in a 450-MHz field sinusoidally amplitude-modulated at 60 Hz. Exposure of the effector cells to the same field prior to adding them to target cells resulted in a similar inhibition of the cytolytic assay (9). This suggested a direct interaction of the field with the function of cytolytic T lymphocytes. The exposed cytolytic T cells recovered their full cytotoxic capacity about 12 hours after termination of exposure. Still more interestingly, a differential susceptibility of the cytolytic T cells was observed with modulation frequencies below 100 Hz. Peak suppression occurred at 60 Hz modulation, with progressively smaller effects at 40, 16 and 3 Hz, while the unmodulated carrier wave (450 MHz) did not affect the cytotoxicity in any way. An identical exposure system and similar conditions of irradiations were used (10) to study the influence of 450-MHz MWs on the activity of certain enzymes in human tonsil lymphocytes irradiated _in vitro_. In these experiments, lymphocytes were kept in the Crawford cell housed in a temperature-controlled chamber (35°C) and irradiated for periods up to 60 minutes. It was found that the activity of cAMP-dependent protein kinase relative to controls remained unaltered by 450-MHz fields modulated at 16 or 60 Hz with exposures of 15, 30 and 60 minutes. On the other hand, total cAMP-independent protein kinase activity fell to less than half of that of the unexposed control levels after 15 and 30 minute exposures. Surprisingly, it returned to control levels during continued exposure for 45 and 60 minutes. The reduced enzyme activity occurred with 16, 40 and 60-Hz modulation frequencies, but not with 3, 6, 80, or 100-Hz modulations.

The biological significance of the large reduction in histone kinase activity of the lymphocyte during exposure to very low and definitely non-thermal (1.0-1.5 mW/cm^2) levels of modulated MW radiation is unknown, but it points out the windowed character of the response regarding both the low modulation frequencies and the time of irradiation.

Adey and his group (8,11, and personal communications) offer a general explanation and hypothesis for the cellular events that occurred after exposure to the weak, modulated MW fields. They presumed, on the basis of numerous experimental studies, that the primary interactions occurred at the cell membrane level, and they concluded that there were three basic steps in the sequence of events at the cell membrane. First, there was a modification of calcium binding that occurred along the membrane surface, this being a highly cooperative step and appearing to be the basis of the amplification that occurred in the response to weak electromagnetic stimuli. The second step was the coupling of the signal across the cell membrane. The electromagnetic waves (depending on their modulation) might have been involved as the vehicle for passing the signal down the strands of helical proteins that span from the outside to the inside of the membrane. The third stage involved coupling to intracellular systems (coupling to the cytoskeleton and activation of intracellular enzymes, either directly at the membrane or indirectly through chemical signaling).

This hypothesis, if proved in more detail and supported by further experiments elucidating principles of the windowing in interaction of weak electromagnetic fields at the subcellular and cellular levels (relation to pulse modulation, time of exposure and frequency of the carrier wave), may change significantly the views on biological effects of MW/RFs, including influence of this radiation on immunocompetent cells.

IMMUNOLOGIC RESPONSE TO LOW-LEVEL SHORT-TERM EXPOSURES *IN VIVO*

Over the past 30 years many reports have appeared (1) dealing with various effects of MW/RF radiation on immunologic functions of irradiated animals. In many earlier cases (1955-1975) the investigators were motivated by the search for possible harmful effects on the immune system, supporting the philosophies of safety standards elaborated in their countries. Thus, when analyzing the literature of the 1960s and early 1970s, one is confronted with different methodological approaches and different conditions of irradiation used by Western and Eastern authors. As a rule, Russian and East European authors searched for any detectable shift in hematologic and/or immune systems after exposure of animals to weak (below 1 mW/cm^2) and very weak (10-100 $\mu W/cm^2$) fields, and related the observed transient shifts to action of MW/RF radiation. West European and American investigators concentrated on power densities leading to subthermal or definitely thermal effects in the organism (10 mW/cm^2 and above). However, the findings due to local or whole-body hyperthermia (even that lasting only a few minutes) were often misinterpreted as being related to the interaction of MW energy with the function of the immune system. In any case, the final interpretation of MW/RF-induced changes in the immune system, and conclusions regarding possible causative relations must consider many variables that may affect the interaction of electromagnetic radiation with the biological system (12). Electromagnetic radiation generally evokes weak, transient, and inconsistent biological effects. In contrast, whole-body hyperthermia is a strong stimulus for the immune system and, as it is well-known in the case of fever, it modulates the function of this system. Thus in experiments involving animals in weak MW/RF fields, sham-irradiated controls should be used, and care should be taken to protect the animals against stresses from handling. In the case of thermogenic MW/RF fields, hyperthermic and stress control groups (positive controls) must be considered. From personal experience and autopsies in numerous laboratories in the Soviet Union, Poland and Czechoslovakia, the present authors can state that up to 1975 the conditions of exposure of the animals, field measurements, and the general handling of the experimental animals were far from being what is presently regarded as acceptable. As a rule, cage-controls were used instead of sham-irradiated controls, and in many experiments generally poor animals were used and were improperly handled. Thus, in our opinion, the earlier Russian and East European investigations involving animals exposed to weak MW/RF fields (6,7,25), including the reported immunological alterations, should be viewed with caution. In the most recent Russian monograph on biological effects of electromagnetic fields (26), immunologic alterations in mice and rats exposed to MW fields at very low power densities (reported repeatedly during 1960-1975) are not presented. Instead, the authors claim that inconsistent and transient immunological alterations may be observed in animals chronically exposed at power densities exceeding 0.5 mW/cm^2, and that clearly demonstrable immunological effects are observed only at thermogenic power densities. The authors also state, although without evidence, that the "high adaptability of the immune system causes the alterations observed after exposure of animals in weak MW/RF fields; they are meaningless from the point of view of safety standards" (26, p.55).

The situation has changed during the last decade. Several laboratories in the USSR and East European countries were equipped with anechoic chambers, modern devices for measurements of power density and SAR, and better handling and care of animals were provided. This is still not the rule everywhere.

In investigations published in 1972-1976 (6,27) transient and reversible increased lymphocyte proliferation and function, as well as activation of single subpopulations of lymphocytes following exposure to subthermal MW/RF fields, were believed to be the most consistent findings. Recent data (Table 2) indicate that immunologic function in a wide range of animal species may be altered by single or short-term (several days) exposure to MW fields; however, the effects are inconsistent and not detrimental for the organism.

Transient enhancement of cell-mediated and humoral immunity, measured with a battery of tests including splenic plaque-forming cells, ability for phagocytosis, and responsiveness to sheep red blood cells (SRBC), were found in mice in a series of experiments performed under well-controlled conditions of irradiation (4 hr/day, 2450 MHz, 10 mW/cm^2) by Roberts and

Steigbeigel (18), and by Ivanoff et al. (28). Liddle et al. (29) reported stimulation of the circulating antibody response (against streptococcal type III polysaccharide) in mice exposed to pulsed (pulse repetition rate about 1000 per second) 9000-MHz MWs at 10 mW/cm^2 (SAR, 4.7 mW/g), 2 hours daily for 5 days, with a concurrent lengthening of the survival time of animals challenged following exposure. More recently, the authors performed a similar experiment at 1 mW/cm^2 (30). They found that the circulating antibody titers for the MW-exposed animals were not significantly different from those of the sham-irradiated animals, and that there were no differences in any of the hemato-immunological parameters analyzed, indicating that the threshold in mice for the response was 1-10 mW/cm^2 (0.47-4.7 mW/g). Similar results were reported recently by Chinese investigators (31). These authors exposed mice to pulsed 3000-MHz MWs (from a radar generator) for 1 hour daily during two consecutive days, and found a fourfold increase of the hemagglutinin titer to SRBC 7 days after termination of exposure, with a return to normal values 22 days after irradiation. In another set of experiments (31) the same investigators searched for a threshold for the observed phenomenon. Mice were irradiated under identical conditions except that 1, 5, 7, and 12 mW/cm^2 (SARs not given) were applied. A maximum increase of the hemagglutinin titer to SRBC was observed after exposures at 7 mW/cm^2, but the titers were also significantly elevated compared with sham-irradiated controls at 5 and 12 mW/cm^2. At 1 mW/cm^2 the hemagglutinin titer did not differ from controls, and thus the authors concluded that for pulsed 3000-MHz MWs, the threshold for the phenomenon was 1-5 mW/cm^2.

In an earlier study, Czerski (14) reported that mice exposed for 6 weeks to 2950-MHz pulsed MWs (from a radar generator) at 0.5 mW/cm^2 (SAR not determined), 12 hr/day, had significantly greater numbers of antibody-producing cells (lymphoblastoids and large lymphocytes) and higher serum hemagglutinin titers following immunization with SRBC. Cage-controls were used, and the facilities available did not permit precise measurements of the field power density in the area of irradiation of the animals. Wiktor-Jedrzejczak et al. (32,33), using facilities available at the Naval Medical Research Institute in Bethesda and at the FDA Bureau of Radiological Health in

Rockville, MD (a rectangular waveguide), exposed mice to 2450-MHz MWs for 30 minutes at a SAR of about 14 mW/g (thermal effects very possible). After one irradiation session, a significant increase in the proportion of B lymphocytes (type of lymphocytes responsible for humoral immunity) was found based on determination of surface receptors (CR, Ig, Fc)(typical for B lymphocytes at various stages of maturity). This was accompanied by enhanced response of murine spleen lymphocytes to the B-cell specific mitogens (lipo-polysaccharides, polyinosinic-polycytydylic acid, and purified protein derivative of tuberculin). In contrast to the other above-cited experiments, Wiktor-Jedrzejczak et al. (32) found a decrease, not an increase, in the primary humoral response to SRBC in mice exposed to a single session of 2450-MHz MWs with a SAR of about 14 mW/g. However, in this case the animals were immunized just prior to the exposure and not after termination of exposures as was the case in the investigations discussed earlier. The authors suggested that the decreased humoral response to SRBC observed in their experiment might have been due to nonspecific stimulation of B lymphocytes by MW exposure to mature before they were activated by the antigen (SRBC). This could have resulted in increased proportion of unresponsive lymphocytes.

The investigations were continued by Sulek et al. (34) using the same facilities. They investigated the kinetics for increased frequency of CR-positive cells (B lymphocytes) in the spleens of mice irradiated as before, and determined the threshold for the observed phenomena. It was found that the frequency of CR-positive cells showed an initial increase 3 days following a single session of 2450-MHz exposure (30 minutes) at a SAR of 12 mW/g, and that it persisted for 5-6 days. A minimum of 15 minutes exposure at a SAR of 11.8 mW/g or 30 minutes at 5 mW/g were needed to cause a significant increase in the number of CR-positive cells in murine spleen. The number of T lymphocytes was unchanged. In contrast, Huang and Mold (35) found a significant increase in number of T lymphocytes without changes in B lymphocytes in mice exposed to 2450-MHz MWs at 15 mW/cm^2 (SAR 10 mW/g) for 30 minutes during 9 days. Schlagel et al. (36) and Smialowicz et al. (37) in two independent investigations re-examined the phenomenon of shifts in murine spleen lymphocyte subpopulations after exposure to 2450-MHz MW fields.

It was found that the observed increase in number of CR-positive and Fc-positive lymphocytes was strain-dependent, and that it occurred only in mice of a certain age (detectable in 16-week old animals, but not in younger animals). The exposure facilities used (waveguide versus anechoic chamber), as well as the environmental conditions were additional sources of variation that might have influenced the appearance of this phenomenon (37).

In fact, all the observed shifts in murine spleen lymphocyte populations were due to thermal stress induced by exposure to MW fields. Evidence for this view has come from studies of Liburdy (38,39), who has found similar shifts in splenic lymphocytes in mice exposed to 2600 MHz at 800 mW/cm^2 (SAR, 5.6 mW/g) that resulted in a 2-3°C increase in rectal temperature. Similar phenomena were observed after administration of glucocorticoids, suggesting that the observed shifts might also have been related to some form of non-specific stress. In another set of experiments Liburdy (39) exposed mice to 2600 MHz MWs for 1 hour at 5 or 25 mW/cm^2 (SARs of 3.8 and 10 mW/g, respectively) and used two positive controls (injection of glucocorticoids and exposure of animals in a 63°C warm-air oven for 1 hour) as well as sham-irradiated mice. Prior to exposure, all animals were injected with ^{51}Cr-labelled syngeneic spleen cells with the aim of examining the distribution of these cells in the lung, liver, spleen and bone marrow. A changed distribution and migration of lymphocytes was observed in mice exposed to 2600-MHz MWs at 25 mW/cm^2 (a 2°C increase in body temperature) and in the glucocorticoid-treated groups, while no changes were found in animals exposed to 5 mW/cm^2 (no detectable increase of body temperature).

Further evidence for an influence of moderate-strength MW/RF-induced brief thermal stress on the function of the immune system was reported in 1983-1985 by Smialowicz et al. (40), Cain, et al. (22,23,41), and Yang et al. (42). Brief exposure to thermal 2450-MHz MW fields resulted in transient suppression of non-specific cell-mediated cytotoxicity (NK cell activity), and a concomitant increase in macrophage phagocytic activity. Smialowicz et al. (40) exposed mice in anechoic chambers to 2450-MHz MW fields at SARs of 3.5, 10.5, or 21 mW/g for 90 minutes, and found a significant reduction of splenic NK cell activity, as measured using *in vivo* and *in vitro* assays, only at the highest SAR. NK activity returned to normal 24 hours following the last exposure at 21 mW/g; treatment of mice with hydrocortisone resulted in a similar suppression of NK activity. Rama Rao et al. (22,23) and Yang et al. (42) exposed hamsters to 2450 MHz at 5-40 mW/cm^2 (SAR of 0.53 mW/g per mW/cm^2). Each irradiation session lasted 1 hour, and the protocol included exposures of 1-5 sessions. Exposure at 25 mW/cm^2 (SAR of 13 mW/g) resulted in an increase of rectal temperature of 3.0-3.5°C at the end of irradiation; the temperature returned to normal 1 hour after termination of exposure. Exposure to 25 mW/cm^2 (single 1-hour session) induced a marked but transient alteration in body temperature, serum glucocorticoid levels, circulating leukocyte profile and NK cell activity. In contrast, after exposure at 15 mW/cm^2 no detectable changes of these parameters were found compared with sham-irradiated controls, and the authors concluded that the suppression of NK activity was related to thermal stress.

In another set of experiments performed under the same conditions of MW exposure, the same authors (22,23) found that exposure at 25 mW/cm^2 (rectal temperature increase, 3.0-3.5°C) resulted in activation of peritoneal macrophages that were significantly more viricidal to vaccinia viruses, compared with sham-irradiated and cage-controls. Moreover, immune macrophage cytotoxicity for virus-infected and non-infected target cells *in vitro* was not suppressed, indicating that peritoneal macrophages were not functionally injured by microwave hyperthermia. The above phenomenon did not occur after exposure at 15 mW/cm^2 (no detectable increase of body temperature).

In summary, the only repeatable effect on the immune system that might be attributed to short-term exposure of experimental animals at non-thermal power densities (SAR below 6-12 mW/g for mice) seems to be the enhanced humoral response to various antigens (SRBC, bacterial polysaccharides). Increased levels of circulating antibodies were found by a few independent groups of investigators, and the threshold for this phenomenon in mice seems to be 1-10 mW/cm^2 -- power densities not related to thermal stress. The mechanisms underlying the increased levels of circulating antibodies are unknown. Recently Rama Rao et al. (41) investigated the effect of a single 1-hour exposure of hamsters to 2450-MHz MWs at 5-15 mW/cm^2 on the IgM antibody response of spleen cells to SRBC, using the direct hemolytic plaque assay. Although most of the observed effects were related to thermal fields, exposure of hamsters at 15 mW/cm^2, not leading to detectable hyperthermia, also resulted in a significant increase of plaque-forming cells. The authors concluded that MW exposure augmented the primary IgM response to SRBC by affecting some early event in the immune response process. Despite the possible mechanisms leading to enhanced humoral response in MW-exposed animals, the biological significance of this phenomenon is uncertain. In fact, short-term exposures in subthermal MW fields act as an adjuvant for various antigens, and thus should be considered beneficial for the organism.

All other reported immunological responses occurring after short-term exposure to MW fields (e.g., shifts in distribution of lymphocytes, changes in number and reactivity of B lymphocytes, suppression of cytotoxic activity of the NK type, activation of macrophages) are related either directly or indirectly to thermal load and/or the concomitant stress with release of adrenal steroids. An interesting possibility based on the results of Rama Rao et al. (41) is that the cause of the immunologic responses observed after short-term thermal MW stress may be the release of endotoxins into the circulation. Endotoxin is a known B-lymphocyte mitogen which causes a polyclonal B-cell stimulation with increased production of antibodies, and which stimulates a number of macrophage functions including their anti-viral, antineoplastic and antibacterial activities. It evokes most of the phenomena observed after short-term MW thermal exposures (single or repeated during a few days). This also may be beneficial for the organism with potential therapeutic use. In fact, mice infected experimentally with herpes or vaccinia viruses and treated with 2-hour sessions of 2450-MHz MW hyperthermia have shown better tolerance of the infections and significantly higher survival of herpetic encephalitis (43). One must, however, remember that prolongation of exposures to moderate MW hyperthermia for several days results in a dramatic shift from stimulation to suppression of immune reactivity, as was clearly demonstrated by our group (4,44,45) in experiments in which mice were exposed to 2450-MHz MWs at

40 mW/cm^2 (SAR of 20 mW/g) 2 hours daily for 1, 4, 7, 10 or 14 days. Stimulation of immune reactions was observed only following 1- and 4-day exposures, and was followed by deep, although reversible, suppression of immune reactivity.

RESPONSE TO LOW-LEVEL LONG-TERM EXPOSURES

Short-term exposures to low-level MW/RFs do not cause harmful effects on the immune system, and thus to search for possible health hazards it is necessary to consider the response to long-term exposures (few weeks to several months for experimental animals, up to several years for humans exposed occupationally or environmentally to MW/RF fields). In cases of long-term exposures, problems of valid control groups and proper conditions of irradiation (environmental control of exposure chambers, avoidance of immobilization, isolation or confinement of animals during irradiation, careful handling of animals, etc.) becomes crucial for obtaining valid results that may be related to the influence of radiation. It must be stressed that the above conditions were not strictly followed in earlier investigations, mainly those performed in the USSR and East European countries, from which came most of the reports of the 1960's and early 1970's indicating immunological responses occurring after long-term exposures to weak MW/RF fields. Thus, despite the fact that there exist numerous publications supporting alteration of function of the immune system after long-term exposure at power densities below 0.5 mW/cm^2 (Table 3), at least some of the observed phenomena seems to be due to different stress situations resulting from normal handling, and from poorly controlled irradiation conditions. Great care must also be taken to avoid incidental bacterial and/or viral infections of the experimental animals exposed for several weeks to daily sessions of irradiation. Infections can easily occur under inadequate breeding conditions, and they are difficult to recognize and diagnose (especially if their course is chronic or subclinical). Undoubtedly, such infections would alter the function of the immune system. We feel that the problem of incidental chronic and/or subclinical infections occurring during investigations of long-term exposures to MW/RF fields is presently underestimated. In most of the long-term experiments performed under acceptable conditions of irradiation and animal care, various effects on immunosuppression have been found (Table 3). The susceptibility of animals to incidental infections should therefore be increased, with consequent further immune-system reactions not related directly to MW exposure.

In earlier experiments (25) Russian investigators found a reduction of circulating antibodies to Salmonella in mice, rabbits, and guinea pigs immunized following several months of daily exposures to 10,000-MHz MWs at 10 mW/cm^2 (SAR not stated). Unfortunately, the conditions of irradiation, time of daily sessions, and even the period of exposure were not described. Most of the acceptable information on immunological responses to low-level long-term exposure of experimental animals to MW/RF radiations have come however from investigations performed during the last decade.

Smialowicz et al. (48) exposed mice to 2450-MHz MWs under far-field conditions for 15 or 30 minutes daily for up to 22 days at 5-35 mW/cm^2 (SAR of 4-25 mW/g), and investigated various immunologic functions, including in vitro mitogen stimulation of isolated splenic lymphocytes, proportion of T and B lymphocytes in the spleen, and the primary humoral response to SRBC. They found no differences in the immunologic parameters compared with sham-irradiated controls, but the period of irradiation was relatively short (up to 22 days). Transient immunosuppression was observed in rats and rabbits exposed during a few months to 2450-MHz MWs in parallel American-Russian investigations (49,50) (Table 3). McRee et al. (51) reported that 1 month after termination of a 6-month exposure to 2450-MHz MWs at SAR of 1.5 mW/g (23 hours daily), spleen cells from rabbits showed a decreased responsiveness to pokeweed mitogen (a specific mitogen for B lymphocytes). Smialowicz (12) commented however, that "although these results are interesting, they are not conclusive and are of questionable value, because only four exposed and four sham-irradiated rabbits were employed. Also, both irradiated and sham-irradiated rabbits were transported from one laboratory to another (University of Washington to Research Triangle Park, NC) between the termination of exposure and spleen cell assay." This is a good example of the importance of personal experience in critical analysis of available published data on biological effects of electromagnetic radiation.

A similar study on rabbits exposed for up to 3 months to weak 2450-MHz MW fields was also reported by Chou et al. (52). Two groups of 16 rabbits were exposed in two experiments of 90 days each at 0.5 and 5 mW/cm^2, respectively. The exposure was preceded by a 2-week adaptation of the animals to the miniature anechoic chambers used for irradiation. Thermographic analysis during MW exposure showed that the SAR ranged from 5.5 mW/g in the head to 7 mW/g in the back at 5 mW/cm^2. Monthly blood samples were taken for hematologic and immunologic examinations, including lymphocyte blast transformation after stimulation with phytohemagglutinin. No significant changes in peripheral blood or in blast-forming activity of blood lymphocytes were noted during the period of observation. Rabbits are rarely used for immunologic studies because they have relatively labile immune systems and it is difficult to evaluate their immune status. Thus the data do not say much about small changes in the function of the immune system.

Recently Guy et al. (53) reported on a 2-year study involving exposure of rats to 2450-MHz MW fields at 0.48 mW/cm^2 (SAR 0.15-0.4 mW/g). Periodically they collected peripheral blood for basic hematologic and immunologic examinations, but were not able to find any consistent changes that might be related to irradiation.

In summary, studies of long-term irradiation of animals in low-level (below 10 mW/cm^2 for 2450-MHz MWs, SAR below 4-6 mW/g) MW/RF fields do not provide convincing evidence for specific response of the immune system to nonionizing radiation. However, slight and transient immunosuppression, explainable in terms of a chronic nonspecific stress reaction, not related directly to interaction with MW/RF energy, is very possible. There is a lack of fully convincing reports on the immune status itself, as opposed to selected parameters, in animals exposed chronically to MW/RF fields of different frequencies and pulse modulations. It is worthwhile to remember the earlier discussed findings of Adey and his group concerning cellular alterations in cultured lymphocytes _in vitro_ caused by exposure to sinusoidal pulse-modulated 450-MHz fields, and the phenomenon of windowing for these alterations. There are no experiments _in vivo_ involving exposure of animals to low-frequency modulated MWs with examination of the immune functions. On the other hand, as discussed below, both the higher susceptibility of animals to chronically exposed bacterial and viral diseases, and the data on acceleration of development of neoplasms in mice exposed for months in non-thermal MW fields (the two phenomena that might result from suppression of immune functions in chronically exposed subjects) emphasize the problem of the response to long-term low-level irradiation in MW/RF fields, and they call for further investigation.

INTEGRATED EVALUATION OF IMMUNITY IN MW/RF-EXPOSED ANIMALS

By "integrated evaluation of immunity" we mean an evaluation of the actual immune status of the host, including its nonspecific and specific mechanisms of resistance and reactivity to typical stimulants of immunity (pathogens, transplants, etc.). As mentioned before, there exists no single test or battery of tests that allows precise measurement of the immune status of the organism. The best procedure is to observe the effects caused by factors known to influence immunologic responses. A typical example of integrated evaluation of immunity is the immunization of the host with an antigen followed by the measurement of the level of circulating antibodies against the antigen. Synthesis of antibodies in the organism requires proper function of different immunocompetent cells, including macrophages, T lymphocytes, B lymphocytes, and plasma cells, as well as numerous control, helper and suppressor mechanisms exerted both by cellular and humoral factors. Although we can observe only the final effect (titer of antibodies), we can reach conclusions about the effectiveness of the whole system of humoral immunity, and concentrate later on an explanation of the function of single elements. Other examples of integrated evaluation of immunity are reaction of animals to transplants, including implantation of neoplastic cells, susceptibility to experimental bacterial and viral infections, and tolerance of factors known to suppress or stimulate immunity (e.g., ionizing radiation and endotoxins).

Surprisingly, in the literature there are only a few investigations where the above models were applied and tested in animals with long-term exposure to electromagnetic radiation. Thus, we still do not know whether long-term irradiation in

MW/RF fields results in immunosuppression, or what the biological significance of the possible suppression would be for the host.

For years we have tested susceptibility to experimental staphylococcal and viral (herpes and vaccinia viruses) infections after long-term exposures to 2450 MHz (continuous wave) at different power densities. In an early study (54) we exposed rabbits to 3000-MHz MW fields for 6 hours daily for 6 or 12 weeks at 3 mW/cm^2 (SAR not determined). After the last exposure the rabbits were infected intravenously with known (sublethal) doses of virulent *Staphylococcus aureus*. Both MW-exposed and control animals survived the infection, but a decreased production of granulocytes and a weaker response of granulopoiesis to the infection were observed in the MW-exposed animals. Unfortunately, at that time (1972-1973) no anechoic chamber was available and the exposures were performed in an open field with an absorbing screen opposite the horn antenna. Thus, measurements of incident power density during exposure were questionable, and measurement of SAR was not possible at all. We reexamined the experimental model a few years later after well-controlled conditions of irradiation and measurement were provided at the Center for Radiobiology and Radiation Safety in Warsaw, Poland. This time we exposed mice to 2450-MHz MWs at 5 and 15 mW/cm^2 (SAR of 2 and 6 mW/g, respectively) 2 hours daily, and irradiated the animals for 6 or 12 weeks, 6 days weekly (45). After termination of the exposure in some of the animals, evaluations were made of phagocytic ability *in vivo* (clearance of labelled staphylococci from blood after intravenous injection of killed microorganisms) and delayed hypersensitivity reaction to oxazolone. Other mice (20 per group) were infected with a lethal dose of virulent *S. aureus* (a strain pathogenic for mice). The dose of staphylococci was titered to provide about 60% survival rate in healthy, unexposed mice, to facilitate possible observation of both elevation and fall of the survival rate. The results are summarized in Figure 2.

Exposure of mice at 5 mW/cm^2 (SAR 2 mW/g) during 12 weeks did not result in detectable differences in phagocytic activity, hypersensitivity to oxazolone, or survival of experimental staphylococcal infections compared with sham-irradiated controls

	Control	5 mW/cm^2		15 mW/cm^2	
		6 weeks	12 weeks	6 weeks	12 weeks
Phagocytosis ($CPM \times 10^3$, labelled staph.)	3572 ± 410	4672 ± 520	3808 ± 376	3885 ± 408	2987 ± 307
Delayed hypersensitivity Oxazolone (ear thickness increment)	21.6	31.6	26.2	22.8	18.2

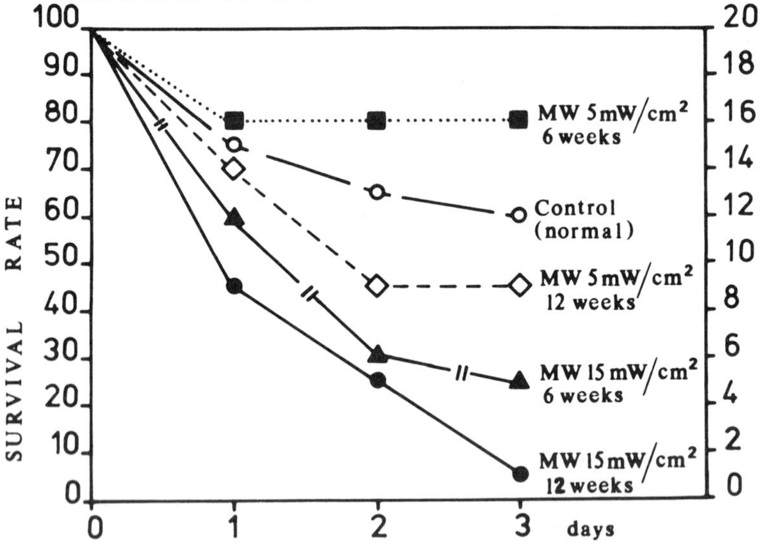

		Microwave-non thermal		Microwave Hyperthermia	
		5 mW/cm^2 6w	15 mW/cm^2 12w	4 days	14 days
Non-specific immunity	granulocytes	↑	↓↓	?	?
	macrophages	no changes	↓	no changes	↓↓
Specific immunity	Delayed hypersensitivity	no changes	no changes	no changes	↓↓↓
	Lymph. prolif.	?	?	no changes	↓↓

FIGURE 2. Summary of immunological effects observed in mice exposed during 6 or 12 weeks to 2450-MHz microwaves at 5 or 15 mW/cm^2 (2 hours daily). Note stimulation of phagocytosis after 6 weeks at 5 mW/cm^2, depression after 12 weeks at 15 mW/cm^2 (top table), lower survival rate of acute staphylococcal infections in mice exposed to 15 mW/cm^2 during 6 and 12 weeks (graph), and differences in immune response in mice exposed to 5 or 15 mW/cm^2 and in those treated with hyperthermia (bottom table) (4).

(Figure 2). In mice exposed for 6 weeks, improved phagocytic ability and reactivity to oxazolone were observed (Figure 2). At 15 mW/cm^2 (SAR of 6 mW/g), no detectable increase of rectal temperature after termination of the 2-hour exposure session, 12-weeks exposure resulted in significant lowering of the survival rate and phagocytic ability _in vivo_ (Figure 2). During further observation (unpublished results) we found that these phenomena were transient and returned to normal values 1 month after termination of the 12-week exposures. In summary, the results indicate that long-term exposure of mice may result in a lowering of antibacterial resistance with weaker phagocytic ability _in vivo_, and the threshold for these phenomena for mice exposed to 2450-MHz MWs is 5-15 mW/cm^2.

SUMMARY OF IMMUNOLOGIC RESPONSE TO MW/RF RADIATION

From the numerous publications on immunologic effects observed in animals exposed to MW and RF fields we have chosen and discussed those that in our opinion, based on many years experience, report experiments performed under acceptable conditions of exposure. We have not discussed the observations of medical and epidemiological investigations of human subjects exposed occupationally to MW and RF radiations. No conclusive results concerning evaluation of the immune status and possible health hazards related to immunologic and/or hematologic findings are possible, especially considering that in the epidemiologic observations only the peripheral blood was analyzed, without more specific tests. There are no screening tests for evaluation of the immune status that may be used in large populations of observed subjects. A recent review of human studies related to occupational MW/RF exposure (55) also does not discuss the immunologic responses.

An overview of the available literature and of our own findings suggests the existence of a biphasic reaction of the immune system to MW/RF radiations -- stimulation of the whole system (mainly of humoral immunity) after a single or a few days exposure, followed by gradual, but transient, suppression of the whole immunity with prolongation of the exposure period (up to several months) and/or increasing power density of the fields. Stimulation and suppression of immunity in MW/RF-exposed animals both seem to be transient and inconsistent phenomena. At low power densities the system recovers soon after the exposure. Thermal effects and the concomitant stress situation also stimulate numerous immune functions and should be viewed as a beneficial factor with potential therapeutic applications. There is some experimental evidence that whole-body MW hyperthermia may be beneficial in the treatment of viral infections.

CANCER-RELATED ASPECTS OF EXPOSURE TO LOW-LEVEL MICROWAVE FIELDS

Carcinogenesis has recently come to be viewed as a multistage process involving both cellular phenomena leading to neoplastic transformation and uncontrolled growth, and the host's antineoplastic systems including immune system surveillance mechanisms, non-specific cytotoxicity (NK cell activity), and interferon production (Figure 3).

The process of cancer development can be divided into three major stages: initiation, which is the occurrence in the cell nucleus of the decoding of oncogenes as a result of the action of an endogenous or exogenous factor, leading to formation of transformed cells; promotion, which is a selective survival of the transformed cells due to factors acting directly on cellular metabolism or membrane function (e.g., phorbol esters) or to factors influencing the host's antineoplastic resistance (e.g. immunosuppressive drugs, stress); and co-carcinogenesis, defined as mechanisms facilitating formation of neoplastic tumors, including their vascularization and spread (Figure 3).

Environmental factors, including MW and RF radiation, may potentially influence the process of carcinogenesis at various steps, either directly (carcinogenic effect) or indirectly, by triggering adaptation mechanisms that in turn may influence the natural antineoplastic resistance of the irradiated host. Potential carcinogenicity has been periodically discussed in relation to MW/RF exposure since 1953, when McLaughin (56) listed various forms of leukemias as one of the possible effects of occupational exposures to radar. More recently, similar suggestions have appeared in the report by Lester and Moore (57), who found significantly higher rate of incidence of cancer mortality in U.S. counties with Air Force bases, compared with

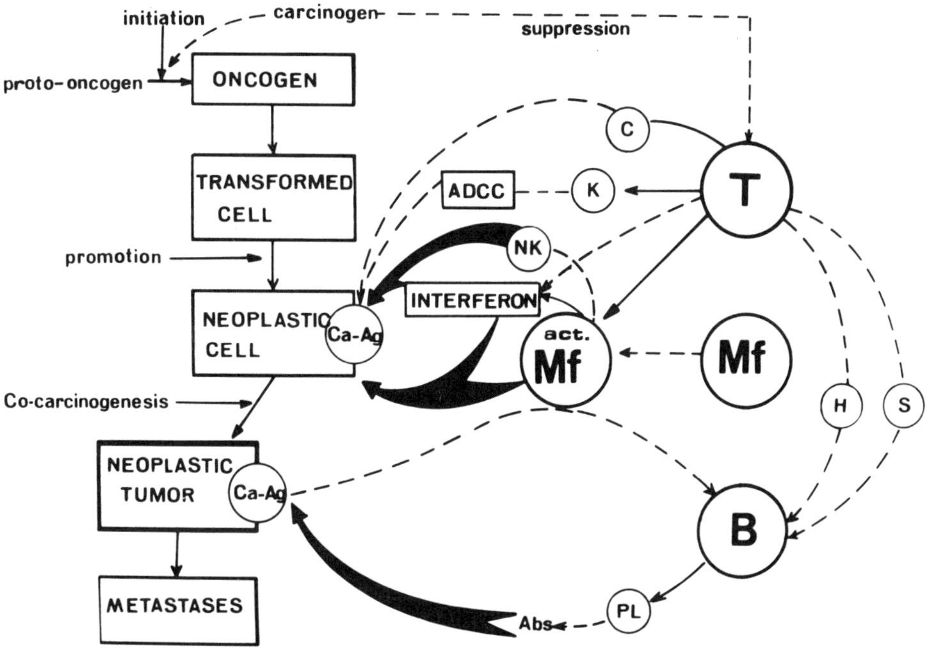

FIGURE 3. Simplified scheme of cancer development and the influence of immune system on carcinogenesis. A three-stage process of cancer development (initiation, promotion, co-carcinogenesis) is influenced by non-specific and specific cell-mediated and humoral reactions that may cope with small number of neoplastic cells and destroy them. Macrophages (Mf), activated macrophages (act.Mf), T and B lymphocytes (T, B), as well as helper (H) and suppressor (S) T lymphocytes cooperate in these reactions. Neoplastic cells can be destroyed by cellular (NK cytotoxicity) and humoral (interferon, antibodies, Abs) mechanisms.

counties without an Air Force base. They related the observed differences to prolonged environmental exposure to MW/RF radiation from radars operating at Air Force bases. However, the above data and suggestions have not been widely accepted, and were criticized after reevaluation (58). Nevertheless, the authors still support their original opinion on the relation of cancer mortality to radar radiation (59).

In a recent review on possible links between RF radiation and carcinogenesis, Kirk (60) stated that the question of RF-carcinogenesis or life shortening remained open, because none of the reports in the literature presented a convincing case for the existence of a significantly higher risk of cancer induction or life shortening in exposed populations. The situation has however changed in the last few years. Publications from our group (61,62) on accelerated development and growth of benzopyrene-induced skin cancer in mice exposed for months to non-thermal power densities of 2450-MHz MWs (10 mW/cm^2), indicated the tumor-promoting (but not carcinogenic) activity of this radiation, and prompted a few groups of investigators in American laboratories to undertake experimental studies in this field. Most of these studies are still in progress, but evidence of co-carcinogenic properties of MWs <u>in vitro</u> (a strong synergy between cancer-promoting phorbol esters and non-thermal MWs in increasing the cell transformation rates) was recently reported (63).

In view of the renewed interest in cancer-related aspects of MW/RF radiation, the recently available information, and the expected progress in this field, it seems desirable to summarize the actual state of knowledge and to stress the further needs and perspectives. Our group presently has several projects on cancer-related aspects of MWs in progress, some of them completed but not yet published, and we will discuss part of this material to give the reader a more complete picture of this intriguing problem.

EXPERIMENTAL OBSERVATIONS

The 1950-1975 literature (64) primarily involves reports of anecdotal and scientifically unsupported cases of neoplasms in workers exposed occupationally to MW/RFs. It does not support the view that these radiations may result in direct or indirect carcinogenic effects, despite the single experimental study of Prausnitz and Susskind (65), who reported an increased frequency of leukosis in mice exposed for 59 weeks to 9270 MHz at 100 mW/cm^2 (thermogenic field) for 4.5 minutes daily (5 days/week). The authors did not specify the type of leukosis, and did not observe the animals after termination of exposures. Considering the conditions of exposure, it seems likely that the observed leukosis was due to heat, which is a known stressor.

Several experimental reports dealing with the effects of MW/RF radiation on development of transplantable, spontaneous, or chemically-induced neoplasms in mice and rats were published during 1975-1985. Preskorn et al. (66) reported that development of tumors following injection of sarcoma cells into 16-day old CFW mice was delayed if the mice were exposed to MWs before birth (_in utero_) on days 11 through 14 of gestation (2450 MHz, 20 minutes daily, SAR of 35 mW/g, thermogenic field). The observation was never confirmed, and thus is difficult to evaluate. Nevertheless, it seems that the slower growth of the sarcoma resulted from stimulation or faster maturation of certain immune functions related to natural antineoplastic resistance (e.g., NK cell activity) in newborn mice after their exposure to MWs during organogenesis. This suggestion is to some degree supported by experiments reported by Smialowicz et al. (48), who exposed rats on day 6 of gestation through 41 days of age (_in utero_ and postnatally) to 2450-MHz MWs at 5 mW/cm^2 (SAR 1-5 mW/g) and found that the exposed young rats had lymphocytes that responded to a significantly greater extent to T- and B-lymphocyte mitogens _in vitro_. A similar increase in lymphocyte responsiveness _in vitro_ was observed by the same authors in young rats exposed pre- and post-natally (as above) to 425-MHz MWs (SAR, 3-7 mW/g)(67).

In experiments performed in our laboratories and reported in 1980-1982 (61,62,68) we demonstrated that daily (2 hour) exposures of BALB/c mice to 2450-MHz MWs at 5 or 15 mW/cm^2 (SARs of 2-3 and 6-8 mW/g, respectively) for 3-6 months, resulted in accelerated appearance and growth of skin neoplasms induced by benzopyrene (Figure 4), suggesting a tumor-promoting activity related to long-term exposure to low-level MW fields. Interestingly, the stress from confinement of unexposed mice used as positive controls, gave a similar acceleration of skin tumor growth as was observed following exposure at 5 mW/cm^2. Exposures at 15 mW/cm^2 resulted in faster appearance and development of tumors compared with both 5 mW/cm^2 and with the controls (61). Exposure of mice to 2450 MHz at the same power densities for 1-3 months, resulted in lowering of the natural antineoplastic resistance of the animals, as measured by the number of lung nodules (neoplastic colonies) formed after intravenous injection of a titered number of viable sarcoma cells

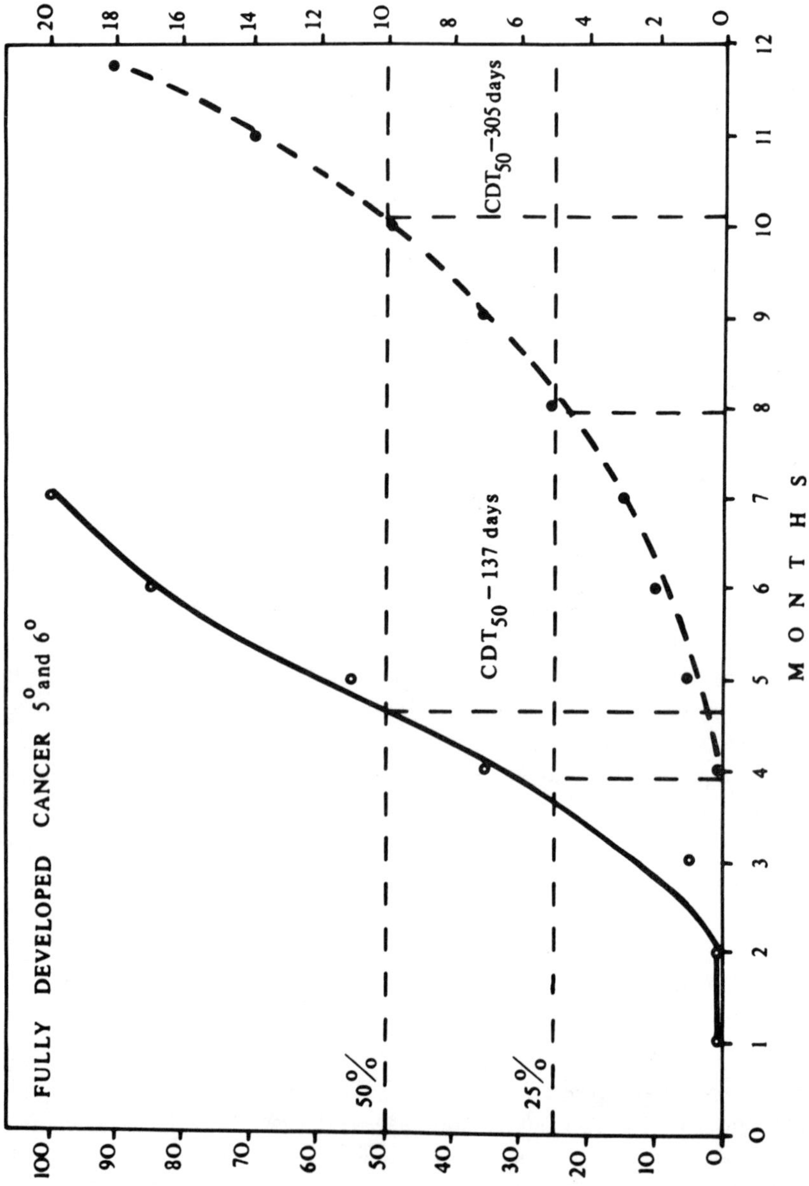
FULLY DEVELOPED CANCER 5° and 6°
50%
25%
CDT50 – 137 days
CDT50 – 305 days
M O N T H S
0 1 2 3 4 5 6 7 8 9 10 11 12
100 90 80 70 60 50 40 30 20 10 0
20 18 16 14 12 10 8 6 4 2 0

FIGURE 4.

(Figure 5). Mechanisms of the natural antineoplastic resistance (Figure 3) can cope with a certain number of implanted (or spontaneously developing) neoplastic cells and destroy them without further consequences for the organism. However, the capacity of the antineoplastic resistance is limited, and if a larger number of neoplastic cells is implanted, the lung nodules (colonies growing from single cells) appear. For BALB/c mice, intravenous injection of 2×10^5 sarcoma L-1 cells results in 1-3 visible colonies on the lung surface in 14 days; 1×10^5 cells do not lead to formation of colonies, while more than 5×10^5 sarcoma cells result in the appearance of numerous nodules. When the number of injected sarcoma cells is fixed (e.g., 2×10^5 cells, as in our experiments), an increase in the number of lung nodules is considered as being due to suppression of natural antineoplastic resistance of the organism (Figure 5). The results obtained after exposure of mice to 2450 MHz for 1-3 months (Figure 5) clearly indicated that irradiation at 15 mW/cm^2 (SAR of 6-8 mW/g) resulted in a significantly increased number of lung nodules. The increase was also significant at 5 mW/cm^2 compared with control animals, but the same effect was found following confinement (over-crowding) of unexposed mice (Oc at Figure 5), a known stressor for mice and used here as a positive control.

Although long-term exposure of mice to 5 mW/cm^2 and 15 mW/cm^2 2450-MHz MWs resulted in acceleration of the appearance and growth of two totally different neoplasms (benzopyrene-induced skin cancer and spontaneous mammary tumors in C_3H/HeA mice), and in the lowering of natural antineoplastic resistance of mice, it cannot be said with any confidence whether the effects are related to a direct action of MWs at the

---

FIGURE 4. Growth curves of 3,4-benzo-alpha-pyrene (BP)-induced skin tumor in mice exposed daily (2 hours) to 2450-MHz radiation at 10 mW/cm^2 (SAR of 4 mW/g) for a whole period of tumor growth. Note earlier appearance and faster growth of tumors in MW-exposed mice compared with sham-irradiated control. MW-exposed, continuous line; control, dotted line; CDT_{50}, cancer development time in 50% of animals.

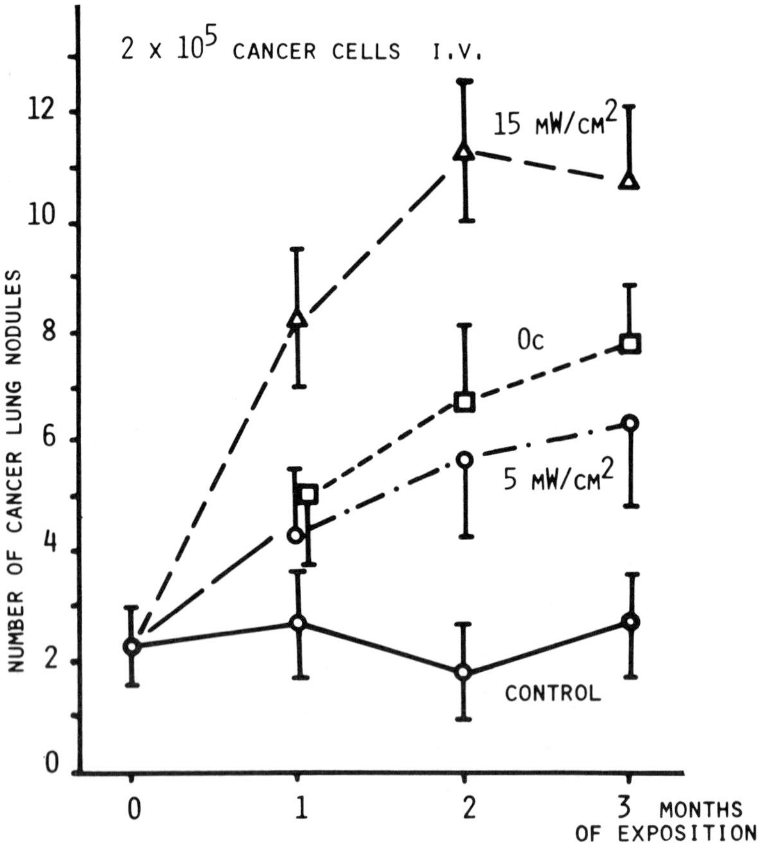

FIGURE 5. Number of lung tumors (following intravenous injection of 2 x 10^5 viable sarcoma cells) in mice exposed during 1, 2 or 3 months to 2450-MHz microwaves (2 hours daily) at 5 or 15 mW/cm^2. Oc, mice treated with nonspecific stress of overcrowding (confinement in cages, without exposure to microwaves), as positive controls; control, sham-irradiated mice (61).

cellular or subcellular levels, a non-specific stress reaction, or to general effects on immune response. We subsequently performed further experiments (not yet published) involving possible mechanisms leading to accelerated development of neoplasms in animals exposed for a long time to low-level MW fields. Because these experiments take much time (a few months of daily exposures, several weeks between application of the carcinogen and the appearance of a chemically-induced tumor), we decided to expose mice only at 10 mW/cm^2 (SAR of 4-5 mW/g). This allowed an increased number of mice in each experimental group. In all experiments the mice were exposed to 2450 MHz continuous wave in an environmentally controlled anechoic chamber for 2 hours daily (5-6 days weekly).

We confirmed acceleration of tumor development in MW-exposed mice after application of two other known carcinogens -- DENA (di-ethyl-nitroso-amine) and methylcholantrene. DENA, when injected intraperitoneally in mice at 10 or 50 mg/kg (single dose), results in the appearance of hepatic tumors after 2-6 months in about 80% of the treated animals. However, this carcinogen is also hepatotoxic and, depending upon dose, it also results in necrosis of hepatocytes. During regeneration of liver tissues, when hepatocytes that survived the toxic shock start to proliferate (3-7 days after injection of DENA), the carcinogenic activity of DENA and the neoplastic transformation of the hepatocytes begins. Thus, in this experiment we studied the survival of DENA-injected MW-exposed and sham-irradiated mice. After autopsy of each cadaver, histologic examination of the liver in search of neoplastic cells was performed. A significantly shorter survival time and earlier appearance of hepatic neoplasms were observed in the exposed animals. A similar acceleration of tumor growth and appearance was noted after administration of methylcholantrene and exposure to the radiation. Mice were injected subcutaneously with a single dose of this carcinogen, leading to development of sarcomas in 3-4 months. In groups of animals exposed 2 hours daily from the day of administration of methylcholantrene the sarcomas appeared in about 2 months, and grew faster. In both experiments we did not treat animals with any known tumor promoters (e.g. phorbol esters). The animals were exposed only to the carcinogen and/or MW radiation. Since we have never observed development of spontaneous neoplasms in mice exposed for a few months in MW fields (in strains not predisposed for spontaneous cancer), and since this phenomenon is not reported in the literature, we conclude that MWs are not carcinogenic, but that they may promote (or enhance the activity of other promoting factors) an already initiated process of carcinogenesis.

In another experiment we tested the reaction of mice to subcarcinogenic doses of 3,4-benzo-alpha-pyrene (BP). In general, the scheme of experiments was similar to that used in our earlier investigations (61). Mice were depilated and the skin on the back was painted every day with 0.01 ml of BP dissolved in a benzene-methanol mixture. Prior to painting, the mice were exposed for 2 hours to 2450-MHz MWs at 10 mW/cm^2. Mice treated with BP and without MW exposures served as controls. The subcarcinogenic dose of BP (the dose leading to the appearance of skin neoplasms in 10-20% of the animals) was established in earlier trials. A daily dose of 100 μg (0.01 ml of 1% solution of BP) resulted in the development of skin neoplasms in almost all mice in 4-10 months, while 3 and 10 μg (0.01 ml of 0.03% and 0.1% BP, respectively), applied twice weekly led to the appearance of neoplasms in about 15% of the treated mice. Thus, a subcarcinogenic dose of BP was established as 10 μg twice weekly, and this dose was applied in mice exposed to daily sessions of 2450 MHz MW exposures. It was found that in MW-exposed mice treated with subcarcinogenic doses of BP, 40-50% of animals exhibited skin neoplasms compared with about 15% in those treated with BP alone. The increased frequency of skin neoplasms in MW-exposed mice treated with subcarcinogenic doses of BP was qualitatively different from the earlier discussed acceleration of appearance and growth of neoplasms induced by full carcinogenic doses of BP and other carcinogens. Although it still must be confirmed, the increased frequency of neoplasms after subcarcinogenic doses of BP and exposure of mice to MWs indicates that long-term exposure in non-thermal MW fields may promote development of neoplasms that normally would not reach the clinically detectable stage, independently of the underlying mechanisms.

In another experiment we investigated intracellular levels of cAMP in murine skin epithelium (scraped) from animals treated with daily doses of 100 μg of BP and/or exposed to 2450-MHz MWs at 10 mW/cm^2 for 2 hours daily. The observations were made during one month of exposure and three experimental groups were used: BP alone, BP and MW exposure, and MW exposure alone, with sham-irradiated mice serving as controls. The results are summarized in Figure 6.

Daily exposures resulted in a fast increase of intracellular level of cAMP and in its reactivity with isoproterenol (IPR)(a substance used widely for testing the beta-adregenic response of membrane receptors). The elevated levels of cAMP

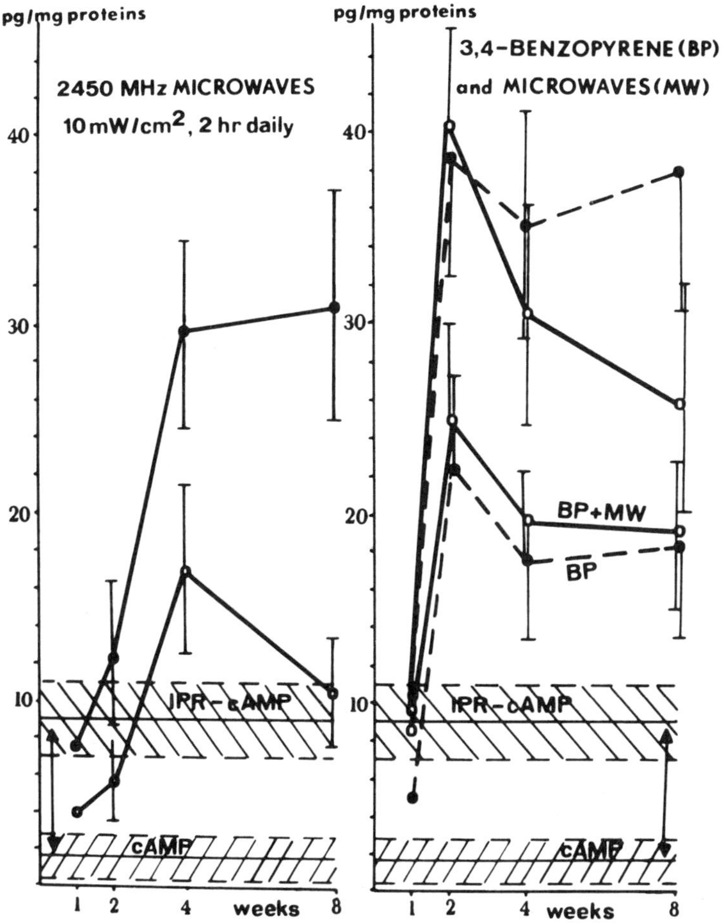

FIGURE 6. Cyclic AMP (cAMP) (RIA kit, Amersham No. TRK-432) in murine epidermis (scraped) in animals exposed to 2450-MHz microwaves (2 hours daily) for one month (left graph), and in animals exposed to daily doses of 100 μg of 3,4-benzo-alpha-pyrene (BP) and to BP and 2450 MHz microwaves (BP + MW). Basic levels of cAMP are represented by lower curves, with normal levels (in healthy mice) and a standard deviation range presented as a bottom line-filled region. The upper curves and upper line-filled regions represent experimental and control data (with standard deviation) of cAMP levels in epidermis after treatment with isoproterenol (IPR), a substance influencing beta-adrenergic receptors and leading to increase of intracellular cAMP. Note increase of cAMP levels and reactivity to IPR in murine epidermis after exposure of animals to 2450-MHz MWs at 10 mW/cm^2, increase after treatment with BP, however without further elevation in mice treated with BP and exposed to microwaves (BP + MW).

and the enhanced reactivity with IPR were clearly demonstrable after 14 days of daily exposure in MW fields (Figure 6). This intriguing observation cannot be explained at present. Levels of cAMP were never before examined in cells or animals exposed to weak MW fields, and there are many known factors that cause an increase of cAMP, including nonspecific stresses. It is accepted that the function of skin is influenced by stress and various reactions from the central nervous system (69). ACTH, adrenal steroids, catecholamines, histamine, and serotonin, released during stress situations, also influence function of skin (70). Thus, the observed increase of cAMP in skin epidermis of mice exposed to MW fields may be due to the concomitant stress reaction. Elevation of intracellular cAMP level in epidermis and enhanced reactivity to IPR were also reported by Murray and Verna (71) for epithelial cells exposed to BP, and by Mufson et al. (72,73) for other carcinogens. It is believed that the reactivity of the cAMP system is caused by a direct influence of carcinogens and/or promotors on cell growth regulatory mechanisms regulated by the cAMP system (74). Surprisingly, MW exposures that alone caused an increase of intracellular cAMP, did not lead to further elevation of this nucleotide when applied to BP-treated mice (Figure 6). Further experiments are needed to explain these observations.

In summary, long-term exposure of mice to 2450-MHz MWs resulted in acceleration of the appearance and growth of tumors initiated by three different carcinogens, and a higher risk of cancer development in animals exposed to subcarcinogenic doses of initiators. The results suggest a tumor-promoting activity of the radiation, but the cause of these phenomena still remains an open question. The experiments were performed only in one animal species and with a single frequency of MWs (2450 MHz). It should be stressed that 2450-MHz MWs is a resonant frequency for mice (maximal absorption of energy), and the situation may be different in larger animals and in human beings. Nothing is known about cancer-related effects of pulse-modulated MW/RF radiation <u>in vivo</u>, especially those sinusoidally and amplitude-modulated at 3-100 Hz that, according to Adey and his group, may directly influence cell membranes and cell metabolism.

An increased number of spontaneous malignant tumors in rats exposed for their life-span to MW fields was recently reported

(53,75). This was the largest single study ever made of the long-term effects of MW exposure. Two hundred Sprague-Dawley rats were maintained under pathogen-free conditions and exposed continuously in a unique circularly-polarized waveguide facility to pulsed 2450-MHz MWs (800 pulses per second with a 10 μsec pulse width) at 0.48 mW/cm^2 (SAR of 0.15-0.4 mW/g). Another 200 rats were maintained in sham-exposure waveguides and served as controls. The presence of neoplasms in spontaneously dying animals from both groups was established on the basis of necropsy and histopathologic examinations (about 40 samples from different organs and tissues were taken from each rat). A total of 192 neoplastic lesions (most of them occurring at 19-30 months) were observed, and there were 83 unique combinations of organ and neoplasm-specific diagnoses. The authors divided the observed lesions into benign neoplasms (115 cases), and primary (23) and metastatic (54) malignancies. The frequency of benign neoplasms did not differ significantly between the exposed and unexposed groups (62 cases versus 53 cases), but differences were noted for primary and metastatic malignancies. In the exposed group, 54 malignancies were found (18 primary and 36 metastatic), while in sham-exposed controls only 23 (5 primary and 18 metastatic), the difference being significant at $P < 0.05$. The authors claimed that (75) "the primary tumors occurred earlier in the exposed group than in the sham-exposed," but they were unable to find a predominance of a specific type of malignancy or tissue of origin. On the basis of neoplastic lesions and diagnoses established for each animal, as listed in the report (75), we tried to reevaluate the frequency of malignancies developing from the hemato-immunologic system (all forms of leukemias and lymphomas, found in different organs). In the MW-exposed animals, 31 cases of hemato-immunologic malignancies were found, while only 19 were found in the sham-exposed animals; the difference was not significant ($\chi^2 = 3.29$, $P = 0.07$), probably due to the relatively small size of groups in terms of morbidity rate from spontaneous neoplasms (200 rats per group). On the other hand, while listing the primary cause of death of their rats, the authors related only 8% mortality in the exposed group and 3% in sham-irradiated controls to neoplasms (their own reevaluation of the published data). This indicates that most of the diagnosed malignancies were established on the basis of histologic examinations rather than on the presence of clinically detectable disease or tumors. It is well known from analysis of mortality data of large groups of older (above 60 years) human subjects who died from various causes other than malignancies, that careful histopathologic examination frequently reveals (up to 40-60% of subjects) small groups of silent neoplastic cells (most frequently in the thyroid gland) that do not have at that moment any clinical significance, and which probably would never develop into clinically detectable neoplasms. A similar situation may also occur in older rats, and we therefore believe that when studying the link between cancer morbidity and MW/RF fields, only clinically detectable malignancies should be taken into account.

The data (75) indicating an increased number of spontaneous malignancies in MW-exposed rats are provocative but far from fully acceptable. The significant difference in frequency of malignancies between the exposed and unexposed group, without differences in any specific type of malignancy or its tissue origin, may be considered only as a trend to increased risk of cancer after prolonged MW exposures. The phenomenon needs further confirmation. For valid conclusions concerning the rate of spontaneous malignancies, the size of the observed groups must be much larger (the order of 1000), and the rates of spontaneous neoplasms in rats must be considered. The data supports the concept of tumor-promoting, but not direct carcinogenic (initiating) activity of MW radiation. In the case of promoting agents the spontaneously-initiated process of cancer development is facilitated, and there are no preferences in organ or type of malignancy. Any group of transformed neoplastic cells that spontaneously appears in any organ or tissue (let us assume even that it appears with the same frequency in MW-exposed and unexposed animals) will be promoted to develop into a detectable neoplasm by MW exposure. Thus, an increased number of malignancies will be found without preference of organ or type.

In summary, the results of studies using different animal models provide only scanty evidence that exposure to low-level MW/RF radiation may in certain cases influence the complicated process of carcinogenesis, with tumor-promoting (direct or indirect) activity possibly being a general phenomenon. The mechanisms leading to these effects are still unknown. The field remains open for good research projects with adequate exposure data, free from possible artifacts.

Another line of search for possible links between weak electromagnetic fields and carcinogenesis is the investigation of cellular/subcellular effects on the process of carcinogenesis _in vitro_. There are presently available valid techniques for neoplastic transformation of cells cultured _in vitro_, including studies on the effect of promoters, and thus it is very surprising that these techniques were applied to bioelectromagnetic studies only very recently. Up to the early 1980s, no laboratory was interested in the influence of MW/RFs on the process of carcinogenesis and promotion _in vitro_ until Balcer-Kubiczek and Harrison (63) reported a strong synergy between phorbol esters (known promoters of carcinogenesis, active _in vitro_ and _in vivo_) and non-thermal MW fields in increasing the transformation rates of cells cultured _in vitro_ and treated formerly with benzopyrene. In 1981, this group started their investigation on the influence of weak 2450-MHz MW fields on carcinogenesis _in vitro_. In the first experiments they were unable to find any effect after exposure of cell cultures to MW fields in combination with BP or x-rays without the addition of promoters (76). In later experiments (63), they investigated the carcinogenic activity of 2450-MHz pulsed MWs (120 pulses per second, with an 83 μsec pulse width) combined with BP or x-rays, using an _in vitro_ assay for malignant transformation in $C_3H/10T1/2$ line of murine embryonic fibroblasts (established cell line). Experiments were performed at low power densities, not leading to elevation of temperature and corresponding to an SAR of 4.4 mW/g. Additional experiments were performed to assess the effect of a non-cytotoxic and non-transforming concentration of the promoter of carcinogenesis (one of the widely applied phorbol esters, TPA) on induction of transformation in cells exposed to MWs and to x-rays (used as a transforming factor). MWs had no effect on transformation induced by BP or x-rays in the absence of tumor promoter (TPA). On the other hand, treatment of cells previously irradiated with MW and x-rays with TPA (0.1 μg/ml) yielded a statistically significant increase in transformation by a factor of 1.6-3.5 compared with the transformation rates of cells irradiated with x-rays alone and treated with TPA. The authors concluded that 2450-MHz MWs can induce latent transformation damage, which can then be revealed by the action of promoters of carcinogenesis (TPA). The results also suggested that the cell membrane might be a sensitive target for the influence of low-level MW fields. It should be stressed that Balcer-Kubiczek and Harrison applied MW radiation modulated at 120 Hz pulses, which was close to the modulation shown by Adey and his group to exert specific effects on cellular membranes independently of the frequency of the carrier waves. It would be interesting to test the Harrison's experimental model with continuous 2450-MHz MWs at different intensities to estimate whether the increase in neoplastic transformation is related to the carrier wave or to its modulation.

As mentioned before, Byus et al. (10) found strong inhibition of histone protein kinase in cultured human lymphocytes exposed to 450-MHz MWs, sinusoidally modulated at 16 Hz, with a dependence of the effect on the modulation frequency. Recently, the same authors (11) extended their studies in ways that appear directly to relate to cancer promotion. They investigated the effect of cancer-promoting phorbol esters (TPA) on the activity of ornithine decarboxylase (an enzyme present in all nucleated cells, being essential for synthesis of polyamines, which in turn, are required for DNA synthesis and cell growth) in cultured hepatoma cells _in vitro_ (Reuber H 35 line). Activity of the above decarboxylase was increased 1.5-fold during exposure (1 mW/cm^2), and the elevated activity persisted for several hours after a 1-hour exposure. Also, the increased activity evoked in the cultured cells by TPA was potentiated by prior exposure to the modulated MWs. The authors concluded that the cell membranes were the site of transductive coupling of MW/RF fields modulated at low frequencies, because the tumor promoter TPA has a specific cell membrane protein kinase receptor (the calcium-phospholipid kinase or protein kinase C), and the synergy between TPA and the modulated MW fields was consistent with this common site of action (11).

On the basis of Balcer-Kubiczek and Harrison's reports, and the above investigations of his own group, Adey (personal communication) recently offered his own concept and initial model of the cancer-promotion process and its influence by MW/RF fields modulated at low frequencies. The promotion appears to relate to a distorted inward stream of signals from the cell membrane to the nucleus (where carcinogenesis was already initiated by other factors) and to intracellular organelles. MW/RFs modulated at low frequencies may in certain cases (depending upon modulation and time of exposure) act synergistically with the action of promoters, activating the same membrane receptors.

HUMAN STUDIES

Before the suggestions of McLaughlin (56) and Lester and Moore (57), who linked increased risk of leukemia with radar exposures, there appeared two letters to the Editor of the New England Journal of Medicine (77,78) suggesting an association of polycythemia vera and leukemias with occupational exposure to a variety of electromagnetic fields, and three letters to the Editors of Lancet (79-81) with similar observations. A higher incidence of cancer in the electronic industry workers was also postulated by Vagero and Olin (82) on the basis of the Swedish Cancer Environment Registry data. All these reports are based on analysis of death certificates and relation of profession of the leukemia victims with probable exposures to unknown intensities of a variety of electromagnetic fields in electronics, electric engineers, radio and television repairmen, and so forth. None of the above analyses meets the criteria for scientifically valid statistics, nevertheless they stress the need for retrospective and prospective studies resolving the questions raised.

The relevant and acceptable study on delayed health effects in U.S. Navy personnel exposed to radar during the Korean War (83) should be mentioned here. No significant differences were found by Robinette et al. (84) and Silverman (85) between the high and low exposure groups for malignant neoplasms as the cause of hospitalization and/or death (from records of Navy and VA hospitals). However, when three sub-groups of the high-exposure group were developed to provide a gradient of potential exposure, a trend appeared for increased number of malignant neoplasms in the subgroup rated as highly exposed. The weak point of this analysis is the fact that only subjects hospitalized in Navy or VA hospitals were analyzed, and only during a certain period of time (1950-1974). These subjects were only a part of the total population of U.S. military personnel that operated in Korea during 1950-1954, and it is difficult, for example, to evaluate which part of the total population was hospitalized or died during 1950-1974.

Very recently we completed a retrospective study on neoplasm morbidity in Polish military career personnel with relation to present and past occupational exposure to RF/MW fields. Since the obtained results, described in detail in a report of limited distribution (86), will be published only in part and with a certain delay, we consider discussion of our data to be desirable here. The total population of career servicemen was analyzed, and a subgroup of personnel exposed occupationally to MW/RF radiation (on the basis of service records) was developed (Figure 7); the E (exposed) group counted about 3% of total population (Figure 8), the rest (97%) was considered as subjects without exposure to MW/RFs (the NE group).

The E group was composed of personnel working on production, repair and use of devices emitting MW/RFs, as well as of those engaged in teaching and research with use of MW/RFs. The accuracy of occupational exposure in the E group was determined on the basis of past and current service records, and on medical records which contained the results of periodic examinations that were introduced in Poland in 1968 for all servicemen exposed to MW/RFs. The extent of daily exposure, power density, frequency, and modulation varied with each individual in the E group. In general, exposure to various types of radar radiation predominated, but exposures to extremely low frequency fields were also noted. For MW radiations, a typical exposure was estimated as 4-8 hours daily at power densities below 0.2 mW/cm^2 ("safety zone," according to rules operating in Poland) with incidental (several minutes) daily expsures at 0.2-1 mW/cm^2. However, mainly in personnel working on production and repair of MW devices, incidents of short-lasting exposures to higher power densities (estimated up to 10-20 mW/cm^2) were reported. These exposures resulted from defying safety rules and were difficult to evaluate. The high-intensity exposures were more frequent in the 1960s, when the safety rules were not yet strictly enforced, but still occurred in the 1970s, despite awareness of the possible health hazards of MW/RF radiation. Thus, in practice, it was not possible to estimate precisely the intensity of MW/RF

Aim of analysis :	MORBIDITY RATE OF NEOPLASMS (1971-80)
Type of analysis :	RETROSPECTIVE
Populations :	MILITARY PERSONNEL (100 %)
Subpopulations :	EXPOSED (MW/RF) (3%) NONEXPOSED (MW/RF) (97%)
Data collections :	NEW CASES OF NEOPLASMS (yearly 1971-80) CUMULATIVE YEARLY MORBIDITY RATE (per 10^5/year)
Subgroups analyzed :	AGE : 20-29; 30-39; 40-49; 50-59
	PERIOD OF EXPOSURE (MW/RF) <2; 2-5; 5-10; 10-15; >15
	LOCALIZATION OF NEOPLAMS 12 + BLOOD & LYMPHATIC T 6
Stastistical methods :	Chi-square 2×2 R×C contingency tables ANOVA (subgroups) $p < 0.01$ $p < 0.05$

FIGURE 7. Retrospective epidemiological study of cancer morbidity in personnel exposed occupationally to microwave and radiofrequency (MW/RF) radiations.

exposures for the whole E group, due to the large individual differences. We divided the exposed subjects into 5 classes; below 2 years, 2-5, 5-10, 10-15, and above 15 years (Figure 7).

Analysis of cancer morbidity was performed for 1971-1980. All cases of neoplasms that were diagnosed during this decade were registered (on the basis of data from all department hospitals and medical commissions with care to avoid duplication of cases) and each victim was classified into the E or NE group based on service records and, if possible, an interview. Only subjects aged 20-59 years at the time of diagnosis (with division into four age groups) were considered. Retiring personnel above 60 years of age were not analyzed because their service records were sometimes doubtful (mainly for those

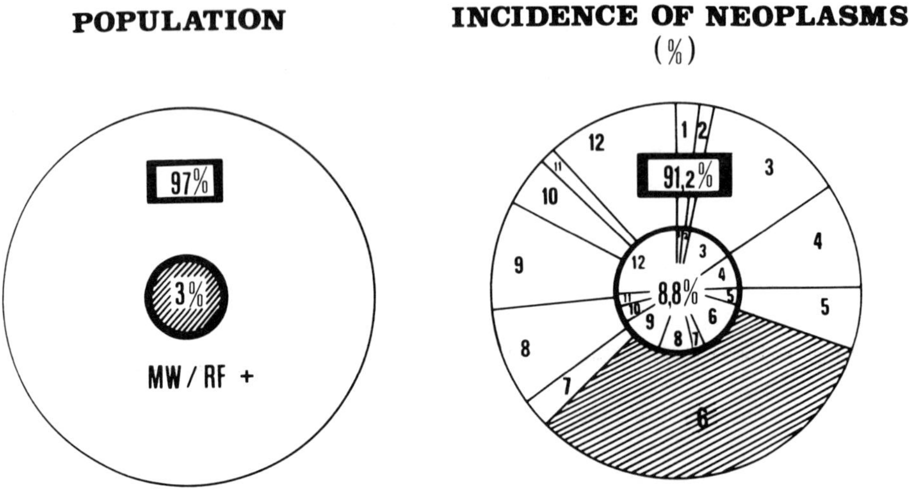

FIGURE 8. The population and incidence of neoplasms in a retrospective epidemiological study of cancer morbidity in personnel exposed occupationally to microwave and/or radiofrequncy (MW/RF+) radiations. Note that the group of personnel exposed to MW/RFs was about 3% of the total population (left circle) and developed 8.8% of all the clinically detectable malignancies found in the total population (right circle). The circle areas marked with numbers 1-12 (right circles) represent frequency of neoplasms in: 1, oral cavity; 2, pharynx and larynx; 3, esophagus, stomach; 4, colo-rectal; 5, liver, pancreas; 6, lungs; 7, bones; 8, skin, including melanoma; 9, kidneys, urogenital tract, prostate; 10, eyes, central nervous system; 11, thyroid gland and other endocrine glands; 12, hematopoietic and lymphatic organs. Note the highest incidence of lung cancer (6) in unexposed group and the highest incidence of hemato-lymphatic malignancies (12) in MW/RF exposed group (right inner circle).

retiring in the early 1970s), and it was not possible to evaluate the occurrence of MW/RF exposures in the far past.

Twelve kinds of neoplasms were differentiated (Figures 8 and 9), and for neoplasms originating in the hemato-lymphatic organs, 6 types of diagnoses were analyzed (Figure 10). From all cases of neoplasms diagnosed during 1971-1980, including their localization and classification in the E or NE group, a cumulative yearly morbidity rate was calculated as: [(total number of neoplasms during 1971-1980)/(mean yearly number of personnel in E or NE group during 1971-1980 x 10)] x 10^5. This gave a morbidity rate of neoplasms expressed as the number of cases per 100,000 subjects per year. The morbidity rate was used for presentation of all results.

The results are summarized in Figures 8-12. In the E group, which was about 3% of the total population, about 8.8% of all the neoplasms appeared (Figure 8). Assuming uniform distribution of neoplasms in the total population, the group E should have accounted for about 3% of all neoplasms (2-4% confidence limits for the size of the population analyzed, $P < 0.05$). This means that the frequency of neoplasms was about 3-fold higher than expected in the E group. The morbidity rate for the NE group was 64.2 cases of various neoplasms per 100,000 per year (at all ages analyzed), while in the E group it was 192.2 cases/100,000/year (Figure 9). Organ localization of the neoplasms (Figures 8 and 9) revealed that the difference between E and NE groups depended mainly on the higher number of neoplasms in esophagus, stomach, colo-rectal region, skin, thyroid gland, and most of all on neoplasms originating from hemato-lymphatic organs. The morbidity rate for all hemato-lymphatic neoplasms was found to be 7.4 cases/100,000/year for the NE group, and 50.8 cases/100,000/year in the E group, the last being about 7 times higher compared with the NE group. Interestingly, no significant difference was found for morbidity from lung cancer, which was the most frequent type of malignancy in the analyzed population. In the NE group (all ages, 20-59 years) the rate for this cancer was 23.6 cases/100,000/year versus 33.2 cases/100,000/year for the E group. However a significantly higher rate for lung cancer was observed in the E group at the ages of 40-49 years. Because hemato-lymphatic malignancies were the most frequent diagnosis in the E group, we analyzed these neoplasms in more detail (Figure 10). It was found that in the E group a higher frequency of lymphatic sarcomas and other lymphomas (but not malignant lymphogranulomatosis), acute lymphoblastic leukemia at a young age, and chronic and acute myelocytic leukemias were found, while the morbidity rate for chronic lymphatic leukemia, as well as for malignant lymphogranulomatosis did not differ for all age groups, but appeared earlier in the E group (at age 40-49, instead of 50-59) (Figure 10).

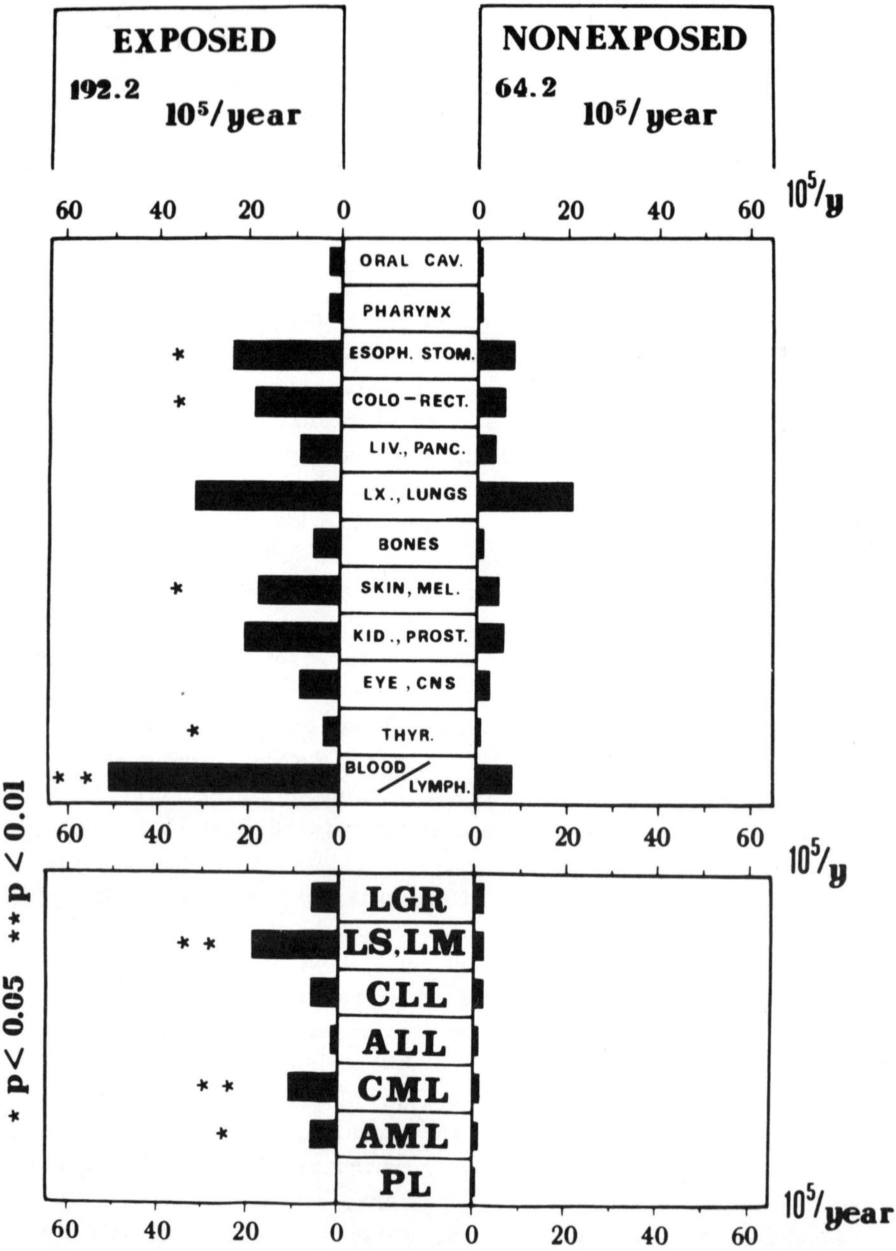

FIGURE 9. Cumulative yearly morbidity rate of neoplasms during 1971-1980 (expressed as a number of new cases per 100,000 subjects per year) for all ages (20-59 years) in exposed (to MW/RF radiations) and non-exposed personnel. Top histograms, organ localization of malignancies; abbreviations equivalent to

Analysis of the morbidity rates for the four developed age groups (20-29, 30-39, 40-49 and 50-59) showed that the largest difference occurred at 40-49 years. In non-exposed personnel, the frequency of neoplasms (all kinds) did not reach 50 cases/100,000/year, while in the E group a rate of about 350 cases/100,000/year was found (Figure 11). In all other age groups the differences in morbidity were also statistically significant, but not as spectacular as that in the 40-49 group (Figure 11). There was also a high correlation of period of exposure to MW/RF fields with the morbidity rate of neoplasms (Figure 12), and with the coefficient of linear correlation for all cases of malignancies, all age groups, and all five classes of period of exposure ($r = 0.87$). The relation of cancer morbidity rate to period of exposure is best seen in the 40-49 year age group (Figure 12), where there were about 70 cases/100,000/year for those working in MW/RF environment for 2-5 years, about 390 cases/100,000/year for those working 5-10 years, and about 450 cases/100,000/year for those working 10-15 years in the fields. A relatively lower morbidity rate in this group (40-49 years) for personnel working above 15 years in the MW/RF environment (about 270 cases/100,000/year) seems to result from two causes. First, most of the personnel at the age of 40-49 had a 5-15 year period of exposure to MW/RFs, and thus the group with exposure during more than 15 years was relatively small in terms of cancer morbidity rates. Second, at age 40-49, many subjects avoided work with MW/RFs and moved to other duties (command, administration) and, although they were still listed in the E group due to past exposures, the exposures no longer continued.

---

numbers 1-12 in Figure 8. Note significant differences in morbidity rate of malignancies in the alimentary tract, skin and malignancies originating from hematopoietic (blood) and lymphatic organs, and no differences in rates for lung cancer (1x, lungs). Bottom histograms, morbidity rate for specific types of malignancies originating from hematopoietic and lymphatic organs: LGR, malignant lymphogranulomatosis; LS, LM, lymphosarcomas and lymphomas; CLL, chronic lymphocytic leukemia; ALL, acute lymphoblastic leukemia; CML, chronic myelocytic leukemia; AML, acute myeloblastic leukemia; PL, plasmocytoma (plasma cell leukemia).

AGE GROUP	Population TOTAL	Population EXPOSED / NON EXPOSED	LGR	Ly Sa Lymp.	CLL	ALL	CML	AML	PL
20–29	3.6	26.3 / 2.7	18.8 / 2.1	/ 0.3		8.8 / 0.3	8.8 /		
30–39	3.8	29.7 / 3.0	/ 0.9	/ 0.3	9.9 / 0.3		19.8 / 1.2	/ 0.3	
40–49	11.8	81.3 / 9.9	11.6 / 2.5	46.5 / 4.2	11.6 / 1.4		/ 0.3	11.6 / 1.8	
50–59	32.7	117.6 / 29.6	/ 3.0	58.8 / 8.9	/ 8.9		29.4 / 1.1	29.4 / 5.9	
TOTAL	8.8	50.8 / 7.4	6.0 / 1.8	18.3 / 2.2	6.1 / 1.3	3.0 / 0.1	12.2 / 0.5	6.1 / 1.1	/ 2.2
χ^2		70.01	2.88	29.85	1.11	14.65	45.32	7.52	0.06
P		< 0.01	NS	< 0.01	NS	< 0.01	< 0.01	< 0.05	NS
RISK FACTOR		6.7		8.3		7.8	9.6	5.5	

FIGURE 10. Morbidity rate of hematopoietic and lymphatic malignancies (number of cases per 100,000 subjects per year) in personnel exposed occupationally to microwave and radiofrequency radiations and in unexposed controls in different groups of age (20–29, 30–39, 40–49, and 50–59 years). LGR, malignant lymphogranulomatosis; Ly Sa, Lymp., lymphosarcomas and other lymphomas; CLL, chronic lymphocytic leukemia; ALL, acute lymphoblastic leukemia; CML, chronic myelocytic leukemia; AML, acute myeloblastic leukemia; PL, plasmocytoma (plasma cell leukemia). Note the largest differences in the 40–49 age group and earlier appearance of malignancies in the MW/RF-exposed group.

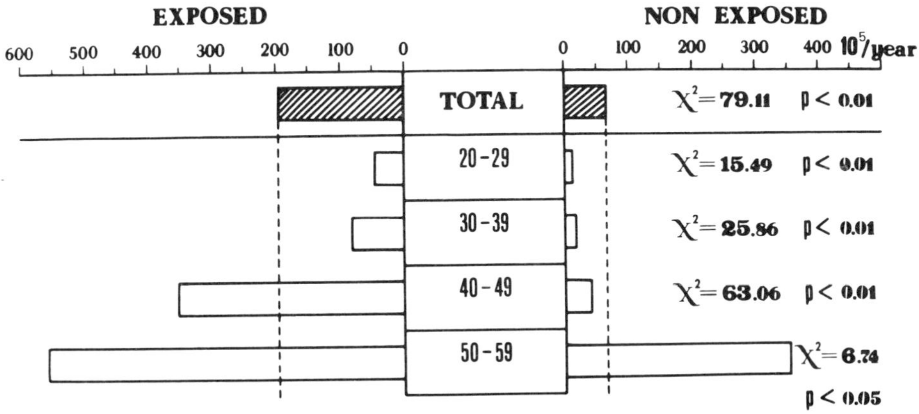

FIGURE 11. Cancer morbidity rates in exposed (to MW/RF radiations) and non-exposed personnel (all types of malignancies) at various age groups (20-29, 30-39, 40-49, 50-59 years). Note the largest differences at the age group of 40-49 years and statistical significance of differences for all age groups.

In summary, from a retrospective study that covered a large and well-controlled population with a known population of subjects, and that had a relatively long period of observation (1971-1980) the following conclusions may be drawn:

- The risk of developing clinically detectable neoplastic disease was about 3 times higher for personnel exposed occupationally to MW/RF radiations. The highest risk appeared for malignancies originating from the hemato-lymphatic systems (morbidity about 7 times higher). Other more frequent neoplasms were located in the alimentary tract and in skin (including melanomas).

- The highest risk factor of cancer morbidity related to occupational exposure to MW/RFs appeared for subjects at the age of 40-49 who had a 5-15 year period of exposure.

- Morbidity rates of neoplasms in personnel exposed occupationally to MW/RFs showed strong correlation with the period of exposure.

- Neoplasms of the same localization and/or type developed earlier (by about 10 years) in personnel exposed occupationally to MW/RFs than in those not working in the MW/RF environment.

AGE GROUP	PERIOD OF EXPOSURE (years) below 2	2-5	5-15	10-15	above 15	TOTAL (=EXPOSED)
20-29	32.3	58.8				44.2
30-39		64.5	96.1	166.7		81.7
40-49		71.4	392.9	454.5	272.7	348.8
50-59			500.6	611.1	666.7	617.6
Total	32.3	62.5	214.3	247.8	478.3	196.2

$\chi^2 = 40.38$

$p < 0.001$

Coefficient of correlation $C = \sqrt{\frac{\chi^2}{\chi^2 + N}} = 0.57$

FIGURE 12. Cancer morbidity rates (for all types of malignancies) in personnel exposed to microwave and/or radiofrequency radiations in relation to age groups (20-29, 30-39, 40-49, 50-59) and period of occupational exposure (below 2 years, 2-5, 5-10, 10-15, and above 15 years). The rates represent number of new cases of malignancies per 100,000 subjects per year. Note the highest relation to period of exposure in the 40-49-year group and significance of the whole table (contingency tables RxC) with coefficients of correlation C = 0.57, and Pearson's coefficient of linear correlation r = 0.87 for total data (not shown).

The above findings are intriguing and disturbing for epidemiologists, medical officers, as well as for society as a whole. It must be emphasized that results of retrospective epidemiologic studies are valid only for the population analyzed and the period of observation covered. Further, despite the correlations found and values of correlation coefficients, they do not provide certain evidence of a causative relationship between the effect and the factor investigated. In our population, we assume that exposure of subjects to other harmful and possibly carcinogenic factors, including smoking and drinking habits, in both the exposed and non-exposed groups were similar and we have no evidence to think differently.

At present we cannot offer a convincing explanation for the observed facts of increased risk of cancer in subjects exposed occupationally to MW/RFs, and do not relate this finding directly to interactions of the radiation with the human organism at any level, until the recently started prospective studies of the same population, planned for 1985-1990, is completed. Nevertheless, the available results point to an urgent need for further epidemiological studies, both retrospective and prospective, in well-controlled populations of people exposed occupationally and/or incidentally to a variety of nonionizing radiation, as well as for elucidation of cancer-related problems in experimental investigations.

REFERENCES

(1) Kinn, J.B. and Postow, E.: Index of Publication on Biological Effects of Electromagnetic Radiation (0-100 GHz), U.S. EPA Report No. 600/9-81-011, Res. Triangle Park, NC, 1981.

(2) Kleinstein, B.H., ed.: Biological Effects of Non-ionizing Electromagnetic Radiation: A Digest of Current Literature, U.S. Dept. Commerce, Rep. NTIA-CR-81-15, Volumes 1-6, NTIS, Springfield, VA, 1982.

(3) Elder, J.A. and Cahill, D.F., eds.: Biological Effects of Radiofrequency Radiation, U.S. EPA Report No. 600/8-83-026 F, Res. Triangle Park, NC, 1984.

(4) Szmigielski, S.: Immunological response of mammals to microwaves, in Biomedical Thermology, C. Alberts and M. Gautherie, eds., Alan R. Liss, New York, 227-246, 1982.

(5) Allis, J.W.: Cellular and subcellular effects, in Biological Effects of Radiofrequency Radiation, J.E. Elder and D.F. Cahill, eds., US EPA Report No. 600/8-83-026 F, Res. Triangle Park, NC, 1-12, 1984.

(6) Baranski, S. and Czerski, P.: Biological Effects of Microwaves, Dowden, Hutchinson and Ross, Stroudsburg, PA, 1976.

(7) Minin, B.A.: Ultra-high Frequencies and Human Safety (in Russian), Energoatomizdat, Moscow, 1974.

(8) Adey, W.R.: Ionic nonequilibrium phenomena, in Tissue Effects of Nonionizing Radiation, K.H. Illinger, ed., ACS Symposium Series, 157:271-297, 1981.

(9) Lyle, D.B., Schechter, P., Adey, W.R. and Lundak, R.L.: Suppression of T-lymphocyte cytotoxicity following exposure to sinusoidally amplitude-modulated fields, Bioelectromagnetics 4:281-292, 1983.

(10) Byus, O.V., Lundak, R.L., Fletcher, R.M. and Adey, W.R.: Alterations in protein kinase activity following exposure of cultured human lymphocytes to modulated microwave fields, Bioelectromagnetics 7:341-351, 1984.

(11) Byus, O.V., Kartun, K., Pieper, S. and Adey, W.R.: Microwaves act at cell membranes alone or in synergy with cancer-promoting phorbol esters to enhance ornithine decarboxylase activity, Bioelectromagnetics 7:1986.

(12) Smialowicz, R.J.: Biological effects of RF radiation: Hematologic and immunologic effects, in Biological Effects of Radiofrequency Radiation, J.A. Elder and D.F. Cahill, eds., U.S. EPA Report No. 600/8-83-026 F, Res. Triangle Park, NC, 1984.

(13) Stolodnik-Baranska, W.: Lymphoblastoid transformation of lymphocytes in vitro after microwave irradiation, Nature 214:102-103, 1967.

(14) Czerski, P.: Microwave effects on the blood-forming system with particular reference to the lymphocyte, Ann. N.Y. Acad. Sci. 247:232-242, 1975.

(15) Smialowicz, R.J.: The effect of microwaves (2450 MHz) on

lymphocyte blast transformation in vitro, in Biological Effects of Electromagnetic Waves, C.C. Johnson and M.L. Shore, eds., Vol. 1, HEW Publ. FDA 77-8010, Rockville, MD, 472-483, 1976.

(16) Hammrick, P.E. and Fox, S.S.: Rat lymphocytes in cell culture exposed to 2450 MHz (CW) microwave radiation, J. Microwave Power 12:125-132, 1977.

(17) Ashman, R.B. and Nahmias, A.J.: Effect of incubation temperature on mitogen responses of lymphocytes from adult peripheral blood and from cord blood, Clin. Exp. Immunol. 33:319-326, 1978.

(18) Roberts, N.J. and Steigbeigel, R.T.: Hyperthermia and human leukocyte function. Effect of response of lymphocytes to mitogens and antigen and bactericidal capacity of monocytes and neutrophils, Infect. Immun. 18:673-684, 1977.

(19) Roszkowski, W., Szmigielski, S., Janiak, M., Wrembel, J.K. and Hryniewicz, W.: Effect of hyperthermia on rabbit macrophages, Immunobiology 157:122-131, 1980.

(20) Lin, J.C. and Peterson, W.D.: Cytological effects of 2450 MHz CW microwave radiation, J. Bioeng. 1:471-478, 1977.

(21) Mayers, C.F. and Habeshaw, J.A.: Depression of phagocytosis - A nonthermal effect of microwave radiation, a potential hazard to health, Int. J. Radiat. Biol. 24:449-461, 1973.

(22) Rama Rao, G., Cain, C.A. and Tompkins, W.A.F.: Effects of microwave exposure on the hamster immune system. II. Activation of peritoneal macrophages, Bioelectromagnetics 4:141-155, 1983.

(23) Rama Rao, G., Cain, C.A. and Tompkins, W.A.F.: Effects of microwave exposure on the hamster immune system. III.

Macrophage resistance to vesicular stomatitis virus infection, Bioelectromagnetics 5:377-388, 1984.

(24) Szmigielski, S.: Effect of 10 cm (3 GHz) electromagnetic radiation (microwaves) on granulocytes _in vitro_, Ann. N.Y. Acad. Sci. 247:275-281, 1975.

(25) Jakovleva, M.E.: Physiologic Mechanisms of Action of Electromagnetic Fields, Izd. Medicina, Moscow, 1973 (in Russian).

(26) Dawidow, B.I., Tichonczuk, W.S. and Antipow, W.W.: Biological Action, Safety and Protection Against Electromagnetic Radiations, Elektroatomizdat, Moscow, 1984 (in Russian).

(27) Smialowicz, R.J.: Hematologic and immunologic effects of nonionizing electromagnetic radiation, Bull. N.Y. Acad. Med. 55:1094-1118, 1979.

(28) Ivanoff, B., Robert, P., Deschaux, P., Pellisier, J.P. and Fontanges, R.: Effect of microwaves on the cellular immune response in Swiss mouse, CR Biology 173:932-941, 1979.

(29) Liddle, C.G., Putnam, J.P., Ali, J.S., Lewis, J.J., Bell, B., West, N.W. and Lewter, O.H.: Alteration of circulating antibody response of mice exposed to 9 GHz pulsed microwaves, Bioelectromagnetics 1:397-408, 1980.

(30) Liddle, C.G., Putnam, J.P., Lewter, O.H., West, M. and Morrow, G.: Circulating antibody response of mice exposed to 9 GHz pulsed microwave radiation, Bioelectromagnetics 7:91-94, 1986.

(31) Chiang-Huai, Shao-Binjie and Wang Xing-Hua: Experimental research in China on the biological effects of microwaves, J. Bioelectricity 4:103-120, 1985.

(32) Wiktor-Jedrzejczak, W., Ahmed, A., Czerski, P., Leach, W.M. and Sell, K.W.: Increase in the frequency of Fc

receptor (FcR)-bearing cells in the mouse spleen following a single exposure of mice to 2450 MHz microwaves, Biomedicine 27:250-252, 1977.

(33) Wiktor-Jedrzejczak, W., Ahmed, A., Sell, K.W., Czerski, P. and Leach, W.M.: Microwaves induce an increase in the frequency of complement-receptor-bearing lymphoid spleen cells in mice, J. Immunol. 118:1499-1502, 1977.

(34) Sulek, K., Schlagel, C.J., Wiktor-Jedrzejczak, W., Ho, H.S., Leach, M.W., Ahmed, J.H. and Woody, J.N.: Biological effects of microwave exposure. I. Threshold conditions for the induction of the increase in complement-receptor-positive (CR+) mouse spleen cells following exposure to 2450 MHz microwaves, Radiat. Res. 83:127-139, 1980.

(35) Huang, A.T. and Mold, N.G.: Immunologic and hematopoietic alterations by 2450 MHz electromagnetic radiation, Bioelectromagnetics 1:177-187, 1980.

(36) Schlagel, C.J., Sulek, K., Ho, H.S., Leach, W.M., Ahmed, A. and Woody, J.N.: Biological effects of microwave exposure. II. Studies on the mechanisms controlling susceptibility to microwave-induced increases in complement-receptor-positive spleen cells, Bioelectromagnetics 1:405-414, 1980.

(37) Smialowicz, R.J., Brugnolotti, P.L. and Riddle, M.: Complement-receptor-positive spleen cells in microwave (2450 MHz) irradiated mice, J. Microwave Power 16:73-77, 1981.

(38) Liburdy, R.P.: Radiofrequency radiation alters the immune system: Modulation of T- and B-lymphocyte levels and cell-mediated immunocompetence, Radiat. Res. 77:34-51, 1979.

(39) Liburdy, R.P.: Radiofrequency radiation alters the immune system. I. Modulation of in vivo lymphocyte circulation, Radiat. Res. 83:66-77, 1980.

(40) Smialowicz, R.J., Rogers, R.R., Garner, R.J., Liddle, M.M, Luebke, R.W. and Rowe, D.G.: Microwaves (2450 MHz) suppress murine natural killer cell activity, Bioelectromagnetics 4:371-381, 1983.

(41) Rama Rao, G., Cain, C.A. and Tompkins, W.A.F.: Effect of microwave exposures on the hamster immune system. IV. Spleen cell IgM hemolytic plaque formation, Bioelectromagnetics 6:41-52, 1985.

(42) Yang, H.K., Cain, C.A., Lockwood, J. and Tompkins, W.A.: Effects of microwave exposure on the hamster immune system. I. Natural killer activity, Bioelectromagnetics 4:123-139, 1983.

(43) Szmigielski, S., Luczak, M., Janiak, M. and Kobus, M.: In vitro and in vivo inhibition of virus multiplication by microwave hyperthermia, Arch. Virol. 53:71-77, 1977.

(44) Roszkowski, W., Wrembel, J.K., Roszkowski, K., Janiak, M. and Szmigielski, S.: Does whole-body hyperthermia therapy involve participation of the immune system?, Int. J. Cancer 25:289-292, 1980.

(45) Szmigielski, S., Roszkowski, W., Kobus, M. and Jeljaszewicz, J.: Modification of acute experimental staphylococcal infections by long-term exposition to non-thermal microwave fields or whole-body microwave hyperthermia, Zbl. Bakt. Hyg. Immunol., Abt. I, Suppl. 10:637-646, 1981.

(46) Ragan, H.A., Phillips, R.D., Buschbom, R.L., Busch, R.H. and Morris, J.E.: Hematologic and immunologic effects of pulsed microwaves in mice, Bioelectromagnetics 4:383-396, 1983.

(47) Sultan, M.F., Cain, C.A. and Tompkins, W.A.F.: Immunological effects of amplitude-modulated radiofrequency radiation: B lymphocyte capping, Bioelectromagnetics 4:157-165, 1983.

(48) Smialowicz, R.J., Kinn, J.B. and Elder, J.A.: Perinatal exposure of rats to 2450 MHz CW microwave radiation. Effects on lymphocytes, Radio Sci. 14:147-153, 1979.

(49) McRee, D.I., Elde, J.A., Gage, M.I., Reiter, L.W. and Guy, A.W.: Effects of non-ionizing radiation on the central nervous system, behaviour and blood: A progress report, Env. Health Perspect. 30:123-132, 1979.

(50) Shandala, M.G., Dumanskij, W.D., Rudnev, M.I., Eskova, L.K. and Los, J.P.: Study of non-ionizing microwave radiation effects upon the central nervous system and behavioural reactions, Env. Health Perspect. 30:115-126, 1979.

(51) McRee, D.I., Faith, R., McDonell, E.E. and Guy, A.W.: Long-term 2450 MHz CW microwave irradiation of rabbits. Evaluation of hematologic and immunologic effects, J. Microwave Power 15:45-52, 1980.

(52) Chou, C.K., Guy, A.W., Borneman, E., Kunz, L.L. and Kramer, P.: Chronic exposure of rabbits to 0.5 and 5 mW/cm^2 2450 MHz CW microwave radiation, Bioelectromagnetics 4:63-77, 1983.

(53) Guy, A.W., Chou, C.K., Kunz, L.L., Crawley, J. and Krupp, J.: Effects of Low-Level Radiofrequency Radiation Exposure on Rats. Vol. 9, Summary. Report No. USA FSAM-TR-85-64, USAF School of Aerospace Med., Brooks Air Force Base, 1985.

(54) Szmigielski, S., Jeljaszewicz, J. and Wiranowska, M.: Acute staphylococcal infections in rabbits irradiated with 3-GHz microwaves, Ann. N.Y. Acad. Sci. 247:305-311, 1975.

(55) Hill, D.: Biological effects of RF radiation - Human studies, in Biological Effects of Radiofrequency Radiation, J.A. Elder and D.F. Cahill, eds., U.S. EPA Report No. 600/8-83-026 F, EPA Res. Triangle Park, NC, 112-124, 1984.

(56) McLaughlin, J.R.: A survey of possible health hazards from

exposure to microwave radiation, Hughes Aircraft Corp., Culver City, CA, 1953.

(57) Lester, J.R. and Moore, D.F.: Cancer mortality and Air Force bases, J. Bioelectricity 1:77-86, 1982.

(58) Polson, P. and Merritt, J.H.: Cancer mortality and Air Force bases: A reevaluation, J. Bioelectricity 4:121-127, 1985.

(59) Lester, J.R.: Reply to "Cancer mortality and Air Force bases: A reevaluation," J. Bioelectricity 4:129-131, 1985.

(60) Kirk, W.P.: Biological effect of radiofrequency radiation - Life span and carcinogenesis, in Biological Effects of Radiofrequency Radiation, J.A. Elder and D.F. Cahill, eds., U.S. EPA Report No. 600/8-83-026 F, Res. Triangle Park, NC, 106-111, 1984.

(61) Szmigielski, S., Szudzinski, A., Pietraszek, A., Bielec, M., Janiak, M. and Wrembel, J.K.: Accelerated development of spontaneous and benzopyrene-induced skin cancer in mice exposed to 2450 MHz microwave radiation, Bioelectromagnetics 3:179-191, 1982.

(62) Szudzinski, A., Pietraszek, A., Janiak, M., Kalczak, M. and Szmigielski, S.: Acceleration of the development of benzopyrene-induced skin cancer in mice by microwave radiation, Arch. Dermatol. Res. 274:303-312, 1982.

(63) Balcer-Kubiczek, E.K. and Harrison, G.: Evidence for microwave carcinogenesis _in vitro_, Carcinogenesis 6:859-864, 1985.

(64) Dwyer, M.J. and Leeper, D.B.: A current literature report on the carcinogenic properties of ionizing and nonionizing radiation. II: Microwave and radiofrequency radiation. DHEW Publ. (NIOSH) No. 78-134, 1978.

(65) Prausnitz, S. and Susskind, C.: Effects of chronic micro-

wave radiation on mice, IRE Trans. Biomed. Electron. 9:104-108, 1962.

(66) Preskorn, S.H., Edwards, D. and Justesen, D.R.: Retarded tumor growth and longer longevity in mice after irradiation by 2450 MHz microwaves, J. Surg. Oncol. 10:483-492, 1978.

(67) Smialowicz, R.J., Weil, C.M., Kinn, J.B. and Elder, J.A.: Exposure of rats to 425 MHz (CW) radiofrequency radiation - Effects on lymphocytes, J. Microwave Power 17:211-221, 1982.

(68) Szmigielski, S., Szudzinski, A., Pietraszek, A. and Bielec, M.: Acceleration of cancer development in mice by long-term exposition to 2450 MHz microwave fields, in Ondes Electromagnetiques and Biologie, A.J. Bertraud and B. Servantie, eds., URSI, Paris, 165-169, 1980.

(69) Teshima, H., Kubo, C. and Kihara, H.: Psychosomatic aspects of skin diseases from the standpoint of immunity, Psychother. Psychosom. 37:165-175, 1982.

(70) David, T.P., Johnson, H.D. and Gehrke, H.: Effect of temperature stress on circulating biogenic amines in bovine, Comp. Biochem. Physiol. 79:369-373, 1984.

(71) Murray, A.W. and Verma, A.K.: The adanyl cyclase system and carcinogenesis: Decreased responsiveness of mouse epidermis to isoproterenol after 3,4,-benzopyrene treatment, Biochem. Biophys. Res. Commun. 54:69-75, 1973.

(72) Mufson, R.A., Simsiman, R.C. and Bontwell, R.K.: The effect of the phorbol ester tumor promoters on the basal and catecholamine-stimulated levels of cyclic adenosine 3'-5'-monophosphate in mouse skin and epidermis <u>in vivo</u>, Cancer Res. 37:665-669, 1977.

(73) Mufson, R.A.: Effects of tumor promoters on cyclic nucleotide metabolism in mouse skin and epidermis <u>in vivo</u>, in Carcinogenesis, Vol. 2. Mechanisms of Tumor Promotion and

Cocarcinogenesis, T.J. Slaga, A. Sivak and R.K. Bontwell, eds., Raven Press, New York, 1978.

(74) Marks, F. and Grimm, W.: Diurnal fluctuation and beta-adrenergic elevation of cyclic AMP in mouse epidermis in vivo, Nature New Biol. 240:178-179, 1972.

(75) Kunz, L.L., Johnson, R.B., Thompson, D., Crowley, J., Chou, C.K. and Guy, A.W.: Effects of Long-Term Low-Level Radiofrequency Radiation Exposure on Rats, Vol. 8. Evaluation of Longevity, Cause of Death and Histopathological Findings, Report USAF School of Aerospace Med. (AFSC), Brooks Air Force Base, TX 78 235-301, 1985.

(76) Balcer-Kubiczek, E.K., Harrison, G.H. and McCulloch, D.: Lack of Effect of 2450 MHz Microwaves on Chemical Carcinogenesis in C3H/10T1/2 Cells, Proc. 7th International Congress of Radiation Res., M. Nijhoff, Rijswijk (NL), 1983.

(77) Friedman, H.L.: Are chronic exposure to microwaves and polycythemia associated?, N. Engl. J. Med. 304:357-358, 1981.

(78) Milham, S.: Mortality from leukemia in workers exposed to electrical and magnetic fields, N. Engl. J. Med. 307:249-250, 1982.

(79) Wright, W.E., Peters, J.M. and Mack, T.M.: Leukemia in workers exposed to electrical and magnetic fields, Lancet 2:1160-1161, 1982.

(80) Coleman, M., Bell, J. and Skeet, R.: Leukemia incidence in electrical workers, Lancet 1:982-983, 1983.

(81) McDowall, M.E.: Leukemia mortality in electrical workers in England and Wales, Lancet 1:246-247, 1983.

(82) Vagero, D. and Olin, R.: Incidence of cancer in the electronic industry: Using the new Swedish Cancer Environ-

ment Registry as a screening instrument, Br. J. Ind. Med. 40:188-192, 1983.

(83) Robinette, C.D. and Silverman, C.: Causes of death following occupational exposure to microwave radiation (radar) 1950-1974, in Proc. Symposium on Biological Effects and Measurements of Radiofrequency and Microwaves, D.G. Hazzard, ed., HEW Publ. FDA 77-8026, Rockville, MD, 1977.

(84) Robinette, C.D., Silverman, C. and Jablon, S.: Effects upon health of occupational exposure to microwave radiation (radar), Am. J. Epidemiol. 112:39-53, 1980.

(85) Silverman, C.: Epidemiologic studies of microwave effects, Proc. IEEE 68:78-84, 1980.

(86) Bielec, M.: Experimental and Epidemiological Investigations on Risk of Cancer in Subjects Exposed for a Long Time to Microwave and Radiofrequency Fields, Report WIHE 85/09 of limited distribution, WIHE, Warsaw, 1985 (in Polish).

(87) Roszkowski, W., Szmigielski, S., Janiak, M. and Wrembel, J.K.: Effect of moderate and intensive hyperthermia on spleen, lymph-node and thymus-derived lymphocytes in vitro, Immunobiology 156:429-437, 1979.

(88) Smith, J.B., Knowlton, R.P. and Agarwal, S.S.: Human lymphocyte responses are enhanced by culture at 40°C, J. Immunol. 121:691-694, 1978.

(89) Wiktor-Jedrzejczak, W., Ahmed, A., Czerski, P. and Leach, W.M.: Immune response of mice to 2450 MHz microwave radiation: Overview of immunology and empiric studies of lymphoid splenic cells, Radio Sci. 12:209-218, 1977.

(90) Smialowicz, R.J., Ali, J.S., Berman, E., Bursian, S.J., Kinn, J.B., Liddle, C.G., Reiter, J.W. and Weil, C.M.: Chronic exposure of rats to 100 MHz (CW) radiofrequency radiation: Assessment of biological effects, Radiat. Res. 86:488-505, 1981.

26
The Biological Effects of Power-Frequency Electric Fields in the Environment

M. G. SHANDALA, YU D. DUMANSKIY, I. S. BEZDOL'NAYA

Marzeev Scientific Research Institute of
General and Communal Hygiene, 50 Popudrenko Str.
Kiev-94, USSR

INTRODUCTION

Electrical energy is the most convenient and well known of the forms of energy used by man. The speed of transmission over large distances, comparatively small amount of power lost during transmission, simplicity of transformation into other types of energy, and availability for immediate utilization render electrical energy indispensable to today's society.

The growth of industry, the use of electrical energy, and the development of electrical power grids have resulted in the increasing exposure of the public to power-frequency electromagnetic fields (EMF) from alternating-current high-voltage powerlines (HVPL). These developments raise the question of the biological significance of fields from HVPLs.

The published material on biological effects of HVPLs involves three basic themes: (1) physical parameters of fields in the vicinity of HVPLs; (2) biological effects of HVPL fields; (3) mechanisms of action of HVPL fields. Each of these areas will be discussed below.

PHYSICAL PARAMETERS OF POWERLINE FIELDS

The electromagnetic environment of HVPLs can be characterized by electric and magnetic fields. Electric fields arise when voltage is applied to the line, and magnetic fields arise

---

Translated from Russian by C.H. Dodge and B.J.B. Dodge, 112 16th St. NE, Washington, DC, 20002.

when current flows through the line. The electric and magnetic fields at 50 Hz can be considered separately with regard to their effects on biological systems.

The magnetic field under a 750-kV HVPL at 1.8 meters above the Earth's surface is 0.30–0.38 gauss (1-33) (0.3 gauss under a 765-kV HVPL (34)), which is comparable to the natural magnetic field of the Earth (0.5 gauss). On the other hand, the electric field under a 765-kV HVPL can reach several thousand volts/meter, while the Earth's natural electric field is 0.13 kV/m (35-40). On this basis, we concluded that effects on organisms are primarily associated with the electric field.

A power-frequency electric field can be considered at any instant as an electrostatic field, such as that created between two charged electrodes. Powerline-electric fields are non-uniform, and they are significantly perturbed by the presence of grounded or metallic structures (41). The field is directly proportional to the voltage of the HVPL. Under standard conditions of measurement and line geometry, the field of a 110-kV line is always lower than that of a 220-kV line, and in turn, the latter is lower than the field of a 330-kV line (14,35,38, 42).

The maximum ground-level electric field from an overhead HVPL occurs midway between the supporting towers, which is the point of maximum line sag. For this reason, and since the towers have a significant shielding effect, the electric field is less in the immediate vicinity of a line tower, compared to other points under the line (Figure 1) (42). Measurements and calculations have established that the strength of the electric field decreases as a function of the lateral distance from a HVPL (Figure 2)(17,24). The field from a 1150-kV powerline has maximum values of 30, 14, 6, and 3.5 kV/m at 10, 30, 40, and 50 m from the line, respectively (33). Broad-leafed trees can produce a screening effect of up to 95% (17,38). The electric field is effectively shielded by typical construction materials (wood, brick, concrete) (17,38).

Electric charges are formed on the body of an individual exposed to a powerline electric field. The magnitude of the charge is determined by the strength of the field, and the dimensions and electrical properties of the body. The potential of a grounded individual is essentially zero, but potentials up to several kilovolts can occur if the body is isolated from ground.

If either a metallic object or an individual is isolated from ground, physical contact between them results in a static

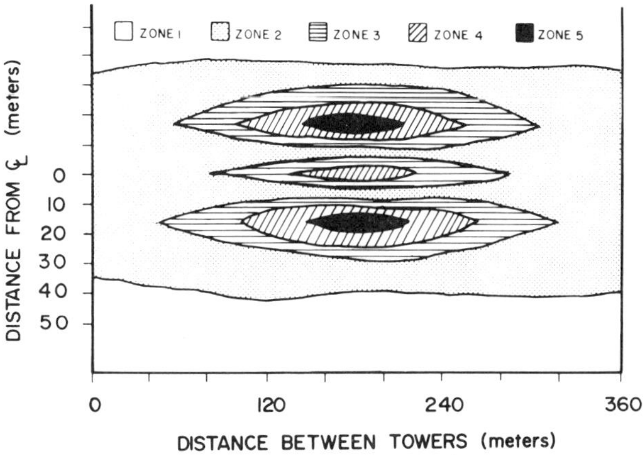

FIGURE 1. Approximate electric-field boundaries near a 750-kV HVPL. Zone (1), < 1 kV/m; (2), 1-5 kV/m; (3), 5-10 kV/m; (4), 10-15 kV/m; (5), > 15 kV/m.

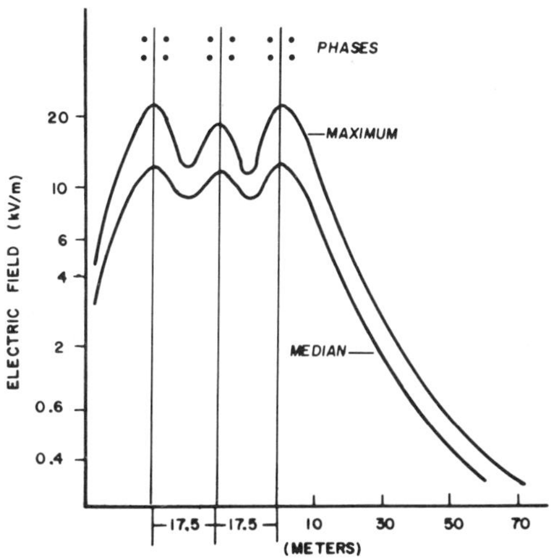

FIGURE 2. Electric field distribution near a 750-kV line.

current that can produce an uncomfortable sensation, particularly at the first moment of contact. A spark discharge is frequently associated with such contact. In the case of contact with large metal objects isolated from ground (pipes, poles) or with metal objects having large areas (roofs), the current passing through the body can reach dangerous levels.

In an individual near a powerline, there is a constant flow of current through the body to ground. If the person is grounded, the current will flow to ground via the contact area. If the body is isolated from ground, the current will flow to ground via a capacitive coupling between the body and ground. The magnitude of the current is essentially the same in both cases (41).

The presence of a human body in the field of a powerline significantly distorts the field in the immediate vicinity of the body (28), and the field intensity may increase by a factor of 8-12 in comparison to the field in the absence of the body (5,44) (Table 1).

In summary, HVPLs are the sources of commercial-frequency electromagnetic fields. The electric component significantly exceeds the natural level of the Earth's electric field, thereby altering the electromagnetic environment in places of human habitation and necessitating the need for assessment and control. Electrical discharges and induced currents are indirect consequences of the action of the electric component of a HVPL. The magnetic component of the field near HVPLs is small, and is not considered here.

BIOLOGICAL EFFECTS OF POWERLINE ELECTRIC FIELDS

INTRODUCTION

Evidence has accumulated in the scientific literature that HVPL electric fields can have effects on the nervous and cardiovascular systems, metabolism, and reproductive function (2,11, 12,18,20,30,31,35,45-50), although there exists some divergence of results (6,51-60).

The first adverse effects of HVPL electric fields in the USSR were observed in the 1960's in the vicinity of 500-kV lines. Investigations were performed on workers in the vicinity

TABLE 1. Electric Field and Induced Current in Parts of a Human Body Situated Under a Powerline. 1, top of head; 2, temple; 3, neck; 4, shoulder; 5, legs, feet; 6, unperturbed field.

LINE VOLTAGE (kV)	ELECTRIC FIELD (kV/m)						INDUCED CURRENT (μA)		
	1	2	3	4	5	6	Head	Body	Legs
225	32.8	21.5	5.6	12.2	5.8	2.5	11.3	23.6	34.9
400	76.0	51.9	12.4	39.6	13.4	5.5	25.6	54.4	80.0
750	135.0	94.1	22.5	72.0	24.4	10.0	46.5	98.9	145.0
1000	173.0	118.0	28.2	90.0	30.4	12.5	58.2	124.0	182.0

of a substation. Subjects in one group (29 men) were exposed for up to two hours to 3-26 kV/m, while those in a second group (25 men) were exposed to a field of similar strength for 5 hrs/day. Medical examinations were conducted before and after exposure to the fields. It was found that the fields influenced thermoregulation, heart-beat, and arterial pressure. The observed effects included functional disturbances of the central nervous system (CNS), including a lengthening of the visual reaction time, and decreased memory and attention span. The functional changes depended on the amount of field exposure (22,26,27).

In a group of 319 workers exposed to 10-32 kV/m, changes in heart and respiration rate, Q and S spikes in the electrocardiogram (ECG), changes in cardiovascular reflexes, and a decrease in work capacity were observed (2,6,31). Reticulocytosis and a qualitative change in neutrophils were also noted (61).

Other investigators reported similar results. People occupationally exposed to electric fields exhibited respiratory distress, bradycardia, and changes in CNS function (8,10,30,62, 63). The latter included hypertension, decreased reaction time to a visual stimulus, and a decrease in memory and attentiveness. In none of these investigations was it possible to establish that the electric fields alone, rather than other physical factors in the powerline environment, were responsible for the observed effects.

The reports stimulated further study by other investigators into the health effects of powerline electric fields. No differences were found beween personnel chronically exposed to 200-400 kV/m in the vicinity of 760 kV powerlines, and control groups (64,65). The occurrence of sudden infant death syndrome among families living near powerlines was attributed to the altered electrical environment (66). Personnel working near 400-kV facilities exhibited headache, sleepiness, nausea, fatigue, and increased blood pressure (67,68). Investigations of the health of male personnel occupationally exposed to powerline electric fields revealed CNS effects, effects on the cardiovascular system, and changes in general physical condition (69,70).

An increased incidence of chromosome damage was reported in personnel working around 400-kV facilities (71). In subsequent studies, a 10% incidence of birth defects was reported in 119 children of substation workers, whereas the incidence was only 2.1% in a group whose parents were not exposed to powerline electric fields (72). Other investigators have reported effects in personnel occupationally exposed to electromagnetic fields (73-76).

The danger of powerline electric fields to human health has been suggested in some studies, but not in others. Some investigators have tried to explain this dichotomy as being a result of a variety of factors including electric fields, low-frequency noise, spark discharges, chemical agents, and variability of social backgrounds (77,78). Clearly, the state of health of workers exposed to powerline electric fields cannot be resolved on the basis of a single factor. The confounding factors must be taken into consideration and evaluated. Experiments on animals have been undertaken in an attempt to resolve the uncertainties inherent in the human studies. The studies can be divided into low (0.1-10 kV/m), medium (10-30 kV/m), and high (greater than 30 kV/m) field-exposure groups.

BIOLOGICAL EFFECTS AT 0.1-10 kV/m

Levels of the trace elements iron, copper, molybdenum, manganese, and nickel in various organs of rats were increased following exposure to 0.05-5 kV/m, 2 hr/day for 4 months (3,4). Following exposure of male rats to 5 kV/m for 5 hrs, a decrease in calcium and magnesium in the liver and thymus was observed, but the concentration of these elements increased in brain, gonads, and the prostate gland (12). In the kidneys only calcium was increased, and in other organs (heart, lungs, stomach, spleen, intestine, adrenals) no changes were observed (12).

An increase in noradrenaline in brain tissue was observed in rats exposed to power-frequency fields of 5.3 kV/m, 15 min/day for 21 days (79). Analogous results were obtained in the plasma of rats exposed to a power-frequency field of 50 kV/m (80).

A decrease in motor activity was reported in animals exposed to 1 kV/m for several days (82). In escape preference tests under the influence of 100 V/m, animals did not perceive the field (83). A change in the latent period of reactions in primates at 7-100 V/m was reported (84-86). Detailed studies of the humoral and cellular aspects of the immune system of rodents under the influence of powerline electric fields of 150-250 V/m did not reveal any changes (87-89).

Many investigations into the biological effects of low-intensity powerline electric fields in chronically exposed animals have been conducted in our laboratory (15,16,18-21,23, 35,36,39,40,42,90-98). The animals were exposed between two conducting plates, one of which carried a potential and the other of which was grounded. The distance between the capacitor plates was scaled to the relationship between a human being and the height of a typical overhead powerline. For rats, the distance between the plates was 430 mm. Each animal was housed in a plastic cage with a metal floor that was placed on the lower grounded plate of the capacitor, thereby grounding the animal.

The effect of 50-Hz electric fields of 100, 500, 1000, 2000, and 5000 V/m on 200 male white rats and 34 rabbits was studied. The respective control groups were housed under identical conditions. The exposure period of 120 days was followed by a recovery period that lasted for 60 days. The animals were evaluated using physiological, biochemical, and hematological indices.

Exposure for 4 months to 5000 V/m caused a decrease in work capacity of the animals to a level of 56% of that of the controls. For rats, after the third and fourth month of exposure there was an increase in reflex latency.

The results indicated a shift in the equilibrium between the processes of stimulation and inhibition in the CNS. These observations were in agreement with the results of measurements of blood cholinesterase activity. There was a decrease in the cholinesterase activity in animals exposed to 1-5 kV/m which persisted into the first month of the recovery period.

Protein metabolism was evaluated based on measurements of metabolites in the urine and blood. Blood nitrogen and uric acid levels were significantly increased following exposure to 5 kV/m in comparison with the controls. The effects disappeared by the end of the recovery period.

Carbohydrate metabolism was evaluated by measuring liver glycogen and blood glucose. At 2-5 kV/m, an increase in blood glucose was observed beginning at the second month of exposure. Liver glycogen was decreased in the exposed animals after four months.

The influence of electric-field exposure on mineral metabolism was evaluated by measuring urinary sodium and potassium ions. At 5000 V/m the levels of both ions was increased.

At 1-5 kV/m, oxygen exchange and phosphorylation in brain mitochondria were altered, indicating a change in the rate of mitochondrial metabolism in the brain cortex. Similar fields affected thyroid function as indicated by radioiodine studies and blood thyroxine measurements. The thyroid effects were confirmed histologically (thickening of epithelial cells, expansion of follicular spaces).

ECG measurements in rabbits revealed a slowing of cardiac rhythm activity and a decrease in the force of the stroke process in the atria of animals exposed to 1 kV/m. Histopathological changes in the myocardium were also observed.

The effect of a 50-Hz field, 1-5 kV/m, on the reproductive function of rats (665 females and 337 males) and their offspring was studied. In the females, an increase in the duration of the estrous cycle and gestational time were seen. In the exposed males, decreased spermatogenesis and sperm concentration, and increased atypical forms were observed. Histopathological changes in the testes also occurred. The offspring of the exposed animals exhibited increased fetal and post-fetal mortality, and a decreased growth rate during the first 3 weeks after birth. The effect occurred in offspring born to exposed females and non-exposed males, but was not observed in offspring of non-exposed females and exposed males.

The experiments led us to conclude that chronic exposure to a 50-Hz electric field, 2-5 kV/m, leads to functional changes in the nervous, cardiovascular, endocrine, and reproductive systems. The observed changes appeared to be related to the duration of exposure and intensity of the field, and returned to normal during a 2-month recovery period. The effects are a result of activation of the body's compensatory mechanisms initiated in response to the application of the fields.

Comparing the results of various investigations, it is possible to discern a general agreement concerning the functional shifts found under the influence of low-intensity power-frequency electric fields (3,4,12,19-21,39,40,95,96,99,100). These are expressed as an increased latent period of the conditioned reflexes, and a decreased activity of the cholinergic system of the brain. There are differences in experimental results related to response thresholds. In some reports the threshold was 500 V/m (15,16,18-21,35,36,39,40,42, 90-97), while thresholds of 50-100 V/m were found in other studies (84-86).

BIOLOGICAL EFFECTS AT 10-30 kV/m

Marino et al. observed a decrease in serum corticosterone in rats exposed to a power-frequency field of 15 kV/m for 30 days (50,101-103). Gann observed cardiovascular changes in dogs exposed to 15 kV/m for 5 hrs (104). Ceretelli and Malaguti (107) reported an increase in blood pressure in dogs under the influence of 10 kV/m. A field of 15 kV/m caused a decrease in heart-rate (108). Other investigators studying the influence of 25 kV/m (8 hr/day for 42 days) on growth, heart-rate, and resistance to experimental infection in rats found that the fields did not influence the immune function or heart-rate (109).

Studies in Sweden (105,106) indicated deleterious changes in the structure of brain-stem cells of rabbits exposed to 14 kV/m. There was a disruption of the integrity of the endoplasmic complex, a decrease in the number of mitochondria, and other structural disruptions of the cell, which indicated serious dysfunctional changes.

An evaluation of the behavior of rats exposed to 25 kV/m revealed that they avoided activity during normally active periods under the influence of the field (110).

In a series of related studies (39,112-114), the effect of intermittent exposure (80 minutes' exposure followed by a 30-minute non-exposure period, with a total daily exposure of 4 hrs) to 10-20 kV/m on white rats was studied. A total of 300 animals were used in the experiment. The experimental animals were exposed for 4 months and were compared with a control group that was housed under identical conditions. A one-month recovery period was provided for both groups. In the exposed rats, changes were seen in blood enzyme activity, glucose, humoral immunity, and in various behavioral measures. The effects were proportional to the intensity and duration of exposure.

Medium intensity powerline electric fields generally produce functional changes, but significant morphological changes can occur (106,107). Inconsistencies have been observed by different investigators, but they can be attributed to the different experimental conditions employed by various investigators. We conclude that, despite these inconsistencies, medium intensity powerline electric fields are a biologically active factor that cause responsive reactions in a number of body systems.

BIOLOGICAL EFFECTS ABOVE 30 kV/m

Investigations of the biological action of high levels of powerline electric fields include essentially all systems of living organisms. However, variable durations of exposure, methods of investigation, and interpretation of results significantly complicate a review of research, and a determination of the biological significance of strong powerline electric fields.

Hackman and Graves (115) observed a sharp increase in blood corticosterone in mice exposed to 50 kV/m, which returned to normal within a day. Opposite results were obtained by Free et al. (116) who exposed rats to 100 kV/m for 30 or 120 days. No differences in corticosterone level were observed between experimental and control groups. Gross (117) analyzed catecholamines in the urine and blood of rats exposed to 100 kV/m. Epinephrine increased after 6 hours' exposure, and the effect persisted for 3 days.

Morphological studies of the brain tissue of chickens (118,119) and rats (120) exposed to 40 kV/m and 100 kV/m did not reveal any changes from normal.

At 100 kV/m, EEG changes were observed after the 5th day of exposure (increase in the amplitude at high frequency). The effect disappeared after 10 exposures. A field of 200 kV/m caused phase changes in the brain bioelectrical activity that the authors believe indicated the development of an inhibition process in the CNS (25).

A 100 kV/m field elicited a decrease in the work capacity of the motor system of rabbits. Injection of novocaine, aminazine, and dibazole in the area of a region of the medulla oblongata decreased the influence of the field. This indicated an influence of the electric field on the CNS. The investigators did not exclude the possibility that electric fields altered energetic processes taking place within muscles (26,27).

The link between powerline electric fields and CNS reactions was confirmed by other investigators. Blanchi (99) reported the occurrence of low-frequency, high-amplitude waves following exposure of mice to 100 kV/m for 30 min. Exposure of rats to 80 kV/m for 16 hrs decreased alpha, beta, and theta in the EEG by 50%. These shifts were thought to be due to hair vibration (195).

Exposure of rats to 100 kV/m for 30 days caused an increase in neuronal excitability (121). An opposite effect was observed in baboons exposed to 300 kV/m (60 Hz), 4 hr/day, for 23-50 days (122). An effect on alertness was noted. It was suggested that animals initially perceived the electric field as a stress factor, and then adapted to it. Notwithstanding the behavioral changes, the authors concluded that there were no harmful effects.

A broad multifaceted investigation was conducted in which mice and rats were exposed to a 60-Hz, 130-kV/m field for 4 months (123). It was concluded that many effects were either directly or indirectly associated with changes in the CNS. Rats avoided 75-100 kV/m (110), while swine preferred a field of 30 kV/m (124).

In our opinion, experiments in which animals are exposed to powerline electric fields while also subjected to other stressors are of interest because this procedure amplifies the effect on the organism. Thus, daily exposure of rats to 100, 200, or 420 kV/m for 7-9 months together with inhalation of mine dust (3-4 mg/m^3) resulted in a presence of silicosis in all experimental animals. The rate of development of silicosis increased as a function of field strength. A difference in sodium and potassium metabolism was also observed (125).

Rabbits exposed to 100 kV/m for 1000 hrs exhibited an increased PQ interval of 19.5% in the ECG (107), and red-blood-cell concentration was decreased (107). A determination of cardiovascular shifts in guinea pigs, mice, and rats was made in response to exposure to a high intensity powerline electric field. The changes in the ECG were similar to those seen in reactions to infectious disease (102,103). Other investigators did not observe changes in heart-rate, blood pressure, or stroke volume upon exposure of anesthetized rabbits to 80 kV/m (126) or 100 kV/m (58,127).

Experiments conducted on rats and rabbits exposed to

50 kV/m for 24 hrs or 70 hrs over 5 days, or 8 hrs/day for 30-1000 days, revealed that the exposed animals exhibited a decreased post-infectious ability to restore normal blood-cell concentrations. Increased white-blood-cell concentrations and decreased red-blood-cell concentrations were observed, leading the authors to conclude that there was a hematopoietic cell stress reaction (128).

Investigations of rats and mice exposed to 100 and 240 kV/m revealed no changes in reproductive capacity, viability, or growth and development (129-134). Likewise, 60-Hz 30-kV/m fields produced no changes in reproductive function on three generations of miniature pigs (135). On the other hand, Graves et al. (136) observed that chicks exposed to powerline electric fields of 40-80 kV/m from 1-22 days after incubation exhibited a significant decrease in motor activity for up to one week after having been removed from the fields.

Some investigators hypothesized that powerline electric fields act as a stressor causing an increase in blood glucocorticoids (121), leukocytosis, and erythrocytic anemia (102,103, 122).

Although most studies have recorded measurable biological shifts under the influence of powerline electric fields of various intensities, the results evidently depend on the experimental conditions (field levels, exposure conditions, methodological approaches, and experimental subjects) as well as on the variable criteria for determining exactly what constitutes a normal or pathological change.

Since low and medium field levels affect the CNS and other body systems, it might be supposed that strong fields would cause more significant biological changes. However, this is not the case. The effects observed at high field levels occur within the limits seen following exposure to low and medium field levels.

Recently, greater attention has been given to the effect of powerline electric fields on humans under controlled laboratory conditions. Experiments conducted on human volunteers exposed to 16 kV/m electric fields for 30 minutes, resulted in altered behavior (27). Extensive investigations by Hauf et al. (52-58) involving volunteers exposed to 1, 15, and 20 kV/m for relatively short exposure periods revealed decreased reaction time, increased white-blood-cell concentration, increased noradrenaline, but no changes in the EEG. The observed changes were considered to be a nonspecific effect of stimulation. Changes in alpha rhythm of the EEGs of 12 volunteers was reported after only 3 minutes of exposure to 40-50 kV/m (140,141).

In 26 volunteers exposed to 20 kV/m for 5 hrs, no changes outside the normal range were found in 20 biochemical indices (142).

Beyer et al. (48) studied the effects of 10 and 20 kV/m on volunteers. One group (20 males) was exposed 6-22 hrs/day, while another group (30 males) was exposed for 12 hours. The typical reaction in all subjects at 10 kV/m was hair vibration. At 20 kV/m, a 2.4% increase in leukocytes was observed.

Schmidt (143) studied a group of volunteers exposed 12 hrs/day for 5 days. No changes were noted at 10 kV/m, but at 20 kV/m there were some hormonal changes (cortisol and insulin) which were interpreted as indicating functional shifts in the mid-brain and hypothalamus.

Medical and biological investigations were conducted in our laboratory to determine the effects on humans exposed to 15 kV/m, 50 Hz (39,113,144-150). The investigation was carried out in a specially designed chamber (5x3x4.5 m) equipped with a high-voltage generator. The upper electrode was connected to the generator, and was bordered with a 5-cm diameter metal tube to minimize corona discharges. The lower electrode (4.5 m separation) was grounded. Uniformity of the field never varied by more than 10%. At any time, two people were in the chamber. The temperature and relative humidity were 23.2 ± 2.5°C and 70.3 ± 5.2%, respectively.

Forty healthy males (20 controls) participated in the study, divided equally between 2 exposure groups (60 and 90 mins/day) and their respective controls. The subjects were adapted to the experimental conditions for 10 days, and then exposed to the field for 20 days, 5 days/week. The recovery period was 10 days. Experimental and control groups were treated identically except for the field exposure.

There were statistically significant changes in the cardiovascular system (heart-rate, blood pressure, ECG) and the EEG (decrease in alpha) beginning after 2 weeks' exposure at 60 mins/day, and 3 weeks' exposure at 90 mins/day. EEG amplitude was not altered by field exposure. Data on brain reaction to rhythmic photostimuli produced results that were statistically significant only in the group exposed for 60 min/day and showed an increase in the stimulated rhythm beginning with the second week of exposure. An investigation of the functional condition of conditioned reflex activity yielded statistically significant changes only in the group exposed to the field for 90 min/day. During the first 3 weeks of exposure, the latent period of response to a robust stimulus was less than in the control group. Various psychological tests involving vigilance and memory revealed no statistically significant differences between experimental and control groups. Similarly, no changes were found in numerous biochemical and hematological indexes.

Thus, this investigation revealed a number of changes in both experimental groups exposed to the electric fields, indicating changes in the balance of inhibitory and excitatory CNS processes (with a predominance of the latter). The changes did not exceed normal physiological levels.

Studies of the sensitivity of 75 male subjects to 50-Hz electric fields were conducted by Cabanes and Gary (151). It was found that 4% could perceive the presence of a 0.35 kV/m field, and 40% could perceive a field of 27 kV/m. In a similar investigation it was found that 5% of the group studied sensed the presence of 1 kV/m, while 50% could sense a field of 7 kV/m (151).

Studies of subjects exposed for 8 hrs/day revealed some differences in endocrine parameters between the experimental and control groups that were within normal physiological limits (153,154).

Almost all studies of human reactions to powerline electric fields under laboratory conditions have revealed changes in the parameters recorded. Inconsistency of some reports can be explained as being a result of different exposure times and field conditions. Also, there is a high variability in thresholds of sensitivity to powerline electric fields (151, 152).

SUMMARY

Most investigations led to the conclusion that powerline electric fields are biologically active factors that elicit measurable reactions from various systems in the organism. Most investigations have recorded effects in exposed organisms, but interpretation of results is often ambiguous because investigators base their evaluation of the biological significance on the occurrence of pathological changes, while others base their evaluation on functional changes.

POSSIBLE MECHANISMS OF THE BIOLOGICAL ACTION OF POWERLINE FIELDS

The main difficulty is that the classical mechanisms involving ionization and thermal effects are not applicable to the powerline situation. However, the existence of data concerning biological effects requires consideration of other, more probable mechanisms.

For low-frequency electric fields, the relaxation of free and bound charges is less than the period of change of the field, leading to a large decrease in the field within a biological system, compared to the external field (72,157). The accumulated ions do not migrate, and only oscillate at an amplitude that is significantly less than the dimensions of a living cell.

Living tissue (possibly excluding bone) can be regarded as an excellent conductor. Consequently, during electric field exposure, an electric current will flow across the body (105). One can compute the magnitude of the internal fields and currents. However, for such subjets as man and animals, exact computations are impractical because of the complexity of shape and internal structure of the body. As a consequence, in studying the current distribution in the body, modeling approaches are widely employed.

The factor having the greatest biological effectiveness is the current density in the body (5). A number of studies have calculated and measured current density with human phantoms (5,13,77,100,139,158-161). It is proposed that an induced current in the body circulates primarily via the vascular network due to the low resistance of blood (137), and that it stimulates the nervous system (162) and alters biochemical reactions (50,102,163).

There are data indicating that electron energy levels in biological systems may be separated by less than 0.01 eV. Thus, it can be supposed that even for very small currents, significant disruption of function can occur (9).

Electric fields can also affect neutral molecules via induced dipole moments, and lead to irreversible changes (164). Other investigators propose that biological effects of powerline electric fields are secondary to vibration of hair and fur, which leads to increased sensitivity and behavioral effects (53-58,137).

CONCLUSIONS

The electric field in the vicinity of powerlines can attain values of several thousand volts/meter. The field decreases sharply with distance from the powerline. Other physical factors present near powerlines are magnetic fields, electrical discharges, and induced current. The presence of a human below a powerline distorts and increases the field in the immediate area of the individual.

Most investigators have concluded that powerline electric fields elicit a measurable reaction in a number of organ systems including the CNS, cardiovascular, blood, and reproductive systems. Some investigators propose that the field elicits regulatory-system reactions based on recorded functional effects. Other investigators consider that a powerline electric field elicits only non-specific or normal physiological reactions that do not pose a threat to human health. There are also reports of no effects of powerline electric fields on biological systems.

Some investigators proposed that the basic active component of a powerline electric field is the induced current. Others have suggested that the biological effects are secondary in nature, and are associated with vibration of hair or fur. As is evident, the questions cannot be resolved at the present time.

Depending on the inclination of one or another group, modeling of fields within laboratories varies considerably. In some cases, electrodes were placed on the surface of the body, while in other cases the field was applied indirectly using capacitor plates. Not surprisingly, it was observed that the fields in each case produced different effects.

In resolving questions about the biological significance of powerline electric fields a number of difficulties are encountered including extrapolation of animal data to humans, and differences in occupational, natural, and laboratory settings. It is extremely difficult to extrapolate data from animal experiments to humans because of differences in anatomic and physiological features. Extrapolation of data based only on field values does not address all problems. In conducting laboratory experiments, it is necessary to also consider ozone, ions, noise and other factors which can play a role in the powerline environment. In evaluating the physiological significance of a powerline electric field, many individual features must be taken into account including state of health, occupational activity, exposure time, and age.

It has been proposed that evaluation of the biological significance of powerline electric fields be based only on observations of pathological changes. But powerline electric fields affect a substantial part of the population including children, old persons, and people who are sick and therefore more sensitive to the influence of environmental factors. Considering these factors and using data from a wide range of experiments (3,4,15-21,23,35-40,45,46,90-98,112-114,144-150, 160), the USSR has developed and promulgated exposure standards for protection of the general public from the action of electric fields produced by overhead HVPLs (138) (Table 2). To protect the public from powerline electric fields "sanitary-protective" zones have been established. Such a zone is a region along a powerline where the electric field strength does not exceed 1 kV/m. For newly planned powerlines and buildings, the sanitary-protective zone along a powerline is established (perpendicular to the powerline) at 20 m for a 330-kV line, 30 m for a 550-kV line, 40 m for a 750-kV line, and 55 m for a 1150-kV line. If the field exceeds these limits, measures for decreasing field strength must be applied. In places of possible human habitation, electric field exposure can be decreased by moving the residence from the powerline area, or by the use of shielding or other means for reducing the electric field strength. Habitable buildings and other structures may be located in sanitary-protective zones of 330-550 kV lines if steps are taken to decrease the electric fields to the recommended safe levels. Such structures may not be located in a

TABLE 2. USSR Exposure Standards for Power-Frequency Electric Fields Emanating from Overhead High-Voltage Powerlines

LOCATION	INTENSITY (kV/m)
Inside homes	0.5
Around residences	1
Populated areas (constructed within 10 years)	5
Roads near powerlines	10
Unpopulated areas (agricultural zones)	15
Difficult terrain (not suitable for farm machinery)	20

750-kV powerline zone.

In conclusion, powerline electric fields alter the electromagnetic environment and have an effect on living organisms. At the present time, there are more questions than answers about the problem of powerline electric field bioeffects. The same is true of the possible mechanisms of action of the magnetic component of the electromagnetic field. Research in this area is very important and many questions deserving attention still await investigation.

REFERENCES

(1) Kalyuzhnyy, V.D. and Lyskov, Yu.M.: Assessment of the physiological effect of an electric field during powerline construction, in Long Distance 750 kV Overhead Line Transmission, Moscow, 1974.

(2) Kalyuzhnyy, V.F. and Mikhailov, M.I.: Influence of a communications line electric field on communications personnel, Elect. Commun. (Elektrosvyaz') 3, 1977.

(3) Kozyarin, I.P.: Effect of a low frequency electric field on the organism, in Hygiene of Populated Places, Kiev, 1976.

(4) Kozyarin, I.P. and Gabovich, V.A.: On the mechanism of interaction of a power frequency electric field with humans and animals, J. Tech. Phys. (Zhurnal Tekhnikoy Fiziki) 5:4, 1978.

(5) Kolesnkov, M.V. and Chukhlovii, V.A.: On the mechanism of interaction of a power frequency electric field with humans and animals, J. Tech. Phys. (Zhurnal Tekhnikoy Fiziki) 15:4, 1978.

(6) Krivova, T.I., Lukovkin, V.V. and Sazonova, T.Ye.: Influence of a power frequency electric field on humans, in Protection from Electromagnetic Fields and Electrical Current in Industry, Moscow, 1973.

(7) Landau, L.D. and Lifshitz, Ye.M.: A Short Course in Theoretical Physics, Science (Nauka), Moscow, Vol. 1, 1969.

(8) Liman, A.D.: Bioelectrical activity and visual light sensitivity in workers under conditions of chronic exposure to electromagnetic fields of various frequency ranges, Dissertation, Khar'kov, 1974.

(9) Manoylov, V.Ye.: Electricity and Man, Energy Publishing House (Ernegoizdat), Leningrad, 1982.

(10) Miller, C.: Measurement of electric fields and currents, in Influence of Electric Fields on People Working around High Voltage Transmission Lines, STsNTI, 1969.

(11) Novikov, V.N.: Effect of an electric field of high voltage on some functions of the organism under conditions of normal and altered reactivity, Dissertation, Moscow, 1971.

(12) Gabovich, R.D., Mikhailyuk, I.A., Kozyarin, I.P. and Shutenko, O.I.: Metabolism and multi-organ distribution

of iron, copper, molybdenum, manganese, and nickel under the influence of electromagnetic fields (UHF), Hygiene and Health (Gig. i San.) 7, 1977.

(13) Guy, A., Davidov, S., Yang, D. and Chou, Ch.K.: Determination of the distribution of electric currents in animals and humans exposed to a homogeneous electric field (60 Hz) of high voltage, in Materials of the 3rd Soviet-American Working Group, NIEHS, Washington, DC, 1982.

(14) Perel'man, L.S., Tikhodeyev, N.N. and Fel'dian, S.D.: Influence of intermediate height lines of medium phase under higher ones on the electric characteristics of 750 kV lines, in Long Range Electric Transmission, Section I, Moscow, 1974.

(15) Popovich, V.M., Prokhvatilo, Ye.V. and Lyubchenko, S.A.: Biological effect of a power frequency (50 Hz) electromagnetic field, in Physico-mathematical and Biological Problems of the Effects of Electromagnetic Fields and Air Ions, Vol. II, Yalta, 1975.

(16) Popovich, V.M. and Kozyarin, I.P.: Influence of power frequency electromagnetic energy on the nervous system of humans and animals, Physician's Work (Vrach. Delo) 6, 1977.

(17) Popovich, V.M. and Dumanskiy, Yu.D.: Hygienic evaluation of an electric field produced by high voltage transmission lines in populated places, in Hygiene of Populated Places, Kiev, 1977.

(18) Prokhvatilo, Ye.V. and Shvayko, I.I.: Changes in thyroid function under the action of a power frequency electric field, in Hygiene of Populated Places, No. 14, Kiev, 1975.

(19) Prokhvatilo, Ye.V.: Reaction of the endocrine system to a 50 Hz power frequency electromagnetic field, Physician's Work (Vrach. Delo) 11, 1976.

(20) Prokhvatilo, Ye.V.: Decreased cardiac capacity under the

influence of a power frequency electromagnetic field, Physician's Work (Vrach. Delo) 7, 1977.

(21) Prokhvatilo, Ye.V. and Andrienko, L.G.: Medical and biological significance of power frequency electromagnetic field, in Protection of Humans from Some Physical Factors in the Environment, Kiev, 1977.

(22) Revnova, N.V.: Influence on the organism of high voltage, power-frequency electric fields, in Hygiene, Occupation, and Biological Effects of Radiofrequency Electromagnetic Fields, Moscow, 1968.

(23) Rudichenko, V.F.: Some aspects of energetic metabolism in animals exposed to power-frequency electromagnetic fields, in Physico-mathematical and Biological Problems of the Effects of Electromagnetic Fields and Air Ions, Vol. II, Yalta, 1975.

(24) Sabarno, R.V., Slonchenko, A.V. and Stepanov, A.R.: Protection Against Power-Frequency Electric Fields, Kiev, 1971.

(25) Sazonova, T.Ye.: Investigation of the biological effect of high-voltage, power-frequency electric fields, in Protection from the Effects of Electromagnetic Fields and Electric Current in Industry, Leningrad, 1963.

(26) Sazonova, T.Ye.: Effect of Low Frequency Electromagnetic Fields on the Work Capacity of the Motor System, Report of Leningrad University (Vestnik Leningradskogo Universiteta), Vol. 3, 1964.

(27) Sazonova, T.Ye.: Effect of a power frequency electric field on the formation of motor dominants, Occupational Hygiene (Gig. Truda) 25, 1971.

(28) Sokovets, I.G.: Calculation of Induced Currents in the Human Body in Non-homogeneous and Homogeneous Electric Fields, Tomsk, 1978.

(29) Solov'yev, N.A.: On the Biological Effect of the Electrical Component of Low-Frequency Electromagnetic Fields, Reports of VNIIMIO (Trudy Inst. VNIIMIO), 6, 1962.

(30) Stroykova, K.V. and Belyayeva, T.I.: Influence of a low-frequency, high-voltage electric field on the level of macro-phosphorus compounds in the skeletal muscles of warm blooded animals, Physiol. J. (Fiziologicheskiy zh.) 43, 1957.

(31) Stroykova, K.V. and Belyayeva, T.I.: Physiological investigation of the effect of a high-voltage, power-frequency electric field on the organism, in Reports of the Laboratory of Electrical Safety, LIOT (Trudy laboratorii elektroobezopasnosti LIOT), Leningrad, 1958.

(32) Tamm, E.Ye.: Basic Theories of Electricity, Science (Nauka), Moscow, 1966.

(33) Tikhodeyev, N.N.: Transmission of Electroenergy Today and Tomorrow, Energy (Energia), 1975.

(34) Kornberg, H.A.: Biological effects of electric fields, RGE-FRA, JSS N 0035-3116, 1976-07, Special Issue.

(35) Dumanskiy, Yu.D., Popovich, V.M. and Prokhvatilo, Ye.V.: Hygienic evaluation of an electromagnetic field produced by high voltage transmission lines, Hygiene and Health (Gig. i. San.) 8, 1976.

(36) Dumanskiy, Yu.D. and Popovich, V.M.: Power frequency electric fields and their hygienic significance, in Problems of Protection Against Electric Current (EMP) in Various Sectors of Agriculture, Kiev, 1976.

(37) Dumanskiy, Yu.D. and Minkh, A.A.: Hygienic evaluation of a power-frequency electromagnetic field in agricultural and populated areas, in Hygiene of Populations and Protection of Health of the Agricultural Populations, Moscow, 1976.

(38) Dumanskiy, Yu.D. and Prokhvatilo, Ye.V.: A power-frequency electromagnetic field as a factor in the environment and its regulation, Hygiene and Health (Gig. i San.) 5, 1979.

(39) Dumanskiy, Yu.D., Tomashevskiy, L.A., Lyubchenko, S.A., et al.: Power-frequency electric field as a significant hygienic factor in the environment, in 10th Ukrainian Session of Hygienists, Thesis of Reports (Tezisy dokla-kov), Kiev, 1981.

(40) Dumanskiy, Yu.D. and Kozyarin, N.P.: Hygienic evaluation of an electrical field produced by high-voltage power-lines in populated areas, Hygiene and Health (Gig. i San.) 11, 1983.

(41) Dolin, P.A.: Fundamentals of Safety Technology in Electrical Facilities, Energy Publishing House (Energoizdat), Moscow, 1984.

(42) Shandala, M.G., Dumanskiy, Yu.D. and Prokhvatilo, Ye.V.: Hygienic basis for power-frequency electric field standards (50 Hz) in populated situations, in Hygiene of the Environment, Kiev, 1979.

(43) Ivanova, Ye.I. and Goncharuk, Ye.P.: Evaluation of hygienic norms under high-voltage powerlines, Energy Publishing House (Energoizdat), Moscow, 8, 1977.

(44) Gavalas, R.J., Walter, D.O., Hammer, J. and Adey, W.R.: Effect of low level, low frequency electric fields on EEG and behavior in _Macaca nemestrine_, Brain Res. 18:491-501, 1970.

(45) Dumanskiy, Yu.D., Los', I.P. and Popovich, V.M.: Effect of an electric field produced by high-voltage lines on humans, in Physico-mathematical and Biological Problems of the Effects of Electromagnetic Fields and Air Ions, Yalta, 1975.

(46) Dumanskiy, Yu.D., Popovich, V.M. and Kozarin, I.P.:

Influence of a low-frequency electromagnetic field (50 Hz) on the functional condition of humans, Hygiene and Health (Gig. i San.) 12, 1977.

(47) Adey, W.R.: A discussion of the functional significance of external and internal electrical fields of low frequency for brain tissue activity, in Functional Significance of Electrical Processes of the Brain, Moscow, 1977.

(48) Beyer, V.M., Brinkmann, K. and Kühne, B.: Einflub elektrischer felder auf lebende Organismen, Elektrigitunwirtschaft, jg 78, Huft 19, 1979.

(49) Eisemann, V.B.: Investigations on the Long-Term Effect of a Small Alternating Current of 50 Hz on the Human Being, Dissertation, Albert-Ludwig University, 1976.

(50) Marino, A.A., Berger, T.J., Mitchell, J.T., Duhacek, B.A. and Becker, R.O.: Electric field effects in selected biologic systems, Ann. N.Y. Acad. Sci., 238:436-444, 1974.

(51) Shterengarts, R.Ya: Problems of occupational hygiene and occupational pathology during high voltage tests of electrotechnical apparatus, Hygiene, Occupation, and Occupational Diseases (Gigiyena truda i prof-zabolevanii), 1, 1985.

(52) Hauf, R.: Effects of 50-Hz alternating fields on man, Electrotechiscka Zeitung, 1974.

(53) Hauf, R.: Effects of electromagnetic fields on man, Electrotechnische Zeitschrift, 1976.

(54) Hauf, R.: Electric and magnetic fields at power frequencies with particular reference to 50 and 60-Hz, in Nonionizing Radiation Protection, Chapter 5, Vol. VIII, M.J. Seuss, ed., World Health Organization, WHO Regional Publications, Copenhagen, Denmark, 1982.

(55) Hauf, R. and Freibing, J.: Medizinische Frogen im Zusem-

menhang mit dem Arbeiten unter Spannung, 3.37d Sonderdruck aus Bulletin SEV/VSR N. 2.3, 1978.

(56) Hauf, R.: Influenkangf no-gietechnischer Felder auf den Henachen, Arbeitsmed. 16, 1981.

(57) Hauf, R.: Influence of alternating electric and magnetic fields on human beings, Rev. Gen. Electr., Special Issue, July, 1976.

(58) Hauf, R.: Einfluss electromagnetisches felder auf den menschen (Effect of electromagnetic fields on human beings), Electrotech. Z. B28, 1976.

(59) Michaelson, S.M.: Human responses to power-frequency exposures, in Biological Effects of Extremely Low Frequency EMF, Proceedings 18th Ann. Hanford Life Sciences Symposium, Richland, Washington, October, 1978.

(60) Michaelson, S.M.: Analysis of studies related to biologic effects and health implications of exposure to power-frequencies, Environ. Profession. 1, 1979.

(61) Danilin, V.A., Voronin, A.K. and Madorskiy, V.A.: Health condition of workers in a high voltage electric field, Occupational Health (Gig. truda) 13, 1969.

(62) Asanova, T.P., Rakov, A.N. and Shcheglova, A.V.: Question of the influence of a strong power-frequency electric field on workers, in Materials of a Scientific Session Commemorating Work of the Institute, 1961-1962, Leningrad, 1963.

(63) Gorpinchenko, I.M.: Physical chemical mechanism of the biological action of electromagnetic fields, Dissertation, 1973.

(64) Kouwenhoven, W.B., Langworthy, O.R., Singewald, M.L. and Knickerbocker, G.G.: Medical evaluation of man working in AC electric fields, IEEE Trans. Power Appar. Syst. PAS-86:506-511, 1967.

(65) Singewald, M.L., Langworthy, O.R. and Kouwenhoven, W.B.: Medical follow-up study of high voltage linemen working in the electric fields, IEEE Trans. Power Appar. Syst. PAS-92, 1973.

(66) Eckert, E.E.: Plötslicher und uner warter Jat im Kleirkindesalter und elektromagnetische Felder, Med. Klin. 71, 1976.

(67) Fole, F.F.: The HVS effect in electric power substations, Medicine by Seguristad del Trabazio, 1973.

(68) Fole, F.F. and Drutrus, E.: New contribution to studies of electromagnetic fields generated by high voltages, Med. V. Seg. del Trab., 1974.

(69) Knave, B., Gamberale, F., Bergstrom, S., Birke, E., Iregren, A., Kolmodin-Hedman, B. and Wennberg, A.: Long-term exposure to electric fields: A cross-sectional epidemiologic investigation of occupationally exposed industrial workers in high voltage substations, Electra 65:41-54, 1979.

(70) Stoops, G.J. and Janischewskyj, W.: Epidemiological Study of Workers Maintaining HV Equipment and Transmission Lines in Ontario. Research Report, Canadian Electrical Association, Montreal, 1979.

(71) Nordstrom, S.E., Birke, E. and Nordenson, I.: Workers in High Voltage Power Stations: A Study of Pregnancies and Chromosome Breaks. Preliminary Report, University of Umea, Umea, Sweden, 1981.

(72) Nordstrom, W., Birke, E. and Gustavsson, L.: Reproductive hazards among workers at high voltage substations, Bioelectromagnetics 4, 1983.

(73) Coleman, M., Bell, J. and Skeet, R.: Leukemia incidence in electrical workers, Lancet 1:982, 1983.

(74) McDowall, M.E.: Leukemia mortality in electric workers in

England and Wales, Lancet 1:8318, 1983.

(75) Milham, S., Jr.: Mortality from leukemia in workers exposed to electrical and magnetic fields, N. Eng. J. Med. 307:249, 1982.

(76) Wright, W.E., Peters, J.M. and Mack, T.M.: Leukaemia in workers exposed to electrical and magnetic fields, Lancet, 1160, 1982.

(77) Bridges, J.E.: Environmental considerations concerning the biological effects of power frequency (50 or 60 Hz) electric fields, IEEE Trans. Power Appar. Syst. PAS-97:19-32, 1978.

(78) Nalboysson, E.: Medical control of men working within electromagnetic fields, Rev. Gen. Electr. Special Issue, July, 1976.

(79) Fisher, G.H., Uhrmann, and Knapp, E.: Übt das netzfrequente Wechselfeld zentrale Wirkungen aus? (Does a 50-Hz alternating field cause central nervous effects?), Zbl. Bakt. Hyg. I. Abt. Orig. B166:381-385, 1978.

(80) Mose, J.R.: Problems of housing quality, Zentralbe. Bakteriol. I. Abt. Orin. B. 166, 1978.

(81) Miller, C.J.: The measurement of electric fields in live line working, IEEE Trans. Power Appar. Syst. PAS-86:493-498, 1967.

(82) Bawin, S.M., Abbot, I., et al.: Effects of 60 Hz environmental electric fields on the central nervous system of laboratory rats, Presented at USVC/URSI Spring Meeting, Seattle, Washington, 1979.

(83) De Lorge, J.: A Psychobiological Study of Rhesus Monkey Exposed to Extremely-Low-Frequency Low-Intensity Magnetic Fields, USN Report NAMRL-1203, NTIS #AD A000078, Springfield, Virginia, 1974.

(84) Gavalas, R.J., Walter, D.O., Hammer, J. and Adey, W.R.: Effect of low level, low frequency electric fields on EEG and behavior in Macaca nemestrine, Brain Res. 18:491-501, 1970.

(85) Gavalas-Medici, R. and Day-Magdaleno, S.R.: Extremely low frequency, weak electric fields affect schedule controlled behavior of monkeys, Nature 261:256-259, 1976.

(86) Kaune, W.T.: Coupling of living organisms to ELF electric and magnetic fields, in Biological and Human Health Effects of Extremely Low Frequency Electromagnetic Fields, American Institute of Biological Sciences, Arlington, Virginia, 1985.

(87) Morris, J.E. and Phillips, R.D.: Effects of 60-Hz electric fields on specific humoral and cellular components of the immune system, Bioelectromagnetics 3:341-347, 1982.

(88) Morris, J.E. and Phillips, R.D.: Erratum. Effects of 60-Hz electric fields on specific humoral and cellular components of the immune system, Bioelectromagnetics 4:293, 1983.

(89) Morris, J.E. and Ragan, N.A.: Immunological studies with 60-Hz electric fields, in Biological Effects of Extremely-Low-Frequency Electromagnetic Fields, 18th Ann. Hanford Life Sciences Symposium, October 16-18, Richland, Washington, CONF-781016, NTIS, Springfield, Virginia, 1979.

(90) Andrienko, L.G.: State of the regenerative function of white rats exposed to an electromagnetic field, in Hygiene of Populated Places, Kiev, 1976.

(91) Andrienko, L.G.: Influence of a power-frequency electric field on reproductive function, in Protection of Humans from Some Physical Factors in the Environment, Kiev, 1977.

(92) Andrienko, L.G.: Influence of a power frequency electric field on regenerative function, Hygiene and Health, 1976.

(93) Dumanskiy, Yu.D. and Tomashevskiy, L.A.: Electromagnetic fields as a hygienically significant factor in the environment, in 9th Ukrainian Session of Hygienists and Occupational Health Medical Personnel, Kiev, 1976.

(94) Dumanskiy, Yu.D. and Andrienko, L.G.: Influence of a power-frequency electric field on spermatogenic function, Physician's Work (Vrach. Delo) 9, 1977.

(95) Dumanskiy, Yu.D. and Andrienko, L.G.: An assessment of adaptive and compensatory reactions during powerline (50 Hz) electromagnetic field exposure, in Adaptation of Humans to Extreme Factors in the Environment, Odessa, 1980.

(96) Dumanskiy, Yu.D. and Andrienko, L.G.: Electric power lines as the origin of electromagnetic radiation exposure and problems of environmental hygiene, in Hygiene of Populated Places, No. 20, 1981.

(97) Tomashevskaya, L.A. and Dumanskiy, Yu.D.: Change in the metabolism of experimental animals as one index of the biological effect of a 50 Hz electric field, Physician's Work (Vrach. Delo) 7, 1981.

(98) Shandala, M.G. and Dumanskiy, Yu.D.: Biological effect and hygienic standard for power frequency electric fields, Knowledge (Znaniye), Kiev, 1977.

(99) Bianchi, D., Cedrini, L., Ceria, F. and Meda, E.: Exposure of mammalians to strong 50 Hz electric fields. Effects on heart's and brain's electrical activity, Arch. Fisiol. 70:33-34, 1973.

(100) Bridges, J.E. and Preache, M.: Biological influences of power frequency electric fields -- A tutorial review from a physical and experimental viewpoint, Proc. IEEE 69:1092-1120, 1981.

(101) Marino, A.A., Berger, T.J., Austin, B.P., Becker, R.O. and

Hart, F.X.: In vivo bioelectrochemical changes associated with exposure to extremely low frequency electric fields, Physiol. Chem. Phys. 9:433-441, 1977.

(102) Marino, A.A. and Becker, R.O.: Biological effects of extremely low frequency electric and magnetic fields: A review, Physiol. Chem. Phys. 9:131-147, 1977.

(103) Marino, A.A., Becker, R.O. and Ulrich, B.: The effect of continuous exposure to low frequency electric fields on three generations of mice: A pilot study, Experientia 32:565-566, 1976.

(104) Gann, D.W.: Biological effects of exposure to high voltage electric fields: Final report, Electric Power Research Institute Report #RP98-02, Electric Power Research Institute, Palo Alto, California, 1976.

(105) Hansson, H.A.: Purkinje nerve cell changes caused by electric fields: Ultrastructural studies on long term effects on rabbits, Med. Biol. (Helsinki), 1981.

(106) Hansson, H.A.: Lamellar bodies in Purkinje nerve cells experimentally induced by electric field, Exp. Brain Res. 216:187-191, 1981.

(107) Cerretelli, P. and Malaguti, C.: Research carried out in Italy by ENEL on the effects of high voltage electric fields, Rev. Gen. Electr. (Special Issue), July, 1976.

(108) Fulton, J.P., Cobb, S., Preble, L., Leone, L. and Forman, E.: Electrical wiring configurations and childhood leukemia in Rhode Island, Am. J. Epidemiol. 111:292-296, 1980.

(109) Margenato, V., Cantone, A. and Viola, D.: Ettettidel campo electtrico di elevata intencita'a 50-Hz sulla resistenze allintenzione sperimentale nel topo, Boll. Soc. Ital. Bill.

(110) Hjeresen, D.L., Kaune, W.T., Decker, J.R. and Phillips,

R.D.: Effects of 60-Hz electric fields on avoidance behavior and activity of rats, Bioelectromagnetics 1:299-312, 1980.

(111) Viola, D., Margenato, V. and Veiesteinas, A.: Effetti del campo elettrico di elevata intensita'a 50-Hz sull etticienza alimentare e Sulla curve di accrescimento del ratto, Boll. Soc. Ital. Biol. Sper. 54, 1978.

(112) Bezdol'naya, I.S.: Influence of a 50-Hz electromagnetic field on conditioned reflexes in white rats, in Hygiene of Populated Places, No. 20, Kiev, 1981.

(113) Shandala, M.G., Dumanskiy, Yu.D., Los', I.P., et al.: Biological effect of a low-frequency electric field (50 Hz), in Materials of the 3rd Soviet-American Working Group on the Problem: Study of the Biological Effect of Physical Factors in the Environment, Kiev, 1981.

(114) Shandala, M.G., Rudnev, M.I., Dumanskiy, Yu.D., et al.: Recommendations for evaluation and regulation of the biological effect of environmental low-intensity microwave radiation, Ministry of Health, Ukrainian SSR, Kiev, 1981.

(115) Hackman, R.M. and Graves, H.B.: Corticosterone levels in mice exposed to high intensity electric fields, Behav. Neurol. Biol. 32:229-234, 1981.

(116) Free, M.J., Kaune, W.T., Phillips, R.D. and Cheng, H.-C.: Endocrinological effects of strong 60-Hz electric fields on rats, Bioelectromagnetics 2:105-121, 1981.

(117) Groza, P., Carmaciu, E. and Bubuianu, E.: Blood and urinary catecholamine variations under the action of high voltage electric field, Physiologie 15:139-144, 1978.

(118) Bankoske, J.W., Mckee, G.W. and Graves, H.B.: Ecological Influence of Electric Fields. Interim report 2. EPRI Research Project 129, EPRI Report #EA-178, Palo Alto, California, 1976.

(119) Carter, J.H. and Graves, H.B.: Effects of High Intensity

AC Electric Fields on the Electroencephalogram and Electrocardiogram of Domestic Chicks: Literature Review and Experimental Results, Pennsylvania State University, University Park, Pennsylvania, 1975.

(120) Phillips, R.D., Chandon, J.H. and Free, M.J.: Biological Effects of 60-Hz Electric Fields on Small Laboratory Animals. Annual Report, DOE Report #HCP/T 1830-3, Department of Energy, Washington, DC, 1978.

(121) Jaffe, R.A., Lassewski, B.L., Carr, D.B. and Phillips, R.D.: Chronic exposure to a 60-Hz electric field: Effects on synaptic transmission and the peripheral nerve function in the rat, Bioelectromagnetics 1:131-147, 1980.

(122) Feldstein, C.S., Gibson, E.G., Polonis, D.D. and Smith, Kh.D.: Effects of 60-Hz electric fields on the behavior of non-human primates, in Materials of the 3rd Soviet-American Working Group, Health, Kiev, 1982.

(123) Phillips, R.D.: Biological effects of electric fields on small laboratory animals, in Materials of the 3rd Soviet-American Working Group, Health (Zdorov'ya), Kiev, 1982.

(124) Hjeresen, D.L., Miller, M.C., Kaune, W.T. and Phillips, R.D.: A behavioral response of swine to 60-Hz electric fields, Bioelectromagnetics 2:443-451, 1982.

(125) Berezyuk, G.S.: Some experimental data on the effects of a high-voltage electric field, in Occupational Hygiene and the Biological Effects of Radiofrequency Electromagnetic Fields, Moscow, 1968.

(126) Veiesteinas, A., Viola, D. and Gussoni, M.: Effto dell'-esposizione and un elevoto gradiente electrico a 50-hz su alcuni parametri cardiocircolatori, Boll. Soc. Ital. Biol. Sper., 1975.

(127) Hilton, D.I. and Phillips, R.D.: Cardiovascular response of rats exposed to 60-Hz electric fields, Bioelectromagnetics 1:55-64, 1980.

(128) Bars, H. and Andre, J.: Biological effects of electric field on rats and rabbits, Rev. Gen. Electr. (Special Issue), July, 1976.

(129) Cerretelli, P., Veiesteinas, A., Margonato, V., Cautone, et al.: 1000 kV project: Research on the biological effects of 50-Hz electric fields in Italy, in Biological Effects of Extremely-Low-Frequency Electromagnetic Fields, 18th Annual Hanford Life Sciences Symposium, Richland, WA, CONF-781016, October 16-18, 1979.

(130) Fam, W.Z.: Long term biological effects of very intense 60-Hz electric fields on mice, IEEE Trans. Biomed. Eng. BME-27:376-381, 1980.

(131) Jaffe, R.A., Lopresti, C.A., Carr, D.B. and Phillips, R.D.: Perinatal exposure to 60-Hz electric fields: Effects on the development of the visual-evoked response in the rat, Bioelectromagnetics 4:327-339, 1983.

(132) Rommereim, D.N., Kaune, W.T., Buschbom, R.L., Phillips, R.D. and Kikov, M.R.: Reproduction and development in rats chronically exposed to 60-Hz fields, in Interaction of Biological Systems with Static and ELF Electric and Magnetic Fields, Proceedings 23rd Ann. Hanford Life Sciences Symposium, L.E. Anderson, R.J. Weigel and B.J. Kelman, eds., DOE Series, NTIS, Springfield, Virginia, 1985.

(133) Seto, J.J.: Pilot Study of 60-Hz Electric Fields Induced Biological Subtle Effects. Final Report, EPRI Report TR-27, Louisiana Power and Light Company, New Orleans, Louisiana, 1979.

(134) Sikov, M.R., Montgomery, L.D., Smith, L.G. and Phillips, R.D.: Studies of prenatal and postnatal development in rat exposed to 60-Hz electric field, Bioelectromagnetics 5:101-112, 1984.

(135) Phillips, R.D.: Study of the biological effect of a 60-Hz electric field on three generations of chronically exposed

miniature pigs, in Materials of the 3rd Soviet-American Working Group, Health (Zdorov'ya), Kiev, 1982.

(136) Graves, H.B., Carter, J.H., Jenkins, T., Kellmel, D. and Poznaniar, D.T.: Responses of domestic chicks to 60-Hz electrostatic fields, in Developmental Toxicology of Energy Related Pollutants, D.D. Nablum, M.R. Sikov, P.L. Hackett and F.D. Andrew, eds., NTIS, Springfield, Virginia, 1978.

(137) Silny, J.: Effect d'influence d'un champ electrique a 50-Hz sur l'organism, Rev. Gen. Electr. 85, 1976.

(138) Shandala, M.G., Dumanskiy, Yu.D., Los, I.P., Akimenko, V.Ya., Bezdol'naya, I.S., et al.: Health standards and guidelines for protecting the public from the effects of an electric field produced by overhead power-frequency electric transmission lines, Moscow Ministry of Health, USSR, 1984.

(139) Barnes, H.C., McElroy, A.J. and Charkow, J.H.: Rational analysis of electric fields in live line working, IEEE Trans. Power Appar. Syst. PAS-86:482-492, 1967.

(140) Waibel, R.: Der Eintluss hiederfreguenter electrischer Felder aut Lebewesend, Dissertation, I.H. Grus, 1975.

(141) Waibel, R. and Schuy, S.: Der Einflub electrischer Felder auf Organismen, Arch. Elektrotechn. (W. Berlin), 1978.

(142) Amon, J.: Einfluß des elektrischen Feldes auf den Elektrolythaushalt, den intermediären Stoffwechsel und den Hormonhaushalt des Menschen, Dissertation, Freiberg, 1977.

(143) Schmidt, U.J.: Biologische Wirkung elettrischer 50-Hz-Felden, Berlin, VDZ, Verley Gmb., 1980.

(144) Bezdol'naya, I.S.: Adaptive reaction during the action of a power-frequency electric field, in Adaptation of Humans to Extreme Environmental Conditions, Thesis of Reports (Tezisy dokl.), Odessa, 1980.

(145) Bezdol'naya, I.S.: Functional condition of the central nervous system of humans during the action of a power-frequency electric field, in Biological Effects of Electromagnetic Fields, Thesis of Reports (Tezisy dokl.), Pushchino, 1983.

(146) Bezdol'naya, I.S.: Reaction of the human central nervous system to a power frequency electric field, in Hygiene of Populated Places, No. 22, 1983.

(147) Shandala, M.G. and Dumanskiy, Yu.D.: Electric transmission lines as a pollution factor in the environment due to low-frequency electromagnetic radiation and its regulation, in Problems of Experimental and Practical Electromagnetobiology, Pushchino, 1983.

(148) Shandala, M.G., Dumanskiy, Yu.D., Tomashevskiy, L.A. and Bezdol'naya, I.S.: Functional condition of humans under the influence of power-frequency electric fields, Physician's Work (Vrach. Delo), No. 6, 1983.

(149) Shandala, M.G., et al.: The biological effects of the low frequency electromagnetic field (50-Hz), USNC-URSY Spring Meeting, Seattle, Washington, 1979.

(150) Shandala, M.G., et al.: Biological effects of industrial frequency (50-Hz) electric fields, Proceedings of US-USSR Workshop, A report by the National Institute of Environmental Health Sciences, 1981.

(151) Cabanes, J. and Gary, C.: La perception direct du champ electrique (Direct perception of the electric field), CIGRE Report #233-08, International Conference on Large High Tension Electric Systems, CIGRE, 1981.

(152) Deno, D.W. and Zaffanella, L.: Electrostatic effects of overhead transmission lines and stations, in Transmission Line Reference Book, 345 kV and Above, 2nd Ed., Electric Power Research Institute, Palo Alto, California, 1982.

(153) Fotopoulos, S.S., Graham, C., Cook, M.R., Koontz, E.,

Gerkovich, M.M. and Cohen, H.D.: Effects of 60-Hz fields on human neuroregulatory immunologic, hematologic, and target organ activity, in Interaction of Biological Systems with Static and ELF Electric and Magnetic Fields, Proceedings of the 23rd Ann. Hanford Life Sciences Symposium, L.E. Anderson, R.J. Weigel and B.J. Kelman, eds., DOE Series, NTIS, Springfield, VA, 1985.

(154) Graham, C., Cohen, H.D., Cook, M.R., Phelps, J., Gerkovich, M. and Fotopoulas, S.S.: A double-blind evaluation of 60-Hz field effects on human performance, physiology and subjective state, in Interaction of Biological Systems with Static and ELF Electric and Magnetic Fields, Proceedings of the 23rd Ann. Hanford Life Sciences Symposium, L.E. Anderson, R.J. Weigel and B.J. Kelman, eds., DOE Series, NTIS, Springfield, VA, 1985.

(155) Kalashnikov, S.G.: Electricity, Science (Nauka), Moscow, 1964.

(156) Feyman, R., Leyton, R. and Sendye, M.: Feyman lecture on physics, Science (Nauka), Moscow, 5, 1966.

(157) Savin, B.Ya. and Stepanskiy, G.A., eds.: Physiology of Man and Animals. Biological Effects of Electromagnetic Radiations, Vol. 22, Moscow, 1978.

(158) Kaune, W.T. and Phillips, R.D.: Comparison of the coupling of grounded humans, swine, and rats to vertical, 60-Hz electric fields, Bioelectromagnetics 1:117-129, 1980.

(159) Kaune, W.T. and Forsythe, W.C.: Current densities measured in human models exposed to 60-Hz electric fields, Bioelectromagnetics, 6:13-32, 1985.

(160) Sheppard, A.R. and Eisenbud, M.: Biological Effects of Electric and Magnetic Fields of Extremely Low Frequencies, New York University Press, New York, 1977.

(161) Spiegel, R.J.: Numerical determination of induced

currents in humans and baboons exposed to 60-Hz electric fields, IEEE Trans. Electromag. Comp. 23:382-390, 1981.

(162) Groze, P., Carmacin, B. and Vacarin, A.: The effect of a high voltage electric field on neurosecretion and on certain diencephalic protein fractions in the rat, Rev. Roum., Morphol., Embryol. et Physiol, Ser. Physiol. 15, 1978.

(163) Wachtel, H., Seaman, R. and Joines, W.: Firing pattern, changes and transmembrane currents produced by extremely low frequency fields in pacemaker neurons, Ann. N.Y. Acad. Sci. 247:46-62, 1975.

(164) Schwan, H.P.: Nonthermal cellular effects of electromagnetic fields: AC-field induced Ponderomotive forces, Brit. J. Cancer 45, 1982.

(165) Kholodov, Yu.A.: Reaction of the nervous system to electromagnetic fields, Science (Nauka), Moscow, 1975.

(166) Durfee, N.K.: Influence of Extremely Low Frequency Electric and Magnetic Fields Upon the Growth, Development and Behavior of Domestic Birds. Biological Effects of High Voltage Fields: An Update, Vol. 1, EPRISA-1123, Project 857-1, Final Report, July, 1979.

(167) Gruner, O.: Psychika a electromagenticke pele, Cesopis Lekeru Ceskych 115, 1976.

(168) Shandala, M.G., Dumanskiy, Yu.D., Prokhvatilo, Ye.V., et al.: Hygienic basis for standards for 50-Hz power-frequency electric fields in populated places, in Hygiene of the Environment, Theses of Reports (Tezisy dokladov), Kiev, 1979.

27
Environmental Electromagnetic Energy and Public Health

ANDREW A. MARINO

Department of Orthopaedic Surgery, Louisiana State University School of Medicine in Shreveport
Shreveport, Louisiana

INTRODUCTION

Electromagnetic fields (EMFs) are real, physical, incorporeal entities that arise from the existence and motion of atomic charges. Man-made EMFs became increasingly common constituents of the general and workplace environments early in the 20th century. Some life-styles and occupations are associated with more than the average amount of exposure to EMFs. People who live near high-voltage powerlines, for example, experience strong electric and magnetic fields that are not usually present in neighborhoods without powerlines. People who use electric blankets similarly experience stronger fields for longer durations compared to the general population. Navy shipboard personnel are exposed to EMFs from many shipboard radars, and in this regard their work environment differs significantly from that of other young men. People living near airports or antenna farms are exposed to radar beams or broadcast radiation, and such residential areas therefore differ from other socio-economically comparable areas with regard to the content of the electromagnetic background. Amateur radio operators experience more EMF exposure than the general population because of their proximity to radiating antennas.

Many patterns of living, working, playing, or resting can be identified with increased intensity and duration of exposure to EMFs. The existence of groups in the general population that experience increased exposure raises the question of whether these groups exhibit an increased incidence or prevalence of disease that is linked to EMF exposure.

In the USA this issue first surfaced in the early 1970's when U.S. powerline engineers heard Soviet reports of adverse health effects among Soviet powerline workers (1). The EMF hazards issue was debated in lengthy hearings in New York, and in two hearings in California, in the context of proposed powerlines. The U.S. Departments of Defense and Energy became involved as the questions widened to include high frequencies. Now, near the second decade of interest in health hazards of environmental EMFs, the data is sufficiently crystallized to permit a realistic assessment of the extent of the problem. This is my purpose.

I begin with an effort to cultivate appreciation for the nature and extent of artificial EMFs in the environment. Next, bioassay studies performed in the laboratory are described. They are important because they convince us that EMFs have physiological significance, and hence have at least a possibility of promoting disease. Equally as important, the bioassay studies tell us of the general nature of EMF-induced bioeffects by keying the EMF studies into a larger and more established literature -- that dealing with chronic stressors. Chronic stressors promote disease, and it is the fact that environmental EMFs are chronic stressors that makes them determinants of human disease. Following my discussion of these points, I describe the reports of the epidemiology of environmental EMFs. It is lamentable that society requires actual evidence of overt disease in innocent people -- dead bodies -- before considering protective steps. But such steps are expensive, and they arguably imperil public order and national security. I do not comment on these arguments, but my views of the public-health significance of environmental EMFs are given in the concluding section.

This Chapter is devoted to an attempt to organize data into a rational framework. I have focused on the reports that are particularly pertinent to the mainstream public-health issues of environmental EMFs. I have not omitted data merely because it does not conform to my (or anyone else's) opinion, and I have tried not to use my opinion to gloss over the absence of data.

EMFs IN THE ENVIRONMENT

The artificial electromagnetic environment of the USA is a

superposition of contributions from many sources having diverse operating characteristics. They include high- and low-power emitters that can be omnidirectional or directional, and that can operate continuously or intermittently. At high frequencies, the general EMF background consists predominantly of the AM radio band (0.535-1.604 MHz) and the FM and television band (54-806 MHz). There are approximately 5000 FM and 5000 AM broadcast stations in the USA, and more than 1200 television stations (2). At any given moment about half the U.S. population is exposed to these sources at levels above 0.005 $\mu W/cm^2$, and about 1% is exposed above 1 $\mu W/cm^2$ (3).

EMFs emanating from the electrical power system (60 Hz in the USA, 50 Hz in Europe and the USSR) constitute most of the artificial low-frequency electromagnetic background. The average background 60-Hz electric field is about 1 V/m, and the average background magnetic field is about 800 μGauss.

EMFs much greater than the background are found in the vicinity of specific sources. The power density from a typical 50,000-watt AM radio station does not decrease below 1 $\mu W/cm^2$ within a radius of about 3280 ft (4). FM radio stations vary in strength and antenna design, but 193 of 2750 such stations in the USA probably have levels exceeding 1000 $\mu W/cm^2$ within 200 ft of the antenna (5). In large urban areas, the elevation necessary for transmission of radio and television signals is sometimes obtained by mounting the antennas atop a tall building. This produces high EMF levels in nearby buildings (6).

When antennas are grouped, they produce relatively intense EMF levels over broad areas. Mount Wilson, California, for example, has 27 radio and television antennas, and they produce strong EMFs in both public and private areas (7). The Sentinel Heights area south of Syracuse, New York contains about a dozen transmitters and they result in essentially ambient levels of about 1 $\mu W/cm^2$ throughout an area of several square miles (8). Only a few of the radiation hotspots have actually been measured. The Environmental Protection Agency (EPA) measured the radiofrequency radiation levels at Honolulu, Hawaii in 1975 (9), and returned 9 years later to make additional measurements (10). They found large fields in homes and businesses near various radio towers. Another hotspot was reported near an antenna farm on Cougar Mountain outside Seattle, Washington,

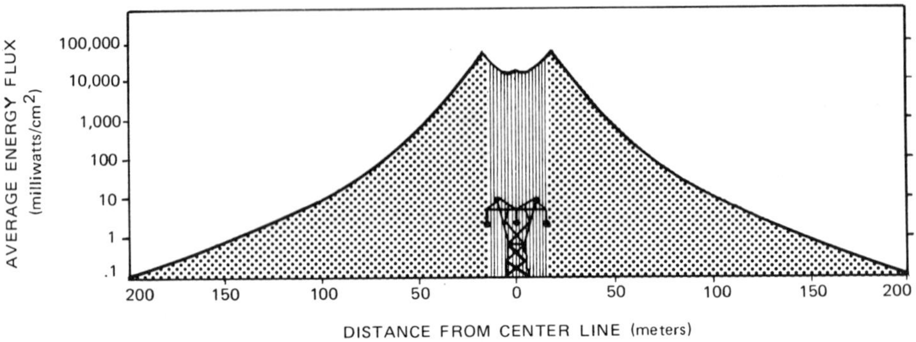

FIGURE 1. Calculated average ground-level energy flux of a typical 765-kV powerline as a function of lateral distance from the center-line (18).

where readings up to 700 $\mu W/cm^2$ were recorded in areas accessible to the public (11). Healy Heights near Portland, Oregon exhibited levels in excess of 100 $\mu W/cm^2$ in public areas and in private homes (12). In Denver, Colorado, measurements in a public area near the antenna farm that services Denver showed levels as high as 1000 $\mu W/cm^2$. Most indoor levels were less than 50 $\mu W/cm^2$, but some were as high as 580 $\mu W/cm^2$ (13).

Because of stray radiation from radar, exposure levels near airports and military bases can be in the range of 10-100 $\mu W/cm^2$ at distances up to 0.5 miles. The Pave Paws 420-450 MHz radar on Cape Cod (1 of 4 such installations in the USA) produces 0.03-3700 $\mu W/cm^2$, depending on location (14).

Microwave-relay antennas are used for long-distance telephone service and for private communications. A 10-ft diameter antenna positioned 100 ft above the ground produces ground level EMFs of approximately 0.03-7.5 $\mu W/cm^2$ within 376 ft of the tower (15). There are several thousand microwave-relay towers in the USA, each with 2 or more antennas.

Mobile communications equipment and hand-held walkie-talkies are relatively low-power sources, but they account for significant exposure levels because the radiating antenna is ordinarily close to the user. A walkie-talkie operating at 165 MHz, with an output of 1.8 watts results in 144-12,000 $\mu W/cm^2$ in the vicinity of the head of the user (16).

Starting and stopping of trains in the Bay Area Rapid Transit System in California produces low-frequency EMFs throughout the entire San Francisco Bay area (17).

Powerlines transport electrical energy, and they are usually built overhead rather than underground. The energy carried by a powerline actually moves through the space that surrounds the wires (18) (Figure 1), and consists of an electric and magnetic field. Various design considerations such as line geometry, phase spacing, and operating voltage materially affect field strength at various lateral distances (19). Electric fields associated with powerlines having different voltages are shown in Figure 2. Figure 3 shows the spatial distribution of the magnetic field of a 345-kV powerline.

Power-frequency magnetic fields measured over 223 km of roads in Quebec showed that the fields were greater than 400 μGauss more than 90% of the time, greater than 1600 μGauss

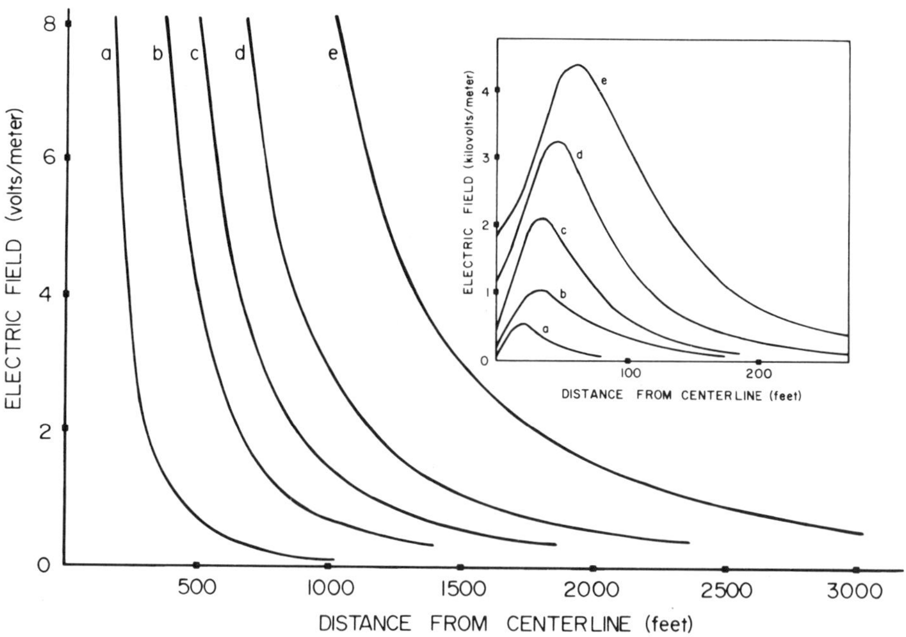

FIGURE 2. Ground-level electric fields of high-voltage powerlines, calculated by the method of images (18). (a), 115 kV; (b), 230 kV; (c), 345 kV; (d), 500 kV; (e), 765 kV.

50% of the time, greater than 5000 μGauss 10% of the time, and greater than 10,000 μGauss 1% of the time (20). The distribution of power-frequency magnetic fields that we measured in a region in the English midlands is shown in Figure 4. Readings were taken at the domiciles of suicide victims and at an equal number of appropriately chosen control addresses (21). A total of 1184 addresses were measured, yielding a mean of about 800 μGauss and a median of about 400 μGauss (range, 10-15,000 μGauss).

An electric field of 2900 V/m was measured from an electric blanket (22). The 60-Hz magnetic field from an electric blanket and from other common electrical devices are shown in Table 1. In addition to line-frequency fields, line-operated devices can produce significant fields at harmonics up to 1 kHz, and beyond (23).

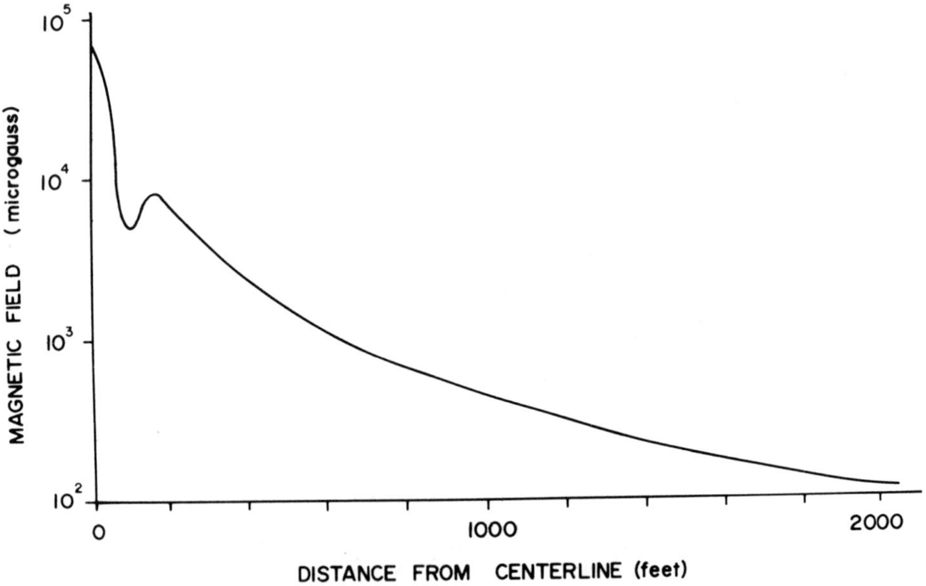

FIGURE 3. Distance dependence of the magnetic field of a twin-circuit vertical 345-kV powerline, calculated by the method of images (18). Each phase (phase relationship 1, 2, 3:3, 1, 2) consists of two conductors (1.08 in, diameter) spaced 1.5 ft apart. The assumed line and phase spacings are 28.5 and 24 ft, respectively. The height of the lowest phase in each line is 63.1 ft.

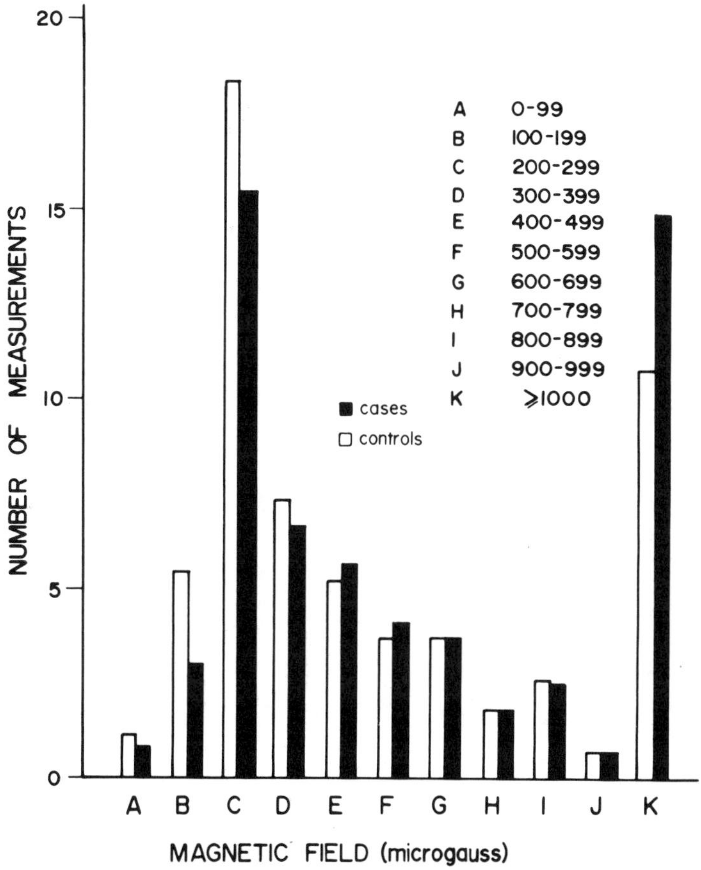

FIGURE 4. Distribution of Power Frequency Magetic Fields in a Study District in the English Midlands. The readings were taken with an FDM-100-10-53 meter (Deno) with the search coil mounted 1 meter above the ground, and 0.5 m from the center of the front door of the residence. Data is taken from a case-control study of the correlation between suicide and power-frequency magnetic field (21). The number of domiciles of suicide cases and controls having a magnetic field within the indicated interval is shown.

TABLE 1. Electric Field at 60 Hz Associated with Common Electrical Appliances. The measurements were made using a magnetic flux probe (Magnetek Model 1846) calibrated by the use of a uniform magnetic field of known magnitude and direction (224).

SOURCE	MAGNETIC FIELD (mGauss) AT INDICATED DISTANCE (cm) FROM SOURCE		
	10	20	30
Electric Blanket* (Northern Style 13, 180 W)	13.8	6.2	2.8
Electric Razor (Remington Model GMF 1)	580	110	40
Electric Wire 18 gauge (5 A)			
Parallel conductors	2.8	1	0.5
Separated conductor	100	55	36
Light Bulb (Sylvania)			
60 W (0.42 A)	1.02	0.56	0.25
100 W (0.75 A)	1.50	0.85	0.46
Electric Iron (GE Catalog 16F66)	44	11	3.7
Electric Typewriter (IBM Selectric II)	2.8	1.2	.56
Electric Motor (Fischer No. 1907501, 1/6 HP)	46	10	3.7
Electric Motor (GE Model 5KC42JG14EX, 1/2 HP)	355	14	6.3
Fluorescent Light (Westinghouse, 15 W (2))	5.2	3.4	2.65
Soldering Iron (Weller Model WTCPN)	185	54	23
Refrigerator (GE Model TBF 21 DW)	4.8	3.6	2.65

*230, 120, 30 mGauss at the surface, 1 cm and 5 cm respectively.

Human exposures to magnetic fields associated with medical imaging, and some occupational activities, have been summarized by Stuchly (24).

Video display terminals produce environmental electric fields of 10 Hz-200 MHz. An electric field in excess of 300 V/m was measured 20 cm from a terminal (25). The electric field exceeded 5 V/m along more than 120° of arc at a radius of 60 cm from the terminal.

Data regarding occupational exposure to EMFs is scanty (26), and that regarding exposure of military servicemen is essentially non-existent.

There are no U.S. federal standards to protect the general population from overexposure to any frequency of nonionizing radiation. The EPA has authority to promulgate high-frequency standards but has not done so. The agency has declared that the sole authority to set standards at low frequencies rests with the several states (none of which has taken any significant action). The Occupational Safety and Health Administration has a 10,000 $\mu W/cm^2$ standard, but the courts have ruled it unenforceable. The Soviet general population exposure standard for radiofrequency and microwave radiation is 10 $\mu W/cm^2$ for 300 MHz-300 GHz, and 3-25 V/m for 30 kHz-300 MHz (27). The standard at low frequencies has been described in the previous Chapter.

LABORATORY STUDIES OF EMF BIOEFFECTS

INTRODUCTION

The threshold public-health question regarding environmental EMFs is whether their presence in human living space is detected by the subjects. The body has a specialized detector, the eye, for receiving EMFs with a narrow frequency range. If human beings have no detector for other environmental EMFs, then the body would remain ignorant of their presence and they could not produce changes in the function of the body -- pathological changes or otherwise. There are different ways to approach this pertinent question. One could search for the detector in model systems, or search for biological effects in animal systems (on the certain inference that if there were an EMF-induced bioeffect, the EMF must have been detected). Historically, only the latter approach has been successful, and it is this data that will be reviewed here.

TRANSIENT EFFECTS IN ANIMALS

Some of the earliest studies were performed in the Soviet Union, and it apparently was Soviet scientists who first recognized the compensatory nature of EMF-induced bioeffects (28). Kholodov studied the effects of magnetic-field exposure on the electroencephalogram (EEG) in rabbits, and found alterations in delta waves and spindles (brief bursts of 8-12 Hz waves) due to 1-3 minutes' exposure at 200-1000 Gauss (29). These reactions usually occurred after a latent period of about 10 seconds, and they lasted at least 30 seconds in about half the animals tested. Sometimes, desynchronization (an abrupt change in the main rhythm) occurred 2-10 seconds after the field was turned on (14% of the cases) or off (24%). Similar changes in EEGs due to EMFs at 50 Hz, and 3 GHz have been reported (28,30).

Lott and McCain measured the total integrated EEG in rats before, during, and after exposure to a DC field of 10 kV/m (31). They found a transient increase associated with either the application or removal of the field, a steady response that persisted during application of the field, and an after-effect. A 640-Hz field, 40 V/m, also increased the total integrated EEG, particularly for readings from the hypothalamic region.

At high frequencies, a different effect on the total integrated electrical activity was observed. Rabbits exposed for 5 minutes to 700-2800 $\mu W/cm^2$, 9.3 GHz exhibited no EEG changes during the exposure period (32). But commencing 10 minutes after exposure, an interval of decreased EEG activity occurred that persisted for up to 15 minutes.

Friedman and Carey measured the corticoid production in monkeys exposed to a 200-Gauss DC magnetic field for 4 hrs/day (33). An increase in urinary corticoids lasting 6 days was observed, followed by a return to baseline despite continued exposure.

Several endocrinological parameters in rats (serum corticoids, pituitary ACTH, and ACTH releasing factor in the hypothalamus) were increased following brief exposure to 10-1000 $\mu W/cm^2$

at 2.4 GHz (34). At 3 GHz, rats exposed to 5-10 $\mu W/cm^2$, 8 hrs/day, had elevated levels of excreted corticoids after 1-3 months of exposure (35). Mice exposed to 25 or 50 kV/m, 60 Hz, exhibited increased corticosterone levels upon application of the field, and the levels returned to baseline despite continued application of the field (36). EMF effects on adrenal tissue _in vitro_ have been reported (37).

The adrenal corticoid response to EMF stimulation is time dependent, and it varies nonlinearly with dose (38). When groups of rats were exposed to 500, 1000, 2000, and 5000 V/m at 50 Hz, the average urine corticoid level of the latter two groups changed similarly during the 4-month exposure period: approximately the same maximum value was achieved in both groups and they exhibited increased corticoid levels as compared to the controls. The 1000-V/m group, however, exhibited lower corticoid levels for the first 2 months of the exposure period followed by a rise above the control level during the last half of the exposure period; at 500 V/m the pattern of corticoid excretion was identical to that of the controls. The biological response was reversible in the sense that when the field was removed, the corticoid level returned to normal within 2 months.

At 3 GHz, 153 $\mu W/cm^2$, an increase in thyroid weight was found after 2 weeks' exposure, but after 5 months' exposure the weights were normal (39). Ossenkopp et al. found that both male and female rats exposed _in utero_ to 0.5 Hz, 0.5-30 Gauss had increased thyroid weights at 105-130 days of age (40).

Fischer et al. exposed rats to 50 and 5300 V/m, 50 Hz, and observed bradycardia at both field strengths beginning 15 minutes after commencement of exposure (41). At the lower field strength an 8% decrease occurred, and it remained statistically significant after 2, 10, 21 and 50 days of continuous exposure. At 5300 V/m the decrease in heart-rate after 15 minutes' exposure was about 16%: it was not seen following 2, 10, or 21 days' exposure, but it was present (about 5%) after 50 days. Bradycardia was also reported in rabbits following exposure to 50 Hz, 1000 V/m (42); the heart-rate decreased by about 9% after 30-60 days. Microwave EMFs also produced bradycardia (43); the effect was seen after 2 weeks' but not after 2 months' exposure.

In preliminary studies, dogs were exposed to 15 kV/m, 60 Hz, for 5 hours to determine whether such exposure altered

TABLE 2. Percent Change in Hematological Parameters in Mice (45). RBC, red blood cell concentration; Hct, hematocrit; Hb, hemoglobin; MCV, mean cell volume; MCH, mean corpuscular hemoglobin; MCHC, mean corpuscular hemoglobin concentration. A, no change in exposure conditions, B, change in exposure condition as indicated. NM, not measured.

		PERCENTAGE CHANGE					
EXPERIMENT	CONDITION	RBC	Hct	Hb	MCV	MCH	MCHC
A							
Male Control	nF ► nF	1.7	2.0	-4.5*	1.0	-6.2*	-6.3*
Female Control	nF ► nF	3.9*	4.1*	1.7	0.2	-5.0*	-5.1*
B							
Male I	F ► nF	-4.7*	-5.1	NM	0	NM	NM
	nF ► F	-5.2*	-4.9	NM	0.2	NM	NM
Male II	F nF	-9.0*	-9.1*	-3.3*	-0.4	5.7*	6.0*
	nF F	-6.5*	-7.0*	-2.4*	-0.7	3.9*	6.1*
Female I	F ► nF	-4.1*	-4.6*	-4.2*	-1.2	0.5	1.2
	nF ► F	-6.4*	-6.7*	-3.4*	-0.5	3.8	4.8
Female II	F ► nF	-5.3*	-6.0*	3.4	-1.2*	8.3*	10.0*
	nF ► F	-7.1*	-9.2*	3.5	-2.3*	11.0*	13.6*

*$P < 0.05$

the physiological response to a controlled hemorrhage (10 ml/kg, over a 3-minute period)(44). The cardiovascular changes at the end of the hemorrhage were: mean arterial pressure fell an average of 5.9 mmHg in the control group and 16 mmHg in the exposed group; arterial pulse pressure fell 0.9 mmHg in the control group and 10.9 mmHg in the exposed group; average heart-rate decreased 9.3 beats per minute in the control group, but increased 57.5 beats per minute in the exposed group. The data was rejected by the sponsor (1).

To study transient responses to EMFs, we looked for changes in hematological parameters of mice due to a 60 Hz electric field of 5 kV/m (45). There were four consecutive experiments, two with males and two with females. In each there were two groups: one for which the control period preceded the exposure period (nF ► F), and one in which the pattern was reversed (F ► nF). On "day 1" of each experiment the mice were divided into the two groups and the electric field was applied to one-half the population. On "day 3" the blood parameters were measured in each mouse and immediately thereafter the exposed and non-exposed groups were interchanged. On "day 5" the blood parameters were measured again and the mice were killed.

Blood was collected from the ophthalmic vessels and it was therefore necessary, before applying the field, to determine the influence of the first blood collection procedure on the values measured after the second such procedure. We measured the blood parameters in two groups of mice, one male and one female, under conditions that were identical in all respects to those employed during the field-exposure portion of the study, and we found that the method of blood collection had a tendency to produce higher RBC, Hct, and MCV values and lower values of Hb, MCH, and MCHC (Table 2).

The results obtained in connection with the application of the electric field are shown in Table 2. In each experiment, RBC on "day 5" was significantly less than on "day 3," regardless of whether the interval between "day 3" and "day 5" was an exposure period or a non-exposure period. A decline in Hct paralleled the RBC changes, but Hb showed no consistent changes. MCV showed a tendency to decrease, but the other computed indices both increased.

The trends in the computed indices, and especially the changes in RBC and Hct, were opposite to those induced by our method of blood collection alone. It follows, therefore, that the applied electric field had a physiological impact. The unique feature of the observed responses is that, for each parameter, a change in the same direction occurred with both the F ► nF and nF ► F groups. An analysis of variance confirmed that in all four experiments there was an effect associated with time but not with the order of field application. This indicated that the animals responded to the change in their electrical environment, not to the electric field itself.

Szmigielski et al. (46) studied the action of an EMF on the granulopoietic reaction in rabbits that had been subjected to an acute staphylococcal infection. Rabbits were exposed to 3000 $\mu W/cm^2$, 3 GHz, 6 hrs/day, for 6 or 12 weeks, and then were infected intravenously with bacteria. Four to six days after infection the 6-weeks exposed animals displayed stronger granulocytosis than did the control animals, but this was reversed by the end of the observation period. These changes were accompanied by a consistent reduction in the bone-marrow reserve pool, and a depressed lysozyme activity. Animals exposed for 3 months displayed consistently depressed granulocytosis after the infection, and both the bone-marrow reserve pool and the blood serum lysozyme activity were lowered during the entire postinfection period. The results were interpreted to mean that the EMF-exposed animals lacked the reserve capacity to adapt to the infection as efficiently as the control animals: fewer granulocytes could be mobilized, and there was a resulting decline in lysozyme activity.

Shandala and Vinogradov studied the effect of an EMF (1-500 $\mu W/cm^2$, 2.4 GHz, for 30 days) on the phagocytic action of neutrophils in peripheral blood of guinea pigs (47). They found that the percent of killed microbes increased following exposure to 1-10 $\mu W/cm^2$, and decreased at 50 and 500 $\mu W/cm^2$; the most pronounced effects occurred at 1 $\mu W/cm^2$. EMF-induced alterations in the complement titer in blood serum were also found. Both immunological indicators returned to normal within two months of the cessation of irradiation.

Shandala et al. reported a significant disturbance in the immunological system of rats exposed intermittently to

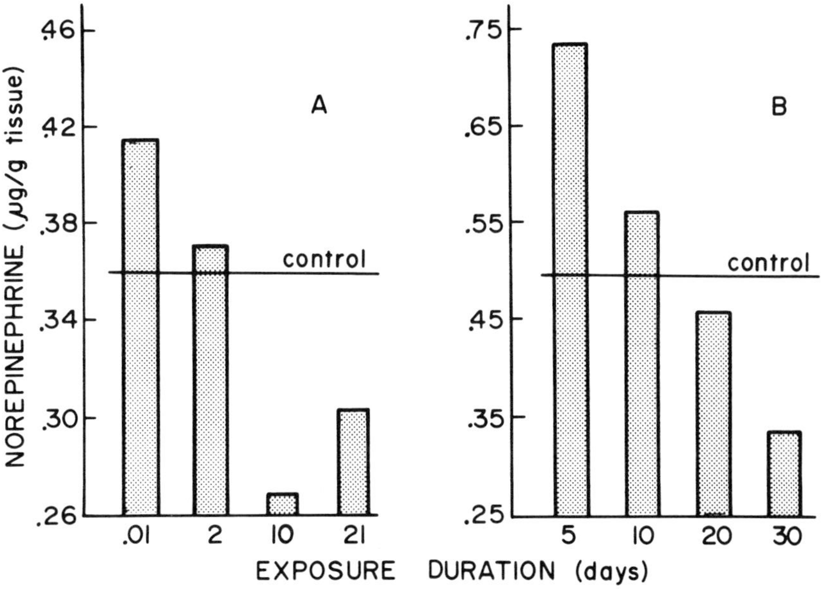

FIGURE 5. Norepinephrine levels in rat brain following exposure to EMFs. A, 5300 V/m, 50 Hz (54); B, 500 $\mu W/cm^2$, 2.4 GHz (55).

500 $\mu W/cm^2$ for 30 days (48); blast cells in peripheral blood, and rosette-forming cells in the spleen and thymus were both affected following EMF exposure.

The threshold for changes in mitochondrial enzymes in murine brain tissue was 100-500 $\mu W/cm^2$ at 3 GHz (49). EMF-drug interactions have been reported (50,51). Merely shielding animals from ambient fields can produce biological changes (52,53).

Fischer et al. found that 50 Hz, 5300 V/m, resulted in an initial rise of norepinephrine in rat brain, and a subsequent decline above the control level (54); Grin observed the same sequence of changes at 2.4 GHz, 500 $\mu W/cm^2$ (55) (Figure 5).

It is apparent from the reported data that all types of EMFs can elicit biological responses. Generally, the responses are transient and the measured parameters return to baseline despite continued application of the EMF.

HUMAN EXPERIMENTS

Soviet scientists had concluded in the 1930's that EMFs altered the human nervous system (56), but human studies were apparently not performed elsewhere until several decades later. In an experimental design in which each subject was exposed to two frequencies in the 2-12 Hz range, at 4 V/m, Hamer found a longer reaction time at the higher frequency (57). Friedman et al. applied magnetic fields of 0.1 and 0.2 Hz to separate groups of male and female subjects, and for both groups found a longer reaction time at the higher frequency compared to the lower frequency (58). Persinger et al. found no difference in the mean reaction time in either males or females due to 0.3-30 V/m, 3-10 Hz, but he reported a significant difference between the sexes in response variability (59). A 60 Hz, 1 Gauss field altered the ability to concentrate (60). All 6 experimental subjects demonstrated a decline in performance (addition of two-digit numbers) in the second test session of the exposure period, and all 6 improved in the first test session of the post-exposure period (60). In contrast, the control subjects showed no consistent changes.

Dumanskiy et al. reported an increase in blood glucose in humans following exposure to 15 kV/m, 50 Hz, 1.5 hrs/day for 6 days (61).

Volunteers confined for up to 7 days were exposed to a 1-Gauss magnetic field, 45 Hz, for 24 hours: they did not know which 24-hour period during their confinement would be chosen for the application of the EMF. Serum triglycerides in 9 of 10 exposed subjects reached a maximum value 1-2 days after EMF exposure; similar trends were not seen in any of the control subjects (62).

A 5 Hz magnetic field (about 10 Gauss) altered the ability of subjects to process and remember verbally delivered information (63).

Wever studied circadian rhythms using an elaborately constructed underground bunker which provided complete isolation from all environmental time cues (64-67). One of the two suites in the bunker was shielded against all electric and magnetic fields of terrestrial or atmospheric origin. In addition, the room contained built in facilities for introducing a range of artificial fields. In the second suite, the Earth's normal fields were continuously present. The rooms were built in such a way that the subjects could not distinguish between them. Subjects were isolated in the bunker for 3-8 weeks, and various circadian rhythms including activity and body temperature were recorded. Wever found that 34 subjects who lived in the non-shielded room had a body temperature rhythm with a mean period of 24.87 ± 0.44 hrs, while 50 subjects who lived in the shielded room had a body temperature rhythm of 25.26 ± 0.85 hrs ($P < 0.01$). In 15 subjects who lived in the shielded room, internal desynchronization occurred. That is, while the body temperature rhythm continued to maintain a circadian rhythm near 25 hours, the period of the activity changed, either lengthening or shortening. Thus, the normal synchronization between the rhythms was destroyed. Internal desynchronization was not observed in the non-shielded room in which the natural fields of the Earth were present.

In another study, an artificial electric field (10 Hz, 2.5 V/m) was switched on and off in a changing temporal sequence. No subject knew when the field was present, and each subject acted as his own control. Wever found that the presence of the EMF reversed the effects found previously. That is, within the field present, the 10 subjects showed lower values of the period of the body temperature rhythm, and in no case did internal desynchronization occur. Moreover, when the field was switched on with the subject in a state of internal desynchronization, the desynchronization was stopped. Wever concluded that the artificial electric field on one hand, and the total of the normal ambient EMFs on the other, had similar influences on circadian rhythms.

Workers in Italian state electric-train substations are exposed to maximum electric and magnetic fields at 50 Hz of 5 kV/m and 150 mG, respectively (68). Substation workers were divided into 4 groups depending on their average weekly exposure to the maximum electric field, namely 0 (controls), 1, 10 and 20 hours/week. The employees in each group were subjected to a clinical examination and laboratory tests. The studies showed a correlation with duration of exposure to the maximum electric field: compared to the controls, the clinical values for most

TABLE 3. Changes in Average Body Weights of Rats Exposed to 45-Hz Vertical Electric Fields (70). The control rats were housed in a field-free environment.

EXPERIMENT	FIELD (V/m)	NO. OF RATS	EXPOSURE TIME (days)	WEIGHT GAIN (gm)
1	25-100	143	36	142 ± 14*
	control	47	36	209 ± 20
2	10-50	47	40	150 ± 19*
	control	16	40	215 ± 11
3	2-10	94	32	131 ± 12*
	control	32	32	166 ± 12
4	0.5-2	32	30	131 ± 11*
	control	32	30	170 ± 11

*$P < 0.001$

parameters in the 1 and 10 hr/week groups were elevated (or depressed), whereas the values in the 20 hr/week group were normal.

In the USA, the rules applicable to human experimentation have become strict, and experiments involving voluntary human exposure to EMFs are presently not being performed.

ADVERSE EFFECTS

Altered spermatogenesis occurred in rats following exposure to 5000 V/m, 50 Hz, for up to 4.5 months (69). After 1.5 months' exposure, the number of atypical sperm cells was significantly greater in the exposed animals (30.7% vs. 15.9%, $P < 0.01$); the percentage of pathological forms increased with the duration of exposure and reached 36.8% after 4.5 months. The exposed rats also produced fewer sperm cells and exhibited a higher ratio of living to dead cells; both effects became significant after 3.5 months.

In a Navy study, exposure to 0.5-100 V/m, 45 Hz consistently produced depression of the body weights of the exposed animals (70) (Table 3). The results of the study were disavowed by the sponsor (but not the investigators), apparently because the control rats had been maintained under Faraday-cage conditions. Low-frequency electric and magnetic fields also produced growth depression in 25-day-old chicks (71).

We performed midshaft fractures on rats, following which half the group was exposed to 5 kV/m, 60 Hz, and half was maintained as a control (72). The extent of bone healing was evaluated at 14 days postfracture on the basis of blind scoring of serial microscopic sections. In two replicate studies, we found a highly significant retardation in healing; the fractures in the exposed rats exhibited the development normally seen in a 10-day fracture. We found no effect on fracture-healing following exposure at 1 kV/m. The adverse effect of a 60-Hz electric field on fracture healing in the rat was confirmed in three replicate studies (73) (the paper must be read carefully before the point is appreciated).

Grissett et al. exposed 30 monkeys to 20 V/m and 2 Gauss at 76 Hz (74). After 1 year, the field-exposed males were significantly heavier than the control males.

When chickens were irradiated for more than 200 days at 0.19-360 $\mu W/cm^2$ (7 GHz), the irradiated animals exhibited a doubled mortality rate, and a deterioration in health (75).

Hansson described histopathological changes in the brains of rabbits that had been exposed to 50-Hz electric fields, 15 kV/m (76). The observed changes included the formation of numerous lamellar bodies in the endoplasmic reticulum in Purkinje cells. Hansson suggested to me that the experiment be repeated using a different animal species. I exposed mice to 9.8 kV/m for 60 days, and Hansson subjected the brains of the exposed and control animals to the same kind of analysis previously performed for the rabbits. The same histopathological changes previously seen in rabbit brain were observed in the brain tissue from the exposed mice (77).

Investigators exposed pregnant rats to 100 $\mu W/cm^2$ at 27.1 MHz for up to 20 days (78). The frequency was chosen because of its widespread industrial use in radiofrequency heaters and heat sealers. One group was exposed for 20 days, and other groups were exposed for 5 and 10 days. The rats exposed for the longest time gained weight the slowest. The rats in the exposed group experienced significantly greater fetal resorption rates. Among their viable offspring, there was a significantly higher incidence of delayed development (higher incidence of incomplete cranial ossification) (78).

IN VITRO STUDIES

In vitro studies have demonstrated EMF-induced changes in growth rate (79,80), respiration (81), metabolism (82,83), membrane receptors (84,85), immune response (86,87), and morphology (88), but the actual mechanism of detection of EMFs has not yet been discovered.

Beyond the phenomenon of detection, the significance of *in vitro* studies with respect to intact organism remains dubious. The point is illustrated by the work of Adey and his colleagues. It was found that EMFs altered behavior in humans

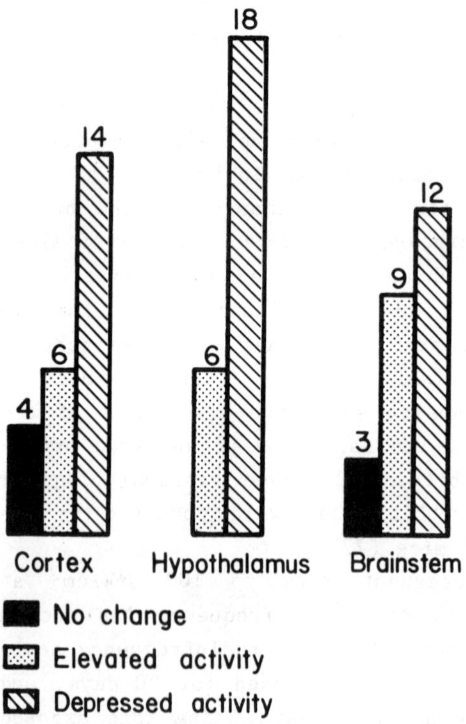

FIGURE 6. EEG response to 100 $\mu W/cm^2$, 3 GHz (95). The numbers indicate rabbits with a given response.

TABLE 4. Biological Effects of Exposure to a 15-kV/m 60-Hz Electric Field (97). The average values were determined following 30 days of continuous exposure. There were no statistically significant changes in water consumption during the first 14 days of any experiment.

EXPERIMENT	NUMBER OF RATS	AVERAGE WATER CONSUMED (ml)	PITUITARY WEIGHT (μg/g)	SERUM CORTICOIDS (μg/100 ml)
1	15 experimental	846 ± 68*	38.7 ± 3.2*	6.8 ± 0.8*
	18 control	940 ± 142	35.2 ± 3.8	8.7 ± 1.2
2	14 experimental	749 ± 80*	43.9 ± 4.1*	7.2 ± 1.5
	20 control	891 ± 93	40.6 ± 3.1	7.6 ± 2.1
3	19 experimental	819 ± 83*	32.9 ± 3.1	
	21 control	890 ± 83	35.2 ± 2.6	
4	16 experimental	901 ± 50*	38.0 ± 2.4	6.0 ± 0.7
	14 control	1054 ± 84	39.0 ± 2.6	6.4 ± 0.6
5	20 experimental	1003 ± 82*	31.4 ± 2.4*	9.1 ± 2.0*
	20 control	1099 ± 117	29.4 ± 2.9	16.3 ± 3.8
6	14 experimental	1143 ± 157	31.2 ± 1.8	9.5 ± 2.0
	16 control	1202 ± 107	30.6 ± 1.8	9.7 ± 4.0

*$P < 0.05$

and monkeys and cats (57,89,90). The idea evolved that EMFs could alter neuronal excitability, if they were in the frequency range of the EEG. An *in vitro* system involving calcium-binding to brain tissue was designed under the belief that the process was important. A complex series of results were then obtained concerning the levels of pre-incubated calcium that were released into solution from live or dead brain tissue (91-93): at 147 MHz, there was an increase when the EMF was modulated at 6-10 Hz, but no increase at 0.5-3 or 25-35 Hz; with EMFs of 6 and 16 Hz, there was a decrease at 10 and 56 V/m, but not at 5 or 100 V/m; there was no change at 1 Hz or 32 Hz, at either 10 or 56 V/m; at 450 MHz, modulated at 16 Hz, there was an increase. Nothing corresponding to these unusual features has been seen in *in vivo* studies.

ROLE OF ENVIRONMENTAL FACTORS

Twenty-four rabbits were exposed to 100 $\mu W/cm^2$, 3 GHz, for 30 minutes, after which their EEG slow-wave activity in different parts of the brain was measured (94). The data (Figure 6) indicated that the rabbits responded differently from one another. If the results of the study had been presented as the average behavior of an experimental compared to control group, it would have been concluded that the fields produced no EEG changes.

Freeman and Carey exposed rabbits to 11-210 Gauss DC and 5-11 Gauss at 0.1-0.2 Hz for up to 60 hrs (95). Four of the 12 exposed rabbits exhibited some histopathological changes which were attributed to exacerbation of a sub-clinical encephalitozoonosis by a stressor effect of the magnetic field (95). Histopathological changes were also observed in the adrenal cortex of 70% of mice chronically exposed to strong static magnetic fields (96). The changes were attributed to the direct stressor effect of the field -- akin to Selye's diseases of adaption (109).

We exposed individually caged rats to 15 kV/m, 60 Hz, for 30 days and measured pituitary weight, serum corticoids, and water consumed (Table 4) (97). In the first replicate, each of the measured indices was significantly different in the treated group. In the sixth replicate none of the indices differed

TABLE 5. Average Glucose Levels in Three Replicate Experiments. (A), reported data analysis (99); (B), analysis of actual data (100) taking into account the 60-Hz background fields (101).

EXPERIMENT	SERUM GLUCOSE LEVELS (mg/dl)					
	Control	2 V/m	10 V/m	20 V/m	50 V/m	100 V/m
(A) 1	281.1 ± 83.8	176.3 ± 74.4	259.4 ± 124.6	218.9 ± 100	286.0 ± 156.9	235.8 ± 49.3
2	210.4 ± 55.4	237.3 ± 62.2	259.4 ± 95.9	241.6 ± 149.1	256.2 ± 118.6	269.4 ± 95.9
3	187.2 ± 34.4	199.0 ± 30.8	199.1 ± 34.3	201.3 ± 40.9	199.1 ± 34.2	232.3 ± 44.0

	SERUM GLUCOSE LEVELS (mg/dl)		STATISTICAL SIGNIFICANCE
	Control + 2 V/m	50 V/m + 100 V/m	
(B) 1	228.7 ± 94.4	260.9 ± 117.2	P = 0.23
2	223.9 ± 59.5	284.2 ± 116.0	P < 0.02
3	193.7 ± 32.7	219.7 ± 42.4	P < 0.01

between the groups, and in replicates 2-5 various combinations of the indices differed between the groups. Numerous sponsors in the USA commissioned attempts to (loosely speaking) replicate this humble study, thereby generating fodder for ten years of argument (1,106,233).

Guinea pigs were exposed to 3 GHz, 10 min/day, for 30 days (98), and both the irradiated and the sham-exposed animals were sampled before and after each daily exposure bout. The sham-exposed group revealed no significant changes, but animals exposed to 25 or 50 $\mu W/cm^2$ exhibited EMF-induced alterations with time dependencies that differed with each animal. For a given exposure duration, the WBC was above the normal level in some animals, and below it in others; as a result, the average values varied little during the study. At 500 $\mu W/cm^2$, however, even on the average there was a pronounced leucopenia and lymphocytosis.

In an Army study, rats were exposed for 28 days to 2, 10, 20, 50 and 100 V/m, 45 Hz, in three replicate experiments, following which complete blood chemistries were performed (99); the serum glucose levels are listed in Table 5A. Although some differences between the control and exposed groups were seen, no linear dose-effect relationship was manifested, and consequently the authors regarded the data as having failed to show a biological effect of the EMF (99). But the 60-Hz electric field in the test cages was 0.18-9.15 V/m (100). In my view, the 2-V/m group should have therefore been considered a control group in relation to the 50-100 V/m exposed groups. When we did this (101), the data revealed increases in serum glucose in each replicate (Table 5B). (This approach to the Army data also suggests the existence of effects on other parameters, including triglycerides.)

By the mid-1970's, no studies had been done to assess the possible impact on successive generations of animals of the continuous presence of a low-frequency EMF; we therefore undertook such a study (102). Initially, mature male and female mice were split into horizontal, vertical, and control groups. Mice in the horizontal group were allowed to mate, gestate, deliver, and rear their offspring in a horizontal 60-Hz electric field of 10 kV/m. At maturity, randomly selected individuals from the first generation were similarly allowed to mate and rear their offspring while being continuously exposed. Randomly selected individuals from the second generation were then mated to produce the third and final generation. A parallel procedure was followed for the vertical group wherein three generations were produced in a 60-Hz vertical electric field of 15 kV/m, and for the control group wherein three generations were produced in the ambient laboratory electric field. In the first and second generations, males and females reared in both the horizontal and vertical electric field were significantly smaller than the controls when weighed at 35 days after birth. In the third generation, the only group whose body weights were significantly affected were the males exposed to the vertical field. In both the second and third generation, a large mortality rate in the vertical-field mice was seen during the 8-35 day postpartum period.

We repeated the multi-generation study at 3.5 kV/m, using an improved exposure system (103). In the first generation, no consistent effect on body weight attributable to the EMF was seen throughout a 63-day observation period. In both the vertical and horizontal groups, however, mortality among newborns was increased. In the vertical-control group 48 animals (about 17%) died between birth and weaning. In the vertical-exposed group, if the electric field wasn't a causative factor, a 17% mortality rate should also have been seen. However, that group exhibited a 31% mortality -- 82 animals died and not the expected 44. Thus, 38 animals, about 16% of those born, failed to live to weaning because of the electric field. A similar result was obtained in the horizontal-exposed group -- about 11% of the animals born failed to live to weaning because of the electric field.

In the second generation, no pattern regarding body weight attributable to the EMF was seen throughout a 108-day observation period. The vertical-exposed group, however, again exhibited a higher mortality; about 6% of the animals alive at weaning failed to live to the final day of observation due to the presence of the EMF. In the third generati n, the exposed animals had higher body weights, particularly in the horizontal-exposed group. At 49 days after birth, the males and females in each exposed group were significantly heavier than their respective controls. At 119 days after birth, only the females in the horizontal-exposed group were significantly heavier, but this was part of a consistent trend for that group. Again we saw an increased mortality in the vertical-exposed group -- 10% of the weaned animals failed to survive to the end because of the electric field.

Following the publication of our first multigeneration study (102), investigators at Battelle Northwest Laboratories, Richland, Washington were commissioned to replicate the work. They first developed an exposure system that was unexcelled with regard to field homogeneity and reproducibility of electrical environment. Every aspect of the animals' physical environment -- light, temperature, humidity, presence of pathogens in the air, air flow, for example -- was rigorously monitored and controlled by automatic equipment. The investigators then constructed two complete exposure facilities: each consisted of a completely characterized exposure unit, an identical unit for sham-irradiation, and a completely controlled environment suitable for housing both units.

The multigeneration study was begun in the first exposure facility, and 3 weeks later a replicate study was begun in the second facility; both replicates were done double blind. At the end of the study, the males and females in the first replicate were statistically significantly smaller than the controls, but in the second replicate they were significantly larger (104). The data appears in the report to the sponsor (not in the literature), and the reader must perform his own statistical analysis to support this conclusion.

ANALYSIS

The reports can be summarized this way (166):

(1) Exposure to EMFs can result in alterations in all body systems, including the nervous, endocrine, cardiovascular, hematological, immune, and reproductive systems;

(2) The effects manifested in each tissue or system are largely independent of the type of electromagnetic field in the sense that common physiological responses are produced by spectrally different electromagnetic fields;

(3) An organism's response to an electromagnetic field is determined by a combination of factors including its

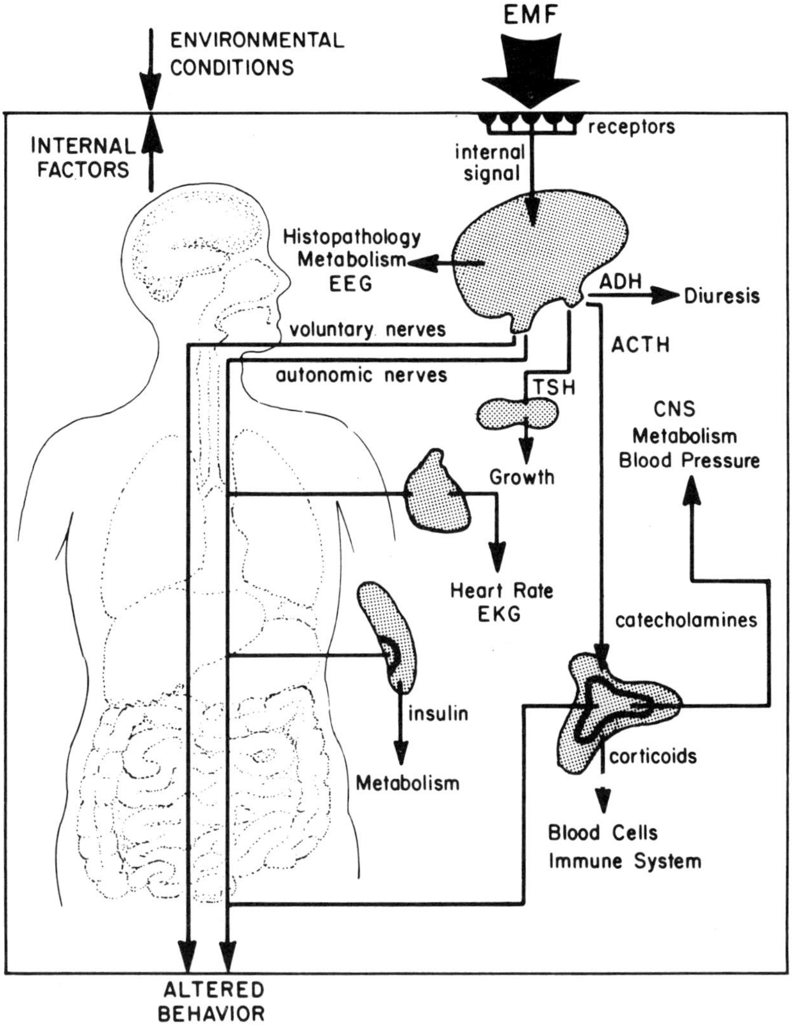

FIGURE 7. Hypothesis regarding the sequence of physiological events underlying the chronic stressor effect of environmental EMFs.

physiological history, genetic predisposition, and the totality of prevailing environmental conditions;

(4) EMF-induced biological effects in animals are best characterized as adaptive or compensatory because the fields present the organism with an environmental factor to which it must accommodate. Linear dose-response relationships are generally not observed.

My hypothesis for the mechanism underlying EMF-induced bioeffects is depicted in Figure 7. The EMF is detected by the peripheral or central nervous system, and the fact of its presence is communicated to the parts of the CNS that initiate and control compensatory responses. The nervous system, both directly and through its rich connections with the endocrine and immune systems modulates the dynamics of the body's various organs and systems in a manner best suited to protect the integrity of the organism. It is the latter transient changes that are detected in most experiments. Their magnitude and direction are simultaneously influenced by other factors in the organism's environment that also impact on the body's compensatory apparatus. Finally, although it is easy to overlook the point, every organism differs from all other organisms of its kind (even though they may look similar) in ways that can affect the outcome of an experiment.

Perhaps the most important lesson of the literature is the guidance that it provides regarding how future studies should be designed. Most studies have been in the nature of bioassays intended to determine the broad parameters of EMF bioeffects. Replicate studies have generally failed to confirm the initial observations in the sense that the specific changes reported in the initial studies were not observed in the replicates. The Battelle attempt (104) to replicate our multi-generation studies (102,103) is a good example. Putative replicate studies are frequently used in argument to deprecate the reliability of the initial reports (105,106). Scientists who come to the EMF bioeffects literature for the first time are consequently presented with a confused picture which perhaps suggests that the area is not worthy of effort. This is not the case, and I hope that the reader will consider my explanation for the essential cause of the apparent confusion.

The precise numerical values that we observed (Table 4) cannot be exactly duplicated because the environmental factors present in our laboratory in Syracuse, New York in 1976 had a significant impact on the specific observations, and those conditions cannot be precisely duplicated (because they were not all catalogued). Similar comments can be made about the work of most of the investigators who first reported EMF bioeffects. If the conditions of an effect cannot be specifically duplicated, the effect cannot be specifically duplicated.

One must recognize exactly what was concluded in most of the original studies. On the basis of the data in Table 4, for example, we concluded that the presence of an electric field was correlated with a change in the average physiological status of the test groups, and therefore that the electric field was detected by the animals. We made no assertion that other investigators would see exactly the same magnitude or direction of change in specific physiological variables. At the time most bioassay studies were performed, the question under consideration was whether any effect occurred, and no investigator then (or now (104)) had the facilities to control for every possible factor that could affect the magnitude or direction of an effect: Again, the question was <u>is</u> there an effect.

Sponsors have not looked beyond the numbers in the Tables, and seen what actually was reported. Having failed to recognize what properly was a candidate for replication, it is not surprising that they failed to recognize that the data from replicate studies was actually consistent with the results of the original work. The assertion of inconsistency or conflict (106,233) is a chimera.

The first decade of sustained laboratory studies of EMF-induced bioeffects is now behind us, and there is no need to repeat the original studies. It is clear to all reasonable investigators that EMFs can affect physiology. What remains is to frame laboratory experiments that have a specific purpose. We have had our pioneers -- now, we need settlers.

Irrespective of the hypothesis depicted in Figure 7, actual laboratory data from numerous experiments shows that simulated environmental EMFs produce biological effects, particularly compensatory reactions. If the data is accepted, we must consider the public-health significance of such reactions.

Even if a reaction to environmental EMFs occurs transiently and then evaporates and can no longer be measured using traditional endpoints, there is a rational basis for expecting EMF-linked pathological changes to occur in exposed subjects. This basis will now be considered.

ENVIRONMENTAL FACTORS AND DISEASE

INTRODUCTION

Many of the physical properties of the fluid matrix of the mammalian body remain remarkably constant despite wide variations in characteristics of the organism's environment. This stable state was termed homeostasis by Cannon (107) who was a student of the physiological mechanisms by which the stability of "this personal, individual sack of salty water, in which each of us lives and moves and has his being" was maintained. His idea was that environmental factors such as temperature or oxygen level constituted a stress on the organism that produced an internal strain which could be pictured as a deviation from a normal level. The CNS and the adrenal gland then act in concert to negate the strain and restore homeostasis. Thus, in an individual undergoing a heat stress, the pores open to facilitate heat loss, whereas a heat-producing response (shivering) occurs in an individual undergoing a cold stress. This use of physical terminology and concepts in a physiological context was an effective descriptive device, even though there existed no means by which to actually measure physiological strain.

In 1936, Selye reported a syndrome in rats produced following exposure to acute agents including cold, surgical injury, muscular exercise, and various injected substances (108). The syndrome consisted of an enlarged adrenal cortex, shrunken thymus, spleen, and lymph nodes, and the appearance of stomach ulcers. He subsequently found that many other acute stimuli produced the syndrome, that it occurred in many species including man, that there existed many other biochemical and physiological indices of the syndrome, and that it was mediated primarily by the anterior pituitary and the adrenal cortex (109). Selye described the syndrome as being a normal physiological response when it was initiated by tolerable levels of the stimuli, but as a pathological process when either the stimulus intensity was too high, or when the organism itself was unable to exhibit a normal response. In these latter cases, diseases of adaption could develop (hypertension, nephrosclerosis, periarteritis nodosa, for example). Selye originally employed the term "stress" as a synonym for the external stimuli or noxious agents, or injected substances that were observed to elicit the syndrome (109,110). Subsequently, he employed the term "stressor" for the external factor, and defined stress in terms of the response of the organism (111). Considered as an internal state variable of the organism, Selye, like Cannon, could provide no means to measure stress, nor even to define it uniquely.

The concept of stress as a special variable -- a variable distinct from quantifiable parameters such as temperature, enzyme concentration, and so forth -- also arose in the context of clinical medicine. Merely from taking a good history, the clinician could observe that various factors in a patient's life had a predisposing influence on the course of the patient's disease. Death of a loved one, loss of a job, an unhappy marital situation, poor diet, or chronic exposure to various physical or chemical factors could exacerbate the clinical course of patients with ulcers, high blood pressure, asthma, and many other diseases. The clinical observation was that the so-called stresses and strains of life could markedly and adversely affect the course of a disease. The life stresses observed by the clinician are historically unrelated to an adaption syndrome.

The idea that human disease is somehow related to stress sprang from these roots. It is obviously an important and popular concept in physiology and medicine. In the laboratory, the adaption syndrome is a useful model for fundamental studies in neuroendocrinology. In clinical medicine, psychosomatic influences on human disease are being increasingly recognized, and explored. Despite this, the concept of stress has led to much confusion, and most modern authorities have recommended that the term itself, or the Cannon/Selye-clinical conception of it be eliminated because, as an organismal state, stress can neither be measured nor defined (112-115). This view seems to me to be the correct one, and will be followed here. As I hope the reader will ultimately agree, the decision to forego an attempt to understand the relationship between environmental factors and disease in terms of an internal state variable called stress is helpful in understanding the influences of environmental EMFs.

Although we shall avoid the term "stress", the term "stressor" is quite useful and will be employed to denote the external stimulus. Specifically, by this term I mean any somatic or neurogenic factor that can elicit an acute adrenal-cortical response. A change in environmental temperature or EMF level, or death of a loved one are examples of stressors, the color of one's socks is an example of a factor that is not (normally) a stressor. The factor can, but need not necessarily, elicit a response. Thus, if a temperature change $\Delta T = T_2 - T_1 = 1°F$ elicits a measurable response in an individual during a reasonable number of independent trials, or in an appreciable percentage of the population of a test species, then that factor can be a stressor. It is not necessary to specify whether T_1 or T_2 are normal or abnormal -- it is enough that they are different. Characterization of a stimulus as a stressor always requires a frame of reference. Thus $\Delta T = 1°F$ may be a stressor for $T_1 = 70°F$, but not for $T_1 = 20°F$. Put another way, a change in temperature can be, but is not always a stressor. Similarly, environmental EMFs can be but are not always stressors (Figure 7).

CHRONIC STRESSORS AND ENDOCRINOLOGICAL ENDPOINTS

Application of acute stressors produces many complicated interrelated, time-dependent changes in adrenal-cortex secretions and other endocrinological endpoints (116-118). These endpoints tend to return to normal levels even with continued application of the stressor. Rats briefly subjected to 85-db noise exhibited a 9-fold increase in serum corticosterone (119). When subjected to the noise for 1 hr, a 4-fold increase was found, and when the stressor was applied continuously for 7 hr, only a 2-fold increase was observed (120).

Rats were placed in an immobilization apparatus that prevented any gross body movements, and plasma corticosterone levels were determined at various times after initiation of confinement (using separate groups of animals at each time point) (121). At 20-200 min after the beginning of confinement, the corticosterone levels in the confined animals were significantly greater than those of the control (not confined) animals. About 24 hrs after initiation of confinement however, the corticosterone levels were identical in the two groups. Adrenal exhaustion could not explain the data (121).

Rats exposed to chronic crowding (0.1 in^2 of floor space per gram of body weight) or a cold stress (6°C) exhibited increased plasma levels of corticosterone after 1 week, but thereafter, at 2 and 4 weeks, the plasma levels were normal (122).

Human subjects exhibit significant adrenal-cortical activation when exposed to real or imagined important changes in their environment (123). Typically, there is a rapid loss of endocrine responsitivity to the initial provocative stimulus, probably reflecting some form of adaption. Studies of plasma cortisol in paramedics, fire fighters, air-traffic controllers, and pilots have failed to demonstrate any long-lasting adrenal-cortical responsitivity (124).

In addition to being inherently transitory, the magnitude of the changes in endocrinological endpoints depends strongly on many factors, only some of which can be satisfactorily quantified. These factors include circadian rhythm and environmental lighting (125), the nature of the stressor (126,127), age (128-130), housing conditions (131), species and strain (132), and psychosomatic factors (123,133).

Since neither man nor animals exhibit altered endocrine activity following repeated or chronic exposure to non-exhausting stressors, it seems clear that endocrinological endpoints are neither the mechanism nor the harbinger for an increased risk of future illness. Despite this, it is important that the reader recognize the importance of endocrinological endpoints in the evaluation of the health implications of stressors such as environmental EMFs. Changes in endocrinological endpoints permit characterization of a factor as a stressor. (They do not, however, permit risk evaluation or even the relative comparison of the magnitude of two different stressors -- say 15 min exposure to 0°C compared to undergoing preparatory procedures in a hospital for coronary bypass surgery.) All chronically applied stressors have a non-zero probability of affecting an individual's risk for disease.

STRESSORS AND DISEASE

Extensive clinical observations have implicated stressors as risk factors for cardiovascular disease (134-137), diabetes (138), depression (139), allergies (140), and cancer (141,142). Although the risk of disease associated with chronic exposure to stressors cannot be expressed in endocrinological terms, it can be studied and measured in the laboratory in other ways.

Animal models have been developed to facilitate controlled exploration of the link between chronic stressors and disease or other adverse physiological consequences. One method involves study of the effect of a chronic stressor on a non-endocrinological endpoint. Parameters involving growth, development, or healing may be sensitive to chronic stressors because they tend to integrate small effects over time. For example, when pregnant rats were subjected to surface illumination (4280 lm/m^2, 35 min/day, on days 14-22 of gestation) the female offspring later experienced fewer conceptions, more spontaneous abortions, longer pregnancies, and fewer viable offspring than did the control rats (143). Beginning at 1-5 kV/m, power-frequency electric fields retarded fracture healing in rats (72,73).

Another approach involves a functional measurement that samples reserve capacity (144). Mice were exposed to a regimen of chronic stressors administered over 2 weeks. The stressors included tail pinch (1 min duration), cold swim (3 min at 4°C), electric shock (30 min), shaker (30 min), water deprivation (24 hr), and isolation (48 hr). The individual stressors were randomly administered over a 2 week period, following which the chronically stressed animals were assigned to one of 3 experimental groups: one group was sacrificed and used for determination of plasma corticosterone levels, a second group was subjected to behavioral testing (gross locomotor activity), and a third group was exposed to an acute noise/light stressor prior to behavioral testing. Following testing, the mice were sacrificed and the corticosterone levels were determined. The corticosterone levels in the chronically stessed mice were normal, but their basal behavioral activity and their behavioral activation response to the acute stressor were each significantly reduced. The corticosterone response in the mice subjected to behavioral testing was significantly greater than that of the comparable control group.

A third approach involves the use of an animal model for a particular disease. The studies involving cancer models illustrate this approach, and will be considered here.

In these studies, the animal is injected or implanted with cancer cells that ordinarily will lead to its death within a reasonable period for laboratory investigation. Parameters that characterize the growth or development of the cancer including growth rate, tumor size, propensity for metastasis, duration of survival, and mortality can be directly measured. The pertinent question in such experiments is: Does the addition of a chronic stressor alter the course of the cancer in the host? We are interested in studies in which the stressor can be reasonably characterized as innocuous because, either by measurement or assumption, it transiently activates the adrenal-cortical system but produces no direct organic damage. (I do not include, for example, full-body exposure to X-rays.) Further, application of the innocuous stressor is continued throughout the period of observation so that we can validly speak of the simultaneous presence of the stressor and the disease. The control for these studies is the subsequent course of parameters that characterize the cancer in a group of animals that do not receive the innocuous stressor.

Female mice were injected with murine sarcoma virus and beginning on the day of inoculation, were subjected to 4 hrs of electric shock (100 10-sec treatments at 2-4 mA, DC) for 3 successive days (145). The shocked animals showed a significant increase in maximum tumor size. Application of a partial body cast altered the incidence of tumors in mice that were inoculated with a sarcoma virus 3 days later (146). Of 24 mice that received the lower inoculum range, only 8 of the control (not casted) but 19 of the casted mice developed tumors. Chronic exposure to microwave EMFs significantly hastened development of benzopyrene-induced skin cancer in mice (147,148).

Mice were injected with fibrosarcoma cells in the hind footpad, and the injected limb or the contralateral limb subsequently were amputated (149). Fourteen days after surgery the mice were killed, and the number of pulmonary metastases was determined. Groups that received the ipsilateral or the contralateral amputation both showed significantly increased pulmonary metastases (149). Similar results were seen in mice that were subjected to tumbling, immobilization or drugs that induced seizures (150). The mice subjected to these stressors exhibited higher numbers of lung colonies of cancer cells following injection of cancer cells.

More complicated experimental designs have yielded even more subtle insights into the nature of chronic stressors. Mice that received inescapable foot shocks exhibited accelerated tumor growth and decreased survival compared to unshocked animals (151). The ability to cope prevented both effects: in animals that received escapable foot shocks, both tumor endpoints were the same as those of the controls. Another report (152) described modification of the antitumor action of an immunomodulator by a stressor. The implanted tumor grew at a known rate in host animals, but did not grow in animals that also received the immunomodulator. If the animals were subjected to a stressor, however, the effect of the immunomodulator was overcome, resulting in tumor growth.

Riley studied the effect of environmental stressors on the latent period for mammary tumors in mice infected with the Bittner oncogenic virus (153). Under usual housing conditions, 80-100% of the female mice developed mammary tumors within 8-18 months after birth. In a normal mouse colony, the animals are exposed to dust, odor, noise, pheromones, and other potential stressors. In a group of mice housed under conditions that reduced exposure to these stressors, the specially-caged group had a median tumor latent period of 566 days, compared to 358 days for the group exposed to the physical stressors (153).

The Riley study brings into focus an inherent uncertainty in stressor studies. Some stressors may have a beneficial effect on an animal's ability to resist development of disease, and the investigator usually does not know which conditions should be associated with a beneficial effect -- he can only establish that some conditions are different than others, and that, speaking anthropomorphically, some conditions are preferable to others. Each of the factors in the colony that Riley labelled "normal" was likely capable of producing a physiological response, and the signs, symptoms, and biochemical indices exhibited by the animals were each the sum of the constituent responses to the various stressors. Removing one factor (say, the dust) or adding a factor (say, doubling the pheromone concentration) might increase or decrease the value of any particular independent variable. Studies involving the effect of caging density of animals on various tumor endpoints are a good example of this complicated interplay. Such studies generally show that isolated animals are more prone to exhibit tumor growth (154-157), but it is not established whether the exacerbation of the cancer should be viewed as associated with the application of a chronic neurogenic stressor (isolation), or the removal of a protective stressor (pheromones, for example).

Immobilization, sound, and electrical shock stressors, each slowed the growth rate of a mammary carcinoma in rats (158). Inhibition of tumor development has also been reported by Newberry et al. When a shock stressor was administered to different groups of rats for 25, 40, or 85 consecutive days, the exposed animals exhibited fewer tumors after 40 or 85 days, compared to the controls (159). Rats subjected to a restraint stressor (5 or 10 hrs/day for 73 days) exhibited significantly reduced average number of tumors (160). The same result was observed when the restraint stress was administered for 12 hrs/day (161).

It seems clear that experimental neoplasms in rats and mice (and other disease states in animal models) are responsive to the environmental conditions of the host. That is, chronic inocuous stressors have a measurable impact on cancer growth. There is, however, no general formulation that permits us to predict outcome in any given set of circumstances. The nature and duration of the stressor, the extent of the exposed organism's control over it, housing conditions, and probably many other factors play an important role in the elaborated response.

MECHANISMS

The physiological details that constitute the link between chronic stresors and disease are peripheral to my purpose here, but they merit some consideration because even a brief discussion increases the plausibility of the link. Electric shocks depressed the cytotoxicity of natural killer (NK) cells in rats (162). A chronic immobilization stressor selectively affected different components of the cellular immune system in rats (163). After 11 days of exposure to the stressor, the number of total T cells was significantly decreased, but after a 12-day recovery period T-cell number was significantly increased. Total leukocytes, T-cell subpopulations, and NK-cell activity were similarly selectively affected by the restraint stress. Okimura et al. also observed suppression of various kinds of cell mediated immunity in mice subjected to a restraint stressor (164).

Stress-induced immunosuppression is not necessarily restricted to the period of exposure to the stressor (165). Rats were exposed to a foot shock, 2-4 hrs/day over a period of 6 months. One month after termination of exposure, the ability of splenic T lymphocytes to respond to a mitogen was suppressed. Thus, the rats apparently had a non-endocrinological memory of their recent history. Even adrenalectomized animals exhibit immunosuppression following chronic exposure to stressors (167, 168).

The effect of a chronic stressor on the immune system can be modulated by psychological factors (169). Rats given electric shocks exhibited immunosuppression (determined by the ability of their lymphocytes to respond to T-cell mitogens). But when the experimental animals were provided with the ability to avoid the stressor (although they received the same intensity, duration, and amount of electric shock as the first group), immunosuppression was avoided. Thus, the neurogenic factor modified the effect of the physical stressor.

For many years it has been recognized that there are nerve endings in the various organs and tissues of the immune system (thymus gland, bone marrow, spleen, and lymph nodes). Recent evidence suggests that the two systems also have an intimate functional relationship in which lymphokines and neuroendocrine peptide hormones can be secreted and detected by the cells of either the immune or neuroendocrine systems (170).

ANALYSIS

Chronic exposure to stressors generally does not alter endocrinological endpoints. Despite this, it is clear that the physiological state of a chronically exposed organism can be differentiated experimentally from that of other similar organisms not exposed to the stressor. Chronic exposure to a stressor can alter an ongoing physiological response such as healing, and it can be manifested by changes in functional measures of reserve capacity. Chronic exposure to stressors can also alter, and worsen, manifestations of the disease state. In all cases, both in the laboratory and in life, an organism is responding to myriad external stimuli of various magnitudes and durations, only some of which are considered and analyzed as stressors. At any instant, each physiological parameter in the organism exhibits a value which is essentially an algebraic sum of the effects associated with each of the factors in the organism's environment. No particular parameter can even theoretically characterize either the state of the organism, nor the quantum of the impact associated with a particular stressor.

Chronic exposure to stressors can exacerbate the disease state, but psychological factors can modulate the response, and sometimes confer a protective benefit. Chronic exposure to stressors also affects the immune system, suggesting that impaired immune surveillance may be the mechanism underlying stressor-related effects.

If we apply the research -- which was conducted mostly on rodents -- to human physiology, what then can we reasonably expect will occur in a human population chronically exposed to an electromagnetic stressor? Since chronic stressors promote disease, and since EMFs are stressors, I conclude that higher disease levels will be found in the exposed population. That is, people who have been exposed to a chronic electromagnetic stressor by virtue of where they live (near a high-voltage powerline), work (electrical trades, military weapons operator), sleep (electrically heated waterbed, electric blankets), or play (amateur radio operators) will exhibit higher disease levels compared to appropriately matched control groups wherein such exposure does not occur.

Becker reached this conclusion almost 15 years ago (171, 172). Its importance can be gauged by the virulence and source of its opposition. He based his view, in 1974, on scanty evidence, instinct and judgment, and consequently could have been wrong. Now, the view is based on voluminous data, and is inescapable.

We have no a priori or experimentally-based reason for expecting that one disease or another will be singled out from among the range of non-traumatic human diseases. Indeed, the generalized clinical experience is that all human diseases can be exacerbated by chronic exposure to stressors. Practically, however, cancer is the likely disease for study in exposed populations because it kills many people, and death certificates often list the decedent's address and occupation. Consequently, analyzable public-health records already exist.

EMFs AND DISEASE

INTRODUCTION

In therapeutic medicine, the physician's goal is the diagnosis and treatment of the patient's disease. In public health, however, the doctor is the government and the patient is the public at large. Consequently, the attitude towards public health varies with the political texture of the society. Concern for the well-being of its citizens has not been a dominant theme in the history of English-speaking people. Malthus viewed public health matters as nature's way of weeding out the genetically inferior individuals. Huxley equated attempts at promoting public health with interference in the survival and propagation of the fittest, and hence an ultimate deterrant to the human species. This attitude became incorporated in state and federal common law in the USA. State health departments were confined primarily to dealing with infectious diseases, filth, and other acute determinants of disease. At the federal level responsibility became split among several federal agencies within the Executive Department of government -- which is to say that their vigor is under the direct political control of the President. As a consequence of this tradition, there has been little governmental (and no industry) interest in electromagnetic-field epidemiology. Despite the general absence of official support however, a broad array of epidemiological studies have been performed.

OCCUPATIONAL EXPOSURE

Zaret reported several small clusters of cancer among occupationally-exposed men (173). These included 2 cases of brain tumors among 18 workers servicing microwave communications equipment, 5 cases of cancer among a group of 17 men who worked on a weapons system involving electromagnetic pulses, and 3 cases of cancer among 8 men employed as repairmen for airborne navigation systems.

During a study of occupational mortality in the state of Washington that involved 438,000 deaths during 1950-1979 (174), Milham found a disproportionately large number of deaths from leukemia among aluminum workers (20 deaths as opposed to the expected mortality of 10.6). Since strong magnetic fields are created as part of the aluminum-manufacturing process, Milham chose job classifications as a surrogate for occupational exposure to EMFs. He found more observed than expected instances of leukemia in 10 of the 11 occupations considered including electricians, aluminum workers, linemen, power-station workers, and electrical engineers (Table 6). Chlorine manufacturing also involves exposure to magnetic fields, but a cohort of 157 exposed men (40-290 Gauss, 1 year or greater) had a normal cancer rate compared with that of Swedish men (175).

Wright et al., used Milham's occupational designations and studied the possible link to leukemia incidence (1972-1979) for white males in Los Angeles County (176). There were no cases in 2 categories, but the number of observed cases exceeded the number of expected cases in 7 of the remaining 9 categories. One of the two occupations that did not show an increase in leukemia (welders and flame-cutters) was the same occupation that did not show an increase in the Milham study. An overall increase in acute leukemia was found (Table 6).

Coleman et al., examined the incidence of leukemia among men who were occupationally exposed to EMFs in southeast England (177). The 10 electrical occupations studied were essentially equivalent to the 11 categories used to classify American workers. They found a 17% excess of leukemias in the electrically-exposed occupations (Table 6). For 8 of the 10 occupations examined, more leukemias were observed than expected.

In studies that only analyze death certificates, an increase in the frequency of one disease may occur as a result of a decrease in the frequency of another disease. Thus, a higher than expected frequency does not necessarily mean that the true relative risk for the disease was increased. More confidence can be placed in the interpretation of the increased frequency

TABLE 6. Leukemia Incidence (Mortality) in Men Occupationally Exposed to Electromagnetic Fields

		LEUKEMIA		ACUTE LEUKEMIA	
INVESTIGATOR	STUDY AREA	Observed	Expected	Observed	Expected
Milham (174)	Washington	(136)*	(92)	(60)*	(36.7)
Wright (176)	Los Angeles	35	27.2	23**	13.3
Coleman (177)	England	113**	96.5	45	35

*P < 0.01
**P < 0.05

as a true indicator of risk if the frequency is greatly increased (by more than a factor of three, for example) or if more than a handful of studies uncover an increased frequency.

Another concern relating to the validity of the EMF-cancer reports involves the occupational factors actually linked with the cancer. Chemicals or other factors might be wholly responsible, thereby acquiting EMFs as etiological factors. The occurrence of a multiplicity of studies (involving different designs and performed in widely separated locations) pointing to the same link is required to overcome this concern. Studies from disparate industries are particularly valuable.

Underground coal miners work in relatively close proximity to electrical power distribution lines that are strung overhead in the mines. In general, coal miners are not at increased risk for leukemia or other cancers (178). In a case-control study comparing 40 coal workers who died from leukemia with 160 controls who died from causes other than cancer or accidents, the odds ratio for chronic leukemia, chronic lymphocytic leukemia, and myelogenous leukemia were 8.2, 6.3, and 4.7, respectively (179). All these increased risk indices were statistically significant.

The incidence of cancer among white males who worked at the Portsmouth Naval Shipyard between 1952 and 1977 was studied in a case-control investigation (180). Fifty-three individuals who died with leukemia were each matched to 4 controls. No associations were found with ionizing radiation or solvent exposure, but significant associations were found for 2 occupations -- electrician (odds ratio 3.0, 95% confidence interval 1.29-6.98), and welder (odds ratio 3.83 for myeloid leukemia, confidence interval 1.28-11.46).

Flodin et al. sought a link between background ionizing radiation and acute myeloid leukemia but, instead, found that electrical technicians, electrical welders and computer and telephone mechanics had a greater than average risk of developing the disease (odds ratio 3.8) (181).

McDowall reported on a case-control study involving 537 deaths in England and Wales (males greater than 15 years of age) that died from acute myeloid leukemia in 1973 (182). The control group included all causes of death except leukemia. They found a consistently increased relative risk for the occupations that involved exposure to EMFs: occupationally-exposed men had a relative risk of acute myeloid leukemia of 2.3 ($P < 0.05$). In a similar case-control study Pearce et al., reported that electrical workers in New Zealand were at increased risk of leukemia (183-184). Wisconsin electrical workers had a normal rate of leukemia mortality between 1963 and 1978, but individual occupations including electrical engineers and radio and telegraph operators exhibited significantly higher rates of leukemia than were expected (185).

In 1971-83, 1,691 deaths occurred in Washington and California among male members of the American Radio Relay League, a group of amateur radio operators. Twenty-four deaths from leukemia were observed, compared to the expected 12.6 (186) ($P < 0.01$). Employment-related exposure did not explain the leukemia excess.

Robinette et al., studied the effects of EMF exposure among Navy servicemen who graduated from training schools and served aboard ships during 1950-1954 (187). One group (20,781 men) consisted of subjects who worked as radiomen, radarmen, and aviation-electricians' mates, and the second group (19,649 men) consisted of aviation, electronics, or fire-control technicians. The men in both groups were heavily exposed, both occupationally (1000 $\mu W/cm^2$) and after work hours (because the ship's radars operated more than 8 hours per day), compared to typical exposure levels experienced by the general public in the USA. The chief difference between the groups was the possibility that some men in the smaller group were occasionally exposed to 100,000 $\mu W/cm^2$. No differences in a variety of mortality and morbidity indices were found by the investigators that could be attributed to the difference in exposure, but there were 16 deaths in the combined groups from eye, brain, and other nervous-system neoplasms. This represented 7.9% of all malignant deaths as compared with the rate of 3.8% for the general population ($P < 0.01$)(188).

The choice of an appropriate control group is a significant inherent difficulty in studies with a control cohort. If the true relative risk for disease is increased in both the case and control groups, as may be true in the Robinette report (187), then the study would actually be only a second-order study of differentially increased relative risk. Such a design has a built-in assumption that relative risk is proportional to a surrogate for magnitude or duration of exposure. If the surrogate is invalid or the effect is not proportional to dose, then the validity of the study is destroyed. Another U.S. government-sponsored study with a dubious control group involved the foreign-service officers at the U.S. Embassy in Moscow who were subjected to microwave EMFs from sources outside the Embassy building (223). The control subjects were foreign-service officers who served in Eastern Europe and who may also have been subjected to microwave EMFs. This realistic possibility undercuts the conclusion (cancer among foreign-service officers in Moscow not related to microwave EMFs) of the study.

Lin et al., studied the relationship between occupation and brain-tumor mortality that occurred among white male Maryland residents between 1969-1982 (188). A total of 951 cases were included, consisting of 370 gliomas, 149 astrocytomas, and 432 brain tumors of unspecified type. Preliminary analysis showed more deaths among occupationally-exposed workers (electricians, electrical engineers, linemen) than expected (from 1970 Maryland Census data): 50 deaths from glioma and astrocytoma were observed (18 expected), 28 deaths from brain tumors of unspecified type (14.7 expected), and both differences were significant ($P < 0.01$). But a possible bias was introduced because the comparison involved mortality data from a 14-year span and occupation prevalance based on only one year. To overcome the problem, a case-control study was performed in which the control group consisted of white males who died from causes other than malignant neoplasms, matched on age and date of death. It was found that patients in occupations involving exposure to electromagnetic fields exhibited more gliomas and astrocytomas than the controls: electricians (13 vs. 10), engineers (18 vs. 6) and utility company employees (19 vs. 11). The patients who died from gliomas or astrocytomas were younger by an average of 5.1 years compared to the controls.

The relative risk for brain cancer among English electrical workers was elevated (189). A Swedish study reported that powerline workers have a slightly elevated risk of leukemia and brain tumors, but that power station operators have normal cancer rates (222).

During 1964-1978, 157 cases of neuroblastoma occurred in children under 15 years of age in Texas for which adequate records could be obtained (190). A control group was formed from randomly selected birth certificates, and data on parental occupation was analyzed based on a system in which occupational exposures were classified according to presumed chemical exposures. Cluster 7 (formed on the basis of presumed moderate exposure to hydrocarbons) was associated with an increased risk of neuroblastoma (odds ratio 3.17). Cluster 7 included occupations that both involved (electricians) and did not involve (printers) exposure to EMFs. When the data was reanalyzed to include only Group 1 occupations (electricians, electric and electronic workers, linemen, welders, and utility employees) the odds ratio for neuroblastoma was 2.14. When Group 1 was expanded to include parents who sold or serviced electrical equipment, the odds ratio was 2.13 ($P = 0.05$). When only electronics workers were evaluated, the odds ratio was 11.75 ($P < 0.05$).

The authors did not specify explicitly whether it was the occupations of the fathers or mothers that were being evaluated. If the link with neuroblastoma actually involved EMFs, it may have resulted from prenatal exposure of the mother. Perhaps the more intriguing possibility for the observed association (190) is that it was linked to EMF exposure of the childrens' fathers. There is some confirmatory support for this possibility: retinoblastoma occurred more often among children whose fathers were radio and television repairmen (191).

During 1968-1975, the annual rate of eye cancer among men who worked in the electrical and electronics industry in England and Wales was consistently greater than that of the general work force (192). None of the other 25 occupational groups showed consistent increases of comparable magnitude.

Vagero and Olin studied whether cancer cases in Sweden reported in 1961-73 contained more cases of cancer among men or women aged 15-64 who were classified as working in the electronics industry (the particular occupations were not specified) (193). They found a 15% excess of cancers among men, and an 8% excess of cancers among women for workers in the electronics industry. Canadian high-voltage powerline workers exhibited more than a three-fold increase in cancer of the intestine ($P < 0.01$) (194).

EXPOSURE IN THE GENERAL ENVIRONMENT

In 1977 Becker reported a cancer cluster among residents of a rural area south of Syracuse, New York that was traversed by high-voltage powerlines and contained 20 antennas (195). The cancer incidence during 1974-1977 was almost double the rate expected in the county as a whole.

In 1979, Wertheimer and Leeper reported the first controlled study of a potential link between EMFs in the general environment and human disease (196). They asked whether children who lived in the greater Denver area, and who died of cancer in 1950-1973, lived near powerlines more commonly than did normal children. Their definitions of both "powerline" and "near" were arbitrary, and were rooted in their idea that any nexus between powerlines and cancer was mediated by the magnetic field of the powerlines. Their definitions were applied to the addresses of both the children who died from cancer, and to the addresses of an appropriately chosen control group. Roughly twice the expected death rates from leukemia, lymphoma, and nervous-system tumors were found in subjects living near powerlines. They performed a similar study among adults who died (or recovered) from cancer in 1967-1977, and again found an association between living near powerlines and cancer (197). Data for both studies is shown in Table 7.

TABLE 7. Incidence of Cancer Reported by Wertheimer and Leeper (16,17). Children, $X^2 = 18.20$, $P < 0.001$. Adults, $X^2 = 7.94$, $P < 0.01$.

	CHILDREN†		ADULTS*	
	Cancer Cases	Controls	Cancer Cases	Controls
Near Powerlines	101	55	438	372
Away from Powerlines	171	217	741	807

† Birth addresses: all cases
* Wiring subcategories combined

Fulton et al. studied the possible association of childhood leukemia with powerlines in Rhode Island (198). The study was similar in some ways to that of Wertheimer and Leeper, and differed mainly in the manner in which nearness was defined. Using their definition, Fulton et al., could not demonstrate a link between childhood leukemia and powerlines. Their data was subsequently reanalyzed by Wertheimer and Leeper, and again an association with powerlines was not found (199).

It is difficult to characterize an ambient electromagnetic environment in engineering terms because of the spatial, spectral, and temporal variations of the fields. In epidemiological studies, this difficulty can sometimes be overcome by relating features of the environment, as opposed to actual EMF measurements, to cancer incidence. But perception of features of an environment is a subjective process, and divisions and classifications may not be made in the same fashion by different groups of investigators. Also, the design and density of powerlines varies significantly throughout the country, and therefore a classification system that indexes exposure in one area may not do so in another area. The validity of the scheme adopted by Wertheimer and Leeper as a surrogate for magnetic-field exposure was independently confirmed (200).

Recently, an exhaustive review and replicate study of the work linking childhood cancer and powerlines (196) was performed under the auspices of public-health and power-company officials in New York, with the result that the conclusion of a link between childhood cancer and powerlines was confirmed. Even more recently, the Electric Power Research Institute commissioned a second replicate study of the first replicate study.

A partial characterization of the EMF environment at two suburban elementary schools in northwest Oregon revealed power-frequency electric fields of 3-4.5 V/m inside the schools, and 3-170 V/m outside one school, and 1-18 V/m outside the other school (202). The magnetic fields inside the schools were less than 1 mGauss, but outside the school they were as high as 75 mGauss at one school and 35 mGauss at the second school. The power density in various frequency ranges between 1 kHz and 15.5 GHz did not differ between the two sites. A cluster of 4 cases of childhood cancer in a fourth grade class occurred within a 9-month period at the school with the higher outdoor power-frequency fields (202). The authors concluded that the measurements indicated that the EMFs were not responsible for the cancer, but the opposite conclusion is equally valid based on the data presented.

In a case-control study that involved 716 tumors (660 malignant, 56 benign) in Stockholm County (1958-1973) in patients 0-18 years of age (203), the authors asked: (1) did more than the expected number of tumors occur in children who lived near 200,000-volt powerlines; (2) did more than the expected number of tumors occur in people who lived in regions with high magnetic fields? Among the tumor cases, they found 32 dwellings at which a 200,000-volt powerline was visible, but only 13 such dwellings were found in the control group ($P < 0.05$). Of the 48 dwellings that exhibited a high magnetic field (3 mG or greater) 34 were tumor cases and 14 controls ($P < 0.05$).

The two major airports in Wichita, Kansas, use radars to control approaches and landings, and their beams blanket the city. Lester and Moore (204) asked whether the geographic pattern of cancer in Wichita was related to the radar fields. The variable topography of Wichita was a complicating factor because the hills interrupted the line-of-sight beam of the radars thereby creating a shield from one or both radars in various parts of the city. A three-tiered measure of exposure was derived consisting of areas with the highest, intermediate, and lowest amounts of radar exposure. It was found that cancer incidence in Wichita residents in 1975-1977 (3,004 cases) was related to the amount of radar exposure ($P < 0.05$) after correcting for age, economic stratification, male/female ratio, and race. Those census tracts with the highest shield showed the lowest cancer incidence, and the tracts with the lowest shield had the highest incidence of cancer.

Counties in the USA that had an Air Force base had a significantly higher cancer mortality during 1950-1969 than did control counties without an Air Force base (205-207).

A cluster of five cases of a rare ovarian tumor diagnosed over a 4-year period was reported in Florida (208). Three potential environmental risk factors were identified in the neighborhood where the children lived: proximity to a major highway, a lead smelter, and powerlines. The children lived 14-592 feet from a 69,000-volt powerline for an average of 7.8 years prior to diagnosis.

Electromagnetic fields have been linked with suicide (21), polycythemia (209), nervous system disorders (210,211), sexual dysfunction (212), reproductive hazards (213), abnormal fetal development (214,215), amyotrophic lateral sclerosis (216), heart disease (217,218), and subjective complaints (219).

ANALYSIS

Leukemia has been linked to many occupations involving electricity. Such workers are exposed to many chemicals including diphenyls, naphthalenes, phenols, epoxys, oils, and solvents, any one or combination of which may have mediated the observed link. But most of the electricity-related occupations listed within the individual studies reported more cancers than expected (25 of 32 occupations described in the studies in Table 6). These occupations, which included electrician, aluminum worker, power-station operator, powerline worker, and telegraph operator, have no discernible chemical factor in common. Furthermore, a link to leukemia has appeared in a group formed on the basis of hobby interests (186), thereby lending more credibility to a non-chemical hypothesis. Increased leukemia has been found in children who lived near powerlines (196). Powerlines can produce ozone, but this seems an unlikely explanation for the observed correlation because there were few addresses near the type of high-voltage powerlines that produce ozone. The evidence of a link between leukemia and EMFs has emerged from many different places: California, Denver, England, Los Angeles, New Zealand, Wales, and Washington.

The only common factor that linked cancer with the subjects' environment was the electromagnetic field. Since the frequency of cancer was increased when the electromagnetic field was added to the environment, the electromagnetic field was a risk factor for cancer.

Although the evidence of a link is reasonably clear, it is unsatisfactory in several salient aspects, the most important of which is that of the quantum of impact associated with specific patterns of exposure. If a child lives in a power-frequency magnetic field of 0.5 Gauss for 5 (10, or 15) years, what is his relative risk for leukemia? Heart disease? To what extent are the clinical signs and symptoms exhibited by the aged who use electric blankets associated with chronic exposure to the fields? To specifically what increased relative risks are individuals who live near airport radars or antenna farms exposed? These answers cannot be distilled from the present literature, but the reason for this inability merits our attention. Conclusive epidemiology (by which I mean the strongest possible statement that can be made linking a disease and a determinant within the intrinsic limitations of epidemiology) occurs in only two situations. Sometimes, the determinant is strongly and uniquely associated with a disease, such as malaria with mosquitos. Conclusive epidemiology can also result when the determinant is weaker but society's desire to know is strong, as manifested by its decision to fund the requisite good-faith research. The link between cancer and cigarette smoking is a good example. EMF epidemiology presently falls in neither category.

Infectious disease epidemiology involves the study of the dynamics of an offending microorganism, and its effects on human health. Good (mankind) and evil (microorganisms) are clearly distinguishable, and we have no difficulty in choosing sides in the conflict. EMF epidemiology is a more complicated matter because the offending fields in the environment are attributable to specific sources and their owners and operators. Thus, unlike the man-against-nature conflict in infectious disease epidemiology, we have a three-cornered conflict. Alas, it is the third corner that has the funds for the requisite studies.

If we choose to force an evaluation of the degree of risk associated with exposure to environmental EMFs from the existing epidemiological literature, then it seems to me that its adumbrations are truly ominous. The design of most epidemiological studies tended to make them relatively insensitive to finding a link between a surrogate for exposure, and disease, because most studies have been done cheaply. An insensitive study detects only the strongest effects. Academic criticism has been directed toward the relatively primitive nature of the epidemiological studies (220), but in my view such criticism is largely misplaced because it is not reasonable to expect cheap studies to provide measures of relative risk or dose/effect. Further, such criticism generally relates to second-order concerns, and not to the basic public-health issues raised by environmental EMFs.

EMFs AND PUBLIC HEALTH

PUBLIC-HEALTH ASPECTS

The prototypical action of stressors is to promote (not initiate) disease, and this is the role that environmental fields play in the incidence and prevalence of human disease. No specific agent can accurately be said to cause any chronic human disease, and EMFs in the environment should therefore be viewed as one of a range of factors that can tax a subject's adaptive capacity. When the total body load of environmental stressors exceeds an individual's capacity, its immune surveillance mechanisms are impaired and disease occurs. Electromagnetic fields are one such disease-promoting factor. No other view in the literature fits all the data.

The most important deficiency in the existing data is our ability to actually measure the risk associated with specific patterns of environmental exposure in comparison to other accepted risk factors for disease such as cigarette smoking, sex, age, cholesterol level, and so forth. Although many aspects of the mechanisms of interaction will ultimately be deduced from relevant animal and tissue studies, such studies are inherently unable to provide data regarding quantification of the relative importance of the risk. Only human beings eat, work, play, sleep, and exercise like human beings.

The molecular mechanism involved in the actual detection by the body of the electromagnetic field remains purely speculative. None of the old theories (166) have born fruit. Recent reports by Liboff and colleagues (220) describe EMF-induced bioeffects whose physical basis may be understood. For now, I believe that it is better to state a general model based on actual data, than to suggest a molecular model based on hypothesis. Consequently, my summary evaluation of the literature is that it establishes that chronic exposure to environmental EMFs taxes adaptive capacity and thereby promotes human disease.

COUNTER-ARGUMENTS

In 1974 (1,225) I first expressed judgments that subsequently evolved into the views expressed here (166,201, 226-229). It was difficult, at that time, to truly accept the idea that such a pervasive entity as the non-thermal electromagnetic field could be harmful. My conclusion was worded delicately, with room for retreat in the face of definitive contrary data. But no data requiring a contrary conclusion has appeared, and I am now convinced that none will appear. Several arguments, however, have been made during the intervening 13 years whose thrust tends to oppose, weaken, or trivialize the conclusion that chronic exposure to environmental EMFs is a risk factor for human disease. The historical details regarding these arguments -- who spoke them, when and where -- is described elsewhere (1,230-232). Here, I intend only to briefly list and describe the principal counter-arguments, and to tell the reader why I have dismissed them. The lesson of the literature concerning the EMF-related health risk is contained in several bodies of data which must be integrated and connected to rationally support the conclusion. One must be aware that prior to the present broad-scale use of EMFs in the environment, there was no serious consideration given to the possible adverse consequences of human exposure. Thus, the present exposure patterns are built on assumptions of safety, and not actual evidence thereof. One must be aware of the laboratory work involving the exposure of numerous animal species to simulated environmental EMFs. If one neglects to consider all the pertinent work, it is possible to promote a prejudice against subsequent epidemiological studies showing apparent links between exposure and disease. One must also be familiar with the literature regarding the disease-promoting nature of chronic stressors. Again, if one does not admit the numerous studies that underlie this concept, then the epidemiological data seems wholly unsupported and slippery. One must be familiar with the human epidemiology of EMFs, and with the inherent limitations of that science -- it forever deals in shades of grey, and never provides an unassailable conclusion. Finally, one must have a realistic notion of how science is done in the USA, and who pays for it. If the individual links in the chain are not connected, then the conclusion cannot be sustained for simple ignorance. But it is not ignorance that I wish to consider under the rubric of counter-arguments. Rather, it is affirmative statements in the literature that run counter to the judgments of risk that I have made.

The argument has been made that there simply are no biological effects due to EMFs, and consequently there can be no hazard associated with exposure to EMFs. This argument was first made in New York in 1975 (1), and subsequently has been repudiated by essentially every investigator in the field. It is, however, of more than historical interest because today, in the bulk of the present litigation in U.S. courts regarding health risks of environmental EMFs, this remains one of the chief arguments advanced by the polluter.

A second argument is that several blue-ribbon government and industry panels have concluded that no significant health risk due to exposure to environmental EMFs has been proved. Such arguments are not bona fide opinion evidence because they are invariably collective judgments of individuals chosen by the polluter (106,233). Such a procedure is simply a matter of the fox being on the jury at a goose's trial (1,234). Blue-ribbon panels make it impossible to pin a given view on a given person, and to ascertain a specific basis for the view.

The biophysical argument is that EMFs can affect biological systems only by stimulating nerves or depositing heat, and that since environmental EMFs do neither, they cannot be a health risk. There is no official U.S. exposure standard for environmental EMFs, but there is an unofficial standard, and it is founded directly on this argument. The proponent of such a view assumes that nothing can happen to a biological system other than via processes that he first accepts as proven. Thus, the argument is invalid because it is unscientific.

It has been argued that power and communications companies have received no reports of illness among people living or working near EMF sources, and that one may visually inspect such premises and observe nothing untoward. The arguments, which appear in essentially all judicial and administrative proceedings in the USA regarding the EMF issue, are both self-serving and irrelevant.

It has been argued that there is much negative literature, and therefore that the reports are in conflict, or are contradictory, and that definitive statements regarding risk cannot be made. The argument is a shibboleth because the negative EMF literature establishes essentially nothing. There is not a single such study that has taught us anything worthwhile about nature. Anyone can drill a hole and fail to strike oil, and the existence of an empty hole is the meanest evidence imaginable that oil does not exist. To weight such studies against actual observations is illogical.

There is one further argument, a powerful one that may prevail. Whatever the magnitude of the risk of environmental EMFs, we must accept it because the alternative would cost too much, disrupt society, and endanger national security. Continued discussion is therefore pointless.

INVOLUNTARY HUMAN EXPERIMENTATION

Power companies, electric-blanket manufacturers, and other organizations produce products that liberate electromagnetic fields into the environment, but they are generally silent about this fact and most subjects are unaware of the presence of the EMFs. Because the various devices directly result in human exposures that are significantly above the ambient, the question arises whether the situation constitutes involuntary human experimentation. In New York, experimentation is defined in terms of physical intervention upon a subject that is not required for the direct benefit of the subject (235). The law provides "no human research may be conducted in this state in the absence of voluntary informed consent subscribed to in writing by the human subject..." (236). U.S. federal regulations governing human research are vastly more detailed, but also require voluntary written informed consent (except in special circumstances) (237). All modern authority opposes involuntary intervention upon a subject (238-240). My opinion is that many present-day exposure patterns, such as living near a powerline, are exactly the kind of physical intervention upon subjects that is proscribed by law and applicable ethical principles. In the USA, civil remedies exist (at least for non-military subjects) to counter such activity, including the law of nuisance, battery, personal injury, and inverse condemnation. Although some of these actions are presently encumbered with significant evidentiary problems for the plaintiff, the problems become fewer with time because of the tide of reports, latches of the defendants, and other factors.

SUMMARY

Chronic exposure to a biological stressor is a risk factor for disease. Laboratory studies show that electromagnetic fields can be biological stressors. Such fields, when present in the environment, are therefore risk factors for disease. The emergence of direct evidence of a link between electromagnetic fields and one class of diseases -- cancer -- has been facilitated by the availability of cancer demographic data, and does not imply that electromagnetic fields have a particular propensity to promote cancer as opposed to heart disease, psychiatric disorders, or other maladies. Controversy, or at least the appearance of controversy, regarding the health risks associated with environmental electromagnetic fields has developed because the emerging scientific picture runs markedly counter to the long-standing interests of some industries and government agencies in unbridled use of the electromagnetic spectrum. The existence of a link between electromagnetic fields in the environment and disease has been established despite the fact that many important details regarding it remain undiscovered.

REFERENCES

(1) Marino, A.A., and Ray, J.: The Electric Wilderness, San Francisco Press, San Francisco, 1986.

(2) An Estimate of the Potential Costs of Guidelines Limiting Public Exposure to Radiofrequency Radiation from Broadcast Sources, EPA 520/1-85-025, PB86-108826, NTIS, Springfield, VA, July, 1985.

(3) Tell, R.A. and Mantiply, E.D.: Population Exposure to VHF and UHF Broadcast Radiation in the United States, ORP/EAD 78-5, U.S. Environmental Protection Agency, Las Vegas, Nevada, 1978.

(4) Tell, R., Mantiply, E., Durney, C. and Massoudi, H.: Electric and magnetic field intensities and associated body currents in man in close proximity to a 50 kW AM standard broadcast station, Presented at Bioelectromagnetics Symposium, Seattle, Washington, 1979.

(5) Tell, R., Lambdin, D., Brown, R. and Mantiply, E.: Electric field strengths in the near vicinity of FM radio broadcast antennas, Presented at IEEE Broadcast Symposium, Washington, DC, 1979.

(6) Tell, R.A. and Hankin, N.H.: Measurements of Radio Frequency Field Intensity in Buildings With Close Proximity to Broadcast Systems, ORP/EAD 78-3, U.S. Environmental Protection Agency, Las Vegas, Nevada, 1978.

(7) Tell, R. and O'Brien, P.J.: An Investigation of Broadcast Radiation Intensities at Mount Wilson, California, ORP/EAD-77-2, U.S. Environmental Protection Agency, Las Vegas, Nevada, 1977.

(8) Cohen, J.: Report to the Town Board of the Town of Onondaga, Onondaga, New York, 1978.

(9) An Analysis of Broadcast Radiation Intervals in Hawaii (Technical Note #ORP/EAD-75-1), Environmental Protection Agency, August, 1975.

(10) Radiofrequency Radiation, Measurement Survey, Honolulu, Hawaii, May 14-25, Environmental Protection Agency, Washington, DC, 1984.

(11) An Investigation of Radiofrequency Radiation Exposure Levels on Cougar Mountain, Issaquah, Washington, May 6-10, Environmental Protection Agency, Washington, DC, 1985.

(12) An Investigation of Radiofrequency Radiation Levels on Healy Heights, Portland, Oregon, July 28-August 1, Environmental Protection Agency, Washington, DC, 1986.

(13) An Investigation of Radiofrequency Radiation Levels on Lookout Mountain, Jefferson County, Colorado, September 22-26, Environmental Protection Agency, Washington, DC, 1986.

(14) Radiofrequency Radiation Surveys for AN/FPS-115 Pave Paws Radar, Cape Cod, AFS, Massachusetts (Report #86-33), September 13-30, 1986.

(15) Hankin, N.: Calculation of Expected Microwave Radiation Exposure Levels at Various Distances From Microwave System Proposed by MCI Corporation in Skaneateles, New York, U.S. Environmental Protection Agency, Washington, D.C., May, 1980.

(16) Lambdin, D.L.: An Investigation of Energy Densities in the Vicinity of Vehicles With Mobile Communications Equipment and Near a Hand-Held Walkie Talkie, ORP/EAD 79-2, U.S. Environmental Protection Agency, Las Vegas, Nevada, 1979.

(17) Ho, A.M., Fraser-Smith, A.C. and Villard, O.G.: Large-amplitude ULF magnetic fields produced by a rapid transit system: Close-range measurements, Radio Sci. 14:1011, 1979.

(18) Hart, F.X. and Marino, A.A.: Energy flux along high-voltage transmission lines, IEEE Trans. Biomed. Eng. BME-24:493-495, 1977.

(19) Reichmanis, M. and Marino, A.A.: Bioelectric considerations in the design of high-voltage powerlines, J. Bioelectricity 1:329-339, 1982.

(20) Heorux, P.: 60-Hz electric and magnetic fields generated by a distribution network, Bioelectromagnetics 8:135-148, 1987.

(21) Perry, F.S., Reichmanis, M., Marino, A.A. and Becker, R.O.: Environmental power-frequency magnetic fields and suicide, Health Physics 41:267-277, 1981.

(22) Florig, H.K., Hoburg, J.F. and Morgan, M.G.: Electric field exposure from electric blankets, IEEE Trans. Power Delivery, 1-10, 1986.

(23) Reiter, R.: On the presence and generation of AC and DC electric fields and small ions in closed rooms as a function of building materials, utilization, and electrical installation, J. Geophys. Res. 90:5936-5944, 1985.

(24) Stuchly, M.: Human exposure to static and time-varying magnetic fields, Health Phys. 51:215-225, 1986.

(25) Marha, K. and Charron, D.: The distribution of pulsed very low frequency electric field around video display terminals, Health Phys. 49:517-521, 1985.

(26) Stuchly, M.A. and Lecuyer, D.L.: Induction heating and operator exposure to electromagnetic fields, Health Phys. 49:693-700, 1985.

(27) USSR Ministry of Health, Interim Health Standards and Regulations on Protecting the General Population from the Effects of Electromagnetic Fields Generated by Radio Transmitting Equipment, #2963-84, Moscow, 1984.

(28) Presman, A.S.: Electromagnetic Fields and Life, Plenum, New York, 1970.

(29) Kholodov, Yu.A.: The Effect of Electromagnetic and Magnetic Fields on the Central Nervous System, N6731733, NTIS, Springfield, VA, 1966.

(30) Bianchi, D., Cedrini, L., Ceria, F., Meda, E. and G.G. Re: Exposure of mammalians to strong 50 Hz electric fields, Arch. Fisiol. 70:33-34, 1973.

(31) Lott, J.R. and McCain, H.B.: Some effects of continuous and pulsating electric fields on brain wave activity in rats, Int. J. Biometeor 17:221-225, 1973.

(32) Goldstein, L. and Sisko, Z.: A quantitative electroencephalographic study of the acute effects of X-band microwaves in rabbits, in Biological Effects and Health Hazards of Microwave Radiation, Polish Medical Publishers, Warsaw, p. 128, 1974.

(33) Friedman, H. and Carey, R.J.: Biomagnetic stressor effects in primates, Physiol. Behav. 9:171-173, 1972.

(34) Novitskiy, A.A., Murashov, B.F., Krasnobaev, P.E. and Markizova, N.F.: Functional state of the hypothalamus-hypophysis-adrenal cortex system as a criterion in setting standards for superhigh frequency electromagnetic radiation, Voen. Med. Zh. 10:53, 1977.

(35) Dumanskiy, Yu.D. and Shandala, M.G.: The biological action and hygenic significance of electromagnetic fields of superhigh and ultrahigh frequencies in densely populated areas, in Biological Effects and Health Hazards of Microwave Radiation, Polish Medical Publishers, Warsaw, 1974.

(36) Hackman, R.M. and Graves, H.B.: Corticosterone levels in mice exposed to high intensity electric fields, Behav. Neural. Biol. 32:201-213, 1981.

(37) Lymangrover, J.R., Keku, E. and Seto, Y.J.: 60-Hz electric field alters the steroidogenic response of rat arenal tissue, in vitro, Life Sci. 32:691-696, 1983.

(38) Prochwatilo, J.W.: Effects of electromagnetic fields of industrial frequency (50 Hz) on the endocrine system, Vrach. Delo. 11:135, 1976.

(39) Demokidova, N.K.: On the biological effects of continuous and intermittent microwave radiation, JPRS 63321, 113, 1973.

(40) Ossenkopp, K.D., Koltek, .W.T. and Persinger, M.A.: Prenatal exposure to an extremely low frequency, low intensity rotating magnetic field and increases in thyroid and testicle weight in rats, Develop. Psycobiol. 5:275, 1972.

(41) Fischer, G., Waibel, R. and Richter, Th.: Influence of line-frequency electric fields on the heart rate of rats, Zbl. Bakt. Hyg., I. Abt. Orig. B 162:374-379, 1976.

(42) Prokhvatilo, Ye.V.: Reduction of functional capacities of the heart following exposure to an electromagnetic field of industrial frequency, JPRS 70101, NTIS, Springfield, VA, p. 76, 1977.

(43) Serdiuk, A.M.: State of the cardiovascular system under the chronic effect of low-intensity electromagnetic fields, JPRS L/5615, NTIS, Springfield, VA, p. 8, 1975.

(44) Gann, D.: Final Report, Electric Power Research Institute Project RP 98-02, Palo Alto, CA, 1976.

(45) Marino, A.A., Cullen, J.M., Reichmanis, M., Becker, R.O. and Hart, F.X.: Sensitivity to change in electrical environment: A new bioelectric effect, Am. J. Physiol. 239 (Regulatory Integrative Comp. Physiol. 8), R424-427, 1980.

(46) Szmigielski, S., Jeljaszewicz, J. and Wiranowski, M.: Acute staphylococcal infections in rabbits irradiated with 3-GHz microwaves, Ann. N.Y. Acad. Sci. 247:305-311, 1975.

(47) Shandala, M.G. and Vinogradov, G.I.: Immunological effects of microwave action, JPRS 72956, NTIS, Springfield, VA, p. 16, 1979.

(48) Shandala, M.G., Dumanskii, U.D., Rudnev, M.I., Ershova, L.K. and Los, I.P.: Study of nonionizing microwave radia-

tion effects upon the central nervous system and behavior reactions, Environ. Health Perspect. 30:115-121, 1979.

(49) Huai, C., Gengdong, Y. and Shuiyan, Z.: Effects of microwave exposure at various power densities on mitochondrial marker enzymes in mouse brains, J. Bioelectricity 3:361-366, 1984.

(50) Quock, R.M., Kouchich, F.J., Ishii, T.K. and Lange, D.G.: Microwave facilitation of methylnaltrexone antagonism of morphine-induced analgesia in mice, J. Bioelectricity 5:35-46, 1986.

(51) Thomas, J.R., Burch, L.S. and Yeandle, S.S.: Microwave radiation and chlordiazepoxide: Synergistic effects on fixed-interval behavior, Science 203:1357-1358, 1979.

(52) Altmann, G. and Soltau, G.: Einfluss luftelektrischer Felder auf das Blut von Meerschweinchen, Z. angew. Bader - u. Klimaheilk 21:28-32, 1974.

(53) Lang, S.: Stoffwechselphysiologische Auswirkungen der Faradayschen Abschirmung und eines kunstlichen luftelektrischen Feldes der Frequenz 10 Hz auf weisse Mause, Arch. Met. Geoph. Biokl., Ser. B. 20:109-122, 1972.

(54) Fischer, G., Udermann, H. and Knapp, E.: Ubt das netzfrequente Wechsefeld zentrale Wirkungen aus?, Zbl. Bakt. Hug., I. Abt. Orig. B 166:381-385, 1978.

(55) Grin, A.N.: Effects of microwaves on catecholamine metabolism in the brain, JPRS 72606, NTIS, Springfield, VA, p. 14, 1978.

(56) Dodge, C.H.: Clinical and hygienic aspects of exposure to electromagnetic fields: A review of Soviet and East European literature, in Biological Effects and Health Implications of Microwave Radiation Symposium Proceedings, S.F. Cleary, ed., BRH, DBE Report #70-2, 140-149, 1970.

(57) Hamer, J.R.: Effects of low-level low-frequency electric fields on human reaction time, Commun. Behav. Biol. 2 Part A:217-222, 1968.

(58) Friedman, H., Becker, R.O. and Bachman, C.H.: Effect of magnetic fields on reaction time performance, Nature 213:949-956, 1967.

(59) Persinger, M.A., Lafreniere, G.F. and Mainprize, D.N.: Human reaction time variability changes from low intensity 3-Hz and 10-Hz electric fields: Interactions with stimulus pattern, sex, and field intensity, Int. J. Biometeor. 19:56-64, 1975.

(60) Gibson, R.S. and Moroney, W.F.: The Effects of Extremely Low Frequency Magnetic Fields on Human Performance, AD A005898, NAMRL-1195, Naval Aerospace Medical Research Laboratory, Pensacola, FL, 1974.

(61) Dumanskiy, Yu.D., Popovich, V.M. and Kozyarin, I.P.: Effects of low-frequency (50-Hz) electromagnetic fields on the functional state of the human body, JPRS L/5615, NTIS, Springfield, VA, p. 33, 1977.

(62) Beischer, D.E., Grissett, J.D. and Mitchell, R.E.: Exposure of Man to Magnetic Fields Alternating at Extremely Low Frequency, AD 770140, NAMRL-1180, Naval Aerospace Medical Research Laboratory, Pensacola, FL, 1973.

(63) Persinger, M.A. and Nolan, M.: Partial amnesia for a narrative following application of theta frequency electromagnetic fields, J. Bioelectricity 4:481-494, 1985.

(64) Wever, R.: The effects of electric fields on circadian rhythmicity in men, Life Sciences and Space Research VIII, North Holland Publishing Co., 177, 1970.

(65) Wever, R.: Influence of electric fields on some parameters of circadian rhythms in man, in Biochronometry, M. Menaker, ed., Washington, DC, 117-132, 1971.

(66) Wever, R.: Human circadian rhythms under the influence of weak electric fields and the different aspects of these studies, Int. J. Biometeor. 17:227-232, 1973.

(67) Wever, R.: ELF - Effects on human circadian rhythms, in ELF and VLF Electromagnetic Field Effects, M.A. Persinger, ed., Plenum Press, New York, 1974.

(68) Baroncelli, P., Battisti, S., Checcucci, A., Comba, P., Grandolfo, M., Serio, A. and Becchia, P.: A health examination of railway high-voltage substation workers exposed to ELF electromagnetic fields, Am. J. Indust. Med. 10:45-55, 1986.

(69) Andrienko, L.G., Dumanskiy, Yu.D., Rouditchenko, V.F. and Meliechko, G.I.: The influence of an electric field of industrial frequency on spermatogenesis, Vrach. Delo. 18, 116, 1977.

(70) Noval, J.J., Sohler, A., Reisberg, R.B., Coyne, H., Straub, K.D. and McKinney, H. Extremely low frequency electric field induced changes in rate of growth and brain and liver enzymes of rats, in Compilation of Navy Sponsored ELF Biomedical and Ecological Research Reports, Vol. 3, AD A035955, Naval Medical Research and Development Command, Bethesda, MD, 1977.

(71) Giarola, A.J. and Krueger, W.F.: Continuous exposure of chicks and rats to electromagnetic fields, IEEE Trans. Microwave Theory Tech. MTT-22:432-437, 1974.

(72) Marino, A.A., Cullen, J.M., Reichmanis, M. and Becker, R.O.: Power frequency electric fields and biological stress: A cause-and-effect relationship, in Biological Effects of Extremely Low Frequency Electromagnetic Fields, U.S. Dept. of Energy, Washington, D.C., p. 258-276, 1979.

(73) McClanahan, B.J. and Phillips, R.D.: The influence of electric field exposure on bone growth and fracture repair in rats, Bioelectromagnetics 4:11-19, 1983.

(74) Grissett, J.D., Kupper, J.L., Kessler, M.J., Brown, R.J., Prettyman, G.D., Cook, L.L. and Griner, T.A.: Exposure of primates for one year to electric and magnetic fields associated with ELF communications systems, USN Report NAMRL-1240, Pensacola, Florida, 1977.

(75) Tanner, J.A. and Romero-Sierra, C.: The effects of chronic exposure to very low intensity microwave radiation on domestic fowl, J. Bioelectricity 1:195-205, 1982.

(76) Hansson, H.: Lamellar bodies and Purkinje nerve cells experimentally induced by electric field, Brain Res. 216:187-191, 1981.

(77) Hansson, H.: (Personal communication), 1984.

(78) Tofani, S., Agnesod, G., Ossola, P., Ferrini, S. and Bussi, R.: Effects of continuous low-level exposure to radiofrequency radiation on intrauterine development in rats, Health Phys. 51:489-499, 1986.

(79) Batkin, S. and Tabrah, F.L.: Effects of alternating magnetic field (12 gauss) on transplanted neuroblastoma, Res. Comm. in Chem. Path. and Pharm. 16:351-362, 1977.

(80) Aarholt, E., Flinn, E.A. and Smith, C.W.: Effects of low frequency magnetic fields on bacterial growth rate, J. Phys. Med. Biol. 26:613-621, 1981.

(81) Goodman, E.M., Greenebaum, B. and Marron, M.T.: Bioeffects of extremely low frequency electromagnetic fields variations with intensity waveform and individual or combined electric and magnetic fields, Radiat. Res. 78:485-501, 1979.

(82) Goodman, R., Bassett, C.A.L. and Henderson, A.S.: Pulsing electromagnetic fields induce cellular transcription, Science 220:1283-1285, 1983.

(83) Liboff, A.R., Williams, T. Jr., Strong, D.M. and Wistar, R. Jr.: Time-varying magnetic fields: Effect on DNA synthesis, Science 223:818-820, 1984.

(84) Lubin, R.A., Cain, C.D., Chen, M.C.-Y., Rosen, D.M. and Adey, W.R.: Effects of electromagnetic stimuli on bone and bone cells in vitro: Inhibition of responses to parathyroid hormone by low-energy low-frequency fields, Proc. Natl. Acad. Sci. (USA) 79:4180-4184, 1982.

(85) Jolley, W.B., Hinsaw, D.B., Knierim, K. and Hinsaw, D.B.: Magnetic field effects on calcium efflux and insulin secretion in isolated islets of Langerhans, Bioelectromagnetics 4:103-106, 1983.

(86) Cantini, M., Cossarizza, A., Bersani, F., Cadossi, R., Ceccherelli, G., Tenconi, R., Gatti, C. and Franceschi, C.: Enhancing effect of low frequency pulsed electromagnetic fields on lectin-induced human lymphocyte proliferation, J. Bioelectricity 5:91-104, 1986.

(87) Conti, P. et al.: Reduced mitogenic stimulation of human lymphocytes by extremely low frequency electromagnetic fields, FEBS Lett. 162:156-160, 1983.

(88) Delgado, J.M.R., Leal, J., Monteagudo, J.L. and Garcia-Gracia, M.G.: Embryological changes induced by weak, extremely low frequency electromagnetic fields, J. Anat. 134:533-551, 1982.

(89) Gavalas, R.J., Walter, D.O., Hamer, J. and Adey, W.R.: Effect of low-level low-frequency electric fields on EEG and behavior in Macaca nemestrina, Brain Res. 18:491-501, 1970.

(90) Bawin, S.M., Gavalas-Medici, R.J. and Adey, W.R.: Effects of modulated very high frequency fields on specific brain rhythms in cats, Brain Res., 365-384, 1973.

(91) Bawin, S.M., Kaczmarek, L.K. and Adey, W.R.: Effects of modulated VHF fields on the central nervous system, Ann. N.Y. Acad. Sci. 247:74-81, 1975.

(92) Bawin, S.M. and Adey, W.R.: Sensitivity of calcium binding in cerebral tissue to weak environmental electric fields oscillating at low frequency, Proc. Natl. Acad. Sci. USA 73:1999-2003, 1976.

(93) Bawin, S.M., Adey, W.R. and Sabbot, I.M.: Ionic factors in release of $^{45}Ca^{2+}$ from chicken cerebral tissue by electromagnetic fields, Proc. Natl. Acad. Sci. USA 75:6314-6318, 1978.

(94) Bychkov, N.S. and Dronov, I.S.: Electroencephalographic data on the effects of very weak microwaves, JPRS 63321, NTIS, Arlington, VA, p. 75, 1973.

(95) Friedman, H. and Carey, R.J.: The effect of magnetic fields upon rabbit brains, Physiol. Behav. 4:539-541, 1969.

(96) Barnothy, M.F. and Sümegi, I.: Effects of the magnetic field on internal organs and the endocrine system of mice, in Biological Effects of Magnetic Fields, M.F. Barnothy, ed., Plenum Press, New York, 103-126, 1969.

(97) Marino, A.A., Berger, T.J., Austin, B.P., Becker, R.O. and Hart, F.X.: _In vivo_ bioelectrochemical changes associated with exposure to extremely low frequency electric fields, Physiol. Chem. Phys. 9:433-441, 1977.

(98) Kartsovnykh, S.A. and Faytelberh-Blank, V.R.: Changes in the peripheral blood of Guinea pigs induced by a three-centimeter electromagnetic field, JPRS 64532, NTIS, Springfield, VA, p. 37, 1974.

(99) Mathewson, N.S., Oosta, G.M., Oliva, S.A., Levin, S.C. and Diamond, S.S.: Influence of 45-Hz electric fields on growth, food and water consumption and blood constituents of rats, Radiat. Res. 79:468-482, 1979.

(100) Mathewson, N.S., Oliva, S.A., Oosta, G.M. and Blasco, A.P.: Extremely Low Frequency (ELF) Vertical Electric Field Exposure of Rats: Irradiation Facility, AD A045080, Technical Note TN77-2, Armed Forces Radiobiology Research Institute, Bethesda, MD, 1977.

(101) Marino, A.A. and Becker,R.O.: Biological effects of extremely low frequency electric and magnetic fields: A review, Physiol. Chem. Phys. 9:131-147, 1977.

(102) Marino, A.A., Becker, R.O. and Ullrich, B.: The effect of continuous exposure to low frequency electric fields on three generations of mice: A pilot study, Experientia 32:565, 1976.

(103) Marino, A.A., Reichmanis, M., Becker, R.O., Ullrich, B. and Cullen, J.M.: Power frequency electric field induces biological changes in successive generations of mice, Experientia 36:309-311, 1980.

(104) Phillips, R.D., Anderson, L.B. and Kaune, W.T.: Biological Effects of High Strength Electric Fields on Small Laboratory Animals, DOE/TIC-10084, U.S. Dept. of Energy, Washington, DC, 1979.

(105) Miller, M.W. and Kaufman, G.E.: High voltage overhead, Environment 20:6-15, 1978.

(106) Graves, H.B., Bracken, T.D., Griffin, J., deLorge, J., Morgan, M.G. and Tenforde, T.S.: Biological effects of 60-Hz power transmission lines, Florida Electric Power Coordinating Group, Tampa, FL, 1985.

(107) Cannon, W.B.: Stresses and strains of homeostasis, Am. J. Med. Sci. 189:1-14, 1935.

(108) Selye, H.: A syndrome produced by diverse noxious agents, Nature 138:32, 1936.

(109) Selye, H.: The general adaptation syndrome and diseases of adaption, J. Clin. Endocrin. 6:117-230, 1946.

(110) Selye, H. and Fortier, C.: Adaptive reactions to stress, in Life Stress and Bodily Disease, H.G. Wolff, S.G. Wolf, Jr. and C.C. Hare, eds., Williams & Wilkins, Baltimore, 1-18, 1950.

(111) Selye, H.: The evolution of the stress concept, Am. Sci. 61:692-699, 1973.

(112) Martin, R.D.: A critical review of the concept of stress in psychosomatic medicine, Perspect. Biol. Med. 27:443-464, 1984.

(113) Hinkle, L.E.: The concept of "stress" in the biological and social sciences, Sci. Med. & Man. 1:31-48, 1973.

(114) Levine, S.: Stress, in Neuroendocrinology and Psychiatric Disorders, G.M. Brown, ed., Raven Press, New York, 145-150, 1984.

(115) Newberry, B.H .: Stress and mammary cancer, in Stress and Cancer, K. Bammer and B.H. Newberry, eds., C.J. Hogrefe, Toronto, 233-264, 1981.

(116) Sapse, A.T.: Stress, cortisol, interferon and "stress" diseases. I. Cortisol as the cause of "stress" diseases, Med. Hypotheses 13:31-44, 1984.

(117) Eliasson, K.: Stress and catecholamines, Acta Med. Scand. 215:197-204, 1984.

(118) Gibbs, D.M.: Vasopressin and oxytocin: Hypothalamic modulators of the stress response: A review, Psychoneuroendocrin. 11:131-140, 1986.

(119) Armario, A., Castellanos, J.M. and Balasch, J.: Orchiectomy does not modify the adrenocortical response to noise stress in the rat, IRCS 10:791, 1982.

(120) Armario, A., Montero, J.L. and Balasch, J.: Sensitivity of corticosterone and some metabolic variables to graded levels of low intensity stresses in adult male rats,

Physiol. Behav. 37:559-561, 1986.

(121) Murison, R.C.C.: Time course of plasma corticosterone under immobilization stress in rats, IRCS 11:20-21, 1983.

(122) Daniels-Severs, A., Goodwin, A., Keil, L.C. and Bernikos-Danellis, J.: Effect of chronic crowding and cold on the pituitary-adrenal system: Responsiveness to an acute stimulus during chronic stress, Pharmacol. 9:348-356, 1973.

(123) Mason, J.W.: A review of psychoendocrine research on the pituitary-adrenal cortical system, Psychosom. Med. 30:576-607, 1968.

(124) Rose, R.M.: Overview of endocrinology of stress, in Neuroendocrinology and Psychiatric Disorder, G.M. Brown, S.H. Koslov and S. Reichlin, eds., Raven Press, New York, 95-122, 1984.

(125) Critchlow, V., Liebelt, R.A., Bar-Sela, M., Mountcastle, W. and Lipscomb, H.S.: Sex difference in resting pituitary-adrenal function in the rat, Am. J. Physiol. 205:807-815, 1963.

(126) Giagnoni, G., Santagostino, A., Senini, R., Fumagalli, P. and Gori, E.: Cold stress in the rat induces parallel changes in plasma and pituitary levels of endorphin and ACTH, Parmacol. Res. Commun. 15:15-21, 1983.

(127) Ganong, W.F.: Neurotransmitter mechanisms underlying stress responses, in Neuroendocrinology and Psychiatric Disorder, G.M. Brown, S.H. Koslov and S. Reichlin, eds., Raven Press, New York, 133-143, 1984.

(128) Walker, C.D., Perrin, M., Vale, W. and Rivier, C.: Ontogeny of the stress response in the rat: Role of the pituitary and the hypothalamus, Endocrinol. 118:1445-1451, 1986.

(129) Talan, M.I., Engel, B.T. and Whitaker, J.R.: A longitudinal study of tolerance to cold stress among C57BL/6J

mice, J. Gerontol. 40:8-14, 1985.

(130) Sapolsky, R.M., Krey, L.C. and McEwen, B.S.: The neuroendocrinology of stress and aging: The glucocorticoid cascade hypothesis, Endocrine Rev. 7:284-301, 1986.

(131) Barrett, A.M. and Stockham, M.A.: The effect of housing conditions and simple experimental procedures upon the corticosterone level in the plasma of rats, J. Endocrin. 26:97-105, 1963.

(132) Seredenin, S.B., Badyshtov, B.A., Nikitina, M.M. and Rozen, V.B.: Changes in plasma corticosterone levels of inbred mice after stress, Byull. Eksp. Biol. Med. 94:36-37, 1982. (Experimental Medicine and Biology, Plenum, 1051-1052, 1983.)

(133) Wolff, H.G: Life stress and bodily disease - A formulation, in Life Stress and Bodily Disease, Williams & Wilkins, Baltimore, 1059-1094, 1950.

(134) Lynch, J.J., Paskewitz, D.A., Gimbel, K.S. and Thomas, S.A.: Psychological aspects of cardiac arrhythmia, Am. Heart J. 93:645-657, 1977.

(135) Jenkins, C.D.: Recent evidence supporting psychologic and social risk factors for coronary disease: Part I, N.E. J. Med. 294:987-994, 1976.

(136) Dimsdale, J.E.: Emotional causes of sudden death, Am. J. Psychiatry 134:1361-1366, 1977.

(137) Eliot, R.S.: Stress and cardiovascular disease, Eur. J. Cardiol. 5:97-104, 1977.

(138) Hinkle, L.E. and Wolf, S.: Studies in diabetes mellitus: Changes in glucose, ketone and water metabolism during stress, in Life Stress and Bodily Disease, Williams & Wilkins, Baltimore, 1950.

(139) Anisman, H. and Zacharko, R.M.: Depression: The predisposing influence of stress, Behav. Brain Sci. 5:89-137, 1982.

(140) Graham, N., Douglas, R.M. and Ryan, P.: Stress and acute respiratory infection, Am. J. Epidemiol. 124:389-401, 1986.

(141) Fox, B.H.: Premorbid psychological factors as related to cancer incidence, J. Behav. Med. 1:45-133, 1978.

(142) Sklar, L.S. and Anisman, H.: Stress and cancer, Psychol. Bull. 89:369-406, 1981.

(143) Herrenkohl, L.R.: Prenatal stress reduces fertility and fecundity in female offspring, Science 206:1097-1099, 1979.

(144) Soblosky, J.S. and Thurmond, J.B.: Biochemical and behavioral correlates of chronic stress: Effects of tricyclic antidepressants, Pharmacol. Biochem. Behav. 24:1361-1368, 1986.

(145) Amkraut, A. and Solomon, G.F.: Stress and murine sarcoma virus (Moloney)-induced tumors, Cancer Res. 32:1428-1433, 1972.

(146) Seifter, E., Rettura, G., Zisblatt, M., Levenson, S.M., Levine, N., Davidson, A. and Seifter, J.: Enhancement of tumor development in physically-stressed mice inoculated with an oncogenic virus, Experentia 29:1379-1382, 1973.

(147) Szmigielski, S., Szudzinski, A., Pietraszek, A., Bielec, M., Janiak, M. and Wremble, J.K. Cocarcinogenic properties of microwave radiation, Bioelectromagnetics 3:179-191, 1982.

(148) Szudzinski, A., Pietraszek, A., Janiak, M., Wrembel, J., Kalczak, M. and Szmigielski, S.: Acceleration of the

development of benzopyrene-induced skin cancer in mice by microwave radiation, Arch. Dermatol. Res. 274:303-312, 1982.

(149) Lundy, J., Lovett, E.J., Wolinsky, S.M. and Conran, P.: Immune impairment and metastatic tumor growth, Cancer 43:945-951, 1979.

(150) van den Brenk, H.A.S., Stone, M.G., Kelly, H. and Sharpington, C.: Lowering of innate resistance of the lungs to the growth of blood-borne cancer cells in states of topical and systemic stress, Br. J. Cancer 33:60-78, 1976.

(151) Sklar, L.S. and Anisman, H.: Stress and coping factors influence tumor growth, Science 205:513-515, 1979.

(152) Turney, T.H., Harmsen, A.G. and Jarpe, M.A.: Modification of the antitumor action of Corynebacterium parvum by stress, Physiol. Behav. 37:555-558, 1986.

(153) Riley, V.: Mouse mammary tumors: Alteration of incidence as apparent function of stress, Science 189:465-467, 1975.

(154) Dechambre, R.-P. and Gosse, C.: Individual versus group caging of mice with grafted tumors, Cancer Res. 33:140-144, 1973.

(155) Andervont, H.B.: Influence of environmental mammary cancer in mice, J. Natl. Cancer Inst. 4:579-581, 1944.

(156) Chouroulinkov, I., Guillon, J.C. and Guerin, M.: Endometrial sarcomas in mice: A survey of 130 cases, J. Natl. Cancer Inst. 42:593-603, 1969.

(157) Lemonde, P.: Influence of fighting on leukemia in mice, Proc. Soc. Exp. Biol. Med. 102:292-295, 1959.

(158) Pradhan, S.N. and Ray, P.: Effects of stress on growth of transplanted and 7,12-dimethylbenz-lanthracene-induced tumors and their modification by psychotrophic drugs, J. Natl. Cancer Inst. 53:1241-1245, 1974.

(159) Newberry, B.H., Frankie, G., Beatty, P.A., Maloney, B.D. and Gilchrist, J.C.: Shock stress and DMBA-induced mammary tumors, Psychosom. Med. 34:295-303, 1972.

(160) Newberry, B.H., Gildow, J., Wogan, J. and Reese, R.L.: Inhibition of Huggins tumors by forced restraint, Psychosom. Med. 38:155-162, 1976.

(161) Newberry, B.H.: Restraint-induced inhibition of 7,12-dimethylbenz[a]anthracene-induced mammary tumors: Relation to stages of tumor development, J. Natl. Cancer Inst. 61:725-729, 1978.

(162) Shavit, Y., Lewis, J.W. and Terman, G.W.: Opioid peptides mediate the suppressive effect of stress on natural killer cell cytotoxicity, Science 223:188-190, 1984.

(163) Steplewski, Z. and Vogel, W.H.: Total leukocytes, T cell subpopulation and natural killer (NK) cell activity in rats exposed to restraint stress, Life Sci. 38:2419-2427, 1986.

(164) Okimura, T., Ogawa, M. and Yamauchi, T.: Stress and immune responses III: Effect of restraint stress on delayed type hypersensitivity (DTH) response, natural killer (NK) activity and phagocytosis in mice, Jap. J. Pharmacol. 41:229-235, 1986.

(165) Odio, M., Goliszek, A., Brodish, A. and Ricardo, M.J. Jr.: Impairment of immune function after cessation of long-term chronic stress, Immunol. Lett. 13:25-31, 1986.

(166) Becker, R.O. and Marino, A.A.: Electromagnetism and Life, SUNY Press, Albany, 1982.

(167) Keller, S.E., Weiss, J.M., Schleifer, S.J., Miller, N.E. and Stein, M.: Stress-induced suppression of immunity in adrenalectomized rats, Science 221:1301-1304, 1983.

(168) Peters, L.J. and Kelly, H.: The influence of stress and stress hormones on the transplantability of a non-immunogenic syngeneic murine tumor, Cancer 39:1482-1488, 1977.

(169) Laudenslager, M.L.: Coping and immunosuppression: Inescapable but not escapable shock suppresses lymphocyte proliferation, Science 221:568-570, 1983.

(170) Blalock, J.E.: The immune system as a sensory organ, J. Immunol. 132:1067-1070, 1984.

(171) Becker, R.O.: Electromagnetic forces and life processes, Technology Review (MIT) 75:32-38, 1972.

(172) Becker, R.O.: Testimony Before N.Y. PSC on Behalf of PSC Staff, October, 1974.

(173) Zaret, M.M.: Potential hazards of Hertzian radiation and tumors, N.Y. State J. Med. 77:146, 1977.

(174) Milham, S., Jr.: Mortality from leukemia in workers exposed to electrical and magnetic fields, N. Eng. J. Med. 307:249, 1982.

(175) Barregard, L., Jarvholm, B. and Ungethum, E.: Cancer among workers exposed to strong static magnetic fields, Lancet, 892, 1985.

(176) Wright, W.E., Peters, J. and Mack, T.: Leukemia in workers exposed to electrical and magnetic fields, Lancet ii:1160, 1982.

(177) Coleman, M., Bell, J. and Skeet, R.: Leukemia incidence in electrical workers, Lancet i:982, 1983.

(178) Rockette, H.: Mortality among coal miners covered by the UNWA Health and Retirement Funds, National Institute for Occupational Safety and Health, Publication 77-155, Washington, DC, 1977.

(179) Gilman, P.A., Ames, R.G. and McCawley, M.A.: Leukemia risk among U.S. white male coal miners, J. Occupat. Med. 27:669-671, 1985.

(180) Stern, F.B., et al.: A case-control study of leukemia at

a Naval nuclear shipyard, Am. J. Epidemiol. 123:980-992, 1986.

(181) Flodin, U., Fredriksson, M., Axelson, O., Persson, B. and Hardell, L.: Background radiation, electrical work and some other exposures associated with acute myeloid leukemia in a case-referent study, Arch. Environ. Health 41:77-84, 1986.

(182) McDowall, M.E.: Leukemia mortality in electrical workers in England and Wales, Lancet i:246, 1983.

(183) Pearce, N.E., Sheppard, R.A., Howard, J.K, Fraser, J. and Lilley, B.M.: Leukemia in electrical workers in New Zealand, Lancet i:811, 1985.

(184) Pearce, N.E., Sheppard, R.A., Howard, J.K, Fraser, J. and Lilley, B.M.: Leukemia among New Zealand agricultural workers, Am. J. Epidemiol. 124:402-409, 1986.

(185) Calle, E.E. and Savitz, D.A.: Leukemia in occupational groups with presumed exposure to electrical and magnetic fields, N. Eng. J. Med., 1475-1476, 1985.

(186) Milham, S., Jr.: Silent keys: Leukemia mortality in amateur radio operators, Lancet i:812, 1985.

(187) Robinette, C.D., Silverman, C. and Jablon, S.: Effects upon health of occupational exposure to microwave radiation (radar), Amer. J. Epidemiol. 112:39-53, 1980.

(188) Lin, R.S., Dischinger, P.C., Condee, J. and Farrell, K.P.: Occupational exposure to electromagnetic fields and the occurrence of brain tumors, J. Occup. Med. 27:413-419, 1985.

(189) Coggon, D., Pannett, B., Osmond, C. and Acheson, E.D.: A survey of cancer and occupation in young and middle-aged men. II. Non-respiratory cancers, Brit. J. Indust. Med. 43:381-386, 1986.

(190) Spitz, M.R. and Johnson, C.C.: Neuroblastoma and parental occupation, Am. J. Epidemiol. 121:924-929, 1985.

(191) Hicks, N., Zack, M., Caldwell, G.G., Fernbach, D.J. and Falletta, J.M.: Childhood cancer and occupational radiation exposure in parents, Cancer 53:1637-1643, 1984.

(192) Swerdlow, A.J.: Epidemiology of eye cancer in adults in England and Wales 1962-1977, Am. J. Epidemiol. 18:294-300, 1983.

(193) Vagero, D. and Olin, R.: Incidence of cancer in the electronics industry: Using the new Swedish Cancer Environment Registry as a screening instrument, Bri. J. Ind. Med. 40:188-192, 1983.

(194) Howe, G.R. and Lindsay, J.P.: A follow-up study of ten-percent sample of the Canadian labor force, J. Nat. Cancer Inst. 70:37-44, 1983.

(195) Becker, R.O.: Microwave radiation, N.Y. State J. Med. 77:2172, 1977.

(196) Wertheimer, N. and Leeper, E.: Electrical wiring configurations and childhood cancer, Am. J. Epidemiol. 109:273-284, 1979.

(197) Wertheimer, N. and Leeper, E.: Adult cancer related to electrical wires near the home, Int. J. Epidemiol. 11:345-355, 1982.

(198) Fulton, J.P., Cobb, S., Preble, L., Leone, L. and Forman, E.: Electrical wiring configurations and childhood leukemia in Rhode Island, Am. J. Epidemiol. 111: 292-296, 1980.

(199) Wertheimer, N. and Leeper, E.: Re: Electrical wiring configurations and childhood leukemia in Rhode Island, Am J. Epidemiol. 112:461-462, 1981.

(200) Barnes, F.S., Wachtel, H., Savitz, D., Fuller, J. and van Feldt, W.: Magnetic fields and wiring configuration, (Abstract), 8th Annual Meeting, BEMS, p. 6, 1986.

(201) Marino, A.A., Berger, T.J., Mitchell, J.T., Duhacek, B.A. and Becker, R.O.: Electric field effects in selected biologic systems, Ann. N.Y. Acad. Sci. 238:436-444, 1974.

(202) Daley, M.L., Morton, W.E., Chartier, V., Zajac, H. and Benitez, H.: Community fear of nonionizing radiation: A field investigation, IEEE Trans. Biomed. Eng. BME32:246-248, 1985.

(203) Tomenius, L., Hellstrom, L. and Enander, B.: Electrical constructions and 50-Hz magnetic field at the dwellings of tumor cases (0-18 years of age) in the county of Stockholm, in Proceedings of the International Symposium of Occupational Health and Safety in Mining and Tunneling, Prague, June 21-25, 1982.

(204) Lester, J.R. and Moore, D.F.: Cancer incidence and electromagnetic radiation, J. Bioelectricity 1:59-76, 1982.

(205) Lester, J.R. and Moore, D.F.: Cancer mortality and Air Force bases, J. Bioelectricity 1:77-82, 1982.

(206) Polson, P.P. and Merritt, J.H.: Cancer mortality and Air Force bases: A reevaluation, J. Bioelectricity 4:121-128, 1985.

(207) Lester, J.R.: Reply to "Cancer mortality and Air Force bases: A reevaluation," J. Bioelectricity 4:129-132, 1985.

(208) Aldrich, T.E., Glorieux, A. and Castro, S.: Florida cluster of five children with endodermal sinus tumors, Oncology 41:233-238, 1984.

(209) Friedman, H.L.: Are chronic exposure to microwaves and polycythemia associated?, N. Eng. J. Med. 304:357-358, 1981.

(210) Sadcikova, M.N.: Clinical manifestations of reactions to microwave irradiation in various occupational groups, in Biologic Effects and Health Hazards of Microwave Radiation, Polish Medical Publishers, Warsaw, 1974.

(211) Siekierzynski, M., Czerski, P., Milczarek, H., Gidynski, A., Czarnecki, C., Dziuk, E. and Jedrzejczak, W.: Health surveillance of personnel occupationally exposed to microwaves. II. Functional disturbances, Aerospace Med. 45:1143-1145, 1974.

(212) Lancranjan, I., Maicanescu, M., Rafaila, E., Klepsch, I. and Popescu, H.I.: Gonadic function in workmen with long-term exposure to microwaves, Health Physics 29:381-383, 1975.

(213) Nordstrom, S., Birke, E. and Gustavsson, L.: Reproductive hazards among workers at high voltage substations, Bioelectromagnetics 4:91-101, 1983.

(214) Wertheimer, N. and Leeper, E.: Fetal development related to the use of electrically-heated waterbeds, Bioelectromagnetics 1:13-22, 1986.

(215) Ericson, A. and Kallen, B.: An epidemiological study of work with video screens and pregnancy outcome: I. A registry study and II. A case-control study, Am. J. Indust. Med. 9:447-457, 459-475, 1986.

(216) Deapen, D.M. and Henderson, B.E.: A case-control study of amyotropic lateral sclerosis, Amer. J. Epidemiol. 123:790-799, 1986.

(217) Hamburger, S., Logue, J.N. and Silverman, P.M.: Occupational exposure to non-ionizing radiation and an association with heart disease: An exploratory study, J. Chronic Dis. 36:791-802, 1983.

(218) Perry, S. and Pearl, L.: Power frequency magnetic field and illness in multi-story blocks, Public Health (England), In Press.

(219) Bini, M., Checcucci, A., Ignesti, A., Millanta, L., Olmi, R., Rubino, N. and Banni, R.: Exposure of workers to intense RF electric fields that leak from plastic sealers, J. Microwave Power 21:33-40, 1986.

(220) Savitz, D.A.: Human studies of carcinogenic, reproductive, and general health effects of ELF fields, in Biological and Human Health Effects of Extremely Low Frequency Electromagnetic Fields, p. 241-270, American Institute of Biological Sciences, Arlington, VA, 1985.

(221) Liboff, A.R., Rozek, R.J., Sherman, M.L., McLeod, B.R. and Smith, S.D.: Ca^{++}-45 cyclotron resonance in human lymphocytes, J. Bioelectricity 6:13-22, 1987.

(222) Tornqvist, S., Norell, S., Ahlbom, A. and Knave, B.: Cancer in the electric power industry, Br. J. Indust. Med. 43:212-213, 1986.

(223) Lilienfeld, A.M.: Foreign Service Health Status Study, Johns Hopkins University, Baltimore, MD, July 31, 1978.

(224) Kotter, F.R. and Misakian, M.: AC transmission line field measurements, US Department of Commerce, National Bureau of Standards, Washington, DC, NBSIR 77-1311, 1981.

(225) Marino, A.A.: Testimony Before the N.Y. PSC on Behalf of the Commission Staff, October, 1974.

(226) Marino, A.A. and Becker, R.O.: High voltage lines: Hazard at a distance, Environment 20:6-15, 1978.

(227) Marino, A.A. and Morris, D.M.: Chronic electromagnetic stressors in the environment: A risk factor in human cancer, J. Environ. Sci. C3:189-219, 1985.

(228) Marino, A.A.: Electromagnetic fields and public health, in Assessments and Viewpoints on the Biological and Human Health Effects of Extremely Low Frequency Electromagnetic Fields, American Institute of Biological Sciences, Arlington, VA, 205-232, 1985.

(229) Marino, A.A.: Health risks from electric power facilities, in Proceedings of International Utility Symposium,

Health Effects of Electric and Magnetic Fields, Ontario Hydro, Toronto, 1986.

(230) Steneck, N.H., Cook, H.J., Vander, A. and Kane, G.L.: The origins of US safety standards for microwave radiation, Science 208:1230-1237, 1980.

(231) Steneck, N.: The Microwave Debate, MIT Press, Cambridge, 1984.

(232) Steneck, N.H.: Risk/Benefit Analysis: The Microwave Case, San Francisco Press, San Francisco, 1982.

(233) Biologic Effects of Electric and Magnetic Fields Associated with the Proposed Project Seafarer, Report of the Committee on Biosphere Effects of Extremely-Low-Frequency Radiation, National Academy of Sciences, Washington, DC, 1977.

(234) Boffey, P.M.: Project Seafarer: Critics attack National Academy's review group, Science 192:1213-1215, 1976.

(235) New York Public Health Law, Section 2441.

(236) New York Public Health Law, Section 2442.

(237) 45 CFR 46 -- Protection of Human Subjects.

(238) Nuremberg Code of 1947.

(239) Helsinki Declarations of 1964 and 1975.

(240) Report of the National Commission for the Protection of Human Subjects of Biomedical and Behavioral Research (The Belmont Report), U.S. Department of Health, Education and Welfare, Washington, DC, 1979.

Index

About the Editor

ANDREW A. MARINO is Associate Professor in the Department of Orthopaedic Surgery at the Louisiana State University Medical School in Shreveport. Formerly, he was research biophysicist for many years at the Syracuse Veterans Administration Hospital in New York, and faculty member at the State University of New York Upstate Medical Center in Syracuse. Dr. Marino is editor of the *Journal of Bioelectricity* (Marcel Dekker, Inc.), and coauthor of *The Electric Wilderness* and *Electromagnetism and Life.* Currently serving as president of the International Society for Bioelectricity, Dr. Marino has written many articles on the subject. He received the B.S. degree from St. Joseph's University in Philadelphia, Pennsylvania, and Ph.D. and J.D. degrees in biophysics and law, respectively, from Syracuse University in New York.